Patho-Epigenetics of Disease

Janos Minarovits • Hans Helmut Niller
Editors

Patho-Epigenetics of Disease

Editors
Janos Minarovits
Microbiological Research Group
National Center for Epidemiology
Budapest, Hungary

Hans Helmut Niller
Institute for Medical Microbiology
and Hygiene at the University
of Regensburg
Regensburg, Germany

ISBN 978-1-4614-3344-6   ISBN 978-1-4614-3345-3 (eBook)
DOI 10.1007/978-1-4614-3345-3
Springer New York Heidelberg Dordrecht London

Library of Congress Control Number: 2012937210

© Springer Science+Business Media New York 2012
This work is subject to copyright. All rights are reserved by the Publisher, whether the whole or part of the material is concerned, specifically the rights of translation, reprinting, reuse of illustrations, recitation, broadcasting, reproduction on microfilms or in any other physical way, and transmission or information storage and retrieval, electronic adaptation, computer software, or by similar or dissimilar methodology now known or hereafter developed. Exempted from this legal reservation are brief excerpts in connection with reviews or scholarly analysis or material supplied specifically for the purpose of being entered and executed on a computer system, for exclusive use by the purchaser of the work. Duplication of this publication or parts thereof is permitted only under the provisions of the Copyright Law of the Publisher's location, in its current version, and permission for use must always be obtained from Springer. Permissions for use may be obtained through RightsLink at the Copyright Clearance Center. Violations are liable to prosecution under the respective Copyright Law.
The use of general descriptive names, registered names, trademarks, service marks, etc. in this publication does not imply, even in the absence of a specific statement, that such names are exempt from the relevant protective laws and regulations and therefore free for general use.
While the advice and information in this book are believed to be true and accurate at the date of publication, neither the authors nor the editors nor the publisher can accept any legal responsibility for any errors or omissions that may be made. The publisher makes no warranty, express or implied, with respect to the material contained herein.

Printed on acid-free paper

Springer is part of Springer Science+Business Media (www.springer.com)

Preface

Epigenetic regulatory mechanisms ensure the heritable transmission of gene expression patterns from one cell generation to another, helping progeny cells to "remember" their proper cellular identity. Thus, epigenetic alterations, including DNA methylation, DNA-associated Polycomb–Trithorax complexes, and histone modifications, provide the means for establishment, maintenance, and programmed alterations of cell type-specific gene expression patterns in multicellular organisms. There is a complex interplay between the various epigenetic processes resulting in the unique transcriptional activity and phenotypic diversity of cells carrying identical or nearly identical DNA sequences. These carefully orchestrated processes can go wrong, however, resulting in epigenetic reprogramming of cells which may manifest in pathological changes, as it was first realized, as far as we know, during the studies of epigenetic alterations in malignant tumors.

Patho-epigenetics is a new discipline dealing with the pathological consequences of dysregulated epigenetic processes (Minarovits 2009). In recent years, epigenetic ideas found their way into many areas of medical and veterinary research, and this new attitude helped to explore many unexpected but most important aspects of disease initiation and progression. This book focuses mainly on patho-epigenetic alterations observed in human diseases, although animal models and veterinary diseases are also touched where relevant. Deciphering how the host cell epigenome is changing in diseased cells and organs may help to develop new tools permitting early and stage-specific detection of pathological alterations. It may lay the foundation of epigenetic therapies, too.

The goal of this book is to bring together recent results of epigenetic research with special attention to disease development. The first chapter by *Walter Doerfler*, senior of epigenetics research in virology, puts epigenetic research into historical context, reviewing the impact of foreign DNA insertion on the epigenotype of the host cell, and vice versa, with all of its vanguard implications for carcinogenesis and evolution.

Next, *Sole Gatto, Maurizio D'Esposito, and Maria R. Matarazzo* describe how distinct mutations of the de novo DNA methyltransferase *DNMT3B* cause ICF syndrome (Immunodeficiency, Centromeric region instability, and Facial anomalies syndrome), a very rare autosomal recessive disease. In addition to describing the consequences of *DNMT3B* loss-of-function mutations, they also present a novel view regarding the setting and maintenance of DNA methylation in mammals and discuss mouse models of maintenance and de novo DNA methyltransferase deficiencies.

DNA stretches rich in methylated cytosine are preferentially bound by methyl-CpG binding proteins implicated in the establishment of repressive chromatin structures that are unfavorable for transcription. *Gaston Caifa, Alan K. Percy, and Lucas Pozzo-Miller* outline how mutations in the *MECP2* gene coding for methyl-CpG-binding protein-2 cause Rett syndrome (RTT), a sporadic neurodevelopmental disorder characterized by cognitive impairment, communication dysfunction, and stereotypic hand movements. They also discuss how the classical view of MECP2 as a global transcriptional repressor changed in the light of new data demonstrating its role in gene activation and alternative splicing. The role of MECP2 and its murine counterpart in brain development is demonstrated in KO and knock-in mouse models, as well as the reversion of many RTT-like symptoms in mice reared in enriched environments.

Glioblastoma multiforme (GBM), the most common malignant brain tumor of adults is the topic of *John K. Wiencke* who argues that the patho-epigenetic background of primary GBM differs from that of secondary GBM arising in patients with a prior history of a non-GBM brain tumor. He emphasizes the importance of isocitrate dehydrogenase mutations resulting in the production of 2-hydroxyglutarate, an "oncometabolite" implicated in the inhibition of TET (ten-eleven-translocation) family enzymes that mediate active DNA demethylation. These processes may result in DNA hypermethylation at hundreds of gene loci in secondary GBMs and a fraction of de novo GBMs.

Epigenetic dysregulation in breast cancer, the most common non-skin malignancy affecting women in the Western world is discussed by *Amanda E. Toland*. She overviews not only the contribution of DNA methylation, Polycomb repressor complexes, and histone modifications to breast carcinogenesis, but in addition to these "classical" epigenetic regulatory mechanisms she elegantly discusses the role of nuclear organization, chromatin looping, microRNAs, and long noncoding RNAs (lncRNAs) as well. These mechanisms are involved in transcriptional or posttranscriptional regulation and they are frequently discussed in an epigenetic context, although they may not contribute to heritable transmission of gene expression patterns.

Hormonal dysbalance is an important factor in mammary carcinogenesis and affects the development and progression of prostate cancer as well, the topic of the review by *Christian Arsov, Wolfgang Goering, and Wolfgang A. Schulz*. They discuss both genetic and epigenetic alterations relevant to prostate carcinogenesis and describe putative epigenetic biomarkers for prostate cancer detection and prognosis. Drugs targeting epigenetic alterations in prostate cancer and their initial clinical trials are overviewed, too.

Epigenetic changes characteristic for lung carcinoma, the leading cause of cancer mortality in the United States, is the topic of *András Kádár and Tibor A. Rauch*.

They give a thorough description of hyper- and hypomethylation events, histone modification changes, and noncoding RNAs associated with lung carcinomas. The pharmacological aspects of DNA methylation-targeted therapy are also dealt with.

Hans Helmut Niller, Ferenc Banati, Eva Ay, and Janos Minarovits discuss the epigenetic alterations frequently observed in virus-associated malignant tumors. They focus on the epigenetic marks of latent gammaherpesvirus genomes related to the host cell-dependent expression of Epstein–Barr virus (EBV) and Kaposi's sarcoma-associated herpesvirus (KSHV) episomes and describe the interactions of the viral oncoproteins with the cellular epigenetic regulatory machinery. Pathoepigenetic processes related to human T-lymphotropic virus, hepatitis B virus, hepatitis C virus, human papillomavirus, Merkel cell polyomavirus, and the associated neoplasms are also discussed.

Both genetic and epigenetic conditions appear to contribute to the pathological forms of aggression, the topic of *Barbara Klausz, József Haller, Áron Tulogdi, and Dóra Zelena.* They describe the relevant animal models and the pioneering studies on the epigenetic effects of steroid hormones, as well as serotonine, vasopressine, and neutrophins, in distinct areas of the brain that could be related to changes in behavior and review the data related to the reduction of aggressive behavior by a histone deacetylase (HDAC) inhibitor.

Neuropsychiatric disorders also became the subjects of intense epigenetic studies. *Dóra Zelena* summarizes the epigenetic changes in schizophrenia (SCZ) and bipolar disorder (BD) which constitute a major public health burden. A series of complex epigenetic alterations were associated with the brain areas implicated in the development of SCZ. In BD, there were less numerous studies, although one of the HDAC inhibitors, valproate, is known to have a mood stabilizing effect.

Epigenetic studies may contribute to the understanding of the etiology and molecular pathogenesis of autoimmune disorders as well. As discussed by *Lorenzo de la Rica* and *Esteban Ballestar*, epigenetic dysregulation may form the basis of as diverse systemic or organ-specific diseases as systemic lupus erythematosus, rheumatoid arthritis, Crohn's disease, ulcerative colitis, autoimmune thyroid diseases, and many others.

Certain genes are expressed only from one of the two parental alleles, and the disturbances of the epigenetic mechanisms regulating this process, called *imprinting*, cause growth restriction or overgrowth and behavior abnormalities. As summarized by *Thomas Eggermann*, imprinting disorders share molecular disturbances including aberrant methylation (epimutation), uniparental disomy, chromosomal aberrations, and point mutations in imprinted genes.

Recently, epigenetic mechanisms were linked to the development of atherosclerosis as well. As overviewed by *Einari Aavik, Mikko Turunen,* and *Seppo Ylä-Herttuala*, dietary preferences may affect human health via epigenetic modifications of chromatin structure. They speculate that a histone methylase may switch the phenotype of smooth muscle cells that may clonally expand in atherosclerotic plaques.

Finally, *Hans Helmut Niller, Ferenc Banati, Eva Ay, and Janos Minarovits* describe the epigenetic changes associated with microbial infections. Certain viruses not associated with neoplasms, most remarkably human immunodeficiency virus (HIV), may affect the epigenotype of the host cell. Bacterial toxins and effector molecules may change histone modification patterns and *Helicobacter pylori*, associated with gastric carcinoma, alters both DNA methylation and histone modification in its target cells. Unicellular protozoa have their own sophisticated epigenetic control systems, but their effects on the epigenome of the host organisms remains to be established.

Budapest, Hungary Janos Minarovits
Regensburg, Germany Hans Helmut Niller

Reference

Minarovits, J. (2009). Microbe-induced epigenetic alterations in host cells: the coming era of patho-epigenetics of microbial infections. A review. *Acta Microbiol Immunol Hung* 56, 1–19.

Contents

1 The Impact of Foreign DNA Integration on Tumor Biology and Evolution via Epigenetic Alterations 1
Walter Doerfler

2 The Role of *DNMT3B* Mutations in the Pathogenesis of ICF Syndrome 15
Sole Gatto, Maurizio D'Esposito, and Maria R. Matarazzo

3 Dysfunction of the Methyl-CpG-Binding Protein MeCP2 in Rett Syndrome 43
Gaston Calfa, Alan K. Percy, and Lucas Pozzo-Miller

4 Epigenetic Alterations in Glioblastoma Multiforme 71
John K. Wiencke

5 Aberrant Epigenetic Regulation in Breast Cancer 91
Amanda Ewart Toland

6 The Impact of Epigenetic Alterations on Diagnosis, Prediction, and Therapy of Prostate Cancer 123
Christian Arsov, Wolfgang Goering, and Wolfgang A. Schulz

7 Epigenetic Reprogramming in Lung Carcinomas 159
András Kádár and Tibor A. Rauch

8 Epigenetic Changes in Virus-Associated Neoplasms 179
Hans Helmut Niller, Ferenc Banati, Eva Ay, and Janos Minarovits

9 Genetic and Epigenetic Determinants of Aggression 227
Barbara Klausz, József Haller, Áron Tulogdi, and Dóra Zelena

10 Co-Regulation and Epigenetic Dysregulation in Schizophrenia and Bipolar Disorder 281
Dóra Zelena

11 Disruption of Epigenetic Mechanisms in Autoimmune Syndromes 349
Lorenzo de la Rica and Esteban Ballestar

12 Imprinting Disorders 379
Thomas Eggermann

13 Epigenetics and Atherosclerosis 397
Einari Aavik, Mikko P. Turunen, and Seppo Ylä-Herttuala

14 Microbe-Induced Epigenetic Alterations 419
Hans Helmut Niller, Ferenc Banati, Eva Ay, and Janos Minarovits

Index 457

Contributors

Einari Aavik, PhD Department of Biotechnology and Molecular Medicine, A. I. Virtanen Institute, University of Eastern Finland, Kuopio, Finland

Christian Arsov, MD Department of Urology, University Hospital, Duesseldorf, Germany

Eva Ay Microbiological Research Group, National Center for Epidemiology, Budapest, Hungary

Esteban Ballestar, PhD Chromatin and Disease Group, Cancer Epigenetics and Biology Programme (PEBC), Bellvitge Biomedical Research Institute (IDIBELL), L'Hospitalet de Llobregat, Barcelona, Spain

Ferenc Banati, PhD Microbiological Research Group, National Center for Epidemiology, Budapest, Hungary

Gaston Calfa, PhD Department of Neurobiology, The University of Alabama at Birmingham, Birmingham, AL, USA

Maurizio D'Esposito, PhD Institute of Genetics and Biophysics "Adriano Buzzati-Traverso"—CNR, Naples, Italy

Lorenzo de la Rica, MSc Chromatin and Disease Group, Cancer Epigenetics and Biology Programme (PEBC), Bellvitge Biomedical Research Institute (IDIBELL), L'Hospitalet de Llobregat, Barcelona, Spain

Walter Doerfler, MD Institute for Virology, Erlangen University, Erlangen, Germany Institute of Genetics, University of Cologne, Cologne, Germany

Thomas Eggermann, PhD Institut für Humangenetik, RWTH Aachen, Aachen, Germany

Sole Gatto Institute of Genetics and Biophysics "Adriano Buzzati-Traverso"—CNR, Naples, Italy

Wolfgang Goering, PhD Department of Urology, Heinrich Heine University, Duesseldorf, Germany

József Haller, PhD, DSc Department of Behavioural Neurobiology, Laboratory of Behavioural and Stress Studies, Institute of Experimental Medicine, Hungarian Academy of Sciences, Budapest, Hungary

András Kádár, MD Section of Molecular Medicine, Department of Orthopedic Surgery, Rush University Medical Center, Chicago, IL, USA

Barbara Klausz Department of Behavioural Neurobiology, Laboratory of Behavioural and Stress Studies, Institute of Experimental Medicine, Hungarian Academy of Sciences, Budapest, Hungary

Maria R. Matarazzo, PhD Institute of Genetics and Biophysics "Adriano Buzzati-Traverso" — CNR, Naples, Italy

Janos Minarovits, MD, MSc Microbiological Research Group, National Center for Epidemiology, Budapest, Hungary

Hans Helmut Niller, MD Institute for Medical Microbiology and Hygiene University of Regensburg, Regensburg, Germany

Alan K. Percy, MD Departments of Pediatrics, Neurology, and Genetics, Civitan International Research Center, Birmingham, AL, USA

Lucas Pozzo-Miller, PhD Department of Neurobiology, The University of Alabama at Birmingham, Birmingham, AL, USA

Tibor A. Rauch, PhD Section of Molecular Medicine, Department of Orthopedic Surgery, Rush University Medical Center, Chicago, IL, USA

Wolfgang A. Schulz, PhD Department of Urology, Heinrich Heine University, Duesseldorf, Germany

Amanda Ewart Toland, PhD Department of Molecular Virology, Immunology and Medical Genetics, Comprehensive Cancer Center, The Ohio State University, Columbus, OH, USA

Áron Tulogdi Department of Behavioural Neurobiology, Laboratory of Behavioural and Stress Studies, Institute of Experimental Medicine, Hungarian Academy of Sciences, Budapest, Hungary

Mikko P. Turunen, PhD Ark Therapeutics, Kuopio, Finland

John K. Wiencke, PhD Department of Neurological Surgery and Epidemiology, University of California San Francisco, Helen Diller Family Cancer Center, San Francisco, CA, USA

Seppo Ylä-Herttuala, PhD, MD, FESC Department of Biotechnology and Molecular Medicine, A. I. Virtanen Institute, University of Eastern Finland, Kuopio, Finland

Gene Therapy Unit, Kuopio University Hospital, Kuopio, Finland

Dóra Zelena, MD, PhD Department of Behavioural Neurobiology, Laboratory of Behavioural and Stress Studies, Institute of Experimental Medicine, Hungarian Academy of Sciences, Budapest, Hungary

Chapter 1
The Impact of Foreign DNA Integration on Tumor Biology and Evolution via Epigenetic Alterations

Walter Doerfler

1.1 Introduction

Throughout evolution and the daily life of all organisms, the ingression of foreign DNA, i.e., of DNA from other organisms, is a continual challenge but also an opportunity for biological systems. The caloric requirements in daily life can be satisfied only by the ingestion of other organisms or their products. In addition, parasites with heterologous genomes find entry into organisms by refined adsorption and penetration mechanisms which have evolved over evolutionary time. In this context, the extensively studied viruses have proved highly adaptable. In the course of frequent encounters between host and parasitic intruder, defense mechanisms have been developed by the victims, the recipients of foreign genetic information. *De novo* methylation and histone modifications of integrated foreign DNA are understood to be essential parts of this ancient cellular defense. The most consequential challenges, both negative and positive, for the genetic stability of all organisms can be seen in the intrusion of foreign genetic information, most frequently in the form of foreign DNA. Retroviral or retrotransposon RNA is frequently converted to recombination-competent DNA. Foreign DNA can utilize the nearly omnipresent mechanisms of genetic recombination and, by this route, access the recipient cells' genomes. There seem to exist elaborate pathways between the ingestion of a foreign organism, e.g., by a unicellular organisms; the liberation and digestion of this foreign genome; and the possible, but probably rare, recombination of remnants of this foreign genome with that of the recipient cells. Much more complex pathways can

---

W. Doerfler, MD (✉)
Institute for Virology, Erlangen University, Erlangen, Germany

Institute of Genetics, University of Cologne, Cologne, Germany
e-mail: walter.doerfler@viro.med.uni-erlangen.de

J. Minarovits and H.H. Niller (eds.), *Patho-Epigenetics of Disease*,
DOI 10.1007/978-1-4614-3345-3_1, © Springer Science+Business Media New York 2012

be envisaged to be operative in multicellular organisms. Although these events and their underlying mechanisms appear to be frequent and likely hold the key to the understanding of important principles in evolution, there have been hardly any systematic studies of these mechanisms. This scarcity of information is all the more astounding as experimental biology has been exploiting these mechanisms in all fields of biotechnology and biomedicine despite the built-in manipulations, as enumerated in a concluding section of this review. A better understanding of precisely these consequences of genome manipulations, however, will help avoid unwanted negative sequelae and guard against unintended hazards in medicine: *Nil nocere* — do not cause harm.

This review addresses the consequences of foreign DNA integration into the genome of the recipient cell, as it were the ultimate and genetically most precarious step in the interaction between a cell and its hostile environment. There is only limited information on this topic, and *de novo* methylation of the transgenome and alterations of methylation and transcription patterns in the targeted genome may be only part of a greater gamut of responses in the host cell. From entry, degradation to eventual fixation, and silencing or activity of foreign DNA, there likely are complex pathways of these events.

1.2 Résumé of Our Previous Work on DNA Methylation

Beginning in the late 1960s, my laboratory has been investigating characteristics of the integration of adenovirus DNA into the genomes of transformed hamster cells or of adenovirus type 12 (Ad12)-induced hamster tumor cells (Doerfler 1968, 2009; Sutter et al. 1978; Doerfler et al. 1983; Knoblauch et al. 1996; Hohlweg et al. 2003). In the course of these studies, we discovered that integrated Ad12 DNA, quite in contrast to virion DNA or free intracellular Ad12 DNA, became *de novo* methylated in specific patterns which probably reflected the selection for transcriptional profiles of the integrated viral genomes to be optimal for the generation of the transformed or tumor phenotype (Sutter et al. 1978; Sutter and Doerfler 1980). As we previously reported, epigenetics plays an important role in the processing of foreign DNA (Doerfler 2008, 2009, 2011):

1. Integrated foreign (adenovirus, bacteriophage λ, or plasmid) DNA becomes *de novo* methylated. Adenovirus DNA inside the virus particle (virion DNA) or free intracellular viral DNA in productively or abortively infected cells remains unmethylated (Günthert et al. 1976; Sutter et al. 1978; Sutter and Doerfler 1980; Orend et al. 1995).
2. *De novo* methylation of artificially introduced foreign genes or genomes is initiated regionally, apparently not at a single CpG dinucleotide (Orend et al. 1995), and spreads from there over extensive genome segments (Toth et al. 1989).
3. Inverse correlations exist between promoter methylation and promoter activity (Sutter and Doerfler 1980; Vardimon et al. 1980).

4. In vitro premethylation of promoters leads to their inactivation upon transfection into mammalian cells or upon microinjection into *Xenopus laevis* oocytes in transient transcription experiments (Vardimon et al. 1982; Kruczek and Doerfler 1983; Doerfler 1983; Langner et al. 1984).
5. The inactivating effect of promoter methylation can be counteracted by transactivators like the E1A genes of adenoviruses or the strong enhancer of human cytomegalovirus (Langner et al. 1986; Weisshaar et al. 1988; Knebel-Mörsdorf et al. 1988).
6. There is evidence for the existence of a DNA methylation memory in the mammalian genome: The reinsertion of the mouse T lymphocyte tyrosine kinase gene at its homologous site on mouse chromosome 14 leads to the reestablishment of the original methylation pattern which is characteristic for this gene. When the same kinase gene was inserted at randomly selected sites by heterologous recombination, completely different methylation patterns were imposed on this transgene (Hertz et al. 1999). The inserted kinase gene had been rendered unmethylated by cloning it in a bacterial plasmid which was propagated in a C-methylation-deficient *Escherichia coli* strain. Apparently, each locus in the genome has its specific DNA methylation memory.
7. In almost completely *de novo* methylated regions of an integrated Ad12 DNA genome, there are singlets, doublets, and even triplets of CpG dinucleotides devoid of methylation (Hochstein et al. 2007). We interpret this punctual lack of methylation as a characteristic of the *de novo* methylation process of unknown functional significance (Doerfler 2011). It is conceivable that a particular chromatin structure and/or the inability of the DNA methyltransferase system to act on certain sequence combinations (Jurkowska et al. 2011) are responsible for this highly selective lack of *de novo* methylation. Moreover, the presence of bona fide or cryptic binding sites for transcription factors may have protected these CpGs from DNA methylation.
8. The insertion of adenoviral, bacterial plasmid, or bacteriophage lambda DNA into established mammalian genomes can cause profound changes of DNA methylation patterns in the recipient genomes at sites remote from the locus of foreign DNA integration (Heller et al. 1995; Doerfler 1995, 1996, 2011; Remus et al. 1999; Muller et al. 2001; Hohlweg et al. 2003 (see below)).
9. There are complex patterns of DNA methylation in the human genome which can be inter-individually preserved (Kochanek et al. 1990, 1991).
10. In the 5′-upstream and promoter regions of the human fragile X mental retardation gene 1 (FMR1), whose continued functionality is crucial for normal human development, we have identified a distinct DNA methylation boundary which is present in all human cell types and cell lines, irrespective of age, developmental stage, and gender. This boundary is also preserved in the homologous region of the mouse genome (Naumann et al. 2009). In individuals with the fragile X syndrome (FRAXA), this distinct DNA methylation boundary has disappeared. As a consequence, the entire region downstream of the boundary, including the

promoter of the FMR1 gene, becomes *de novo* methylated and the promoter inactivated. The ensuing lack of the FMR1 protein is considered to be the cause of FRAXA. However, the methylation boundary is stable in premutation carriers with a CGG repeat expansion >50 but <200 repeats and in high-functioning males (HFM) with CGG repeat lengths of up to 400 (Naumann et al., manuscript). In HFMs, the FMR1 promoter remains active since the intact methylation boundary safeguards the promoter against methylation and inactivation.

11. In the *de novo* methylated FMR1 promoter region in FRAXA genomes, there are also islets of adjacently located unmethylated CpG dinucleotides (Naumann et al. 2009) reminiscent of the situation in integrated Ad12 genomes (Hochstein et al. 2007), perhaps as a hallmark of *de novo* methylated DNA in mammalian cells. When the same FMR1 promoter sequence is in vitro methylated by a bacterial DNA methyltransferase, these characteristic unmethylated islets are not observed (Naumann et al. 2009). This finding lends credence to the notion that a specific chromatin structure or the presence of a nucleotide sequence precluding access of the mammalian DNA methyltransferase system might be involved in the exclusion of certain CpG dinucleotides from *de novo* methylation in an otherwise heavily *de novo* methylated genome segment.

1.3 Manipulations of the Genome Can Lead to Extensive Alterations of Its Epigenetic Profile

The integration of foreign DNA into established genomes is obviously a frequent and biologically significant event. Viral genomes, particularly those of oncogenic viruses, have been extensively studied for their ability to recombine with and integrate into the genomes of their host cells. Moreover, the presence of retroviral and retrotransposon DNA in today's genomes documents that foreign genomes had had frequent encounters with and successful entries into the genomes of our ancestors on an evolutionary time scale. What are the possible consequences of foreign DNA insertions?

1. The integrated genome can be recognized as foreign and become *de novo* methylated in specific patterns and inactivated or remain unmethylated and active.
2. The nucleotide sequence at the site of insertion could be interrupted or partially deleted. Genes at such sites would be deleted or mutated, a phenomenon referred to as *site-directed mutagenesis*. As genes, e.g., in the human genome comprise only 1–2% of the total nucleotide sequence, mutations due to foreign DNA insertions are expected to be infrequent, aside from the counter-selective value of such mutations and the then compromised survival of the affected cells or organisms.
3. The foreign DNA insert could also act as a promoter/enhancer for an endogenous gene.

This review will discuss the topic of epigenetic alterations upon insertion of foreign DNA in the host genomes. The concept underlying the work reviewed in this section is premised on the following observations:

1. In our studies on integrated adenovirus (Ad) genomes, we noted in more than 100 Ad-transformed or Ad12-induced tumor cells that the persisting viral genomes were invariably integrated into the host genome. Multiple copies of Ad12 DNA were integrated in most instances at one chromosomal site, but the sites of integration were randomly selected and were different from tumor to tumor (Doerfler et al. 1983; Knoblauch et al. 1996; Hilger-Eversheim and Doerfler 1997; Hohlweg et al. 2003). There was no evidence for site-specific integration of Ad12 DNA in hamster cells. However, short patch-like sequence homologies between the termini of Ad12 DNA and cellular DNA were frequently found at the junctions (Doerfler et al. 1983; Knoblauch et al. 1996).
2. In several other well-studied systems of viral oncogenesis, the viral genomes had also become immortalized by integration into the host genomes, e.g., in SV40- (Pipas 2009), retrovirus- (Skalka and Katz 2005), HPV- (Pett and Coleman 2007), or HBV-induced (Neuveut et al. 2010) tumors. Conventionally, one had argued that the integrated state served to ascertain the permanent fixation of the foreign genome in the recipient cell and to facilitate the continued expression of viral "oncogenes." These earlier ideas did not incorporate the possibility that the insertion of foreign DNA by itself and the ensuing transcriptional and structural alterations in the host cell genome played an important role in viral oncogenesis which might well surpass or at least equal that of the viral "oncogene" products.
3. One of the unresolved puzzles in molecular genetics remains the role of the repetitive DNA sequences including the biological functions of the retroviral and retrotransposon DNA segments which are estimated to comprise up to 40% of a mammalian genome (Lander et al. 2001). Again, the focus on the transcripts and presumptive gene products of these repetitive sequences may have obstructed the alternative view on the impact which the insertion event itself might have had on the epigenetic structure and function of the genome targeted by the retro-transposition events. The impact of the integration event itself has likely altered the methylation patterns of remote parts of the recipient genome and its tran-scriptional profiles. Subsequently, the prevailing environmental conditions would have determined the survival value of cells carrying genomes with thus funda-mentally altered transcription patterns.

In the 1990s, our laboratory (Heller et al. 1995; Doerfler 1995, 1996; Remus et al. 1999; Muller et al. 2001) has started to investigate possible alterations of methylation and transcription patterns in mammalian cells transgenic for:

1. Ad12 genomes
2. DNA of bacteriophage lambda
3. Bacterial plasmid DNA

Initially hamster and mouse cells and most recently the human cell line HCT116 have been used in these studies:

1. *Ad12-transformed hamster cell lines.* The genome of the Ad12-transformed hamster cell line T637 carries in an integrated state an estimated 10–15 (almost) complete Ad12 genomes, each measuring 34.125 kb This cell line was established by in vitro transformation of BHK21 cells with human Ad12 (Strohl et al. 1967). The results of restriction and FISH (fluorescent in situ hybridization) experiments indicated that the integrated viral genomes were located at one chromosomal site (Stabel et al. 1980; Knoblauch et al. 1996; Schroer et al. 1997; Hochstein et al. 2007). When the extent of DNA methylation in the retrotransposon sequences IAP (intracisternal A particles) (Kuff and Lueders 1988) with an estimated abundance of 900 copies per diploid genome was determined in T637 cells and compared to that in the parent BHK21 cell line, a marked increase in DNA methylation was noted (Heller et al. 1995). In this study, methylation patterns were assessed by cleaving the entire genomic DNA of the cell line with HpaII or MspI or with HhaI and subsequently hybridizing the DNA on Southern blots to cloned segments from the IAP genome. Moreover, since the IAP genomes are distributed over a large number of hamster chromosomes (Heller et al. 1995; Meyer zu Altenschildesche et al. 1996), and the alterations of DNA methylation had been quite extensive, the effect of the inserted foreign (Ad12) DNA on authentic cellular (IAP) methylation patterns had to be in trans and had to involve genome-wide regions remote from the site of foreign DNA insertion. It was conceivable that repetitive, possibly retrotransposon, sequences were particularly prone to such induced changes of their methylation patterns, but unique genome sequences, like MHC classes I and II, Ig Cμ I and II segments, as well as the serine protease and cytochrome P450 genes were also affected (Heller et al. 1995). It has been suggested that the site of foreign DNA integration might determine location and extent of these epigenetic alterations. It is likely that Ad12 DNA integrates early after the infection of hamster cells (Doerfler 1968). Similar alterations of cellular methylation patterns were not observed in BHK21 cells infected with Ad12 (Heller et al. 1995). These findings argued against the possibility that Ad12 transcripts or their products might have been responsible for the epigenetic effects on the recipient hamster cell genome. Ad12 transcription in BHK21 and in Ad12-transformed cells is limited to some of the early viral genes, and the infection is completely abortive. Late Ad12 genes are not transcribed, and viral DNA is not replicated in hamster cells (Doerfler 1969, 2009; Hochstein et al. 2008). The stability of the altered cellular methylation patterns was not dependent on the continued presence of the Ad12 transgenome since in the TR3 revertant of T637 cells (Groneberg et al. 1978), which had lost all Ad12 genomes, the altered epigenetic profiles of the IAP sequences remained stable (Heller et al. 1995). These altered epigenetic profiles therefore are thought to be stable and play a major causal role in viral oncogenic transformation.

2. *Hamster cells transgenic for the DNA of bacteriophage lambda.* Since the Ad12-transformed hamster cell line T637 resembled tumor cells (Strohl et al. 1967),

the epigenetic alterations in DNA methylation patterns could have been due to the oncogenic phenotype. We therefore extended the study on epigenetic alterations due to foreign DNA insertions to hamster cells rendered transgenic for the genome of bacteriophage lambda (Remus et al. 1999). Again, multiple lambda DNA molecules were found integrated at a single chromosomal site, and these phage genomes became also *de novo* methylated. As indicator sequences for changes in DNA methylation, IAP segments of the hamster genome in transgenic cells were analyzed by HpaII and MspI restriction followed by Southern blot hybridization as well as by bisulfite sequencing. Both methods revealed an increase in DNA methylation in the IAP sequences in the transgenic cells in several independently isolated hamster cell clones transgenic for bacteriophage lambda DNA (Remus et al. 1999). In 66 control clones of non-transgenic hamster cells, such changes in IAP methylation profiles were not observed by HpaII and MspI restriction analyses. Hence, we considered it unlikely that the observed alterations in methylation patterns had already preexisted in some of the hamster cells prior to having become transgenic by transfection and fixation of lambda DNA. We concluded that the fixation of Ad12 DNA as well as of bacterial plasmid or lambda DNA (Heller et al. 1995; Remus et al. 1999) caused considerable changes in the DNA methylation patterns of the recipient genome. The mechanism(s) responsible for these trans-effects are not understood.

3. *Studies using subtractive hybridization methods* (Muller et al. 2001). A wide scope of cellular DNA segments and genes was analyzed in non-transgenic vs. transgenic hamster cells by applying the technique of methylation-sensitive representational difference analysis (MSRDA) which was based on a subtractive hybridization protocol (Ushijima et al. 1997). By applying this procedure, one selected against DNA segments that were heavily methylated and hence rarely cleaved by the methylation-sensitive restriction endonuclease HpaII. The use of this protocol led to the isolation of several cellular DNA segments which were indeed more heavily methylated in bacteriophage lambda DNA-transgenic hamster cell lines (Muller et al. 2001). In addition, by targeting the suppressive subtractive hybridization technique to cDNA preparations from non-transgenic and Ad12-transformed or lambda DNA-transgenic hamster cells, several cellular genes with altered transcription patterns were cloned from Ad12-transformed or lambda DNA-transgenic hamster cells. Many of the DNA segments with altered methylation and cDNA fragments derived from genes with altered transcription patterns were identified by their nucleotide sequences (Muller et al. 2001). In control experiments, no differences in gene expression or DNA methylation patterns were detectable among individual non-transgenic BHK21 cell clones. These data further support the notion that the insertion of foreign DNA into an established mammalian genome can lead to alterations in cellular DNA methylation and transcription patterns.

4. *Human cell line HCT 116 carrying integrated bacterial plasmid DNA (pEGFP-FMR1).* Currently, we are investigating another transgenic cell system to assess the generality of the phenomenon of epigenetic alterations in the genomes of cells transgenic for foreign DNA. The human cell line HCT116 was transfected

with the 5.688 kb bacterial plasmid pEGFP-FMR1, and cells transgenic for the plasmid were selected for kanamycin resistance. The continued presence of the plasmid in the transformed cells was also ascertained by PCR using plasmid-specific primers. Preliminary data (S. Weber and W. Doerfler, unpublished experiments) indicate that several cell clones of non-transgenic HCT116 cells have very similar, if not identical, transcription profiles. In plasmid-transgenic cells, however, transcription patterns appear to be markedly altered. We are aware of the limitations of these studies as we have chosen a human tumor cell line in culture. However, this choice assures a constant source of a genetically homogeneous cell type which can be easily cloned. Since the tumor cell phenotype may create problems in the interpretation of data, we plan to control this problem by using primary human cells in similar studies.

A critical evaluation of the present state of our own work on the consequences of foreign DNA integration reveals a number of unresolved questions:

1. It is presently unknown how general the described observations apply to mammalian or plant systems. We can offer several, as we believe, well-documented observations and invite other researchers to expand them in their fields of interest.
2. The mechanisms by which the insertion of foreign DNA into an established genome could elicit trans-effects on remote segments in the recipient genome are not understood.

1.4 Impact of (Repetitive) Foreign DNA Insertions on Evolution

About 40% of the human genome is derived from retroviral or retrotransposon sequences (Lander et al. 2001). They have been preserved for long evolutionary periods, and some of them are actively transcribed. Their function in present-day genomes is only partly understood. Large intergenic noncoding RNAs (lincRNAs) are considered to function as transcriptional regulators and to be sensors of transcriptional responses and epigenetic events, possibly by impacting on the activity of histone-modifying enzymes and/or DNA methyltransferases (Yang et al. 2011). There are interesting considerations ranging from the contribution of promoter or regulatory systems to serving as foreign effectors on adjacent endogenous functional signals. However, one obvious possibility has so far escaped consideration. In my model, it is not merely the presence and presumptive genetic functions of these sequences or their mutagenic actions at the sites of insertion which deserve foremost evaluation but rather the alterations of cellular DNA methylation and histone modification patterns which had been elicited at an evolutionarily distant time of impact of foreign DNA insertions on the recipient genome.

Epigenetic *cis*- and/or *trans*-effects on patterns of DNA methylation and/or histone modifications, mainly at sites remote from the location of insertion, appear less obvious and have been underrepresented in the ongoing discussion about the possible functions of retroviral and retrotransposon sequences in the human or other mammalian genomes. Under this new premise, the major role of retroviral and retrotransposon sequences is their impact at the time of insertion on the epigenetic system of the genome by causing far-reaching activity changes.

I submit that induced epigenetic alterations due to the integration of foreign DNA have played an important role in evolution by altering the functional profile of the affected cell without causing mutations by insertional mutagenesis. According to this hypothesis, viruses with the potential of integration into the host genome, i.e., essentially all DNA viruses and all retroviruses, have served and continue to serve as a major driving force in evolution. Retroviral and retrotransposon sequences present in huge copy numbers could thus have made major functional contributions to the evolution of mammalian and other genomes. Of course, these epigenetic effects are not limited to viral genomes since any foreign DNA gaining access to the nucleus of a cell has the potential of becoming inserted and of eliciting similar epigenetic effects on the host. We have investigated the possibility to what extent trace amounts of foreign DNA ingested with the food supply might also gain access to cells in the organism and then become subject to the same cellular mechanisms as viral DNA molecules (Schubbert et al. 1997). In a speculative mode: Although the persistence of food-ingested DNA is transient, occasionally, this DNA insertion might have consequences similar to those of viral DNA in rare cases and might then contribute to oncogenesis.

A related, equally important problem has not yet been studied in sufficient detail: To what extent is the transcriptional activity of a cell affected by the mere introduction of foreign DNA into a cell by any of the experimentally used transfection methods? By the cell, transfection with foreign DNA will be registered as a highly invasive event. Since these transfection methods are routinely applied in all areas of biomedicine, a detailed study of this problem appears timely.

1.5 Role in Tumor Disease

In addition to the long-term evolutionary effects of epigenetic mechanisms, in an individual's life span as well, the insertion of foreign, e.g., viral DNA might fundamentally alter the transcriptional profile of cells by epigenetic changes and thus modulate the cell's growth control towards the pathway of disturbed growth supervision and tumor disease. The integration of viral DNA and its consequences might thus make a fundamental contribution to oncogenesis. In tumor cells, the total transcriptional profile has been altered, and it is likely that these alterations have been elicited by epigenetic effects. Activity changes of a single "oncogene" or "tumor suppressor gene" may be merely part of much more general alterations of cellular transcription profiles that occurred during the oncogenic transformation process.

1.6 HIV Infections and AIDS

Epigenetic alterations in HIV-infected cells isolated directly from AIDS patients have not yet been systematically investigated. As retroviral DNA integrates randomly at many sites in the recipient genomes, a plethora of epigenetic alterations are suspected to ensue. Would these changes play a role in the pathogenesis of the disease? In virus research in general, studies on epigenetic effects of virus infection on the methylation and transcription patterns of the infected cell have just begun (Grafodatskaya et al. 2010; Alberter et al. 2011; Leonard et al. 2011).

1.7 A Caveat Towards Manipulations on Existing Eukaryotic Genomes

The evidence reviewed here indicates that it is important to consider the possibility that foreign DNA insertions in mammalian and other genomes can alter epigenetic parameters and therefore transcriptional profiles in the manipulated cells. As genome manipulations have taken center stage in many branches of biomedical research, it will be prudent to investigate closely the possibly far-reaching consequences of these experimental procedures. These implications will be important for several fields of biomedicine:

- Gene therapeutic regimens
- Knockout and knockin experiments
- Induced pluripotent stem cells
- Tumor biology
- Virus infections
- Pathogenesis of HIV infections and AIDS
- Generation of transgenic cells and organisms
- Gene-manipulated organisms

1.8 Future Perspectives

It was the goal of this review to introduce and provide evidence for a new hypothesis about one of the globally relevant consequences of foreign DNA insertions into established genomes. This hypothesis is based on the observation that, upon the integration of Ad12, bacteriophage lambda, or plasmid DNA into the genomes of mammalian cells in culture, genome-wide alterations of DNA methylation and transcription patterns can be documented. I have suggested in a more general way of argumentation that these findings can be extended to the interpretation of foreign DNA insertions in tumor disease—e.g., caused by oncogenic viruses—and in evolution. In the latter context, I have not intended to reiterate or summarize cogent hypotheses of the past (Ohno 1970, 1999; Brosius 1999; Kidwell and Lisch 2001; Bowen and Jordan 2002; Kazazian 2004; Medstrand et al. 2005; Moyes et al. 2007; Romanish et al. 2010; Rebollo et al. 2011). There, it has been proposed that the integrated retroviral or retrotransposon genomes have contributed to evolution by introducing new genetic information, by insertional mutagenesis, or by adding regulatory elements in functionally decisive genome locations. In marked contrast, the novel hypothesis elaborated here concentrates on the impact of epigenetic alterations in the wake of foreign DNA insertions. These insertion events, possibly dependent on size of the transgenome and site of its integration, have the potential to alter methylation and transcription patterns genome wide. Depending on the environmental conditions prevalent at the time of impact, cells thus fundamentally altered in their transcriptional profiles are then selected for or against survival under the environmental conditions prevalent at the time and place. The genetic stability of this alteration is of course secured by the continued presence of the transgenome. There is evidence that, at least on the time scale of years, persistence of the transgenome may not be absolutely required to ascertain stability of the altered methylation profiles (Heller et al. 1995). This aspect, however, requires further experimental work.

There are additional important features in which this hypothesis requires future critical evaluation:

1. The mechanisms by which the trans-effects of foreign DNA insertions are elicited are completely unknown.
2. Moreover, it will be important to investigate whether and how the site of foreign DNA integration relates to the locations in which methylation and transcription patterns are altered.
3. It remains to be studied in detail how histone modifications are affected by the insertion of foreign DNA.
4. How generally valid are these observations? So far, only a small number of experimental systems have been analyzed; only cells growing in culture were included in our analyses. This limitation was intentional since proof of principle had to be sought first.
5. The use of much more complicated technology and careful controls will be required in relevant studies in animal or plant systems.
6. Future work will be focused on experiments directed towards the problems just enumerated. It will also be our intent to alert biomedical researchers to the potential complexities of widely applied genome manipulations in the interpretation of their results garnered under such premises.

Acknowledgments The author's current work is supported by the Deutsche Forschungsgemeinschaft (DFG), Bonn, Germany, and the Thyssen Foundation, Köln, Germany. W. D. is guest professor in the Institute for Clinical and Molecular Virology, Erlangen University Medical School. Previous projects carried out in the Institute of Genetics in Cologne were made possible, at different times, by grants from the DFG (SFB 74 and 274), from the Center for Molecular Medicine Cologne (CMMC, TP13), and from the Thyssen, Sander, and Humboldt Foundations.

Note of Acknowledgment: This article has been published in EPIGENOMICS (volume 4/1, 41–49, 2012) and is reproduced here with the permission of Future Medicine Ltd, Unitec House, 2 Albert Place, London, N3 1QB, UK.

References

Alberter, B., Vogel, B., Lengenfelder, D., Full, F. & Ensser, A. (2011). Genome-wide histone acetylation profiling of Herpesvirus saimiri in human T cells upon induction with a histone deacetylase inhibitor. *J Virol* 85, 5456–5464.

Bowen, N. J. & Jordan, I. K. (2002). Transposable elements and the evolution of eukaryotic complexity. *Curr Issues Mol Biol* 4, 65–76.

Brosius, J. (1999). Genomes were forged by massive bombardments with retroelements and retrosequences. *Genetica* 107, 209–238.

Doerfler, W. (1968). The fate of the DNA of adenovirus type 12 in baby hamster kidney cells. *Proc Natl Acad Sci U S A* 60, 636–643.

Doerfler, W. (1969). Nonproductive infection of baby hamster kidney cells (BHK21) with adenovirus type 12. *Virology* 38, 587–606.

Doerfler, W. (1983). DNA methylation and gene activity. *Annu Rev Biochem* 52, 93–124.

Doerfler, W. (1995). The insertion of foreign DNA into mammalian genomes and its consequences: a concept in oncogenesis. *Adv Cancer Res* 66, 313–344.

Doerfler, W. (1996). A new concept in (adenoviral) oncogenesis: integration of foreign DNA and its consequences. *Biochim Biophys Acta* 1288, F79–F99.

Doerfler, W. (2008). In pursuit of the first recognized epigenetic signal--DNA methylation: a 1976 to 2008 synopsis. *Epigenetics* 3, 125–133.

Doerfler, W. (2009). Epigenetic mechanisms in human adenovirus type 12 oncogenesis. *Semin Cancer Biol* 19, 136–143.

Doerfler, W. (2011). Epigenetic consequences of foreign DNA insertions: de novo methylation and global alterations of methylation patterns in recipient genomes. *Rev Med Virol* 21, 336–346.

Doerfler, W., Gahlmann, R., Stabel, S., Deuring, R., Lichtenberg, U., Schulz, M., Eick, D. & Leisten, R. (1983). On the mechanism of recombination between adenoviral and cellular DNAs: the structure of junction sites. *Curr Top Microbiol Immunol* 109, 193–228.

Grafodatskaya, D., Choufani, S., Ferreira, J. C., Butcher, D. T., Lou, Y., Zhao, C., Scherer, S. W. & Weksberg, R. (2010). EBV transformation and cell culturing destabilizes DNA methylation in human lymphoblastoid cell lines. *Genomics* 95, 73–83.

Groneberg, J., Sutter, D., Soboll, H. & Doerfler, W. (1978). Morphological revertants of adenovirus type 12-transformed hamster cells. *J Gen Virol* 40, 635–645.

Günthert, U., Schweiger, M., Stupp, M. & Doerfler, W. (1976). DNA methylation in adenovirus, adenovirus-transformed cells, and host cells. *Proc Natl Acad Sci U S A* 73, 3923–3927.

Heller, H., Kämmer, C., Wilgenbus, P. & Doerfler, W. (1995). Chromosomal insertion of foreign (adenovirus type 12, plasmid, or bacteriophage lambda) DNA is associated with enhanced methylation of cellular DNA segments. *Proc Natl Acad Sci U S A* 92, 5515–5519.

Hertz, J. M., Schell, G. & Doerfler, W. (1999). Factors affecting de novo methylation of foreign DNA in mouse embryonic stem cells. *J Biol Chem* 274, 24232–24240.

Hilger-Eversheim, K. & Doerfler, W. (1997). Clonal origin of adenovirus type 12-induced hamster tumors: nonspecific chromosomal integration sites of viral DNA. *Cancer Res* 57, 3001–3009.

Hochstein, N., Muiznieks, I., Mangel, L., Brondke, H. & Doerfler, W. (2007). Epigenetic status of an adenovirus type 12 transgenome upon long-term cultivation in hamster cells. *J Virol* 81, 5349–5361.

Hochstein, N., Webb, D., Hösel, M., Seidel, W., Auerochs, S. & Doerfler, W. (2008). Human CAR gene expression in nonpermissive hamster cells boosts entry of type 12 adenovirions and nuclear import of viral DNA. *J Virol* 82, 4159–4163.

Hohlweg, U., Hösel, M., Dorn, A., Webb, D., Hilger-Eversheim, K., Remus, R., Schmitz, B., Buettner, R., Schramme, A., Corzilius, L., Niemann, A. & Doerfler, W. (2003). Intraperitoneal dissemination of Ad12-induced undifferentiated neuroectodermal hamster tumors: de novo methylation and transcription patterns of integrated viral and of cellular genes. *Virus Res* 98, 45–56.

Jurkowska, R. Z., Ceccaldi, A., Zhang, Y., Arimondo, P. B. & Jeltsch, A. (2011). DNA methyltransferase assays. *Methods Mol Biol* 791, 157–177.

Kazazian, H. H. (2004). Mobile elements: drivers of genome evolution. *Science* 303, 1626–1632.

Kidwell, M. G. & Lisch, D. R. (2001). Perspective: transposable elements, parasitic DNA, and genome evolution. *Evolution* 55, 1–24.

Knebel-Mörsdorf, D., Achten, S., Langner, K. D., Rüger, R., Fleckenstein, B. & Doerfler, W. (1988). Reactivation of the methylation-inhibited late E2A promoter of adenovirus type 2 by a strong enhancer of human cytomegalovirus. *Virology* 166, 166–174.

Knoblauch, M., Schroer, J., Schmitz, B. & Doerfler, W. (1996). The structure of adenovirus type 12 DNA integration sites in the hamster cell genome. *J Virol* 70, 3788–3796.

Kochanek, S., Radbruch, A., Tesch, H., Renz, D. & Doerfler, W. (1991). DNA methylation profiles in the human genes for tumor necrosis factors alpha and beta in subpopulations of leukocytes and in leukemias. *Proc Natl Acad Sci U S A* 88, 5759–5763.

Kochanek, S., Toth, M., Dehmel, A., Renz, D. & Doerfler, W. (1990). Interindividual concordance of methylation profiles in human genes for tumor necrosis factors alpha and beta. *Proc Natl Acad Sci U S A* 87, 8830–8834.

Kruczek, I. & Doerfler, W. (1983). Expression of the chloramphenicol acetyltransferase gene in mammalian cells under the control of adenovirus type 12 promoters: effect of promoter methylation on gene expression. *Proc Natl Acad Sci U S A* 80, 7586–7590.

Kuff, E. L. & Lueders, K. K. (1988). The intracisternal A-particle gene family: structure and functional aspects. *Adv Cancer Res* 51, 183–276.

Lander, E. S., Linton, L. M., Birren, B., Nusbaum, C., Zody, M. C., Baldwin, J., Devon, K., Dewar, K., Doyle, M., FitzHugh, W., et al., International Human Genome Sequencing Consortium (2001). Initial sequencing and analysis of the human genome. *Nature* 409, 860–921.

Langner, K. D., Vardimon, L., Renz, D. & Doerfler, W. (1984). DNA methylation of three 5′ C-C-G-G 3′ sites in the promoter and 5′ region inactivate the E2a gene of adenovirus type 2. *Proc Natl Acad Sci U S A* 81, 2950–2954.

Langner, K. D., Weyer, U. & Doerfler, W. (1986). Trans effect of the E1 region of adenoviruses on the expression of a prokaryotic gene in mammalian cells: resistance to 5′ -CCGG- 3′ methylation. *Proc Natl Acad Sci U S A* 83, 1598–1602.

Leonard, S., Wei, W., Anderton, J., Vockerodt, M., Rowe, M., Murray, P. G. & Woodman, C. B. (2011). Epigenetic and transcriptional changes which follow Epstein-Barr virus infection of germinal center B cells and their relevance to the pathogenesis of Hodgkin's lymphoma. *J Virol* 85, 9568–9577.

Medstrand, P., van de Lagemaat, L. N., Dunn, C. A., Landry, J. R., Svenback, D. & Mager, D. L. (2005). Impact of transposable elements on the evolution of mammalian gene regulation. *Cytogenet Genome Res* 110, 342–352.

Meyer zu Altenschildesche, G., Heller, H., Wilgenbus, P., Tjia, S. T. & Doerfler, W. (1996). Chromosomal distribution of the hamster intracisternal A-particle (IAP) retrotransposons. *Chromosoma* 104, 341–344.

Moyes, D., Griffiths, D. J. & Venables, P. J. (2007). Insertional polymorphisms: a new lease of life for endogenous retroviruses in human disease. *Trends Genet* 23, 326–333.

Müller, K., Heller, H. & Doerfler, W. (2001). Foreign DNA integration. Genome-wide perturbations of methylation and transcription in the recipient genomes. *J Biol Chem* 276, 14271–14278.

Naumann, A., Hochstein, N., Weber, S., Fanning, E. & Doerfler, W. (2009). A distinct DNA-methylation boundary in the 5′- upstream sequence of the FMR1 promoter binds nuclear proteins and is lost in fragile X syndrome. *Am J Hum Genet* 85, 606–616.

Naumann, A., Kraus, C., Hoogeveen, A., Fanning, E. & Doerfler, W. (2011). The FMR1 methylation boundary safeguards against methylation spreading and is (partly) stable upon foreign DNA insertions in trans. Manuscript.

Neuveut, C., Wei, Y. & Buendia, M. A. (2010). Mechanisms of HBV-related hepatocarcinogenesis. *J Hepatol* 52, 594–604.

Ohno, S. (1970). *Evolution by gene duplication.* Berlin, Heidelberg, New York: Springer.

Ohno, S. (1999). Gene duplication and the uniqueness of vertebrate genomes circa 1970–1999. *Semin Cell Dev Biol* 10, 517–522.

Orend, G., Knoblauch, M., Kammer, C., Tjia, S. T., Schmitz, B., Linkwitz, A., Meyer zu Altenschildesche, G., Maas, J. & Doerfler, W. (1995). The initiation of de novo methylation of foreign DNA integrated into a mammalian genome is not exclusively targeted by nucleotide sequence. *J Virol* 69, 1226–1242.

Pett, M. & Coleman, N. (2007). Integration of high-risk human papillomavirus: a key event in cervical carcinogenesis? *J Pathol* 212, 356–367.

Pipas, J. M. (2009). SV40: Cell transformation and tumorigenesis. *Virology* 384, 294–303.

Rebollo, R., Karimi, M. M., Bilenky, M., Gagnier, L., Miceli-Royer, K., Zhang, Y., Goyal, P., Keane, T. M., Jones, S., Hirst, M., Lorincz, M. C. & Mager, D. L. (2011). Retrotransposon-induced heterochromatin spreading in the mouse revealed by insertional polymorphisms. *PLoS Genet* 7, e1002301.

Remus, R., Kämmer, C., Heller, H., Schmitz, B., Schell, G. & Doerfler, W. (1999). Insertion of foreign DNA into an established mammalian genome can alter the methylation of cellular DNA sequences. *J Virol* 73, 1010–1022.

Romanish, M. T., Cohen, C. J. & Mager, D. L. (2010). Potential mechanisms of endogenous retroviral-mediated genomic instability in human cancer. *Semin Cancer Biol* 20, 246–253.

Schroer, J., Hölker, I. & Doerfler, W. (1997). Adenovirus type 12 DNA firmly associates with mammalian chromosomes early after virus infection or after DNA transfer by the addition of DNA to the cell culture medium. *J Virol* 71, 7923–7932.

Schubbert, R., Renz, D., Schmitz, B. & Doerfler, W. (1997). Foreign (M13) DNA ingested by mice reaches peripheral leukocytes, spleen, and liver via the intestinal wall mucosa and can be covalently linked to mouse DNA. *Proc Natl Acad Sci U S A* 94, 961–966.

Skalka, A. M. & Katz, R. A. (2005). Retroviral DNA integration and the DNA damage response. *Cell Death Differ* 12 Suppl 1, 971–978.

Stabel, S., Doerfler, W. & Friis, R. R. (1980). Integration sites of adenovirus type 12 DNA in transformed hamster cells and hamster tumor cells. *J Virol* 36, 22–40.

Strohl, W. A., Rabson, A. S. & Rouse, H. (1967). Adenovirus tumorigenesis: role of the viral genome in determining tumor morphology. *Science* 156, 1631–1633.

Sutter, D. & Doerfler, W. (1980). Methylation of integrated adenovirus type 12 DNA sequences in transformed cells is inversely correlated with viral gene expression. *Proc Natl Acad Sci U S A* 77, 253–256.

Sutter, D., Westphal, M. & Doerfler, W. (1978). Patterns of integration of viral DNA sequences in the genomes of adenovirus type 12-transformed hamster cells. *Cell* 14, 569–585.

Toth, M., Lichtenberg, U. & Doerfler, W. (1989). Genomic sequencing reveals a 5-methylcytosine-free domain in active promoters and the spreading of preimposed methylation patterns. *Proc Natl Acad Sci U S A* 86, 3728–3732.

Ushijima, T., Morimura, K., Hosoya, Y., Okonogi, H., Tatematsu, M., Sugimura, T. & Nagao, M. (1997). Establishment of methylation-sensitive-representational difference analysis and isolation of hypo- and hypermethylated genomic fragments in mouse liver tumors. *Proc Natl Acad Sci U S A* 94, 2284–2289.

Vardimon, L., Kressmann, A., Cedar, H., Maechler, M. & Doerfler, W. (1982). Expression of a cloned adenovirus gene is inhibited by in vitro methylation. *Proc Natl Acad Sci U S A* 79, 1073–1077.

Vardimon, L., Neumann, R., Kuhlmann, I., Sutter, D. & Doerfler, W. (1980). DNA methylation and viral gene expression in adenovirus-transformed and -infected cells. *Nucleic Acids Res* 8, 2461–2473.

Weisshaar, B., Langner, K. D., Jüttermann, R., Müller, U., Zock, C., Klimkait, T. & Doerfler, W. (1988). Reactivation of the methylation-inactivated late E2A promoter of adenovirus type 2 by E1A (13 S) functions. *J Mol Biol* 202, 255–270.

Yang, L., Lin, C. & Rosenfeld, M. G. (2011). A lincRNA switch for embryonic stem cell fate. *Cell Res* 21(12):1646–1648.

Chapter 2
The Role of *DNMT3B* Mutations in the Pathogenesis of ICF Syndrome

Sole Gatto, Maurizio D'Esposito, and Maria R. Matarazzo

2.1 Introduction

Immunodeficiency, centromere instability, facial abnormalities (ICF) syndrome was firstly described during the 1978 Symposium of the European Society of Human Genetics, where two independent cases of immunodeficiency associated with chromosome multibranching were reported by Tiepolo and colleagues, and one of them was published in detail (Tiepolo et al. 1979). In these patients, combined immunodeficiency was associated with facial anomalies and instability of the centromeric regions of chromosomes 1, 9, and 16 (Maraschio et al. 1988). In one case, centromeric instability was reported initially only in chromosome 1, even though further analysis revealed the involvement of chromosomes 9 and 16 as well.

Lately, patients with similar chromosomal and clinical features are being recruited, and the picture of the complex phenotype related to this rare human disease is still being drawn. A little more than 60 patients have been reported worldwide since its first description. Most patients come from Europe. However, ICF patients are of multiple origin, and some cases of consanguinity have been reported (Wijmenga et al. 2000), although most cases are not familial. Recently, the number of diagnosed, nonfamilial cases in Europe and Japan is largely increased, suggesting that this disease is underdiagnosed. This would explain why in some countries, only a few cases have been reported (Carpenter et al. 1988).

Although the disease has been described more than 20 years ago, ICF syndrome pathogenesis is far from being completely clarified. It has been classified as belonging to the group of chromatin diseases, and a growing number of molecular aspects have been subject of intense investigations over the years. However, only recently, the technological breakthroughs linked to the next generation sequencing make it

---

S. Gatto • M. D'Esposito, PhD • M.R. Matarazzo, PhD (✉)
Institute of Genetics and Biophysics "Adriano Buzzati-Traverso"—CNR, Naples, Italy
e-mail: gatto@igb.cnr.it; maria.matarazzo@igb.cnr.it

J. Minarovits and H.H. Niller (eds.), *Patho-Epigenetics of Disease*,
DOI 10.1007/978-1-4614-3345-3_2, © Springer Science+Business Media New York 2012

possible to undertake large-scale epigenomic studies, which promise to clarify the obscure aspects of chromatin disease pathogenesis.

Indeed, the detailed study of the epigenetic maps would be of enormous use, not only in basic, but also in applied research, and would be relevant for focusing pharmacological research on the most promising epigenetic targets.

2.1.1 ICF Syndrome: A Chromatin Disorder of Gene Silencing

2.1.1.1 Clinical Manifestation and Diagnosis

The ICF syndrome (OMIM #242860) is a very rare autosomal recessive disease that severely damages the immune system of the affected subjects and exhibits a diffuse hypomethylation of specific heterochromatic regions of the DNA (Tiepolo et al. 1979; Maraschio et al. 1988).

ICF patients are mostly diagnosed during childhood due to recurrent infections, the characteristic symptom of the syndrome. In the blood biochemical analysis, they all show a combined immunodeficiency with reduction or absence of serum immunoglobulins of all subtypes (in different combinations) with a normal number of B and T cells. In detail, B cells from the peripheral blood of these patients present characteristics of anergy and are mainly naive, with absence of memory or plasma cells, leading to the malfunctioning of their immune response. However, those cells can secrete immunoglobulin if stimulated in vitro, suggesting that their lack of function can be probably the cause of some other impairment in the upstream pathway of antigen response. On the other hand, T cells have a deficient proliferation response to antigen exposure that probably leads to the impairment of B cells activation (Blanco-Betancourt et al. 2004). ICF patients, thus, are prone to recurrent severe respiratory and gastrointestinal infections that often cause death at young age.

The hallmark of this syndrome lays in the karyotype of the affected subjects, where chromosomes 1, 9, and 16 show evident decondensation of juxtacentromeric heterochromatin causing chromosome breaks and rearrangements in radial structures only in phytohemagglutinin-stimulated peripheral blood lymphocytes. The molecular basis of this phenomenon has mainly been addressed to the loss of DNA methylation within classical satellites (Sat 2 and 3) at the juxtacentromeric heterochromatin of the long arms of chromosomes 1, 16, sometimes 9 and Y in males (Fig. 2.1a). In the latter case, chromosomal abnormalities have never been found. DNA hypomethylation is also present in the nonsatellite repeats NBL2 on acrocentric chromosomes and D4Z4 in the subtelomeres of the long arms of chromosomes 4 and 10 (Jeanpierre et al. 1993; Kondo et al. 2000; Tuck-Muller et al. 2000). Additional hypomethylation, localized in the α-satellite repeats of the centromeres, is found only in a subset of patients (Miniou et al. 1997; Jiang et al. 2005) (Fig. 2.1c). This DNA hypomethylation is present in all analyzed cell types, but it gives rise to rearrangements only in lymphoblasts, probably playing a specific role in the onset of the immunologic phenotype (Jeanpierre et al. 1993).

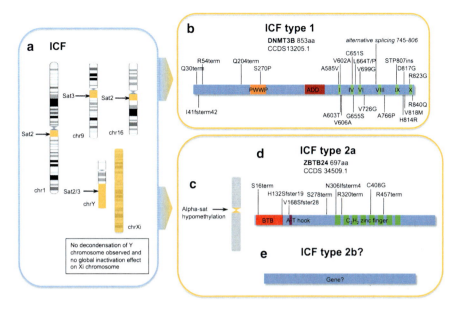

Fig. 2.1 ICF syndrome molecular features. (**a**) Hypomethylation of juxtacentromeric heterochromatin of chromosomes 1, 9, 16, and Y and of the inactive X chromosome. Regions of interest are marked in yellow. (**b**) Mutations in the DNMT3B gene causing ICF type 1. In green are the active sites of the catalytic domain. (**c**) Alpha satellite of centromeric heterochromatin is hypomethylated only in ICF type 2 on all chromosomes. (**d**) Mutations in ZBTB24 are mostly nonsense and represent the hallmark of ICF type 2a. (**e**) ICF type 2b has yet to be well characterized. It can be only defined as neither type 1 nor 2

These anomalies in DNA methylation and chromosomal instability could very easily suggest a correlation of ICF cells with cancer cells, where satellite heterochromatin hypomethylation has often been noticed (Ehrlich 2002), but in ICF patients, an onset of malignancies has been reported in very few cases. Two patients were affected by hematological malignancies, namely, myelodysplastic syndrome and Hodgkin lymphoma (Schuetz et al. 2007; Hagleitner et al. 2008), and more recently, angiosarcoma, a nonhematological malignancy, was reported in one ICF case (van den Brand et al. 2011). The reason why the epigenetic defects in ICF cells do not often lead to malignancies has not been addressed, but probably the young age at the death of those patients does not leave enough time for their organisms to develop cancer.

To complete the heterogeneous picture of the ICF phenotype, only some patients show facial anomalies as epicanthic folds, hypertelorism, low-set ears, and a flat nasal bridge. All the other symptoms have an even more reduced penetrance, being present only in few individuals. Mental retardation, neurological problems, and a delay of motor development are occasionally present, all in a highly variable range. Few ICF patients present congenital defects, hematological abnormalities, or malignancies (see Hagleitner et al. 2008 for a complete description of the range of phenotypes).

For these patients, the most common therapy consists in repeated intravenous infusions of immunoglobulins during their whole lifetime. Interestingly, in 2007, Gennery et al. succeeded in correcting the humoral and cellular immunological defect of three ICF patients through hematopoietic stem cell transplantation, encountering the development of autoimmune phenomena in two of them as only sequelae (Gennery et al. 2007). It is the only documented case about the restoration of a healthier immune condition and growth improvement for such individuals.

2.1.1.2   Genetics: ICF Type 1 and ICF Type 2

ICF syndrome is an extremely rare disease. So far, around 60 patients have been reported worldwide, and they have been classified in two distinct disease classes, ICF types 1 and 2, due to their genetic and epigenetic features (ICF1 and ICF2, Hansen et al. 1999; Jiang et al. 2005, Fig. 2.1).

Both classes present the same clinical phenotype, and until early 2011, their distinction criteria were the presence of mutations in the DNA methyltransferase 3B gene (DNMT3B) for ICF1 and hypomethylation of alpha satellites in centromeric heterochromatin for ICF2 patients (Jiang et al. 2005). Recently, de Greef et al. (2011) identified several mutations in the zinc-finger- and BTB (bric-a-bric, tramtrack, broad complex)-domain-containing 24 (ZBTB24) gene at 6q21 highly associated to ICF phenotype in seven out of eleven studied ICF2 patients, and another one has been found in three further ICF siblings (Chouery et al. 2011) (Fig. 2.1d). With this finding, a new interpretation of ICF2 is defined, assigning it a genetic etiology and splitting the ICF type 2 in two subcategories, where alpha satellite hypomethylation is present, but ZBTB24 can either be mutated or not, indicating that probably there could be another affected gene in a correlated pathway causing the same phenotype yet to be discovered (Fig. 2.1e).

ICF1 subjects present biallelic mutations in the *DNMT3B* gene at chromosomal locus 20q11.2, all leading to the hypofunctioning of the protein. Twenty three mutations have been reported until now, and they are listed in Fig. 2.1b (Jiang et al. 2005; Hagleitner et al. 2008). DNMT3B is a *de novo* DNA methyltransferase with multiple domains for DNA binding, for interacting with histones and regulatory proteins (ADD, PWWP, see below), and for transferring the methyl group from S-adenosyl methionine (SAM) to the cytosine in the CpG dinucleotide. Its mutations are mainly missense and mostly concentrated in the C-terminal portion where they partially affect the catalytic function of the protein. All the major mutations, like the nonsense ones, appear in the N-terminal regulatory part of the protein and are always found as compound heterozygous, as the complete loss of function in the homozygous state is probably incompatible with life, analogous to the situation in mice (see Sect. 1.2.3).

ZBTB24 (also known as ZNF450, BIF1, or PATZ2) is a member of the ZBTB family of transcriptional factors with a preeminent role in hematopoiesis. It has a highly conserved amino-terminal BTB/POZ dimerization domain, an AT-hook motif, and eight C_2H_2 zinc fingers, which are mediators of its functions, allowing ZBTB24 protein to dimerize, to interact with transcriptional repressor complexes

(BTB-POZ, C_2H_2 zinc fingers), and to bind DNA (AT hook) (Edgar et al. 2005; de Greef et al. 2011). Mutations of this protein in ICF2 are always biallelic and mostly nonsense, leading to the loss of function of the protein (Fig. 2.1d). Up to now, eight mutations have been identified, only one missense, and only two of ten mutated patients are compound heterozygous, with the rest being homozygous.

Both DNMT3B and ZBTB24 are ubiquitously expressed and apparently have different functions in the cell, but mutations in both lead to the same phenotype. The effects of DNMT3B mutations have been studied more in depth, and more information is available on their pathogenic effects, while, due to the only recent discovery of ZBTB24 mutations, their pathogenic mechanisms are still obscure.

2.1.2 DNMT3B Mutations: How Do They Produce the ICF Phenotype?

2.1.2.1 Setting and Maintenance of DNA Methylation in Mammals: A Novel View

In mammals, DNA methylation represents a key layer of the transmitted epigenetic information mostly correlated with transcriptional gene silencing. Cytosine methylation is required for embryonic development during which it plays a critical role in maintaining genomic integrity and regulating gene expression programs (Li 2002; Bird 2002; Mohn and Schubeler 2009). X chromosome inactivation, genomic imprinting, and the control of lineage specificity and pluripotency programs all represent processes for which proper DNA methylation is essential (Oda et al. 2006; Mohn et al. 2008; Borgel et al. 2010). Aberrant establishment of DNA methylation patterns is associated with several human disorders including chromatin diseases (Matarazzo et al. 2009), imprinting syndromes (Hirasawa and Feil 2010), psychiatric and neurodevelopmental defects, and immunological diseases (Portela and Esteller 2010). It also contributes both to the initiation and to the progression of various cancers (Jones 2002; Scarano et al. 2005).

In the mammalian genome, DNA methylation occurs predominantly at the CpG dinucleotides and only occasionally at non-CpG sites. However, only certain CpG sites are methylated, resulting in the generation of a tissue- and cell-type-specific pattern of methylation. Genomes of plants and fungi show a more extensive methylation at different sites, for example, at CHG and CHH sites, where H is A, C, or T (Vanyushin and Ashapkin 2011). CHG and CHH methylation has recently been found in human stem cells and seems to be enriched in gene bodies directly connected with gene expression and to be depleted in promoters and enhancers (Lister et al. 2009). The levels of non-CpG methylation decrease during differentiation and are restored in induced pluripotent stem cells, suggesting a key function in establishment and maintenance of the pluripotent state (Laurent et al. 2010).

Additionally, 5-hydroxymethylcytosine (5-hmC), which arises from the oxidation of the methyl group of 5-hmC, has recently been discovered in the mammalian genome (Kriaucionis and Heintz 2009; Tahiliani et al. 2009). So far, 5-hmC has been reported in Purkinje neurons and the brain, but it seems not to be present in cancer cells (Kriaucionis and Heintz 2009). Mechanisms and biological roles of non-CpG methylation and 5-hydroxymethylation remain unclear.

In the mammalian genome, CpGs are normally underrepresented, being usually quite rare. However, they can be found at a frequency closer to the statistical expectation in specific genomic regions, termed CpG islands (Gardiner-Garden and Frommer 1987). These are defined as regions of more than 200 bases with a $G+C$ content of at least 50% and a ratio of observed to statistically expected CpG frequencies of at least 0.6. CpG islands are found in promoter regions of about 70% of all human genes and are usually unmethylated in normal cells. However, about 6% of them become methylated in a tissue-specific program during early development or differentiation (Straussman et al. 2009).

DNA methylation does not occur exclusively at CpG islands. Regions of lower CpG density lying in close proximity (~2 kb) of CpG islands, defined as CpG island *shores,* are methylated when associated with transcriptional inactivation. Most of the tissue-specific DNA methylation seems to occur not at CpG islands but at CpG island shores (Doi et al. 2009) which are also conserved between human and mouse. Furthermore, 70% of the differentially methylated regions during reprogramming are associated with CpG island shores (Ji et al. 2010).

DNA methylation is less frequently correlated with permissive transcription, and in that case, it occurs at gene bodies. Gene body methylation is common in housekeeping genes (Hellman and Chess 2007), and it is thought to be related to elongation efficiency and prevention of spurious initiations of transcription (Zilberman et al. 2007).

DNA methylation and DNA methylation–related protein complexes not only control gene expression in *cis*, but also act in *trans*, being involved in nuclear architecture and influencing the establishment and the positioning of specific chromosome territories. The hypomethylated and aberrantly expressed *SYBL1* gene in ICF B cells shows a repositioning that alters its localization within the Xi and Y chromosome territories, indicating impaired regulatory properties of higher order nuclear organization (Matarazzo et al. 2007a).

The enzymes responsible for DNA methylation patterns are grouped in a family of cytosine C5-DNA methyltransferases (DNMTs) which act by transferring a methyl group from the universal methyl group donor, S-adenosyl-L-methionine (SAM), onto DNA (Bestor 2000; Jurkowska et al. 2011). In mammals, three enzymatically active members of the DNMT family have been reported (DNMT1, 3A, and 3B) and one related regulatory protein, DNMT3L, which lacks catalytic activity.

DNMT3A and DNMT3B have been considered as mainly devoted to the *de novo* methylation, being responsible for establishing the pattern of DNA methylation during embryonic development, whereas DNMT1, with preferential activity for hemimethylated DNA, acts mainly as maintenance methyltransferase. Null mutations of the three DNA methyltransferases are lethal in mice (Li et al. 1992; Okano et al. 1999), clearly demonstrating that DNA methylation is essential for mammalian survival (see Sect. 1.2.3).

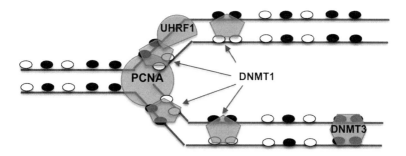

Fig. 2.2 Up-to-date model explaining the maintenance of DNA methylation patterns after replication. DNMT1 localizes at the replication fork, and its methyltransferase activity on hemimethylated cytosines is promoted through its interaction with PCNA and UHRF1 proteins. DNMT3 enzymes actively participate also in the maintenance process of heavily methylated regions, ensuring methylation at CpG sites which are missed by DNMT1

The *de novo* DNMTs are highly expressed in embryonic tissue and stem (ES) cells and become downregulated in differentiated cells (Esteller 2007). Both DNMT3A and DNMT3B are stably associated with chromatin containing methylated DNA (Jeong et al. 2009) and localize to pericentromeric heterochromatin (Hansen et al. 1999). DNMT3L is essential for the establishment of genomic imprints in oocytes and for the silencing of dispersed repeated sequences in male germ cells, despite being catalytically inactive (Bourc'his et al. 2001). Accordingly, DNMT3L is expressed during gametogenesis when genomic imprinting takes place and in embryonic stages (Bourc'his and Bestor 2004). It acts as a stimulatory factor for DNMT3A and DNMT3B and interacts with them, being co-localized in the nucleus (Chen et al. 2005; Holz-Schietinger and Reich 2010).

The maintenance methyltransferase, DNMT1, shows a 30- to 40-fold preference for hemimethylated DNA (Jeltsch 2006). It is the most abundant DNMT in the cell and is localized at DNA replication foci during the S phase of the cell cycle; it is mostly required to methylate hemimethylated sites that are produced during semiconservative DNA replication. However, it also has *de novo* DNMT activity; in this latter function, DNMT1 might support DNMT3A and DNMT3B by using hemimethylated CpG sites produced by the DNMT3 enzymes as substrates (Fatemi et al. 2002).

In the cell, the affinity of DNMT1 to newly replicated DNA is promoted by its interaction with a known component of the DNA replication machinery, the proliferating cell nuclear antigen (PCNA), serving as platform for tethering it to the replication fork (Chuang et al. 1997). The ubiquitin-like plant homeodomain (PHD) and RING finger domain–containing protein 1 (UHRF1) could perform a similar function, recruiting DNMT1 to hemimethylated DNA. Dnmt1 shows a strong preferential binding to hemimethylated CpGs due to its SET- and RING-associated (SRA) domain or the PHD (Bostick et al. 2007; Achour et al. 2008).

However, the distinction of functions between *de novo* and maintenance methylation is not always so clear, and several observations suggested an active involvement of DNMT3 enzymes in the preservation of DNA methylation after DNA replication, especially in densely methylated or repetitive sequences. Accordingly, a revised and updated model has recently been proposed (Fig. 2.2). This model still sustains the idea that the bulk of DNA methylation in replicating cells would be maintained by DNMT1 together with UHRF1 and PCNA. However, it also proposes that DNMT3A and DNMT3B which have been shown to anchor strongly to nucleosomes containing methylated DNA contribute to the maintenance of methylation at heterochromatic regions, *de novo* methylating the sites missed by DNMT1 at the replication fork (Jones and Liang 2009).

In addition, a fundamental problem of the classical model of maintenance methylation is that the inherent preference of DNMT1 for hemimethylated DNA over unmethylated CpG sites observed in vitro cannot be sufficient to copy a site-specific pattern of DNA methylation of the approximately 50 million CpG sites present in the human genome. As is also emerging from the genome-wide methylome studies, the novel view is that "maintenance DNA methylation" implies the preservation of average levels of DNA methylation at certain regions rather than the accurate copy of individual CpG sites. That would be sufficient to ensure the inheritance of the epigenetic information in a stable manner.

2.1.2.2 Dnmt3s Target Specificity: Is it a Matter of the Neighbors?

One of the most intriguing questions in the DNA methylation field is how the DNA methylation machinery is directed to target sequences in the genome and which is the specific contribution of DNMT3A and 3B in this process. The clearly distinct phenotypes of null mutant mice for the two enzymes demonstrate that they have partially nonoverlapping biological functions despite high sequence identity and related biochemical properties. Indeed, it has become clear that the methylation pattern of human DNA might at least in part be influenced by the flanking sequence preferences of *de novo* methyltransferases (Lin et al. 2002; Handa and Jeltsch 2005). DNMT3A displays strong preference for a CpG site surrounded by purine bases at the 5′-end of the CpG site, whereas pyrimidines are favored at its 3′-end (Handa and Jeltsch 2005). Additionally, it was shown that selection for high or low efficiency sites is mediated by the base composition at the −2 and +2 positions flanking the CpG site for DNMT3A and at the −1 and +1 positions for DNMT3B. This inherent preference reproducibly leads to the formation of specific methylation patterns characterized by up to 34-fold variations in the efficiency of DNA methylation at individual sites (Wienholz et al. 2010). All these results suggest that flanking sequence preference of DNMTs has contributed to the shaping of the CpG island composition in the human genome.

However, it is difficult to believe that DNA sequence alone could be sufficient to shape the human methylome because all cells share the same DNA sequence and they still carry specific methylation patterns. It seems possible that sequence preferences of DNMT3A and 3B might be important in the overall *de novo* methylation, whereas for locus-specific changes, other factors are dominant.

Besides sequence preferences, several mechanisms have been proposed to explain the targeting of DNMTs to specific genomic regions, mainly suggesting the interaction of DNMT3A and 3B with other epigenetic factors (Ooi et al. 2007; Tachibana et al. 2008; Jeong et al. 2009; Esteve et al. 2009).

Fig. 2.3 Models for the targeting of DNMT3s to specific genomic regions. (**a**) DNMT3s might be selectively attracted by specific histone marks, such as unmethylated H3K4, through interaction with additional protein domains. Alternatively, the recruitment of DNMT3 to specific sequences might be mediated by proteins recognizing and interacting with a consensus sequence in the genome, such as (**b**) transcription factors, or (**c**) with chromatin marks

One hypothesis is that the DNA methylation machinery might be attracted to selected genomic regions carrying specific histone marks (Fig. 2.3a). Recent data have reported that DNMTs can directly read histone modifications through their N-terminal domains and apparently could be recruited to the nucleosomes containing unmethylated H3K4 (Ooi et al. 2007; Otani et al. 2009; Zhang et al. 2010). Because methylation of H3K4 is a chromatin mark associated with transcribed genes, the absence of this modification in specific regions could be read as a signal for their inactivation, whereas its presence could reject DNA methyltransferases. Moreover, targeting of DNA methylation by H3K36me3 is consistent with many studies, indicating that this histone mark accumulates in the bodies of active genes (Vakoc et al. 2006; Barski et al. 2007). Indeed, it has been described that expressed genes show high DNA methylation levels in the gene bodies, whereas inactive genes do not (Weber et al. 2007; Ball et al. 2009). Besides, further results suggest that DNA methylation and H3K36 methylation might play a role in regulating the splicing, with exons having increased levels of both H3K36me3 and DNA methylation, compared to introns (Kolasinska-Zwierz et al. 2009).

In addition, delivery of DNMTs to target genes through the interaction with sequence-specific transcription factors or chromatin-interacting proteins has already been demonstrated in several examples (Fig. 2.3b, c). DNMT3A has been reported to interact with several transcription factors, such as PU.1 (Suzuki et al. 2006), Myc (Brenner et al. 2005), and p53 (Fuks et al. 2001; Wang et al. 2005). Additionally, the mammalian H3K9/H3K27 histone methyl transferase (HMT), G9a, is required for the recruitment of *de novo* DNMTs to gene promoters during mouse ES cells differentiation (Feldman et al. 2006), whereas EZH2 (enhancer of zeste homologue 2), an H3K27-specific HMT, is involved in the recruitment of DNMT3A and 3B in cancer cells (Vire et al. 2006). Finally, histone deacetylases (HDACs) and heterochromatin protein 1 (HP1) directly interact with DNMTs, and it has been suggested that they participate in the delivery of DNMTs to silenced chromatin regions (Fuks et al. 2003). Overall, this suggests that the targeting of DNMTs by DNA- or chromatin-binding proteins is a widespread and general mechanism for the generation of specific DNA methylation patterns within a cell.

Certain genomic regions could be protected from *de novo* methylation through the binding of other proteins or sequestering of *de novo* enzymes. In support of this model, a general reduction of DNA methylation at protein–DNA interaction sites has recently been observed in a methylome analysis (Lister et al. 2009). Chromatin remodeling might decide which genomic regions are subject to DNA methylation or demethylation because DNMTs are connected with the problem of accessibility of nucleosome-linked DNA which is particularly relevant for highly condensed heterochromatin. It has been shown that the methyltransferases activity on nucleosomal templates is much weaker than on naked DNA (Gowher et al. 2005), and that they preferentially methylate linker DNA sequences (Zhang et al. 2010). The support of ATP-dependent chromatin remodeling complexes is therefore required to make the methylation of previously condensed DNA efficient. Accordingly, DNMT3A is able to interact with members of chromatin remodeling complexes such as the SWI/SNF-BRM complex, the Mi-2/NuRD complex (Datta et al. 2005), and lymphoid-specific helicase (Lsh) (Zhu et al. 2006). However, the biological significance of these interactions remains unknown.

Moreover, siRNA-mediated, RNA-directed DNA methylation (RdDM) has also been thought to occur. In plants, RdDM is a stepwise process initiated by double-stranded RNAs recruiting DNMTs to specific regions including gene promoters and repetitive sequences (Matzke and Birchler 2005; Mosher and Melnyk 2010). Although some of the RdDM components are conserved in mammals, it is still unclear whether similar processes are involved in regulating DNA methylation in animals.

Whatever the mechanism, cooperation between DNMTs and factors that allow them to access specialized chromosomal regions might be particularly relevant for heterochromatic regions, which contain highly compacted chromatin.

2.1.2.3 Mouse Models for DNMT Deficiency

Knockout models of the Dnmt family members helped to elucidate their specific biological functions. Inactivation of the Dnmt1 gene results in extensive demethylation of genomic sequences even if a remaining DNA methylation (5–10%) is present. Mice lacking Dnmt1 are not viable and die before gastrulation (Li et al. 1992). Mouse embryonic stem cells deficient in Dnmt1, although viable, die when induced to differentiate, and even human cells lacking DNMT1 cannot survive, via apoptosis-related mechanisms (Brown and Robertson 2007). Given that Dnmt1 mutants are not viable, efforts have been made to study its function in specific tissues and organs. Fan and coworkers produced a conditional gene deletion of Dnmt1 in neural progenitor cells (NPCs), which results in DNA hypomethylation and precocious astroglial differentiation. The molecular effect is a developmentally regulated demethylation of astrocyte marker genes, as well as genes encoding the crucial components of the JAK-STAT pathway, with a role in gliogenesis (Fan et al. 2005).

Dnmt3a$^{-/-}$ mice are normal at birth and die around 4 weeks, whereas Dnmt3b$^{-/-}$ mice do not develop to term. Careful examination of mutant embryos reveals a normal development before E9.5 dpc, with embryonic lethality evident at E13.5 (Ueda et al. 2006). However, double mutants reveal more severe abnormalities, such as lack of somites and lack of development just after gastrulation (Okano et al. 1999). Results in null mutant mice support the hypothesis that ICF mutations do not cause a complete loss of function of both alleles because of their milder phenotype.

Two studies on mouse models carrying the same mutations in DNMT3B as in ICF patients show that this protein is indispensable for embryonic development. The defects of these mice mostly recapitulate the phenotypes observed in ICF syndrome.

A study from Ueda and coworkers analyzed molecular alterations and functions of different mutant Dnmt3b proteins, constructed by the introduction of specific mutations into mouse Dnmt3B cDNA: specifically, two mutants bearing specific mutations were produced. These mutants were viable, but the majority of them died within 24 h after birth (Ueda et al. 2006). ICF mice show facial anomalies similar to those observed in ICF patients: *DNMT3B* mutations lead to immune defects mainly due to apoptosis of thymocytes. It is worth to note that DNMT3B is not the only DNA methylation–related gene affecting lymphocyte survival: in fact, a similar role is suggested by the phenotype of $Lsh^{-/-}$ mice (Dennis et al. 2001).

A second Dnmt3B model has been recently derived (Velasco et al. 2010) with the production of compound heterozygous mice (mEx3/mEx24). A detailed morphological and molecular analysis unveiled a number of alterations, previously unrecognized. Notably, there are posterior axial transformations, which transform a thoracic vertebra into a lumbar one. These alterations often derive from the altered expression of Hox genes, a gene family representing the vertebrate counterpart of Drosophila homeotic genes. Accordingly, altered expression of Hoxa11 and Hoxa13 are detected in mEx3/mEx24 Dnmt3B mice (Velasco et al. 2010).

Microarray analysis in thymus of wild-type compared to mEx3/mEx24 mice revealed 25 deregulated genes, 17 of them expressed in testis. Among them, five are germ line specific and characterized by the presence of E2F6 binding sites (Velasco et al. 2010). Differently from genes analyzed in ICF syndrome (Jin et al. 2008; Gatto et al. 2010), all regulatory regions of these genes are hypomethylated in mutant mice. However, loss of methylation is not the result of an altered localization of Dnmt3B, but rather of its altered activity. ChIP and co-immunoprecipitation experiments revealed that E2F6 binding is crucial for the maintenance of DNA methylation and subsequent silencing of a subset of genes (Velasco et al. 2010). Noticeable is that $E2F6^{-/-}$ is characterized by similar homeotic transformations to those observed in mEx3/mEx24 Dnmt3B mutants (Storre et al. 2002).

The targeted disruption of the third member of the Dnmt3 family, DNA methyltransferase 3-like (Dnmt3L), caused azoospermia in homozygous males, and heterozygous progeny of homozygous females died before midgestation (Bourc'his et al. 2001). A defect in the establishment of the germ line–specific methylation imprints was first observed in these mice. Despite lacking methyltransferase activity, Dnmt3L is essential for the establishment of maternal methylation imprints during oogenesis probably by stimulation of Dnmt3A. The defect was specific to imprinted regions, and global genome methylation levels were not affected (Bourc'his et al. 2001). Recent reports indicate that it has a similar role in the establishment of paternal imprints.

A more complete view of the effects of Dnmt deletions on genomic imprinting has been reached with the use of conditional knockout experiments. Germ line–specific gene knockout studies showed that Dnmt3A is essential for the establishment of the maternal methylation imprints C57BL/6J (Kaneda et al. 2004). The conditional disruption of Dnmt3A and Dnmt3B in germ cells, and their preservation in somatic cells, led to very different phenotypes between these two genes. Offspring from Dnmt3A conditional mutant females die in utero and lack methylation and allele-specific expression at all maternally imprinted loci examined. Dnmt3A conditional mutant males show impaired spermatogenesis and lack methylation at two of three paternally imprinted loci examined in spermatogonia. It is worth to note that the phenotype of Dnmt3A conditional mutants is almost indistinguishable from that of Dnmt3L knockout mice. By contrast, Dnmt3B conditional mutants and their offspring show no apparent phenotype.

The relationships between the various Dnmts have been analyzed through multiple DNMT knockout mice. Single inactivated Dnmt3A or 3B mutants still retain DNA methylation activities capable of *de novo* methylation of specific substrates, such as infecting retroviruses (Okano et al. 1999). However, double mutants completely lack *de novo* methylation activity, as evaluated by the analysis of endogenous C-type retrovirus and iAP particles: methylation of these elements is normal in DNMT3A$^{-/-}$ mice, slightly reduced in Dnmt3B$^{-/-}$ mice, but highly undermethylated in the double Dnmt3A$^{-/-}$, Dnmt3B$^{-/-}$ mice. These latter results demonstrate (1) the absolute need of these genes for *de novo* methylation and (2) the redundant activities of Dnmt3A and 3B.

Dnmt1 and Dnmt3A are expressed in postmitotic neurons, but their function in the central nervous system is unclear. Fan and coworkers (Feng et al. 2010) revealed that only their forebrain-specific double knockout (DKO) has effects on specific functions, such as deficits in learning and memory. DKO neurons showed deregulated expression of genes that are known to contribute to synaptic plasticity. A significant decrease in DNA methylation is observed in neurons from double mutated animals.

The recently produced triple KO (Dnmt1, 3A, and 3B mutant; TKO) unveiled unknown aspects about the role of DNA methylation in development (Sakaue et al. 2010). This chapter suggests that the cell response to epigenomic stress, such as the complete absence of DNA methylation, is quite different in different tissues. In fact, while very few TKO cells have been found in the proper embryo, they can be found in extraembryonic tissues, such as placenta and yolk sac, thus indicating that these latter tissues can survive and proliferate even in the absence of DNA methylation.

2.1.2.4 Functional Domains and Enzymatic Properties of DNMT3 Enzymes

The DNMT3 enzymes contain an N-terminal regulatory region and a C-terminal catalytic domain harboring the conserved methyltransferase motifs. The N-terminal part contains two domains: a cysteine-rich region, referred to as an ATRX/DNMT3/

DNMT3L (ADD) domain, also known as the PHD, and a PWWP domain (proline–tryptophan motif).

The ADD domain is a zinc-finger domain that constitutes a platform mediating the interaction of DNMT3 enzymes with many proteins, such as transcription factors and histone-modifying proteins. Additionally, the ADD domains of DNMT3 enzymes have recently been shown to interact specifically with the N-terminal portion of histone H3 tails unmodified at lysine 4, the binding being disrupted by the methylation of H3K4 (Ooi et al. 2007; Otani et al. 2009; Zhang et al. 2010).

The PWWP domain of DNMT3A and DNMT3B is a scarcely conserved region of roughly 150 amino acids, including a conserved proline–tryptophan motif. The PWWP domains of DNMT3A and DNMT3B are required for the targeting of the methyltransferase to pericentromeric heterochromatin, (Bachman et al. 2001; Chen et al. 2004), although the mechanism of this targeting remains unknown. It has been recently shown that the PWWP domain of DNMT3A specifically recognizes the H3K36me3 mark and that this interaction increases the activity of DNMT3A to methylate nucleosomal DNA, suggesting that it can contribute to the targeting of DNA methylation by DNMT3A.

The catalytic domains of DNMT3A and 3B share more than 80% sequence identity and contain several highly conserved motifs important for their enzymatic catalysis (motifs IV and VI), DNA binding (motif IX), and S-adenosylmethionine cofactor binding (motifs I and X) (Jurkowska et al. 2011).

Recent studies indicate that the formation of homo- and hetero-oligomers is a critical feature of the regulation of the DNMT3s (recently reviewed by Jurkowska et al. 2011). DNMT3A forms large homo-oligomers due to the presence of two nonequivalent interaction surfaces in the catalytic domain, one hydrophobic FF interface (characterized by the interaction of two phenylalanine residues) and one polar RD interface (characterized by a hydrogen bonding network between arginine and aspartate residues; Fig. 2.4a) (Jia et al. 2007; Cheng and Blumenthal 2008). In fact, a hetero-homo-dimer consists of two DNMT3L molecules at the edges and two DNMT3A molecules in the center of the tetramer. The FF interface creates the DNMT3A/3L contact. In contrast, the RD interface mediates the central 3A/3A interaction in the 3A/3L tetramer and creates the DNA binding site. The interaction of 3A with 3L through the FF interface presumably influences the structure of DNMT3A, directly intervening at the key catalytic or SAM binding residues, which might explain the stimulatory effect that DNMT3L exerts on DNMT3A function. The hetero-homo-dimer structure of the DNMT3A/3L complex was also verified in solution (Jurkowska et al. 2008).

Since DNMT3A and DNMT3B catalytic domains are conserved, DNMT3B likely exhibits two similar interfaces, one of them interacting with DNMT3L, thereby accounting for the large soluble complexes, in which the wild-type protein elutes.

2.1.2.5 Mutations Perturbing DNMT3B Function at Multiple Levels

Mutations in the human DNMT3B, causing ICF type 1 (called ICF for simplicity from here on) syndrome, are mostly scattered across the catalytic domain of DNMT3B,

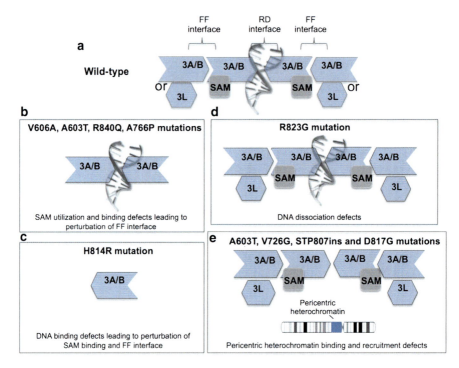

Fig. 2.4 Proposed models representing biochemical and localization defects in *DNMT3B*-mediated methylation. (**a**) Wild-type homo- or heterotetrameric complex of DNMT3A/B and the stimulating protein DNMT3L interacting through the FF interface. The RD interface mediates the self-oligomerization. (**b**) ICF variants deficient in SAM utilization and binding, proposed to compromise FF interface through internal destabilization, thereby accounting for the frequently observed dimeric form of these variants. (**c**) ICF variant whose mutated residue is predicted to be located at interface RD, and proposed to perturb the DNA binding, that is then translated into a lack of SAM binding and into a resulting destabilization of the FF interface. This hypothesis would explain the fact that this variant is mainly found as a monomer. (**d**) ICF variant with reduced ability to dissociate from DNA and to scan for new target sites. (**e**) ICF variants showing perturbation of protein localization to the pericentromeric heterochromatin within the nucleus

suggesting that they impair the catalytic function of the protein (Hagleitner et al. 2008). Indeed, most ICF mutants studied show significantly reduced activity that is less than 10% of wild-type levels (Gowher and Jeltsch 2002; Xie et al. 2006). However, the precise nature of the catalytic mechanism perturbed by these mutations has been unknown for a long time.

More recently, specific ICF mutations provide the scientist with a tool to study the catalytic mechanism of DNMT3B in vitro and, in particular, to characterize residues involved in various aspects of the reaction, including DNA dissociation, SAM binding, catalytic transfer, and the regulation of protein oligomerization (Moarefi and Chedin 2011). Six ICF-associated DNMT3B variants, corresponding to V606A, A766P, H814R, R823G, R840Q, and C651S are shown to cause a broad spectrum of biochemical defects in DNMT3B function. On the other hand, the results revealed that catalysis by DNMT3B is much more complex than previously expected.

In fact, it appeared that DNMT3B is distinct from DNMT3A in that it has a specific and restricted substrate-binding order, requiring the presence of DNA to achieve a stable SAM binding and methyl transfer. This finding suggests that, while highly homologous to each other, both enzymes might act in a different manner.

Nevertheless, the identification of the crystal structure of the DNMT3A:DNMT3L complex allowed to make more precise predictions on the impact that specific ICF variants produce on DNMT3B catalytic function. Indeed, the R840Q and A766P mutations were previously suggested to make the protein unable to interact with and to be stimulated by DNMT3L (Xie et al. 2006). Instead, located far away from the DNMT3L interaction surface and unlikely to perturb it, they appear to be defective in SAM utilization and binding and in undergoing normal stimulation (Fig. 2.4b). Therefore, all the studied ICF variants to date appear to impair DNMT3B methyltransferase activity but do not affect DNMT3L stimulation. This indicates that DNMT3L is probably not involved in the ICF pathogenesis which is consistent with the fact that DNMT3B and DNMT3L mouse knockouts show nonoverlapping phenotypes (Okano et al. 1999; Bourc'his et al. 2001; Kaneda et al. 2004).

Furthermore, the study of the three variants V606A, A603T, and H814R provided evidence for the functional coupling of the SAM binding and oligomerization phases of the reaction. Interestingly, the SAM binding–defective mutants (V606A and A603T) show oligomerization defects, although the protein–protein interface is unaffected (Fig. 2.4b). Similarly, the H814R variant which is supposed to reduce the formation of homo-oligomers being located at interface RD is unable to bind SAM. It is interesting to note that H814R variant proteins exist mainly as monomer, indicating that the FF interface is also perturbed (Fig. 2.4c).

A proposed network of stabilizing polar interactions associating the SAM molecule bound to the FF interface may provide a mechanistic link between the otherwise distant SAM binding pocket and the interaction surface. In the case of the H814R mutant, it has been suggested that the compromised RD interface would cause an incapacity to bind SAM and this would result in the further destabilization of the FF interface, thereby explaining the mostly monomeric elution profile of the H814R.

In that context, it is surprising that several DNMT3B variants with a destabilized FF interface still appear to undergo normal stimulation by DNMT3L, as demonstrated by the fold stimulation. This indicates that DNMT3L may not need to be involved in extended protein–protein interactions to provide its effects or, alternatively, that its interaction mode is such that it can suit a weakened FF interface.

DNA dissociation and scanning for a new target site are also events that may be disturbed in the mutated DNMT3B protein (Fig. 2.4d). For instance, the R823G variant is catalytically active, and the initial rates of methylation at early phases appear almost identical with that of the wild-type protein. However, the R823G variant becomes catalytically impaired later after several turnovers. A slightly increased DNA binding pattern compared to that of the wild-type protein would be consistent with the proposed DNA dissociation defect.

Overall, these results provide new insights into the manner by which oligomerization, SAM binding, and DNA binding are coupled and contribute to the regulation of DNMT3B catalytic activity. Given the conservation of sequences between DNMT3A and DNMT3B and considering similarities in the behavior of known interface mutants between both enzymes, it is likely that such coupling is a conserved feature of the DNMT3 family of enzymes.

In this light, other ICF mutations could reduce the enzymatic activity of DNMT3B in several alternative ways: perturbing overall stability (V699G and V726G), altering configuration of the active-site loop (G663S and L664T) directly or indirectly by affecting the interface with DNMT3L, impairing binding with the methyl donor SAM (A585V), or affecting the proposed RD interface (V818M, D817G, and H814R).

Within the six conserved catalytic motifs, the motif IV invariably includes a proline–cysteine dipeptide. The conserved cysteine functions as the nucleophilic thiolate responsible for attacking the carbon 6 position of the pyrimidine ring, thus forming a covalent methyltransferase DNA intermediate. It seems that this cysteine is absolutely critical for catalytic function. Indeed, its substitution with a serine (C651S) makes the mutated protein completely unable to catalyze the transfer of methyl groups independently of DNA, SAM concentrations, and the presence or absence of DNMT3L.

Some loss-of-function mutations affect DNMT3B activity via different mechanisms: some by altering protein localization and others by disrupting protein–protein interactions. The ICF mouse model demonstrated that four of the ICF mutations, A609T, V732G, STP813, and D823G (corresponding to the A603T, V726G, STP807ins, and D817G human variants), exhibit changes in localization patterns when compared with wild-type DNMT3b (Fig. 2.4e).

A609T does not show accumulation in pericentromeric heterochromatin anymore, suggesting that DNMT3b hetero- or homodimerization may be required for targeting DNMT3b to pericentric heterochromatin. V732G, STP813, and D823G show diffuse patterns in the nucleus, indicating that these mutations impair the association of DNMT3b with nuclear foci. Although these foci remain not clearly identified, it is possible that these structures represent heterochromatin regions, which usually contain repetitive DNA sequences, including satellite repeats. Failure to target DNMT3b to heterochromatin may thus contribute to demethylation of satellite DNA.

Interesting relationships between DNA methyltransferases and protein mediating chromatin condensation have been reported. Thus, it is reasonable to assume that some mutations may directly or indirectly affect these interactions.

2.1.2.6 Epigenetic Network: Relationship of DNMT3B Perturbation to the ICF Phenotype

Since *de novo* DNMT3B has been pointed out to be the causative gene for the ICF syndrome some years ago, many efforts have been spent in dissecting the pathogenesis of this syndrome starting from the loss of DNA methylation. In ICF1 patients with impaired DNMT3B function, no global decrease of methylation is observed,

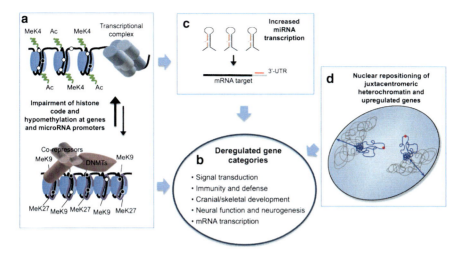

Fig. 2.5 Insights into ICF syndrome pathogenesis. (**a**) Histone modification machinery and DNA methylation in an open (*top*) and in a closed conformation (*bottom*). Green waves stand for histone H3 tail acetylated or trimethylated at K4. *Purple lines* stand for histone H3 tail trimethylated in K9 or K27. *White* and *black circles* are unmethylated and methylated CpGs. In ICF cells, some genes pass from the "closed" to the "open" conformation due to altered DNA methylation and histone code. (**b**) Functional categories enriched in deregulated genes in ICF B cells. (**c**) Epigenetic mechanisms controlling transcription of microRNAs are deregulated in ICF syndrome affecting the regulation of their downstream targets. (**d**) A cell nucleus is represented. Chr1 is in gray and juxtacentromeric chromatin is the blue mass that, in ICF, relocalizes toward the inner part of the nucleus, marked by *blue arrows*. The red dots represent CNN3 and RGS1 genes that move far from the juxtacentromeric chromatin to be expressed in ICF cells

but an abnormal methylation pattern has been proposed for 7% of the genome, specifically in constitutive and facultative heterochromatin (Miniou et al. 1994). This would suggest that the overall level of 5-methylcytosine seems to be only slightly reduced in ICF cells. However, it cannot be ruled out that the approach employed to examine overall levels of DNA methylation in the genome at low resolution has failed to reveal subtle differences in local DNA methylation patterns.

Classical satellites in the juxtacentromeric chromatin of chromosomes 1, 16, 9, and Y are significantly hypomethylated, together with the whole inactivated X chromosome (Xi) and other nonsatellite repeats, like D4Z4 and NBL2. Moreover, the DNA methylation at a limited number of specific CpG islands (Jin et al. 2008; Gatto et al. 2010) and in active gene bodies (Aran et al. 2011) looks impaired (Fig. 2.5a).

The fact that the DNMT3B function in ICF cells is only impaired and not completely disrupted or deleted might account for the limited DNA hypomethylation effect. The partial overlap with the other DNMTs, 1 and 3A, with whom 3B interacts physically (Kim et al. 2002; Chen et al. 2003), might also play a role.

Not all the hypomethylated regions, though, have been directly correlated to a straight effect in the cell, for example, on gene functions or chromatin structure. Actually, juxtacentromeric chromatin of chromosomes 1 and 16 (sometimes 9, but never Y) is decondensed only in mitogen-stimulated lymphocytes but not in other tissues (Jeanpierre et al. 1993), leading to the formation of multiradial structures, micronuclei (Stacey et al. 1995), bridges, and breakages. The formation of such structures might be due to the defective interaction of Dnmt3B with several components of the condensin complex, KIF4A (chromokinesin homolog), and CENP-C (constitutive centromere protein) due to its mutations in ICF syndrome. DNA hypomethylation in juxtacentromeric chromatin in chromosomes 1 and 16, full of large sites of repetitive DNA, probably drives this process, impairing DNMT3B ability to stabilize those proteins during the onset of mitosis in pericentromeric chromatin (Geiman et al. 2004a; Gopalakrishnan et al. 2009).

Conversely, chromosomes Xi and Y do not show structural or functional anomalies as it would be expected from the general loss of methylation observed, but the inactivation process is conserved (Hansen 2003), and the global profile of histone modifications remains unchanged (Gartler et al. 2004). Two X-linked genes, SYBL1 and G6PD, escape the normal process of X inactivation in ICF cells and are biallelically expressed due to the loss of promoter CpG island methylation and to the increase of activating histone modifications, like acetylated H3 and H4. SYBL1, belonging to the pseudoautosomal region 2 (PAR2), becomes biallelically expressed also in male cells (Hansen et al. 2000). Also methyl-binding proteins and Polycomb complex PCR2 are involved in the maintenance of long-term silencing of X- and Y-inactivated alleles of the pseudoautosomal gene SYBL1 (Matarazzo et al. 2007b), and mutations in DNMT3B disturb the repressive activity of those proteins. Another gene from the same region, SPRY3, has instead been shown to be unaffected in response to DNA methylation changes and being dependent on chromatin remodeling proteins binding and histone modifications (De Bonis et al. 2006).

In addition, the subtelomeric regions of most chromosomes in ICF lymphoblastoid and fibroblast cells show heterogeneous hypomethylation, and the telomeres are abnormally short. These anomalies have been associated to advanced replication timing and increased transcription of telomeric-repeat-containing RNA from telomeres (TERRA or TelRNA) (Yehezkel et al. 2008). Negative loop mediated by these molecules which normally would act to reinforce the telomeric heterochromatin fails in ICF cells (Deng et al. 2010).

Not only telomeric RNA and Xi gene transcription is deregulated in ICF cells, but also several genes and regulatory RNAs are differentially expressed in mutated immortalized B cells (Ehrlich et al. 2001; Jin et al. 2008; Gatto et al. 2010). Many of these genes are importantly involved in immune functions and in the development of the neurological system, the two mostly impaired pathways in ICF (Fig. 2.5b). Around 75% of dysregulated genes are associated to a CpG island, and their promoters are hypomethylated only in half of them, specifically in cases where they are already scarcely methylated in normal cells (Jin et al. 2008). Also, miRNA expression is disrupted in ICF cells and they similarly present associated CpG islands in 70% of cases. CpG islands associated with upregulated miRNAs, however, never show any differences in DNA methylation, but their differential expression in most cases has an opposite correlation with the expression of their predicted targets (Fig. 2.5c) (Gatto et al. 2010). In any case, though, both upregulated genes and miRNAs show an impairment of histone modification patterns, with reduction of silencing markers like H3K27me3 and H3K9me3 and an increase of transcriptional activators, such as H3K4me3 and H4ac. This is according with the fact that DNMT3B activity is crucial not only for its methyltransferase activity but also for its interactions with other chromatin remodeling proteins which are not effective in the presence of ICF mutations (Fig. 2.5a) (Ehrlich et al. 2001; Jin et al. 2008; Gatto et al. 2010). In fact, DNMT3B associates with four chromatin-associated enzymatic activities common to transcriptionally repressed, heterochromatic regions of the genome: DNA methyltransferase, HDAC, ATPase, and histone methyl transferase (HMT) activities. DNMT3B co-localizes in the heterochromatin and interacts with HDACs 1 and 2, the corepressor SIN3A, the ATP-dependent chromatin remodeling enzyme hSNF2H, HP1 proteins, and Suv39h1 (Geiman et al. 2004b). Particularly, the PcG protein EZH2 (Enhancer of Zeste homolog 2) in the Polycomb repressive complexes 2 and 3 (PRC2/3) interacts with all the DNMTs and is required as a recruitment platform for their proper methyltransferase activity (Geiman et al. 2004a). The so far analyzed ICF mutations in DNMT3B do not seem to affect protein–protein interactions between DNMT3B and HDAC, ATPase, and HMT in enzymatic assays in vitro, leaving all their enzymatic activities unaltered (Geiman et al. 2004b). Anyway, it is known that SUZ12 and EZH2 binding is lost or impaired on promoters of upregulated genes (Jin et al. 2008) and on the promoter of one inactivation-escaping gene (SYBL1) in ICF cells (Matarazzo et al. 2002), showing the existence of a functional connection between Dnmts and the Polycomb group complex. Moreover, in most ICF G2 nuclei, HP1 proteins are concentrated and co-localize with the hypomethylated satellite heterochromatin of chromosomes 1qh and 16qh forming an aberrant giant nuclear body (NB) together with SP100, SUMO-1, and other proteins from the promyelocytic leukemia (PML) nuclear bodies. This bright focus of HP1 protein could reflect the difficulty to condense rapidly the heterochromatin during the transition from G2 to M phase, as a consequence of the hypomethylation of satellite sequences, leaving the complex accumulating at those sites (Luciani et al. 2005) and leading to chromosome decondensation and formation of the aberrant structures in ICF lymphoblasts.

Furthermore, heterochromatin from chromosome 1 in interphase ICF nuclei (both ICF type 1 and 2) looks to be much more concentrated in a dense spot compared to control nuclei and repositioned in a more internal location compared to the external nuclear rim (Fig. 2.5d). The same alteration is not found in chromosomes 16 and 9 (Jefferson et al. 2010; Dupont et al. 2011). Besides that, classical satellite repeats are reduced in number in two ICF patients compared to related controls as proposed also from (Luciani et al. 2005), but this difference does not correlate directly with heterochromatin configuration (Jefferson et al. 2010). Probably, DNA demethylation, protein assembly, and satellite length polymorphisms all concur to the development of the phenotype through a concerted action.

Thus, nuclear architecture appears to be crucial in this pathology as one of the potential driving mechanisms for gene deregulation and cellular reorganization. In fact, it has been observed that the PAR2 genes on the Xi and Y chromosomes escaping inactivation in ICF cells loop out of their own chromosome territories (CTs), like transcriptionally active genes normally do (Matarazzo et al. 2007a). This finding supports the possibility of an impaired regulatory effect in *trans* of higher order nuclear organization. The altered CT organization that encompasses much larger regions than the restricted hypomethylated area could explain the failure to detect DNA hypomethylation of genes despite their dysregulated expression in this disease. Intriguingly, it has been demonstrated also that CNN3 and RGS1, two upregulated genes from chromosome 1 in ICF cells (type 1 and 2), display no change in DNA methylation in their promoters; are less associated to juxtacentromeric heterochromatin of chromosome 1 in ICF, unlike normal cells; and are in general more frequently associated with the nuclear periphery compared to the other analyzed genes. This could mean that those genes are normally repressed by the trans-action of chromosome 1 heterochromatin, but in ICF cells, this tridimensional regulation is missing, thereby causing their repositioning into the nuclear center and their abnormal expression (Fig. 2.5d).

As a conclusion, we can affirm that there are probably several interconnected mechanisms leading to the immunodeficiency through impairment of gene expression in B and T cells. First of all, gene promoters are regulated by DNA methylation, histone modifications, miRNA targeting, and binding of other regulatory proteins, like chromatin remodeling proteins. All these fine-tuners are altered subsequently to DNMT3B mutations. Besides that, hypomethylation of juxtacentromeric heterochromatin has been proven to alter the nuclear organization of the chromosomes and affect *in trans* the expression of such regulators or key genes and miRNAs through delocalization, altered binding of chromatin complexes, and length polymorphisms of repetitive satellites.

2.2 Conclusions and Perspectives

The number of human diseases resulting from mutations in the component of chromatin or in enzymes that modify chromatin structure is considerable. The difficulty in approaching this type of pathologies is that genes with affected expression are not linked to the underlying mutation within an obvious impaired molecular pathway. This is because the epigenetic regulatory network is complex and the levels at which chromatin structure can be modified are numerous.

ICF syndrome can be a highly informative model for this class of human genetic disorders, not only for those showing inheritance of aberrant patterns of genomic DNA methylation, such as the X-linked alpha-thalassemia/mental retardation syndrome (ATR-X) and the facioscapulohumeral muscular dystrophy (FSHD). Since DNMT3B, with its ability to recruit several chromatin proteins which are partners of methyl-CpG binding proteins, plays a central role in the DNA methylation–mediated control of the gene expression, its study may provide useful insights on Rett syndrome (mutated in MeCP2).

Understanding these diseases as disorders of chromatin will inform us either about chromatin-based regulatory mechanisms or about ways to design rational approaches aiming to treat these disorders.

Together with transcriptomics and proteomics, large-scale analysis of genomic DNA methylation (Brown et al. 2007; Dunn et al. 2007) is likely to play increasingly essential roles in understanding normal and abnormal human development and physiology. Indeed, it will be helpful in contributing to the deciphering of DNA methylation disturbances in ICF cells.

The experimental investigation of DNMTs is providing a promptly growing picture of catalytic mechanisms, interacting proteins, and chromatin target sites. However, understanding the molecular basis for establishing, maintaining, and perturbing DNA methylation patterns will require a much better understanding of the connection between structure and function in the DNMT proteins than is currently in our hands.

References

Achour, M., Jacq, X., Ronde, P., Alhosin, M., Charlot, C., Chataigneau, T., Jeanblanc, M., Macaluso, M., Giordano, A., Hughes, A. D., Schini-Kerth, V. B. & Bronner, C. (2008). The interaction of the SRA domain of ICBP90 with a novel domain of DNMT1 is involved in the regulation of VEGF gene expression. *Oncogene* 27, 2187–2197.

Aran, D., Toperoff, G., Rosenberg, M. & Hellman, A. (2011). Replication timing-related and gene body-specific methylation of active human genes. *Hum Mol Genet* 20, 670–680.

Bachman, K. E., Rountree, M. R. & Baylin, S. B. (2001). Dnmt3a and Dnmt3b are transcriptional repressors that exhibit unique localization properties to heterochromatin. *J Biol Chem* 276, 32282–32287.

Ball, M. P., Li, J. B., Gao, Y., Lee, J. H., LeProust, E. M., Park, I. H., Xie, B., Daley, G. Q. & Church, G. M. (2009). Targeted and genome-scale strategies reveal gene-body methylation signatures in human cells. *Nat Biotechnol* 27, 361–368.

Barski, A., Cuddapah, S., Cui, K., Roh, T. Y., Schones, D. E., Wang, Z., Wei, G., Chepelev, I. & Zhao, K. (2007). High-resolution profiling of histone methylations in the human genome. *Cell* 129, 823–837.

Bestor, T. H. (2000). The DNA methyltransferases of mammals. *Hum Mol Genet* 9, 2395–2402.

Bird, A. (2002). DNA methylation patterns and epigenetic memory. *Genes Dev* 16, 6–21.

Blanco-Betancourt, C. E., Moncla, A., Milili, M., Jiang, Y. L., Viegas-Pequignot, E. M., Roquelaure, B., Thuret, I. & Schiff, C. (2004). Defective B-cell-negative selection and terminal differentiation in the ICF syndrome. *Blood* 103, 2683–2690.

Borgel, J., Guibert, S., Li, Y., Chiba, H., Schubeler, D., Sasaki, H., Forne, T. & Weber, M. (2010). Targets and dynamics of promoter DNA methylation during early mouse development. *Nat Genet* 42, 1093–1100.

Bostick, M., Kim, J. K., Esteve, P. O., Clark, A., Pradhan, S. & Jacobsen, S. E. (2007). UHRF1 plays a role in maintaining DNA methylation in mammalian cells. *Science* 317, 1760–1764.

Bourc'his, D. & Bestor, T. H. (2004). Meiotic catastrophe and retrotransposon reactivation in male germ cells lacking Dnmt3L. *Nature* 431, 96–99.

Bourc'his, D., Xu, G. L., Lin, C. S., Bollman, B. & Bestor, T. H. (2001). Dnmt3L and the establishment of maternal genomic imprints. *Science* 294, 2536–2539.

Brenner, C., Deplus, R., Didelot, C., Loriot, A., Vire, E., De Smet, C., Gutierrez, A., Danovi, D., Bernard, D., Boon, T., Pelicci, P. G., Amati, B., Kouzarides, T., de Launoit, Y., Di Croce, L. & Fuks, F. (2005). Myc represses transcription through recruitment of DNA methyltransferase corepressor. *EMBO J* 24, 336–346.

Brown, K. D. & Robertson, K. D. (2007). DNMT1 knockout delivers a strong blow to genome stability and cell viability. *Nat Genet* 39, 289–290.

Brown, S. E., Fraga, M. F., Weaver, I. C., Berdasco, M. & Szyf, M. (2007). Variations in DNA methylation patterns during the cell cycle of HeLa cells. *Epigenetics* 2, 54–65.

Carpenter, N. J., Filipovich, A., Blaese, R. M., Carey, T. L. & Berkel, A. I. (1988). Variable immunodeficiency with abnormal condensation of the heterochromatin of chromosomes 1, 9, and 16. *J Pediatr* 112, 757–760.

Chen, T., Tsujimoto, N. & Li, E. (2004). The PWWP domain of Dnmt3a and Dnmt3b is required for directing DNA methylation to the major satellite repeats at pericentric heterochromatin. *Mol Cell Biol* 24, 9048–9058.

Chen, T., Ueda, Y., Dodge, J. E., Wang, Z. & Li, E. (2003). Establishment and maintenance of genomic methylation patterns in mouse embryonic stem cells by Dnmt3a and Dnmt3b. *Mol Cell Biol* 23, 5594–5605.

Chen, Z. X., Mann, J. R., Hsieh, C. L., Riggs, A. D. & Chedin, F. (2005). Physical and functional interactions between the human DNMT3L protein and members of the de novo methyltransferase family. *J Cell Biochem* 95, 902–917.

Cheng, X. & Blumenthal, R. M. (2008). Mammalian DNA methyltransferases: a structural perspective. *Structure* 16, 341–350.

Chouery, E., Abou-Ghoch, J., Corbani, S., El Ali, N., Korban, R., Salem, N., Castro, C., Klayme, S., Azoury-Abou, R. M., Khoury-Matar, R., Debo, G., Germanos-Haddad, M., Delague, V., Lefranc, G. & Megarbane, A. (2011). A novel deletion in ZBTB24 in a Lebanese family with immunodeficiency, centromeric instability, and facial anomalies syndrome type 2. *Clin Genet.* doi: 10.1111/j.1399-0004.2011.01783.x.

Chuang, L. S., Ian, H. I., Koh, T. W., Ng, H. H., Xu, G. & Li, B. F. (1997). Human DNA-(cytosine-5) methyltransferase-PCNA complex as a target for p21WAF1. *Science* 277, 1996–2000.

Datta, J., Majumder, S., Bai, S., Ghoshal, K., Kutay, H., Smith, D. S., Crabb, J. W. & Jacob, S. T. (2005). Physical and functional interaction of DNA methyltransferase 3A with Mbd3 and Brg1 in mouse lymphosarcoma cells. *Cancer Res* 65, 10891–10900.

De Bonis, M. L., Cerase, A., Matarazzo, M. R., Ferraro, M., Strazzullo, M., Hansen, R. S., Chiurazzi, P., Neri, G. & D'Esposito, M. (2006). Maintenance of X- and Y-inactivation of the pseudoautosomal (PAR2) gene SPRY3 is independent from DNA methylation and associated to multiple layers of epigenetic modifications. *Hum Mol Genet* 15, 1123–1132.

de Greef, J. C., Wang, J., Balog, J., den Dunnen, J. T., Frants, R. R., Straasheijm, K. R., Aytekin, C., van der, B. M., Duprez, L., Ferster, A., Gennery, A. R., Gimelli, G., Reisli, I., Schuetz, C., Schulz, A., Smeets, D. F., Sznajer, Y., Wijmenga, C., van Eggermond, M. C., van Ostaijen-Ten Dam MM, Lankester, A. C., van Tol, M. J., van den Elsen, P. J., Weemaes, C. M. & van der Maarel, S. M. (2011). Mutations in ZBTB24 are associated with immunodeficiency, centromeric instability, and facial anomalies syndrome type 2. *Am J Hum Genet* 88, 796–804.

Deng, Z., Campbell, A. E. & Lieberman, P. M. (2010). TERRA, CpG methylation and telomere heterochromatin: lessons from ICF syndrome cells. *Cell Cycle* 9, 69–74.

Dennis, K., Fan, T., Geiman, T., Yan, Q. & Muegge, K. (2001). Lsh, a member of the SNF2 family, is required for genome-wide methylation. *Genes Dev* 15, 2940–2944.

Doi, A., Park, I. H., Wen, B., Murakami, P., Aryee, M. J., Irizarry, R., Herb, B., Ladd-Acosta, C., Rho, J., Loewer, S., Miller, J., Schlaeger, T., Daley, G. Q. & Feinberg, A. P. (2009). Differential methylation of tissue- and cancer-specific CpG island shores distinguishes human induced pluripotent stem cells, embryonic stem cells and fibroblasts. *Nat Genet* 41, 1350–1353.

Dunn, J. J., McCorkle, S. R., Everett, L. & Anderson, C. W. (2007). Paired-end genomic signature tags: a method for the functional analysis of genomes and epigenomes. *Genet Eng (N Y)* 28, 159–173.

Dupont, C., Guimiot, F., Perrin, L., Marey, I., Smiljkovski, D., Le Tessier, D., Lebugle, C., Baumann, C., Bourdoncle, P., Tabet, A. C., Aboura, A., Benzacken, B. & Dupont, J. M. (2011). 3D position of pericentromeric heterochromatin within the nucleus of a patient with ICF syndrome. *Clin Genet.* doi: 10.1111/j.1399-0004.2011.01697.

Edgar, A. J., Dover, S. L., Lodrick, M. N., McKay, I. J., Hughes, F. J. & Turner, W. (2005). Bone morphogenetic protein-2 induces expression of murine zinc finger transcription factor ZNF450. *J Cell Biochem* 94, 202–215.

Ehrlich, M. (2002). DNA hypomethylation, cancer, the immunodeficiency, centromeric region instability, facial anomalies syndrome and chromosomal rearrangements. *J Nutr* 132, 2424S–2429S.

Ehrlich, M., Buchanan, K. L., Tsien, F., Jiang, G., Sun, B., Uicker, W., Weemaes, C. M., Smeets, D., Sperling, K., Belohradsky, B. H., Tommerup, N., Misek, D. E., Rouillard, J. M., Kuick, R. & Hanash, S. M. (2001). DNA methyltransferase 3B mutations linked to the ICF syndrome cause dysregulation of lymphogenesis genes. *Hum Mol Genet* 10, 2917–2931.

Esteller, M. (2007). Epigenetic gene silencing in cancer: the DNA hypermethylome. *Hum Mol Genet* 16, R50–R59.

Esteve, P. O., Chin, H. G., Benner, J., Feehery, G. R., Samaranayake, M., Horwitz, G. A., Jacobsen, S. E. & Pradhan, S. (2009). Regulation of DNMT1 stability through SET7-mediated lysine methylation in mammalian cells. *Proc Natl Acad Sci U S A* 106, 5076–5081.

Fan, G., Martinowich, K., Chin, M. H., He, F., Fouse, S. D., Hutnick, L., Hattori, D., Ge, W., Shen, Y., Wu, H., ten Hoeve, J., Shuai, K. & Sun, Y. E. (2005). DNA methylation controls the timing of astrogliogenesis through regulation of JAK-STAT signaling. *Development* 132, 3345–3356.

Fatemi, M., Hermann, A., Gowher, H. & Jeltsch, A. (2002). Dnmt3a and Dnmt1 functionally cooperate during de novo methylation of DNA. *Eur J Biochem* 269, 4981–4984.

Feldman, N., Gerson, A., Fang, J., Li, E., Zhang, Y., Shinkai, Y., Cedar, H. & Bergman, Y. (2006). G9a-mediated irreversible epigenetic inactivation of Oct-3/4 during early embryogenesis. *Nat Cell Biol* 8, 188–194.

Feng, J., Zhou, Y., Campbell, S. L., Le, T., Li, E., Sweatt, J. D., Silva, A. J. & Fan, G. (2010). Dnmt1 and Dnmt3a maintain DNA methylation and regulate synaptic function in adult forebrain neurons. *Nat Neurosci* 13, 423–430.

Fuks, F., Burgers, W. A., Godin, N., Kasai, M. & Kouzarides, T. (2001). Dnmt3a binds deacetylases and is recruited by a sequence-specific repressor to silence transcription. *EMBO J* 20, 2536–2544.

Fuks, F., Hurd, P. J., Deplus, R. & Kouzarides, T. (2003). The DNA methyltransferases associate with HP1 and the SUV39H1 histone methyltransferase. *Nucleic Acids Res* 31, 2305–2312.

Gardiner-Garden, M. & Frommer, M. (1987). CpG islands in vertebrate genomes. *J Mol Biol* 196, 261–282.

Gartler, S. M., Varadarajan, K. R., Luo, P., Canfield, T. K., Traynor, J., Francke, U. & Hansen, R. S. (2004). Normal histone modifications on the inactive X chromosome in ICF and Rett syndrome cells: implications for methyl-CpG binding proteins. *BMC Biol* 2:21.

Gatto, S., Della, R. F., Cimmino, A., Strazzullo, M., Fabbri, M., Mutarelli, M., Ferraro, L., Weisz, A., D'Esposito, M. & Matarazzo, M. R. (2010). Epigenetic alteration of microRNAs in DNMT3B-mutated patients of ICF syndrome. *Epigenetics* 5, 427–443.

Geiman, T. M., Sankpal, U. T., Robertson, A. K., Chen, Y., Mazumdar, M., Heale, J. T., Schmiesing, J. A., Kim, W., Yokomori, K., Zhao, Y. & Robertson, K. D. (2004a). Isolation and characterization of a novel DNA methyltransferase complex linking DNMT3B with components of the mitotic chromosome condensation machinery. *Nucleic Acids Res* 32, 2716–2729.

Geiman, T. M., Sankpal, U. T., Robertson, A. K., Zhao, Y., Zhao, Y. & Robertson, K. D. (2004b). DNMT3B interacts with hSNF2H chromatin remodeling enzyme, HDACs 1 and 2, and components of the histone methylation system. *Biochem Biophys Res Commun* 318, 544–555.

Gennery, A. R., Slatter, M. A., Bredius, R. G., Hagleitner, M. M., Weemaes, C., Cant, A. J. & Lankester, A. C. (2007). Hematopoietic stem cell transplantation corrects the immunologic abnormalities associated with immunodeficiency-centromeric instability-facial dysmorphism syndrome. *Pediatrics* 120, e1341–e1344.

Gopalakrishnan, S., Sullivan, B. A., Trazzi, S., Della, V. G. & Robertson, K. D. (2009). DNMT3B interacts with constitutive centromere protein CENP-C to modulate DNA methylation and the histone code at centromeric regions. *Hum Mol Genet* 18, 3178–3193.

Gowher, H. & Jeltsch, A. (2002). Molecular enzymology of the catalytic domains of the Dnmt3a and Dnmt3b DNA methyltransferases. *J Biol Chem* 277, 20409–20414.

Gowher, H., Liebert, K., Hermann, A., Xu, G. & Jeltsch, A. (2005). Mechanism of stimulation of catalytic activity of Dnmt3A and Dnmt3B DNA-(cytosine-C5)-methyltransferases by Dnmt3L. *J Biol Chem* 280, 13341–13348.

Hagleitner, M. M., Lankester, A., Maraschio, P., Hulten, M., Fryns, J. P., Schuetz, C., Gimelli, G., Davies, E. G., Gennery, A., Belohradsky, B. H., de Groot, R., Gerritsen, E. J., Mattina, T.,

Howard, P. J., Fasth, A., Reisli, I., Furthner, D., Slatter, M. A., Cant, A. J., Cazzola, G., van Dijken, P. J., van Deuren, M., de Greef, J. C., van der Maarel, S. M. & Weemaes, C. M. (2008). Clinical spectrum of immunodeficiency, centromeric instability and facial dysmorphism (ICF syndrome). *J Med Genet* 45, 93–99.

Handa, V. & Jeltsch, A. (2005). Profound flanking sequence preference of Dnmt3a and Dnmt3b mammalian DNA methyltransferases shape the human epigenome. *J Mol Biol* 348, 1103–1112.

Hansen, R. S. (2003). X inactivation-specific methylation of LINE-1 elements by DNMT3B: implications for the Lyon repeat hypothesis. *Hum Mol Genet* 12, 2559–2567.

Hansen, R. S., Stoger, R., Wijmenga, C., Stanek, A. M., Canfield, T. K., Luo, P., Matarazzo, M. R., D'Esposito, M., Feil, R., Gimelli, G., Weemaes, C. M., Laird, C. D. & Gartler, S. M. (2000). Escape from gene silencing in ICF syndrome: evidence for advanced replication time as a major determinant. *Hum Mol Genet* 9, 2575–2587.

Hansen, R. S., Wijmenga, C., Luo, P., Stanek, A. M., Canfield, T. K., Weemaes, C. M. & Gartler, S. M. (1999). The DNMT3B DNA methyltransferase gene is mutated in the ICF immunodeficiency syndrome. *Proc Natl Acad Sci U S A* 96, 14412–14417.

Hellman, A. & Chess, A. (2007). Gene body-specific methylation on the active X chromosome. *Science* 315, 1141–1143.

Hirasawa, R. & Feil, R. (2010). Genomic imprinting and human disease. *Essays Biochem* 48, 187–200.

Holz-Schietinger, C. & Reich, N. O. (2010). The inherent processivity of the human de novo methyltransferase 3A (DNMT3A) is enhanced by DNMT3L. *J Biol Chem* 285, 29091–29100.

Jeanpierre, M., Turleau, C., Aurias, A., Prieur, M., Ledeist, F., Fischer, A. & Viegas-Pequignot, E. (1993). An embryonic-like methylation pattern of classical satellite DNA is observed in ICF syndrome. *Hum Mol Genet* 2, 731–735.

Jefferson, A., Colella, S., Moralli, D., Wilson, N., Yusuf, M., Gimelli, G., Ragoussis, J. & Volpi, E. V. (2010). Altered intra-nuclear organisation of heterochromatin and genes in ICF syndrome. *PLoS ONE* 5, e11364.

Jeltsch, A. (2006). On the enzymatic properties of Dnmt1: specificity, processivity, mechanism of linear diffusion and allosteric regulation of the enzyme. *Epigenetics* 1, 63–66.

Jeong, S., Liang, G., Sharma, S., Lin, J. C., Choi, S. H., Han, H., Yoo, C. B., Egger, G., Yang, A. S. & Jones, P. A. (2009). Selective anchoring of DNA methyltransferases 3A and 3B to nucleosomes containing methylated DNA. *Mol Cell Biol* 29, 5366–5376.

Ji, H., Ehrlich, L. I., Seita, J., Murakami, P., Doi, A., Lindau, P., Lee, H., Aryee, M. J., Irizarry, R. A., Kim, K., Rossi, D. J., Inlay, M. A., Serwold, T., Karsunky, H., Ho, L., Daley, G. Q., Weissman, I. L. & Feinberg, A. P. (2010). Comprehensive methylome map of lineage commitment from haematopoietic progenitors. *Nature* 467, 338–342.

Jia, D., Jurkowska, R. Z., Zhang, X., Jeltsch, A. & Cheng, X. (2007). Structure of Dnmt3a bound to Dnmt3L suggests a model for de novo DNA methylation. *Nature* 449, 248–251.

Jiang, Y. L., Rigolet, M., Bourc'his, D., Nigon, F., Bokesoy, I., Fryns, J. P., Hulten, M., Jonveaux, P., Maraschio, P., Megarbane, A., Moncla, A. & Viegas-Pequignot, E. (2005). DNMT3B mutations and DNA methylation defect define two types of ICF syndrome. *Hum Mutat* 25, 56–63.

Jin, B., Tao, Q., Peng, J., Soo, H. M., Wu, W., Ying, J., Fields, C. R., Delmas, A. L., Liu, X., Qiu, J. & Robertson, K. D. (2008). DNA methyltransferase 3B (DNMT3B) mutations in ICF syndrome lead to altered epigenetic modifications and aberrant expression of genes regulating development, neurogenesis and immune function. *Hum Mol Genet* 17, 690–709.

Jones, P. A. (2002). DNA methylation and cancer. *Oncogene* 21, 5358–5360.

Jones, P. A. & Liang, G. (2009). Rethinking how DNA methylation patterns are maintained. *Nat Rev Genet* 10, 805–811.

Jurkowska, R. Z., Anspach, N., Urbanke, C., Jia, D., Reinhardt, R., Nellen, W., Cheng, X. & Jeltsch, A. (2008). Formation of nucleoprotein filaments by mammalian DNA methyltransferase Dnmt3a in complex with regulator Dnmt3L. *Nucleic Acids Res* 36, 6656–6663.

Jurkowska, R. Z., Jurkowski, T. P. & Jeltsch, A. (2011). Structure and function of mammalian DNA methyltransferases. *Chembiochem* 12, 206–222.

Kaneda, M., Okano, M., Hata, K., Sado, T., Tsujimoto, N., Li, E. & Sasaki, H. (2004). Essential role for de novo DNA methyltransferase Dnmt3a in paternal and maternal imprinting. *Nature* 429, 900–903.

Kim, G. D., Ni, J., Kelesoglu, N., Roberts, R. J. & Pradhan, S. (2002). Co-operation and communication between the human maintenance and de novo DNA (cytosine-5) methyltransferases. *EMBO J* 21, 4183–4195.

Kolasinska-Zwierz, P., Down, T., Latorre, I., Liu, T., Liu, X. S. & Ahringer, J. (2009). Differential chromatin marking of introns and expressed exons by H3K36me3. *Nat Genet* 41, 376–381.

Kondo, T., Bobek, M. P., Kuick, R., Lamb, B., Zhu, X., Narayan, A., Bourc'his, D., Viegas-Pequignot, E., Ehrlich, M. & Hanash, S. M. (2000). Whole-genome methylation scan in ICF syndrome: hypomethylation of non-satellite DNA repeats D4Z4 and NBL2. *Hum Mol Genet* 9, 597–604.

Kriaucionis, S. & Heintz, N. (2009). The nuclear DNA base 5-hydroxymethylcytosine is present in Purkinje neurons and the brain. *Science* 324, 929–930.

Laurent, L., Wong, E., Li, G., Huynh, T., Tsirigos, A., Ong, C. T., Low, H. M., Kin Sung, K. W., Rigoutsos, I., Loring, J. & Wei, C. L. (2010). Dynamic changes in the human methylome during differentiation. *Genome Res* 20, 320–331.

Li, E. (2002). Chromatin modification and epigenetic reprogramming in mammalian development. *Nat Rev Genet* 3, 662–673.

Li, E., Bestor, T. H. & Jaenisch, R. (1992). Targeted mutation of the DNA methyltransferase gene results in embryonic lethality. *Cell* 69, 915–926.

Lin, I. G., Han, L., Taghva, A., O'Brien, L. E. & Hsieh, C. L. (2002). Murine de novo methyltransferase Dnmt3a demonstrates strand asymmetry and site preference in the methylation of DNA in vitro. *Mol Cell Biol* 22, 704–723.

Lister, R., Pelizzola, M., Dowen, R. H., Hawkins, R. D., Hon, G., Tonti-Filippini, J., Nery, J. R., Lee, L., Ye, Z., Ngo, Q. M., Edsall, L., Antosiewicz-Bourget, J., Stewart, R., Ruotti, V., Millar, A. H., Thomson, J. A., Ren, B. & Ecker, J. R. (2009). Human DNA methylomes at base resolution show widespread epigenomic differences. *Nature* 19, 315–322.

Luciani, J. J., Depetris, D., Missirian, C., Mignon-Ravix, C., Metzler-Guillemain, C., Megarbane, A., Moncla, A. & Mattei, M. G. (2005). Subcellular distribution of HP1 proteins is altered in ICF syndrome. *Eur J Hum Genet* 13, 41–51.

Maraschio, P., Zuffardi, O., Dalla, F. T. & Tiepolo, L. (1988). Immunodeficiency, centromeric heterochromatin instability of chromosomes 1, 9, and 16, and facial anomalies: the ICF syndrome. *J Med Genet* 25, 173–180.

Matarazzo, M. R., Boyle, S., D'Esposito, M. & Bickmore, W. A. (2007a). Chromosome territory reorganization in a human disease with altered DNA methylation. *Proc Natl Acad Sci U S A* 104, 16546–16551.

Matarazzo, M. R., De Bonis, M. L., Gregory, R. I., Vacca, M., Hansen, R. S., Mercadante, G., D'Urso, M., Feil, R. & D'Esposito, M. (2002). Allelic inactivation of the pseudoautosomal gene SYBL1 is controlled by epigenetic mechanisms common to the X and Y chromosomes. *Hum Mol Genet* 11, 3191–3198.

Matarazzo, M. R., De Bonis, M. L., Strazzullo, M., Cerase, A., Ferraro, M., Vastarelli, P., Ballestar, E., Esteller, M., Kudo, S. & D'Esposito, M. (2007b). Multiple binding of methyl-CpG and polycomb proteins in long-term gene silencing events. *J Cell Physiol* 210, 711–719.

Matarazzo, M. R., De Bonis, M. L., Vacca, M., Della, R. F. & D'Esposito, M. (2009). Lessons from two human chromatin diseases, ICF syndrome and Rett syndrome. *Int J Biochem Cell Biol* 41, 117–126.

Matzke, M. A. & Birchler, J. A. (2005). RNAi-mediated pathways in the nucleus. *Nat Rev Genet* 6, 24–35.

Miniou, P., Jeanpierre, M., Blanquet, V., Sibella, V., Bonneau, D., Herbelin, C., Fischer, A., Niveleau, A. & Viegas-Pequignot, E. (1994). Abnormal methylation pattern in constitutive and facultative (X inactive chromosome) heterochromatin of ICF patients. *Hum Mol Genet* 3, 2093–2102.

Miniou, P., Jeanpierre, M., Bourc'his, D., Coutinho Barbosa, A. C., Blanquet, V. & Viegas-Pequignot, E. (1997). alpha-satellite DNA methylation in normal individuals and in ICF patients: heterogeneous methylation of constitutive heterochromatin in adult and fetal tissues. *Hum Genet* 99, 738–745.

Moarefi, A. H. & Chedin, F. (2011). ICF syndrome mutations cause a broad spectrum of biochemical defects in DNMT3B-mediated de novo DNA methylation. *J Mol Biol* 409, 758–772.

Mohn, F. & Schubeler, D. (2009). Genetics and epigenetics: stability and plasticity during cellular differentiation. *Trends Genet* 25, 129–136.

Mohn, F., Weber, M., Rebhan, M., Roloff, T. C., Richter, J., Stadler, M. B., Bibel, M. & Schubeler, D. (2008). Lineage-specific polycomb targets and de novo DNA methylation define restriction and potential of neuronal progenitors. *Mol Cell* 30, 755–766.

Mosher, R. A. & Melnyk, C. W. (2010). siRNAs and DNA methylation: seedy epigenetics. *Trends Plant Sci* 15, 204–210.

Oda, M., Yamagiwa, A., Yamamoto, S., Nakayama, T., Tsumura, A., Sasaki, H., Nakao, K., Li, E. & Okano, M. (2006). DNA methylation regulates long-range gene silencing of an X-linked homeobox gene cluster in a lineage-specific manner. *Genes Dev* 20, 3382–3394.

Okano, M., Bell, D. W., Haber, D. A. & Li, E. (1999). DNA methyltransferases Dnmt3a and Dnmt3b are essential for de novo methylation and mammalian development. *Cell* 99, 247–257.

Ooi, S. K., Qiu, C., Bernstein, E., Li, K., Jia, D., Yang, Z., Erdjument-Bromage, H., Tempst, P., Lin, S. P., Allis, C. D., Cheng, X. & Bestor, T. H. (2007). DNMT3L connects unmethylated lysine 4 of histone H3 to de novo methylation of DNA. *Nature* 448, 714–717.

Otani, J., Nankumo, T., Arita, K., Inamoto, S., Ariyoshi, M. & Shirakawa, M. (2009). Structural basis for recognition of H3K4 methylation status by the DNA methyltransferase 3A ATRX-DNMT3-DNMT3L domain. *EMBO Rep* 10, 1235–1241.

Portela, A. & Esteller, M. (2010). Epigenetic modifications and human disease. *Nat Biotechnol* 28, 1057–1068.

Sakaue, M., Ohta, H., Kumaki, Y., Oda, M., Sakaide, Y., Matsuoka, C., Yamagiwa, A., Niwa, H., Wakayama, T. & Okano, M. (2010). DNA methylation is dispensable for the growth and survival of the extraembryonic lineages. *Curr Biol* 20, 1452–1457.

Scarano, M. I., Strazzullo, M., Matarazzo, M. R. & D'Esposito, M. (2005). DNA methylation 40 years later: Its role in human health and disease. *J Cell Physiol* 204, 21–35.

Schuetz, C., Barbi, G., Barth, T. F., Hoenig, M., Schulz, A., Moeller, P., Smeets, D., de Greef, J. C., van der Maarel, S. M., Vogel, W., Debatin, K. M. & Friedrich, W. (2007). ICF syndrome: high variability of the chromosomal phenotype and association with classical Hodgkin lymphoma. *Am J Med Genet A* 143A, 2052–2057.

Stacey, M., Bennett, M. S. & Hulten, M. (1995). FISH analysis on spontaneously arising micronuclei in the ICF syndrome. *J Med Genet* 32, 502–508.

Storre, J., Elsasser, H. P., Fuchs, M., Ullmann, D., Livingston, D. M. & Gaubatz, S. (2002). Homeotic transformations of the axial skeleton that accompany a targeted deletion of E2f6. *EMBO Rep* 3, 695–700.

Straussman, R., Nejman, D., Roberts, D., Steinfeld, I., Blum, B., Benvenisty, N., Simon, I., Yakhini, Z. & Cedar, H. (2009). Developmental programming of CpG island methylation profiles in the human genome. *Nat Struct Mol Biol* 16, 564–571.

Suzuki, M., Yamada, T., Kihara-Negishi, F., Sakurai, T., Hara, E., Tenen, D. G., Hozumi, N. & Oikawa, T. (2006). Site-specific DNA methylation by a complex of PU.1 and Dnmt3a/b. *Oncogene* 25, 2477–2488.

Tachibana, M., Matsumura, Y., Fukuda, M., Kimura, H. & Shinkai, Y. (2008). G9a/GLP complexes independently mediate H3K9 and DNA methylation to silence transcription. *EMBO J* 27, 2681–2690.

Tahiliani, M., Koh, K. P., Shen, Y., Pastor, W. A., Bandukwala, H., Brudno, Y., Agarwal, S., Iyer, L. M., Liu, D. R., Aravind, L. & Rao, A. (2009). Conversion of 5-methylcytosine to 5-hydroxymethylcytosine in mammalian DNA by MLL partner TET1. *Science* 324, 930–935.

Tiepolo, L., Maraschio, P., Gimelli, G., Cuoco, C., Gargani, G. F. & Romano, C. (1979). Multibranched chromosomes 1, 9, and 16 in a patient with combined IgA and IgE deficiency. *Hum Genet* 51, 127–137.

Tuck-Muller, C. M., Narayan, A., Tsien, F., Smeets, D. F., Sawyer, J., Fiala, E. S., Sohn, O. S. & Ehrlich, M. (2000). DNA hypomethylation and unusual chromosome instability in cell lines from ICF syndrome patients. *Cytogenet Cell Genet* 89, 121–128.

Ueda, Y., Okano, M., Williams, C., Chen, T., Georgopoulos, K. & Li, E. (2006). Roles for Dnmt3b in mammalian development: a mouse model for the ICF syndrome. *Development* 133, 1183–1192.

Vakoc, C. R., Sachdeva, M. M., Wang, H. & Blobel, G. A. (2006). Profile of histone lysine methylation across transcribed mammalian chromatin. *Mol Cell Biol* 26, 9185–9195.

van den Brand, M., Flucke, U. E., Bult, P., Weemaes, C. M. & van Deuren, M. (2011). Angiosarcoma in a patient with immunodeficiency, centromeric region instability, facial anomalies (ICF) syndrome. *Am J Med Genet A* 155A, 622–625.

Vanyushin, B. F. & Ashapkin, V. V. (2011). DNA methylation in higher plants: past, present and future. *Biochim Biophys Acta* 1809, 360–368.

Velasco, G., Hube, F., Rollin, J., Neuillet, D., Philippe, C., Bouzinba-Segard, H., Galvani, A., Viegas-Pequignot, E. & Francastel, C. (2010). Dnmt3b recruitment through E2F6 transcriptional repressor mediates germ-line gene silencing in murine somatic tissues. *Proc Natl Acad Sci U S A* 107, 9281–9286.

Vire, E., Brenner, C., Deplus, R., Blanchon, L., Fraga, M., Didelot, C., Morey, L., Van Eynde, A., Bernard, D., Vanderwinden, J. M., Bollen, M., Esteller, M., Di Croce, L., de Launoit, Y. & Fuks, F. (2006). The Polycomb group protein EZH2 directly controls DNA methylation. *Nature* 439, 871–874.

Wang, Y. A., Kamarova, Y., Shen, K. C., Jiang, Z., Hahn, M. J., Wang, Y. & Brooks, S. C. (2005). DNA methyltransferase-3a interacts with p53 and represses p53-mediated gene expression. *Cancer Biol Ther* 4, 1138–1143.

Weber, M., Hellmann, I., Stadler, M. B., Ramos, L., Paabo, S., Rebhan, M. & Schubeler, D. (2007). Distribution, silencing potential and evolutionary impact of promoter DNA methylation in the human genome. *Nat Genet* 39, 457–466.

Wienholz, B. L., Kareta, M. S., Moarefi, A. H., Gordon, C. A., Ginno, P. A. & Chedin, F. (2010). DNMT3L modulates significant and distinct flanking sequence preference for DNA methylation by DNMT3A and DNMT3B in vivo. *PLoS Genet* 6, e1001106.

Wijmenga, C., Hansen, R. S., Gimelli, G., Bjorck, E. J., Davies, E. G., Valentine, D., Belohradsky, B. H., van Dongen, J. J., Smeets, D. F., van den Heuvel, L. P., Luyten, J. A., Strengman, E., Weemaes, C. & Pearson, P. L. (2000). Genetic variation in ICF syndrome: evidence for genetic heterogeneity. *Hum Mutat* 16, 509–517.

Xie, Z. H., Huang, Y. N., Chen, Z. X., Riggs, A. D., Ding, J. P., Gowher, H., Jeltsch, A., Sasaki, H., Hata, K. & Xu, G. L. (2006). Mutations in DNA methyltransferase DNMT3B in ICF syndrome affect its regulation by DNMT3L. *Hum Mol Genet* 15, 1375–1385.

Yehezkel, S., Segev, Y., Viegas-Pequignot, E., Skorecki, K. & Selig, S. (2008). Hypomethylation of subtelomeric regions in ICF syndrome is associated with abnormally short telomeres and enhanced transcription from telomeric regions. *Hum Mol Genet* 17, 2776–2789.

Zhang, Y., Jurkowska, R., Soeroes, S., Rajavelu, A., Dhayalan, A., Bock, I., Rathert, P., Brandt, O., Reinhardt, R., Fischle, W. & Jeltsch, A. (2010). Chromatin methylation activity of Dnmt3a and Dnmt3a/3L is guided by interaction of the ADD domain with the histone H3 tail. *Nucleic Acids Res* 38, 4246–4253.

Zhu, H., Geiman, T. M., Xi, S., Jiang, Q., Schmidtmann, A., Chen, T., Li, E. & Muegge, K. (2006). Lsh is involved in de novo methylation of DNA. *EMBO J* 25, 335–345.

Zilberman, D., Gehring, M., Tran, R. K., Ballinger, T. & Henikoff, S. (2007). Genome-wide analysis of Arabidopsis thaliana DNA methylation uncovers an interdependence between methylation and transcription. *Nat Genet* 39, 61–69.

Chapter 3
Dysfunction of the Methyl-CpG-Binding Protein MeCP2 in Rett Syndrome

Gaston Calfa, Alan K. Percy, and Lucas Pozzo-Miller

3.1 Introduction

Rett syndrome (RTT; Online Mendelian Inheritance in Man #312750; http://www. ncbi.nlm.nih.gov/omim/), first recognized and described in German by Andreas Rett (1966), is a neurodevelopmental disorder predominantly occurring in females. Almost 20 years later, Hagberg and colleagues presented the first description of RTT in the English language, leading to worldwide diagnosis in all ethnic and racial groups (Hagberg et al. 1983). Currently, the incidence of RTT is estimated to be approximately 1:10,000 female births (Laurvick et al. 2006; Chahrour and Zoghbi 2007). Early studies proposing a genetic basis for RTT were later confirmed by Zoghbi and colleagues with the identification of mutations in the *MECP2* gene located at chromosome Xq28 (Amir et al. 1999). *MECP2* encodes methyl-CpG-binding protein-2, a transcription factor that binds methylated DNA and is ubiquitously expressed in mammalian tissues (Lewis et al. 1992). Following the identification of loss-of-function mutations in the *MECP2* gene in RTT individuals, research efforts expanded rapidly on an international scale with comprehensive clinical investigations into the complex array of medical and behavioral issues. In addition, intensive laboratory-based studies have been spurred by the availability of human autopsy tissue and experimental animal models, such as deletions of the endogenous *Mecp2* gene (knockout), insertions of premature STOP codons or RTT-associated mutations (knock-in) common in the human *MECP2* gene, and overexpression of Mecp2 to model the newly identified *MECP2* duplication syndrome (Meins et al. 2005; Friez et al. 2006; Ramocki et al. 2010).

G. Calfa • L. Pozzo-Miller(✉)
Department of Neurobiology, The University of Alabama at Birmingham,
Birmingham, AL, USA
e-mail: lucaspm@uab.edu

A.K. Percy
Departments of Pediatrics, Neurology, and Genetics, Civitan International
Research Center, Birmingham, AL, USA

J. Minarovits and H.H. Niller (eds.), *Patho-Epigenetics of Disease*,
DOI 10.1007/978-1-4614-3345-3_3, © Springer Science+Business Media New York 2012

RTT is a sporadic condition in >99% of the cases, with the risk of familial recurrence being extremely low. Indeed, *MECP2* mutations appear to be spontaneous transitions occurring in the paternal germline (Girard et al. 2001; Trappe et al. 2001), explaining in part the paucity of males with RTT or carrying *MECP2* mutations. The clinical hallmarks of RTT include a period of apparently normal early development, followed by a plateau or stagnation in development and a subsequent frank regression. It is during this period that both visual and aural contact is impaired leading to an initial diagnosis of autism. The convergence of clinical presentations in RTT and autism in association with *MECP2* mutations represents an intriguing link between these disorders and other neurodevelopmental conditions with an established genetic basis and clinical features consistent with autism spectrum disorders, such as Fragile X syndrome and Down syndrome (Kaufmann and Moser 2000). This chapter reviews the features of RTT and its genetic bases, as well as the role of *MECP2* in neurodevelopment at the clinical as well as molecular and cellular levels, exploring potential neurobiological mechanisms shared with other autism spectrum disorders.

3.2 Clinical Features of RTT

RTT is a neurodevelopmental disorder characterized by profound cognitive impairment, communication dysfunction, stereotypic hand movements, breathing irregularities, and pervasive growth failure. The diagnosis of RTT is based on consensus criteria, originally developed in 1984, revised in 1988 and 2002, and recently updated in 2010 (Table 3.1). Current criteria recognize both classic and variant forms of RTT, differentiated by meeting all or only some of the consensus criteria. In a RTT natural history study, 85% have classic and 15% variant RTT; however, whether classic or variant, regression is universal (Percy et al. 2007). Initial development appears to be normal during early infancy, after which acquired language, socialization, and fine motor skills are lost either partially or completely. In many instances, early development is accompanied by a reduction in muscle tone (hypotonia). The majority (85%) of RTT individuals show abnormal deceleration in the rate of head growth as early as 3 months of age, with more than 50% of those falling below the 2nd percentile (microcephaly). For those who begin walking (~80%), gait becomes dyspraxic, broad-based, and ataxic. At the time of regression or shortly thereafter, the stereotypic hand movements evident only while awake become so predominant that they may preclude any purposeful hand use. These stereotypies consist of hand-wringing, hand-patting or clasping, and hand-mouthing, either together in the midline or independently, and in some may also be noted around the mouth or in the lower extremities. Each RTT girl has her own repertoire of stereotypies that in most will evolve over time.

During the regression phase, eye contact is markedly reduced or lost, as is their response to spoken language. It is at this point that the diagnosis of autism is often suggested. In some, this phase is accompanied by periods of inconsolable crying or

3 Dysfunction of the Methyl-CpG-Binding Protein MeCP2 in Rett Syndrome

Table 3.1 Rett syndrome consensus criteria (2002 and 2010)

Typical RTT consensus criteria, 2002	
Necessary criteria	Apparently normal prenatal and perinatal history
	Psychomotor development largely normal throughout life span
	Normal head circumference at birth
	Postnatal deceleration of head growth in the majority
	Loss of achieved purposeful hand skill between 6 and 18 months of age
	Stereotypic hand movements, such as hand-wringing, hand-patting or clasping, and hand-mouthing
	Emerging social withdrawal, communication dysfunction
	Impaired (dyspraxic) or falling locomotion
Exclusion criteria	Organomegaly or other signs of storage disease
	Retinopathy, optic atrophy, or cataract
	Evidence of perinatal or postnatal brain damage
	Existence of an identifiable metabolic disorder
	Acquired neurological disorders resulting from severe infections or head trauma
Supportive criteria	Awake disturbances of breathing
	Bruxism
	Impaired sleep pattern
	Abnormal muscle tone
	Peripheral vasomotor disturbances
	Scoliosis/kyphosis
	Growth retardation
	Hypotrophic small and cold feet; small thin hands
Atypical RTT consensus criteria 2002	
Inclusion criteria	Meet at least 3 of 6 main criteria
	Meet at least 5 of 10 supportive criteria
Main criteria	Absence or reduction of hand skills
	Reduction or loss of babble speech
	Monotonous pattern to hand stereotypies
	Reduction or loss of communication skills
	Deceleration of head growth from first years of life
	Rett syndrome disease profile: a regression stage followed by a recovery of interaction contrasting with slow neuromotor regression
Supportive criteria	Breathing irregularities
	Bloating/air swallowing
	Bruxism, harsh sounding type
	Abnormal locomotion
	Scoliosis/kyphosis
	Lower limb amyotrophy
	Cold, purplish feet, usually growth impaired
	Sleep disturbances including night screaming outbursts
	Laughing/screaming spells
	Diminished response to pain

(continued)

Table 3.1 (continued)

RTT diagnostic criteria 2010

	Consider diagnosis when postnatal deceleration of head growth is observed
Required inclusion criteria for typical RTT	A period of regression followed by recovery or stabilization
	Meet at least 2 out of 4 main criteria
	Meet at least 5 out of 11 supportive criteria
Main criteria	Partial or complete loss of acquired purposeful hand skills
	Partial or complete loss of acquired spoken language
	Impaired (dyspraxic) gait or absence of gait ability
	Stereotypic hand movements, such as hand wringing/squeezing, clapping/tapping, mouthing, and washing/rubbing automatisms
Exclusion criteria for typical RTT	Brain injury secondary to trauma (peri- or postnatal), neurometabolic disease, or severe infection causing neurological symptoms
	Grossly abnormal psychomotor development in first 6 months of life
Supportive criteria for atypical RTT	Breathing disturbances when awake
	Bruxism when awake
	Impaired sleep pattern
	Abnormal muscle tone
	Peripheral vasomotor disturbances
	Scoliosis/kyphosis
	Growth retardation
	Small cold hands and feet
	Inappropriate laughing/screaming spells
	Diminished response to pain
	Intense eye communication—"eye pointing"

Table 3.2 Temporal profile of classic Rett syndrome

Apparently normal early development
Arrest of developmental progress at 6–18 months
Frank regression of social contact, language, and finger skills
Improved social contact and eye gaze by age 5; gradual slowing of motor functions in adulthood

irritability that may last for days or even months. Thereafter, a period of stabilization is typical, with markedly improved socialization and eye gaze that affords the opportunity to develop communication skills, often through augmentative communication strategies ranging from choice cards to gaze-directed computer technologies. As such, a clear temporal profile of RTT has now been recognized (Table 3.2). A stabilization phase of their socialization and interpersonal interactions continues throughout life, which allows families to develop an enduring pattern of interaction and schools to engage in effective therapeutic strategies. However, during later years, particularly beginning after the teens, hypotonia is replaced by increased tone or rigidity, dystonia may be noted, and purposeful motor functions become slow. Along with slowing of motor functions, hand stereotypies may also diminish or even disappear.

Despite a preponderance of neurological and neurodevelopmental features, the medical issues associated with RTT are systemic (Table 3.3), and thus, the major clinical problems relate to growth, nutrition, and the gastrointestinal system.

3 Dysfunction of the Methyl-CpG-Binding Protein MeCP2 in Rett Syndrome

Table 3.3 Medical issues in Rett syndrome

Longevity
Growth and nutrition
Epilepsy
Gastrointestinal dysfunction
Anxiety
Scoliosis
Breathing irregularities
Sleep
Cardiac conduction
Sexual maturation

As with head growth, increases in length and weight slow during the first 6–18 months, such that both parameters fall below the 5th percentiles by adolescence. Maintaining adequate weight gain is indeed challenging for many parents. It does appear that during childhood, individuals with RTT require increased caloric intake on a per kilogram basis compared to age-matched peers; this is further compounded by poor chewing and swallowing, gastrointestinal reflux, and chronic constipation. Therefore, it is often necessary to provide high-calorie supplements or to institute gastrostomy feedings for delivery of adequate calories.

Other important medical manifestations of RTT include anxiety, sleep dysfunction, scoliosis, breathing irregularities, and epilepsy. Anxiety may manifest as difficulty adjusting to new environments or situations or could simply be reluctance to transition from one floor surface to another. Disrupted sleep is common, often associated with gastrointestinal issues such as reflux or constipation, but also without apparent causes; at times, RTT girls are simply awake in the night playing (Young et al. 2007). Cardiac conduction abnormalities in the form of a prolonged interval between the Q and T waves of the electrocardiogram have been described, requiring annual monitoring (Ellaway et al. 1999). Scoliosis is present as early as age 4 and is noted in >85% by mid-teens, often being progressive and ultimately requiring corrective surgery in 10–12% of cases (Percy et al. 2010). Breathing irregularities such as breath holding, hyperventilation, or both are recurrent during waking hours (Katz et al. 2009) and may interfere with feeding and thus the maintenance of an adequate nutrition. Sexual maturation is normal, making RTT girls a highly vulnerable group that requires appropriate monitoring by the parents and caregivers.

The longevity of RTT individuals is considerably greater now than when this disorder was first recognized in the 1960s. The original cohort of Andreas Rett survived into the mid-20s, while current estimates of longevity, based on the North American database, indicate that median survival now extends to 50–55 years of age (Kirby et al. 2010). This improvement likely reflects a number of factors, including greater ascertainment to include more mildly affected individuals, and improvements in medical management and nutrition, as our understanding of RTT has advanced over the last 40 years.

One of the most vexing medical issues relates to adverse behaviors that range from inconsolable crying to self-abuse in the form of head-banging or self-biting to outbursts of aggression towards others as in hitting, biting, or scratching or damaging objects or furniture. While these may be regarded simply as adverse behaviors, it is crucial to consider the broad variety of medical issues that could produce pain or discomfort and thus represent the explanation for these behaviors. In fact, RTT individuals do not respond appropriately to pain: either their response to pain is delayed or blunted or they do not know how to communicate they are suffering pain. Therefore, when acute behavioral changes occur, a thorough investigation of possible underlying causes is essential. These may include ear or sinus infections, bone fractures, gastrointestinal dysfunction, urinary tract infection or kidney stones, or ovarian cysts. Recently, gallbladder dysfunction has emerged as a common cause of acute changes in behavior in RTT individuals. In many instances, gallstones are not the issue but rather the lack of gallbladder function. Thus, even though stones are not revealed by either ultrasound or radiological techniques, a functional radionuclide test is required, i.e., hepatobiliary imino-diacetic acid (HIDA) scan.

The presentation of paroxysmal or epileptic episodes is particularly challenging because many of the seizure-like behaviors, such as partial and generalized convulsive or silent (i.e., absence) seizures, lack a clear-cut electrographic signature, and do not respond to antiepileptic agents. Indeed, video-EEG monitoring is generally required to differentiate epileptic events as seizure and nonseizure events (Glaze et al. 1998). The EEG in RTT individuals is characterized by focal, multifocal, and generalized epileptiform abnormalities, as well as rhythmic slow (theta) activity primarily in the frontal-central regions (Glaze 2005). Across the life span, more than 85% of RTT individuals will have seizures, but at any given time, this figure is closer to 30% (Glaze et al. 2010). Seizures rarely begin after the teenage years and tend to abate in adulthood. The EEG, which typically demonstrates marked background slowing and multifocal epileptiform waves after age 2–3 years, may no longer have this epileptiform pattern after age 20. Lastly, RTT individuals also show abnormally large somatosensory evoked potentials (Yoshikawa et al. 1991; Kimura et al. 1992).

3.3 RTT Neuropathology

Gross anatomy and cellular morphology in autopsy brains from RTT individuals reveals very consistent and distinct features (Table 3.4) (Armstrong 2005). First and foremost, the absence of any recognizable pattern of neuronal or glial cell atrophy, degeneration or death, gliosis, demyelination, or neuronal migration defects, as well as the lack of disease progression, is critical to differentiate RTT from a neurodegenerative disorder (Jellinger et al. 1988; Reiss et al. 1993).

Table 3.4 Brain morphology in Rett syndrome

Reduced brain weight
Reduced volume of specific regions: frontal, temporal, and caudate
Reduced melanin pigmentation, especially substantia nigra
Small neurons, simplified dendrites, and low dendritic spine density
No recognizable disease progression or degeneration of neurons or glial cells

Reduced brain and neuronal size with increased cell density is consistently observed in several brain regions, including cerebral cortex, hypothalamus, and the hippocampal formation (Bauman et al. 1995a, b). In addition, biopsies of nasal epithelium revealed far fewer terminally differentiated olfactory receptor neurons and significantly greater numbers of immature neurons in RTT individuals compared to unaffected controls (Ronnett et al. 2003). Furthermore, the size and complexity of dendritic trees was reduced in pyramidal cells of the frontal and motor cortices and of the subiculum (Armstrong et al. 1995, 1998), while levels of microtubule-associated protein-2 (MAP-2), a protein involved in microtubule stabilization, were lower throughout the neocortex of RTT autopsy material (Kaufmann et al. 1995, 2000). In addition, autoradiography in frontal cortex and basal ganglia of autopsy RTT brains revealed complex, age-related abnormalities in the density of neurotransmitter receptors, such as excitatory NMDA-(N-methyl-D-aspartate), AMPA-(α-amino-3-hydroxy-5-methyl-4-isoxazolepropionic acid), kainate- and metabotropic-type glutamate receptors (GluRs) as well as inhibitory GABA receptors (Blue et al. 1999a, b). Furthermore, 1H spectroscopy at 4.1 T revealed that the ratio of glutamate to N-acetylaspartate was elevated in gray matter of RTT individuals compared to their unaffected siblings while unchanged in white matter (Pan et al. 1999). Finally, the density of dendritic spines is reduced in pyramidal neurons of the frontal cortex and CA1 region of the hippocampal formation (Jellinger et al. 1988; Belichenko et al. 1994; Chapleau et al. 2009a), which is consistent with a reduced expression of cyclooxygenase, a protein enriched in dendritic spines (Kaufmann et al. 1997). This so-called spine dysgenesis phenotype (Purpura 1974) is common to other neurodevelopmental disorders, including Down syndrome, autism, Angelman syndrome, and Fragile X syndrome (Armstrong et al. 1998; Kaufmann and Moser 2000; Fiala et al. 2002). Such striking commonality across a spectrum of disorders, most with distinct molecular mechanisms, suggests a fundamental linkage through common pathways of neurobiological development responsible for cognitive performance (Chapleau et al. 2009b).

3.4 Links to Autism Spectrum Disorders

RTT is classified among the autism spectrum disorders, and indeed, individuals with RTT have a period during which autistic features are prominent, namely, during their regression. As such, RTT joins a growing list of neurodevelopmental disorders of genetic origin in which autism is a major clinical feature (Table 3.5). Despite this association, clear phenotypic differences exist between autism and RTT, most notably gender, head growth, and incidence, as well as defined genetic etiologies (Table 3.6). Furthermore, regression is universal in RTT but only in a subset of those individuals with autism. In addition, eye contact and socialization improve significantly in RTT following the initial period of regression with autistic features. Other distinguishing characteristics include poor to absent fine motor skills, dyspraxic or absent gait, and breathing irregularities in RTT individuals.

Table 3.5 Neurodevelopmental disorders of genetic origin associated with autism

Fragile X syndrome
Down syndrome
Angelman syndrome
Prader–Willi syndrome
Smith–Magenis syndrome
Neurofibromatosis, NF-1
Tuberous sclerosis
Rett syndrome

Table 3.6 Comparative features of autism and Rett syndrome

Autism	Rett syndrome
Affects primarily boys	Affects primarily girls
Initial period of normal development	Initial period of normal development
Hand-flapping	Hand-wringing, clasping
Accelerated head growth	Decelerated head growth
Abnormal social interaction	Abnormal social interaction
Impaired language/communication	Impaired language/communication
Restricted stereotyped behavior/activities	restricted stereotyped behavior/activities
75% cognitively impaired	100% cognitively impaired
1:100–1:500 births	1:10,000 female births
Genetic basis: undefined	Genetic basis: *MECP2* gene

Table 3.7 Molecular convergence of autism and Rett syndrome (based on Swanberg et al. 2009)

EGR2/MeCP2 regulation disrupted in both conditions
MECP2 expression decreased in autism cerebral cortex
EGR2 decreased in autism/RTT cerebral cortex
EGR2 decreased in cerebral cortex of *Mecp2* null mice
EGR2 has predicted binding site in *MECP2* promoter region
MeCP2 family binds EGR2 enhancer region
EGR2/MeCP2 increase coordinately in mouse and human cerebral cortex

Beyond reduced *MECP2* expression in the cerebral cortex of individuals with autism (Samaco et al. 2004) and the autistic features of boys with *MECP2* duplications (Ramocki et al., 2009), a potential linkage between RTT and autism is *EGR2* (early growth response-2), a member of the activity-dependent immediate-early genes (IEG). *EGR2* encodes a DNA-binding protein with a zinc finger domain that is postulated to participate in forebrain development and synaptic plasticity (Chavrier et al. 1988; Herdegen et al. 1993; Williams et al. 2000). Egr2 is the only member of the EGR/IEG family restricted to neurons in the CNS (Herdegen et al. 1993), with an intriguing cytoplasmic and nuclear localization (Mack et al. 1992). Protein levels of Egr2 are low in the cerebral cortex of individuals with autism and RTT, while Egr2 and MeCP2 protein levels increase in parallel in human and mouse cerebral cortex (Swanberg et al. 2009) (Table 3.7).

Furthermore, Angelman syndrome, an imprinted disorder caused by maternal 15q11-q13 or *UBE3A* deficiency, has phenotypic and genetic overlap with autism and RTT. The similarities include significant defects in expression of *UBE3A/E6AP* (ubiquitin protein ligase E3A) and *GABRB3* (ß3 subunit of GABA$_A$ receptors) in two different *Mecp2*-deficient mouse strains and autopsy brains from individuals with RTT, Angelman syndrome, and autism (Samaco et al. 2005). These results suggest an overlapping pathway of gene dysregulation within 15q11-q13 in RTT, Angelman, and autism.

3.5 Gene Function: MeCP2

MeCP2 is a member of a family of proteins that bind regions of DNA enriched with methylated CpG regions, i.e., cytosine and guanine nucleosides separated by a phosphate group (Hendrich and Bird 1998). Two major functional domains characterize this family of proteins: the methyl-CpG-binding domain (MBD) and the transcriptional repression domain (TRD) (Nan et al. 1993, 1997). The best and first characterized function of MeCP2 is to repress gene transcription by recruiting corepressor and the mSin3a/histone deacetylase complex (HDAC) and altering the structure of genomic DNA (Nan et al. 1998). This classical view of MeCP2 as a global transcriptional repressor exclusively has been questioned by the recent realization that out of all the genes misregulated in the hypothalamus of both *MECP2*-overexpressing and *Mecp2* knockout mice (see below), the majority (~85%) was activated in the overexpressing mice and downregulated in the knockout mice, which indicates that MeCP2 has a broader gene transcription role than originally thought (Chahrour et al. 2008). In addition, MeCP2 interacts with the RNA-binding protein Y box-binding protein 1 and regulates splicing of reporter minigenes, which may explain the aberrant alternative splicing patterns observed in *MeCP2*[308/Y] mice (Young et al. 2005), which carry a premature STOP codon and express a truncated nonfunctional MeCP2 protein (see below) (Shahbazian et al. 2002).

The human *MECP2* gene has four exons (Reichwald et al. 2000) from which two protein isoforms differing in their N-termini are expressed by alternative splicing: MECP2-e1 (previously identified as MECP2B/MECP2α), the most abundant isoform, and MECP2-e2 (previously identified as MECP2A/MECP2β) (Kriaucionis and Bird 2004; Mnatzakanian et al. 2004). Total protein and mRNA expression from the mouse *Mecp2* gene (without differentiating between isoforms) is widely distributed throughout the developing and adult brain (Shahbazian et al. 2002; Jung et al. 2003; Kishi and Macklis 2004). More recently, brain-region-specific splicing of the *Mecp2* gene was observed during mouse brain development: *Mecp2-e2* mRNA was enriched in the dorsal thalamus and layer V of the cerebral cortex, while more *Mecp2-e1* transcript was detected in the hypothalamus than in the thalamus between postnatal days 1 and 21 (Dragich et al. 2007).

So far, more than 250 different *MECP2* mutations have been identified in individuals with RTT. However, eight common point mutations (R106W, R133C, T158M, R168X, R255X, R270X, R294X, R306C) account for about 65% of those with RTT, while large deletions involving one or more exons and 3' deletions account for another 15–18% of RTT individuals (RettBASE: IRSF *MECP2* Variation Database; http://mecp2.chw.edu.au/). While the vast majority of individuals expressing a *MECP2* mutation do fulfill the diagnostic criteria for typical RTT (Table 3.1), significant phenotypic variation is associated with such mutations (Bebbington et al. 2008; Neul et al. 2008). These include individuals meeting variant atypical RTT criteria, female carriers who may be normal or have mild learning or cognitive disabilities, and more significantly include individuals with prominent behavioral phenotypes, such as autism and obsessive-compulsive and aggressive behaviors in associations with moderate to severe cognitive delay. Individuals meeting the RTT variant criteria, such as preserved speech but with a significant delayed onset (up to 10 years of age or more) are noted in association with milder involvement, whereas early onset seizures and congenital onset are noted with more severe clinical features (Bebbington et al. 2008; Glaze et al. 2010). Inasmuch as RTT is a sporadic condition with extremely low risk of familial recurrence, female carriers represent less than 3% of the total participants in the RTT natural history study (Percy et al. 2007). However, this group is likely to be underrepresented because they would not be recognized if it were not for an affected child or sibling.

A possible explanation for the wide phenotypic variability in clinical presentations is skewing of the normally expected random X chromosome inactivation (XCI), whereby the inactivation of one of the X chromosomes in every female cell typically permits the expression of genes from the active X chromosome in the adult. As such, individuals with normal function or only mild involvement would represent cases of wild-type MECP2 expression in the majority of cells. However, it is important to mention that this explanation may account for only ~20% of RTT cases related to variances in severity (Archer et al. 2006).

Among males, the most common *MECP2* dysfunction is associated with duplications of the Xq28 region, which include variable numbers of other genes. However, the principal clinical features in these male individuals appear to relate to the overexpression of *MECP2* because overexpressing the human *MECP2* gene in mice caused significant neurological deficits (see below) (Collins et al. 2004). Thus, a *MECP2* duplication syndrome has been defined (Meins et al. 2005; Friez et al. 2006; Ramocki et al. 2010). Male individuals carrying *MECP2* mutations generally have nonspecific cognitive delay with or without progressive motor problems or a severe early infantile encephalopathy (Kankirawatana et al. 2006), as well as clear autistic features (del Gaudio et al. 2006). As with carrier females, these males are likely underrepresented, unless a sibling has been identified with RTT or a progressive encephalopathy. In addition, typical RTT features have been described in a small number of males with either somatic mosaicism or in combination with Klinefelter syndrome (47XXY) because in both instances, two cell populations of X chromosomes would exist, similar to females with RTT.

3.6 Model Systems: Mecp2 Knockout, Mecp2 Mutant, and MeCP2 Knock-In Mice

To advance the basic knowledge of the role of Mecp2 on brain development and function, as well as to understand the molecular and cellular bases of RTT pathophysiology, several different mouse models were generated based on targeted manipulations of the endogenous *Mecp2* gene or targeted introduction of the human *MECP2* gene, either wild-type or carrying RTT-associated mutations. The similarities between the following mouse models to the clinical presentation in RTT individuals have been recently reviewed (Calfa et al. 2011).

1. *Mecp2 knockout mice.* The first report of constitutive *Mecp2* deletion in mice described embryonic lethality (Tate et al. 1996), which led to the generation of mice using the Cre recombinase-loxP system of conditional deletion in selected tissues at desired times (Sauer 1998). Embryonic lethality was prevented by breeding mice that have *Mecp2* exons 3 and 4 flanked by loxP sites (i.e., "floxed" or *Mecp2*[2lox]) with "Cre deleter" mice that ubiquitously express a Cre transgene in the X chromosome (Schwenk et al. 1995). The resulting progeny (*Mecp2*[tm1.1Bird], "Bird strain") carry deletions spanning *Mecp2* exons 3 and 4 starting in early embryonic development (Guy et al. 2001). These mice showed no signs of Mecp2 expression using antibodies directed against either N- or C-termini of the protein (Guy et al. 2001; Braunschweig et al. 2004). In addition, brain-specific targeting to the neuron and glial lineage was achieved by breeding floxed *Mecp2* mice with mice expressing a *Nestin-Cre* transgene, which is highly expressed in neural precursor cells beginning at E12 (Tronche et al. 1999). However, it should be noted that the mice with *Mecp2* exons flanked by loxP sites used in the generation of these knockouts (Bird *Mecp2*[2lox/Y]) already show ~50% lower levels of Mecp2 protein than wild-type controls (without Cre-mediated recombination), and thus should be considered "*Mecp2* hypomorphs" (see below) (Samaco et al. 2008).

   A recent report described the generation of a *Mecp2* knockout mouse line by removal of the coding MBD sequence (entire exon 3 and part of exon 4) and introduction of a nonfunctional splicing site at the 5′ end of the gene sequence encoding the TRD, preventing splicing and transcription of downstream *Mecp2* sequence for the TRD, the C-terminal domain and the 3′ UTR (*Mecp2*[tm1Tam]) (Pelka et al. 2006). These mice showed no signs of Mecp2 protein expression using antibodies directed against either N- or C-termini and developed a postnatal behavioral phenotype that resembles RTT.

2. *Mecp2 mutant mice expressing a truncated protein:* Brain-specific deletions of *Mecp2* exon 3, which encodes 116 amino acids including most of the MBD, led to the expression of a truncated protein in the neuron and astrocyte lineage beginning at E12 using a *Nestin-Cre* transgene (Chen et al. 2001) (*Mecp2*[2lox/X]; *Nestin-Cre*). Because these mice express a truncated Mecp2 protein lacking the MBD, but with an intact C-terminus—which includes the nuclear localization signal and potentially the TRD and other downstream domains (Chen et al. 2001;

Braunschweig et al. 2004; Luikenhuis et al. 2004)—they should be considered mutant mice rather than "knockout" or "null" mice. Therefore, the consequences of the expression of a mutant protein of unknown function should be taken into account. In addition, it is unknown whether the introduction of the loxP sites flanking $Mecp2$ exon 3 in Jaenisch $Mecp2^{2lox}$ mice (before Cre-mediated loxP deletion) had consequences on Mecp2 protein expression similar to that detected in Bird $Mecp2^{2lox/Y}$ mice (Samaco et al. 2008). It should be noted that most studies of this so-called Jaenisch strain use mice derived by germline recombination ($Mecp2^{1lox}$), where the mutations in $Mecp2$ are no longer brain-specific (Chen et al. 2001).

The introduction of a premature STOP codon in the mouse $Mecp2$ gene led to the expression of a protein truncated at amino acid 308 ($Mecp2^{308}$) (Shahbazian et al. 2002). These mice express a truncated protein with the MBD and a portion of the TRD elements still intact, suggesting residual protein function, as thought to be the case in RTT patients presenting with milder disease features.

3. *Cell-type-specific Mecp2 deletions or mutations.* The Cre/loxP recombination system has been used to generate the following conditional $Mecp2$ knockout or mutant mice:

   (a) Breeding Jaenisch $Mecp2^{2lox}$ mice with mice expressing Cre under control of the $CamkII$ promoter ($CamkII$-Cre^{93} line) (Minichiello et al. 1999) allowed the $Mecp2$ mutations to be selectively expressed in forebrain excitatory neurons after approximately postnatal day 20 (Chen et al. 2001). It should be noted that these mice express a milder behavioral phenotype with a delayed onset, compared to that caused by more widespread deletion (Guy et al. 2001) or truncation (Chen et al. 2001) of Mecp2 using the *Nestin-Cre* transgene.

   (b) Breeding Bird $Mecp2^{2lox/Y}$ mice with mice expressing Cre under control of the $Sim1$ promoter (Balthasar et al. 2005) yielded cell-type-specific $Mecp2$ deletions in hypothalamic neurons (Fyffe et al. 2008).

   (c) Breeding Bird $Mecp2^{2lox/Y}$ mice with mice expressing Cre in tyrosine hydroxylase (TH) neurons (Lindeberg et al. 2004) generated another cell-type-specific $Mecp2$ deletion (Samaco et al. 2009).

   (d) Breeding Bird $Mecp2^{2lox/Y}$ mice with mice expressing Cre in PC12 *ets* factor 1 (PET1)-expressing neurons (Scott et al. 2005) generated another cell-type-specific $Mecp2$ deletion (Samaco et al. 2009).

   (e) Lastly, infusions of *Cre*-expressing lentiviruses into specific brain areas of Jaenisch $Mecp2^{2lox}$ mice yielded useful mouse models of RTT amenable for behavioral studies without confounding issues of brain development (Adachi et al. 2009).

4. *Mice expressing reduced levels of Mecp2.* The introduction of loxP sites flanking $Mecp2$ exons 3 and 4 for the generation of conditional knockout mice of the Bird knockout mouse line (Guy et al. 2001) may be the reason for a ~50% reduction in the expression of Mecp2 protein in male $Mecp2^{2lox/Y}$ compared to wild-type littermates (Samaco et al. 2008). Interestingly, these mice express a delayed and

milder behavioral phenotype similar to other *Mecp2*-deficient mice, which may originate from reduced expression levels of wild-type Mecp2 protein.

An alternative approach to reduce the expression of endogenous Mecp2 is by knock-down with small interfering RNAs (Hamilton and Baulcombe 1999; Elbashir et al. 2001). Intraventricular injections of lentiviruses that express a *Mecp2*-targeted short hairpin RNA (shRNA) in 1-day-old rat pups reduced *Mecp2* mRNA levels and caused subtle but transient sensory-motor impairments (Jin et al. 2008). So far, the in utero transfection of shRNA constructs to knock-down Mecp2 expression in sparsely distributed cortical pyramidal neurons in mice has been used to map intracortical connectivity by glutamate uncaging and laser scanning photostimulation in brain slices (Wood et al. 2009).

5. *Mice overexpressing wild-type full-length MeCP2.* The introduction of the human *MECP2* gene under control of its entire regulatory promoter using the P1-derived artificial chromosome (PAC) led to a ~2-fold increase in expression levels compared to the endogenous mouse *Mecp2* (MeCP2^{Tg1} mice) (Collins et al. 2004). Another mouse line overexpressed a Tau-MeCP2 fusion protein selectively in postmitotic neurons from the *Tau* locus in homozygous *Tau* knockout mice (Luikenhuis et al. 2004).

6. *Knock-in mice carrying RTT-associated MECP2 mutations.* So far, only two mouse strains carrying RTT-associated mutations have been generated: the R168X mouse (*Mecp2^{R168X}*) (Lawson-Yuen et al. 2007) and the MeCP2 A140V mouse (Jentarra et al. 2010). In theory, female heterozygous mice with mosaic expression of RTT-associated mutations in *MECP2* would represent the closest experimental animal model of the human disease. The next closest experimental model would be human neurons derived from induced pluripotent stem (iPS) cells obtained from individual RTT patients (e.g., Hotta et al. 2009), with the obvious limitation of being a dissociated culture system allowing only studies at the molecular and cellular level, and not network or behavioral studies.

The most important feature of all these genetically manipulated mice is that all present some behavioral features that correlate well with specific clinical symptoms observed in RTT individuals, although no single mouse line truly mimics the human disease (reviewed by Ricceri et al. 2008; Tao et al. 2009).

It should be stressed that most studies use male hemizygous mice (i.e., *Mecp2$^{-/y}$*) because they consistently develop a severe and characteristic behavioral phenotype much earlier than female heterozygous mice (i.e., *Mecp2$^{-/+}$*), which express a mosaic pattern of wild-type and mutant cells due to XCI. However, XCI is not uniform in female heterozygous mice from the *Mecp2$^{tm1.1Jae}$* and *Mecp2$^{308/X}$* mutant lines, or from the *Mecp2$^{tm1.1Bird}$* knockout strains, being skewed towards the wild-type *Mecp2* allele (Braunschweig et al. 2004; Young and Zoghbi 2004). Intriguingly, wild-type cells in *Mecp2$^{tm1.1Jae}$* mutant and *Mecp2$^{tm1.1Bird}$* knockout mice express lower levels of Mecp2 protein than in wild-type mice (Braunschweig et al. 2004). Therefore, the delayed appearance and variability of the behavioral phenotypes observed in *Mecp2$^{-/+}$* heterozygous female mice could be caused by the combination of mosaic expression of mutant *Mecp2*, the degree of XCI unbalance, and reduced Mecp2 levels in wild-type cells. To simplify the analyses by reducing the contribution of these contributing factors, *Mecp2^{-/y}* male hemizygous mice are used as a more homogenous population, which seem more amenable for experimental work in the laboratory. There are, however, a number of studies that compare RTT-like phenotypes between *Mecp2^{-/+}* heterozygous female and *Mecp2^{-/y}* hemizygous male mice and their wild-type littermates, and all agree that female heterozygous mice display a delayed onset, milder phenotype than male hemizygous mice (e.g., Metcalf et al. 2006; Stearns et al. 2007; Bissonnette and Knopp 2008; Jugloff et al. 2008; Kondo et al. 2008; Ward et al. 2008; Belichenko et al. 2009a; D'Cruz et al. 2010; Isoda et al. 2010; Lonetti et al. 2010; Roux et al. 2010).

Another critical point is that only one study characterized the consequences of the complete deletion of the Mecp2 protein without potential deleterious effects of genetic engineering (e.g., *neo* cassettes) (Pelka et al. 2006) since at least one of the mouse lines commonly called "*Mecp2* null" or "*Mecp2* knockout" in fact expresses an internally deleted protein detectable by western blotting and immunohistochemistry (Chen et al. 2001; Braunschweig et al. 2004; Luikenhuis et al. 2004). Despite these limitations, Mecp2-based mouse models represent useful experimental models of RTT in which to test novel therapeutic approaches before moving to the clinic.

3.7 Reversal of Behavioral and Cellular Impairments in MeCP2-Based Mouse Models of RTT

The experimental reversal of behavioral and synaptic impairments in several models of neurodevelopmental disorders by pharmacological approaches in adult animals has raised hope for similar interventions in humans after the onset of neurological symptoms (Ehninger et al. 2008b). For example, the behavioral impairment in a mouse model of Down syndrome (Ts65Dn) caused by an excitatory/inhibitory imbalance of synaptic function in the hippocampus can be reverted by $GABA_AR$ antagonists in symptomatic animals (Fernandez et al. 2007; Rueda et al. 2008). A mouse model of neurofibromatosis-1 (*Nf1^{+/-}*) also has higher levels of inhibition than their wild-type littermates (but comparable excitation), which can be reversed by decreasing Ras/MAPK signaling (Costa et al. 2002; Li et al. 2005). Also, mice that model tuberous sclerosis (*Tsc2^{+/-}*) improve after treatment with inhibitors of the mTOR/Akt signaling cascade (Meikle et al. 2008; Zeng et al. 2008; Ehninger et al. 2008a). In addition, a mouse model of Rubinstein–Taybi syndrome (*CBP^{+/-}*) improves after treatment with inhibitors of either phosphodiesterase 4 (Bourtchouladze et al. 2003) or histone deacetylases (Alarcon et al. 2004). Using a genetic manipulation, most neurological deficits in a mouse model of Fragile X syndrome (*Fmr1^{-/-}*) are prevented after breeding them with heterozygous *mGluR5* mice (Dolen et al. 2007). Likewise, reducing aCaMKII inhibitory phosphorylation in a mouse model of Angelman syndrome (*Ube3a* mutants) by crossing them with mice harboring a targeted aCaMKII mutation (T305V/T306A) prevents the development of Angelman syndrome-like behavioral deficits (van Woerden et al. 2007).

As we reviewed in the preceding sections, the neuropathology in RTT individuals and *Mecp2*-deficient mice is subtle, including reduced neuronal complexity and dendritic spine density rather than severe neuronal degeneration (Belichenko et al. 1994, 2009a, b; Belichenko and Dahlstrom 1995; Armstrong 2002; Kishi and Macklis 2004; Chapleau et al. 2009a), thus raising the possibility that some specific deficits in RTT individuals may be reversible (Cobb et al. 2010).

Several experimental approaches have been tested for the reversal of behavioral impairments in symptomatic *Mecp2*-deficient mice, four based on gene expression manipulations and two on pharmacological treatments, in addition to a dietary supplementation and behavioral interventions in the form of rearing in enriched environments:

1. To selectively increase the expression of full-length wild-type Mecp2 in post-mitotic neurons of *Mecp2* mutant mice, they were crossed with a mouse line overexpressing a Tau-MeCP2 fusion protein from the *Tau* locus in homozygous *Tau* knockout mice. The resulting offspring showed improved body and brain weights, as well as locomotor activity and fertility, compared to *Mecp2* mutants, and seemed indistinguishable from their wild-type littermates (Luikenhuis et al. 2004).

2. Inducible and neuron-specific expression of human *MECP2* in either *Mecp2* knockout mice or $Mecp2^{308/Y}$ mice was achieved by using tetracycline-inducible *MECP2* under control of either the *CamkII* or the *Eno2* promoters. Despite the presence of specific patterns of transgene expression, most behavioral impairments in $Mecp2^{308/Y}$ mice (i.e., dowel test, suspended wire, and accelerating rotarod) were not improved by neuron-specific *MECP2* transgene expression. Similarly, neuron-specific *MECP2* transgene expression failed to extend the longevity or prevent the tremors and breathing irregularities of *Mecp2* knockout mice, with a subtle effect on locomotor activity in their home cages (Alvarez-Saavedra et al. 2007). The conclusion of these studies was that either the levels of MeCP2 achieved were insufficient or not in the relevant brain regions. Alternatively, these results may reflect the critical role of proper Mecp2 expression in glial cells and its non-cell autonomous consequences on neuronal structure and function (see preceding section).

3. *Mecp2* overexpression in postmitotic neurons of *Mecp2* knockout mice was achieved by using tetracycline-inducible *Mecp2-e2* cDNA under control of the *CamkII* promoter. Females of this rescue line showed improved rearing activity, overall mobility, and rotarod performance compared to female $Mecp2^{-/+}$ mice, reaching levels of performance comparable to wild-type littermates (Jugloff et al. 2008).

4. The reactivation of the endogenous *Mecp2* gene under control of its own promoter and regulatory elements was achieved by silencing it with a *lox-Stop* cassette, which can be removed by transgene expression of a fusion protein between Cre recombinase and a modified estrogen receptor (*Cre-ER*). The Cre-ER protein remains in the cytoplasm, unless exposed to the estrogen analog tamoxifen, which induces its nuclear translocation. Male mice of this strain

($Mecp2^{lox\text{-}Stop/y}$) developed RTT-like symptoms at 16 weeks of age and survived for ~11 weeks, being comparable to $Mecp2$ knockout mice. Tamoxifen injections in $Mecp2^{lox\text{-}Stop/y}$ mice with advanced symptoms (between 7 and 17 weeks of age) led to milder symptoms and extended life span, demonstrating that reactivation of the endogenous $Mecp2$ gene reverses established symptoms. Similar behavioral results were obtained in female $Mecp2^{lox\text{-}Stop/+}$ mice after treatment with tamoxifen. Furthermore, the impairment in high frequency- and TBS-induced LTP in area CA1 of slices from symptomatic female $Mecp2^{lox\text{-}Stop/+}$ was completely prevented by tamoxifen treatment (Guy et al. 2007).

5. A conditional $Mecp2$ transgene that can be activated by Cre-mediated deletion of a loxP-STOP-loxP cassette was used in $Mecp2$ mutant mice to show that increasing Mecp2 levels as late as 2–4 weeks of age prevented the onset of RTT-like symptoms. The mouse $Mecp2e2$ cDNA was placed downstream of a loxP-flanked STOP cassette, which in turn was downstream of the synthetic CAGGS promoter/enhancer/intron. Mice carrying this rescue transgene were crossed with $Mecp2$ mutant mice, and the resulting offspring bred with Cre deleter mice. When Cre was driven by the $Nestin$ promoter (E12 in neural precursors of neurons and glia), or by Tau promoter (postmitotic neurons), the life span of the rescue mice increased to more than 8 months (compared to 10–12 weeks in $Mecp2$ mutants). Activation of the $Mecp2$ transgene by the $CamkII$ promoter also extended the life span, but to a lesser extent. Nocturnal locomotor activity was also improved in all the lines of rescue mice, albeit more efficiently in those where Cre expression was driven earlier and in most neurons, i.e., Nest-Cre and Tau-Cre. Finally, rescue mice lacked the decrease in brain weight and neuronal soma size in hippocampus and cerebral cortex characteristic of $Mecp2$ mutant mice (Giacometti et al. 2007).

6. $Bdnf$, the gene coding for brain-derived neurotrophic factor, was one of the first $Mecp2$ targets to be identified, binding to its promoter region (Martinowich et al. 2003; Chen et al. 2003). The initial interpretation of Mecp2 as a transcriptional repressor of $Bdnf$ was later confronted with the observations that $Mecp2$ mutant mice express lower levels of BDNF mRNA and protein in the cerebral cortex and cerebellum than wild-type controls and that conditional postnatal deletion of $Bdnf$ in the forebrain, parts of midbrain and hindbrain of $Mecp2$ mutant mice exacerbated the onset of their locomotor dysfunction and shortened their longevity, two consistent RTT-associated impairments in these mice (Chang et al. 2006). Furthermore, a microarray study comparing hypothalamic samples from $Mecp2$-deficient and MeCP2^{Tg1}-overexpressing mice found that BDNF mRNA levels were lower in the absence of $Mecp2$ and higher when MeCP2 levels were doubled (Chahrour et al. 2008). Considering the well-established role of BDNF on synaptic transmission and plasticity (Black 1999; Poo 2001; Tyler et al. 2002; Lu 2003; Chapleau et al. 2009b), restoring proper levels of BDNF in Mecp2-deficient brains is an attractive therapeutic strategy.

To selectively overexpress $BDNF$ in postmitotic forebrain neurons of $Mecp2$ mutant mice, they were first bred with a mouse line that expresses Cre recombinase under control of the $CamkII$ promoter (cre93 transgenic line) (Rios et al. 2001). The resulting mice ($Mecp2^{+/-}$; cre93) were then crossed with mice carrying a human *BDNF* transgene under regulation of the synthetic CAGGS promoter/enhancer/intron followed by a loxP-STOP-loxP cassette. In the presence of Cre in postmitotic forebrain neurons, the STOP cassette is removed, resulting in the activation of the *BDNF* transgene. The resulting overexpression of *BDNF* in postnatal forebrain neurons in a *Mecp2*-deficient background extended the life span and prevented a locomotor defect (hypoactivity) as well as an electrophysiological deficit (low spike firing frequency in cortical layer V pyramidal neurons) consistently observed in *Mecp2* mutant mice (Chang et al. 2006). In support of these observations, an in vitro study showed that overexpression of *Bdnf* in primary hippocampal cultures rescued the dendritic phenotype caused by either shRNA-mediated *Mecp2* knockdown or expression of RTT-associated *MECP2* mutations (Larimore et al. 2009). Altogether, these studies indicate that BDNF levels can be targeted for therapeutic interventions to alleviate RTT symptoms and are the bases of two pharmacological approaches and the beneficial effect of environmental enrichment (see below).

7. The inability of BDNF to cross the blood-brain barrier has hampered its use as a therapeutic agent in several neurological disorders. AMPAkines are a family of allosteric modulators of AMPA-type glutamate receptors known to enhance BDNF mRNA and protein levels (Lauterborn et al. 2000; Lynch and Gall 2006). In support of their use to ameliorate RTT symptoms by elevating BDNF levels, the breathing pattern irregularities in *Mecp2* mutant mice are alleviated by a 10-day treatment with the AMPAkine CX546 (Ogier et al. 2007). Consistently, direct application of recombinant BDNF to brainstem slices from *Mecp2* mutant mice reversed their synaptic dysfunction phenotype (Kline et al. 2010). Intriguingly, cultured neurons from *Mecp2* knockout mice are able to release more BDNF than wild-type cells, despite showing lower BDNF expression levels. Such hypersecretion phenotype was also observed for catecholamine release from chromaffin cells (Wang et al. 2006). The parsimonious interpretation of these observations is that *Mecp2* null neurons may eventually exhaust their pool of releasable BDNF.

8. Supporting the potential use of "BDNF-mimetic" trophic factors to reverse the RTT-like impairments in Jaenisch *Mecp2* mutant mice, a 2-week treatment with an active tri-peptide fragment of insulin-like growth factor-1 (IGF-1) extended the life span, improved locomotor function, ameliorated breathing patterns, reduced heart rate irregularity, and increased brain weight. Indeed, the IGF-1 receptor activates intracellular pathways common to those induced by BDNF signaling through its TrkB receptor (i.e., PI3K/Akt and MAPK) (Zheng and Quirion 2004). Furthermore, IGF-1 partially restored dendritic spine density in pyramidal neurons of layer V in motor cortex, the amplitude of spontaneous EPSCs in pyramidal neurons of sensorimotor cortex, the cortical expression of the synaptic scaffolding protein PSD-95, and stabilized cortical plasticity in *Mecp2* mutant mice to wild-type levels (Tropea et al. 2009).

9. Daily injections of desipramine, a selective inhibitor of the norepinephrine transporter used to increase extracellular levels of this neurotransmitter, improved respiratory rhythm, the number of tyrosine hydroxylase-expressing neurons in the brainstem, and longevity in *Mecp2* knockout mice (Roux et al. 2007).

10. Dietary choline supplementation improved motor coordination and locomotor activity in male *Mecp2* mutant mice and enhanced grip strength in female *Mecp2* mutant mice (Nag and Berger-Sweeney 2007). Increased NGF protein levels in the striatum (Nag et al. 2008) and of N-acetylaspartate content, as measured by NMR spectroscopy (Ward et al. 2009), suggest improved neuronal proliferation and survival after choline supplementation in *Mecp2* mutant mice.

11. In several neurological disorders, environmental enrichment has beneficial effects on various behavioral and cellular phenotypes, including increased levels of BDNF expression (van Praag et al. 2000; Nithianantharajah and Hannan 2006). Potentially related to the ability of either *Bdnf* overexpression (Chang et al. 2006) or increased BDNF levels after AMPAkine treatment (Ogier et al. 2007) to improve RTT-like symptoms in *Mecp2*-deficient mice, rearing them in enriched environments also ameliorated some of their behavioral and synaptic phenotypes. For example, the cerebellar and hippocampal-/amygdala-based learning deficits, as well as the reduced motor dexterity and decreased anxiety levels characteristic of heterozygous *Mecp2*[tm1Tam] females, were prevented by their housing in larger-sized home cages with nesting material; a variety of objects with different textures, shapes, and sizes; and running wheels starting at 4 weeks of age (Kondo et al. 2008). Similarly, housing male mice of the Jaenisch *Mecp2* mutant line in enriched environments starting at weaning (postnatal day 21) improved their locomotor activity, but not motor coordination or contextual or cued fear conditioning. Curiously, MRI revealed a reduction in ventricular volume after environmental enrichment in both *Mecp2* mutant and wild-type littermates, without changes in total brain volume. Together with the known reduction in brain size in *Mecp2*-deficient mice, this observation suggests that environmental enrichment selectively increased grey and white matter (Nag et al. 2009).

Intriguingly, more robust effects were obtained when environmental enrichment began earlier in the development of the pups and included their dams, which displayed enhanced maternal care behaviors. Rearing Jaenisch *Mecp2* mutant mice and their dams from postnatal day 10 in enriched environments led to improved motor coordination and motor learning compared to control *Mecp2* mutant mice kept in standard housing conditions. In addition, environmental enrichment prevented the deficit of TBS-induced LTP in layer II/III of slices from primary somatosensory cortex typical of *Mecp2* mutant mice kept in standard housing. Despite not being different between wild-type and *Mecp2* mutant mice, the density of spine synapses in layer III of the primary sensory cortex, as well as of parallel fiber-Purkinje cell synapses in the molecular layer of the cerebellum, was higher in mice reared in enriched environments than in standard housing, suggesting that *Mecp2*-deficient neurons are still capable of structural plasticity. Consistent with previous reports in rats and mice, this protocol of early environmental intervention increased the levels of BDNF mRNA and protein in the cerebral cortex of both wild-type and *Mecp2* mutant mice (Lonetti et al. 2010).

Altogether, these successful therapeutic approaches that reversed many RTT-like symptoms in both Jaenisch *Mecp2* mutant and Bird *Mecp2* knockout mouse lines provide further support to the potential pharmacological reversal of neurodevelopmental disorders in adults (Ehninger et al. 2008b).

3.8 Summary

The wide range of clinical symptoms in RTT individuals and of phenotypes in MeCP2-based mouse models was initially thought to originate from unbalanced XCI. Ample evidence in mouse models now indicates that the role of MeCP2 in neuronal development and function is very different across various brain regions, suggesting that dysfunction in specific neuronal populations due to differential distribution of mutant MeCP2 also contributes to phenotypic variability. In addition, the recently uncovered role of *Mecp2* dysfunction in glial cells leading to neuronal pathology cannot be overlooked, raising more questions regarding the primary deficits that initiate the cascade of events leading to clinical symptoms. Also, the realization that MeCP2 acts as both a repressor and activator of potentially thousands of genes has increased the complexity of this once thought simple monogenetic disorder. Finally, none of the experimental approaches tested in MeCP2-based mouse models fully reversed their RTT-like phenotypes, suggesting additional molecular and cellular deficits. In spite of these seemingly overwhelming limitations in our state of knowledge, we should remind ourselves that we have learned more in the last decade since the discovery that MeCP2 mutations cause RTT than in the preceding 30 years from the first description by Andreas Rett. Following this trajectory, it is likely that rational therapies grounded on basic scientific knowledge will be available for RTT individuals within the next decade.

Acknowledgements We thank Dr. Carolyn Schanen for useful comments on the manuscript. This is supported by NIH grants from NINDS (NS40593, NS057780, NS-065027) and NICHD (U54 grant HD061222 and IDDRC grant HD38985), IRSF, and the Civitan Foundation. The authors acknowledge the gracious participation and provision of information by families in the Rare Disease Natural History Study for which Dr. Mary Lou Oster-Granite, Health Scientist Administrator at NICHD, provided invaluable guidance, support, and encouragement.

References

Adachi, M., Autry, A. E., Covington, H. E., III & Monteggia, L. M. (2009). MeCP2-mediated transcription repression in the basolateral amygdala may underlie heightened anxiety in a mouse model of Rett syndrome. *J Neurosci* 29, 4218–4227.

Alarcon, J. M., Malleret, G., Touzani, K., Vronskaya, S., Ishii, S., Kandel, E. R. & Barco, A. (2004). Chromatin acetylation, memory, and LTP are impaired in CBP+/- mice: a model for the cognitive deficit in Rubinstein-Taybi syndrome and its amelioration. *Neuron* 42, 947–959.

Alvarez-Saavedra, M., Saez, M. A., Kang, D., Zoghbi, H. Y. & Young, J. I. (2007). Cell-specific expression of wild-type MeCP2 in mouse models of Rett syndrome yields insight about pathogenesis. *Hum Mol Genet* 16, 2315–2325.

Amir, R. E., Van, d. V., I, Wan, M., Tran, C. Q., Francke, U. & Zoghbi, H. Y. (1999). Rett syndrome is caused by mutations in X-linked MECP2, encoding methyl-CpG-binding protein 2. *Nat Genet* 23, 185–188.

Archer, H. L., Whatley, S. D., Evans, J. C., Ravine, D., Huppke, P., Kerr, A., Bunyan, D., Kerr, B., Sweeney, E., Davies, S. J., Reardon, W., Horn, J., MacDermot, K. D., Smith, R.

A., Magee, A., Donaldson, A., Crow, Y., Hermon, G., Miedzybrodzka, Z., Cooper, D. N., Lazarou, L., Butler, R., Sampson, J., Pilz, D. T., Laccone, F. & Clarke, A. J. (2006). Gross rearrangements of the MECP2 gene are found in both classical and atypical Rett syndrome patients. *J Med Genet* 43, 451–456.

Armstrong, D., Dunn, J. K., Antalffy, B. & Trivedi, R. (1995). Selective dendritic alterations in the cortex of Rett syndrome. *J Neuropathol Exp Neurol* 54, 195–201.

Armstrong, D. D. (2002). Neuropathology of Rett syndrome. *Ment Retard Dev Disabil Res Rev* 8, 72–76.

Armstrong, D. D. (2005). Neuropathology of Rett syndrome. *J Child Neurol* 20, 747–753.

Armstrong, D. D., Dunn, K. & Antalffy, B. (1998). Decreased dendritic branching in frontal, motor and limbic cortex in Rett syndrome compared with trisomy 21. *J Neuropathol Exp Neurol* 57, 1013–1017.

Balthasar, N., Dalgaard, L. T., Lee, C. E., Yu, J., Funahashi, H., Williams, T., Ferreira, M., Tang, V., McGovern, R. A., Kenny, C. D., Christiansen, L. M., Edelstein, E., Choi, B., Boss, O., Aschkenasi, C., Zhang, C. Y., Mountjoy, K., Kishi, T., Elmquist, J. K. & Lowell, B. B. (2005). Divergence of melanocortin pathways in the control of food intake and energy expenditure. *Cell* 123, 493–505.

Bauman, M. L., Kemper, T. L. & Arin, D. M. (1995a). Microscopic observations of the brain in Rett syndrome. *Neuropediatrics* 26, 105–108.

Bauman, M. L., Kemper, T. L. & Arin, D. M. (1995b). Pervasive neuroanatomic abnormalities of the brain in three cases of Rett's syndrome. *Neurology* 45, 1581–1586.

Bebbington, A., Anderson, A., Ravine, D., Fyfe, S., Pineda, M., de Klerk, N., Ben Zeev, B., Yatawara, N., Percy, A., Kaufmann, W. E. & Leonard, H. (2008). Investigating genotype-phenotype relationships in Rett syndrome using an international data set. *Neurology* 70, 868–875.

Belichenko, N. P., Belichenko, P. V. & Mobley, W. C. (2009a). Evidence for both neuronal cell autonomous and nonautonomous effects of methyl-CpG-binding protein 2 in the cerebral cortex of female mice with Mecp2 mutation. *Neurobiol Dis* 34, 71–77.

Belichenko, P. V. & Dahlstrom, A. (1995). Studies on the 3-dimensional architecture of dendritic spines and varicosities in human cortex by confocal laser scanning microscopy and Lucifer yellow microinjections. *J Neurosci Methods* 57, 55–61.

Belichenko, P. V., Oldfors, A., Hagberg, B. & Dahlstrom, A. (1994). Rett syndrome: 3-D confocal microscopy of cortical pyramidal dendrites and afferents. *Neuroreport* 5, 1509–1513.

Belichenko, P. V., Wright, E. E., Belichenko, N. P., Masliah, E., Li, H. H., Mobley, W. C. & Francke, U. (2009b). Widespread changes in dendritic and axonal morphology in Mecp2-mutant mouse models of Rett syndrome: evidence for disruption of neuronal networks. *J Comp Neurol* 514, 240–258.

Bissonnette, J. M. & Knopp, S. J. (2008). Effect of inspired oxygen on periodic breathing in methy-CpG-binding protein 2 (Mecp2) deficient mice. *J Appl Physiol* 104, 198–204.

Black, I. B. (1999). Trophic regulation of synaptic plasticity. *J Neurobiol* 41, 108–118.

Blue, M. E., Naidu, S. & Johnston, M. V. (1999a). Altered development of glutamate and GABA receptors in the basal ganglia of girls with Rett syndrome. *Exp Neurol* 156, 345–352.

Blue, M. E., Naidu, S. & Johnston, M. V. (1999b). Development of amino acid receptors in frontal cortex from girls with Rett syndrome. *Ann Neurol* 45, 541–545.

Bourtchouladze, R., Lidge, R., Catapano, R., Stanley, J., Gossweiler, S., Romashko, D., Scott, R. & Tully, T. (2003). A mouse model of Rubinstein-Taybi syndrome: defective long-term memory is ameliorated by inhibitors of phosphodiesterase 4. *Proc Natl Acad Sci U S A* 100, 10518–10522.

Braunschweig, D., Simcox, T., Samaco, R. C. & LaSalle, J. M. (2004). X-Chromosome inactivation ratios affect wild-type MeCP2 expression within mosaic Rett syndrome and Mecp2-/+ mouse brain. *Hum Mol Genet* 13, 1275–1286.

Calfa, G., Percy, A. K. & Pozzo-Miller, L. (2011). Experimental models of Rett syndrome based on Mecp2 dysfunction. *Exp Biol Med (Maywood)* 236, 3–19.

Chahrour, M., Jung, S. Y., Shaw, C., Zhou, X., Wong, S. T., Qin, J. & Zoghbi, H. Y. (2008). MeCP2, a key contributor to neurological disease, activates and represses transcription. *Science* 320, 1224–1229.

3 Dysfunction of the Methyl-CpG-Binding Protein MeCP2 in Rett Syndrome

Chahrour, M. & Zoghbi, H. Y. (2007). The story of Rett syndrome: from clinic to neurobiology. *Neuron* 56, 422–437.

Chang, Q., Khare, G., Dani, V., Nelson, S. & Jaenisch, R. (2006). The disease progression of Mecp2 mutant mice is affected by the level of BDNF expression. *Neuron* 49, 341–348.

Chapleau, C. A., Calfa, G. D., Lane, M. C., Albertson, A. J., Larimore, J. L., Kudo, S., Armstrong, D. L., Percy, A. K. & Pozzo-Miller, L. (2009a). Dendritic spine pathologies in hippocampal pyramidal neurons from Rett syndrome brain and after expression of Rett-associated MECP2 mutations. *Neurobiol Dis* 35, 219–233.

Chapleau, C. A., Larimore, J. L., Theibert, A. & Pozzo-Miller, L. (2009b). Modulation of dendritic spine development and plasticity by BDNF and vesicular trafficking: fundamental roles in neurodevelopmental disorders associated with mental retardation and autism. *J Neurodev Disord* 1, 185–196.

Chavrier, P., Zerial, M., Lemaire, P., Almendral, J., Bravo, R. & Charnay, P. (1988). A gene encoding a protein with zinc fingers is activated during G0/G1 transition in cultured cells. *EMBO J* 7, 29–35.

Chen, R. Z., Akbarian, S., Tudor, M. & Jaenisch, R. (2001). Deficiency of methyl-CpG binding protein-2 in CNS neurons results in a Rett-like phenotype in mice. *Nat Genet* 27, 327–331.

Chen, W. G., Chang, Q., Lin, Y., Meissner, A., West, A. E., Griffith, E. C., Jaenisch, R. & Greenberg, M. E. (2003). Derepression of BDNF transcription involves calcium-dependent phosphorylation of MeCP2. *Science* 302, 885–889.

Cobb, S., Guy, J. & Bird, A. (2010). Reversibility of functional deficits in experimental models of Rett syndrome. *Biochem Soc Trans* 38, 498–506.

Collins, A. L., Levenson, J. M., Vilaythong, A. P., Richman, R., Armstrong, D. L., Noebels, J. L., Sweatt, J. D. & Zoghbi, H. Y. (2004). Mild overexpression of MeCP2 causes a progressive neurological disorder in mice. *Hum Mol Genet* 13, 2679–2689.

Costa, R. M., Federov, N. B., Kogan, J. H., Murphy, G. G., Stern, J., Ohno, M., Kucherlapati, R., Jacks, T. & Silva, A. J. (2002). Mechanism for the learning deficits in a mouse model of neurofibromatosis type 1. *Nature* 415, 526–530.

D'Cruz, J. A., Wu, C., Zahid, T., El Hayek, Y., Zhang, L. & Eubanks, J. H. (2010). Alterations of cortical and hippocampal EEG activity in MeCP2-deficient mice. *Neurobiol Dis* 38, 8–16.

del Gaudio, D., Fang, P., Scaglia, F., Ward, P. A., Craigen, W. J., Glaze, D. G., Neul, J. L., Patel, A., Lee, J. A., Irons, M., Berry, S. A., Pursley, A. A., Grebe, T. A., Freedenberg, D., Martin, R. A., Hsich, G. E., Khera, J. R., Friedman, N. R., Zoghbi, H. Y., Eng, C. M., Lupski, J. R., Beaudet, A. L., Cheung, S. W. & Roa, B. B. (2006). Increased MECP2 gene copy number as the result of genomic duplication in neurodevelopmentally delayed males. *Genet Med* 8, 784–792.

Dolen, G., Osterweil, E., Rao, B. S., Smith, G. B., Auerbach, B. D., Chattarji, S. & Bear, M. F. (2007). Correction of fragile X syndrome in mice. *Neuron* 56, 955–962.

Dragich, J. M., Kim, Y. H., Arnold, A. P. & Schanen, N. C. (2007). Differential distribution of the MeCP2 splice variants in the postnatal mouse brain. *J Comp Neurol* 501, 526–542.

Ehninger, D., Han, S., Shilyansky, C., Zhou, Y., Li, W., Kwiatkowski, D. J., Ramesh, V. & Silva, A. J. (2008a). Reversal of learning deficits in a Tsc2+/- mouse model of tuberous sclerosis. *Nat Med* 14, 843–848.

Ehninger, D., Li, W., Fox, K., Stryker, M. P. & Silva, A. J. (2008b). Reversing neurodevelopmental disorders in adults. *Neuron* 60, 950–960.

Elbashir, S. M., Harborth, J., Lendeckel, W., Yalcin, A., Weber, K. & Tuschl, T. (2001). Duplexes of 21-nucleotide RNAs mediate RNA interference in cultured mammalian cells. *Nature* 411, 494–498.

Ellaway, C. J., Sholler, G., Leonard, H. & Christodoulou, J. (1999). Prolonged QT interval in Rett syndrome. *Arch Dis Child* 80, 470–472.

Fernandez, F., Morishita, W., Zuniga, E., Nguyen, J., Blank, M., Malenka, R. C. & Garner, C. C. (2007). Pharmacotherapy for cognitive impairment in a mouse model of Down syndrome. *Nat Neurosci* 10, 411–413.

Fiala, J. C., Spacek, J. & Harris, K. M. (2002). Dendritic spine pathology: cause or consequence of neurological disorders? *Brain Res Brain Res Rev* 39, 29–54.

Friez, M. J., Jones, J. R., Clarkson, K., Lubs, H., Abuelo, D., Bier, J. A., Pai, S., Simensen, R., Williams, C., Giampietro, P. F., Schwartz, C. E. & Stevenson, R. E. (2006). Recurrent infections, hypotonia, and mental retardation caused by duplication of MECP2 and adjacent region in Xq28. *Pediatrics* 118, e1687–e1695.

Fyffe, S. L., Neul, J. L., Samaco, R. C., Chao, H. T., Ben Shachar, S., Moretti, P., McGill, B. E., Goulding, E. H., Sullivan, E., Tecott, L. H. & Zoghbi, H. Y. (2008). Deletion of Mecp2 in Sim1-expressing neurons reveals a critical role for MeCP2 in feeding behavior, aggression, and the response to stress. *Neuron* 59, 947–958.

Giacometti, E., Luikenhuis, S., Beard, C. & Jaenisch, R. (2007). Partial rescue of MeCP2 deficiency by postnatal activation of MeCP2. *Proc Natl Acad Sci U S A* 104, 1931–1936.

Girard, M., Couvert, P., Carrie, A., Tardieu, M., Chelly, J., Beldjord, C. & Bienvenu, T. (2001). Parental origin of de novo MECP2 mutations in Rett syndrome. *Eur J Hum Genet* 9, 231–236.

Glaze, D. G. (2005). Neurophysiology of Rett syndrome. *J Child Neurol* 20, 740–746.

Glaze, D. G., Percy, A. K., Skinner, S., Motil, K. J., Neul, J. L., Barrish, J. O., Lane, J. B., Geerts, S. P., Annese, F., Graham, J., McNair, L. & Lee, H. S. (2010). Epilepsy and the natural history of Rett syndrome. *Neurology* 74, 909–912.

Glaze, D. G., Schultz, R. J. & Frost, J. D. (1998). Rett syndrome: characterization of seizures versus non-seizures. *Electroencephalogr Clin Neurophysiol* 106, 79–83.

Guy, J., Gan, J., Selfridge, J., Cobb, S. & Bird, A. (2007). Reversal of neurological defects in a mouse model of Rett syndrome. *Science* 315, 1143–1147.

Guy, J., Hendrich, B., Holmes, M., Martin, J. E. & Bird, A. (2001). A mouse Mecp2-null mutation causes neurological symptoms that mimic Rett syndrome. *Nat Genet* 27, 322–326.

Hagberg, B., Aicardi, J., Dias, K. & Ramos, O. (1983). A progressive syndrome of autism, dementia, ataxia, and loss of purposeful hand use in girls: Rett's syndrome: report of 35 cases. *Ann Neurol* 14, 471–479.

Hamilton, A. J. & Baulcombe, D. C. (1999). A species of small antisense RNA in posttranscriptional gene silencing in plants. *Science* 286, 950–952.

Hendrich, B. & Bird, A. (1998). Identification and characterization of a family of mammalian methyl-CpG binding proteins. *Mol Cell Biol* 18, 6538–6547.

Herdegen, T., Kiessling, M., Bele, S., Bravo, R., Zimmermann, M. & Gass, P. (1993). The KROX-20 transcription factor in the rat central and peripheral nervous systems: novel expression pattern of an immediate early gene-encoded protein. *Neuroscience* 57, 41–52.

Hotta, A., Cheung, A. Y., Farra, N., Vijayaragavan, K., Seguin, C. A., Draper, J. S., Pasceri, P., Maksakova, I. A., Mager, D. L., Rossant, J., Bhatia, M. & Ellis, J. (2009). Isolation of human iPS cells using EOS lentiviral vectors to select for pluripotency. *Nat Methods* 6, 370–376.

Isoda, K., Morimoto, M., Matsui, F., Hasegawa, T., Tozawa, T., Morioka, S., Chiyonobu, T., Nishimura, A., Yoshimoto, K. & Hosoi, H. (2010). Postnatal changes in serotonergic innervation to the hippocampus of methyl-CpG-binding protein 2-null mice. *Neuroscience* 165, 1254–1260.

Jellinger, K., Armstrong, D., Zoghbi, H. Y. & Percy, A. K. (1988). Neuropathology of Rett syndrome. *Acta Neuropathol* 76, 142–158.

Jentarra, G. M., Olfers, S. L., Rice, S. G., Srivastava, N., Homanics, G. E., Blue, M., Naidu, S. & Narayanan, V. (2010). Abnormalities of cell packing density and dendritic complexity in the MeCP2 A140V mouse model of Rett syndrome/X-linked mental retardation. *BMC Neurosci* 11:19.

Jin, J., Bao, X., Wang, H., Pan, H., Zhang, Y. & Wu, X. (2008). RNAi-induced down-regulation of Mecp2 expression in the rat brain. *Int J Dev Neurosci* 26, 457–465.

Jugloff, D. G., Vandamme, K., Logan, R., Visanji, N. P., Brotchie, J. M. & Eubanks, J. H. (2008). Targeted delivery of an Mecp2 transgene to forebrain neurons improves the behavior of female Mecp2-deficient mice. *Hum Mol Genet* 17, 1386–1396.

Jung, B. P., Jugloff, D. G., Zhang, G., Logan, R., Brown, S. & Eubanks, J. H. (2003). The expression of methyl CpG binding factor MeCP2 correlates with cellular differentiation in the developing rat brain and in cultured cells. *J Neurobiol* 55, 86–96.

Kankirawatana, P., Leonard, H., Ellaway, C., Scurlock, J., Mansour, A., Makris, C. M., Dure, L. S., Friez, M., Lane, J., Kiraly-Borri, C., Fabian, V., Davis, M., Jackson, J., Christodoulou, J., Kaufmann, W. E., Ravine, D. & Percy, A. K. (2006). Early progressive encephalopathy in boys and MECP2 mutations. *Neurology* 67, 164–166.

Katz, D. M., Dutschmann, M., Ramirez, J. M. & Hilaire, G. (2009). Breathing disorders in Rett syndrome: progressive neurochemical dysfunction in the respiratory network after birth. *Respir Physiol Neurobiol* 168, 101–108.

Kaufmann, W. E., MacDonald, S. M. & Altamura, C. R. (2000). Dendritic cytoskeletal protein expression in mental retardation: an immunohistochemical study of the neocortex in Rett syndrome. *Cereb Cortex* 10, 992–1004.

Kaufmann, W. E. & Moser, H. W. (2000). Dendritic anomalies in disorders associated with mental retardation. *Cereb Cortex* 10, 981–991.

Kaufmann, W. E., Naidu, S. & Budden, S. (1995). Abnormal expression of microtubule-associated protein 2 (MAP-2) in neocortex in Rett syndrome. *Neuropediatrics* 26, 109–113.

Kaufmann, W. E., Worley, P. F., Taylor, C. V., Bremer, M. & Isakson, P. C. (1997). Cyclooxygenase-2 expression during rat neocortical development and in Rett syndrome. *Brain Dev* 19, 25–34.

Kimura, K., Nomura, Y. & Segawa, M. (1992). Middle and short latency somatosensory evoked potentials (SEPm, SEPs) in the Rett syndrome: chronological changes of cortical and subcortical involvements. *Brain Dev* 14 Suppl, S37–S42.

Kirby, R. S., Lane, J. B., Childers, J., Skinner, S. A., Annese, F., Barrish, J. O., Glaze, D. G., Macleod, P. & Percy, A. K. (2010). Longevity in Rett syndrome: analysis of the North American Database. *J Pediatr* 156, 135–138.

Kishi, N. & Macklis, J. D. (2004). MECP2 is progressively expressed in post-migratory neurons and is involved in neuronal maturation rather than cell fate decisions. *Mol Cell Neurosci* 27, 306–321.

Kline, D. D., Ogier, M., Kunze, D. L. & Katz, D. M. (2010). Exogenous brain-derived neurotrophic factor rescues synaptic dysfunction in Mecp2-null mice. *J Neurosci* 30, 5303–5310.

Kondo, M., Gray, L. J., Pelka, G. J., Christodoulou, J., Tam, P. P. & Hannan, A. J. (2008). Environmental enrichment ameliorates a motor coordination deficit in a mouse model of Rett syndrome--Mecp2 gene dosage effects and BDNF expression. *Eur J Neurosci* 27, 3342–3350.

Kriaucionis, S. & Bird, A. (2004). The major form of MeCP2 has a novel N-terminus generated by alternative splicing. *Nucleic Acids Res* 32, 1818–1823.

Larimore, J. L., Chapleau, C. A., Kudo, S., Theibert, A., Percy, A. K. & Pozzo-Miller, L. (2009). Bdnf overexpression in hippocampal neurons prevents dendritic atrophy caused by Rett-associated MECP2 mutations. *Neurobiol Dis* 34, 199–211.

Laurvick, C. L., de Klerk, N., Bower, C., Christodoulou, J., Ravine, D., Ellaway, C., Williamson, S. & Leonard, H. (2006). Rett syndrome in Australia: a review of the epidemiology. *J Pediatr* 148, 347–352.

Lauterborn, J. C., Lynch, G., Vanderklish, P., Arai, A. & Gall, C. M. (2000). Positive modulation of AMPA receptors increases neurotrophin expression by hippocampal and cortical neurons. *J Neurosci* 20, 8–21.

Lawson-Yuen, A., Liu, D., Han, L., Jiang, Z. I., Tsai, G. E., Basu, A. C., Picker, J., Feng, J. & Coyle, J. T. (2007). Ube3a mRNA and protein expression are not decreased in Mecp2R168X mutant mice. *Brain Res* 1180, 1–6.

Lewis, J. D., Meehan, R. R., Henzel, W. J., Maurer-Fogy, I., Jeppesen, P., Klein, F. & Bird, A. (1992). Purification, sequence, and cellular localization of a novel chromosomal protein that binds to methylated DNA. *Cell* 69, 905–914.

Li, W., Cui, Y., Kushner, S. A., Brown, R. A., Jentsch, J. D., Frankland, P. W., Cannon, T. D. & Silva, A. J. (2005). The HMG-CoA reductase inhibitor lovastatin reverses the learning and attention deficits in a mouse model of neurofibromatosis type 1. *Curr Biol* 15, 1961–1967.

Lindeberg, J., Usoskin, D., Bengtsson, H., Gustafsson, A., Kylberg, A., Soderstrom, S. & Ebendal, T. (2004). Transgenic expression of Cre recombinase from the tyrosine hydroxylase locus. *Genesis* 40, 67–73.

Lonetti, G., Angelucci, A., Morando, L., Boggio, E. M., Giustetto, M. & Pizzorusso, T. (2010). Early environmental enrichment moderates the behavioral and synaptic phenotype of MeCP2 null mice. *Biol Psychiatry* 67, 657–665.

Lu, B. (2003). BDNF and activity-dependent synaptic modulation. *Learn Mem* 10, 86–98.

Luikenhuis, S., Giacometti, E., Beard, C. F. & Jaenisch, R. (2004). Expression of MeCP2 in postmitotic neurons rescues Rett syndrome in mice. *Proc Natl Acad Sci U S A* 101, 6033–6038.

Lynch, G. & Gall, C. M. (2006). Ampakines and the threefold path to cognitive enhancement. *Trends Neurosci* 29, 554–562.

Mack, K. J., Cortner, J., Mack, P. & Farnham, P. J. (1992). krox 20 messenger RNA and protein expression in the adult central nervous system. *Brain Res Mol Brain Res* 14, 117–123.

Martinowich, K., Hattori, D., Wu, H., Fouse, S., He, F., Hu, Y., Fan, G. & Sun, Y. E. (2003). DNA methylation-related chromatin remodeling in activity-dependent BDNF gene regulation. *Science* 302, 890–893.

Meikle, L., Pollizzi, K., Egnor, A., Kramvis, I., Lane, H., Sahin, M. & Kwiatkowski, D. J. (2008). Response of a neuronal model of tuberous sclerosis to mammalian target of rapamycin (mTOR) inhibitors: effects on mTORC1 and Akt signaling lead to improved survival and function. *J Neurosci* 28, 5422–5432.

Meins, M., Lehmann, J., Gerresheim, F., Herchenbach, J., Hagedorn, M., Hameister, K. & Epplen, J. T. (2005). Submicroscopic duplication in Xq28 causes increased expression of the MECP2 gene in a boy with severe mental retardation and features of Rett syndrome. *J Med Genet* 42, e12.

Metcalf, B. M., Mullaney, B. C., Johnston, M. V. & Blue, M. E. (2006). Temporal shift in methyl-CpG binding protein 2 expression in a mouse model of Rett syndrome. *Neuroscience* 139, 1449–1460.

Minichiello, L., Korte, M., Wolfer, D., Kuhn, R., Unsicker, K., Cestari, V., Rossi-Arnaud, C., Lipp, H. P., Bonhoeffer, T. & Klein, R. (1999). Essential role for TrkB receptors in hippocampus-mediated learning. *Neuron* 24, 401–414.

Mnatzakanian, G. N., Lohi, H., Munteanu, I., Alfred, S. E., Yamada, T., MacLeod, P. J., Jones, J. R., Scherer, S. W., Schanen, N. C., Friez, M. J., Vincent, J. B. & Minassian, B. A. (2004). A previously unidentified MECP2 open reading frame defines a new protein isoform relevant to Rett syndrome. *Nat Genet* 36, 339–341.

Nag, N. & Berger-Sweeney, J. E. (2007). Postnatal dietary choline supplementation alters behavior in a mouse model of Rett syndrome. *Neurobiol Dis* 26, 473–480.

Nag, N., Mellott, T. J. & Berger-Sweeney, J. E. (2008). Effects of postnatal dietary choline supplementation on motor regional brain volume and growth factor expression in a mouse model of Rett syndrome. *Brain Res* 1237, 101–109.

Nag, N., Moriuchi, J. M., Peitzman, C. G., Ward, B. C., Kolodny, N. H. & Berger-Sweeney, J. E. (2009). Environmental enrichment alters locomotor behaviour and ventricular volume in Mecp2 1lox mice. *Behav Brain Res* 196, 44–48.

Nan, X., Campoy, F. J. & Bird, A. (1997). MeCP2 is a transcriptional repressor with abundant binding sites in genomic chromatin. *Cell* 88, 471–481.

Nan, X., Cross, S. & Bird, A. (1998). Gene silencing by methyl-CpG-binding proteins. *Novartis Found Symp* 214, 6–16.

Nan, X., Meehan, R. R. & Bird, A. (1993). Dissection of the methyl-CpG binding domain from the chromosomal protein MeCP2. *Nucleic Acids Res* 21, 4886–4892.

Neul, J. L., Fang, P., Barrish, J., Lane, J., Caeg, E. B., Smith, E. O., Zoghbi, H., Percy, A. & Glaze, D. G. (2008). Specific mutations in methyl-CpG-binding protein 2 confer different severity in Rett syndrome. *Neurology* 70, 1313–1321.

Nithianantharajah, J. & Hannan, A. J. (2006). Enriched environments, experience-dependent plasticity and disorders of the nervous system. *Nat Rev Neurosci* 7, 697–709.

Ogier, M., Wang, H., Hong, E., Wang, Q., Greenberg, M. E. & Katz, D. M. (2007). Brain-derived neurotrophic factor expression and respiratory function improve after ampakine treatment in a mouse model of Rett syndrome. *J Neurosci* 27, 10912–10917.

Pan, J. W., Lane, J. B., Hetherington, H. & Percy, A. K. (1999). Rett syndrome: 1H spectroscopic imaging at 4.1 Tesla. *J Child Neurol* 14, 524–528.

Pelka, G. J., Watson, C. M., Radziewic, T., Hayward, M., Lahooti, H., Christodoulou, J. & Tam, P. P. (2006). Mecp2 deficiency is associated with learning and cognitive deficits and altered gene activity in the hippocampal region of mice. *Brain* 129, 887–898.

Percy, A. K., Lane, J. B., Childers, J., Skinner, S., Annese, F., Barrish, J., Caeg, E., Glaze, D. G. & Macleod, P. (2007). Rett syndrome: North American database. *J Child Neurol* 22, 1338–1341.

Percy, A. K., Lee, H. S., Neul, J. L., Lane, J. B., Skinner, S. A., Geerts, S. P., Annese, F., Graham, J., McNair, L., Motil, K. J., Barrish, J. O. & Glaze, D. G. (2010). Profiling scoliosis in Rett syndrome. *Pediatr Res* 67, 435–439.

Poo, M. M. (2001). Neurotrophins as synaptic modulators. *Nat Rev Neurosci* 2, 24–32.

Purpura, D. P. (1974). Dendritic spine "dysgenesis" and mental retardation. *Science* 186, 1126–1128.

Ramocki, M. B., Peters, S. U., Tavyev, Y. J., Zhang, F., Carvalho, C. M., Schaaf, C. P., Richman, R., Fang, P., Glaze, D. G., Lupski, J. R. & Zoghbi, H. Y. (2009). Autism and other neuropsychiatric symptoms are prevalent in individuals with MeCP2 duplication syndrome. *Ann Neurol* 66, 771–782.

Ramocki, M. B., Tavyev, Y. J. & Peters, S. U. (2010). The MECP2 duplication syndrome. *Am J Med Genet A* 152A, 1079–1088.

Reichwald, K., Thiesen, J., Wiehe, T., Weitzel, J., Poustka, W. A., Rosenthal, A., Platzer, M., Stratling, W. H. & Kioschis, P. (2000). Comparative sequence analysis of the MECP2-locus in human and mouse reveals new transcribed regions. *Mamm Genome* 11, 182–190.

Reiss, A. L., Faruque, F., Naidu, S., Abrams, M., Beaty, T., Bryan, R. N. & Moser, H. (1993). Neuroanatomy of Rett syndrome: a volumetric imaging study. *Ann Neurol* 34, 227–234.

Rett, A. (1966). Über ein eigenartiges hirnatrophisches Syndrom bei Hyperammonämie im Kindesalter [On a unusual brain atrophy syndrome in hyperammonemia in childhood]. *Wien Med Wochenschr* 116, 723–726.

Ricceri, L., De Filippis, B. & Laviola, G. (2008). Mouse models of Rett syndrome: from behavioural phenotyping to preclinical evaluation of new therapeutic approaches. *Behav Pharmacol* 19, 501–517.

Rios, M., Fan, G., Fekete, C., Kelly, J., Bates, B., Kuehn, R., Lechan, R. M. & Jaenisch, R. (2001). Conditional deletion of brain-derived neurotrophic factor in the postnatal brain leads to obesity and hyperactivity. *Mol Endocrinol* 15, 1748–1757.

Ronnett, G. V., Leopold, D., Cai, X., Hoffbuhr, K. C., Moses, L., Hoffman, E. P. & Naidu, S. (2003). Olfactory biopsies demonstrate a defect in neuronal development in Rett's syndrome. *Ann Neurol* 54, 206–218.

Roux, J. C., Dura, E., Moncla, A., Mancini, J. & Villard, L. (2007). Treatment with desipramine improves breathing and survival in a mouse model for Rett syndrome. *Eur J Neurosci* 25, 1915–1922.

Roux, J. C., Panayotis, N., Dura, E. & Villard, L. (2010). Progressive noradrenergic deficits in the locus coeruleus of Mecp2 deficient mice. *J Neurosci Res* 88, 1500–1509.

Rueda, N., Florez, J. & Martinez-Cue, C. (2008). Chronic pentylenetetrazole but not donepezil treatment rescues spatial cognition in Ts65Dn mice, a model for Down syndrome. *Neurosci Lett* 433, 22–27.

Samaco, R. C., Fryer, J. D., Ren, J., Fyffe, S., Chao, H. T., Sun, Y., Greer, J. J., Zoghbi, H. Y. & Neul, J. L. (2008). A partial loss of function allele of methyl-CpG-binding protein 2 predicts a human neurodevelopmental syndrome. *Hum Mol Genet* 17, 1718–1727.

Samaco, R. C., Hogart, A. & LaSalle, J. M. (2005). Epigenetic overlap in autism-spectrum neurodevelopmental disorders: MECP2 deficiency causes reduced expression of UBE3A and GABRB3. *Hum Mol Genet* 14, 483–492.

Samaco, R. C., Mandel-Brehm, C., Chao, H. T., Ward, C. S., Fyffe-Maricich, S. L., Ren, J., Hyland, K., Thaller, C., Maricich, S. M., Humphreys, P., Greer, J. J., Percy, A., Glaze, D. G., Zoghbi, H. Y. & Neul, J. L. (2009). Loss of MeCP2 in aminergic neurons causes cell-autonomous defects in neurotransmitter synthesis and specific behavioral abnormalities. *Proc Natl Acad Sci U S A* 106, 21966–21971.

Samaco, R. C., Nagarajan, R. P., Braunschweig, D. & LaSalle, J. M. (2004). Multiple pathways regulate MeCP2 expression in normal brain development and exhibit defects in autism-spectrum disorders. *Hum Mol Genet* 13, 629–639.

Sauer, B. (1998). Inducible gene targeting in mice using the Cre/lox system. *Methods* 14, 381–392.

Schwenk, F., Baron, U. & Rajewsky, K. (1995). A cre-transgenic mouse strain for the ubiquitous deletion of loxP-flanked gene segments including deletion in germ cells. *Nucleic Acids Res* 23, 5080–5081.

Scott, M. M., Wylie, C. J., Lerch, J. K., Murphy, R., Lobur, K., Herlitze, S., Jiang, W., Conlon, R. A., Strowbridge, B. W. & Deneris, E. S. (2005). A genetic approach to access serotonin neurons for in vivo and in vitro studies. *Proc Natl Acad Sci U S A* 102, 16472–16477.

Shahbazian, M., Young, J., Yuva-Paylor, L., Spencer, C., Antalffy, B., Noebels, J., Armstrong, D., Paylor, R. & Zoghbi, H. (2002). Mice with truncated MeCP2 recapitulate many Rett syndrome features and display hyperacetylation of histone H3. *Neuron* 35, 243–254.

Stearns, N. A., Schaevitz, L. R., Bowling, H., Nag, N., Berger, U. V. & Berger-Sweeney, J. (2007). Behavioral and anatomical abnormalities in Mecp2 mutant mice: a model for Rett syndrome. *Neuroscience* 146, 907–921.

Swanberg, S. E., Nagarajan, R. P., Peddada, S., Yasui, D. H. & LaSalle, J. M. (2009). Reciprocal co-regulation of EGR2 and MECP2 is disrupted in Rett syndrome and autism. *Hum Mol Genet* 18, 525–534.

Tao, J., Wu, H. & Sun, Y. E. (2009). Deciphering Rett syndrome with mouse genetics, epigenomics, and human neurons. *Int Rev Neurobiol* 89, 147–160.

Tate, P., Skarnes, W. & Bird, A. (1996). The methyl-CpG binding protein MeCP2 is essential for embryonic development in the mouse. *Nat Genet* 12, 205–208.

Trappe, R., Laccone, F., Cobilanschi, J., Meins, M., Huppke, P., Hanefeld, F. & Engel, W. (2001). MECP2 mutations in sporadic cases of Rett syndrome are almost exclusively of paternal origin. *Am J Hum Genet* 68, 1093–1101.

Tronche, F., Kellendonk, C., Kretz, O., Gass, P., Anlag, K., Orban, P. C., Bock, R., Klein, R. & Schutz, G. (1999). Disruption of the glucocorticoid receptor gene in the nervous system results in reduced anxiety. *Nat Genet* 23, 99–103.

Tropea, D., Giacometti, E., Wilson, N. R., Beard, C., McCurry, C., Fu, D. D., Flannery, R., Jaenisch, R. & Sur, M. (2009). Partial reversal of Rett Syndrome-like symptoms in MeCP2 mutant mice. *Proc Natl Acad Sci U S A* 106, 2029–2034.

Tyler, W. J., Alonso, M., Bramham, C. R. & Pozzo-Miller, L. D. (2002). From acquisition to consolidation: on the role of brain-derived neurotrophic factor signaling in hippocampal-dependent learning. *Learn Mem* 9, 224–237.

van Praag, H., Kempermann, G. & Gage, F. H. (2000). Neural consequences of environmental enrichment. *Nat Rev Neurosci* 1, 191–198.

van Woerden, G. M., Harris, K. D., Hojjati, M. R., Gustin, R. M., Qiu, S., de Avila, F. R., Jiang, Y. H., Elgersma, Y. & Weeber, E. J. (2007). Rescue of neurological deficits in a mouse model for Angelman syndrome by reduction of alphaCaMKII inhibitory phosphorylation. *Nat Neurosci* 10, 280–282.

Wang, H., Chan, S. A., Ogier, M., Hellard, D., Wang, Q., Smith, C. & Katz, D. M. (2006). Dysregulation of brain-derived neurotrophic factor expression and neurosecretory function in Mecp2 null mice. *J Neurosci* 26, 10911–10915.

Ward, B. C., Agarwal, S., Wang, K., Berger-Sweeney, J. & Kolodny, N. H. (2008). Longitudinal brain MRI study in a mouse model of Rett Syndrome and the effects of choline. *Neurobiol Dis* 31, 110–119.

Ward, B. C., Kolodny, N. H., Nag, N. & Berger-Sweeney, J. E. (2009). Neurochemical changes in a mouse model of Rett syndrome: changes over time and in response to perinatal choline nutritional supplementation. *J Neurochem* 108, 361–371.

Williams, J. M., Beckmann, A. M., Mason-Parker, S. E., Abraham, W. C., Wilce, P. A. & Tate, W. P. (2000). Sequential increase in Egr-1 and AP-1 DNA binding activity in the dentate gyrus following the induction of long-term potentiation. *Brain Res Mol Brain Res* 77, 258–266.

Wood, L., Gray, N. W., Zhou, Z., Greenberg, M. E. & Shepherd, G. M. (2009). Synaptic circuit abnormalities of motor-frontal layer 2/3 pyramidal neurons in an RNA interference model of methyl-CpG-binding protein 2 deficiency. *J Neurosci* 29, 12440–12448.

Yoshikawa, H., Kaga, M., Suzuki, H., Sakuragawa, N. & Arima, M. (1991). Giant somatosensory evoked potentials in the Rett syndrome. *Brain Dev* 13, 36–39.

Young, D., Nagarajan, L., de Klerk, N., Jacoby, P., Ellaway, C. & Leonard, H. (2007). Sleep problems in Rett syndrome. *Brain Dev* 29, 609–616.

Young, J. I., Hong, E. P., Castle, J. C., Crespo-Barreto, J., Bowman, A. B., Rose, M. F., Kang, D., Richman, R., Johnson, J. M., Berget, S. & Zoghbi, H. Y. (2005). Regulation of RNA splicing by the methylation-dependent transcriptional repressor methyl-CpG binding protein 2. *Proc Natl Acad Sci U S A* 102, 17551–17558.

Young, J. I. & Zoghbi, H. Y. (2004). X-chromosome inactivation patterns are unbalanced and affect the phenotypic outcome in a mouse model of rett syndrome. *Am J Hum Genet* 74, 511–520.

Zeng, L. H., Xu, L., Gutmann, D. H. & Wong, M. (2008). Rapamycin prevents epilepsy in a mouse model of tuberous sclerosis complex. *Ann Neurol* 63, 444–453.

Zheng, W. H. & Quirion, R. (2004). Comparative signaling pathways of insulin-like growth factor-1 and brain-derived neurotrophic factor in hippocampal neurons and the role of the PI3 kinase pathway in cell survival. *J Neurochem* 89, 844–852.

Chapter 4
Epigenetic Alterations in Glioblastoma Multiforme

John K. Wiencke

Abbreviations

CNS	Central nervous system
CpG	Cytosine-phosphate-guanine dinucleotide
DNMT	DNA methyltransferase
EGFR	Epithelial growth factor receptor
GBM	Glioblastoma multiforme WHO grade IV astrocytoma
sGBM	Secondary glioblastoma multiforme
G-Cimp	Glioma-CpG island methylator phenotype
GFAP	Glial fibrillary acid protein
HDAC	Histone deacetylase
PTEN	Phosphatase tensin analogue deleted on chromosome 10
MSP	Methylation-specific polymerase chain reaction
MGMT	O6-Methylguanine DNA methyltransferase
qMSP	Quantitative methylation-specific polymerase chain reaction
SAHA	Suberoylanilide hydroxamic acid
STAT	Signal transducer and activator of transcription
TET	Enzymes of the ten-eleven-translocation family
TP53	Tumor suppressor protein p53
WHO	World Health Organization

J.K. Wiencke, PhD (✉)
Department of Neurological Surgery and Epidemiology, University of California
San Francisco, Helen Diller Family Cancer Center, San Francisco, CA, USA
e-mail: John.Wiencke@ucsf.edu

J. Minarovits and H.H. Niller (eds.), *Patho-Epigenetics of Disease*,
DOI 10.1007/978-1-4614-3345-3_4, © Springer Science+Business Media New York 2012

4.1 Overview: The Epigenome and CNS Malignancy

Here the focus is on the current state of science regarding epigenetic alterations in GBM. Some examples are provided from our laboratory to illustrate epigenetic alterations in GBM. However, the field of cancer epigenetics is rapidly evolving, and many developments have not yet been addressed in this disease specifically. This report will not cover many findings in the field, and other reviews are cited throughout this chapter to provide the reader with perspective on the potential for further advances in understanding the epigenetic origins of cancer and shaping new therapeutic paradigms (Costello et al. 2009; Rodriguez-Paredes and Esteller 2011; Tsai and Baylin 2011). The epigenetic landscape is highly complex with many hundreds of proteins being involved in coordinating the expression of genes in a cell and developmental stage appropriate context. To date, the most widely studied epigenetic phenomena in GBM are those that historically have been the most easy to study, i.e., DNA methylation, which is a highly stable chemical modification. On the other hand, protein modifications of chromatin are highly complex, labile, and require intact chromatin and highly specific antibodies or sophisticated chemical analytic approaches and thus pose a more formidable experimental challenge to the researcher (Bernstein et al. 2010). One point that should be kept in mind from the outset with regards to DNA methylation is that this phenomenon is a normal mechanism in CNS development (Kim and Rosenfeld 2010). Viewing early reports on GBM or other cancers, one could be left with the impression that DNA methylation of gene promoters was purely a cancer-associated feature. Although inappropriate DNA methylation is a cancer-associated feature, an ongoing problem for epigenetic research in cancer is defining the "normal" epigenetic state for the cell type that is represented in a malignant tissue. Thus, the problem of "cell of origin" is very prominent in GBM as in most cancer cell types (Clarke et al. 2006). The appropriate reference for epigenetic alterations in a tumor should be the cell type that has given rise to the transformed cell, which in the case of GBM is unknown (Clarke et al. 2006; Palm and Schwamborn 2010). The stem cell theory of GBM posits that an early progenitor cell population gives rise to these tumors (Germano et al. 2010; Palm and Schwamborn 2010; Tabatabai and Weller 2011). Recent studies demonstrate the problems with this theory and its limitations in predicting the clinical behavior of GBM (Chen et al. 2010).

A relevant example of gene promoter DNA methylation as being a normal and highly conserved mechanism in CNS development is provided by the work of Takizawa et al. (2001) and others (Barresi et al. 1999; Condorelli et al. 1999; Shimozaki et al. 2005) showing that astrocyte differentiation in the fetal brain from neural precursors normally involves a cycle of demethylation and re-methylation of specific STAT3 binding sites within the glial fibrillary acid protein (GFAP) gene at critical times in embryogenesis. Precursors undergoing re-methylation have blocked the signal transducer and activator of transcription 3 (STAT3) binding site and contain a transcriptionally silenced GFAP gene. These cells become neurons. Those cells with STAT3 sites that are unmethylated retain signaling and associated cytokine responsiveness and evolve into astrocytes. GFAP is aberrantly expressed in many astrocytomas and oligodendrogliomas (Chumbalkar et al. 2005; Mokhtari et al. 2005). Others have concluded as well that DNA methylation is one of the key mechanisms regulating the timing of glial cell differentiation (Martinowich et al. 2003). No doubt the tremendous advances in epigenetic and stem cell technology will dominate and reshape our views of normal CNS development and new therapeutic applications in GBM going forward (Ahmed et al. 2010; Ahmed and Lesniak 2011; Baylin and Jones 2011).

4.2 Glioblastoma Multiforme: Description and Incidence

In 2010 in the USA, 22,020 individuals will be diagnosed with a primary malignant brain tumor (the vast majority of which are gliomas) and 13,140 will die from the disease (Jemal et al. 2010). Glioma is further subdivided into different subtypes with the most aggressive form glioblastoma multiforme (GBM) accounting for about 50–60% of all adult gliomas. The World Health Organization (WHO) classification of glioma, based on the presumed cell origin, distinguishes astrocytic, oligodendrocytic, and mixed gliomas (Louis et al. 2007). The grading system is based on the presence of the following criteria: increased cellular density, nuclear atypia, mitosis, vascular proliferation, and necrosis. GBM corresponds to grade IV astrocytoma and demonstrates all of the above histopathological features; GBM tumors are usually fatal within 12–18 months. Hallmarks of the GBM tumor phenotype are extensive vascular proliferation (angio- and vasculogenesis), cellular heterogeneity, hypoxia, high cellular motility, and diffusely infiltrating growth patterns. Few consistent risk factors for GBM have been identified aside from ionizing radiation exposures and rare genetic cancer syndromes (Schwartzbaum et al. 2006). Recent studies suggest moderate but reproducible risks for GBM associated with immune factors and common polymorphisms in the *CDKN2B, RTEL,* and *hTERT* genes (Wiemels et al. 2009; Wrensch et al. 2009).

4.3 Primary and Secondary GBM: Clinical and Genetic Characteristics

Two subsets of GBM have been recognized for many years based on the clinical and genetic characteristics of the tumors (Ohgaki and Kleihues 2007) (Table 4.1). The most common type is denoted primary or de novo GBM. De novo GBMs occur in older patients without a prior clinical history of a primary brain tumor. The primary GBM genome frequently contains amplifications of the EGFR locus, chromosomal deletion of the CDKN2A locus, and mutations in the PTEN gene. Thus, primary GBM contains genetic abnormalities in well-established cell growth/proliferation and cell cycle control pathways. TP53 mutations are not common in primary GBM.

Table 4.1 Clinical and tumor genetic features distinguishing primary and secondary glioblastoma multiforme (GBM) (Ohgaki and Kleihues, 2007[a]; Kloosterhof et al., 2011[b])

Clinical history	Primary GBM	Secondary GBM
Sex ratio (M/F)	1.4	0.8
Mean age at diagnosis	55	40
TP53 mutation[a]	10	65%
EGFR amplification[a]	40%	0%
MDM2 amplification[a]	5%	0%
CDKN2A deletion[a]	35%	5%
PTEN mutation[a]	30%	5%
IDH1/2 mutation[b]	5%	83%

In contrast, sGBM are those tumors arising in patients with a positive prior history of a non-GBM brain tumor. sGBM patients are typically younger than primary GBM patients and are less likely to be male. In addition, sGBM often contain TP53 mutations. Very recently, additional genetic and epigenetic characteristics have been discovered (discussed below) that further reinforce the idea that there exist unique pathogenetic pathways for primary and sGBM subtypes.

4.4 Distinct Epigenetic Alterations in Primary Versus sGBM: Early Clues to a Hypermethylation Phenotype

Clinically, the development of GBM following the presentation of a lower grade glioma is distinguished from the more common de novo presentation of GBM that arises without a documented earlier malignancy. As indicated above, sGBM are known to display different genomic alterations (Maher et al. 2006) and different transcriptional signatures compared to de novo GBM tumors (Tso et al. 2006). More recently studies of DNA methylation that compared de novo with sGBM identified individual gene targets that were more hypermethylated in sGBM compared with de novo GBM. These studies preceded the transition to genome-wide methylation arrays. These early studies utilized methylation-specific PCR (MSP) and quantitative methylation-specific PCR (qMSP) that amplify individual genes. MSP is a qualitative estimate of DNA methylation. Typically, bisulfite converted DNA is used as a template and an endpoint of either a methylated or unmethylated PCR product is visualized on an agarose gel. The disadvantage of this qualitative approach is that a positive response may not differentiate a minor cell fraction harboring methylation compared to one that is uniformly methylated. More recently, the MSP procedure has been modified to be more quantitative using methods such as MethylLight (Eads et al. 2000; Trinh et al. 2001; Campan et al. 2009). These qMSP methods provide a more stringent and reproducible endpoint of DNA methylation that can be compared across studies more readily than the subjective MSP. Below is a summary of studies that have identified specific methylation events that differentiate primary from sGBM. sGBM are recognized by their frequent high levels of CpG methylation within promoters of many genes.

4.5 IDH Mutation and the DNA Hypermethylator Phenotype

Following on the evidence that many genes were coordinately hypermethylated in sGBM and some de novo GBM came the groundbreaking discoveries of mutations within genes involved in glucose and energy metabolism. These developments have dramatically affected our view that primary and sGBM pathogeneses are driven by pathogenetically distinct mechanisms. In 2008, a genome-wide sequencing study identified mutations in the genes encoding the isocitrate dehydrogenase (IDH1 and IDH2) enzymes in low-grade gliomas and several GBM tumors (Parsons et al. 2008). These were not germ line mutations but acquired somatic alterations. IDH1 is a cytoplasmic and peroxisomal enzyme whereas IDH2 is found in mitochondria; both proteins catalyze the oxidative decarboxylation of isocitrate into α-ketoglutarate. Overall, *IDH* mutations were found among 12% of the GBM tumors examined. *IDH1* mutations were highly localized within codon 132 of the *IDH1* gene and in the analogous codon 172 in *IDH2*. The most common mutations (>90% of all *IDH* mutations) produce arginine to histidine amino acid substitutions (R132H) in *IDH1*. *IDH* mutations were most frequent in non-GBM gliomas, among younger GBM patients and most sGBM patients, which as noted develop GBM through a prior low-grade glioma. The presence of *IDH1* mutations were also associated with increased survival times compared with GBM patients whose tumors did not contain *IDH* mutations. Many large clinical series have now been published on patients from the United States, Europe, and Japan that define the incidence of *IDH* mutations in GBM and their associations with other genetic abnormalities and clinical outcomes (Gravendeel et al. 2009; Nobusawa et al. 2009; Sanson et al. 2009; Sonoda et al. 2009; Yan et al. 2009; Verhaak et al. 2010). Interestingly, IDH mutations are not found among non-glioma brain tumors or other cancers with the exception of acute myeloid leukemia (Bleeker et al. 2009; Kang et al. 2009; Mardis et al. 2009; Park et al. 2009) and rarely in prostate cancer. *IDH1* mutations result in a diminished enzymatic activity towards the native substrate, isocitrate, and in addition the mutant IDH1 catalyzes the formation of 2-hydroxyglutarate (2-HG) from α-ketoglutarate (Zhao et al. 2009; Dang et al. 2009). The product 2-HG has been termed an "oncometabolite" (Xu et al. 2011). Great attention has been given to the effects of accumulating high intracellular levels of 2-HG in *IDH* mutant glioma cells (5–35 mM), and it is now thought that this abnormal product leads to the disruption of the epigenome and to characteristic DNA hypermethylation alterations. These results have refocused the attention of the scientific community on metabolic factors as fundamental drivers in tumorigenesis and have rekindled interest in examining the so call "Warburg effect." The Warburg effect (Warburg 1923) is the observation that most cancer cells predominantly produce energy by a high rate of glycolysis followed by lactic acid fermentation in the cytosol, rather than by a comparatively low rate of glycolysis followed by oxidation of pyruvate in mitochondria like most normal cells. Such altered energy metabolism has been considered a consequence not a cause of malignant transformation in modern oncology; however, IDH mutations in glioma and AML now point to a more primary role in cancer pathogenesis (Diaz-Ruiz et al. 2011; Weljie and Jirik 2011).

Table 4.2 Genes differentially methylated in secondary versus primary glioblastoma

Gene name	Primary GBM%	Secondary GBM%	Study method	Reference
MGMT	36, 38	75, 45	MSP qMSP	(Nakamura et al., 2001a)
				(Zheng et al., 2011)
PTEN	9, 11	82, 82	MSP qMSP	Wiencke et al. (2007)
				Zheng et al. (2011)
RASSF1A	63	100	qMSP	Zheng et al. (2011))
RFX1	73	100	qMSP	Zheng et al. (2011)
EMP3	17, 16	89, 70	Sequence qMSP	Kunitz et al. (2007)
				Zheng et al. (2011)
TMS1	68	90	qMSP	Zheng et al. (2011)
ZNF342	13	60	qMSP	Zheng et al. (2011)
SOCS1	61	90	qMSP	Zheng et al. (2011)
P14ARF	6	31	MSP	Ohgaki and Kleihues (2007)
RB1	14	43	MSP	Nakamura et al. (2001b)
TIMP-3	28	71	MSP	Nakamura et al. (2005)

qMSP quantitative methylation-specific PCR. For comparison, the percentages represent the positive methylation-positive tumors; a quantitative threshold was applied in qMSP methods; positive for MSP is subjective and refers to a visible band being observable on an agarose gel

As indicated above (Table 4.2), the presence of a pattern of coordinated hypermethylated genes in sGBM and lower grade gliomas had been noted for several years before the discovery of the *IDH1* mutations. With the advent of DNA methylation arrays, it became practical to assess the methylation status of hundreds of genes simultaneously in sGBM and de novo GBM and relate genome wide methylation events to the *IDH* mutation. Our group assessed DNA methylation and *IDH* using an approach that compared glioma tumor methylation at 1,500 gene associated CpG sites in 805 genes with nonmalignant brain tissue (Christensen et al. 2011). Figure 4.1a shows that *IDH* mutant gliomas contained predominantly hypermethylation events (compared with nonmalignant brain tissue), whereas *IDH* wild-type glioma was characterized by DNA hypomethylation and relatively little hypermethylation. Using a statistical clustering procedure called "recursive partitioning mixture models" (Houseman et al. 2008), our glioma methylation data produced homogeneous methylation groups with one predominant tumor group that contained almost all of the IDH mutations (Fig. 4.1c). The linkage between uniform hypermethylation of genes and IDH mutation was dramatic. Furthermore, the *IDH* mutant glioma patients showed much longer survival times compared with patient with wild-type (nonmutant) *IDH* tumors (Fig. 4.1d).

A component of the glioma-CpG island phenotype is the DNA repair enzyme O6-methylguanine-DNA methyltransferase (MGMT). Methylation of MGMT was earlier identified as a prognostic factor in glioma patients receiving temozolomide, which is a DNA alkylating agent (Hegi et al. 2005, 2008). Because of the close correlation of *IDH* mutation with *MGMT* methylation, it is possible that much of the previously reported survival benefits associated with *MGMT* could have been due to *IDH* mutation. The independent value of *MGMT* methylation and the most valid technique to assess MGMT status in tumors are the subjects of some controversy and current research (Ohka et al. 2011).

4 Epigenetic Alterations in Glioblastoma Multiforme

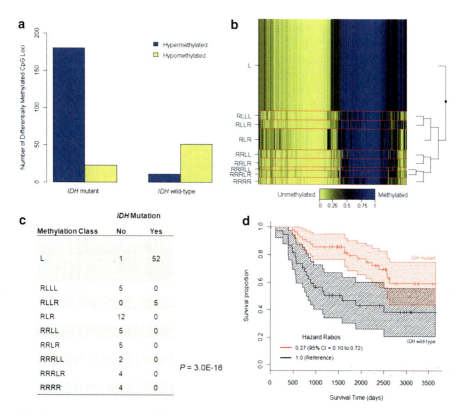

Fig. 4.1 Association between *IDH* mutation and methylation phenotype in gliomas. (**a**) The number of statistically significantly differentially hyper- and hypomethylated loci ($Q < .05$ and $|Db| > 0.2$), are plotted by tumor *IDH* mutation status. (**b**) Recursively partitioned mixture model (RPMM) of glioma samples with both methylation and mutation data (n = 95). Methylation profile classes are stacked in rows separated by **red lines, class height** corresponds to the number of samples in each class. Class methylation at each CpG locus (**columns**) is the mean methylation for all samples in a class where **blue** designates methylated CpG loci (average $\beta = 1$), and **yellow** designates unmethylated CpG loci (average $\beta = 0$). To the right of the RPMM is the clustering dendrogram. (**c**) Methylation-class-specific *IDH* mutation status (Fisher $P = 3.0 \times 10216$). (**d**) Kaplan–Meier survival probability strata for *IDH* mutant (**red**, n = 57) and *IDH* wild-type (**black**, n = 38) tumors, **tick marks** are censored observations and **banding patterns** represent 95% confidence intervals. (Reprinted with permission from Christensen et al. 2011)

4.6 Mechanistic Links Between IDH Mutation and DNA Hypermethylation in GBM: The Role of 5-Hydroxymethylcytosine (the Sixth Base)

Historically, the loss (removal) of 5-methylcytosine (5mC) in the mammalian genome was considered a "passive" process mediated by a failure of DNA methyltransferase DNMT1 enzyme to remethylate hemi-methylated CpG sites following DNA synthesis (Huang et al. 2010).

Demethylation and maintenance of 5m-free regions in regulatory regions of actively transcribed genes is critical during cell differentiation and to fix cell type-specific gene expression. An active demethylation mechanism has been elusive to scientists until very recently, although this pathway was predicted based on studies of demethylation that occur at the earliest stages of development (e.g., paternal gene demethylation in the newly fertilized zygote). The elements of an active DNA demethylation process in mammalian cells have now at least been partially revealed and importantly have also been shown to play an important role in cellular differentiation and tumorigenesis including gliomagenesis. Central in this process are the TET family of proteins. TET enzymes are implicated in 5mC demethylation by catalyzing the hydroxylation of 5mC that yields 5-hydroxy-mC (5hmC), which is then an intermediate in a base excision process that effectively returns a methylated CpG site to an unmethylated state. The human ten-eleven-translocation (TET) family consists of three members, namely, TET1 (10q21), TET2 (4q24), and TET3 (2p13) (Tan and Manley 2009). TET enzymes are oxoglutarate Fe (II)-dependent dioxygenases; moreover, the human dioxygenases are involved in a wide range of biological functions including histone demethylation, hypoxic response, and DNA repair (Mohr et al. 2011). TET1 was the first identified member of the family and was discovered as a fusion partner of the mixed lineage leukemia gene in the translocation t (10;11)(q22;q23) (Ono et al. 2002; Burmeister et al. 2009; Burmeister 2010). TET1 catalyzes the conversion of 5mC to 5hmC (Tahiliani et al. 2009), and depletion of the enzyme leads to a decrease in cellular 5hmC concentrations. Maintaining stem cell features in mouse embryonic stem cells has further been shown to be dependent on Tet1 function (Ito et al. 2010). TET2 is frequently found to be mutated in myeloid malignancies but not in GBM.

Initial studies uncovered high levels of 5hmC in brain and neural tissues (Kriaucionis and Heintz 2009), which are largely nondividing cell populations. More recently, malignant tissues have been found to have relatively low levels of this modified base, and this is correlated with DNA hypermethylation (Li and Liu 2011). A crucial aspect of this work is the close link between 5hmC and cell type-specific gene expression. High-throughput sequencing of 5hmC containing DNA in mouse embryonic stem cells revealed that both 5hmC and 5mC were enriched within transcribed regions, particularly exons. Importantly, however, only 5hmC was enriched at transcription start sites and within the 5' untranslated regions of genes (Pastor et al. 2011). 5hmC was also more enriched in gene expression enhancers compared with 5mC; this latter observation supports a strong connection between 5hmC and regulatory elements. Mapping of 5hmC in the human brain genome revealed high concentrations in TSSs and active gene bodies of genes involved in nerve cell-specific functions (Jin et al. 2011). Taken together, the evidence thus far indicates that 5hmC plays an important role in regulation of DNA methylation, chromatin remodeling, and gene expression in a tissue-, cell-, or organ-specific manner (Tan and Manley 2009; Dahl et al. 2011; Guo et al. 2011; Mohr et al. 2011). A current model is that glioma progenitors containing IDH mutations accumulate high concentrations of 2-HG, which leads to inhibition of TET enzymes that leads to an accumulation of 5mC within promoters of many genes and perhaps in repetitive CpG sites as well (see below).

4.7 DNA Hypomethylation: A Common Feature of the De Novo GBM Genome

Even a few years ago, the phenotype of cancer with respect to DNA methylation was characterized as suffering genome-wide loss of CpG methylation while at the same time accumulating gene associated DNA hypermethylation events. This view has been modified by recent developments including the application of powerful genome-wide DNA methylation scanning methylation arrays. To summarize briefly, the current view holds that DNA hypermethylation of gene regulatory regions (promoters) is a normal developmental occurrence for some genes in a tissue appropriate context (Christensen et al. 2009). Also, DNA hypermethylation of actively transcribed gene regions (gene bodies) is now also recognized as part of the normal epigenome (Maunakea et al. 2010), particularly for genes with alternative promoter sites. During cellular differentiation, DNA methylation may occur in waves and be followed by demethylation and subsequent re-methylation events. The demethylation events may be mediated by specific DNA demethylases (see below). Loss of DNA methylation in both gene bodies, promoters, and in repetitive sequences, which constitute about 45% of the genome, is a common and aberrant feature of the transformed cell likely contributing to the transformation process. There is very convincing evidence in animal models and human cancer cell systems that loss of DNMT activity and resulting genomic (Holm et al. 2005) hypomethylation increases genomic instability and promotes tumorigenesis and mutagenesis (Chen et al. 1998; Gaudet et al. 2003; Eden et al. 2003; Karpf and Matsui 2005). Within genic regions, hypomethylation can lead to inappropriate gene expression. The classic example of this are the increased melanoma antigen gene (MAGE) family expression in GBM and other cancers associated with demethylation of DNA on the X chromosome (Scarcella et al. 1999; Liu et al. 2004).

Most of the CpG methylation of DNA within the genome exists within repetitive elements (Gama-Sosa et al. 1983a, b). One example of these is the long interspersed nucleotide elements such as LINE1 elements, which comprise nearly 18% of the human genome (Prak and Kazazian 2000). The loss of CpG methylation can be assessed using PCR or chemical analytical methodologies (Malzkorn et al. 2011). Briefly, chemical measures typically assess global 5mC levels. PCR-based methods are also useful to assess global methylation and technically more accessible to most laboratories. For example, our group employed a quantitative bisulfite pyrosequencing PCR method that amplifies LINE elements to assess global hypomethylation (Bollati et al. 2007). In Fig. 4.2a is shown comparison of LINE1 methylation levels in GBM compared with other glioma subtypes and nonmalignant brain tissue. Overall, de novo GBM showed lower LINE1 methylation but also displayed a wide range of values. Non-GBM glioma had much higher levels of global DNA methylation and even higher LINE1 scores compared with nonmalignant brain tissue. These latter results suggest the accumulation of DNA methylation events associated with *IDH* mutation as most of the non-GBM gliomas were shown to contain *IDH* mutations. Looking at gene methylation, we also showed a pattern of hypermethylation within gene regions that was associated with higher LINE1 methylation and hence a positive correlation of gene-specific hypermethylation with global methylation (Fig. 4.2b).

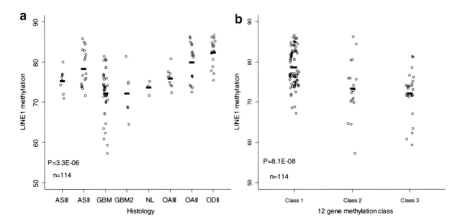

Fig. 4.2 LINE1 DNA methylation values of primary glioma tissues. (**a**) Mean LINE1 scores stratified by histopathological subtype; NL indicates nonmalignant brain tissues. (**b**) Mean LINE1 scores stratified by methylation class. All glioma tumors were clustered using RPMM procedure that yielded three methylation classes. Class I contains IDH mutant tumors and sGBM. (Reprinted with permission from Zheng et al., 2011)

4.8 GBM Therapy Targeting Epigenetic Pathways

One often cited advantage of targeting epigenetic alterations in cancer is that they are potentially reversible in contrast to genetic alterations. While this is true in theory, much work needs to be done to take advantage of this property in GBM therapy. To date, the experience with epigenetic therapies in GBM patients is very limited. Inhibition of the DNMT and HDAC enzymes have been explored using in vitro cell systems, animal models, and in only a few clinical trials.

Nonetheless, epigenetic therapies have the potential to target a wide range of chromatin modifications and DNA methylation that regulate gene expression and ultimately affect tumor cell survival. The amino acid tails of the histone proteins are marked by a large number of modifications that affect gene expression; these include: acetylation, methylation, phosphorylation, poly-ADP ribosylation, ubiquitinylation, sumoylation, carbonylation, and glycosylation (Kelly et al. 2002; de Ruijter et al. 2003; Bolden et al. 2006). Blocking the deacetylation of histone proteins via histone deacetylase inhibitor action has been explored most extensively using in vitro cell systems, animal models, and human trials (Bolden et al. 2006). HDAC inhibitors have proven antitumor activity by inducing cell apoptosis, blocking cell proliferation, and promoting differentiation. A fundamental early observation on HDAC inhibition was to show that these drugs could preferentially affect the transcription of a small subset of genes (Van Lint et al. 1996). Eighteen different HDAC enzymes comprising three classes have been identified in humans. These enzymes differ in their location within the cell and their homology to yeast counterparts. Inhibitors of the HDACs can be divided into six classes based on their chemical structures. These include short chain fatty acids such as valproic acid and

4 Epigenetic Alterations in Glioblastoma Multiforme

butyrate derivatives, hydroxamic acids such as trichostatin, SAHA and AQ824, cyclic tetrapeptides such as trapoxin, and benzamides such as MS-275. Understanding the multiplicity of HDAC protein targets and the specificity of HDAC inhibitors towards them remains a major challenge to scientists. In GBM cells, mutations in several HDAC genes have been discovered as well as in related genes encoding the histone demethylases JMJD1A, JMJD1B, and histone methyltransferase (Parsons et al. 2008). There is intense interest in this latter class of chromatin modifiers. Two classes of histone lysine demethylases remove histone methylation. Lysine demethylase 1 (KDM1, also known as LSD1) is a flavin adenine dinucleotide-containing enzyme that removes mono-/di-methylation. The Jumonji C-terminal domain (JmjC) family of histone demethylases uses Fe(2+) and α-ketoglutarate as cofactors to remove all methylation states. Further implicating these epigenetic regulators in gliomagenesis are reports showing greatly reduced expression of class I, II, and IV HDAC genes in GBM cells compared to low-grade astrocytoma or nonmalignant brain tissues (Lucio-Eterovic et al. 2008) while increased expression of HDAC3 was observed in another study (Liby et al. 2006). Using small interfering RNAs, investigators showed that HDAC4 specifically participates in the repression of p21(WAF1/Cip1) through Sp1/Sp3– in GBM cell lines in vitro and tumors in vivo. SAHA is a small-molecule inhibitor of most human class I and class II HDACs that binds directly at the enzyme's active site. Preclinical studies showed that SAHA demonstrated antitumor activity towards glioma cell lines in vitro and in glioma orthotopic xenografts in vivo (Eyupoglu et al. 2005; Ugur et al. 2007; Yin et al. 2007). Exposure of glioma cells to SAHA increased the expression of markers of apoptosis such as DR5, tumor necrosis factor, p21Waf1, and p27Kip1; also, animal studies indicate that SAHA effectively crossed the blood-brain barrier (Bolden et al. 2006; Yin et al. 2007; Ellis et al. 2009). A phase II study of 66 recurrent GBM patients treated with SAHA reported that 9 out of 52 patents were progression-free at 6 months and acceptable toxicity of the agent (Galanis et al. 2009).

Recently, SAHA was shown to inhibit the expression of an important oncogene Enhancer of Zeste 2 (EZH2) in experiment GBM cell models. EZH2 is the catalytic subunit of the polycomb repressive complex 2 (PRC2) and is responsible for the methylation of the lysine 27 position of histone H3 (H3K27) (Cao et al. 2002). H3K27me3 is a chromatin mark associated with repressing critical genes in stem cells and during early phases of cellular differentiation. It is a component of the specialized chromatin state, termed the "bivalent domain," that marks cell differentiation genes that are silenced in embryonic stem cells (Bernstein et al. 2006). H3K27me3 and the reactivation of PCR repression is also a feature of cellular transformation (Valk-Lingbeek et al. 2004). EZH2 is non-detectable in normal nonmalignant brain tissues but overexpressed in all human glioma histopathological subtypes compared with nonmalignant brain tissue and dramatically overexpressed in GBM compared with lower grade gliomas (Orzan et al. 2011; Zheng et al. 2011). These observations support the idea that EZH2, a stem cell epigenetic regulator, is a marker of the most aggressive forms of glioma and a favorable target for therapeutic intervention. Illustrating this potential is a study that isolated glioma stem cells from 66 primary tumors of different grades that were treated with the SAHA, which led to a repression of EZH2 expression and to increased expression of genes that were targets of PRC-related gene repression (Orzan et al. 2011).

Combining HDAC inhibitors with other agents may achieve improved antitumor response. One study of phenylbutyrate (PB) in patients with recurrent glioma reported only a 5% overall response rate in 23 treated patients (Phuphanich et al. 2005). Because in general the response rates for HDAC inhibitors as monotherapies have been low, efforts have been made to achieve more favorable outcomes by combining these HDAC inhibitors with radiation therapy (Camphausen et al. 2004, 2005; Kim et al. 2004; Chinnaiyan et al. 2005). PB was shown to in induce p21-independent cytostasis and enhance radiation sensitivity in TP53 mutant human GBM cells in vitro (Lopez et al. 2007). Combining therapies that target genetic subgroups of GBM will require consideration of the individual tumor characteristics, and these can be limitations to such approaches. As indicated in Table 4.1, only about 10% of primary GBMs harbor TP53 mutations whereas they occur in about 60% of the less common sGBM cases. Another limitation noted in human studies is that the pharmacology of PB appears to be affected by concomitant administration of P450-inducing anticonvulsants (Phuphanich et al. 2005). The identification of *IDH* mutations in glioma has renewed interest in glucose and energy metabolism in gliomagenesis and has spurred new research into targeting metabolic pathways in conjunction with epigenetic targets. Studies are now focusing on combining HDAC drug treatments with glycolysis inhibitors such as the glucose analogue 2-deoxy-D-glucose. Glucose metabolism and ATP production are inhibited by 2-deoxy-D-glucose. One study reported that HDAC inhibitors trichostatin A, sodium butyrate, and LAQ824 combined with 2-deoxy-D-glucose induced a strong apoptotic response in brain tumor cell lines in a TP53-independent manner (Egler et al. 2008). Further study is required to test combined HDAC and 2-deoxy-D-glucose in cells with different *IDH* mutation profiles.

4.9 Perspective and Future Challenges

The dramatic advances in genome science and technology have ushered in a new and exciting era of research and many potential applications in neuro-oncology. Applying the new technologies to characterize the GBM epigenome and integrating it with patient care remains one of the greatest challenges. Epigenetic alterations and the genetic and metabolic factors that drive these abnormalities are being revealed. Several outstanding questions remain regarding the epigenetics of GBM:

- The observed abnormalities in DNA hypermethylation for the first time are being related to genetic mutations affecting common metabolic pathways and interference with developmentally regulated demethylation. The further elucidation of the targeting of specific genes by the TET system and design of drugs to target the aberrant metabolic profile hold great promise for new therapies.
- The recent genome-wide association studies and epidemiological investigations of immune factors suggest important heritable risk factors for GBM

4 Epigenetic Alterations in Glioblastoma Multiforme

(Wrensch et al. 2009). One future challenge is integrating this population level research with tumor epigenetics. Prevention of GBM if it could be achieved would be superior to treating the disease.

- The dysregulation of CpG methylation in de novo GBM that leads to DNA hypomethylation is largely unexplained. One possibility is that germ line or acquired gene alterations may compromise the methyl donor metabolism in GBM cells (Cadieux et al. 2006). Developing therapies that correct this demethylation or target it pharmacologically for therapeutic outcome deserve greater attention.
- The histone code modifications of GBM require much greater study especially using fresh tumor explants and advanced genome-wide array technologies. The histone code is still being refined, and important modifications of these concepts (Vakoc et al. 2005; Wiencke et al. 2008; Blahnik et al. 2011) require continual updating and flexibility in defining the paradigm.

Acknowledgments I acknowledge the important contributions of Shichun Zheng, Ashley Smith, Karl T. Kelsey, Brock Christensen, Carmen Marsit, E. Andres Houseman, Heather Nelson, Margaret Wrensch, Joe Wiemels, Jennette D. Sison, Terri Rice, Lucie McCoy, and Joe Patoka in the data presented from our publications. JKW was supported by NIEHS R01ES06717, R01CA126831, P50CA092725, and RO1CA52689.

References

Ahmed, A. U., Alexiades, N. G. & Lesniak, M. S. (2010). The use of neural stem cells in cancer gene therapy: predicting the path to the clinic. *Curr Opin Mol Ther* 12, 546–552.

Ahmed, A. U. & Lesniak, M. S. (2011). Glioblastoma multiforme: can neural stem cells deliver the therapeutic payload and fulfill the clinical promise? *Expert Rev Neurother* 11, 775–777.

Barresi, V., Condorelli, D. F. & Giuffrida Stella, A. M. (1999). GFAP gene methylation in different neural cell types from rat brain. *Int J Dev Neurosci* 17, 821–828.

Baylin, S. B. & Jones, P. A. (2011). A decade of exploring the cancer epigenome - biological and translational implications. *Nat Rev Cancer* 11, 726–734.

Bernstein, B. E., Mikkelsen, T. S., Xie, X., Kamal, M., Huebert, D. J., Cuff, J., Fry, B., Meissner, A., Wernig, M., Plath, K., Jaenisch, R., Wagschal, A., Feil, R., Schreiber, S. L. & Lander, E. S. (2006). A bivalent chromatin structure marks key developmental genes in embryonic stem cells. *Cell* 125, 315–326.

Bernstein, B. E., Stamatoyannopoulos, J. A., Costello, J. F., Ren, B., Milosavljevic, A., Meissner, A., Kellis, M., Marra, M. A., Beaudet, A. L., Ecker, J. R., Farnham, P. J., Hirst, M., Lander, E. S., Mikkelsen, T. S. & Thomson, J. A. (2010). The NIH Roadmap Epigenomics Mapping Consortium. *Nat Biotechnol* 28, 1045–1048.

Blahnik, K. R., Dou, L., Echipare, L., Iyengar, S., O'Geen, H., Sanchez, E., Zhao, Y., Marra, M. A., Hirst, M., Costello, J. F., Korf, I. & Farnham, P. J. (2011). Characterization of the contradictory chromatin signatures at the 3' exons of zinc finger genes. *PLoS ONE* 6, e17121.

Bleeker, F. E., Lamba, S., Leenstra, S., Troost, D., Hulsebos, T., Vandertop, W. P., Frattini, M., Molinari, F., Knowles, M., Cerrato, A., Rodolfo, M., Scarpa, A., Felicioni, L., Buttitta, F., Malatesta, S., Marchetti, A. & Bardelli, A. (2009). IDH1 mutations at residue p.R132 (IDH1(R132)) occur frequently in high-grade gliomas but not in other solid tumors. *Hum Mutat* 30, 7–11.

Bolden, J. E., Peart, M. J. & Johnstone, R. W. (2006). Anticancer activities of histone deacetylase inhibitors. *Nat Rev Drug Discov* 5, 769–784.

Bollati, V., Baccarelli, A., Hou, L., Bonzini, M., Fustinoni, S., Cavallo, D., Byun, H. M., Jiang, J., Marinelli, B., Pesatori, A. C., Bertazzi, P. A. & Yang, A. S. (2007). Changes in DNA methylation patterns in subjects exposed to low-dose benzene. *Cancer Res* 67, 876–880.

Burmeister, T. (2010). MLL: exploring the methylome. *Blood* 115, 4627–4628.

Burmeister, T., Meyer, C., Schwartz, S., Hofmann, J., Molkentin, M., Kowarz, E., Schneider, B., Raff, T., Reinhardt, R., Gokbuget, N., Hoelzer, D., Thiel, E. & Marschalek, R. (2009). The MLL recombinome of adult CD10-negative B-cell precursor acute lymphoblastic leukemia: results from the GMALL study group. *Blood* 113, 4011–4015.

Cadieux, B., Ching, T. T., VandenBerg, S. R. & Costello, J. F. (2006). Genome-wide hypomethylation in human glioblastomas associated with specific copy number alteration, methylenetetrahydrofolate reductase allele status, and increased proliferation. *Cancer Res* 66, 8469–8476.

Campan, M., Weisenberger, D. J., Trinh, B. & Laird, P. W. (2009). MethyLight. *Methods Mol Biol* 507, 325–337.

Camphausen, K., Burgan, W., Cerra, M., Oswald, K. A., Trepel, J. B., Lee, M. J. & Tofilon, P. J. (2004). Enhanced radiation-induced cell killing and prolongation of gammaH2AX foci expression by the histone deacetylase inhibitor MS-275. *Cancer Res* 64, 316–321.

Camphausen, K., Cerna, D., Scott, T., Sproull, M., Burgan, W. E., Cerra, M. A., Fine, H. & Tofilon, P. J. (2005). Enhancement of in vitro and in vivo tumor cell radiosensitivity by valproic acid. *Int J Cancer* 114, 380–386.

Cao, R., Wang, L., Wang, H., Xia, L., Erdjument-Bromage, H., Tempst, P., Jones, R. S. & Zhang, Y. (2002). Role of histone H3 lysine 27 methylation in Polycomb-group silencing. *Science* 298, 1039–1043.

Chen, R., Nishimura, M. C., Bumbaca, S. M., Kharbanda, S., Forrest, W. F., Kasman, I. M., Greve, J. M., Soriano, R. H., Gilmour, L. L., Rivers, C. S., Modrusan, Z., Nacu, S., Guerrero, S., Edgar, K. A., Wallin, J. J., Lamszus, K., Westphal, M., Heim, S., James, C. D., VandenBerg, S. R., Costello, J. F., Moorefield, S., Cowdrey, C. J., Prados, M. & Phillips, H. S. (2010). A hierarchy of self-renewing tumor-initiating cell types in glioblastoma. *Cancer Cell* 17, 362–375.

Chen, R. Z., Pettersson, U., Beard, C., Jackson-Grusby, L. & Jaenisch, R. (1998). DNA hypomethylation leads to elevated mutation rates. *Nature* 395, 89–93.

Chinnaiyan, P., Vallabhaneni, G., Armstrong, E., Huang, S. M. & Harari, P. M. (2005). Modulation of radiation response by histone deacetylase inhibition. *Int J Radiat Oncol Biol Phys* 62, 223–229.

Christensen, B. C., Houseman, E. A., Marsit, C. J., Zheng, S., Wrensch, M. R., Wiemels, J. L., Nelson, H. H., Karagas, M. R., Padbury, J. F., Bueno, R., Sugarbaker, D. J., Yeh, R. F., Wiencke, J. K. & Kelsey, K. T. (2009). Aging and environmental exposures alter tissue-specific DNA methylation dependent upon CpG island context. *PLoS Genet* 5, e1000602.

Christensen, B. C., Smith, A. A., Zheng, S., Koestler, D. C., Houseman, E. A., Marsit, C. J., Wiemels, J. L., Nelson, H. H., Karagas, M. R., Wrensch, M. R., Kelsey, K. T. & Wiencke, J. K. (2011). DNA methylation, isocitrate dehydrogenase mutation, and survival in glioma. *J Natl Cancer Inst* 103, 143–153.

Chumbalkar, V. C., Subhashini, C., Dhople, V. M., Sundaram, C. S., Jagannadham, M. V., Kumar, K. N., Srinivas, P. N., Mythili, R., Rao, M. K., Kulkarni, M. J., Hegde, S., Hegde, A. S., Samual, C., Santosh, V., Singh, L. & Sirdeshmukh, R. (2005). Differential protein expression in human gliomas and molecular insights. *Proteomics* 5, 1167–1177.

Clarke, M. F., Dick, J. E., Dirks, P. B., Eaves, C. J., Jamieson, C. H., Jones, D. L., Visvader, J., Weissman, I. L. & Wahl, G. M. (2006). Cancer stem cells--perspectives on current status and future directions: AACR Workshop on cancer stem cells. *Cancer Res* 66, 9339–9344.

Condorelli, D. F., Nicoletti, V. G., Barresi, V., Conticello, S. G., Caruso, A., Tendi, E. A. & Giuffrida Stella, A. M. (1999). Structural features of the rat GFAP gene and identification of a novel alternative transcript. *J Neurosci Res* 56, 219–228.

Costello, J. F., Krzywinski, M. & Marra, M. A. (2009). A first look at entire human methylomes. *Nat Biotechnol* 27, 1130–1132.

Dahl, C., Gronbaek, K. & Guldberg, P. (2011). Advances in DNA methylation: 5-hydroxymethylcytosine revisited. *Clin Chim Acta* 412, 831–836.

Dang, L., White, D. W., Gross, S., Bennett, B. D., Bittinger, M. A., Driggers, E. M., Fantin, V. R., Jang, H. G., Jin, S., Keenan, M. C., Marks, K. M., Prins, R. M., Ward, P. S., Yen, K. E., Liau, L. M., Rabinowitz, J. D., Cantley, L. C., Thompson, C. B., Vander Heiden, M. G. & Su, S. M. (2009). Cancer-associated IDH1 mutations produce 2-hydroxyglutarate. *Nature* 462, 739–744.

de Ruijter, A. J., van Gennip, A. H., Caron, H. N., Kemp, S. & van Kuilenburg, A. B. (2003). Histone deacetylases (HDACs): characterization of the classical HDAC family. *Biochem J* 370, 737–749.

Diaz-Ruiz, R., Rigoulet, M. & Devin, A. (2011). The Warburg and Crabtree effects: On the origin of cancer cell energy metabolism and of yeast glucose repression. *Biochim Biophys Acta* 1807, 568–576.

Eads, C. A., Danenberg, K. D., Kawakami, K., Saltz, L. B., Blake, C., Shibata, D., Danenberg, P. V. & Laird, P. W. (2000). MethyLight: a high-throughput assay to measure DNA methylation. *Nucleic Acids Res* 28:E32.

Eden, A., Gaudet, F., Waghmare, A. & Jaenisch, R. (2003). Chromosomal instability and tumors promoted by DNA hypomethylation. *Science* 300, 455.

Egler, V., Korur, S., Failly, M., Boulay, J. L., Imber, R., Lino, M. M. & Merlo, A. (2008). Histone deacetylase inhibition and blockade of the glycolytic pathway synergistically induce glioblastoma cell death. *Clin Cancer Res* 14, 3132–3140.

Ellis, L., Atadja, P. W. & Johnstone, R. W. (2009). Epigenetics in cancer: targeting chromatin modifications. *Mol Cancer Ther* 8, 1409–1420.

Eyupoglu, I. Y., Hahnen, E., Buslei, R., Siebzehnrubl, F. A., Savaskan, N. E., Luders, M., Trankle, C., Wick, W., Weller, M., Fahlbusch, R. & Blumcke, I. (2005). Suberoylanilide hydroxamic acid (SAHA) has potent anti-glioma properties in vitro, ex vivo and in vivo. *J Neurochem* 93, 992–999.

Galanis, E., Jaeckle, K. A., Maurer, M. J., Reid, J. M., Ames, M. M., Hardwick, J. S., Reilly, J. F., Loboda, A., Nebozhyn, M., Fantin, V. R., Richon, V. M., Scheithauer, B., Giannini, C., Flynn, P. J., Moore, D. F., Zwiebel, J. & Buckner, J. C. (2009). Phase II trial of vorinostat in recurrent glioblastoma multiforme: a north central cancer treatment group study. *J Clin Oncol* 27, 2052–2058.

Gama-Sosa, M. A., Slagel, V. A., Trewyn, R. W., Oxenhandler, R., Kuo, K. C., Gehrke, C. W. & Ehrlich, M. (1983a). The 5-methylcytosine content of DNA from human tumors. *Nucleic Acids Res* 11, 6883–6894.

Gama-Sosa, M. A., Wang, R. Y., Kuo, K. C., Gehrke, C. W. & Ehrlich, M. (1983b). The 5-methylcytosine content of highly repeated sequences in human DNA. *Nucleic Acids Res* 11, 3087–3095.

Gaudet, F., Hodgson, J. G., Eden, A., Jackson-Grusby, L., Dausman, J., Gray, J. W., Leonhardt, H. & Jaenisch, R. (2003). Induction of tumors in mice by genomic hypomethylation. *Science* 300, 489–492.

Germano, I., Swiss, V. & Casaccia, P. (2010). Primary brain tumors, neural stem cell, and brain tumor cancer cells: where is the link? *Neuropharmacology* 58, 903–910.

Gravendeel, L. A., Kouwenhoven, M. C., Gevaert, O., de Rooi, J. J., Stubbs, A. P., Duijm, J. E., Daemen, A., Bleeker, F. E., Bralten, L. B., Kloosterhof, N. K., De Moor, B., Eilers, P. H., van der Spek, P. J., Kros, J. M., Sillevis Smitt, P. A., van den Bent, M. J. & French, P. J. (2009). Intrinsic gene expression profiles of gliomas are a better predictor of survival than histology. *Cancer Res* 69, 9065–9072.

Guo, J. U., Su, Y., Zhong, C., Ming, G. L. & Song, H. (2011). Emerging roles of TET proteins and 5-hydroxymethylcytosines in active DNA demethylation and beyond. *Cell Cycle* 10, 2662–2668.

Hegi, M. E., Diserens, A. C., Gorlia, T., Hamou, M. F., de Tribolet, N., Weller, M., Kros, J. M., Hainfellner, J. A., Mason, W., Mariani, L., Bromberg, J. E., Hau, P., Mirimanoff, R. O., Cairncross, J. G., Janzer, R. C. & Stupp, R. (2005). MGMT gene silencing and benefit from temozolomide in glioblastoma. *N Engl J Med* 352, 997–1003.

Hegi, M. E., Liu, L., Herman, J. G., Stupp, R., Wick, W., Weller, M., Mehta, M. P. & Gilbert, M. R. (2008). Correlation of O6-methylguanine methyltransferase (MGMT) promoter methylation with clinical outcomes in glioblastoma and clinical strategies to modulate MGMT activity. *J Clin Oncol* 26, 4189–4199.

Holm, T. M., Jackson-Grusby, L., Brambrink, T., Yamada, Y., Rideout, W. M. & Jaenisch, R. (2005). Global loss of imprinting leads to widespread tumorigenesis in adult mice. *Cancer Cell* 8, 275–285.

Houseman, E. A., Christensen, B. C., Yeh, R. F., Marsit, C. J., Karagas, M. R., Wrensch, M., Nelson, H. H., Wiemels, J., Zheng, S., Wiencke, J. K. & Kelsey, K. T. (2008). Model-based clustering of DNA methylation array data: a recursive-partitioning algorithm for high-dimensional data arising as a mixture of beta distributions. *BMC Bioinformatics* 9:365.

Huang, Y., Pastor, W. A., Shen, Y., Tahiliani, M., Liu, D. R. & Rao, A. (2010). The behaviour of 5-hydroxymethylcytosine in bisulfite sequencing. *PLoS ONE* 5, e8888.

Ito, S., D'Alessio, A. C., Taranova, O. V., Hong, K., Sowers, L. C. & Zhang, Y. (2010). Role of Tet proteins in 5mC to 5hmC conversion, ES-cell self-renewal and inner cell mass specification. *Nature* 466, 1129–1133.

Jemal, A., Siegel, R., Xu, J. & Ward, E. (2010). Cancer statistics, 2010. *CA Cancer J Clin* 60, 277–300.

Jin, S. G., Wu, X., Li, A. X. & Pfeifer, G. P. (2011). Genomic mapping of 5-hydroxymethylcytosine in the human brain. *Nucleic Acids Res* 39, 5015–5024.

Kang, M. R., Kim, M. S., Oh, J. E., Kim, Y. R., Song, S. Y., Seo, S. I., Lee, J. Y., Yoo, N. J. & Lee, S. H. (2009). Mutational analysis of IDH1 codon 132 in glioblastomas and other common cancers. *Int J Cancer* 125, 353–355.

Karpf, A. R. & Matsui, S. (2005). Genetic disruption of cytosine DNA methyltransferase enzymes induces chromosomal instability in human cancer cells. *Cancer Res* 65, 8635–8639.

Kelly, W. K., O'Connor, O. A. & Marks, P. A. (2002). Histone deacetylase inhibitors: from target to clinical trials. *Expert Opin Investig Drugs* 11, 1695–1713.

Kim, H. J. & Rosenfeld, M. G. (2010). Epigenetic control of stem cell fate to neurons and glia. *Arch Pharm Res* 33, 1467–1473.

Kim, J. H., Shin, J. H. & Kim, I. H. (2004). Susceptibility and radiosensitization of human glioblastoma cells to trichostatin A, a histone deacetylase inhibitor. *Int J Radiat Oncol Biol Phys* 59, 1174–1180.

Kloosterhof, N. K., Bralten, L. B., Dubbink, H. J., French, P. J. & van den Bent, M. J. (2011). Isocitrate dehydrogenase-1 mutations: a fundamentally new understanding of diffuse glioma? *Lancet Oncol* 12, 83–91.

Kriaucionis, S. & Heintz, N. (2009). The nuclear DNA base 5-hydroxymethylcytosine is present in Purkinje neurons and the brain. *Science* 324, 929–930.

Kunitz, A., Wolter, M., van den Boom J., Felsberg, J., Tews, B., Hahn, M., Benner, A., Sabel, M., Lichter, P., Reifenberger, G., von Deimling, A. & Hartmann, C. (2007). DNA hypermethylation and aberrant expression of the EMP3 gene at 19q13.3 in Human Gliomas. *Brain Pathol* 17, 363–370.

Li, W. & Liu, M. (2011). Distribution of 5-hydroxymethylcytosine in different human tissues. *J Nucleic Acids* 2011:870726.

Liby, P., Kostrouchova, M., Pohludka, M., Yilma, P., Hrabal, P., Sikora, J., Brozova, E., Kostrouchova, M., Rall, J. E. & Kostrouch, Z. (2006). Elevated and deregulated expression of HDAC3 in human astrocytic glial tumours. *Folia Biol (Praha)* 52, 21–33.

Liu, G., Ying, H., Zeng, G., Wheeler, C. J., Black, K. L. & Yu, J. S. (2004). HER-2, gp100, and MAGE-1 are expressed in human glioblastoma and recognized by cytotoxic T cells. *Cancer Res* 64, 4980–4986.

Lopez, C. A., Feng, F. Y., Herman, J. M., Nyati, M. K., Lawrence, T. S. & Ljungman, M. (2007). Phenylbutyrate sensitizes human glioblastoma cells lacking wild-type p53 function to ionizing radiation. *Int J Radiat Oncol Biol Phys* 69, 214–220.

Louis, D. N., Ohgaki, H., Wiestler, O. D., Cavenee, W. K., Burger, P. C., Jouvet, A., Scheithauer, B. W. & Kleihues, P. (2007). The 2007 WHO classification of tumours of the central nervous system. *Acta Neuropathol* 114, 97–109.

Lucio-Eterovic, A. K., Cortez, M. A., Valera, E. T., Motta, F. J., Queiroz, R. G., Machado, H. R., Carlotti, C. G., Jr., Neder, L., Scrideli, C. A. & Tone, L. G. (2008). Differential expression of 12 histone deacetylase (HDAC) genes in astrocytomas and normal brain tissue: class II and IV are hypoexpressed in glioblastomas. *BMC Cancer* 8:243.

Maher, E. A., Brennan, C., Wen, P. Y., Durso, L., Ligon, K. L., Richardson, A., Khatry, D., Feng, B., Sinha, R., Louis, D. N., Quackenbush, J., Black, P. M., Chin, L. & DePinho, R. A. (2006). Marked genomic differences characterize primary and secondary glioblastoma subtypes and identify two distinct molecular and clinical secondary glioblastoma entities. *Cancer Res* 66, 11502–11513.

Malzkorn, B., Wolter, M., Riemenschneider, M. J. & Reifenberger, G. (2011). Unraveling the glioma epigenome-from molecular mechanisms to novel biomarkers and therapeutic targets. *Brain Pathol* 21, 619–632.

Mardis, E. R., Ding, L., Dooling, D. J., Larson, D. E., McLellan, M. D., Chen, K., Koboldt, D. C., Fulton, R. S., Delehaunty, K. D., McGrath, S. D., Fulton, L. A., Locke, D. P., Magrini, V. J., Abbott, R. M., Vickery, T. L., Reed, J. S., Robinson, J. S., Wylie, T., Smith, S. M., Carmichael, L., Eldred, J. M., Harris, C. C., Walker, J., Peck, J. B., Du, F., Dukes, A. F., Sanderson, G. E., Brummett, A. M., Clark, E., McMichael, J. F., Meyer, R. J., Schindler, J. K., Pohl, C. S., Wallis, J. W., Shi, X., Lin, L., Schmidt, H., Tang, Y., Haipek, C., Wiechert, M. E., Ivy, J. V., Kalicki, J., Elliott, G., Ries, R. E., Payton, J. E., Westervelt, P., Tomasson, M. H., Watson, M. A., Baty, J., Heath, S., Shannon, W. D., Nagarajan, R., Link, D. C., Walter, M. J., Graubert, T. A., DiPersio, J. F., Wilson, R. K. & Ley, T. J. (2009). Recurring mutations found by sequencing an acute myeloid leukemia genome. *N Engl J Med* 361, 1058–1066.

Martinowich, K., Hattori, D., Wu, H., Fouse, S., He, F., Hu, Y., Fan, G. & Sun, Y. E. (2003). DNA methylation-related chromatin remodeling in activity-dependent BDNF gene regulation. *Science* 302, 890–893.

Maunakea, A. K., Nagarajan, R. P., Bilenky, M., Ballinger, T. J., D'Souza, C., Fouse, S. D., Johnson, B. E., Hong, C., Nielsen, C., Zhao, Y., Turecki, G., Delaney, A., Varhol, R., Thiessen, N., Shchors, K., Heine, V. M., Rowitch, D. H., Xing, X., Fiore, C., Schillebeeckx, M., Jones, S. J., Haussler, D., Marra, M. A., Hirst, M., Wang, T. & Costello, J. F. (2010). Conserved role of intragenic DNA methylation in regulating alternative promoters. *Nature* 466, 253–257.

Mohr, F., Dohner, K., Buske, C. & Rawat, V. P. (2011). TET genes: new players in DNA demethylation and important determinants for stemness. *Exp Hematol* 39, 272–281.

Mokhtari, K., Paris, S., Aguirre-Cruz, L., Privat, N., Criniere, E., Marie, Y., Hauw, J. J., Kujas, M., Rowitch, D., Hoang-Xuan, K., Delattre, J. Y. & Sanson, M. (2005). Olig2 expression, GFAP, p53 and 1p loss analysis contribute to glioma subclassification. *Neuropathol Appl Neurobiol* 31, 62–69.

Nakamura, M., Ishida, E., Shimada, K., Kishi, M., Nakase, H., Sakaki, T. & Konishi, N. (2005). Frequent LOH on 22q12.3 and TIMP-3 inactivation occur in the progression to secondary glioblastomas. *Lab Invest* 85, 165–175.

Nakamura, M., Yonekawa, Y., Kleihues, P. & Ohgaki, H. (2001b). Promoter hypermethylation of the RB1 gene in glioblastomas. *Lab Invest* 81, 77–82.

Nobusawa, S., Watanabe, T., Kleihues, P. & Ohgaki, H. (2009). IDH1 mutations as molecular signature and predictive factor of secondary glioblastomas. *Clin Cancer Res* 15, 6002–6007.

Ohgaki, H. & Kleihues, P. (2007). Genetic pathways to primary and secondary glioblastoma. *Am J Pathol* 170, 1445–1453.

Ohka, F., Natsume, A., Motomura, K., Kishida, Y., Kondo, Y., Abe, T., Nakasu, Y., Namba, H., Wakai, K., Fukui, T., Momota, H., Iwami, K., Kinjo, S., Ito, M., Fujii, M. & Wakabayashi, T. (2011). The global DNA methylation surrogate LINE-1 methylation is correlated with MGMT promoter methylation and is a better prognostic factor for glioma. *PLoS ONE* 6, e23332.

Ono, R., Taki, T., Taketani, T., Taniwaki, M., Kobayashi, H. & Hayashi, Y. (2002). LCX, leukemia-associated protein with a CXXC domain, is fused to MLL in acute myeloid leukemia with trilineage dysplasia having t(10;11)(q22;q23). *Cancer Res* 62, 4075–4080.

Orzan, F., Pellegatta, S., Poliani, P. L., Pisati, F., Caldera, V., Menghi, F., Kapetis, D., Marras, C., Schiffer, D. & Finocchiaro, G. (2011). Enhancer of Zeste 2 (EZH2) is up-regulated in malignant gliomas and in glioma stem-like cells. *Neuropathol Appl Neurobiol* 37, 381–394.

Palm, T. & Schwamborn, J. C. (2010). Brain tumor stem cells. *Biol Chem* 391, 607–617.

Park, S. W., Chung, N. G., Han, J. Y., Eom, H. S., Lee, J. Y., Yoo, N. J. & Lee, S. H. (2009). Absence of IDH2 codon 172 mutation in common human cancers. *Int J Cancer* 125, 2485–2486.

Parsons, D. W., Jones, S., Zhang, X., Lin, J. C., Leary, R. J., Angenendt, P., Mankoo, P., Carter, H., Siu, I. M., Gallia, G. L., Olivi, A., McLendon, R., Rasheed, B. A., Keir, S., Nikolskaya, T., Nikolsky, Y., Busam, D. A., Tekleab, H., Diaz, L. A., Jr., Hartigan, J., Smith, D. R., Strausberg, R. L., Marie, S. K., Shinjo, S. M., Yan, H., Riggins, G. J., Bigner, D. D., Karchin, R., Papadopoulos, N., Parmigiani, G., Vogelstein, B., Velculescu, V. E. & Kinzler, K. W. (2008). An integrated genomic analysis of human glioblastoma multiforme. *Science* 321, 1807–1812.

Pastor, W. A., Pape, U. J., Huang, Y., Henderson, H. R., Lister, R., Ko, M., McLoughlin, E. M., Brudno, Y., Mahapatra, S., Kapranov, P., Tahiliani, M., Daley, G. Q., Liu, X. S., Ecker, J. R., Milos, P. M., Agarwal, S. & Rao, A. (2011). Genome-wide mapping of 5-hydroxymethylcytosine in embryonic stem cells. *Nature* 473, 394–397.

Phuphanich, S., Baker, S. D., Grossman, S. A., Carson, K. A., Gilbert, M. R., Fisher, J. D. & Carducci, M. A. (2005). Oral sodium phenylbutyrate in patients with recurrent malignant gliomas: a dose escalation and pharmacologic study. *Neuro Oncol* 7, 177–182.

Prak, E. T. & Kazazian, H. H. (2000). Mobile elements and the human genome. *Nat Rev Genet* 1, 134–144.

Rodriguez-Paredes, M. & Esteller, M. (2011). Cancer epigenetics reaches mainstream oncology. *Nat Med* 17, 330–339.

Sanson, M., Marie, Y., Paris, S., Idbaih, A., Laffaire, J., Ducray, F., El Hallani, S., Boisselier, B., Mokhtari, K., Hoang-Xuan, K. & Delattre, J. Y. (2009). Isocitrate dehydrogenase 1 codon 132 mutation is an important prognostic biomarker in gliomas. *J Clin Oncol* 27, 4150–4154.

Scarcella, D. L., Chow, C. W., Gonzales, M. F., Economou, C., Brasseur, F. & Ashley, D. M. (1999). Expression of MAGE and GAGE in high-grade brain tumors: a potential target for specific immunotherapy and diagnostic markers. *Clin Cancer Res* 5, 335–341.

Schwartzbaum, J. A., Fisher, J. L., Aldape, K. D. & Wrensch, M. (2006). Epidemiology and molecular pathology of glioma. *Nat Clin Pract Neurol* 2, 494–503.

Shimozaki, K., Namihira, M., Nakashima, K. & Taga, T. (2005). Stage- and site-specific DNA demethylation during neural cell development from embryonic stem cells. *J Neurochem* 93, 432–439.

Sonoda, Y., Kumabe, T., Nakamura, T., Saito, R., Kanamori, M., Yamashita, Y., Suzuki, H. & Tominaga, T. (2009). Analysis of IDH1 and IDH2 mutations in Japanese glioma patients. *Cancer Sci* 100, 1996–1998.

Tabatabai, G. & Weller, M. (2011). Glioblastoma stem cells. *Cell Tissue Res* 343, 459–465.

Tahiliani, M., Koh, K. P., Shen, Y., Pastor, W. A., Bandukwala, H., Brudno, Y., Agarwal, S., Iyer, L. M., Liu, D. R., Aravind, L. & Rao, A. (2009). Conversion of 5-methylcytosine to 5-hydroxymethylcytosine in mammalian DNA by MLL partner TET1. *Science* 324, 930–935.

Takizawa, T., Nakashima, K., Namihira, M., Ochiai, W., Uemura, A., Yanagisawa, M., Fujita, N., Nakao, M. & Taga, T. (2001). DNA methylation is a critical cell-intrinsic determinant of astrocyte differentiation in the fetal brain. *Dev Cell* 1, 749–758.

Tan, A. Y. & Manley, J. L. (2009). The TET family of proteins: functions and roles in disease. *J Mol Cell Biol* 1, 82–92.

Trinh, B. N., Long, T. I. & Laird, P. W. (2001). DNA methylation analysis by MethyLight technology. *Methods* 25, 456–462.

Tsai, H. C. & Baylin, S. B. (2011). Cancer epigenetics: linking basic biology to clinical medicine. *Cell Res* 21, 502–517.

Tso, C. L., Freije, W. A., Day, A., Chen, Z., Merriman, B., Perlina, A., Lee, Y., Dia, E. Q., Yoshimoto, K., Mischel, P. S., Liau, L. M., Cloughesy, T. F. & Nelson, S. F. (2006). Distinct transcription profiles of primary and secondary glioblastoma subgroups. *Cancer Res* 66, 159–167.

Ugur, H. C., Ramakrishna, N., Bello, L., Menon, L. G., Kim, S. K., Black, P. M. & Carroll, R. S. (2007). Continuous intracranial administration of suberoylanilide hydroxamic acid (SAHA) inhibits tumor growth in an orthotopic glioma model. *J Neurooncol* 83, 267–275.

Vakoc, C. R., Mandat, S. A., Olenchock, B. A. & Blobel, G. A. (2005). Histone H3 lysine 9 methylation and HP1gamma are associated with transcription elongation through mammalian chromatin. *Mol Cell* 19, 381–391.

Valk-Lingbeek, M. E., Bruggeman, S. W. & van Lohuizen, M. (2004). Stem cells and cancer; the polycomb connection. *Cell* 118, 409–418.

Van Lint, C., Emiliani, S. & Verdin, E. (1996). The expression of a small fraction of cellular genes is changed in response to histone hyperacetylation. *Gene Expr* 5, 245–253.

Verhaak, R. G., Hoadley, K. A., Purdom, E., Wang, V., Qi, Y., Wilkerson, M. D., Miller, C. R., Ding, L., Golub, T., Mesirov, J. P., Alexe, G., Lawrence, M., O'Kelly, M., Tamayo, P., Weir, B. A., Gabriel, S., Winckler, W., Gupta, S., Jakkula, L., Feiler, H. S., Hodgson, J. G., James, C. D., Sarkaria, J. N., Brennan, C., Kahn, A., Spellman, P. T., Wilson, R. K., Speed, T. P., Gray, J. W., Meyerson, M., Getz, G., Perou, C. M. & Hayes, D. N. (2010). Integrated genomic analysis identifies clinically relevant subtypes of glioblastoma characterized by abnormalities in PDGFRA, IDH1, EGFR, and NF1. *Cancer Cell* 17, 98–110.

Warburg, O. (1923). Versuche an überlebendem Carcinom-Gewebe (Methoden). *Biochem Zeitschr* 142, 317–333.

Weljie, A. M. & Jirik, F. R. (2011). Hypoxia-induced metabolic shifts in cancer cells: moving beyond the Warburg effect. *Int J Biochem Cell Biol* 43, 981–989.

Wiemels, J. L., Wilson, D., Patil, C., Patoka, J., McCoy, L., Rice, T., Schwartzbaum, J., Heimberger, A., Sampson, J. H., Chang, S., Prados, M., Wiencke, J. K. & Wrensch, M. (2009). IgE, allergy, and risk of glioma: update from the San Francisco Bay Area Adult Glioma Study in the temozolomide era. *Int J Cancer* 125, 680–687.

Wiencke, J. K., Zheng, S., Jelluma, N., Tihan, T., Vandenberg, S., Tamguney, T., Baumber, R., Parsons, R., Lamborn, K. R., Berger, M. S., Wrensch, M. R., Haas-Kogan, D. A. & Stokoe, D. (2007). Methylation of the PTEN promoter defines low-grade gliomas and secondary glioblastoma. *Neuro Oncol* 9, 271–279.

Wiencke, J. K., Zheng, S., Morrison, Z. & Yeh, R. F. (2008). Differentially expressed genes are marked by histone 3 lysine 9 trimethylation in human cancer cells. *Oncogene* 27, 2412–2421.

Wrensch, M., Jenkins, R. B., Chang, J. S., Yeh, R. F., Xiao, Y., Decker, P. A., Ballman, K. V., Berger, M., Buckner, J. C., Chang, S., Giannini, C., Halder, C., Kollmeyer, T. M., Kosel, M. L., LaChance, D. H., McCoy, L., O'Neill, B. P., Patoka, J., Pico, A. R., Prados, M., Quesenberry, C., Rice, T., Rynearson, A. L., Smirnov, I., Tihan, T., Wiemels, J., Yang, P. & Wiencke, J. K. (2009). Variants in the CDKN2B and RTEL1 regions are associated with high-grade glioma susceptibility. *Nat Genet* 41, 905–908.

Xu, W., Yang, H., Liu, Y., Yang, Y., Wang, P., Kim, S. H., Ito, S., Yang, C., Wang, P., Xiao, M. T., Liu, L. X., Jiang, W. Q., Liu, J., Zhang, J. Y., Wang, B., Frye, S., Zhang, Y., Xu, Y. H., Lei, Q. Y., Guan, K. L., Zhao, S. M. & Xiong, Y. (2011). Oncometabolite 2-hydroxyglutarate is a competitive inhibitor of alpha-ketoglutarate-dependent dioxygenases. *Cancer Cell* 19, 17–30.

Yan, H., Parsons, D. W., Jin, G., McLendon, R., Rasheed, B. A., Yuan, W., Kos, I., Batinic-Haberle, I., Jones, S., Riggins, G. J., Friedman, H., Friedman, A., Reardon, D., Herndon, J., Kinzler, K. W., Velculescu, V. E., Vogelstein, B. & Bigner, D. D. (2009). IDH1 and IDH2 mutations in gliomas. *N Engl J Med* 360, 765–773.

Yin, D., Ong, J. M., Hu, J., Desmond, J. C., Kawamata, N., Konda, B. M., Black, K. L. & Koeffler, H. P. (2007). Suberoylanilide hydroxamic acid, a histone deacetylase inhibitor: effects on gene expression and growth of glioma cells in vitro and in vivo. *Clin Cancer Res* 13, 1045–1052.

Zhao, S., Lin, Y., Xu, W., Jiang, W., Zha, Z., Wang, P., Yu, W., Li, Z., Gong, L., Peng, Y., Ding, J., Lei, Q., Guan, K. L. & Xiong, Y. (2009). Glioma-derived mutations in IDH1 dominantly inhibit IDH1 catalytic activity and induce HIF-1alpha. *Science* 324, 261–265.

Zheng, S., Houseman, E. A., Morrison, Z., Wrensch, M. R., Patoka, J. S., Ramos, C., Haas-Kogan, D. A., McBride, S., Marsit, C. J., Christensen, B. C., Nelson, H. H., Stokoe, D., Wiemels, J. L., Chang, S. M., Prados, M. D., Tihan, T., VandenBerg, S. R., Kelsey, K. T., Berger, M. S. & Wiencke, J. K. (2011). DNA hypermethylation profiles associated with glioma subtypes and EZH2 and IGFBP2 mRNA expression. *Neuro Oncol* 13, 280–289.

Chapter 5
Aberrant Epigenetic Regulation in Breast Cancer

Amanda Ewart Toland

5.1 Introduction

Breast cancer is the most common nonskin malignancy affecting women in the Western world. It is diagnosed in approximately 200,000 women and 1,900 men in the United States each year and is estimated to affect 1.3 million individuals worldwide (www.cancer.org; American Cancer Society). Breast cancer is not one disease but a heterogeneous group of cancers which can arise from both lobular and ductal epithelial cells. Several subtypes of breast cancer have been characterized based on molecular and histological profiles; these vary in clinical presentation, prognosis, and available therapies. Breast cancer, like other cancers, is caused by changes in gene regulation and function which lead to abnormal cell growth and spread. Genes important in suppressing tumor growth, known as tumor suppressor genes, frequently have inactivating mutations or are deleted in breast tumors. Oncogenes, genes which act to promote cell growth, may have activating mutations or may have extra copies in breast cancer cells. In addition to mutations which activate or inactivate cancer-related genes and alterations to copy numbers of genes, there is increasing evidence to suggest that epigenetics is playing a key role in breast cancer tumorigenesis. Epigenetics refers to heritable factors that influence gene expression and regulation which are independent of the DNA sequence. Deregulation of normal epigenetic processes during aging and by environmental factors has been associated with breast cancer development. Several mechanisms of epigenetic regulation of genes exist. CpG methylation of gene promoters and introns, histone modifications leading to chromatin remodeling and accessibility to transcription factors, methylation of cytosines in gene bodies, and microRNA expression are all epigenetic processes reported to be altered during breast carcinogenesis (Fig. 5.1). As technological advancement of identification of

---

A.E. Toland, PhD (✉)
Department of Molecular Virology, Immunology and Medical Genetics,
Comprehensive Cancer Center, The Ohio State University, Columbus, OH, USA
e-mail: Amanda.Toland@osumc.edu

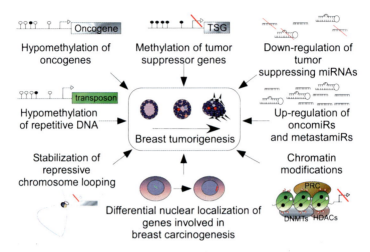

Fig. 5.1 Epigenetic regulation in breast cancer. The main perturbations in epigenetic regulation occurring during breast carcinogenesis are illustrated. TSG, tumor suppressor gene; miRNAs, microRNAs; DNMTs, DNA methyltransferases; HDACs, histone deaceteylases; PRC, polycomb repressor complexes

epigenetic changes across the genome has improved our understanding, the role of epigenetic dysregulation leading to breast cancer has accelerated.

This chapter will highlight the role of epigenetics during normal breast development and will outline basic epigenetic processes which are perturbed during the development of breast cancer. In addition, the role of epigenetics in classifying breast cancers, in determining prognosis, and as potential therapeutic targets will be discussed. What is clear from studies to date is that epigenetics plays a critical role in breast tumorigenesis and that new whole-genome approaches will shed additional insight into the relationship between aberrant epigenetic regulation and the development of breast cancer.

5.2 Epigenetics

Epigenetics refers to heritable factors which influence the phenotype of a cell or organism and are not dependent upon DNA sequence. Most epigenetic mechanisms impact gene regulation and expression and are stable across multiple cell divisions and, in some cases, across generations. Unlike adaptive mutations which may take generations before they allow for better survival, a cell can adapt more rapidly to environmental stimuli through changes in epigenetic regulation. Identifying the normal epigenetic events of the breast and abnormal epigenetic changes that take place during breast tumorigenesis are not only important in understanding how breast cancer develops but can highlight biomarkers for early diagnosis, response to and resistance from therapy, and can themselves be targets for treatment. There are several epigenetic mechanisms by which normal gene regulation can be perturbed.

The best studied in breast cancer are DNA methylation, chromatin modifications, and small noncoding RNAs. Other less-described epigenetic phenomenon, such as nuclear localization and chromatin loops, shows some evidence of being altered in breast tumors. There are additional epigenetic mechanisms such as paramutations, but as these do not yet have an established role in human cancer, these will not be described.

5.3 Methylation

DNA methylation is the addition of a methyl group to a cytosine which is typically located 5' to a guanine (CpG). During methylation, a methyl group from S-adenosylmethionine (SAM) is transferred by DNA methyltransferases (DNMTs) to the C5 position of cytosine. DNMT1 is responsible for maintenance of the methylation state of the genome during replication (Hermann et al. 2004). DNMT3a and 3b establish new methylation on cytosines not previously methylated (Chedin 2011). Traditionally, methylation studies have focused on CpG islands, cytosine- and guanine-rich regions, found in promoter regions of about 70% of genes and near retrotransposable elements and repetitive DNA such as Alu and LINE-1 sequences (Widschwendter and Jones 2002). Generally, methylation of CpGs within a promoter region is associated with silencing of gene transcription, and hypomethylation is associated with gene expression. In normal tissues, CpG islands in promoter regions are mainly unmethylated, but CpG islands in repetitive DNA regions tend to be highly methylated (Jintaridth and Mutirangura 2010). This pattern of methylation ensures that important genes are expressed but that regions housing retrotransposable and retroviral elements are repressed. Recent studies show that cytosines in introns and non-CpG-rich regions are methylated and may play a role in how genes are expressed (Colaneri et al. 2011; Cotton et al. 2011). There is a strong link between DNA methylation status in gene promoters and in concordant occupancy of repressive histone modifications (like H3K27me3) resulting in active or inactive chromatin states which will be discussed later (Ballestar and Esteller 2005).

5.3.1 Aberrant Methylation Patterns in Cancers

Gene-specific promoter hypermethylation and global genomic DNA hypomethylation of repetitive sequences of the genome are hallmarks of several cancers (Ehrlich 2000). Promoter hypermethylation generally leads to repression of gene expression; several tumor suppressor genes important in breast cancer show increased methylation and subsequent decreased expression. Hypomethylation of the genome is an early event in many tumors (Ehrlich 2009) and is thought to reactivate gene expression including oncogenes. Hypomethylation has also been postulated to be one phenomenon leading to chromosomal instability in cancer cells (Rodriguez et al. 2006), possibly via hypomethylation of viral genes or retrotransposable elements whose activation is associated with genomic instability (Daskalos et al. 2009).

5.3.2 Breast Cancer Genes Showing Aberrant Methylation

When methylation was first recognized to be a critical regulator of gene expression, researchers proposed that aberrant methylation might be occurring in tumorigenesis (Feinberg and Vogelstein 1983). In the first studies to test this hypothesis, researchers identified the genes with aberrant methylation in breast cancer using a candidate gene approach. Techniques for determining whether a specific gene or promoter of interest is methylated include bisulfite sequencing, methylation-specific PCR, restriction landmark genomic scanning, methylation-specific restriction digests (COBRA), mass spectrometry (Sequenom EpiTyper), Pyrosequencing, and methylation arrays (Fraga and Esteller 2002; Fazzari and Greally 2010; Nair et al. 2011). Over 150 genes have been shown to be hypermethylated in breast tumors or breast cancer cell lines (Hinshelwood and Clark 2008; Huang and Esteller 2010); many of these genes have roles in processes important in breast carcinogenesis like cell signaling, apoptosis, tissue invasion and metastasis, cell cycle, angiogenesis, and hormone signaling. It is also important to point out that genes from multiple pathways shown to be important in breast tumorigenesis are silenced or show altered activation by CpG promoter methylation (Huang and Esteller 2010). Pathways perturbed by methylation include cell cycle, DNA repair, angiogenesis, hormone receptors, cellular invasion, and tumor suppression. A much smaller number of genes with tumor-promoting functions show hypomethylation in breast cancer (Scelfo et al. 2002; Hinshelwood and Clark 2008).

The number of genes that are hypermethylated in breast tumors is too long to review in detail, and there are several reviews which list many of them (Agrawal et al. 2007; Huang and Esteller 2010; Jovanovic et al. 2010). However, there are a handful of classic examples of genes that show hypermethylation in tumors and which have important roles in breast cancer. These illustrate the importance of promoter methylation of specific genes during breast cancer development.

5.3.2.1 Estrogen Receptor Alpha

One of the standard means of characterizing breast tumors for therapeutic and prognosis reasons is analysis of estrogen receptor (ER) staining. There are two estrogen receptors, α and β. ER-α is encoded by estrogen receptor alpha (*ESR1*) and is the receptor which appears to be more clinically relevant. Tumors that are positive for ER-α (ER+) are more likely to have better prognosis and are amenable to therapies such as tamoxifen which target the estrogen receptor pathway. About 30–40% of breast tumors lack ER-α expression and are denoted as ER– (ER "minus"). ER– tumors have a poorer prognosis and can be part of a subgroup of tumors called triple negative tumors which are also negative for progesterone receptor (PR) staining and HER2/Neu amplification. Depending on the study, the amount of measured *ESR1* promoter methylation varies in ER– tumors (Kim et al. 2004), but promoter methylation appears to be more frequent in triple negative tumors. One study found that

80% of sporadic triple negative breast cancers in Chinese women show silencing of the *ESR1* gene by methylation (Jing et al. 2011). There is also evidence suggesting that demethylating agents can restore ER-α expression in a subset of tumors (Lapidus et al. 1996). Despite the different data suggesting that methylation is an important means of regulation of ER-α, some studies show weak correlation between methylation patterning and ER-α expression (Gaudet et al. 2009).

5.3.2.2 Ras-Associated Domain Family Member 1 Gene

Ras-associated domain family member 1 gene (*RASSF1A*) is one of seven unique transcripts (*A–G*) encoded from the *RASSF1* gene. The promoter for *RASSF1A* is in a CpG island and is one of the most frequently hypermethylated promoters described in cancer. Approximately 60–77% of breast cancers show *RASSF1A* promoter hypermethylation, but hypermethylation is very rare in normal tissues of individuals who do not have cancer. RASSF1A is a Ras effector and has a role in inducing apoptosis providing a link between silencing of the gene and cancer development. Methylated *RASSF1A* in sera from breast cancer patients is associated with metastasis, tumor size, and mortality and has been identified as a marker for response to tamoxifen treatment (Muller et al. 2003; Fiegl et al. 2005). Thus, methylation of this gene may be a useful biomarker for prognosis and response to therapy.

5.3.2.3 Breast Cancer 1

Germline mutations of breast cancer 1 (*BRCA1*) are found in roughly 30% of individuals with a strong personal and family history of breast and ovarian cancer. Breast tumors from individuals with *BRCA1* mutations are frequently triple negative. *BRCA1* is well known for its role as a tumor suppressor gene and a gene that can predispose individuals to hereditary breast and ovarian cancer, but it also has a direct connection to methylation. In its role as a transcription factor, BRCA1 transcriptionally activates *DNMT1*, one of the methyltransferases which maintains the methylation status of the genome. When *BRCA1* is inactivated or silenced through methylation, the levels of DNMT1 decrease and the genome becomes hypomethylated with an open chromatin confirmation structure (Shukla et al. 2010). Between 11 and 31% of primary sporadic breast tumors show methylation of CpGs in the *BRCA1* promoter (Catteau and Morris 2002); however, the *BRCA1* promoter is much more frequently methylated in triple negative/basal like-breast tumors strengthening a connection between loss of *BRCA1* and the development of these aggressive tumors (Stefansson et al. 2011). Like *RASSF1A*, methylation of *BRCA1* is associated with prognosis; individuals with tumors positive for *BRCA1* methylation have an increased mortality (Xu et al. 2010). *BRCA1* methylation is also associated with low RB expression and high p16 levels which correlate with poorer outcomes (Stefansson et al. 2011). Consistent with these findings, tumors with methylated *BRCA1* are more likely to be invasive, have nodal involvement, and be larger in size.

5.3.3 Genome-Wide Methylation Patterning

Epigenetic studies assessing methylation status of genes important in breast cancer has moved from the analysis of candidate genes to microarrays which can interrogate hundreds to thousands of specific promoter regions to whole genome sequencing-based methods which can potentially assess all methylated sites in the genome (Fouse et al. 2010). In methylation arrays, probes for promoter regions, tiling probes of nonrepetitive regions of the genome, or probes for specific CpG are contained on the arrays. The number of sites assayed can range from hundreds to nearly 30,000. One of the differences between array-based methods is how the DNA for probing the arrays is isolated. Methylated DNA immunoprecipitation (MeDIP) is an approach which utilizes immunoprecipitation of methylated DNA using antibodies against 5-mC to pull down DNA to hybridize to the array. A related approach is to use antibodies against methyl-CpG-binding domain proteins (MBD). Differential methylation hybridization (DMH) utilizes methylation-specific digests prior to hybridization on an array (Yan et al. 2006). Similar approaches include methylated CpG island amplification microarray (MCAM), HpaII tiny fragment enrichment by ligation-mediated PCR (HELP), and comprehensive high-throughput arrays for relative methylation (CHARM). Following the use of these and other related approaches, enriched methylated DNA is hybridized to arrays. Array-based studies of methylation patterns have led to the identification of genes previously unrecognized to show methylation in breast cancer. In addition, these studies have identified large blocks of DNA which are contiguously hypomethylated in multiple tumor types relative to normal cells (Hansen et al. 2011). Microarrays have led to a vast increase in knowledge of the epigenome; however, they may be biased based on the choice of probes to be included or by false results generated by polymorphisms. In addition, arrays require prior knowledge of sequences of interest.

More recently, methylation has been assessed across the whole genome by high-throughput next-generation sequencing which is not biased by proximity to promoters or by inclusion in CpG islands. As there are an estimated 29 million CpG sites in the genome, only 7% of which are within CpG islands; genomic sequencing has the potential to greatly add to our knowledge of genomic methylation (Rollins et al. 2006). These large-scale sequencing efforts are just beginning to be published but confirm, in general, that there are specific regions near genes that show increased methylation during breast cancer development and there are repetitive regions which show decreased methylation (Sun et al. 2011). In addition, the genome-wide sequencing methods have also led to the identification of intragenic regions and non-CpG residues which show methylation in embryonic stem cells (Lister et al. 2009).

There are multiple methods used to prepare DNA for next-generation sequencing (Hirst and Marra 2010; Huang and Esteller 2010). Two, methylC-seq and reduced representation bisulfite sequencing, are bisulfite-based methods and require treatment of DNA to convert methylated cytosines to uracil using bisulfite treatment. In reduced representation bisulfite sequencing, the genome is reduced by digesting DNA with methylation-insensitive enzymes and selecting fragments of a certain size which are then treated with bisulfite and sequenced. The bisulfite-based methods provide greater resolution as it is known exactly which base pairs are methylated. In addition to these strategies, there are antibody-based methods using antibodies to enrich methylated DNA, and these include MeDIP sequencing and methylated DNA-binding domain sequencing. The enrichment methods are slightly cheaper as less of the genome is sequenced, but they provide less resolution as it is not known exactly which residue is methylated, just the general region. Methylcytosines which do not occur in a CpG context are unable to be captured by array-based capture methods. However, they can be detected by bisulfite conversion followed by genome sequencing (Lister et al. 2009).

5.4 Chromatin

Chromatin, a DNA-protein complex, is responsible for packaging DNA and determining its potential to be transcribed. Chromatin consists of nucleosomes, 147 bp of DNA wrapped around a histone protein octamer. The octamer is composed of four histone proteins, H2A, H2B, H3, and H4, each in two copies (Kornberg and Lorch 1999). Chromatin exists in two states, euchromatin, also called open or non-condensed, and heterochromatin, or closed. Genes in open (noncondensed) chromatin are able to be transcribed, and genes in heterochromatin (highly compacted) are transcriptionally silenced. Chromatin states are regulated by a myriad of histone posttranslational modifications that include acetylation, methylation, ubiquitination, phosphorylation, and sumoylation which mainly occur on residues of amino-terminal tails of the histones (Fullgrabe et al. 2011). Acetylation is one of the main means by which chromatin structure is regulated. Two opposing classes of enzymes, histone acetyltransferases (HATs) and histone deacetylates (HDACs), help to maintain histone acetylation levels by adding or removing acetyl groups from lysines, respectively. Methylation of histones occurs mainly on lysine and arginine residues. Lysines can be mono-, di-, or trimethylated, and arginines may be mono-, symmetrically, or asymmetrically dimethylated (Zhang and Reinberg 2001). Together, the combination of histone modifications, sometimes called the "histone code," determines its conformational state (Strahl and Allis 2000).

Traditionally, promoter methylation and the acetylation state of chromatin are thought to be separate epigenetic mechanisms controlling gene expression. More recently, studies have linked the two of them together mechanistically (Ballestar and Esteller 2005) (Fig. 5.2). The methylation state of a promoter helps to signal whether a region should become silenced at the chromatin level, but several proteins and protein complexes, including HDACs, histone methyltransferases (HMTs), methyl-CpG-binding proteins (MBPs), and polycomb repressor complexes (PRCs), are key for this process. During the process of gene silencing, methylated cytosines (5-mC) recruit methyl-CpG-binding proteins to methylated DNA which then attract nucleosome remodeling complexes, HDACs, and other proteins which modify histone tails and result in a closed chromatin configuration (Lopez-Serra and Esteller 2008).

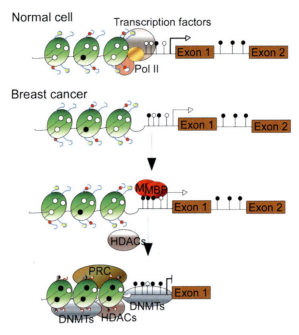

Fig. 5.2 The link between DNA methylation and chromatin silencing. In a normal cell, tumor suppressor genes are transcriptionally active, there is little promoter methylation (*small open circles*), the chromatin is in an open state, and transcription factors can bind. In breast cancer cells, there is an alteration of transcriptional regulation, and promoter methylation occurs. This leads to decreased transcription and the recruitment of methyl CpG-binding proteins (MBP) and histone deacetylases (HDACs). HDACs in combination with polycomb repressor complexes (PRC) further alter the chromatin state by histone deacetylation and histone H3K9 and/or H3K27 methylation which lead to gene silencing (Alford et al. 2012)

5.5 Methods Used to Measure Histone Modifications

Changes in chromatin modifications for candidate genes of interest are typically measured using chromatin-immunoprecipitation (ChIP) which is a method by which cross-linked DNA and protein are immunoprecipitated using an antibody specific for the histone modification of interest (Lo and Sukumar 2008). The DNA precipitated with the chromatin is then amplified using promoter-specific PCR primers. ChIP has been modified for profiling on a genome-wide basis. ChIP-chip uses chromatin-IP for a specific protein combined with high-density genomic microarrays to identify DNA sequences where the protein is binding. ChIP-SAGE combines ChIP with SAGE sequence analysis to quantify and map binding sites for specifically modified histones. ChIP-seq is the most recent method and involves massively parallel signature sequencing of all DNA pulled down by ChIP. ChIP-seq is the most quantitative of the three and is not biased for the specific parts of the genome being assessed like ChIP-on-chip. Candidate gene- and array-based methods have led to the identification of histone marks that are important in breast cancer, but no published ChIP-seq data for specific histone-specific modifications in breast cancer is yet available. Histone modifications can also be measured by mass spectrometry mapping.

5.5.1 Chromatin Modification in Breast Cancer

In addition to studies looking at methylation of genes as a biomarker for breast cancer, modifications to histones have also been assessed. Histone modifications of specific residues are associated with active versus repressed gene transcription; unlike CpG dinucleotides, methylation of histones can be associated with both repression and activation. Lysine acetylation (H3K9ac, H3K18ac, and H4K12ac), lysine trimethylation (H3K4me3), and arginine dimethylation (H4R3met2) are associated with transcriptional activation (Pokholok et al. 2005). Repressed chromatin is typically characterized by lysine methylation (H3K9me2, H3K9me3, H4K20me3), and often, a combination of methylation at multiple resides (H3K9me3, H3K27me3, and K4K20me3) occurs (reviewed by Bannister and Kouzarides 2011). In tumors, the combination of loss of acetylation of histone H4 at lysine 16 (H4K16Ac) with trimethylation of histone H4 at lysine 20 (H4K20me3) is associated with hypomethylation as well as increased sensitivity to chemotherapy (Fraga et al. 2005b). The combination of DNA methylation and repressive histone codes doubly assures the silencing of target genes. Several histone markers including those for histone lysine acetylation, lysine methylation, and arginine methylation have been evaluated in breast tumors. In one study of 880 breast tumors, approximately 80% of tumors showed low or absent histone lysine acetylation of H4K16ac (Elsheikh et al. 2009). Luminal breast cancers showed an extremely high correlation with high relative levels of global histone acetylation and methylation. Breast cancer subtypes conferring a poorer prognosis, including triple negative/basal-like tumors and HER2 positive tumors, had moderate to low levels of lysine acetylation (H3K9ac, H3K18ac, H4K12ac), lysine methylation (H3K4me2, H4K20met3), and arginine methylation (H4R3me2). Together, the data indicate that specific histone modifications strongly correlate with clinical phenotypes and breast cancer subtypes.

5.5.1.1 Polycomb Repressor Complex 1 and 2

Polycomb group proteins (PcG) were originally identified in Drosophila and were shown to play a key role in development via the regulation of *HOX* genes. There are two main polycomb repressive complexes (PRC1 and PRC2) which are composed of multiple proteins, many of which show deregulation during cancer. The PRCs are associated with transcriptional repression via their impact on chromatin remodeling. PRC1 is considered to be the complex important in maintenance of silencing. PRC2, 3, and 4 are complexes associated with initiation of the silencing. In general, the complexes are organized by the recruitment of HDACs by different isoforms of embryonic ectoderm development (EED) which leads to the deacetylation of histones. This is thought to be followed by the trimethylation of H3K27 by the PcG enhancer of zeste homolog 2 (EZH2) which allows for the binding of PRC1 to PRC2 (Lin et al. 2011; Tsang and Cheng 2011). *EZH2* expression is increased in a variety of cancers including prostate, bladder, gastric, lung, hepatocellular, and breast where its expression is associated with invasive cancer, metastasis, and poor survival (Chase and Cross 2011). Breast tumors lacking *BRCA1* expression have higher *EZH2* expression. Studies show a possible synthetic lethality between the two; knockdown of *EZH2* in *BRCA1*-deficient cells led to increased killing ability using a PRC2 inhibitor (Puppe et al. 2009). The mechanisms by which PRCs are recruited to promoter regions are still being defined.

5.6 Nuclear Organization and Chromatin Looping

In addition to DNA methylation and chromatin conformation, the position of DNA in the nucleus or in relationship to promoter and enhancers has been associated with gene regulation. Whereas there has not been as much study in these epigenetic mechanisms, there is support of their importance in breast malignancies.

5.6.1 Nuclear Organization

Another way in which gene expression is thought to be regulated epigenetically is through its position in the nucleus. Chromosomes have long been recognized to be positioned in specific territories in the nucleus (Sengupta et al. 2008; Heride et al. 2010). The epigenetic regulation of chromatin states may be modulated by position-effect variegation, an untested hypothesis first generated from observations in Drosophila (Henikoff 2008). In position-effect variegation, when a gene is juxtaposed next to heterochromatin, the gene becomes heritably silenced. More recently, it has been proposed that chromosomes occupy specific regions of the nucleus which impact their transcriptional activity. The periphery of the nucleus is associated with repression, but there may be "microdomains" which are excluded from repression (Deniaud and Bickmore 2009). One unanswered question is if heterochromatin or silenced genes are more likely to be located at the periphery of the nucleus or if the nuclear positioning itself impacts gene regulation. One hypothesis of the positional effect is that proteins related to chromatin silencing may be more likely to be tethered to the nuclear membrane; another theory is that there are fewer transcription factors in the periphery (Deniaud and Bickmore 2009). Few studies have looked at changes of nuclear organization in cancer, but in Hodgkin's lymphoma, nuclear remodeling has been shown to occur which may be due in part to aneuploidy and chromosomal rearrangements (Guffei et al. 2010). In breast cancer, eight genes, including *ERBB2* (the gene for HER2), have been shown to be differentially positioned in the nucleus of invasive breast cancers relative to normal breast tissue (Meaburn et al. 2009). The effect of differential nuclear location and expression was not tested; further research is needed to understand the significance of these findings for the regulation of genes in cancer.

5.6.2 Chromatin Looping

In addition to chromatin structure effects on the regulation of nearby genes, chromatin looping between distal sites on a chromosome and between chromosomes has been discovered as another means of epigenetic regulation. These long-distance physical interactions between enhancers/repressors and promoters can be identified by chromatin conformation capture (3C) and variations on 3C (e.g., 4, 5C) which can be applied across the genome (Dostie et al. 2006; Simonis et al. 2006). Evidence of the importance of long-distance regulation in normal breast tissue has been established. During lactation, lactogenic hormones induce the interaction of the β-casein promoter and its upstream enhancer (Kabotyanski et al. 2009). In normal breast cells in vitro, estrogen induces long-range epigenetic silencing through long DNA loops which bring together promoter regions with estrogen binding sites. In normal cells, the repressive effect of estrogen on these specific loci is transient, but in tumor cells, the repression is long term. These data suggest that epigenetic silencing in tumors can take place between loci which are separated by long distances (Hsu et al. 2010). Another example of the importance of chromatin looping in breast cancer is the connection between $1,25(OH)_2D3$ (vitamin D) and *CYP24*, a vitamin D-responsive gene encoding a vitamin D hydroxylase. Activated vitamin D has been shown to promote apoptosis and have an antiproliferative role in breast cancer cells (Welsh 1994; Flanagan et al. 2003). Vitamin D-responsive genes like *CYP24* contain vitamin D receptor (VDR) binding sites in their regulatory regions. In a malignant breast cancer cell line, MCF7, there are a higher number of activated VDR binding sites in the CYP24 promoter region compared to normal mammary epithelial cells because chromosomal looping in the tumor cells brings the distal sites in proximity to the *CYP24* promoter (Matilainen et al. 2010).

5.7 MicroRNAs and Other Noncoding RNA

MicroRNAs (miRNAs) are small 18–21 bp RNAs which regulate gene expression and protein translation. They were first discovered in plants and Drosophila but have since been found to play key roles in carcinogenesis of every cancer assessed thus far. miRNAs act to regulate gene expression of specific mRNAs by targeting mRNAs for degradation or preventing their translation. The discovery of miRNAs has led to an increased interest in the identification of additional miRNAs and the study of the role of small RNAs in normal biological processes. In addition to miRNAs, there are many other noncoding RNAs such as long noncoding RNAs (lncRNAs), small nucleolar RNAs (snoRNAs), piwi-interacting RNAs (piRNAs), promoter-associated small RNAs (pasRNAs), transcription initiation RNAs (tiRNAs), and endogenous small interfering RNAs (siRNAs) (Taft et al. 2010). LncRNAs are RNAs which are generally greater than 200 bp in size. They can regulate gene expression via modulation of chromatin modifications and transcriptional interference via antisense transcription; a few studies show a role for lncRNAs in breast cancer (Taft et al. 2010). piRNAs associate with Argonaute proteins and appear to be important in transposon defense via silencing of active transposons. One study to date shows aberrant expression of piRNAs in breast cancer (Cheng et al. 2011). As the rate of identification and characterization of new types of noncoding RNAs does not seem to be slowing down, it is expected that additional types of regulatory RNAs will be identified. While still under investigation, it is likely that at least some of these other ncRNAs will prove to be important in tumorigenesis and breast cancer. miRNAs have been the most studied of the small noncoding RNAs in breast cancer.

5.7.1 miRNAs in Breast Cancer

There is an ever-growing list of miRNAs associated with breast cancer. miRNAs have been identified as being important in breast cancer by gene expression studies comparing breast tumors to normal breast samples. miRNAs showing aberrant expression in breast cancers include *miR-21*, *miR-27a*, *miR-17*, *miR-155*, *miR-10b*, *miR-125*, *miR-145*, *miR-195*, and *miR-497* (Le Quesne and Caldas 2010; Shi et al. 2010). In addition to these, several miRNA families have been implicated in breast cancer such as the *miR-17-92* cluster and the *miR-200*, *miR-10*, and *let-7* families (Le Quesne and Caldas 2010). Together, these and additional miRNAs have been found to regulate genes important in a variety of cancer-related phenotypes including angiogenesis, cell invasion, migration, apoptosis, and cell cycle. As this field is relatively new, it is likely that additional miRNAs will be implicated in breast cancer pathogenesis.

Some miRNAs act as tumor suppressors by downregulating genes important in cancer progression while others act more as oncogenes (oncomirs) by downregulating genes important in tumor suppression. miRNAs can show cancer-specific up- or downregulation. *miR-205* is an example of a gene which is upregulated in several cancer types but shows decreased expression in breast cancer (Fletcher et al. 2008; Markou et al. 2008; Iorio et al. 2009). In breast cancer, *miR-205* regulates genes important for epithelial to mesenchymal transition and angiogenesis (Shi et al. 2010). These studies suggest that a miRNA can act as a tumor suppressor or as an oncogene depending on cell type and context-specific gene expression patterns. Thus, it will be important if anti-miR therapies are developed that off-target effects impacting other diseases and tissues are carefully assessed.

5.7.2 Epigenetic Regulation of miRNAs

miRNAs are transcriptionally regulated much in the same manner as mRNAs; thus, some miRNAs show evidence of epigenetic regulation. miRNAs have an additional layer of regulation as they are also regulated at the posttranscriptional level by enzymes important in processing the long transcribed forms into the shorter pre- and mature forms. About half of miRNA genes reside in CpG islands (Weber et al. 2007), and many miRNA promoters regions show aberrant promoter methylation in breast cancer. One miRNA, *miR-9,* shows both up- and downregulation in breast cancer in different studies. In late-stage cancers, *miR-9* is upregulated which leads to downregulation of E-cadherin, activation of beta-catenin, and promotion of breast cancer metastasis (Ma et al. 2010). Interestingly, in a subset of breast cancers, *miR-9-1* is downregulated via promoter methylation; the hypermethylation appears to occur early in preinvasive ductal cancers (Lehmann et al. 2008). From these studies, it appears that *miR-9* is downregulated in early stages of tumorigenesis or nonaggressive breast tumors but is upregulated in aggressive tumors. Another miRNA that can be regulated epigenetically is *miR-34*. *miR-34* is transcriptionally regulated by p53. It targets genes involved in apoptosis and cell proliferation. About one quarter of breast cancers show methylation of the *miR-34* promoter and decreased *miR-34* expression even in the presence of activated p53 (Lodygin et al. 2008). Some miRNAs themselves impact epigenetic regulation. *miR-29C* targets DNMTs *DNMT3a* and *DNMT3b* (Nguyen et al. 2011). Downregulation of *miR-29C* results in hypermethylation of tumor suppressor genes in other cancers; aberrant expression of *miR-29* has not yet been associated with breast cancer. These examples illustrate the importance of miRNA-mediated regulation during breast tumorigenesis.

5.7.3 LncRNAs

A subset of lncRNAs associated with PRC2 and other chromatin-modifying complexes suggest a key role of these RNAs in epigenetic regulation (Khalil et al. 2009). Only a few lncRNAs identified thus far show evidence of involvement in breast cancer. LncRNA *LSINCT5* is a 2.5 kb RNA which is overexpressed in ovarian and breast cancers and has a role in proliferation (Silva et al. 2011). Genome sequencing of the transcriptome in a lobular breast cancer led to the identification of two novel noncoding RNAs, *SRA1* and *MALAT1*, which were differentially expressed in multiple breast cancer samples assessed (Guffanti et al. 2009). Interestingly, *SRA1* might have both a protein-coding and noncoding transcripts which impact regulation of the estrogen signaling pathway (Leygue 2007). *HOTAIR*, another lncRNA with a role in *HOX* gene regulation, is associated with metastatic breast cancer (Gupta et al. 2010). It is likely that additional lncRNAs and other small RNAs will be found to be important in breast tumorigenesis.

5.8 The Role of Epigenetics in Normal Breast Development and Breast Stem Cells

5.8.1 Development

The mammary gland develops in several stages throughout a female's lifetime. Initially, mammary gland development begins during embryogenesis, but it undergoes significant changes during puberty, pregnancy, lactation, and postlactation and involution. Epigenetic changes occur during each of the major developmental stages of the breast. One example of the importance of epigenetics during breast development is lactation. During lactation, a number of milk-producing genes such as *β-casein* and *WAP*, which are normally silenced at other developmental stages and in all other tissues, are activated by lactogenic hormones. Studies show that these hormones induce more open chromatin by inducing histone modifications leading to differential expression of genes important in lactation (Rijnkels et al. 2010). Some of the genes transcriptionally regulated by gene silencing through methylation and chromatin remodeling show similar expression patterns in fetal breast development and breast cancer.

In addition to differential regulation of key regulatory and lactation genes, noncoding regulatory RNAs show differential regulation during breast development. miRNAs such as *miR-25* and *mir-17-92* which are associated with basal-like breast cancer are expressed in the normal breast during puberty and gestation when a great deal of remodeling and proliferation occur (Avril-Sassen et al. 2009). Other small RNAs also are important in breast development. A noncoding RNA, pregnancy-induced noncoding RNA (*PINC*), is expressed in the regressed terminal ductal lobular unit-like structures of the parous mammary gland. These cells have stem-cell-like features including the ability to proliferate and form all the cells of a mammary gland; they also survive the massive cell death that occurs during mammary gland involution (Ginger et al. 2006). In the mouse, expression of *Pinc* is induced during pregnancy and mammary gland involution and appears to be epigenetically regulated (Rijnkels et al. 2010). Whereas the role of *PINC* in the various stages of breast development is not established, *PINC* is proposed to impart a parity-induced "epigenetic imprint" such that the mammary cells show different expression patterning than pre-parity.

5.8.2 Stem Cells

Mammary stem cells are a long-lived population of cells that have the ability to create all the cells of the breast architecture including luminal and myoepithelial cells. In addition to breast cell pluripotency, these cells have self-renewal properties. Unique to other organs which undergo little dynamic change in adulthood, the breast undergoes extensive remodeling during pregnancy, lactation, and involution. Stem cells play a key role in breast remodeling. Because breast stem cells are long lived and slow dividing, it has been proposed that mammary stem and/or early progenitor cells are the primary targets for transformation into cancer cells. Key genes for the maintenance of stem cells and for the differentiation of stem cells into different cell lineages show evidence of epigenetic regulation (Bloushtain-Qimron et al. 2009). In addition, undifferentiated stem cells show higher global demethylation, and more differentiated progenitor cells show higher degrees of methylation. Transcription factors associated with stem cell function exhibit similar methylation patterns in stem cells as breast tumor cells (Bloushtain-Qimron et al. 2008). In addition to methylation and chromatin regulation of stem cells, multiple miRNAs show differential regulation in stem cells compared to differentiated cells.

5.8.3 Breast Cancer Stem Cells

There are two main hypotheses for the progression and development of a breast tumor (Bombonati and Sgroi 2011). One hypothesis, the sporadic clonal evolution model, proposes that any breast epithelial cell is the target of random mutations and epigenetic alterations. Cells that contain the most advantageous mix of tumor predisposing changes will be selected for and have the capability of contributing to tumor development. The alternative hypothesis, the cancer stem cell model, suggests that only stem and/or breast progenitor cells can initiate and maintain a tumor. As the stem cells are long lived, it is thought that they have a longer time in which to accumulate mutations (Wicha et al. 2006). A combination of these two possibilities in which stem cells are important for initiation and/or progression of a tumor and clonal evolution of particular cellular populations help to drive carcinogenesis is also plausible.

Cancer stem cells, also known as tumor-initiating or stem-like cells, are a subpopulation of tumor cells which retain stem cell abilities to differentiate into multiple cell lineages and are able to self-renew. Breast cancer stem cells were first identified in 2003 as $CD44^{+}/CD24^{-/low}$ cells (Al Hajj et al. 2003) that also express *ALDH1*. In addition, the IL-6/JAK2/STAT3 pathway appears to be preferentially active in breast cancer stem cells (Marotta et al. 2011). Epigenetic studies of breast cancer stem cells indicate that there is increased hypomethylation of several genes in the JAK-STAT pathway in precursor/stem cells compared to non-stem cell-like cells (Hernandez-Vargas et al. 2011) suggesting a key role of epigenetics in activation/regulation of breast cancer stem cells in addition to the epigenetic regulation of normal mammary stem cells.

5.9 Environmental Epigenetic Influences on Breast Cancer

Cancer is a genetic disease caused by changes in how genes are expressed or how they function. Environmental factors can impact gene regulation and function by direct mutation or by inducing epigenetic alterations of the genome. Several environmental agents associated with breast cancer may induce important epigenetic alterations. Carcinogens are substances or exposures, such as gamma radiation, which cause an increased incidence of tumors. Many carcinogens have been shown to damage DNA directly leading to inactivation of tumor suppressor genes or activation of onco-genes. Some carcinogens do not directly mutate DNA, but act by perturbing normal epigenetic signaling. Carcinogens and environmental exposures that lead to an increase of breast cancer risk through epigenetic signaling have been identified through epidemiological studies. These include alcohol, bisphenol A (BpA), and diethylstilbestrol (DES).

5.9.1 Xenoestrogens

Naturally occurring estrogens and estrogen mimics found in the environment may both impact the methylation status of genes. In utero and earlylife exposures to xenoestrogens have been proposed to alter epigenetic programming of stem cells which lead to increased risk of cancer in adulthood. In vitro, breast stem and pro-genitor cells are targets of estrogen imprinting, alterations to the epigenome which specifically increase hypermethylation, following exposure to estrogens (Cheng et al. 2008). In vitro studies showed that 0.5% of CpG islands become hypermethy-lated in epithelial cells following estrogen exposure (Cheng et al. 2008). Targets include polycomb protein-regulated genes and miRNAs (Hsu et al. 2009). Studies are ongoing to fully assess the impact that xenoestrogens have on cancer risk and the epigenome, but data thus far supports a role of these in breast cancer risk.

5.9.2 Lifetime

Age is one of the greatest risk factors for breast cancer with two thirds of breast cancers diagnosed in women over the age of 55 (www.cancer.org). Epigenetic changes in DNA methylation patterning are age-related. Twin studies show that young monozygotic twin pairs have a higher concordance of methylation status than older twin pairs (Fraga et al. 2005a). Comparison of global methylation patterns in the young babies compared to centenarians shows significant epigenetic differences associated with age. Babies and children tend to have a higher degree of intragenic methylation and hypomethylation of genomic repeats and housekeeping genes. As a person ages, CpG islands become more methylated and loci not in CpG islands lose methylation (Christensen et al. 2010). DNA hypomethylation of repetitive DNA is particularly associated with older age (Bollati et al. 2009; Jintaridth and Mutirangura 2010). The mechanisms leading to age-related epigenetic changes are not known.

5.9.3 Alcohol

Alcohol has been associated with an increased risk of breast cancer in some studies (Hamajima et al. 2002; Key et al. 2006; Allen et al. 2009). One potential mechanism of the role of alcohol in breast cancer development is through its effect on folate absorption in the intestine. Folate plays a key role in the process of methylation by donating a methyl group for homocysteine conversion to methionine during one-carbon metabolism. Consistent with this hypothesis is a correlation between alcohol and methylation status in breast tumors. In one study, increasing levels of alcohol consumption were associated with decreased methylation at the genome level in breast tumors (Christensen et al. 2010).

5.9.4 Endocrine-Disrupting Chemicals

BpA is an endocrine-disrupting chemical found in polycarbonate plastics such as reusable water bottles, the coating in food cans, and baby bottles. Cell studies and evidence from animal models suggest that exposure to BpA during development is a risk factor for breast cancer in adulthood (reviewed by Vandenberg et al. 2009). Animal studies indicate that BpA exposure causes an increase in the number of terminal end buds that persist in the mammary gland which is where most breast cancers are thought to arise (Markey et al. 2001). Several studies have shown the impact of BpA on methylation patterning of mice exposed to BpA in utero (Dolinoy et al. 2007; Rosenfeld 2010; Bromer et al. 2010). Studies of human breast epithelial cells in vitro show that BpA can induce epigenetic changes including silencing of genes, such as *ERS1*, the gene for ER-α (Weng et al. 2010). A related compound to BpA is DES. DES was given to pregnant women from the 1940s to the 1970s to prevent miscarriages. The incidence of breast cancer is higher in daughters of women exposed to DES in utero compared to offspring of women who were not exposed to DES (Palmer et al. 2006). Animal studies of DES show changes in methylation patterns in mice exposed to DES compared to nonexposed control mice (Li et al. 2003). DES treatment also results in decreased *DNMT* expression (Sato et al. 2009). Interestingly, some of aberrant methylation patterns in DES-exposed mice appear to persist for more than one generation (Newbold et al. 2006).

5.9.5 Transgenerational Effects

In animal model studies, exposure of a mouse to an environmental agent (like DES) can induce epigenetic changes which persist in subsequent generations and increase their risk of cancer (Fleming et al. 2008). The mechanism for this has not been fully established in animal models and has not been extended more than one generation in humans, but this newly described type of transgenerational epigenetic inheritance may explain why identification of genetic risk factors for familial breast cancer has been difficult.

5.10 Epigenetics and Clinical Outcome

5.10.1 Epigenetic Events as Biomarkers for Breast Cancer

Promoter methylation occurs as an early event in breast cancer, making methylation of specific gene promoters a good target to identify biomarkers to be used for early detection. Several groups have looked for aberrant mRNAs or methylation patterns in serum, nipple aspirates, and breast lavage cells which could serve as biomarkers for detecting breast cancer. Despite some tantalizing preliminary results with high sensitivity and specificity, none of these studies have been robustly reproduced (Suijkerbuijk et al. 2011). In addition to looking at the methylation status of specific candidate genes in order to predict breast cancer status, some groups have assessed overall methylation patterning. One study showed a higher degree of leukocyte genomic hypomethylation in breast cancer cases compared to matched controls (Choi et al. 2009); however, this finding has not yet been replicated in a larger series. More recently, serum and plasma miRNAs have been evaluated as breast cancer biomarkers. Very preliminary studies identified 31 miRNAs that showed differential expression between serum from cancer patients and matched controls (Zhao et al. 2010). As techniques for detecting small numbers of circulating cancer cells get better, identification of early epigenetic events of cancers likely to progress and metastasize will improve diagnosis and treatment.

5.10.2 Classification of Breast Cancer Subtypes

In 2000, Perou et al. showed that mRNA expression profiling can separate breast cancers into multiple subtypes (Perou et al. 2000) with differences in outcome (Sorlie et al. 2001). Breast cancer subtypes include basal-like, luminal A, luminal B, HER2-enriched, claudin-low, and normal breast-like; each subtype has general clinical characteristics, prognosis, and treatment options which differ from the others (Prat and Perou 2011). miRNA expression profiles can also distinguish subtypes of breast cancer and proliferation profiles (Enerly et al. 2011). Now that more complete datasets are available, more sophisticated bioinformatics approaches have generated network and pathway correlations of epigenetic events occurring during breast carcinogenesis. Global DNA methylation profiling can also separate breast tumors into subgroups of tumors which share some overlap and some distinctions from the subgroups defined by gene expression analysis (Bae et al. 2004; Ronneberg et al. 2011). In addition to being able to identify breast cancer subtypes from women with nonhereditary breast cancers, methylation profiling can distinguish tumors from women with germline *BRCA1* and *BRCA2* mutations and non-*BRCA* familial breast cancer (*BRCA*-X). A study using MeDIP on tiling microarrays with probes for over 25,000 promoter regions led to the identification of 822 genes showing differential methylation between the groups (Flanagan et al. 2010). These data suggest that germline mutations may influence the methylation patterns that occur during tumorigenesis.

Few studies have looked specifically at epigenetic changes occurring in male breast cancer. Two small studies looked at miRNA expression in male breast cancer. A few miRNAs were differentially expressed, some of which had not been found to show aberrant expression in female breast tumors (Fassan et al. 2009; Lehmann et al. 2010). These early data suggest gender-specific differences in breast cancer methylation patterning.

5.10.3 Prognosis and Progression

Breast cancer likely arises through a multistep process. The steps in the progression of normal cells to ductal carcinoma are thought to be epithelial atypia, followed by atypical ductal hyperplasia, ductal carcinoma in situ (DCIS), invasive ductal carcinoma, and metastasis. Lobular carcinoma has a similar progression series. As detection methods get better and more women are being screened, more women are being diagnosed with early-stage DCIS. Not all DCIS tumors are predicted to become aggressive and require significant therapeutic interventions. One clinical problem in breast cancer is predicting which early-stage cancers will progress and therefore require more invasive interventions. Currently, all DCIS lesions are treated to some degree. The use of epigenetic signatures to identify those DCIS tumors which are more likely to advance is an area of active research. Several studies have compared epigenetic patterning in early and advanced cancers to determine their utility in predicting progression; however, many of the studies have found no significant differences in methylation patterns between DCIS and invasive cancers (Moelans et al. 2011; Park et al. 2011). miRNA profiling of normal breast epithelium from reduction mammoplasty patients with paired normal and DCIS samples has led to the identification of miRNAs that are expressed in these early cancers but have not tied these into progression (Hannafon et al. 2011). Whereas these studies are promising, they have yet to be validated in larger clinical series.

Another way in which methylation profiling may have clinical utility is in assessing response to therapy. A few studies have examined the methylation status of genes known to be methylated in breast cancer for a correlation with treatment response (Dejeux et al. 2010; Dietrich et al. 2010). Methylation of *ABCB1* and *GSTP1* promoters correlated with good prognosis and response to doxorubicin treatment. Conversely, sensitivity to doxorubicin correlates with expression of the *TG2* gene; methylation corresponded to resistance (Ai et al. 2008). It is possible that methylation status of key genes may be used in the future for determining the best therapy for treatment.

5.10.4 Metastasis

Metastasis is the spread of cancer cells from the originating site to distant sites in the body where the cells colonize and form new tumor foci. Breast cancers that metastasize generally have poorer outcomes with approximately 40% of women surviving 5 years or more after a diagnosis with local metastatic disease. Women with stage IV breast cancer, which is cancer that has disseminated throughout the body, have 5-year survival rates of less than 15% (www.cancer.org). As shown from these grim statistics, it is very important to develop therapies that can treat metastatic breast cancer and predict which breast cancers are likely to metastasize so that more aggressive therapies can be implemented prior to spread.

Several studies have assessed whether differential epigenetic regulation of genes plays a role in breast cancer metastasis (Rodenhiser et al. 2008; Dietrich et al. 2010; Fang et al. 2011). A study assessed 202 gene promoters to identify biomarkers predictive of distant metastasis in lymph-node-negative breast cancers. Of the 202 promoter regions, 37 showed some differential methylation and six were highly correlated with metastatic potential (Dietrich et al. 2010). Of these, cysteine dioxygenase 1 (*CDO1*) was the best predictor of metastasis. Other candidate genes showing hypermethylation associated with metastasis include *CDH1* (Shinozaki et al. 2005). Using a whole-genome approach in breast cancer cell lines, 2,209 hypermethylated and 1,235 hypomethylated genes were found to be different between a cell line with metastatic potential and its derivative cell line. Promoter methylation was more pronounced in genes involved in cell signaling and cellular movement/migration and epithelial to mesenchymal transition (Rodenhiser et al. 2008).

Interestingly, breast cancers that show extensive hypermethylation of a large number of genes are associated with a low risk of metastasis, and the converse appears to be true (Fang et al. 2011). Hypermethylated breast tumors are described as having a breast CpG island methylator phenotype or B-CIMP tumors. Genes silenced by methylation in the B-CIMP tumors include those that were previously known to be associated with outcome and those with epithelial to mesenchymal transition, a phenotype associated with metastasis. B-CIMP tumors were also more likely to be positive for ER and progesterone receptor (PR) expression, two proteins known to be correlated with better outcome. It is notable that some genes which in previous studies correlate with poor prognosis when methylated were not methylated in B-CIMP tumors. If the B-CIMP methylator phenotype occurs early enough during breast cancer development, it could be used as a marker to identify tumors more likely to result in poor outcomes.

Metastamir is a term used to describe miRNAs that play a role in metastasis. Metastamirs can be divided into two categories: those that suppress metastasis and are downregulated in aggressive breast tumors and those that promote metastasis and are upregulated in tumors that spread. *miR-126*, *miR-31*, *miR-146a/b*, *miR-200*, and *miR-335* are thought to suppress metastasis in breast cancer by targeting genes which have roles in migration or epithelial to mesenchymal transition (EMT) (Hurst et al. 2009). In mouse models, reexpression of *miR-31* in metastatic breast tumors

5   Aberrant Epigenetic Regulation in Breast Cancer

leads to regression of the metastatic lesion and increases survival (Valastyan et al. 2011). *miR-200a* and *200c* are frequently downregulated in tumors including breast tumors. These *miR-200* family members also are important in the regulation of EMT by targeting *ZEB1* and *ZEB2*, mediators of EMT (Radisky 2011). miRNAs that are upregulated in breast metastases include *miR-373*, *miR-520c*, *miR-21*, *miR-143*, and *miR-182* (Hurst et al. 2009). *miR-9* targets E-cadherin, a protein which is essential for maintaining the epithelial nature of cells (Ma et al. 2010; Enerly et al. 2011).

Other noncoding RNAs have been shown to be important in breast cancer metastasis. *HOTAIR*, a long noncoding RNA which maps to the *HOXC* locus, shows increased expression in breast tumors and metastases and may be a good predictor of metastasis. *HOTAIR* binds to the polycomb repressive complex 2 (PRC2 complex) and promotes binding of this complex to specific promoter regions. PRC2 is a methylase responsible for H3K27 methylation, and its binding to promoter regions leads to epigenetic silencing of metastasis suppressor genes (Gupta et al. 2010). In addition to noncoding RNAs, proteins involved in chromatin structure and the PRCs have been shown to play important roles in breast cancer metastasis. For example, high *EZH2* expression is associated with aggressive breast cancer subtypes and an increased risk of distant metastases (Alford et al. 2012).

5.10.5   Field Cancerization

Field cancerization is when normal-appearing cells surrounding cancer cells also contain similar mutations or aberrant epigenetic patterning as the tumor cells. Environmental agents can induce alterations in cells which lead to selective growth advantage and expansion resulting in a field effect of cells surrounding a tumor which contain the same epigenetic aberration. Field cancerization is well-documented for breast cancer (Heaphy et al. 2009). In one study, methylation of specific tumor suppressor genes persisted up to 4 cm away from the primary tumor (Yan et al. 2006). In some cases, genes such as *CYP26A1*, *KCNAB1*, and *SNCA* which were methylated in the primary tumor exhibited methylation of nearby normal tissue 70% of the time. In another study, normal tissue adjacent to breast tumors had significantly higher methylation of the *RASSF1A* and *APC* genes compared to unrelated normal breast tissue (Van der Auwera et al. 2010). These findings have implications for breast tumors treated surgically because normal-appearing tissue may harbor premalignant epigenetic alterations which may make the cell at increased risk of developing into a tumor cell.

5.11   Therapeutic Interventions

The evidence is strong that epigenetic alterations influence the initiation, growth, and outcome of breast cancers. Thus, there has been considerable effort to identify therapeutic agents to reverse the breast cancer-related epigenetic events. Many of these have been developed or are being tested in a variety of tumor types. As this is an active area of research, we will highlight only a few key drugs in trial or being used clinically. Despite their promise, these drugs have weaknesses. One of potential pitfalls to general epigenetic drugs is off-target effects. For example, given that hypomethylation of the gene also appears to play a role in cancer initiation or progression, then agents which induce hypomethylation may have unintended consequences for subsequent cancer risk or development of therapy resistance.

5.11.1 DNA Methyltransferase Inhibitors

DNMTS are the primary enzymes responsible for DNA methylation. A variety of tumor suppressor genes are methylated in breast cancers. Thus, the first FDA-approved therapeutics specific for targeting epigenetic events in cancer were the DNMT inhibitors 5-azacytdine (Vidaza™) and 5-aza-2′-deoxycytidine (decitabine). These agents work by their incorporation into DNA in place of the normal cytosine during DNA replication. Unfortunately, these compounds are very unstable in solution. Another more recent DNMT inhibitor is zebularine which has the advantage of being more stable in solution (Cheng et al. 2004). As high levels of zebularine are required for efficacy, the search continues for "the perfect" DNMT inhibitor.

5.11.2 HDAC Inhibitors

HDACs are responsible for removing acetyl groups and promoting repression. HDAC inhibitors are being developed for the treatment of breast cancer, in the hopes of leading to reactivation of key tumor suppressor genes. Vorinostat (suberoylanilide hydroxamic acid, SAHA) is an HDAC inhibitor that has been used in combination with commonly used breast cancer drugs paclitaxel and bevacizumab. Numerous other inhibitors of class I and class II HDACs, such as entinostat (MS-275) and belinostat, are in preclinical or early clinical trials (Lo and Sukumar 2008). HDAC inhibitors have been shown to induce tumor suppressor genes such as *p21* and apoptotic genes such as *BAX*, *BIM*, and *TRAIL* (Bolden et al. 2006). Whereas the data look promising for many, it is too soon to tell how well these will work for breast cancer.

As many breast tumors refractory to treatment are estrogen receptor-negative tumors, one approach to therapy is to use HDAC inhibitors or demethylating agents to trigger reexpression of the ER so that the tumors will be responsive to therapies which work in ER-positive tumors. Tamixofen-resistant breast cancer cell lines treated with both DNMT and HDAC inhibitors become sensitive to tamoxifen treatment (Sharma et al. 2006). It also appears that treatment of breast cancers with HER2 amplification with HDAC inhibitors enhances the tumor response to trastuzumab (Herceptin), an antibody therapy targeted against HER2/Neu (Fuino et al. 2003). Thus, combination therapy of HDAC inhibitors with other drugs may increase

5 Aberrant Epigenetic Regulation in Breast Cancer

the efficacy of other therapies. Perturbation of the epigenetic state of tumors may induce tumors which are resistant to particular therapies to become sensitive.

5.11.3 In the Pipeline

Several addition lines of research are being conducted to identify agents which can perturb epigenetic alternations which occur in breast tumors. The identification of inhibitors of PcG complexes for therapeutic uses is an active area of research. 3-Deazaneplanocin A (DZNep) is a PRC2 inhibitor which has been used for in vitro studies of breast cancer and shows enhancement of apoptosis and decreased proliferation in breast cancer cell lines (Tan et al. 2007). Another area of development is the identification of miRNA agonists and delivery systems for miRNAs (Sioud 2009; Kanwar et al. 2011). miRNA therapy has inherent difficulties such as lack of tissue specificity, off-target effects, poor cellular uptake, and lack of optimal delivery systems (Wu et al. 2011). If some of these problems can be solved and delivery methods optimized, miRNAs-based therapy as part of combination therapy has the potential to be quite powerful for the treatment of breast cancer.

5.12 Conclusions

In the past 15 years, we have made significant leaps of our knowledge of the role of epigenetic events in breast cancer development. Aberrant promoter methylation, chromatin modifications, and small noncoding RNAs impact the initiation, growth, and metastasis of breast cancer cells. It is likely that as technology advances our understanding of the role of additional epigenetic mechanisms, such as paramutation and transgenerational epigenetic marks, and complex epigenetic regulation, such as interchromosomal chromatin interactions, on breast cancer development will lead to new insight of this disease and how to prevent and treat it. The use of early epigenetic changes in breast cells as potential biomarkers for early diagnosis and prognosis is promising but has yet to be clinically validated. Exploiting the tumor cell's abnormal epigenetic state is a promising area for breast cancer therapy. This is an avenue which has yet to reach full success in the clinic. Demethylating agents, HDACs, or miRNA inhibitors may work best in combination with existing chemo- or immunotherapeutic agents and may be able to help to sensitize the cells to standard therapies. What is clear from the body of work amassed to date is that epigenetics plays a critical role in breast cancer pathogenesis by deregulating important genes in the tumor-related pathways as well as impacting stability of the genome.

Acknowledgments This work was funded in part by the National Cancer Institute/National Institutes of Health grant (R01 CA134461) and the Ohio State University Comprehensive Cancer Center.

References

Agrawal, A., Murphy, R. F. & Agrawal, D. K. (2007). DNA methylation in breast and colorectal cancers. *Mod Pathol* 20, 711–721.

Ai, L., Kim, W. J., Demircan, B., Dyer, L. M., Bray, K. J., Skehan, R. R., Massoll, N. A. & Brown, K. D. (2008). The transglutaminase 2 gene (TGM2), a potential molecular marker for chemotherapeutic drug sensitivity, is epigenetically silenced in breast cancer. *Carcinogenesis* 29, 510–518.

Al Hajj, M., Wicha, M. S., Benito-Hernandez, A., Morrison, S. J. & Clarke, M. F. (2003). Prospective identification of tumorigenic breast cancer cells. *Proc Natl Acad Sci U S A* 100, 3983–3988.

Alford, S. H., Toy, K., Merajver, S. D. & Kleer, C. G. (2012). Increased risk for distant metastasis in patients with familial early-stage breast cancer and high EZH2 expression. *Breast Cancer Res Treat* 132, 429–437.

Allen, N. E., Beral, V., Casabonne, D., Kan, S. W., Reeves, G. K., Brown, A. & Green, J. (2009). Moderate alcohol intake and cancer incidence in women. *J Natl Cancer Inst* 101, 296–305.

Avril-Sassen, S., Goldstein, L. D., Stingl, J., Blenkiron, C., Le Quesne, J., Spiteri, I., Karagavriilidou, K., Watson, C. J., Tavare, S., Miska, E. A. & Caldas, C. (2009). Characterisation of microRNA expression in post-natal mouse mammary gland development. *BMC Genomics* 10:548.

Bae, Y. K., Brown, A., Garrett, E., Bornman, D., Fackler, M. J., Sukumar, S., Herman, J. G. & Gabrielson, E. (2004). Hypermethylation in histologically distinct classes of breast cancer. *Clin Cancer Res* 10, 5998–6005.

Ballestar, E. & Esteller, M. (2005). Methyl-CpG-binding proteins in cancer: blaming the DNA methylation messenger. *Biochem Cell Biol* 83, 374–384.

Bannister, A. J. & Kouzarides, T. (2011). Regulation of chromatin by histone modifications. *Cell Res* 21, 381–395.

Bloushtain-Qimron, N., Yao, J., Shipitsin, M., Maruyama, R. & Polyak, K. (2009). Epigenetic patterns of embryonic and adult stem cells. *Cell Cycle* 8, 809–817.

Bloushtain-Qimron, N., Yao, J., Snyder, E. L., Shipitsin, M., Campbell, L. L., Mani, S. A., Hu, M., Chen, H., Ustyansky, V., Antosiewicz, J. E., Argani, P., Halushka, M. K., Thomson, J. A., Pharoah, P., Porgador, A., Sukumar, S., Parsons, R., Richardson, A. L., Stampfer, M. R., Gelman, R. S., Nikolskaya, T., Nikolsky, Y. & Polyak, K. (2008). Cell type-specific DNA methylation patterns in the human breast. *Proc Natl Acad Sci U S A* 105, 14076–14081.

Bolden, J. E., Peart, M. J. & Johnstone, R. W. (2006). Anticancer activities of histone deacetylase inhibitors. *Nat Rev Drug Discov* 5, 769–784.

Bollati, V., Schwartz, J., Wright, R., Litonjua, A., Tarantini, L., Suh, H., Sparrow, D., Vokonas, P. & Baccarelli, A. (2009). Decline in genomic DNA methylation through aging in a cohort of elderly subjects. *Mech Ageing Dev* 130, 234–239.

Bombonati, A. & Sgroi, D. C. (2011). The molecular pathology of breast cancer progression. *J Pathol* 223, 307–317.

Bromer, J. G., Zhou, Y., Taylor, M. B., Doherty, L. & Taylor, H. S. (2010). Bisphenol-A exposure in utero leads to epigenetic alterations in the developmental programming of uterine estrogen response. *FASEB J* 24, 2273–2280.

Catteau, A. & Morris, J. R. (2002). BRCA1 methylation: a significant role in tumour development? *Semin Cancer Biol* 12, 359–371.

Chase, A. & Cross, N. C. (2011). Aberrations of EZH2 in cancer. *Clin Cancer Res* 17, 2613–2618.

Chedin, F. (2011). The DNMT3 family of mammalian de novo DNA methyltransferases. *Prog Mol Biol Transl Sci* 101, 255–285.

Cheng, A. S., Culhane, A. C., Chan, M. W., Venkataramu, C. R., Ehrich, M., Nasir, A., Rodriguez, B. A., Liu, J., Yan, P. S., Quackenbush, J., Nephew, K. P., Yeatman, T. J. & Huang, T. H. (2008). Epithelial progeny of estrogen-exposed breast progenitor cells display a cancer-like methylome. *Cancer Res* 68, 1786–1796.

Cheng, J., Guo, J. M., Xiao, B. X., Miao, Y., Jiang, Z., Zhou, H. & Li, Q. N. (2011). piRNA, the new non-coding RNA, is aberrantly expressed in human cancer cells. *Clin Chim Acta* 412, 1621–1625.

Cheng, J. C., Yoo, C. B., Weisenberger, D. J., Chuang, J., Wozniak, C., Liang, G., Marquez, V. E., Greer, S., Orntoft, T. F., Thykjaer, T. & Jones, P. A. (2004). Preferential response of cancer cells to zebularine. *Cancer Cell* 6, 151–158.

Choi, J. Y., James, S. R., Link, P. A., McCann, S. E., Hong, C. C., Davis, W., Nesline, M. K., Ambrosone, C. B. & Karpf, A. R. (2009). Association between global DNA hypomethylation in leukocytes and risk of breast cancer. *Carcinogenesis* 30, 1889–1897.

Christensen, B. C., Kelsey, K. T., Zheng, S., Houseman, E. A., Marsit, C. J., Wrensch, M. R., Wiemels, J. L., Nelson, H. H., Karagas, M. R., Kushi, L. H., Kwan, M. L. & Wiencke, J. K. (2010). Breast cancer DNA methylation profiles are associated with tumor size and alcohol and folate intake. *PLoS Genet* 6, e1001043.

Colaneri, A., Staffa, N., Fargo, D. C., Gao, Y., Wang, T., Peddada, S. D. & Birnbaumer, L. (2011). Expanded methyl-sensitive cut counting reveals hypomethylation as an epigenetic state that highlights functional sequences of the genome. *Proc Natl Acad Sci U S A* 108, 9715–9720.

Cotton, A. M., Lam, L., Affleck, J. G., Wilson, I. M., Penaherrera, M. S., McFadden, D. E., Kobor, M. S., Lam, W. L., Robinson, W. P. & Brown, C. J. (2011). Chromosome-wide DNA methylation analysis predicts human tissue-specific X inactivation. *Hum Genet* 130, 187–201.

Daskalos, A., Nikolaidis, G., Xinarianos, G., Savvari, P., Cassidy, A., Zakopoulou, R., Kotsinas, A., Gorgoulis, V., Field, J. K. & Liloglou, T. (2009). Hypomethylation of retrotransposable elements correlates with genomic instability in non-small cell lung cancer. *Int J Cancer* 124, 81–87.

Dejeux, E., Ronneberg, J. A., Solvang, H., Bukholm, I., Geisler, S., Aas, T., Gut, I. G., Borresen-Dale, A. L., Lonning, P. E., Kristensen, V. N. & Tost, J. (2010). DNA methylation profiling in doxorubicin treated primary locally advanced breast tumours identifies novel genes associated with survival and treatment response. *Mol Cancer* 9:68.

Deniaud, E. & Bickmore, W. A. (2009). Transcription and the nuclear periphery: edge of darkness? *Curr Opin Genet Dev* 19, 187–191.

Dietrich, D., Krispin, M., Dietrich, J., Fassbender, A., Lewin, J., Harbeck, N., Schmitt, M., Eppenberger-Castori, S., Vuaroqueaux, V., Spyratos, F., Foekens, J. A., Lesche, R. & Martens, J. W. (2010). CDO1 promoter methylation is a biomarker for outcome prediction of anthracycline treated, estrogen receptor-positive, lymph node-positive breast cancer patients. *BMC Cancer* 10:247.

Dolinoy, D. C., Huang, D. & Jirtle, R. L. (2007). Maternal nutrient supplementation counteracts bisphenol A-induced DNA hypomethylation in early development. *Proc Natl Acad Sci U S A* 104, 13056–13061.

Dostie, J., Richmond, T. A., Arnaout, R. A., Selzer, R. R., Lee, W. L., Honan, T. A., Rubio, E. D., Krumm, A., Lamb, J., Nusbaum, C., Green, R. D. & Dekker, J. (2006). Chromosome Conformation Capture Carbon Copy (5C): a massively parallel solution for mapping interactions between genomic elements. *Genome Res* 16, 1299–1309.

Ehrlich, M. (2000). DNA hypomethylation and cancer. In *DNA Alterations in Cancer*, pp. 273–291. Edited by M. Ehrlich. Westborough: Eaton Publishing.

Ehrlich, M. (2009). DNA hypomethylation in cancer cells. *Epigenomics* 1, 239–259.

Elsheikh, S. E., Green, A. R., Rakha, E. A., Powe, D. G., Ahmed, R. A., Collins, H. M., Soria, D., Garibaldi, J. M., Paish, C. E., Ammar, A. A., Grainge, M. J., Ball, G. R., Abdelghany, M. K., Martinez-Pomares, L., Heery, D. M. & Ellis, I. O. (2009). Global histone modifications in breast cancer correlate with tumor phenotypes, prognostic factors, and patient outcome. *Cancer Res* 69, 3802–3809.

Enerly, E., Steinfeld, I., Kleivi, K., Leivonen, S. K., Aure, M. R., Russnes, H. G., Ronneberg, J. A., Johnsen, H., Navon, R., Rodland, E., Makela, R., Naume, B., Perala, M., Kallioniemi, O., Kristensen, V. N., Yakhini, Z. & Borresen-Dale, A. L. (2011). miRNA-mRNA integrated analysis reveals roles for miRNAs in primary breast tumors. *PLoS ONE* 6, e16915.

Fang, F., Turcan, S., Rimner, A., Kaufman, A., Giri, D., Morris, L. G., Shen, R., Seshan, V., Mo, Q., Heguy, A., Baylin, S. B., Ahuja, N., Viale, A., Massague, J., Norton, L., Vahdat, L. T., Moynahan, M. E. & Chan, T. A. (2011). Breast cancer methylomes establish an epigenomic foundation for metastasis. *Sci Transl Med* 3, 75ra25.

Fassan, M., Baffa, R., Palazzo, J. P., Lloyd, J., Crosariol, M., Liu, C. G., Volinia, S., Alder, H., Rugge, M., Croce, C. M. & Rosenberg, A. (2009). MicroRNA expression profiling of male breast cancer. *Breast Cancer Res* 11:R58.

Fazzari, M. J. & Greally, J. M. (2010). Introduction to epigenomics and epigenome-wide analysis. *Methods Mol Biol* 620, 243–265.

Feinberg, A. P. & Vogelstein, B. (1983). Hypomethylation distinguishes genes of some human cancers from their normal counterparts. *Nature* 301, 89–92.

Fiegl, H., Millinger, S., Mueller-Holzner, E., Marth, C., Ensinger, C., Berger, A., Klocker, H., Goebel, G. & Widschwendter, M. (2005). Circulating tumor-specific DNA: a marker for monitoring efficacy of adjuvant therapy in cancer patients. *Cancer Res* 65, 1141–1145.

Flanagan, J. M., Cocciardi, S., Waddell, N., Johnstone, C. N., Marsh, A., Henderson, S., Simpson, P., da Silva, L., Khanna, K., Lakhani, S., Boshoff, C. & Chenevix-Trench, G. (2010). DNA methylome of familial breast cancer identifies distinct profiles defined by mutation status. *Am J Hum Genet* 86, 420–433.

Flanagan, L., Packman, K., Juba, B., O'Neill, S., Tenniswood, M. & Welsh, J. (2003). Efficacy of Vitamin D compounds to modulate estrogen receptor negative breast cancer growth and invasion. *J Steroid Biochem Mol Biol* 84, 181–192.

Fleming, J. L., Huang, T. H. & Toland, A. E. (2008). The role of parental and grandparental epigenetic alterations in familial cancer risk. *Cancer Res* 68, 9116–9121.

Fletcher, A. M., Heaford, A. C. & Trask, D. K. (2008). Detection of metastatic head and neck squamous cell carcinoma using the relative expression of tissue-specific mir-205. *Transl Oncol* 1, 202–208.

Fouse, S. D., Nagarajan, R. P. & Costello, J. F. (2010). Genome-scale DNA methylation analysis. *Epigenomics* 2, 105–117.

Fraga, M. F., Ballestar, E., Paz, M. F., Ropero, S., Setien, F., Ballestar, M. L., Heine-Suner, D., Cigudosa, J. C., Urioste, M., Benitez, J., Boix-Chornet, M., Sanchez-Aguilera, A., Ling, C., Carlsson, E., Poulsen, P., Vaag, A., Stephan, Z., Spector, T. D., Wu, Y. Z., Plass, C. & Esteller, M. (2005a). Epigenetic differences arise during the lifetime of monozygotic twins. *Proc Natl Acad Sci U S A* 102, 10604–10609.

Fraga, M. F., Ballestar, E., Villar-Garea, A., Boix-Chornet, M., Espada, J., Schotta, G., Bonaldi, T., Haydon, C., Ropero, S., Petrie, K., Iyer, N. G., Perez-Rosado, A., Calvo, E., Lopez, J. A., Cano, A., Calasanz, M. J., Colomer, D., Piris, M. A., Ahn, N., Imhof, A., Caldas, C., Jenuwein, T. & Esteller, M. (2005b). Loss of acetylation at Lys16 and trimethylation at Lys20 of histone H4 is a common hallmark of human cancer. *Nat Genet* 37, 391–400.

Fraga, M. F. & Esteller, M. (2002). DNA methylation: a profile of methods and applications. *Biotechniques* 33, 632–639.

Fuino, L., Bali, P., Wittmann, S., Donapaty, S., Guo, F., Yamaguchi, H., Wang, H. G., Atadja, P. & Bhalla, K. (2003). Histone deacetylase inhibitor LAQ824 down-regulates Her-2 and sensitizes human breast cancer cells to trastuzumab, taxotere, gemcitabine, and epothilone B. *Mol Cancer Ther* 2, 971–984.

Fullgrabe, J., Kavanagh, E. & Joseph, B. (2011). Histone onco-modifications. *Oncogene* 30, 3391–3403.

Gaudet, M. M., Campan, M., Figueroa, J. D., Yang, X. R., Lissowska, J., Peplonska, B., Brinton, L. A., Rimm, D. L., Laird, P. W., Garcia-Closas, M. & Sherman, M. E. (2009). DNA hypermethylation of ESR1 and PGR in breast cancer: pathologic and epidemiologic associations. *Cancer Epidemiol Biomarkers Prev* 18, 3036–3043.

Ginger, M. R., Shore, A. N., Contreras, A., Rijnkels, M., Miller, J., Gonzalez-Rimbau, M. F. & Rosen, J. M. (2006). A noncoding RNA is potential marker of cell fate during mammary gland development. *Proc Natl Acad Sci U S A* 103, 5781–5786.

Guffanti, A., Iacono, M., Pelucchi, P., Kim, N., Soldà, G., Croft, L. J., Taft, R. J., Rizzi, E., Askarian-Amiri, M., Bonnal, R. J., Callari, M., Mignone, F., Pesole, G., Bertalot, G., Bernardi, L. R., Albertini, A., Lee, C., Mattick, J. S., Zucchi, I., & De Bellis, G. A. (2009). A transcriptional sketch of a primary human breast cancer by 454 sequencing. BMC Genomics 10, 163.

Guffei, A., Sarkar, R., Klewes, L., Righolt, C., Knecht, H. & Mai, S. (2010). Dynamic chromosomal rearrangements in Hodgkin's lymphoma are due to ongoing three-dimensional nuclear remodeling and breakage-bridge-fusion cycles. *Haematologica* 95, 2038–2046.

Gupta, R. A., Shah, N., Wang, K. C., Kim, J., Horlings, H. M., Wong, D. J., Tsai, M. C., Hung, T., Argani, P., Rinn, J. L., Wang, Y., Brzoska, P., Kong, B., Li, R., West, R. B., van de Vijver, M. J., Sukumar, S. & Chang, H. Y. (2010). Long non-coding RNA HOTAIR reprograms chromatin state to promote cancer metastasis. *Nature* 464, 1071–1076.

Hamajima, N., Hirose, K., Tajima, K., Rohan, T., Calle, E. E., Heath, C. W., Jr., Coates, R. J., Liff, J. M., Talamini, R., Chantarakul, N., Koetsawang, S., Rachawat, D., Morabia, A., Schuman, L., Stewart, W., Szklo, M., Bain, C., Schofield, F., Siskind, V., Band, P., Coldman, A. J., Gallagher, R. P., Hislop, T. G., Yang, P., Kolonel, L. M., Nomura, A. M., Hu, J., Johnson, K. C., Mao, Y., De Sanjose, S., Lee, N., Marchbanks, P., Ory, H. W., Peterson, H. B., Wilson, H. G., Wingo, P. A., Ebeling, K., Kunde, D., Nishan, P., Hopper, J. L., Colditz, G., Gajalanski, V., Martin, N., Pardthaisong, T., Silpisornkosol, S., Theetranont, C., Boosiri, B., Chutivongse, S., Jimakorn, P., Virutamasen, P., Wongsrichanalai, C., Ewertz, M., Adami, H. O., Bergkvist, L., Magnusson, C., Persson, I., Chang-Claude, J., Paul, C., Skegg, D. C., Spears, G. F., Boyle, P., Evstifeeva, T., Daling, J. R., Hutchinson, W. B., Malone, K., Noonan, E. A., Stanford, J. L., Thomas, D. B., Weiss, N. S., White, E., Andrieu, N., Bremond, A., Clavel, F., Gairard, B., Lansac, J., Piana, L., Renaud, R., Izquierdo, A., Viladiu, P., Cuevas, H. R., Ontiveros, P., Palet, A., Salazar, S. B., Aristizabel, N., Cuadros, A., Tryggvadottir, L., Tulinius, H., Bachelot, A., Le, M. G., Peto, J., Franceschi, S., Lubin, F., Modan, B., Ron, E., Wax, Y., Friedman, G. D., Hiatt, R. A., Levi, F., Bishop, T., Kosmelj, K., Primic-Zakelj, M., Ravnihar, B., Stare, J., Beeson, W. L., Fraser, G., Bullbrook, R. D., Cuzick, J., Duffy, S. W., Fentiman, I. S., Hayward, J. L., Wang, D. Y., McMichael, A. J., McPherson, K., Hanson, R. L., Leske, M. C., Mahoney, M. C., Nasca, P. C., Varma, A. O., Weinstein, A. L., Moller, T. R., Olsson, H., Ranstam, J., Goldbohm, R. A., van den Brandt, P. A., Apelo, R. A., Baens, J., de, l. C., Jr., Javier, B., Lacaya, L. B., Ngelangel, C. A., La Vecchia, C., Negri, E., Marubini, E., Ferraroni, M., Gerber, M., Richardson, S., Segala, C., Gatei, D., Kenya, P., Kungu, A., Mati, J. G., Brinton, L. A., Hoover, R., Schairer, C., Spirtas, R., Lee, H. P., Rookus, M. A., van Leeuwen, F. E., Schoenberg, J. A., McCredie, M., Gammon, M. D., Clarke, E. A., Jones, L., Neil, A., Vessey, M., Yeates, D., Appleby, P., Banks, E., Beral, V., Bull, D., Crossley, B., Goodill, A., Green, J., Hermon, C., Key, T., Langston, N., Lewis, C., Reeves, G., Collins, R., Doll, R., Peto, R., Mabuchi, K., Preston, D., Hannaford, P., Kay, C., Rosero-Bixby, L., Gao, Y. T., Jin, F., Yuan, J. M., Wei, H. Y., Yun, T., Zhiheng, C., Berry, G., Cooper, B. J., Jelihovsky, T., MacLennan, R., Shearman, R., Wang, Q. S., Baines, C. J., Miller, A. B., Wall, C., Lund, E., Stalsberg, H., Shu, X. O., Zheng, W., Katsouyanni, K., Trichopoulou, A., Trichopoulos, D., Dabancens, A., Martinez, L., Molina, R., Salas, O., Alexander, F. E., Anderson, K., Folsom, A. R., Hulka, B. S., Bernstein, L., Enger, S., Haile, R. W., Paganini-Hill, A., Pike, M. C., Ross, R. K., Ursin, G., Yu, M. C., Longnecker, M. P., Newcomb, P., Bergkvist, L., Kalache, A., Farley, T. M., Holck, S. & Meirik, O. (2002). Alcohol, tobacco and breast cancer--collaborative reanalysis of individual data from 53 epidemiological studies, including 58,515 women with breast cancer and 95,067 women without the disease. *Br J Cancer* 87, 1234–1245.

Hannafon, B. N., Sebastiani, P., de Las, M. A., Lu, J. & Rosenberg, C. L. (2011). Expression of microRNA and their gene targets are dysregulated in preinvasive breast cancer. *Breast Cancer Res* 13, R24.

Hansen, K. D., Timp, W., Bravo, H. C., Sabunciyan, S., Langmead, B., McDonald, O. G., Wen, B., Wu, H., Liu, Y., Diep, D., Briem, E., Zhang, K., Irizarry, R. A. & Feinberg, A. P. (2011). Increased methylation variation in epigenetic domains across cancer types. *Nat Genet* 43, 768–775.

Heaphy, C. M., Griffith, J. K. & Bisoffi, M. (2009). Mammary field cancerization: molecular evidence and clinical importance. *Breast Cancer Res Treat* 118, 229–239.

Henikoff, S. (2008). Nucleosome destabilization in the epigenetic regulation of gene expression. *Nat Rev Genet* 9, 15–26.

Heride, C., Ricoul, M., Kieu, K., von Hase, J., Guillemot, V., Cremer, C., Dubrana, K. & Sabatier, L. (2010). Distance between homologous chromosomes results from chromosome positioning constraints. *J Cell Sci* 123, 4063–4075.

Hermann, A., Goyal, R. & Jeltsch, A. (2004). The Dnmt1 DNA-(cytosine-C5)-methyltransferase methylates DNA processively with high preference for hemimethylated target sites. *J Biol Chem* 279, 48350–48359.

Hernandez-Vargas, H., Ouzounova, M., Calvez-Kelm, F., Lambert, M. P., McKay-Chopin, S., Tavtigian, S. V., Puisieux, A., Matar, C. & Herceg, Z. (2011). Methylome analysis reveals Jak-STAT pathway deregulation in putative breast cancer stem cells. *Epigenetics* 6, 428–439.

Hinshelwood, R. A. & Clark, S. J. (2008). Breast cancer epigenetics: normal human mammary epithelial cells as a model system. *J Mol Med (Berl)* 86, 1315–1328.

Hirst, M. & Marra, M. A. (2010). Next generation sequencing based approaches to epigenomics. *Brief Funct Genomics* 9, 455–465.

Hsu, P. Y., Deatherage, D. E., Rodriguez, B. A., Liyanarachchi, S., Weng, Y. I., Zuo, T., Liu, J., Cheng, A. S. & Huang, T. H. (2009). Xenoestrogen-induced epigenetic repression of microRNA-9-3 in breast epithelial cells. *Cancer Res* 69, 5936–5945.

Hsu, P. Y., Hsu, H. K., Singer, G. A., Yan, P. S., Rodriguez, B. A., Liu, J. C., Weng, Y. I., Deatherage, D. E., Chen, Z., Pereira, J. S., Lopez, R., Russo, J., Wang, Q., Lamartiniere, C. A., Nephew, K. P. & Huang, T. H. (2010). Estrogen-mediated epigenetic repression of large chromosomal regions through DNA looping. *Genome Res* 20, 733–744.

Huang, T. H. & Esteller, M. (2010). Chromatin remodeling in mammary gland differentiation and breast tumorigenesis. *Cold Spring Harb Perspect Biol* 2, a004515.

Hurst, D. R., Edmonds, M. D. & Welch, D. R. (2009). Metastamir: the field of metastasis-regulatory microRNA is spreading. *Cancer Res* 69, 7495–7498.

Iorio, M. V., Casalini, P., Piovan, C., Di Leva, G., Merlo, A., Triulzi, T., Menard, S., Croce, C. M. & Tagliabue, E. (2009). microRNA-205 regulates HER3 in human breast cancer. *Cancer Res* 69, 2195–2200.

Jing, M. X., Mao, X. Y., Li, C., Wei, J., Liu, C. & Jin, F. (2011). Estrogen receptor-alpha promoter methylation in sporadic basal-like breast cancer of Chinese women. *Tumour Biol* 32, 713–719.

Jintaridth, P. & Mutirangura, A. (2010). Distinctive patterns of age-dependent hypomethylation in interspersed repetitive sequences. *Physiol Genomics* 41, 194–200.

Jovanovic, J., Ronneberg, J. A., Tost, J. & Kristensen, V. (2010). The epigenetics of breast cancer. *Mol Oncol* 4, 242–254.

Kabotyanski, E. B., Rijnkels, M., Freeman-Zadrowski, C., Buser, A. C., Edwards, D. P. & Rosen, J. M. (2009). Lactogenic hormonal induction of long distance interactions between beta-casein gene regulatory elements. *J Biol Chem* 284, 22815–22824.

Kanwar, J. R., Mahidhara, G. & Kanwar, R. K. (2011). Antiangiogenic therapy using nanotechnological-based delivery system. *Drug Discov Today* 16, 188–202.

Key, J., Hodgson, S., Omar, R. Z., Jensen, T. K., Thompson, S. G., Boobis, A. R., Davies, D. S. & Elliott, P. (2006). Meta-analysis of studies of alcohol and breast cancer with consideration of the methodological issues. *Cancer Causes Control* 17, 759–770.

Khalil, A. M., Guttman, M., Huarte, M., Garber, M., Raj, A., Rivea, M. D., Thomas, K., Presser, A., Bernstein, B. E., van Oudenaarden, A., Regev, A., Lander, E. S. & Rinn, J. L. (2009). Many human large intergenic noncoding RNAs associate with chromatin-modifying complexes and affect gene expression. *Proc Natl Acad Sci U S A* 106, 11667–11672.

Kim, S. J., Kim, T. W., Lee, S. Y., Park, S. J., Kim, H. S., Chung, K. W., Lee, E. S. & Kang, H. S. (2004). CpG methylation of the ERalpha and ERbeta genes in breast cancer. *Int J Mol Med* 14, 289–293.

Kornberg, R. D. & Lorch, Y. (1999). Twenty-five years of the nucleosome, fundamental particle of the eukaryote chromosome. *Cell* 98, 285–294.

Lapidus, R. G., Ferguson, A. T., Ottaviano, Y. L., Parl, F. F., Smith, H. S., Weitzman, S. A., Baylin, S. B., Issa, J. P. & Davidson, N. E. (1996). Methylation of estrogen and progesterone receptor gene 5′ CpG islands correlates with lack of estrogen and progesterone receptor gene expression in breast tumors. *Clin Cancer Res* 2, 805–810.

Le Quesne, J. & Caldas, C. (2010). Micro-RNAs and breast cancer. *Mol Oncol* 4, 230–241.

Lehmann, U., Hasemeier, B., Christgen, M., Muller, M., Romermann, D., Langer, F. & Kreipe, H. (2008). Epigenetic inactivation of microRNA gene hsa-mir-9-1 in human breast cancer. *J Pathol* 214, 17–24.

Lehmann, U., Streichert, T., Otto, B., Albat, C., Hasemeier, B., Christgen, H., Schipper, E., Hille, U., Kreipe, H. H. & Langer, F. (2010). Identification of differentially expressed microRNAs in human male breast cancer. *BMC Cancer* 10:109.

Leygue, E. (2007). Steroid receptor RNA activator (SRA1): unusual bifaceted gene products with suspected relevance to breast cancer. *Nucl Recept Signal* 5, e006.

Li, S., Hansman, R., Newbold, R., Davis, B., McLachlan, J. A. & Barrett, J. C. (2003). Neonatal diethylstilbestrol exposure induces persistent elevation of c-fos expression and hypomethylation in its exon-4 in mouse uterus. *Mol Carcinog* 38, 78–84.

Lin, Y. W., Chen, H. M. & Fang, J. Y. (2011). Gene silencing by the Polycomb group proteins and associations with cancer. *Cancer Invest* 29, 187–195.

Lister, R., Pelizzola, M., Dowen, R. H., Hawkins, R. D., Hon, G., Tonti-Filippini, J., Nery, J. R., Lee, L., Ye, Z., Ngo, Q. M., Edsall, L., Antosiewicz-Bourget, J., Stewart, R., Ruotti, V., Millar, A. H., Thomson, J. A., Ren, B. & Ecker, J. R. (2009). Human DNA methylomes at base resolution show widespread epigenomic differences. *Nature* 19, 315–322.

Lo, P. K. & Sukumar, S. (2008). Epigenomics and breast cancer. *Pharmacogenomics* 9, 1879–1902.

Lodygin, D., Tarasov, V., Epanchintsev, A., Berking, C., Knyazeva, T., Korner, H., Knyazev, P., Diebold, J. & Hermeking, H. (2008). Inactivation of miR-34a by aberrant CpG methylation in multiple types of cancer. *Cell Cycle* 7, 2591–2600.

Lopez-Serra, L. & Esteller, M. (2008). Proteins that bind methylated DNA and human cancer: reading the wrong words. *Br J Cancer* 98, 1881–1885.

Ma, L., Young, J., Prabhala, H., Pan, E., Mestdagh, P., Muth, D., Teruya-Feldstein, J., Reinhardt, F., Onder, T. T., Valastyan, S., Westermann, F., Speleman, F., Vandesompele, J. & Weinberg, R. A. (2010). miR-9, a MYC/MYCN-activated microRNA, regulates E-cadherin and cancer metastasis. *Nat Cell Biol* 12, 247–256.

Markey, C. M., Luque, E. H., Munoz, D. T., Sonnenschein, C. & Soto, A. M. (2001). In utero exposure to bisphenol A alters the development and tissue organization of the mouse mammary gland. *Biol Reprod* 65, 1215–1223.

Markou, A., Tsaroucha, E. G., Kaklamanis, L., Fotinou, M., Georgoulias, V. & Lianidou, E. S. (2008). Prognostic value of mature microRNA-21 and microRNA-205 overexpression in non-small cell lung cancer by quantitative real-time RT-PCR. *Clin Chem* 54, 1696–1704.

Marotta, L. L., Almendro, V., Marusyk, A., Shipitsin, M., Schemme, J., Walker, S. R., Bloushtain-Qimron, N., Kim, J. J., Choudhury, S. A., Maruyama, R., Wu, Z., Gonen, M., Mulvey, L. A., Bessarabova, M. O., Huh, S. J., Silver, S. J., Kim, S. Y., Park, S. Y., Lee, H. E., Anderson, K. S., Richardson, A. L., Nikolskaya, T., Nikolsky, Y., Liu, X. S., Root, D. E., Hahn, W. C., Frank, D. A. & Polyak, K. (2011). The JAK2/STAT3 signaling pathway is required for growth of CD44CD24 stem cell-like breast cancer cells in human tumors. *J Clin Invest* 121, 2723–2735.

Matilainen, J. M., Malinen, M., Turunen, M. M., Carlberg, C. & Vaisanen, S. (2010). The number of vitamin D receptor binding sites defines the different vitamin D responsiveness of the CYP24 gene in malignant and normal mammary cells. *J Biol Chem* 285, 24174–24183.

Meaburn, K. J., Gudla, P. R., Khan, S., Lockett, S. J. & Misteli, T. (2009). Disease-specific gene repositioning in breast cancer. *J Cell Biol* 187, 801–812.

Moelans, C. B., Verschuur-Maes, A. H. & van Diest, P. J. (2011). Frequent promoter hypermethylation of BRCA2, CDH13, MSH6, PAX5, PAX6 and WT1 in ductal carcinoma in situ and invasive breast cancer. *J Pathol* 225, 222–231.

Muller, H. M., Widschwendter, A., Fiegl, H., Ivarsson, L., Goebel, G., Perkmann, E., Marth, C. & Widschwendter, M. (2003). DNA methylation in serum of breast cancer patients: an independent prognostic marker. *Cancer Res* 63, 7641–7645.

Nair, S. S., Coolen, M. W., Stirzaker, C., Song, J. Z., Statham, A. L., Strbenac, D., Robinson, M. W. & Clark, S. J. (2011). Comparison of methyl-DNA immunoprecipitation (MeDIP) and methyl-CpG binding domain (MBD) protein capture for genome-wide DNA methylation analysis reveal CpG sequence coverage bias. *Epigenetics* 6, 34–44.

Newbold, R. R., Padilla-Banks, E. & Jefferson, W. N. (2006). Adverse effects of the model environmental estrogen diethylstilbestrol are transmitted to subsequent generations. *Endocrinology* 147, S11–S17.

Nguyen, T., Kuo, C., Nicholl, M. B., Sim, M. S., Turner, R. R., Morton, D. L. & Hoon, D. S. (2011). Downregulation of microRNA-29c is associated with hypermethylation of tumor-related genes and disease outcome in cutaneous melanoma. *Epigenetics* 6, 388–394.

Palmer, J. R., Wise, L. A., Hatch, E. E., Troisi, R., Titus-Ernstoff, L., Strohsnitter, W., Kaufman, R., Herbst, A. L., Noller, K. L., Hyer, M. & Hoover, R. N. (2006). Prenatal diethylstilbestrol exposure and risk of breast cancer. *Cancer Epidemiol Biomarkers Prev* 15, 1509–1514.

Park, S. Y., Kwon, H. J., Lee, H. E., Ryu, H. S., Kim, S. W., Kim, J. H., Kim, I. A., Jung, N., Cho, N. Y. & Kang, G. H. (2011). Promoter CpG island hypermethylation during breast cancer progression. *Virchows Arch* 458, 73–84.

Perou, C. M., Sorlie, T., Eisen, M. B., van de, R. M., Jeffrey, S. S., Rees, C. A., Pollack, J. R., Ross, D. T., Johnsen, H., Akslen, L. A., Fluge, O., Pergamenschikov, A., Williams, C., Zhu, S. X., Lonning, P. E., Borresen-Dale, A. L., Brown, P. O. & Botstein, D. (2000). Molecular portraits of human breast tumours. *Nature* 406, 747–752.

Pokholok, D. K., Harbison, C. T., Levine, S., Cole, M., Hannett, N. M., Lee, T. I., Bell, G. W., Walker, K., Rolfe, P. A., Herbolsheimer, E., Zeitlinger, J., Lewitter, F., Gifford, D. K. & Young, R. A. (2005). Genome-wide map of nucleosome acetylation and methylation in yeast. *Cell* 122, 517–527.

Prat, A. & Perou, C. M. (2011). Deconstructing the molecular portraits of breast cancer. *Mol Oncol* 5, 5–23.

Puppe, J., Drost, R., Liu, X., Joosse, S. A., Evers, B., Cornelissen-Steijger, P., Nederlof, P., Yu, Q., Jonkers, J., van Lohuizen, M. & Pietersen, A. M. (2009). BRCA1-deficient mammary tumor cells are dependent on EZH2 expression and sensitive to Polycomb Repressive Complex 2-inhibitor 3-deazaneplanocin A. *Breast Cancer Res* 11:R63.

Radisky, D. C. (2011). miR-200c at the nexus of epithelial-mesenchymal transition, resistance to apoptosis, and the breast cancer stem cell phenotype. *Breast Cancer Res* 13, 110.

Rijnkels, M., Kabotyanski, E., Montazer-Torbati, M. B., Beauvais, C. H., Vassetzky, Y., Rosen, J. M. & Devinoy, E. (2010). The epigenetic landscape of mammary gland development and functional differentiation. *J Mammary Gland Biol Neoplasia* 15, 85–100.

Rodenhiser, D. I., Andrews, J., Kennette, W., Sadikovic, B., Mendlowitz, A., Tuck, A. B. & Chambers, A. F. (2008). Epigenetic mapping and functional analysis in a breast cancer metastasis model using whole-genome promoter tiling microarrays. *Breast Cancer Res* 10:R62.

Rodriguez, J., Frigola, J., Vendrell, E., Risques, R. A., Fraga, M. F., Morales, C., Moreno, V., Esteller, M., Capella, G., Ribas, M. & Peinado, M. A. (2006). Chromosomal instability correlates with genome-wide DNA demethylation in human primary colorectal cancers. *Cancer Res* 66, 8462–9468.

Rollins, R. A., Haghighi, F., Edwards, J. R., Das, R., Zhang, M. Q., Ju, J. & Bestor, T. H. (2006). Large-scale structure of genomic methylation patterns. *Genome Res* 16, 157–163.

Ronneberg, J. A., Fleischer, T., Solvang, H. K., Nordgard, S. H., Edvardsen, H., Potapenko, I., Nebdal, D., Daviaud, C., Gut, I., Bukholm, I., Naume, B., Borresen-Dale, A. L., Tost, J. & Kristensen, V. (2011). Methylation profiling with a panel of cancer related genes: association

5 Aberrant Epigenetic Regulation in Breast Cancer

with estrogen receptor, TP53 mutation status and expression subtypes in sporadic breast cancer. *Mol Oncol* 5, 61–76.

Rosenfeld, C. S. (2010). Animal models to study environmental epigenetics. *Biol Reprod* 82, 473–488.

Sato, K., Fukata, H., Kogo, Y., Ohgane, J., Shiota, K. & Mori, C. (2009). Neonatal exposure to diethylstilbestrol alters expression of DNA methyltransferases and methylation of genomic DNA in the mouse uterus. *Endocr J* 56, 131–139.

Scelfo, R. A., Schwienbacher, C., Veronese, A., Gramantieri, L., Bolondi, L., Querzoli, P., Nenci, I., Calin, G. A., Angioni, A., Barbanti-Brodano, G. & Negrini, M. (2002). Loss of methylation at chromosome 11p15.5 is common in human adult tumors. *Oncogene* 21, 2564–2572.

Sengupta, K., Camps, J., Mathews, P., Barenboim-Stapleton, L., Nguyen, Q. T., Difilippantonio, M. J. & Ried, T. (2008). Position of human chromosomes is conserved in mouse nuclei indicating a species-independent mechanism for maintaining genome organization. *Chromosoma* 117, 499–509.

Sharma, D., Saxena, N. K., Davidson, N. E. & Vertino, P. M. (2006). Restoration of tamoxifen sensitivity in estrogen receptor-negative breast cancer cells: tamoxifen-bound reactivated ER recruits distinctive corepressor complexes. *Cancer Res* 66, 6370–6378.

Shi, M., Liu, D., Duan, H., Shen, B. & Guo, N. (2010). Metastasis-related miRNAs, active players in breast cancer invasion, and metastasis. *Cancer Metastasis Rev* 29, 785–799.

Shinozaki, M., Hoon, D. S., Giuliano, A. E., Hansen, N. M., Wang, H. J., Turner, R. & Taback, B. (2005). Distinct hypermethylation profile of primary breast cancer is associated with sentinel lymph node metastasis. *Clin Cancer Res* 11, 2156–2162.

Shukla, V., Coumoul, X., Lahusen, T., Wang, R. H., Xu, X., Vassilopoulos, A., Xiao, C., Lee, M. H., Man, Y. G., Ouchi, M., Ouchi, T. & Deng, C. X. (2010). BRCA1 affects global DNA methylation through regulation of DNMT1. *Cell Res* 20, 1201–1215.

Silva, J. M., Boczek, N. J., Berres, M. W., Ma, X. & Smith, D. I. (2011). LSINCT5 is over expressed in breast and ovarian cancer and affects cellular proliferation. *RNA Biol* 8, 496–505.

Simonis, M., Klous, P., Splinter, E., Moshkin, Y., Willemsen, R., de Wit, E., van Steensel, B. & de Laat, W. (2006). Nuclear organization of active and inactive chromatin domains uncovered by chromosome conformation capture-on-chip (4 C). *Nat Genet* 38, 1348–1354.

Sioud, M. (2009). Targeted delivery of antisense oligonucleotides and siRNAs into mammalian cells. *Methods Mol Biol* 487, 61–82.

Sorlie, T., Perou, C. M., Tibshirani, R., Aas, T., Geisler, S., Johnsen, H., Hastie, T., Eisen, M. B., van de, R. M., Jeffrey, S. S., Thorsen, T., Quist, H., Matese, J. C., Brown, P. O., Botstein, D., Eystein, L. P. & Borresen-Dale, A. L. (2001). Gene expression patterns of breast carcinomas distinguish tumor subclasses with clinical implications. *Proc Natl Acad Sci U S A* 98, 10869–10874.

Stefansson, O. A., Jonasson, J. G., Olafsdottir, K., Hilmarsdottir, H., Olafsdottir, G., Esteller, M., Johannsson, O. T. & Eyfjord, J. E. (2011). CpG island hypermethylation of BRCA1 and loss of pRb as co-occurring events in basal/triple-negative breast cancer. *Epigenetics* 6, 638–649.

Strahl, B. D. & Allis, C. D. (2000). The language of covalent histone modifications. *Nature* 403, 41–45.

Suijkerbuijk, K. P., van Diest, P. J. & van der, W. E. (2011). Improving early breast cancer detection: focus on methylation. *Ann Oncol* 22, 24–29.

Sun, Z., Asmann, Y. W., Kalari, K. R., Bot, B., Eckel-Passow, J. E., Baker, T. R., Carr, J. M., Khrebtukova, I., Luo, S., Zhang, L., Schroth, G. P., Perez, E. A. & Thompson, E. A. (2011). Integrated analysis of gene expression, CpG island methylation, and gene copy number in breast cancer cells by deep sequencing. *PLoS ONE* 6, e17490.

Taft, R. J., Pang, K. C., Mercer, T. R., Dinger, M. & Mattick, J. S. (2010). Non-coding RNAs: regulators of disease. *J Pathol* 220, 126–139.

Tan, J., Yang, X., Zhuang, L., Jiang, X., Chen, W., Lee, P. L., Karuturi, R. K., Tan, P. B., Liu, E. T. & Yu, Q. (2007). Pharmacologic disruption of Polycomb-repressive complex 2-mediated gene repression selectively induces apoptosis in cancer cells. *Genes Dev* 21, 1050–1063.

Tsang, D. P. & Cheng, A. S. (2011). Epigenetic regulation of signaling pathways in cancer: role of the histone methyltransferase EZH2. *J Gastroenterol Hepatol* 26, 19–27.

Valastyan, S., Chang, A., Benaich, N., Reinhardt, F. & Weinberg, R. A. (2011). Activation of miR-31 function in already-established metastases elicits metastatic regression. *Genes Dev* 25, 646–659.

Van der Auwera, I., Bovie, C., Svensson, C., Trinh, X. B., Limame, R., van Dam, P., van Laere, S. J., van Marck, E. A., Dirix, L. Y. & Vermeulen, P. B. (2010). Quantitative methylation profiling in tumor and matched morphologically normal tissues from breast cancer patients. *BMC Cancer* 10:97.

Vandenberg, L. N., Maffini, M. V., Sonnenschein, C., Rubin, B. S. & Soto, A. M. (2009). Bisphenol-A and the great divide: a review of controversies in the field of endocrine disruption. *Endocr Rev* 30, 75–95.

Weber, B., Stresemann, C., Brueckner, B. & Lyko, F. (2007). Methylation of human microRNA genes in normal and neoplastic cells. *Cell Cycle* 6, 1001–1005.

Welsh, J. (1994). Induction of apoptosis in breast cancer cells in response to vitamin D and antiestrogens. *Biochem Cell Biol* 72, 537–545.

Weng, Y. I., Hsu, P. Y., Liyanarachchi, S., Liu, J., Deatherage, D. E., Huang, Y. W., Zuo, T., Rodriguez, B., Lin, C. H., Cheng, A. L. & Huang, T. H. (2010). Epigenetic influences of low-dose bisphenol A in primary human breast epithelial cells. *Toxicol Appl Pharmacol* 248, 111–121.

Wicha, M. S., Liu, S. & Dontu, G. (2006). Cancer stem cells: an old idea--a paradigm shift. *Cancer Res* 66, 1883–1890.

Widschwendter, M. & Jones, P. A. (2002). DNA methylation and breast carcinogenesis. *Oncogene* 21, 5462–5482.

Wu, Y., Crawford, M., Yu, B., Mao, Y., Nana-Sinkam, S. P. & Lee, L. J. (2011). MicroRNA delivery by cationic lipoplexes for lung cancer therapy. *Mol Pharm* 8, 1381–1389.

Xu, X., Gammon, M. D., Zhang, Y., Cho, Y. H., Wetmur, J. G., Bradshaw, P. T., Garbowski, G., Hibshoosh, H., Teitelbaum, S. L., Neugut, A. I., Santella, R. M. & Chen, J. (2010). Gene promoter methylation is associated with increased mortality among women with breast cancer. *Breast Cancer Res Treat* 121, 685–692.

Yan, P. S., Venkataramu, C., Ibrahim, A., Liu, J. C., Shen, R. Z., Diaz, N. M., Centeno, B., Weber, F., Leu, Y. W., Shapiro, C. L., Eng, C., Yeatman, T. J. & Huang, T. H. (2006). Mapping geographic zones of cancer risk with epigenetic biomarkers in normal breast tissue. *Clin Cancer Res* 12, 6626–6636.

Zhang, Y. & Reinberg, D. (2001). Transcription regulation by histone methylation: interplay between different covalent modifications of the core histone tails. *Genes Dev* 15, 2343–2360.

Zhao, H., Shen, J., Medico, L., Wang, D., Ambrosone, C. B. & Liu, S. (2010). A pilot study of circulating miRNAs as potential biomarkers of early stage breast cancer. *PLoS ONE* 5, e13735.

Chapter 6
The Impact of Epigenetic Alterations on Diagnosis, Prediction, and Therapy of Prostate Cancer

Christian Arsov, Wolfgang Goering, and Wolfgang A. Schulz

6.1 Introduction

6.1.1 The Challenge of Prostate Cancer

In men, prostate cancer is the most commonly diagnosed malignancy in Western industrialized countries, accounting for approximately 900,000 new cases worldwide every year. Nevertheless, prostate cancer represents only the third common cause of death from cancer in Western industrialized countries (Jemal et al. 2011). In other regions, especially in Asia, incidence and mortality are much lower. Generally, the discrepancy between incidence and mortality reflects a broad range of disease phenotypes ranging from aggressive and lethal metastatic cancers to clinically insignificant indolent tumors that will not cause symptoms, least limit the life expectancy of elderly men.

The widespread use of prostate-specific antigen (PSA) as a method for early detection has contributed to an increased incidence and a shift of detected cancers toward earlier stages and lower grades (Smith et al. 1996). While early detection based on PSA may increase the chance of curative treatments, it comes along with a considerable risk of overtreatment (Schroder et al. 2009; Hugosson et al. 2010). This is due to the limited ability of the currently used clinical parameters like PSA level, Gleason score of biopsies, and clinical tumor stage to discriminate between clinically significant and insignificant prostate cancers. Ideally, patients with clinically insignificant cancer should be followed up by active surveillance, whereas patients

---

C. Arsov, MD
Department of Urology, University Hospital, Duesseldorf, Germany

W. Goering, PhD • W.A. Schulz, PhD (✉)
Department of Urology, Heinrich Heine University, Moorenstrasse 5,
Duesseldorf 40225, Germany
e-mail: wolfgang.schulz@uni-duesseldorf.de

J. Minarovits and H.H. Niller (eds.), *Patho-Epigenetics of Disease*,
DOI 10.1007/978-1-4614-3345-3_6, © Springer Science+Business Media New York 2012

with potentially lethal tumors should be treated by surgery (radical prostatectomy) or radiotherapy to remove their cancers. However, while about 50% of diagnosed prostate cancers theoretically fulfill criteria for clinically insignificant cancers (Cooperberg et al. 2007), a significant proportion of these patients present with progression of their disease under active surveillance and have to undergo secondary curative treatment (Klotz et al. 2010). Moreover, after 15 years, up to 14.6% of patients with initial low-risk prostate cancers will die from cancer in spite of primary curative treatment by surgery (Bill-Axelson et al. 2011). These patients might benefit from a more aggressive treatment, e.g., by means of combined surgery plus radiotherapy.

An even greater challenge is that recurrent disease after treatment with curative intent is commonly incurable. Many recurrent tumors can be contained by androgen deprivation therapy (ADT) for a certain period. However, eventually all tumors become resistant to ADT, resulting in a condition now designated as castrate-resistant prostate cancer (CRPC). After its emergence, life expectancy is limited, and further treatments like chemotherapy with taxanes or second-generation antiandrogenic compounds only prolong survival by a few months (Tannock et al. 2004; de Bono et al. 2010, 2011).

On that background, molecular analyses of prostate cancer pathogenesis aim to provide tools for early detection of aggressive cancers and prediction of individual prognosis as well as alternative treatments based on an improved understanding of the pathomechanisms. Over the last years, important insights have been gained into genetic alterations in prostate cancer, which will be summarized briefly below. In addition, a large number of diverse epigenetic alterations have been linked to initiation, progression, and clinical behavior of prostate cancer. These will be described in the greater part of this chapter, with an emphasis on their potential clinical application.

6.1.2 Genetic Alterations in Prostate Cancer

Two crucial alterations commonly found in prostate cancer are chromosomal translocations leading to overexpression and oncogenic activation of ETS family transcription factors and overactivity of the phosphatidylinositol 3-kinase (PI3K) signaling pathway (Majumder and Sellers 2005; Tomlins et al. 2005). Both interact with androgen signaling in a mutual fashion (Hermans et al. 2006; Jiao et al. 2007; Yu et al. 2010; Carver et al. 2011).

The most common chromosomal rearrangements in about 40% of prostate cancers juxtapose the regulatory region of the protease gene TMPRSS2 and the larger part of the coding region of the ETS family factor ERG. In the simplest case, the fusion gene is created by the deletion of several-Mbp DNA normally separating the two genes at chromosome 21q (Perner et al. 2006). TMPRSS2 transcription is prostate specific and regulated by androgens (Lin et al. 1999). Therefore, the gene fusion

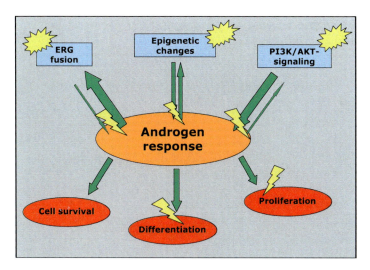

Fig. 6.1 Pathogenetic alterations in prostate cancer. Pathogenetic mechanisms in prostate cancer. Mutations or epimutations are marked by *exploding stars*. Consequences are marked by *bolts*

leads to overexpression of ERG in an androgen-dependent manner. In many other cases, analogous translocations are observed, typically fusing other ETS factors, such as ETV-1, to promoters expressed in an androgen-dependent or at least prostate-specific fashion. The oncogenic ETS factors stimulate cell proliferation and invasion and impede differentiation, not least by distorting androgen action.

Activation of the PI3K/Akt signaling pathway in prostate cancer is brought about by various mechanisms, including increased activity of growth factors. Arguably the most severe defect in regulation of the pathway is loss of the negative regulator PTEN (Majumder and Sellers 2005). Homozygous deletions of this gene located at 10q are observed in 10–30% of prostate cancers, especially at advanced stages (Sircar et al. 2009). PTEN acts as a phosphatase to reverse the reaction catalyzed by PI3K. In contrast, INPP enzymes antagonize PI3K by limiting the concentration of its substrate. Expression of INPP4B is lost in many prostate cancers, in particular in metastatic tumors, probably due to epigenetic mechanisms. In the normal prostate, INPP4B may act as a feedback inhibitor for PI3K signaling that is induced by androgens (Hodgson et al. 2011). This is only one example of crosstalk between the two signaling pathways disturbed in prostate cancer. Others include activation of the androgen receptor (AR) through phosphorylation by Akt, direct inhibition of the receptor by PTEN, and transcriptional repression of PTEN by the AR. Finally, the effects of ERG fusion genes, PI3K signal activation, and altered androgen response on prostate differentiation are compounded by epigenetic dysregulation, in particular by overexpression of the H3K27 histone methyltransferase (HMT) EZH2 and altered activity of its associated polycomb repressive complex 2 (PRC2) (Fig. 6.1).

6.2 Epigenetic Alterations as Biomarkers for Prostate Cancer Detection

6.2.1 Rationale for Epigenetic Alterations as Biomarkers for Prostate Cancer Detection

At present, early detection of prostate cancer is based on measurements of serum PSA concentration. These immunoassays provide a relatively high sensitivity, but limited specificity resulting in a large number of false-positive results, especially in unselected populations. A more specific diagnostic biomarker would ideally be based on molecular changes crucially involved in prostate carcinoma initiation. Furthermore, an optimal biomarker assay for initial prostate cancer screening should use body fluids like urine or peripheral blood accessible by noninvasive or minimally invasive means. A definitive diagnosis of prostate cancer is currently made by histological investigation of biopsies, which can be difficult because of sampling errors and ambiguous histology. A molecular biomarker of prostate cancer might also be helpful in such cases.

Epigenetics-based biomarkers appear to satisfy these requirements for several reasons. First, many epigenetic alterations are highly prevalent, and some arise at early stages, indicating that they are associated with tumor initiation. Second, DNA methylation in particular is very stable, and altered DNA methylation can be detected in body fluids as well as tissues (Hoque 2009). Third, sensitive techniques like quantitative real-time methylation-specific PCR, which are now widely employed, are capable of detecting methylated alleles even in the presence of a 10,000-fold excess of unmethylated alleles (Eads et al. 2000). Finally, the development of high-throughput technologies enables the simultaneous screening of multiple samples at several loci (Costa et al. 2007; Schulz and Goering 2011).

Several epigenetics-based biomarkers including alterations of DNA methylation, histone modifications, and microRNA (miRNAs) expression have been suggested for prostate cancer early detection. Of these, DNA methylation alterations are by far the best characterized.

6.2.2 Modifications of DNA Methylation as Biomarkers for Prostate Cancer Detection

Currently, more than 50 genes have been reported to be hypermethylated in prostate cancers (Li 2007; Nelson et al. 2007). Their number is rapidly expanding due to novel techniques allowing genome-wide screening for methylation alterations (Kim et al. 2011; Kobayashi et al. 2011) and may eventually run up to many hundreds. While not all of these genes are consistently affected in a majority of the cases, some genes are hypermethylated at remarkably high frequencies, as compiled by Park (2010) (Table 6.1). The best characterized epigenetic change in that respect is

Table 6.1 Selection of genes frequently hypermethylated in prostate cancer tissues

Gene	Common name	Frequency methylated	References
APC	Adenomatous polyposis coli	27% (27/101)	Maruyama et al. (2002)
		90% (66/73)	Yegnasubramanian et al. (2004)
		57% (21/37)	Kang et al. (2004)
		100% (118/118)	Jeronimo et al. (2004)
		78% (88/113)	Florl et al. (2004)
		82% (59/72)	Tokumaru et al. (2004)
		64% (109/170)	Enokida et al. (2005)
		83% (44/53)	Bastian et al. (2005)
		73% (131/179)	Cho et al. (2007)
		27% (21/79)	Henrique et al. (2007)
		83% (65/78)	Bastian et al. (2007a)
		40% (182/459)	Richiardi et al. (2009)
GSTP1	Glutathione S-transferase π	100% (20/20)	Lee et al. (1994)
		91% (52/57)	Lee et al. (1997)
		75% (24/32)	Santourlidis et al. (1999)
		91% (63/69)	Goessl et al. (2000)
		94% (16/17)	Jeronimo et al. (2001)
		79% (22/28)	Cairns et al. (2001)
		85% (89/105)	Jeronimo et al. (2002b)
		36% (36/101)	Maruyama et al. (2002)
		75% (24/32)	Konishi et al. (2002)
		58% (7/12)	Gonzalgo et al. (2003)
		71% (43/61)	Harden et al. (2003b)
		88% (96/109)	Yamanaka et al. (2003)
		84% (99/118)	Woodson et al. (2003)
		100% (18/18)	Kollermann et al. (2003)
		95% (69/73)	Yegnasubramanian et al. (2004)
		87% (32/37)	Kang et al. (2004)
		95% (112/118)	Jeronimo et al. (2004)
		72% (58/81)	Singal et al. (2004)
		79% (89/113)	Florl et al. (2004)
MDR1	Multidrug resistance	88% (64/73)	Yegnasubramanian et al. (2004)
		55% (97/177)	Enokida et al. (2004)
		100% (53/53)	Bastian et al. (2005)
		49% (89/164)	Enokida et al. (2006)
		95% (124/130)	Yegnasubramanian et al. (2006)
		51% (91/179)	Cho et al. (2007)
PTGS2	Prostaglandin-endoperoxide synthase 2	88% (64/73)	Yegnasubramanian et al. (2004)
		71% (38/53)	Bastian et al. (2005)
		65% (51/78)	Bastian et al. (2007a)
		68% (54/80)	Ellinger et al. (2008)
RARB2	Retinoic acid receptor β	79% (11/14)	Nakayama et al. (2001)
		53% (54/101)	Maruyama et al. (2002)
		78% (85/109)	Yamanaka et al. (2003)
		84% (42/50)	Zhang et al. (2004)
		70% (79/113)	Florl et al. (2004)
		40% (32/81)	Singal et al. (2004)
RARRES1 (TIG1)	Retinoic acid receptor responder protein 1 (Tazarotene-induced gene 1)	55% (17/31)	Tokumaru et al. (2003)
		53% (26/50)	Zhang et al. (2004)
		70% (43/61)	Topaloglu et al. (2004)
		70% (125/179)	Cho et al. (2007)
		96% (77/80)	Ellinger et al. (2008)
RASSF1A	Ras association domain family 1 isoform A	71% (37/52)	Liu et al. (2002)
		53% (54/101)	Maruyama et al. (2002)
		99% (117/118)	Jeronimo et al. (2004)
		49% (40/81)	Singal et al. (2004)
		78% (88/113)	Florl et al. (2004)
		96% (70/73)	Yegnasubramanian et al. (2004)
		84% (31/37)	Kang et al. (2004)
		74% (97/131)	Kawamoto et al. (2007)
RPRM	Reprimo	59% (47/80)	Ellinger et al. (2008)

Adapted from Park (2010)

hypermethylation of *GSTP1*, a gene located at 11q13, first described in 1994 (Lee et al. 1994). The *GSTP1* gene encodes glutathione transferase π which functions as a detoxifying enzyme by conjugating chemically reactive electrophiles to glutathione (Pickett and Lu 1989; Hayes and Pulford 1995). Thus, GSTP1 is important for detoxification and inactivation of potential carcinogens. In prostate cancer, hypermethylation of the *GSTP1* promoter region is extensive throughout the CpG island and associated with loss of expression of the enzyme (Millar et al. 1999; Lin et al. 2001b). *GSTP1* promoter hypermethylation is one of the earliest events reported in prostatic carcinogenesis. It occurs in 75–100% of prostate cancers and in up to 70% of high-grade prostatic intraepithelial neoplasia (PIN) precursor lesions (Brooks et al. 1998; Lin et al. 2001b; Kang et al. 2004). Since high-grade PIN lesions are the most common precursor of prostate cancer (Epstein 2009), it has been speculated that inactivation of *GSTP1* by hypermethylation might serve as one initiating event in cancer development, e.g., by increasing susceptibility to carcinogens (Perry et al. 2006). In contrast, restoration of GSTP1 expression in established prostate cancers has no effect on tumor growth in vivo (Lin et al. 2001a). Therefore, *GSTP1* is certainly not a classical tumor suppressor gene, and it is not clear whether its hypermethylation is functionally relevant for prostate carcinogenesis or rather reflects the altered state of differentiation of prostate cancer cells (Schulz and Hatina 2006).

GSTP1 hypermethylation can be detected in body fluids—such as urine, blood, and ejaculate—and tissues. Because *GSTP1* hypermethylation is infrequent in benign prostate tissues (Lee et al. 1994; Millar et al. 1999) and less frequent in other malignancies of the genitourinary tract (Esteller et al. 1998), it is a promising diagnostic biomarker for prostate cancer. A large number of studies have evaluated its sensitivity and specificity for the detection of prostate cancer in different specimens using various techniques, especially conventional and quantitative methylation-specific PCR. Not surprisingly, sensitivity and specificity are high in tissue samples with reported ranges of 70.5–94% and 96.8–100%, respectively (Goessl et al. 2000; Cairns et al. 2001; Jeronimo et al. 2001, 2002a; Harden et al. 2003a). Sensitivity in body fluids is however relatively poor and inferior to that of PSA. In blood, urine, and ejaculate, the sensitivity of *GSTP1* hypermethylation assays was reported as 13–72%, while specificity remained high at 93–100% (Goessl et al. 2000; Cairns et al. 2001; Jeronimo et al. 2001, 2002a; Harden et al. 2003a). Sensitivity in urine samples can be increased by prostatic massage prior to urine collection (Goessl et al. 2001). It is worth noting that in all cited studies, the number of subjects was <100 and that a high proportion of subjects with locally advanced or metastatic disease were enrolled. Therefore, data from these studies cannot be straightforwardly transferred to populations encountered during screening for prostate cancer. A minor additional complication is that *GSTP1* hypermethylation is not completely specific for cancer of the prostate. Its prevalence in renal cancer is about 20%, and it is also observed in 9% of lung cancer and 4% of colorectal cancer (Esteller et al. 1998). For these reasons, determination of *GSTP1* methylation status in body fluids is not suitable as a single screening test for diagnostic purposes. Its strength lies instead in its high specificity, which could help avoiding false-positive test results in high-prevalence diseases like prostate cancer, because of the inverse correlation

6 The Impact of Epigenetic Alterations on Diagnosis, Prediction...

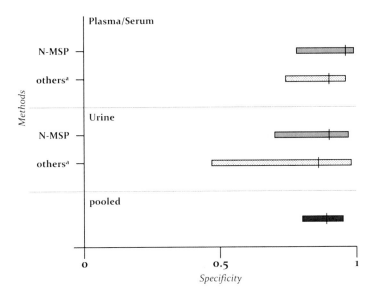

Fig. 6.2 Specificity of GSTP1 methylation testing in plasma and urine samples. Graphic representation of the data by Wu et al. (2011). *N-MSP* nonquantitative methylation-specific PCR. [a]Other methods include quantitative methylation-specific PCR, methylation-sensitive restriction endonuclease-qPCR, and bisulfite sequencing

between negative predictive value and prevalence at given specificity. Therefore, several authors suggested multigene methylation testing (see below) to increase sensitivity without compromising specificity.

An obvious option is to sequentially apply PSA testing and *GSTP1* promoter hypermethylation assays, combining their respective high sensitivity and specificity (Wu et al. 2011). A meta-analysis of the sensitivity and specificity for prostate cancer detection in body fluids in 22 studies with a total of 1,635 samples from prostate cancer patients and 573 samples from normal controls confirmed modest sensitivity (52%; 95% CI: 0.40–0.64) and high specificity (89%; 95% CI: 0.80–0.95), as illustrated in Figs. 6.2 and 6.3. The authors (Wu et al. 2011) proposed a sequence in which only patients with suspicious PSA values subsequently undergo *GSTP1* hypermethylation measurement in body fluids and only those with a positive *GSTP1* methylation test would be subjected to prostate biopsy. This procedure should significantly reduce the frequency of unnecessary prostate biopsies, suspected to amount up to 75% of patients with suspicious PSA (Schroder et al. 2009). This proposal should be confirmed in a controlled prospective clinical setting.

Additional genes demonstrated to be hypermethylated in prostate cancers but rarely in normal prostate tissues at comparable frequencies to *GSTP1* include *APC*, *RARB2*, *RASSF1A*, *RARRES1* (*TIG1*) encoding a carboxypeptidase inhibitor, *RPRM* (*reprimo*) a downstream target of p53, *PTGS2* encoding the cyclooxygenase-2

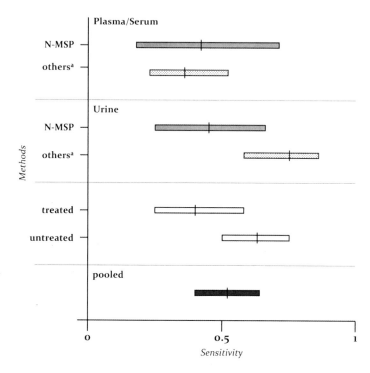

Fig. 6.3 Sensitivity of GSTP1 methylation test in plasma and urine samples. Graphic representation of the data by Wu et al. (2011). *N-MSP* nonquantitative methylation-specific PCR. [a]Other methods include quantitative methylation-specific PCR, methylation-sensitive restriction endonuclease-qPCR, and bisulfite sequencing. Treated or untreated means that the samples were collected either after treatments or before treatments

(COX-2), and *MDR1*, encoding a P-glycoprotein involved in multidrug resistance (Table 6.1). Among these, *APC*, *RARB2*, and *RASSF1A* are reasonable tumor suppressor candidates. *APC* is the most frequently mutated gene in colorectal cancers, but it has not been demonstrated that its hypermethylation in prostate cancer leads to transcriptional silencing. *RARB2* methylation is associated with gene silencing (Nakayama et al. 2001). The gene product is a retinoic acid receptor regulating differentiation of several tissues, but its function in prostate cancer is unknown. RASSF1A is a negative regulator of RAS signaling that is inactivated by promoter hypermethylation or allelic loss in many cancers, but no evidence for a specific function in prostate cancers has been brought forward. At this point, it is therefore not known, which genes affected by the coordinate hypermethylation associated with the early stages of prostate cancer are essential targets of inactivation and which are bystanders involved by a common mechanism.

Promoter hypermethylation of several genes listed in Table 6.1 is not only similarly frequent but also highly correlated with each other (Florl et al. 2004;

Yegnasubramanian et al. 2004). It has therefore been proposed to occur as part of an "epigenetic catastrophe" at early stage of prostate carcinogenesis during the progression of high-grade PIN lesions towards invasive prostate cancers (Yegnasubramanian et al. 2004). Because most methylation changes occurring during this epigenetic catastrophe will be maintained throughout disease progression, measurement of methylation of several of these genes ought to provide a robust tool for diagnostic purposes. From an analysis of prostate carcinoma tissues, our group has suggested four genes for this purpose, namely, *GSTP1*, *RASSF1A*, *RARB2*, and *APC* (Florl et al. 2004).

In body fluids, the most specific single gene methylation biomarker remains *GSTP1*. In urine samples, sensitivity was 29–62.1% for *RARB2*, 36–51% for *APC*, and 77.9% for *RASSF1A* methylation (Roupret et al. 2007; Vener et al. 2008). The four-gene set tested in urine samples after prostate massage demonstrated high sensitivity and accuracy of 86% and 89%, respectively (Roupret et al. 2007). Reducing the number of genes went along with a significant decrease of sensitivity. In urine samples, the combined use of *GSTP1/APC/RARB2* or *GSTP1/APC* demonstrated sensitivities of only 55% and 51%, respectively (Harden et al. 2003b; Vener et al. 2008; Baden et al. 2009). Hoque et al. 2005 evaluated urine sediment DNA from patients with prostate cancer and normal controls for aberrant methylation of nine gene promoters. From these, a combination of only four genes including *GSTP1*, *MGMT*, *ARF*, and *p16INK4A* (*CDKN2A*) achieved a sensitivity of 87% and a specificity of 100%. This result is a bit surprising in that methylation of the three latter genes was rare in most other studies. A combination of *TIG1* (*RARRES1*) methylation combined with either *GSTP1*, *APC*, *RARB2*, or *EDNRB* achieved high sensitivities of up to 100%, while specificity remained high (Ellinger et al. 2008).

In general, while sensitivity increases with the number of separate assays, there is an inverse correlation between the number of tests combined and specificity. Therefore, for practical applications, the total number of genes selected for a panel should be restricted to an optimum. Since methylation of *GSTP1* and many other genes is so highly correlated, combined assays may tend to identify or fail to detect the same cases. One of the most important practical questions in this respect is, therefore, whether the limited sensitivity of methylation assays in body fluids is due to a lack of DNA from tumors or to a subset of cancers lacking methylation of *GSTP1* and correlated genes.

Lack of tumor DNA is certainly relevant in blood samples because sensitivities are generally lower than in urine. For instance, in cell-free serum DNA, a four-gene panel (*GSTP1*, *PTGS2*, *RPRM*, *PTGS2*) significantly increased detection rates compared to *GSTP1* singly. Specificity was still high (92%), but sensitivity decreased to only 42–47% (Ellinger et al. 2008). Thus, multigene testing in blood samples is probably inferior to urine specimens for prostate cancer detection. Conceivably, though, the appearance of methylated DNA in serum signifies a stage progression of prostate cancer. In that case, positive assays might possess prognostic value. However, neither urine-based nor serum-based multigene panels have been validated in prospective clinical trials for prostate cancer detection or prognosis.

6.3 Epigenetic Alterations as Prognostic Biomarkers in Prostate Cancer

Because current clinical prognostic parameters discriminate insufficiently between indolent and aggressive prostate cancers, there is a strong need for novel prognostic biomarkers. Since epigenetic dysregulation not only is involved in initiation of prostate cancer but also contributes to disease progression, it should also provide biomarkers for individual prognosis or even prediction of the responses to different therapy modalities. In principle, two types of biomarkers have been explored for these purposes. Most straightforwardly, biomarkers for prognostic purposes could be derived from epigenetic alterations restricted to metastatic cancer or CRPC. However, genetic or epigenetic markers associated with cancer initiation can be used under two conditions. One favorable instance is the association of a change with a subgroup of the disease with particular prognosis or response to therapy. For instance, *MLH1* hypermethylation is associated with a "molecular" subtype of colon cancer, and *MGMT* hypermethylation predicts responsiveness to temozolomide in glioblastoma. In prostate cancer, it is hotly debated whether molecular subgroups can be defined by *TMPRSS2–ERG* fusions. Prognostic biomarkers may also be developed from an initial event aggravating with tumor progression, as revealed by quantitative measurements. In prostate cancer, PSA serum levels are routinely used in this fashion. In the development of epigenetic biomarkers for prostate cancer, both approaches have been followed.

6.3.1 Modifications of DNA Methylation as Prognostic Biomarkers in Prostate Cancer

Both genome-wide hypomethylation evident as a decrease of total 5-methylcytosine (5meC) content and local hypermethylation of certain CpG islands are associated with progression and poor prognosis in prostate cancer (Li 2007; Nelson et al. 2007; Schulz and Hoffmann 2009).

Global DNA hypomethylation was the first reported epigenetic alteration in cancer. It is thought to cause activation of oncogenes and to induce chromosomal instability (Hoffmann and Schulz 2005; Sharma et al. 2010). Although good correlative evidence has been reported for both effects, a causative relation has not been conclusively established. Hypomethylation affects especially retroelements like LINE-1, ALUs, and HERVs, which are densely methylated in normal cells. Global hypomethylation sets in at different stages and can be associated with cancer initiation or cancer progression, depending on the cancer type (Hoffmann and Schulz 2005). In prostate cancers, hypomethylation is generally associated with tumor progression. Decreased overall 5meC content and widespread hypomethylation of retroelements are regularly found in advanced cancer stages (Florl et al. 2004; Yegnasubramanian et al. 2008). It is not yet known to which extent hypomethylation reactivates silenced retroelements. Data available so far suggests that the relationship is not straightforward (Goering et al. 2011b). For example, LINE-1 hypomethylation is observed in one half of primary prostate cancers, and its

6 The Impact of Epigenetic Alterations on Diagnosis, Prediction...

Table 6.2 DNA hypermethylation events suggested as prognostic biomarkers in prostate cancer in recent years

Gene	Common name	Correlation	References
HOXD3	Homeobox D3	Biochemical recurrence Gleason score Tumor stage	Kron et al. (2010), Liu et al. (2011)
CCDN2	Cyclin-D2	Gleason score Tumor stage	Henrique et al. (2006)
TIG1	Tazarotene-induced gene 1	Gleason score Tumor stage	Ellinger et al. (2008)
CD44		Biochemical recurrence	Woodson et al. (2006)
GPR7	G protein-coupled receptor	Biochemical recurrence	Cottrell et al. (2007)
ABHD9 (EPHX3)	Abhydrolase domain containing 9 (epoxide hydrolase 3)	Biochemical recurrence	Cottrell et al. (2007)
Bcl-2	B cell lymphoma	Advanced tumor stages	Carvalho et al. (2010)
RBP1	Retinol binding protein 1 cellular	Gleason score Tumor stage	Cho et al. (2007)
ASC	Apoptosis-associated speck-like protein containing a CARD	Gleason score Tumor stage	Cho et al. (2007)
THBS1	Thrombospondin 1	Gleason score Tumor stage	Cho et al. (2007)
CDKN2A	Cyclin-dependent kinase inhibitor 2A	Gleason score	Verdoodt et al. (2011)
SFN	Stratifin	Gleason score	Vasiljevic et al. (2011)
SLIT2	Slit homolog 2	Gleason score	Vasiljevic et al. (2011)
SERPINB5	Serpin peptidase inhibitor, clade B (ovalbumin), member 5	Gleason score	Vasiljevic et al. (2011)

frequency increases in patients with progression to lymph node metastases (Santourlidis et al. 1999; Yegnasubramanian et al. 2008). Two small series investigated the relation of global 5meC content as measured by immunohistochemistry to outcome. They confirmed a pronounced decrease in global methylation as compared to normal prostate tissues, but the extent of hypomethylation did not correlate with recurrence (Brothman et al. 2005; Yang et al. 2011). However, precise quantification of methylcytosine by immunochemistry is difficult.

While the functional effects of genome-wide hypomethylation are not well understood, hypermethylation of gene promoters commonly silences the corresponding genes (Park 2010). An exception to the rule is *MCAM* (melanoma cell adhesion molecule). Nevertheless, *MCAM* hypermethylation is associated with higher tumor stages and higher Gleason scores (Liu et al. 2008). Hypermethylated genes that have been proposed to discriminate between aggressive and insignificant prostate cancer include *SOCS3, DcR1 and DcR2 (TNFRSF10C and TNFRSF10D), PITX2, MCAM, APC, CDH1, EDNRB, GSTP1, MDR1, MT1G, PTGS2, RARB2, RASSF1A, RUNX3,* and others (Li 2007; Nelson et al. 2007; Jeronimo et al. 2011; Goering et al. 2011a). Table 6.2 lists additional candidates that have been proposed since 2006. The products of these genes are involved in DNA repair, apoptosis, cell cycle control, steroid hormone response, and metastasis, but—as discussed above for genes hypermethylated at early stages—the functional consequences of their hypermethylation have only rarely been elucidated. Moreover, few studies linking hypermethylation of particular genes and clinical course of disease have been replicated independently. These may owe to technical differences. Significantly, the DNA methylation pattern at silenced genes is often not homogeneous, and the results of methylation analyses depend on the chosen method and on which CpG sites are examined (Mikeska et al. 2010; Goering et al. 2011a).

Extensive data link hypermethylation of the paired-like homeodomain transcription factor 2 (*PITX2*) to poor prognosis in prostate cancer and other cancers (Hartmann et al. 2009; Weiss et al. 2009; Banez et al. 2010). In prostate cancer, *PITX2* promoter hypermethylation is a strong and independent prognostic marker of biochemical recurrence in patients with localized prostate cancer treated by radical prostatectomy. *PITX2* promoter hypermethylation is associated with a dramatically diminished gene expression that also significantly correlates with poor prognosis. Interestingly, *PITX2* hypermethylation is more pronounced in cases with ERG overexpression (Vinarskaja et al. 2011). Similar findings were previously reported for *PTGS2*. *PTGS2* hypermethylation independently predicts biochemical recurrence and metastases after radical prostatectomy, but no multicenter confirmation study has yet been performed (Yegnasubramanian et al. 2004). The suppressor of cytokine signaling (*SOCS3*) is hypermethylated in about 40% of cases, strongly correlating with higher Gleason scores and a worse prognosis. Tissue samples exhibiting *SOCS3* DNA hypermethylation display weaker mRNA and protein expression (Pierconti et al. 2011).

Despite such promising results, up to now, epigenetics-derived prognostic biomarkers for prostate cancer have not entered into clinical routine for several reasons. First, most studies had a retrospective design with small numbers of cases. Therefore, a selection bias cannot be excluded. Second, due to the use of different assays, data are inconsistent. Third, biochemical recurrence in terms of PSA increase was chosen as the endpoint in most studies. However, biochemical recurrence is a rather poor surrogate parameter for "hard" endpoints like occurrence of metastases or death from disease. For instance, in patients with initial curative treatment, the risk of dying from prostate cancer is at most 9.9% within 10 years after biochemical recurrence (Boorjian et al. 2011). Finally, a prognostic biomarker has little value in clinical practice if it does not result in a choice of different treatment. Whether treatment choice can be based on a prognostic biomarker must be evaluated in prospective and randomized trials with hard endpoints. The way an epigenetics-based biomarker could enter into clinical routine is illustrated by *PITX2* hypermethylation. By means of quantitative methylation-specific real-time PCR, *PITX2* hypermethylation was found to be predictive for biochemical recurrence after radical prostatectomy in a retrospective multicenter trial including more than 600 patients (Weiss et al. 2009). Promisingly, *PITX2* methylation was predictive especially in intermediate-risk carcinomas. In a next step, these results were validated and confirmed in a distinct set of more than 400 patients using a standardized microarray assay (Banez et al. 2010). While these data are convincing, this biomarker has still limited value for clinical decisions as it indicates simply that the patient has an increased risk for biochemical recurrence after prostatectomy has already been done. Therefore, in a next step, the standardized *PITX2* methylation assay should to be further evaluated in a prospective randomized trial dividing *PITX2* methylation-positive patients in two treatment arms, e.g., an experimental arm with adjuvant radiotherapy or androgen deprivation therapy and a comparator arm without adjuvant therapy. Superiority of the experimental arm in terms of hard endpoints like overall survival would then allow using this biomarker for prediction and treatment decisions. As an example, in breast cancer, determination of expression of the proteases uPA and PAI-1, albeit not strictly an epigenetic marker, is established to make decisions in favor of or against adjuvant chemotherapy (Harbeck et al. 2007).

Analogous to multigene testing for diagnostic biomarkers, a multigene methylation analysis including the genes *GSTP1*, *APC*, and *MDR1* has been proposed for prognostic purposes (Enokida et al. 2005). A methylation score for these genes based on investigation of 170 cancers and 69 benign tissues significantly correlated with tumor stage, Gleason score, high preoperative PSA, and advanced pathologic features. The methylation score discriminated between organ confined and locally advanced prostate cancers with a sensitivity of 72.1% and a specificity of 67.8%. Even if the results can be replicated, it is doubtful whether these percentages would be sufficient for routine clinical use. The combined methylation profile of *CD44* and *PTGS2* was also reported to be an independent predictor of biochemical recurrence (Phe et al. 2010). Our own group made the intriguing observation that intensification of methylation at the *RARRES1* gene at 3q was associated with appearance of hypermethylation at the neighboring paralogous gene *LXN* as well as adverse prognosis (Kloth et al. 2012).

As in the above examples, most DNA methylation-based prognostic markers have been evaluated in tissues from prostatectomy specimens. In clinical practice, it would be desirable to predict prognosis prior to prostatectomy or radiotherapy, especially in order to identify indolent tumors. However, studies evaluating respective prognostic biomarkers in body fluids are scarce and inconsistent. For instance, one study demonstrated that biochemical recurrence within 2 years of radical prostatectomy can be predicted by *GSTP1* methylation status in cell-free serum samples (Bastian et al. 2007b). This could indicate that emergence of hypermethylated DNA from the tumor in serum is associated with progression to an incurable state. In contrast, in another study, neither hypermethylation at single genes including *GSTP1*, *TIG1*, *PTGS2*, and *reprimo* nor the combination of multiple gene sites correlated with pathological stage, Gleason grade, or biochemical recurrence following radical prostatectomy (Ellinger et al. 2008). Studies evaluating DNA methylation modifications in urine samples for prognostic purposes have not been published yet.

6.3.2 Alterations of Histone Modifications and Histone-Modifying Enzymes as Prognostic Biomarkers in Prostate Cancer

The EZH2 HMT activity of the PRC2, consisting of EZH2, EED, SUZ12, and further subunits, mediates gene silencing by trimethylation of H3 lysine 27 (H3K27me3) (Crea et al. 2011). EZH2 is overexpressed in prostate cancers, increasing gradually with tumor progression (Varambally et al. 2002). Because PRC2 components are overexpressed in metastatic prostate cancers and CRPC (Kuzmichev et al. 2005) and are likely required for cancer stem cell renewal, they are plausible prognostic markers of prostate cancer (Varambally et al. 2002). Indeed, a microarray-derived signature of 14 genes repressed by PRC2 predicted overall survival and recurrence in two distinct prostate cancer cohorts (Yu et al. 2007).

Many alterations of other enzymes involved in histone methylation (HMT), histone acetylation (histone acetyltransferases, HAT), histone deacetylases (HDAC), or histone demethylases (HDM) have been linked to prostate cancers (Cooper and Foster 2009). In addition, global changes in histone modifications are observed (Seligson et al. 2005; Bianco-Miotto et al. 2010; Ellinger et al. 2010). It is not yet known by which mechanisms such global changes in histone modifications arise and how they relate to the altered expression of individual histone-modifying enzymes.

Since evaluation of histone modifications requires tissue samples, they will presumably be more applicable as prognostic or predictive biomarkers than for prostate cancer detection. Moreover, biological changes in histone modifications seem to be associated with prostate cancer tumor progression rather than initiation. To date, there are a relatively small number of studies on this topic. For instance, lower H3K4me2 and H3K18 acetylation have been reported to predict a higher risk of prostate cancer recurrence (Seligson et al. 2005). Another study has claimed precisely the opposite, namely, that high H3K4me2 and H3K18 acetylation levels are associated with an increased risk of recurrence (Bianco-Miotto et al. 2010). A third report (Ellinger et al. 2010) described acetylation of H3 and H4 as well as H3K4me1, H3K9me2, and H3K9me3 as significantly reduced in prostate cancers compared to benign prostate tissues. The combination of H3 acetylation and H3K9me2 levels allowed discrimination between prostate cancers and benign prostate tissues with a sensitivity of >78% and a specificity of >92%. Moreover, increased H3K4me1 was suggested as a predictor of biochemical recurrence, and increased H3K4me1, H3K4me2, and H3K4me3 levels were linked to CRPC. The discrepancies in these reports underline that this field is still in its infancy, pointing to technical difficulties in quantifying histone modifications in stained tissue sections. In addition to technical improvements, a better understanding of the mechanisms underlying the changes may be essential to alleviate these difficulties.

6.4 Epigenetic Alterations as Therapeutic Targets in Prostate Cancer

Epigenetic alterations are very likely as essential as chromosomal alterations and base mutations for the initiation and progression of prostate cancer. However, in contrast to genetic changes, they are potentially reversible. Therefore, inhibitors of enzymes maintaining the aberrant epigenome of cancer cells may reverse the pathologic state of the genome. In the most straightforward instance, dysregulated epigenetic regulator proteins may silence genes limiting proliferation or inducing apoptosis of cancer cells. Inhibition of such regulators may restore transcription of silenced genes to block tumor growth. A number of novel compounds have been designed based on that rationale. The following discussion will emphasize therapeutic drugs targeting epigenetic alterations which have already made the transition to clinical trials.

6.4.1 Therapeutic Targeting of DNA Methylation

The human genome contains five DNA methyltransferase (DNMT) genes, the major DNA methyltransferases DNMT1, DNMT3A, DNMT3B, as well as DNMT2 and DNMTL, which encode proteins with distinct functions (Goll and Bestor 2005). To which extent and by which mechanisms these proteins become dysregulated during prostate carcinogenesis has not been fully resolved (Patra et al. 2002; Hoffmann et al. 2007; Kobayashi et al. 2011). Both quantitative changes in DNMT gene expression as well as aberrant enzyme targeting might be involved (Stresemann et al. 2006). There is good evidence that both the maintenance methyltransferase DNMT1 as well as the de novo methyltransferases DNMT3A and DNMT3B are required for establishment and maintenance of promoter hypermethylation and gene silencing (Vire et al. 2006). It is much less clear how the generally observed increases in DNMT expression fit with the overall decrease in methylcytosine content during prostate cancer progression.

Inhibitors of DNMTs can be divided into nucleoside and nonnucleoside inhibitors. The best characterized DNMT nucleoside inhibitors are 5-azacytidine (Vidaza®) and 5-aza-2'-deoxycytidine (decitabine). They become incorporated into DNA during replication, react with, and sequester DNMTs which results in diminished methylation of cytosines (Perry et al. 2010). Their potential to restore expression of silenced genes has been broadly demonstrated (Bender et al. 1998; Lin et al. 2001a), and they are widely used for experiments in vitro and in animal models. In the clinic, 5-azacytidine and 5-aza-2'-deoxycytidine significantly improve overall survival in myelodysplastic syndrome (Fenaux et al. 2009; Lubbert et al. 2011) and have been approved for this use by the US Food and Drug Administration (FDA). In CRPC patients, 5-azacytidine and 5-aza-2'-deoxycytidine were evaluated in phase II trials, but their antitumor effects proved modest. 5-Aza-2'-deoxycytidine led to stable disease in 17% of patients, and median time to progression was only 10 weeks (Thibault et al. 1998). The same poor efficacy in chemonaive CRPC patients was observed for 5-azacytidine. The treatment attained prolongation of PSA doubling time to ≥ 3 months in 56% of the patients. Methylation of LINE-1 retroelements in blood cells was used as a surrogate parameter for efficacy and showed a trend toward decreasing. However, no patient reached reduction of PSA serum levels by $\geq 50\%$, a commonly used indicator of chemotherapy efficacy in prostate cancer, and median clinical time to progression was just 12.4 weeks (Sonpavde et al. 2011). Unfortunately, nucleoside analogs have a complex mode of action and do not only inhibit DNMTs and DNA methylation (Yang et al. 2010). These additional unspecific effects but also specific effects on hematopoietic differentiation (Santos et al. 2010) may explain their relatively high toxicity in clinical trials. In prostate cancer, it is questionable to which extent the drugs are actually incorporated into the genome of the tumor cells because of a relatively low proliferative fraction. Another phase I/II trial evaluating the combination of 5-azacytidine with a cytotoxic drug (docetaxel) is not yet finished.

As compared to 5-azacytidine and 5-aza-2'-deoxycytidine, another cytidine analog, zebularine, shows minimal acute toxic effects (Cheng et al. 2003) and has fewer side effects in long-term therapy (Yoo et al. 2008). It is chemically more stable and can be administered orally. Currently, the compound is being evaluated in preclinical studies for myeloid malignancies and selected carcinomas, but has not yet been tried for the treatment of prostate cancer.

Nonnucleoside DNMT inhibitors like procaine and procainamide might be safer than nucleoside analogs due to their lower mutagenic potential (Stresemann et al. 2006). Other than nucleoside analogs, their mechanism of action does not require incorporation into DNA. They instead target DNMTs directly and can therefore be specific for individual DNMT isoenzymes (Perry et al. 2010). For example, procaine, in common use as a local anesthetic drug, is thought to selectively inhibit DNMT1 (Lin et al. 2001a). So far, nonnucleoside analogs have been much less efficacious in inhibiting DNMT and in reactivating silenced genes than nucleoside analogs (Chuang et al. 2005). Procaine and the chemically related procainamide, clinically used for the treatment of cardiac arrhythmia, have not been evaluated in clinical trials probably due to their main pharmacologic effects. RG108 (*N*-phthalyl-L-tryptophan), a small molecule DNMT inhibitor designed by virtual screening, could constitute an improvement. It selectively inhibits DNMT1 and is capable of restoring silenced tumor suppressors (Brueckner et al. 2005; Braun et al. 2010).

Epigallocatechin 3-gallate (EGCG3) is the major phenol in green tea. Consumption of large amounts of green tea has been linked to a lowered risk of prostate cancer by strong epidemiological data and is ascribed to EGCG3 (Bettuzzi et al. 2006). Many mechanisms have been proposed for the anticancer activity of this compound, especially a function as an antioxidant (Yang et al. 2002). In addition, EGCG3 is a potent inhibitor of certain *S*-adenosylmethionine-dependent methyltransferases, including catechol-*O*-methyltransferase and DNMTs. By inhibiting DNMT, EGCG3 reactivates genes silenced by methylation in cancer cell lines (Fang et al. 2003) and restores cell cycle inhibitors such as p21[CIP1] and p16[INK4a] (Yu et al. 2008; Nandakumar et al. 2011). It is still unclear to which extent any cancer-preventive activity of EGCG3 in prostate cancers is due to DNMT inhibition (Fang et al. 2003). Recently, a placebo-controlled phase I/II study has been initiated evaluating the short-term effects of daily EGCG (800 mg) administration during the interval between prostate biopsy and radical prostatectomy in men with localized prostate cancer (http://www.clinicaltrials.gov). Another placebo-controlled double-blind phase II trial evaluates the ability of EGCG to prevent progression towards invasive carcinoma in patients with high-grade PIN lesions (http://www.clinicaltrials.gov).

Disulfiram inhibits acetaldehyde dehydrogenase and thus the conversion of acetaldehyde to acetic acid. The accumulation of acetaldehyde induces hypersensitivity

6   The Impact of Epigenetic Alterations on Diagnosis, Prediction...   139

to alcohol, and disulfiram is therefore used for treatment of chronic alcoholism (Johnson 2008). Recently, it has been demonstrated that disulfiram inhibits DNMT1 in prostate cancer cell lines in a dose-dependent fashion. It decreases global genomic methylcytosine content, reverses promoter hypermethylation, and increases expression of the tumor suppressor genes *APC* and *RARB2*. No effect on PSA expression was observed. Growth of prostate cancer cell lines was inhibited, and growth of prostate cancer xenografts was attenuated (Lin et al. 2011). Evaluation of these effects in human patients would now be necessary.

In summary, although effective in prostate cancer cell and animal models, nucleoside DNMTs inhibitors have failed in clinical trials due to lack of efficacy and to high toxicity. Nonnucleoside inhibitors have a more favorable toxic profile which however goes along with generally weaker antitumor effect. Therefore, they have not achieved the transition to clinical trials. Obviously, there is a strong need for novel specially designed inhibitors of DNMTs with improved toxic and therapeutic profiles.

6.4.2   Therapeutic Targeting of Histone Acetylation by HDAC Inhibitors

In humans, 18 HDACs have been identified which are classified into four classes (I–IV) (Bertrand 2010). The class III HDACs are the NAD^+-dependent sirtuins; HDAC1–11 constitute the classes I, II, and IV. Of note, neither all sirtuins nor all HDACs are nuclear enzymes deacetylating histones and other chromatin proteins (Horio et al. 2011). Both sirtuins, especially SIRT1 (Hoffmann et al. 2007; Huffman et al. 2007), and HDACs (Abbas and Gupta 2008) have been implicated in prostate cancer. While drugs and natural compounds targeting sirtuins are under investigation for use in cancer treatment (Aljada et al. 2010; Balcerczyk and Pirola 2010), most studies have dealt with HDAC inhibitors. This section focuses on drugs that have been employed in humans and a few additional agents with strong potential of becoming future drugs (Table 6.3).

In prostate cancers, HDACs are thought to promote in particular disease progression. As compared to benign prostate tissues, HDAC1 exhibits two- to fourfold enhanced activity and expression in carcinoma (Patra et al. 2001). Since HDACs regulate several tumor suppressor genes as part of transcriptional corepressor complexes (Abbas and Gupta 2008), they are regarded as attractive targets for inhibitory compounds. HDAC inhibitors can be subdivided into seven classes (Perry et al. 2010), namely, short-chain fatty acids, hydroxamates, cyclic tetrapetides, benzamides, hydroxamate-tethered phenylbutyrate derivatives, trifluoromethyl ketones, plus miscellaneous compounds. Of these, the first five have demonstrated antitumor effects in prostate cancers.

Suberoylanilide hydroxamic acid (SAHA, vorinostat, Zolinza®) belongs to the hydroxamate class of HDAC inhibitors. Vorinostat inhibits HDAC classes I, II, and IV, but not sirtuins (Xu et al. 2007). It selectively downregulates the class II HDAC7

Table 6.3 Inhibitors of histone deacetylases (HDACs) in clinical trials for treatment of prostate cancer

Drug	Regimen	Number of subjects	Phase	Status	References
Vorinostat	400 mg orally daily	27	II	Completed	Bradley et al. (2009)
Vorinostat plus docetaxel	Dose escalation (vorinostat 100, 100, 200, and 200 mg plus docetaxel 50, 60, 60, and 75 mg/m^2)	12	I	Completed	Schneider et al. (2012)
Vorinostat plus doxorubicin	Dose escalation (vorinostat 400, 600, 800, and 1,000 mg days 1–3 followed by doxorubicin 20 mg/m^2)	32	I	Completed	Munster et al. (2009)
Vorinostat plus temsirolimus	Vorinostat days 1–14 and temsirolimus on days 1, 8, and 15	29 (Planned)	I	Not yet recruiting	http://www.clinicaltrials.gov
Panobinostat plus docetaxel	Arm I Panobinostat 20 mg on days 1, 3, and 5 for 2 consecutive weeks followed by 1-week break Arm II Panobinostat 20 mg on days 1, 3, and 5 for 2 consecutive weeks followed by 1-week break plus docetaxel 75 mg/m^2 every 21 days	16	I	Completed	Rathkopf et al. (2010)
Panobinostat plus bicalutamide	Arm I Panobinostat 20 mg triweekly 2 of 3 weeks plus bicalutamide 50 mg daily Arm II Panobinostat 30 mg triweekly 2 of 3 weeks plus bicalutamide 50 mg daily Arm III Panobinostat 40 mg triweekly 2 of 3 weeks plus bicalutamide 50 mg daily	78 (Planned)	I/II	Recruiting	Ferrari et al. (2011)
Panobinostat	Panobinostat mono	35	II	Completed (outstanding results)	http://www.clinicaltrials.gov
Belinostat	Dose escalation	100	I	Ongoing (recruiting finished)	http://www.clinicaltrials.gov
Romidepsin	Romidepsin 13 mg/m^2 on days 1, 8, and 15 every 28 days	35	II	Completed	Molife et al. (2010)
Entinostat	Dose escalation entinostat mono	100	I	Completed (outstanding results)	http://www.clinicaltrials.gov

by binding to its catalytic site with little or no effect on expression of other class I or II HDACs (Dokmanovic et al. 2007). HDAC7 is linked to initiation of apoptosis (Bakin and Jung 2004). In preclinical prostate cancer models, vorinostat led to cell death via downregulation of HDAC7 (Butler et al. 2000; Rokhlin et al. 2006) and enhanced sensitivity of prostate cancer cell lines to radiation therapy (Chinnaiyan et al. 2005). In a metastatic mouse model, it reduced tumor growth in bones in up to a third of the animals (Pratap et al. 2010). Furthermore, vorinostat downregulates expression of the androgen receptor (AR) as well as of PSA, and it synergizes with the AR antagonist bicalutamide (Marrocco et al. 2007).

In humans, vorinostat is currently the most investigated HDAC inhibitor and is approved for treatment of cutaneous T cell lymphoma (Jeronimo et al. 2011). Following proof of preclinical antitumor activity in prostate cancer, it was evaluated in clinical trials. Mature results from a phase II trial investigating its use as a single agent (400 mg orally daily) after prior chemotherapy were disappointing. All patients were taken off therapy before 6 months mainly due to disease progression (48%) or toxicity (41%). Efficacy was poor with stable disease as the best response in 7% of the patients and lack of PSA response in more than half. Additionally, 48% of patients experienced grade 3 or 4 toxicities (Bradley et al. 2009). Addition of vorinostat to docetaxel is also poorly tolerated (Schneider et al. 2012). A combination of vorinostat and doxorubicin (an anthracycline) in a phase I trial seems to be better tolerated and showed partial responses. However, further studies are needed to evaluate this combination (Munster et al. 2009). A study adding vorinostat to temsirolimus (mTOR inhibitor) will be opened for recruitment shortly. The rationale for this study design is a synergistic antitumor effect exerted by combined mTOR and HDAC inhibition in several prostate cancer cell lines (Wedel et al. 2011a, b).

Panobinostat is another hydroxamic acid pan-HDAC inhibitor with potent inhibitory activity at low concentrations against all class I, II, and IV HDACs. In vitro, it is at least tenfold more potent than vorinostat (Atadja 2009). In preclinical models, panobinostat elicits more cell death in AR-positive prostate cancer cells as compared to those lacking an AR. In PC-3 xenografts, it inhibited tumor angiogenesis (Qian et al. 2006b). The latter two effects might be explained by an inhibitory effect of panobinostat on class II HDACs like HDAC6 that primarily deacetylate nonhistones. Thus, panobinostat is assumed to mediate acetylation and degradation of hypoxia-inducible factor 1α (HIF-1α) (Qian et al. 2006a). Deacetylation of the chaperone heat shock protein 90 (Hsp90) by HDAC6 increases its ability to bind ATP enhancing its function as a chaperone. This function is crucial for the stability of wild-type as well as mutated AR and for stability and function of oncogenic tyrosine and serine–threonine kinases like HER2 and Akt (Fang et al. 1996; Solit et al. 2003; Atadja 2009). Interestingly, the combination of panobinostat with docetaxel delays progression to CRPC (Shao et al. 2008), suggesting that panobinostat might help to overcome resistance to chemotherapy which is common in CRPC. This conclusion is supported by a small phase I study comparing panobinostat alone vs. docetaxel plus panobinostat in CRPC patients. While panobinostat as a single agent exhibited no effects, the combination panobinostat/docetaxel achieved PSA reductions $\geq 50\%$ in 63% of patients. Importantly, responses were even observed in patients who had previously progressed on docetaxel (Rathkopf et al. 2010). Additionally, panobinostat seems to be capable to restore the sensitivity of CRPC patients to bicalutamide (Ferrari et al. 2011). A phase II study of panobinostat as single agent in CRPC patients has finished recruitment, and results are awaited.

Belinostat is a third hydroxamic acid pan-HDAC inhibitor investigated in human patients. In an orthotopic prostate cancer tumor model, it inhibited tumor growth by up to 43% and prevented lung metastases. Belinostat increases the expression of tissue inhibitor of metalloproteinase-1 (TIMP-1) by inhibition of HDAC3, upregulates p21[CIP1], and downregulates ERG (Qian et al. 2008). While its evaluation in hematological malignancies is more advanced, a clinical trial of various solid tumors including prostate cancer has begun.

A recent addition to the hydroxamate-based pan-HDAC inhibitor arsenal is CG200745. Its clinical evaluation is at an early phase. In the carcinoma cell lines LNCaP, DU145, and PC-3, it inhibits deacetylation of histone H3 and shows synergistic cytotoxicity with docetaxel. The combination decreased expression levels of antiapoptotic Bcl-2 family members and reduced tumor growth in DU145 xenograft models (Hwang et al. 2011).

Romidepsin, also known as FK228, is a cyclic tetrapeptide derived from *Chromobacterium violaceum*. Its mode of action is considered comparable to that of vorinostat, including increased Hsp90 acetylation via inhibition of HDAC6. In preclinical prostate cancer models, it demonstrates antitumor efficacy as a single agent and enhances the activity of cytotoxic agents like docetaxel and gemcitabine (Kanzaki et al. 2007; Zhang et al. 2007). In humans, romidepsin is approved for treatment of cutaneous T cell lymphoma. However, in prostate cancer patients, response to this drug was weak (5.7%), and the toxicity turned out to be relatively high, leading especially to nausea, vomiting, fatigue, and anorexia (Molife et al. 2010).

No member of the other HDAC inhibitor classes has been investigated in advanced clinical trials, but some will be discussed briefly. Valproic acid belongs to the short-chain fatty acids class of HDAC inhibitors. It is an anticonvulsant drug known to induce anomalies during pregnancy. Valproic acid inhibits class I and II HDACs and causes hyperacetylation of the N-terminal tails of histones H3 and H4 in vitro and in vivo (Duenas-Gonzalez et al. 2008). Its activity in prostate cancers is explained by several mechanisms including p21[CIP1] induction, downregulation of proliferating cell nuclear antigen (PCNA), and antiangiogenic effects (Shabbeer et al. 2007). Valproic acid acts synergistically with the mTOR inhibitor RAD001 and enhances radiosensitivity (Chen et al. 2011; Wedel et al. 2011a). However, there is some concern that valproic acid might induce neuroendocrine differentiation in prostate cancers which is associated with poor prognosis (Valentini et al. 2007; Sidana et al. 2011). No clinical trials of valproic acid in the treatment of prostate cancer are ongoing.

Entinostat (MS-275) is a benzamide-based inhibitor of class I HDAC1 and HDAC3 (Lemoine and Younes 2010). In prostate cancers, it enhances histone hyperacetylation, restores retinoid sensitivity, and sensitizes to radiation (Abbas and Gupta 2008). It is currently investigated in a small phase I trial in different solid tumors including prostate cancers.

OSU-HDAC42 is a novel phenylbutyrate HDAC inhibitor which causes histone H3 acetylation as well as acetylation of nonhistones like α-tubulin, indicative of

HDAC6 inhibition. It upregulates p21[CIP1] and modulates several regulators of cell survival such as Akt, Bcl-x$_L$, Bax, and survivin. In the transgenic adenocarcinoma of the mouse prostate (TRAMP) model, it decreased the severity of PIN lesions and blocked progression to invasive cancers (Sargeant et al. 2008). Therefore, it might be investigated for prevention of prostate cancer in men with PIN lesions.

In summary, several HDAC inhibitors have demonstrated strong antitumor activity in preclinical prostate cancer models. While some HDAC inhibitors such as vorinostat and romidepsin are already approved for treatment of hematological malignancies, HDAC inhibitors as single agents have shown no or at best modest effects in prostate cancer clinical trials. Unfortunately, poor efficacy went along with relative high toxicity. This is probably due to the low specificity of all inhibitors for individual HDACs. Unfortunately, the function and importance of individual HDACs for prostate cancer progression have not been resolved in sufficient detail. It is therefore difficult to decide which HDACs might constitute optimal targets for therapy. Promisingly, though, many HDAC inhibitors seem to be able to sensitize prostate cancers to radiation and to overcome resistance to chemotherapy. This makes them attractive for combination strategies with classical therapeutics. This application will however require a more favorable toxicity profile, which should be a must for future HDAC inhibitors.

6.4.3 Therapeutic Targeting of Histone Modifications by Histone Methyltransferase Inhibitors

Targeting alterations of histone methylation in cancer is a theoretically highly attractive concept. However, inhibitors of HMTs are still far away from application, even from clinical trials. As an example, 3-deazaneplanocin (DZNep) inhibits the histone H3K27 HMT activity of EZH2 and disrupts the PRC2 complex (Crea et al. 2011). However, DZNep is not a selective inhibitor of EZH2 and H3K27me3 methylation. Instead, it inhibits the turnover of the common methyltransferase reaction product S-adenosylhomocysteine, leading to a global inhibition of histone methylation and loss of both repressive and active histone marks (Miranda et al. 2009). Nevertheless, the first preclinical results with DZNep are encouraging. The compound reactivated several epigenetically silenced genes and inhibited growth of the prostate cancer cell lines LNCaP and DU145 (Miranda et al. 2009; Crea et al. 2011). Evidently, more selective drugs targeting histone methylases hold great promise for prostate cancer therapy.

6.4.4 Therapeutic Targeting of Histone Modifications by Histone Demethylase Inhibitors

Like inhibitors of HMTs, inhibitors of HDMs are still at early stages of development. HDMs can be classified into amine-oxidase-related enzymes that remove mono-/dimethylation and Jumonji C-terminal domain (JmjC)-containing enzymes that remove all methylation states (Hou and Yu 2010). Lysine-specific demethylase (LSD1, formerly named KDM1) and KDM2A are representatives of these two classes. Members of both classes are overexpressed in prostate cancer (Kahl et al. 2006; Xiang et al. 2007).

In particular, LSD1 interacts with the AR and stimulates AR-dependent transcription by removing repressive histone marks. As a coactivator of the AR, it demethylates especially H3K9 supporting activation of AR-dependent genes. The LSD1 inhibitor paragyline blocks demethylation of mono- and dimethyl H3K9, but not demethylation of other histone methylation states. In LNCaP cells, it reduced PSA transcription induced by androgen (Metzger et al. 2005). Since LSD1 overexpression has been reported to occur mainly in CRPC, inhibitors may find application in the treatment of advanced stage prostate cancers.

6.5 Conclusions and Perspectives

Prostate cancer is the most common malignancy among men in Western industrialized countries. Several epigenetic alterations have now conclusively been linked to initiation and progression of this disease, although the function of many others remains to be elucidated. Well-established alterations include genome-wide hypomethylation, local promoter hypermethylation, and alterations of histone modifications, which in a few instances can be traced to dysregulated enzymes involved in epigenetic modification of DNA and histones. Epigenetic alterations are suitable for diagnostic purposes as well as prognosis and, in a next step, will provide targets for therapeutic approaches. Because of its stability and the availability of robust and sensitive assays, DNA methylation may be particularly well suited for the development of biomarkers.

Similar properties as a biomarker are ascribed to miRNAs. Research on miRNAs in prostate cancer has begun quite recently, but is rapidly expanding. It is therefore impossible to give a timely review here. In brief, more than 50 miRNAs are presently considered to be abnormally expressed in prostate cancers as compared to benign prostate tissues, but only few of them have been experimentally evaluated for their functional role (Pang et al. 2010). Many efforts have been made to identify a miRNA signature for prostate cancer, but the data are still inconsistent due to widely differing techniques and lack of standardization (Coppola et al. 2010). Published studies report accuracies for prostate cancer detection of single miRNAs in a range between 50 and 80%, but the combination of already two miRNAs (miR-205 and miR-183) increased correct classification to 84% (Schaefer et al. 2010). MiRNAs originating from prostate cancers were also present in plasma as well as in serum in a stable form protected from endogenous RNase activity, independent of circulating tumor cells (Mitchell et al. 2008). Indeed, Moltzahn et al. (2011) propose a new diagnostic miRNA signature in sera from prostate cancer patients that included different miRNAs from those previously identified in tissues. Similarly, several specific miRNA signatures have been correlated with pathological features or patient outcome, such as tumor stage or Gleason score (Schaefer et al. 2010), lymph node involvement (Martens-Uzunova et al. 2011), biochemical recurrence (Leite et al. 2011), and clinical recurrence (Spahn et al. 2010). However, by far, not all findings from these and other studies are concordant, so verification and prospective evaluation of the suggested miRNA biomarkers will be necessary.

The increasing knowledge of miRNA function in carcinogenesis might moreover open therapeutical strategies by means of miRNA replacement or drugs targeting miRNAs. In prostate cancer, several groups have demonstrated that miRNA replacement by water-soluble collagen molecules (atelocollagen) or viruses (herpes simplex virus-1, adenoviruses) is feasible. Promising results indicate antitumor effects in terms of tumor reduction and attenuated growth of metastases in animal models (Wedel et al. 2011b).

In spite of such progress, a number of obstacles need to be overcome. In the field of diagnostic and prognostic markers, standardized assays are a critical requirement. Clearly, there will be no acceptance for broad application of epigenetic markers until preclinical data are confirmed in prospective multicenter trials. The main difficulties in exploiting epigenetic alterations for therapeutic purposes appear to relate to the nonspecific nature of the current available compounds, which may underlie the unsatisfactory outcome of weak effects and simultaneous high toxicity observed with these drugs. More research into the mechanisms causing epigenetic alterations is warranted.

Hopefully, many of these obstacles will be resolved in the near future by emerging whole genome-wide analyses of epigenetic alterations which will provide a basis for detailed insights into methylation changes in individual cancers, their relations to chromosomal and other genetic alterations, and their functional importance.

References

Abbas, A. & Gupta, S. (2008). The role of histone deacetylases in prostate cancer. *Epigenetics* 3, 300–309.

Aljada, A., Dong, L. & Mousa, S. A. (2010). Sirtuin-targeting drugs: Mechanisms of action and potential therapeutic applications. *Curr Opin Investig Drugs* 11, 1158–1168.

Atadja, P. (2009). Development of the pan-DAC inhibitor panobinostat (LBH589): successes and challenges. *Cancer Lett* 280, 233–241.

Baden, J., Green, G., Painter, J., Curtin, K., Markiewicz, J., Jones, J., Astacio, T., Canning, S., Quijano, J., Guinto, W., Leibovich, B. C., Nelson, J. B., Vargo, J., Wang, Y. & Wuxiong, C. (2009). Multicenter evaluation of an investigational prostate cancer methylation assay. *J Urol* 182, 1186–1193.

Bakin, R. E. & Jung, M. O. (2004). Cytoplasmic sequestration of HDAC7 from mitochondrial and nuclear compartments upon initiation of apoptosis. *J Biol Chem* 279, 51218–51225.

Balcerczyk, A. & Pirola, L. (2010). Therapeutic potential of activators and inhibitors of sirtuins. *Biofactors* 36, 383–393.

Banez, L. L., Sun, L., van Leenders, G. J., Wheeler, T. M., Bangma, C. H., Freedland, S. J., Ittmann, M. M., Lark, A. L., Madden, J. F., Hartman, A., Weiss, G. & Castanos-Velez, E. (2010). Multicenter clinical validation of PITX2 methylation as a prostate specific antigen recurrence predictor in patients with post-radical prostatectomy prostate cancer. *J Urol* 184, 149–156.

Bastian, P. J., Ellinger, J., Heukamp, L. C., Kahl, P., Muller, S. C. & von Rucker, A. (2007a). Prognostic value of CpG island hypermethylation at PTGS2, RAR-beta, EDNRB, and other gene loci in patients undergoing radical prostatectomy. *Eur Urol* 51, 665–674.

Bastian, P. J., Ellinger, J., Wellmann, A., Wernert, N., Heukamp, L. C., Muller, S. C. & von Ruecker, A. (2005). Diagnostic and prognostic information in prostate cancer with the help of a small set of hypermethylated gene loci. *Clin Cancer Res* 11, 4097–4106.

Bastian, P. J., Palapattu, G. S., Yegnasubramanian, S., Lin, X., Rogers, C. G., Mangold, L. A., Trock, B., Eisenberger, M., Partin, A. W. & Nelson, W. G. (2007b). Prognostic value of preoperative serum cell-free circulating DNA in men with prostate cancer undergoing radical prostatectomy. *Clin Cancer Res* 13, 5361–5367.

Bender, C. M., Pao, M. M. & Jones, P. A. (1998). Inhibition of DNA methylation by 5-aza-2′-deoxycytidine suppresses the growth of human tumor cell lines. *Cancer Res* 58, 95–101.

Bertrand, P. (2010). Inside HDAC with HDAC inhibitors. *Eur J Med Chem* 45, 2095–2116.

Bettuzzi, S., Brausi, M., Rizzi, F., Castagnetti, G., Peracchia, G. & Corti, A. (2006). Chemoprevention of human prostate cancer by oral administration of green tea catechins in volunteers with high-grade prostate intraepithelial neoplasia: a preliminary report from a one-year proof-of-principle study. *Cancer Res* 66, 1234–1240.

Bianco-Miotto, T., Chiam, K., Buchanan, G., Jindal, S., Day, T. K., Thomas, M., Pickering, M. A., O'Loughlin, M. A., Ryan, N. K., Raymond, W. A., Horvath, L. G., Kench, J. G., Stricker, P. D., Marshall, V. R., Sutherland, R. L., Henshall, S. M., Gerald, W. L., Scher, H. I., Risbridger, G. P., Clements, J. A., Butler, L. M., Tilley, W. D., Horsfall, D. J. & Ricciardelli, C. (2010). Global levels of specific histone modifications and an epigenetic gene signature predict prostate cancer progression and development. *Cancer Epidemiol Biomarkers Prev* 19, 2611–2622.

Bill-Axelson, A., Holmberg, L., Ruutu, M., Garmo, H., Stark, J. R., Busch, C., Nordling, S., Haggman, M., Andersson, S. O., Bratell, S., Spangberg, A., Palmgren, J., Steineck, G., Adami, H. O. & Johansson, J. E. (2011). Radical prostatectomy versus watchful waiting in early prostate cancer. *N Engl J Med* 364, 1708–1717.

Boorjian, S. A., Thompson, R. H., Tollefson, M. K., Rangel, L. J., Bergstralh, E. J., Blute, M. L. & Karnes, R. J. (2011). Long-term risk of clinical progression after biochemical recurrence following radical prostatectomy: the impact of time from surgery to recurrence. *Eur Urol* 59, 893–899.

Bradley, D., Rathkopf, D., Dunn, R., Stadler, W. M., Liu, G., Smith, D. C., Pili, R., Zwiebel, J., Scher, H. & Hussain, M. (2009). Vorinostat in advanced prostate cancer patients progressing on prior chemotherapy (National Cancer Institute Trial 6862): trial results and interleukin-6 analysis: a study by the Department of Defense Prostate Cancer Clinical Trial Consortium and University of Chicago Phase 2 Consortium. *Cancer* 115, 5541–5549.

Braun, J., Boittiaux, I., Tilborg, A., Lambert, D. & Wouters, J. (2010). The dicyclo-hexyl-amine salt of RG108 (N-phthalyl-l-tryptophan), a potential epigenetic modulator. *Acta Crystallogr Sect E Struct Rep Online* 66, o3175–o3176.

Brooks, J. D., Weinstein, M., Lin, X., Sun, Y., Pin, S. S., Bova, G. S., Epstein, J. I., Isaacs, W. B. & Nelson, W. G. (1998). CG island methylation changes near the GSTP1 gene in prostatic intraepithelial neoplasia. *Cancer Epidemiol Biomarkers Prev* 7, 531–536.

Brothman, A. R., Swanson, G., Maxwell, T. M., Cui, J., Murphy, K. J., Herrick, J., Speights, V. O., Isaac, J. & Rohr, L. R. (2005). Global hypomethylation is common in prostate cancer cells: a quantitative predictor for clinical outcome? *Cancer Genet Cytogenet* 156, 31–36.

Brueckner, B., Garcia, B. R., Siedlecki, P., Musch, T., Kliem, H. C., Zielenkiewicz, P., Suhai, S., Wiessler, M. & Lyko, F. (2005). Epigenetic reactivation of tumor suppressor genes by a novel small-molecule inhibitor of human DNA methyltransferases. *Cancer Res* 65, 6305–6311.

Butler, L. M., Agus, D. B., Scher, H. I., Higgins, B., Rose, A., Cordon-Cardo, C., Thaler, H. T., Rifkind, R. A., Marks, P. A. & Richon, V. M. (2000). Suberoylanilide hydroxamic acid, an inhibitor of histone deacetylase, suppresses the growth of prostate cancer cells in vitro and in vivo. *Cancer Res* 60, 5165–5170.

6 The Impact of Epigenetic Alterations on Diagnosis, Prediction…   147

Cairns, P., Esteller, M., Herman, J. G., Schoenberg, M., Jeronimo, C., Sanchez-Cespedes, M., Chow, N. H., Grasso, M., Wu, L., Westra, W. B. & Sidransky, D. (2001). Molecular detection of prostate cancer in urine by GSTP1 hypermethylation. *Clin Cancer Res* 7, 2727–2730.

Carvalho, J. R., Filipe, L., Costa, V. L., Ribeiro, F. R., Martins, A. T., Teixeira, M. R., Jeronimo, C. & Henrique, R. (2010). Detailed analysis of expression and promoter methylation status of apoptosis-related genes in prostate cancer. *Apoptosis* 15, 956–965.

Carver, B. S., Chapinski, C., Wongvipat, J., Hieronymus, H., Chen, Y., Chandarlapaty, S., Arora, V. K., Le, C., Koutcher, J., Scher, H., Scardino, P. T., Rosen, N. & Sawyers, C. L. (2011). Reciprocal feedback regulation of PI3K and androgen receptor signaling in PTEN-deficient prostate cancer. *Cancer Cell* 19, 575–586.

Chen, X., Wong, J. Y., Wong, P. & Radany, E. H. (2011). Low-dose valproic acid enhances radiosensitivity of prostate cancer through acetylated p53-dependent modulation of mitochondrial membrane potential and apoptosis. *Mol Cancer Res* 9, 448–461.

Cheng, J. C., Matsen, C. B., Gonzales, F. A., Ye, W., Greer, S., Marquez, V. E., Jones, P. A. & Selker, E. U. (2003). Inhibition of DNA methylation and reactivation of silenced genes by zebularine. *J Natl Cancer Inst* 95, 399–409.

Chinnaiyan, P., Vallabhaneni, G., Armstrong, E., Huang, S. M. & Harari, P. M. (2005). Modulation of radiation response by histone deacetylase inhibition. *Int J Radiat Oncol Biol Phys* 62, 223–229.

Cho, N. Y., Kim, B. H., Choi, M., Yoo, E. J., Moon, K. C., Cho, Y. M., Kim, D. & Kang, G. H. (2007). Hypermethylation of CpG island loci and hypomethylation of LINE-1 and Alu repeats in prostate adenocarcinoma and their relationship to clinicopathological features. *J Pathol* 211, 269–277.

Chuang, J. C., Yoo, C. B., Kwan, J. M., Li, T. W., Liang, G., Yang, A. S. & Jones, P. A. (2005). Comparison of biological effects of non-nucleoside DNA methylation inhibitors versus 5-aza-2′-deoxycytidine. *Mol Cancer Ther* 4, 1515–1520.

Cooper, C. S. & Foster, C. S. (2009). Concepts of epigenetics in prostate cancer development. *Br J Cancer* 100, 240–245.

Cooperberg, M. R., Broering, J. M., Kantoff, P. W. & Carroll, P. R. (2007). Contemporary trends in low risk prostate cancer: risk assessment and treatment. *J Urol* 178, S14–S19.

Coppola, V., De Maria, R. & Bonci, D. (2010). MicroRNAs and prostate cancer. *Endocr Relat Cancer* 17, F1–F17.

Costa, V. L., Henrique, R. & Jeronimo, C. (2007). Epigenetic markers for molecular detection of prostate cancer. *Dis Markers* 23, 31–41.

Cottrell, S., Jung, K., Kristiansen, G., Eltze, E., Semjonow, A., Ittmann, M., Hartmann, A., Stamey, T., Haefliger, C. & Weiss, G. (2007). Discovery and validation of 3 novel DNA methylation markers of prostate cancer prognosis. *J Urol* 177, 1753–1758.

Crea, F., Hurt, E. M., Mathews, L. A., Cabarcas, S. M., Sun, L., Marquez, V. E., Danesi, R. & Farrar, W. L. (2011). Pharmacologic disruption of Polycomb Repressive Complex 2 inhibits tumorigenicity and tumor progression in prostate cancer. *Mol Cancer* 10:40.

de Bono, J. S., Logothetis, C. J., Molina, A., Fizazi, K., North, S., Chu, L., Chi, K. N., Jones, R. J., Goodman, O. B., Jr., Saad, F., Staffurth, J. N., Mainwaring, P., Harland, S., Flaig, T. W., Hutson, T. E., Cheng, T., Patterson, H., Hainsworth, J. D., Ryan, C. J., Sternberg, C. N., Ellard, S. L., Flechon, A., Saleh, M., Scholz, M., Efstathiou, E., Zivi, A., Bianchini, D., Loriot, Y., Chieffo, N., Kheoh, T., Haqq, C. M. & Scher, H. I. (2011). Abiraterone and increased survival in metastatic prostate cancer. *N Engl J Med* 364, 1995–2005.

de Bono, J. S., Oudard, S., Ozguroglu, M., Hansen, S., Machiels, J. P., Kocak, I., Gravis, G., Bodrogi, I., Mackenzie, M. J., Shen, L., Roessner, M., Gupta, S. & Sartor, A. O. (2010). Prednisone plus cabazitaxel or mitoxantrone for metastatic castration-resistant prostate cancer progressing after docetaxel treatment: a randomised open-label trial. *Lancet* 376, 1147–1154.

Dokmanovic, M., Perez, G., Xu, W., Ngo, L., Clarke, C., Parmigiani, R. B. & Marks, P. A. (2007). Histone deacetylase inhibitors selectively suppress expression of HDAC7. *Mol Cancer Ther* 6, 2525–2534.

Duenas-Gonzalez, A., Candelaria, M., Perez-Plascencia, C., Perez-Cardenas, E., Cruz-Hernandez, E. & Herrera, L. A. (2008). Valproic acid as epigenetic cancer drug: preclinical, clinical and transcriptional effects on solid tumors. *Cancer Treat Rev* 34, 206–222.

Eads, C. A., Danenberg, K. D., Kawakami, K., Saltz, L. B., Blake, C., Shibata, D., Danenberg, P. V. & Laird, P. W. (2000). MethyLight: a high-throughput assay to measure DNA methylation. *Nucleic Acids Res* 28:E32.

Ellinger, J., Bastian, P. J., Jurgan, T., Biermann, K., Kahl, P., Heukamp, L. C., Wernert, N., Muller, S. C. & von Ruecker, A. (2008). CpG island hypermethylation at multiple gene sites in diagnosis and prognosis of prostate cancer. *Urology* 71, 161–167.

Ellinger, J., Kahl, P., von der, G. J., Rogenhofer, S., Heukamp, L. C., Gutgemann, I., Walter, B., Hofstadter, F., Buttner, R., Muller, S. C., Bastian, P. J. & von Ruecker, A. (2010). Global levels of histone modifications predict prostate cancer recurrence. *Prostate* 70, 61–69.

Enokida, H., Shiina, H., Igawa, M., Ogishima, T., Kawakami, T., Bassett, W. W., Anast, J. W., Li, L. C., Urakami, S., Terashima, M., Verma, M., Kawahara, M., Nakagawa, M., Kane, C. J., Carroll, P. R. & Dahiya, R. (2004). CpG hypermethylation of MDR1 gene contributes to the pathogenesis and progression of human prostate cancer. *Cancer Res* 64, 5956–5962.

Enokida, H., Shiina, H., Urakami, S., Igawa, M., Ogishima, T., Li, L. C., Kawahara, M., Nakagawa, M., Kane, C. J., Carroll, P. R. & Dahiya, R. (2005). Multigene methylation analysis for detection and staging of prostate cancer. *Clin Cancer Res* 11, 6582–6588.

Enokida, H., Shiina, H., Urakami, S., Terashima, M., Ogishima, T., Li, L. C., Kawahara, M., Nakagawa, M., Kane, C. J., Carroll, P. R., Igawa, M. & Dahiya, R. (2006). Smoking influences aberrant CpG hypermethylation of multiple genes in human prostate carcinoma. *Cancer* 106, 79–86.

Epstein, J. I. (2009). Precursor lesions to prostatic adenocarcinoma. *Virchows Arch* 454, 1–16.

Esteller, M., Corn, P. G., Urena, J. M., Gabrielson, E., Baylin, S. B. & Herman, J. G. (1998). Inactivation of glutathione S-transferase P1 gene by promoter hypermethylation in human neoplasia. *Cancer Res* 58, 4515–4518.

Fang, M. Z., Wang, Y., Ai, N., Hou, Z., Sun, Y., Lu, H., Welsh, W. & Yang, C. S. (2003). Tea polyphenol (-)-epigallocatechin-3-gallate inhibits DNA methyltransferase and reactivates methylation-silenced genes in cancer cell lines. *Cancer Res* 63, 7563–7570.

Fang, Y., Fliss, A. E., Robins, D. M. & Caplan, A. J. (1996). Hsp90 regulates androgen receptor hormone binding affinity in vivo. *J Biol Chem* 271, 28697–28702.

Fenaux, P., Mufti, G. J., Hellstrom-Lindberg, E., Santini, V., Finelli, C., Giagounidis, A., Schoch, R., Gattermann, N., Sanz, G., List, A., Gore, S. D., Seymour, J. F., Bennett, J. M., Byrd, J., Backstrom, J., Zimmerman, L., McKenzie, D., Beach, C. & Silverman, L. R. (2009). Efficacy of azacitidine compared with that of conventional care regimens in the treatment of higher-risk myelodysplastic syndromes: a randomised, open-label, phase III study. *Lancet Oncol* 10, 223–232.

Ferrari, A. C., Stein, M. N., Alumkal, J. J., Gomez-Pinillos, A., Catamero, D. D., Mayer, T. M., Collins, F., Beer, T. M. & DiPaola, R. S. (2011). A phase I/II randomized study of panobinostat and bicalutamide in castration-resistant prostate cancer (CRPC) patients progressing on second-line hormone therapy. *J Clin Oncol* 29 suppl 7, abstract 156.

Florl, A. R., Steinhoff, C., Muller, M., Seifert, H. H., Hader, C., Engers, R., Ackermann, R. & Schulz, W. A. (2004). Coordinate hypermethylation at specific genes in prostate carcinoma precedes LINE-1 hypomethylation. *Br J Cancer* 91, 985–994.

Goering, W., Kloth, M. & Schulz, W. A. (2011a). DNA Methylation Changes in Prostate Cancer. In *Cancer Epigenetics*. Edited by R. Dumitrescu & V. Mukesh. New York: Humana Press.

Goering, W., Ribarska, T. & Schulz, W. A. (2011b). Selective changes of retroelement expression in human prostate cancer. *Carcinogenesis* 32, 1484–1492.

Goessl, C., Krause, H., Muller, M., Heicappell, R., Schrader, M., Sachsinger, J. & Miller, K. (2000). Fluorescent methylation-specific polymerase chain reaction for DNA-based detection of prostate cancer in bodily fluids. *Cancer Res* 60, 5941–5945.

Goessl, C., Muller, M., Heicappell, R., Krause, H., Straub, B., Schrader, M. & Miller, K. (2001). DNA-based detection of prostate cancer in urine after prostatic massage. *Urology* 58, 335–338.

Goll, M. G. & Bestor, T. H. (2005). Eukaryotic cytosine methyltransferases. *Annu Rev Biochem* 74, 481–514.

Gonzalgo, M. L., Pavlovich, C. P., Lee, S. M. & Nelson, W. G. (2003). Prostate cancer detection by GSTP1 methylation analysis of postbiopsy urine specimens. *Clin Cancer Res* 9, 2673–2677.

Harbeck, N., Schmitt, M., Paepke, S., Allgayer, H., Kates, R. E. (2007). Tumor-associated proteolytic factors uPA and PAI-1: critical appraisal of their clinical relevance in breast cancer and their integration into decision-support algorithms. *Crit Rev Clin Lab* 44, 179–201.

Harden, S. V., Guo, Z., Epstein, J. I. & Sidransky, D. (2003a). Quantitative GSTP1 methylation clearly distinguishes benign prostatic tissue and limited prostate adenocarcinoma. *J Urol* 169, 1138–1142.

Harden, S. V., Sanderson, H., Goodman, S. N., Partin, A. A., Walsh, P. C., Epstein, J. I. & Sidransky, D. (2003b). Quantitative GSTP1 methylation and the detection of prostate adenocarcinoma in sextant biopsies. *J Natl Cancer Inst* 95, 1634–1637.

Hartmann, O., Spyratos, F., Harbeck, N., Dietrich, D., Fassbender, A., Schmitt, M., Eppenberger-Castori, S., Vuaroqueaux, V., Lerebours, F., Welzel, K., Maier, S., Plum, A., Niemann, S., Foekens, J. A., Lesche, R. & Martens, J. W. (2009). DNA methylation markers predict outcome in node-positive, estrogen receptor-positive breast cancer with adjuvant anthracycline-based chemotherapy. *Clin Cancer Res* 15, 315–323.

Hayes, J. D. & Pulford, D. J. (1995). The glutathione S-transferase supergene family: regulation of GST and the contribution of the isoenzymes to cancer chemoprotection and drug resistance. *Crit Rev Biochem Mol Biol* 30, 445–600.

Henrique, R., Costa, V. L., Cerveira, N., Carvalho, A. L., Hoque, M. O., Ribeiro, F. R., Oliveira, J., Teixeira, M. R., Sidransky, D. & Jeronimo, C. (2006). Hypermethylation of Cyclin D2 is associated with loss of mRNA expression and tumor development in prostate cancer. *J Mol Med (Berl)* 84, 911–918.

Henrique, R., Ribeiro, F. R., Fonseca, D., Hoque, M. O., Carvalho, A. L., Costa, V. L., Pinto, M., Oliveira, J., Teixeira, M. R., Sidransky, D. & Jeronimo, C. (2007). High promoter methylation levels of APC predict poor prognosis in sextant biopsies from prostate cancer patients. *Clin Cancer Res* 13, 6122–6129.

Hermans, K. G., van Marion, R., van Dekken, H., Jenster, G., van Weerden, W. M. & Trapman, J. (2006). TMPRSS2:ERG fusion by translocation or interstitial deletion is highly relevant in androgen-dependent prostate cancer, but is bypassed in late-stage androgen receptor-negative prostate cancer. *Cancer Res* 66, 10658–10663.

Hodgson, M. C., Shao, L. J., Frolov, A., Li, R., Peterson, L. E., Ayala, G., Ittmann, M. M., Weigel, N. L. & Agoulnik, I. U. (2011). Decreased expression and androgen regulation of the tumor suppressor gene INPP4B in prostate cancer. *Cancer Res* 71, 572–582.

Hoffmann, M. J., Engers, R., Florl, A. R., Otte, A. P., Muller, M. & Schulz, W. A. (2007). Expression changes in EZH2, but not in BMI-1, SIRT1, DNMT1 or DNMT3B are associated with DNA methylation changes in prostate cancer. *Cancer Biol Ther* 6, 1403–1412.

Hoffmann, M. J. & Schulz, W. A. (2005). Causes and consequences of DNA hypomethylation in human cancer. *Biochem Cell Biol* 83, 296–321.

Hoque, M. O. (2009). DNA methylation changes in prostate cancer: current developments and future clinical implementation. *Expert Rev Mol Diagn* 9, 243–257.

Hoque, M. O., Topaloglu, O., Begum, S., Henrique, R., Rosenbaum, E., Van Criekinge, W., Westra, W. H. & Sidransky, D. (2005). Quantitative methylation-specific polymerase chain reaction gene patterns in urine sediment distinguish prostate cancer patients from control subjects. *J Clin Oncol* 23, 6569–6575.

Horio, Y., Hayashi, T., Kuno, A. & Kunimoto, R. (2011). Cellular and molecular effects of sirtuins in health and disease. *Clin Sci (Lond)* 121, 191–203.

Hou, H. & Yu, H. (2010). Structural insights into histone lysine demethylation. *Curr Opin Struct Biol* 20, 739–748.

Huffman, D. M., Grizzle, W. E., Bamman, M. M., Kim, J. S., Eltoum, I. A., Elgavish, A. & Nagy, T. R. (2007). SIRT1 is significantly elevated in mouse and human prostate cancer. *Cancer Res* 67, 6612–6618.

Hugosson, J., Carlsson, S., Aus, G., Bergdahl, S., Khatami, A., Lodding, P., Pihl, C. G., Stranne, J., Holmberg, E. & Lilja, H. (2010). Mortality results from the Goteborg randomised population-based prostate-cancer screening trial. *Lancet Oncol* 11, 725–732.

Hwang, J. J., Kim, Y. S., Kim, T., Kim, M. J., Jeong, I. G., Lee, J. H., Choi, J., Jang, S., Ro, S. & Kim, C. S. (2011). A novel histone deacetylase inhibitor, CG200745, potentiates anticancer effect of docetaxel in prostate cancer via decreasing Mcl-1 and Bcl-(XL). *Invest New Drugs.* 2011 Jul 20. [Epub ahead of print].

Jemal, A., Bray, F., Center, M. M., Ferlay, J., Ward, E. & Forman, D. (2011). Global cancer statistics. *CA Cancer J Clin* 61, 69–90.

Jeronimo, C., Bastian, P. J., Bjartell, A., Carbone, G. M., Catto, J. W., Clark, S. J., Henrique, R., Nelson, W. G. & Shariat, S. F. (2011). Epigenetics in prostate cancer: biologic and clinical relevance. *Eur Urol* 60, 753–766.

Jeronimo, C., Henrique, R., Hoque, M. O., Mambo, E., Ribeiro, F. R., Varzim, G., Oliveira, J., Teixeira, M. R., Lopes, C. & Sidransky, D. (2004). A quantitative promoter methylation profile of prostate cancer. *Clin Cancer Res* 10, 8472–8478.

Jeronimo, C., Usadel, H., Henrique, R., Oliveira, J., Lopes, C., Nelson, W. G. & Sidransky, D. (2001). Quantitation of GSTP1 methylation in non-neoplastic prostatic tissue and organ-confined prostate adenocarcinoma. *J Natl Cancer Inst* 93, 1747–1752.

Jeronimo, C., Usadel, H., Henrique, R., Silva, C., Oliveira, J., Lopes, C. & Sidransky, D. (2002a). Quantitative GSTP1 hypermethylation in bodily fluids of patients with prostate cancer. *Urology* 60, 1131–1135.

Jeronimo, C., Varzim, G., Henrique, R., Oliveira, J., Bento, M. J., Silva, C., Lopes, C. & Sidransky, D. (2002b). I105V polymorphism and promoter methylation of the GSTP1 gene in prostate adenocarcinoma. *Cancer Epidemiol Biomarkers Prev* 11, 445–450.

Jiao, J., Wang, S., Qiao, R., Vivanco, I., Watson, P. A., Sawyers, C. L. & Wu, H. (2007). Murine cell lines derived from Pten null prostate cancer show the critical role of PTEN in hormone refractory prostate cancer development. *Cancer Res* 67, 6083–6091.

Johnson, B. A. (2008). Update on neuropharmacological treatments for alcoholism: scientific basis and clinical findings. *Biochem Pharmacol* 75, 34–56.

Kahl, P., Gullotti, L., Heukamp, L. C., Wolf, S., Friedrichs, N., Vorreuther, R., Solleder, G., Bastian, P. J., Ellinger, J., Metzger, E., Schule, R. & Buettner, R. (2006). Androgen receptor coactivators lysine-specific histone demethylase 1 and four and a half LIM domain protein 2 predict risk of prostate cancer recurrence. *Cancer Res* 66, 11341–11347.

Kang, G. H., Lee, S., Lee, H. J. & Hwang, K. S. (2004). Aberrant CpG island hypermethylation of multiple genes in prostate cancer and prostatic intraepithelial neoplasia. *J Pathol* 202, 233–240.

Kanzaki, M., Kakinuma, H., Kumazawa, T., Inoue, T., Saito, M., Narita, S., Yuasa, T., Tsuchiya, N. & Habuchi, T. (2007). Low concentrations of the histone deacetylase inhibitor, depsipeptide, enhance the effects of gemcitabine and docetaxel in hormone refractory prostate cancer cells. *Oncol Rep* 17, 761–767.

Kawamoto, K., Okino, S. T., Place, R. F., Urakami, S., Hirata, H., Kikuno, N., Kawakami, T., Tanaka, Y., Pookot, D., Chen, Z., Majid, S., Enokida, H., Nakagawa, M. & Dahiya, R. (2007). Epigenetic modifications of RASSF1A gene through chromatin remodeling in prostate cancer. *Clin Cancer Res* 13, 2541–2548.

Kim, J. H., Dhanasekaran, S. M., Prensner, J. R., Cao, X., Robinson, D., Kalyana-Sundaram, S., Huang, C., Shankar, S., Jing, X., Iyer, M., Hu, M., Sam, L., Grasso, C., Maher, C. A., Palanisamy, N., Mehra, R., Kominsky, H. D., Siddiqui, J., Yu, J., Qin, Z. S. & Chinnaiyan, A. M. (2011). Deep sequencing reveals distinct patterns of DNA methylation in prostate cancer. *Genome Res* 21, 1028–1041.

Kloth, M., Goering, W., Arsov, C. & Schulz, W. A. (2012). Intensification of DNA methylation at 3q25.32 during progression of prostate cancer. Submitted for publication.

Klotz, L., Zhang, L., Lam, A., Nam, R., Mamedov, A. & Loblaw, A. (2010). Clinical results of long-term follow-up of a large, active surveillance cohort with localized prostate cancer. *J Clin Oncol* 28, 126–131.

Kobayashi, Y., Absher, D. M., Gulzar, Z. G., Young, S. R., McKenney, J. K., Peehl, D. M., Brooks, J. D., Myers, R. M. & Sherlock, G. (2011). DNA methylation profiling reveals novel biomarkers and important roles for DNA methyltransferases in prostate cancer. *Genome Res* 21, 1017–1027.

Kollermann, J., Muller, M., Goessl, C., Krause, H., Helpap, B., Pantel, K. & Miller, K. (2003). Methylation-specific PCR for DNA-based detection of occult tumor cells in lymph nodes of prostate cancer patients. *Eur Urol* 44, 533–538.

Konishi, N., Nakamura, M., Kishi, M., Nishimine, M., Ishida, E. & Shimada, K. (2002). DNA hypermethylation status of multiple genes in prostate adenocarcinomas. *Jpn J Cancer Res* 93, 767–773.

Kron, K. J., Liu, L., Pethe, V. V., Demetrashvili, N., Nesbitt, M. E., Trachtenberg, J., Ozcelik, H., Fleshner, N. E., Briollais, L., van der Kwast, T. H. & Bapat, B. (2010). DNA methylation of HOXD3 as a marker of prostate cancer progression. *Lab Invest* 90, 1060–1067.

Kuzmichev, A., Margueron, R., Vaquero, A., Preissner, T. S., Scher, M., Kirmizis, A., Ouyang, X., Brockdorff, N., Abate-Shen, C., Farnham, P. & Reinberg, D. (2005). Composition and histone substrates of polycomb repressive group complexes change during cellular differentiation. *Proc Natl Acad Sci U S A* 102, 1859–1864.

Lee, W. H., Isaacs, W. B., Bova, G. S. & Nelson, W. G. (1997). CG island methylation changes near the GSTP1 gene in prostatic carcinoma cells detected using the polymerase chain reaction: a new prostate cancer biomarker. *Cancer Epidemiol Biomarkers Prev* 6, 443–450.

Lee, W. H., Morton, R. A., Epstein, J. I., Brooks, J. D., Campbell, P. A., Bova, G. S., Hsieh, W. S., Isaacs, W. B. & Nelson, W. G. (1994). Cytidine methylation of regulatory sequences near the pi-class glutathione S-transferase gene accompanies human prostatic carcinogenesis. *Proc Natl Acad Sci U S A* 91, 11733–11737.

Leite, K. R., Tomiyama, A., Reis, S. T., Sousa-Canavez, J. M., Sanudo, A., Dall'Oglio, M. F., Camara-Lopes, L. H. & Srougi, M. (2011). MicroRNA-100 expression is independently related to biochemical recurrence of prostate cancer. *J Urol* 185, 1118–1122.

Lemoine, M. & Younes, A. (2010). Histone deacetylase inhibitors in the treatment of lymphoma. *Discov Med* 10, 462–470.

Li, L. C. (2007). Epigenetics of prostate cancer. *Front Biosci* 12, 3377–3397.

Lin, B., Ferguson, C., White, J. T., Wang, S., Vessella, R., True, L. D., Hood, L. & Nelson, P. S. (1999). Prostate-localized and androgen-regulated expression of the membrane-bound serine protease TMPRSS2. *Cancer Res* 59, 4180–4184.

Lin, J., Haffner, M. C., Zhang, Y., Lee, B. H., Brennen, W. N., Britton, J., Kachhap, S. K., Shim, J. S., Liu, J. O., Nelson, W. G., Yegnasubramanian, S. & Carducci, M. A. (2011). Disulfiram is a DNA demethylating agent and inhibits prostate cancer cell growth. *Prostate* 71, 333–343.

Lin, X., Asgari, K., Putzi, M. J., Gage, W. R., Yu, X., Cornblatt, B. S., Kumar, A., Piantadosi, S., DeWeese, T. L., De Marzo, A. M. & Nelson, W. G. (2001a). Reversal of GSTP1 CpG island hypermethylation and reactivation of pi-class glutathione S-transferase (GSTP1) expression in human prostate cancer cells by treatment with procainamide. *Cancer Res* 61, 8611–8616.

Lin, X., Tascilar, M., Lee, W. H., Vles, W. J., Lee, B. H., Veeraswamy, R., Asgari, K., Freije, D., van Rees, B., Gage, W. R., Bova, G. S., Isaacs, W. B., Brooks, J. D., DeWeese, T. L., De Marzo, A. M. & Nelson, W. G. (2001b). GSTP1 CpG island hypermethylation is responsible for the absence of GSTP1 expression in human prostate cancer cells. *Am J Pathol* 159, 1815–1826.

Liu, J. W., Nagpal, J. K., Jeronimo, C., Lee, J. E., Henrique, R., Kim, M. S., Ostrow, K. L., Yamashita, K., van Criekinge, V., Wu, G., Moon, C. S., Trink, B. & Sidransky, D. (2008). Hypermethylation of MCAM gene is associated with advanced tumor stage in prostate cancer. *Prostate* 68, 418–426.

Liu, L., Kron, K. J., Pethe, V. V., Demetrashvili, N., Nesbitt, M. E., Trachtenberg, J., Ozcelik, H., Fleshner, N. E., Briollais, L., van der Kwast, T. H. & Bapat, B. (2011). Association of tissue promoter methylation levels of APC, TGFbeta2, HOXD3 and RASSF1A with prostate cancer progression. *Int J Cancer* 129, 2454–2462.

Liu, L., Yoon, J. H., Dammann, R. & Pfeifer, G. P. (2002). Frequent hypermethylation of the RASSF1A gene in prostate cancer. *Oncogene* 21, 6835–6840.

Lubbert, M., Suciu, S., Baila, L., Ruter, B. H., Platzbecker, U., Giagounidis, A., Selleslag, D., Labar, B., Germing, U., Salih, H. R., Beeldens, F., Muus, P., Pfluger, K. H., Coens, C., Hagemeijer, A., Eckart, S. H., Ganser, A., Aul, C., de Witte, T. & Wijermans, P. W. (2011). Low-dose decitabine versus best supportive care in elderly patients with intermediate- or high-risk myelodysplastic syndrome (MDS) ineligible for intensive chemotherapy: final results of the randomized phase III study of the European Organisation for Research and Treatment of Cancer Leukemia Group and the German MDS Study Group. *J Clin Oncol* 29, 1987–1996.

Majumder, P. K. & Sellers, W. R. (2005). Akt-regulated pathways in prostate cancer. *Oncogene* 24, 7465–7474.

Marrocco, D. L., Tilley, W. D., Bianco-Miotto, T., Evdokiou, A., Scher, H. I., Rifkind, R. A., Marks, P. A., Richon, V. M. & Butler, L. M. (2007). Suberoylanilide hydroxamic acid (vorinostat) represses androgen receptor expression and acts synergistically with an androgen receptor antagonist to inhibit prostate cancer cell proliferation. *Mol Cancer Ther* 6, 51–60.

Martens-Uzunova, E. S., Jalava, S. E., Dits, N. F., van Leenders, G. J., Moller, S., Trapman, J., Bangma, C. H., Litman, T., Visakorpi, T. & Jenster, G. (2011). Diagnostic and prognostic signatures from the small non-coding RNA transcriptome in prostate cancer. *Oncogene* (2012) 31, 978–991.

Maruyama, R., Toyooka, S., Toyooka, K. O., Virmani, A. K., Zochbauer-Muller, S., Farinas, A. J., Minna, J. D., McConnell, J., Frenkel, E. P. & Gazdar, A. F. (2002). Aberrant promoter methylation profile of prostate cancers and its relationship to clinicopathological features. *Clin Cancer Res* 8, 514–519.

Metzger, E., Wissmann, M., Yin, N., Muller, J. M., , R., Peters, A. H., Gunther, T., Buettner, R. & Schule, R. (2005). LSD1 demethylates repressive histone marks to promote androgen-receptor-dependent transcription. *Nature* 437, 436–439.

Mikeska, T., Candiloro, I. L. M. & Dobrovic, A. (2010). The implications of heterogeneous DNA methylation for the accurate quantification of methylation. *Epigenomics* 2, 561–573.

Millar, D. S., Ow, K. K., Paul, C. L., Russell, P. J., Molloy, P. L. & Clark, S. J. (1999). Detailed methylation analysis of the glutathione S-transferase pi (GSTP1) gene in prostate cancer. *Oncogene* 18, 1313–1324.

Miranda, T. B., Cortez, C. C., Yoo, C. B., Liang, G., Abe, M., Kelly, T. K., Marquez, V. E. & Jones, P. A. (2009). DZNep is a global histone methylation inhibitor that reactivates developmental genes not silenced by DNA methylation. *Mol Cancer Ther* 8, 1579–1588.

Mitchell, P. S., Parkin, R. K., Kroh, E. M., Fritz, B. R., Wyman, S. K., Pogosova-Agadjanyan, E. L., Peterson, A., Noteboom, J., O'Briant, K. C., Allen, A., Lin, D. W., Urban, N., Drescher, C. W., Knudsen, B. S., Stirewalt, D. L., Gentleman, R., Vessella, R. L., Nelson, P. S., Martin, D. B. & Tewari, M. (2008). Circulating microRNAs as stable blood-based markers for cancer detection. *Proc Natl Acad Sci U S A* 105, 10513–10518.

Molife, L. R., Attard, G., Fong, P. C., Karavasilis, V., Reid, A. H., Patterson, S., Riggs, C. E., Jr., Higano, C., Stadler, W. M., McCulloch, W., Dearnaley, D., Parker, C. & de Bono, J. S. (2010). Phase II, two-stage, single-arm trial of the histone deacetylase inhibitor (HDACi) romidepsin in metastatic castration-resistant prostate cancer (CRPC). *Ann Oncol* 21, 109–113.

Moltzahn, F., Olshen, A. B., Baehner, L., Peek, A., Fong, L., Stoppler, H., Simko, J., Hilton, J. F., Carroll, P. & Blelloch, R. (2011). Microfluidic-based multiplex qRT-PCR identifies diagnostic and prognostic microRNA signatures in the sera of prostate cancer patients. *Cancer Res* 71, 550–560.

Munster, P. N., Marchion, D., Thomas, S., Egorin, M., Minton, S., Springett, G., Lee, J. H., Simon, G., Chiappori, A., Sullivan, D. & Daud, A. (2009). Phase I trial of vorinostat and doxorubicin in solid tumours: histone deacetylase 2 expression as a predictive marker. *Br J Cancer* 101, 1044–1050.

Nakayama, T., Watanabe, M., Yamanaka, M., Hirokawa, Y., Suzuki, H., Ito, H., Yatani, R. & Shiraishi, T. (2001). The role of epigenetic modifications in retinoic acid receptor beta2 gene expression in human prostate cancers. *Lab Invest* 81, 1049–1057.

Nandakumar, V., Vaid, M. & Katiyar, S. K. (2011). (-)-Epigallocatechin-3-gallate reactivates silenced tumor suppressor genes, Cip1/p21 and p16INK4a, by reducing DNA methylation and increasing histones acetylation in human skin cancer cells. *Carcinogenesis* 32, 537–544.

Nelson, W. G., Yegnasubramanian, S., Agoston, A. T., Bastian, P. J., Lee, B. H., Nakayama, M. & De Marzo, A. M. (2007). Abnormal DNA methylation, epigenetics, and prostate cancer. *Front Biosci* 12, 4254–4266.

Pang, Y., Young, C. Y. & Yuan, H. (2010). MicroRNAs and prostate cancer. *Acta Biochim Biophys Sin (Shanghai)* 42, 363–369.

Park, J. Y. (2010). Promoter hypermethylation in prostate cancer. *Cancer Control* 17, 245–255.

Patra, S. K., Patra, A. & Dahiya, R. (2001). Histone deacetylase and DNA methyltransferase in human prostate cancer. *Biochem Biophys Res Commun* 287, 705–713.

Patra, S. K., Patra, A., Zhao, H. & Dahiya, R. (2002). DNA methyltransferase and demethylase in human prostate cancer. *Mol Carcinog* 33, 163–171.

Perner, S., Demichelis, F., Beroukhim, R., Schmidt, F. H., Mosquera, J. M., Setlur, S., Tchinda, J., Tomlins, S. A., Hofer, M. D., Pienta, K. G., Kuefer, R., Vessella, R., Sun, X. W., Meyerson, M., Lee, C., Sellers, W. R., Chinnaiyan, A. M. & Rubin, M. A. (2006). TMPRSS2:ERG fusion-associated deletions provide insight into the heterogeneity of prostate cancer. *Cancer Res* 66, 8337–8341.

Perry, A. S., Foley, R., Woodson, K. & Lawler, M. (2006). The emerging roles of DNA methylation in the clinical management of prostate cancer. *Endocr Relat Cancer* 13, 357–377.

Perry, A. S., Watson, R. W., Lawler, M. & Hollywood, D. (2010). The epigenome as a therapeutic target in prostate cancer. *Nat Rev Urol* 7, 668–680.

Phe, V., Cussenot, O. & Roupret, M. (2010). Methylated genes as potential biomarkers in prostate cancer. *BJU Int* 105, 1364–1370.

Pickett, C. B. & Lu, A. Y. (1989). Glutathione S-transferases: gene structure, regulation, and biological function. *Annu Rev Biochem* 58, 743–764.

Pierconti, F., Martini, M., Pinto, F., Cenci, T., Capodimonti, S., Calarco, A., Bassi, P. F. & Larocca, L. M. (2011). Epigenetic silencing of SOCS3 identifies a subset of prostate cancer with an aggressive behavior. *Prostate* 71, 318–325.

Pratap, J., Akech, J., Wixted, J. J., Szabo, G., Hussain, S., McGee-Lawrence, M. E., Li, X., Bedard, K., Dhillon, R. J., van Wijnen, A. J., Stein, J. L., Stein, G. S., Westendorf, J. J. & Lian, J. B. (2010). The histone deacetylase inhibitor, vorinostat, reduces tumor growth at the metastatic bone site and associated osteolysis, but promotes normal bone loss. *Mol Cancer Ther* 9, 3210–3220.

Qian, D. Z., Kachhap, S. K., Collis, S. J., Verheul, H. M., Carducci, M. A., Atadja, P. & Pili, R. (2006a). Class II histone deacetylases are associated with VHL-independent regulation of hypoxia-inducible factor 1 alpha. *Cancer Res* 66, 8814–8821.

Qian, D. Z., Kato, Y., Shabbeer, S., Wei, Y., Verheul, H. M., Salumbides, B., Sanni, T., Atadja, P. & Pili, R. (2006b). Targeting tumor angiogenesis with histone deacetylase inhibitors: the hydroxamic acid derivative LBH589. *Clin Cancer Res* 12, 634–642.

Qian, X., Ara, G., Mills, E., LaRochelle, W. J., Lichenstein, H. S. & Jeffers, M. (2008). Activity of the histone deacetylase inhibitor belinostat (PXD101) in preclinical models of prostate cancer. *Int J Cancer* 122, 1400–1410.

Rathkopf, D., Wong, B. Y., Ross, R. W., Anand, A., Tanaka, E., Woo, M. M., Hu, J., Dzik-Jurasz, A., Yang, W. & Scher, H. I. (2010). A phase I study of oral panobinostat alone and in combination with docetaxel in patients with castration-resistant prostate cancer. *Cancer Chemother Pharmacol* 66, 181–189.

Richiardi, L., Fiano, V., Vizzini, L., De Marco, L., Delsedime, L., Akre, O., Tos, A. G. & Merletti, F. (2009). Promoter methylation in APC, RUNX3, and GSTP1 and mortality in prostate cancer patients. *J Clin Oncol* 27, 3161–3168.

Rokhlin, O. W., Glover, R. B., Guseva, N. V., Taghiyev, A. F., Kohlgraf, K. G. & Cohen, M. B. (2006). Mechanisms of cell death induced by histone deacetylase inhibitors in androgen receptor-positive prostate cancer cells. *Mol Cancer Res* 4, 113–123.

Roupret, M., Hupertan, V., Yates, D. R., Catto, J. W., Rehman, I., Meuth, M., Ricci, S., Lacave, R., Cancel-Tassin, G., de la Taille A., Rozet, F., Cathelineau, X., Vallancien, G., Hamdy, F. C. & Cussenot, O. (2007). Molecular detection of localized prostate cancer using quantitative methylation-specific PCR on urinary cells obtained following prostate massage. *Clin Cancer Res* 13, 1720–1725.

Santos, F. P., Kantarjian, H., Garcia-Manero, G., Issa, J. P. & Ravandi, F. (2010). Decitabine in the treatment of myelodysplastic syndromes. *Expert Rev Anticancer Ther* 10, 9–22.

Santourlidis, S., Florl, A., Ackermann, R., Wirtz, H. C. & Schulz, W. A. (1999). High frequency of alterations in DNA methylation in adenocarcinoma of the prostate. *Prostate* 39, 166–174.

Sargeant, A. M., Rengel, R. C., Kulp, S. K., Klein, R. D., Clinton, S. K., Wang, Y. C. & Chen, C. S. (2008). OSU-HDAC42, a histone deacetylase inhibitor, blocks prostate tumor progression in the transgenic adenocarcinoma of the mouse prostate model. *Cancer Res* 68, 3999–4009.

Schaefer, A., Jung, M., Mollenkopf, H. J., Wagner, I., Stephan, C., Jentzmik, F., Miller, K., Lein, M., Kristiansen, G. & Jung, K. (2010). Diagnostic and prognostic implications of microRNA profiling in prostate carcinoma. *Int J Cancer* 126, 1166–1176.

Schneider, B. J., Kalemkerian, G. P., Bradley, D., Smith, D. C., Egorin, M. J., Daignault, S., Dunn, R. & Hussain, M. (2012). Phase I study of vorinostat (suberoylanilide hydroxamic acid, NSC 701852) in combination with docetaxel in patients with advanced and relapsed solid malignancies. *Invest New Drugs* 30, 249–257. Epub 2010 Aug 5.

Schroder, F. H., Hugosson, J., Roobol, M. J., Tammela, T. L., Ciatto, S., Nelen, V., Kwiatkowski, M., Lujan, M., Lilja, H., Zappa, M., Denis, L. J., Recker, F., Berenguer, A., Maattanen, L., Bangma, C. H., Aus, G., Villers, A., Rebillard, X., van der Kwast T., Blijenberg, B. G., Moss, S. M., de Koning, H. J. & Auvinen, A. (2009). Screening and prostate-cancer mortality in a randomized European study. *N Engl J Med* 360, 1320–1328.

Schulz, W. A. & Goering, W. (2011). Eagles report: Developing cancer biomarkers from genome-wide DNA methylation analyses. *World J Clin Oncol* 2, 1–7.

Schulz, W. A. & Hatina, J. (2006). Epigenetics of prostate cancer: beyond DNA methylation. *J Cell Mol Med* 10, 100–125.

Schulz, W. A. & Hoffmann, M. J. (2009). Epigenetic mechanisms in the biology of prostate cancer. *Semin Cancer Biol* 19, 172–180.

Seligson, D. B., Horvath, S., Shi, T., Yu, H., Tze, S., Grunstein, M. & Kurdistani, S. K. (2005). Global histone modification patterns predict risk of prostate cancer recurrence. *Nature* 435, 1262–1266.

Shabbeer, S., Kortenhorst, M. S., Kachhap, S., Galloway, N., Rodriguez, R. & Carducci, M. A. (2007). Multiple Molecular pathways explain the anti-proliferative effect of valproic acid on prostate cancer cells in vitro and in vivo. *Prostate* 67, 1099–1110.

Shao, W., Growney, J. & O'Connor, G. (2008). Efficacy of panobinostat (LBH589) in prostate cancer cell models: Targeting the androgen receptor in hormone-refractory prostate cancer (HRPC). *ASCO Genitourinary Cancers Symposium* abstract 216.

Sharma, S., Kelly, T. K. & Jones, P. A. (2010). Epigenetics in cancer. *Carcinogenesis* 31, 27–36.

Sidana, A., Wang, M., Chowdhury, W. H., Toubaji, A., Shabbeer, S., Netto, G., Carducci, M., Lupold, S. E. & Rodriguez, R. (2011). Does valproic acid induce neuroendocrine differentiation in prostate cancer? *J Biomed Biotechnol* 2011:607480.

Singal, R., Ferdinand, L., Reis, I. M. & Schlesselman, J. J. (2004). Methylation of multiple genes in prostate cancer and the relationship with clinicopathological features of disease. *Oncol Rep* 12, 631–637.

Sircar, K., Yoshimoto, M., Monzon, F. A., Koumakpayi, I. H., Katz, R. L., Khanna, A., Alvarez, K., Chen, G., Darnel, A. D., Aprikian, A. G., Saad, F., Bismar, T. A. & Squire, J. A. (2009). PTEN genomic deletion is associated with p-Akt and AR signalling in poorer outcome, hormone refractory prostate cancer. *J Pathol* 218, 505–513.

6 The Impact of Epigenetic Alterations on Diagnosis, Prediction... 155

Smith, D. S., Catalona, W. J. & Herschman, J. D. (1996). Longitudinal screening for prostate cancer with prostate-specific antigen. *JAMA* 276, 1309–1315.

Solit, D. B., Scher, H. I. & Rosen, N. (2003). Hsp90 as a therapeutic target in prostate cancer. *Semin Oncol* 30, 709–716.

Sonpavde, G., Aparicio, A. M., Zhan, F., North, B., Delaune, R., Garbo, L. E., Rousey, S. R., Weinstein, R. E., Xiao, L., Boehm, K. A., Asmar, L., Fleming, M. T., Galsky, M. D., Berry, W. R. & Von Hoff, D. D. (2011). Azacitidine favorably modulates PSA kinetics correlating with plasma DNA LINE-1 hypomethylation in men with chemonaive castration-resistant prostate cancer. *Urol Oncol* 29, 682–689. Epub 2009 Dec 3.

Spahn, M., Kneitz, S., Scholz, C. J., Stenger, N., Rudiger, T., Strobel, P., Riedmiller, H. & Kneitz, B. (2010). Expression of microRNA-221 is progressively reduced in aggressive prostate cancer and metastasis and predicts clinical recurrence. *Int J Cancer* 127, 394–403.

Stresemann, C., Brueckner, B., Musch, T., Stopper, H. & Lyko, F. (2006). Functional diversity of DNA methyltransferase inhibitors in human cancer cell lines. *Cancer Res* 66, 2794–2800.

Tannock, I. F., de Wit, R., Berry, W. R., Horti, J., Pluzanska, A., Chi, K. N., Oudard, S., Theodore, C., James, N. D., Turesson, I., Rosenthal, M. A. & Eisenberger, M. A. (2004). Docetaxel plus prednisone or mitoxantrone plus prednisone for advanced prostate cancer. *N Engl J Med* 351, 1502–1512.

Thibault, A., Figg, W. D., Bergan, R. C., Lush, R. M., Myers, C. E., Tompkins, A., Reed, E. & Samid, D. (1998). A phase II study of 5-aza-2′deoxycytidine (decitabine) in hormone independent metastatic (D2) prostate cancer. *Tumori* 84, 87–89.

Tokumaru, Y., Harden, S. V., Sun, D. I., Yamashita, K., Epstein, J. I. & Sidransky, D. (2004). Optimal use of a panel of methylation markers with GSTP1 hypermethylation in the diagnosis of prostate adenocarcinoma. *Clin Cancer Res* 10, 5518–5522.

Tokumaru, Y., Sun, D. I., Nomoto, S., Yamashita, K. & Sidransky, D. (2003). Re: Is TIG1 a new tumor suppressor in prostate cancer? *J Natl Cancer Inst* 95, 919–920.

Tomlins, S. A., Rhodes, D. R., Perner, S., Dhanasekaran, S. M., Mehra, R., Sun, X. W., Varambally, S., Cao, X., Tchinda, J., Kuefer, R., Lee, C., Montie, J. E., Shah, R. B., Pienta, K. J., Rubin, M. A. & Chinnaiyan, A. M. (2005). Recurrent fusion of TMPRSS2 and ETS transcription factor genes in prostate cancer. *Science* 310, 644–648.

Topaloglu, O., Hoque, M. O., Tokumaru, Y., Lee, J., Ratovitski, E., Sidransky, D. & Moon, C. S. (2004). Detection of promoter hypermethylation of multiple genes in the tumor and bronchoalveolar lavage of patients with lung cancer. *Clin Cancer Res* 10, 2284–2288.

Valentini, A., Biancolella, M., Amati, F., Gravina, P., Miano, R., Chillemi, G., Farcomeni, A., Bueno, S., Vespasiani, G., Desideri, A., Federici, G., Novelli, G. & Bernardini, S. (2007). Valproic acid induces neuroendocrine differentiation and UGT2B7 up-regulation in human prostate carcinoma cell line. *Drug Metab Dispos* 35, 968–972.

Varambally, S., Dhanasekaran, S. M., Zhou, M., Barrette, T. R., Kumar-Sinha, C., Sanda, M. G., Ghosh, D., Pienta, K. J., Sewalt, R. G., Otte, A. P., Rubin, M. A. & Chinnaiyan, A. M. (2002). The polycomb group protein EZH2 is involved in progression of prostate cancer. *Nature* 419, 624–629.

Vasiljevic, N., Wu, K., Brentnall, A. R., Kim, D. C., Thorat, M. A., Kudahetti, S. C., Mao, X., Xue, L., Yu, Y., Shaw, G. L., Beltran, L., Lu, Y. J., Berney, D. M., Cuzick, J. & Lorincz, A. T. (2011). Absolute quantitation of DNA methylation of 28 candidate genes in prostate cancer using pyrosequencing. *Dis Markers* 30, 151–161.

Vener, T., Derecho, C., Baden, J., Wang, H., Rajpurohit, Y., Skelton, J., Mehrotra, J., Varde, S., Chowdary, D., Stallings, W., Leibovich, B., Robin, H., Pelzer, A., Schafer, G., Auprich, M., Mannweiler, S., Amersdorfer, P. & Mazumder, A. (2008). Development of a multiplexed urine assay for prostate cancer diagnosis. *Clin Chem* 54, 874–882.

Verdoodt, B., Sommerer, F., Palisaar, R. J., Noldus, J., Vogt, M., Nambiar, S., Tannapfel, A., Mirmohammadsadegh, A. & Neid, M. (2011). Inverse association of p16(INK4a) and p14(ARF) methylation of the CDKN2a locus in different Gleason scores of prostate cancer. *Prostate Cancer Prostatic Dis* 14, 295–301.

Vinarskaja, A., Schulz, W. A., Ingenwerth, M., Hader, C. & Arsov, C. (2011). Association of PITX2 mRNA down-regulation in prostate cancer with promoter hypermethylation and poor prognosis. *Urol Oncol.* 2011 Jul 29. [Epub ahead of print].

Vire, E., Brenner, C., Deplus, R., Blanchon, L., Fraga, M., Didelot, C., Morey, L., Van Eynde, A., Bernard, D., Vanderwinden, J. M., Bollen, M., Esteller, M., Di Croce, L., de Launoit, Y. & Fuks, F. (2006). The Polycomb group protein EZH2 directly controls DNA methylation. *Nature* 439, 871–874.

Wedel, S., Hudak, L., Seibel, J. M., Juengel, E., Tsaur, I., Wiesner, C., Haferkamp, A. & Blaheta, R. A. (2011a). Inhibitory effects of the HDAC inhibitor valproic acid on prostate cancer growth are enhanced by simultaneous application of the mTOR inhibitor RAD001. *Life Sci* 88, 418–424.

Wedel, S., Hudak, L., Seibel, J. M., Makarevic, J., Juengel, E., Tsaur, I., Wiesner, C., Haferkamp, A. & Blaheta, R. A. (2011b). Impact of combined HDAC and mTOR inhibition on adhesion, migration and invasion of prostate cancer cells. *Clin Exp Metastasis* 28, 479–491.

Weiss, G., Cottrell, S., Distler, J., Schatz, P., Kristiansen, G., Ittmann, M., Haefliger, C., Lesche, R., Hartmann, A., Corman, J. & Wheeler, T. (2009). DNA methylation of the PITX2 gene promoter region is a strong independent prognostic marker of biochemical recurrence in patients with prostate cancer after radical prostatectomy. *J Urol* 181, 1678–1685.

Woodson, K., O'Reilly, K. J., Ward, D. E., Walter, J., Hanson, J., Walk, E. L. & Tangrea, J. A. (2006). CD44 and PTGS2 methylation are independent prognostic markers for biochemical recurrence among prostate cancer patients with clinically localized disease. *Epigenetics* 1, 183–186.

Woodson, K., Tangrea, J. A., Pollak, M., Copeland, T. D., Taylor, P. R., Virtamo, J. & Albanes, D. (2003). Serum insulin-like growth factor I: tumor marker or etiologic factor? A prospective study of prostate cancer among Finnish men. *Cancer Res* 63, 3991–3994.

Wu, T., Giovannucci, E., Welge, J., Mallick, P., Tang, W. Y. & Ho, S. M. (2011). Measurement of GSTP1 promoter methylation in body fluids may complement PSA screening: a meta-analysis. *Br J Cancer* 105, 65–73.

Xiang, Y., Zhu, Z., Han, G., Ye, X., Xu, B., Peng, Z., Ma, Y., Yu, Y., Lin, H., Chen, A. P. & Chen, C. D. (2007). JARID1B is a histone H3 lysine 4 demethylase up-regulated in prostate cancer. *Proc Natl Acad Sci U S A* 104, 19226–19231.

Xu, W. S., Parmigiani, R. B. & Marks, P. A. (2007). Histone deacetylase inhibitors: molecular mechanisms of action. *Oncogene* 26, 5541–5552.

Yamanaka, M., Watanabe, M., Yamada, Y., Takagi, A., Murata, T., Takahashi, H., Suzuki, H., Ito, H., Tsukino, H., Katoh, T., Sugimura, Y. & Shiraishi, T. (2003). Altered methylation of multiple genes in carcinogenesis of the prostate. *Int J Cancer* 106, 382–387.

Yang, B., Sun, H., Lin, W., Hou, W., Li, H., Zhang, L., Li, F., Gu, Y., Song, Y., Li, Q. & Zhang, F. (2011). Evaluation of global DNA hypomethylation in human prostate cancer and prostatic intraepithelial neoplasm tissues by immunohistochemistry. *Urol Oncol.* 2011 Jun 23. [Epub ahead of print].

Yang, C. S., Maliakal, P. & Meng, X. (2002). Inhibition of carcinogenesis by tea. *Annu Rev Pharmacol Toxicol* 42, 25–54.

Yang, X., Lay, F., Han, H. & Jones, P. A. (2010). Targeting DNA methylation for epigenetic therapy. *Trends Pharmacol Sci* 31, 536–546.

Yegnasubramanian, S., Haffner, M. C., Zhang, Y., Gurel, B., Cornish, T. C., Wu, Z., Irizarry, R. A., Morgan, J., Hicks, J., DeWeese, T. L., Isaacs, W. B., Bova, G. S., De Marzo, A. M. & Nelson, W. G. (2008). DNA hypomethylation arises later in prostate cancer progression than CpG island hypermethylation and contributes to metastatic tumor heterogeneity. *Cancer Res* 68, 8954–8967.

Yegnasubramanian, S., Kowalski, J., Gonzalgo, M. L., Zahurak, M., Piantadosi, S., Walsh, P. C., Bova, G. S., De Marzo, A. M., Isaacs, W. B. & Nelson, W. G. (2004). Hypermethylation of CpG islands in primary and metastatic human prostate cancer. *Cancer Res* 64, 1975–1986.

Yegnasubramanian, S., Lin, X., Haffner, M. C., DeMarzo, A. M. & Nelson, W. G. (2006). Combination of methylated-DNA precipitation and methylation-sensitive restriction enzymes

(COMPARE-MS) for the rapid, sensitive and quantitative detection of DNA methylation. *Nucleic Acids Res* 34, e19.

Yoo, C. B., Chuang, J. C., Byun, H. M., Egger, G., Yang, A. S., Dubeau, L., Long, T., Laird, P. W., Marquez, V. E. & Jones, P. A. (2008). Long-term epigenetic therapy with oral zebularine has minimal side effects and prevents intestinal tumors in mice. *Cancer Prev Res (Phila)* 1, 233–240.

Yu, A. F., Shen, J. Z., Chen, Z. Z., Fan, L. P. & Lin, F. A. (2008). [Demethylation and transcription of p16 gene in malignant lymphoma cell line CA46 induced by EGCG]. *Zhongguo Shi Yan Xue Ye Xue Za Zhi* 16, 1073–1078.

Yu, J., Mani, R. S., Cao, Q., Brenner, C. J., Cao, X., Wang, X., Wu, L., Li, J., Hu, M., Gong, Y., Cheng, H., Laxman, B., Vellaichamy, A., Shankar, S., Li, Y., Dhanasekaran, S. M., Morey, R., Barrette, T., Lonigro, R. J., Tomlins, S. A., Varambally, S., Qin, Z. S. & Chinnaiyan, A. M. (2010). An integrated network of androgen receptor, polycomb, and TMPRSS2-ERG gene fusions in prostate cancer progression. *Cancer Cell* 17, 443–454.

Yu, J., Rhodes, D. R., omlins, S. A., ao, X., hen, G., ehra, R., ang, X., hosh, D., hah, R. B., arambally, S., ienta, K. J. & hinnaiyan, A. M. (2007). A polycomb repression signature in metastatic prostate cancer predicts cancer outcome. *Cancer Res* 67, 10657–10663.

Zhang, J., Liu, L. & Pfeifer, G. P. (2004). Methylation of the retinoid response gene TIG1 in prostate cancer correlates with methylation of the retinoic acid receptor beta gene. *Oncogene* 23, 2241–2249.

Zhang, Z., Stanfield, J., Frenkel, E., Kabbani, W., & Hsieh, J. T. (2007). Enhanced therapeutic effect on androgen-independent prostate cancer by depsipeptide (FK228), a histone deacetylase inhibitor, in combination with docetaxel. *Urology* 70, 396–401.

Chapter 7
Epigenetic Reprogramming in Lung Carcinomas

András Kádár and Tibor A. Rauch

7.1 Lung Cancers and Associated Epigenetic Changes

Lung carcinoma is the leading cause of cancer mortality in the United States. It accounts for 30% of all deaths from cancer, and 1.5 million annual deaths from lung cancer are projected worldwide by 2012. The >85% mortality rate associated with lung carcinomas is in part related to suboptimal therapeutic strategies and lack of an efficient screening approach for early detection. The relevance of a screening method is emphasized by the fact that the 5-year survival rate of lung cancer (14%) is at least six times lower than that of breast (85%) and prostate (93%) cancers for which appropriate screening tests exist. The benefit of early detection is evident in patients with stage I tumors, where surgical resection is the preferred treatment option and the recurrence within 5 years is <35%. Lung cancers are divided into small cell (SCLC) and non-small cell lung carcinomas (NSCLC). NSCLCs are further classified on the basis of histological appearance into three subtypes: squamous cell carcinoma (SCC), adenocarcinoma (ADC), and large cell carcinoma (LCC).

The development of lung cancer is a multifactorial process; both genetic and epigenetic events are involved in carcinogenesis. Tumor suppressor genes are frequent targets of these processes, and the downregulation of them ultimately leads to uncontrolled cell proliferation. In general, both alleles of a tumor suppressor gene need to be inactivated by genetic alterations such as deletion, inversion, or point mutation. As a possible alternative mechanism for gene inactivation, transcriptional silencing can be achieved by various epigenetic changes including de novo DNA methylation, histone modifications, and aberrant expression of noncoding RNAs (ncRNAs). Disease-associated epigenetic alterations are not restricted to gene silencing events; aberrant activation/reactivation of genes are also included.

A. Kádár • T.A. Rauch (✉)
Section of Molecular Medicine, Department of Orthopedic Surgery,
Rush University Medical Center, 1735 West Harrison Street, Chicago, IL 60612, USA
e-mail: andras.kadar.md@gmail.com; tibor_rauch@rush.edu

J. Minarovits and H.H. Niller (eds.), *Patho-Epigenetics of Disease*,
DOI 10.1007/978-1-4614-3345-3_7, © Springer Science+Business Media New York 2012

The highly dynamic epigenome actively contributes to the regulation of transcriptional activity of the cell and actually defines the identity of the cell. Epigenetic reprogramming of cells can be the outcome of miswritten, misinterpreted, or miserased modifications of the chromatin signals (Chi et al. 2010). Perturbations in the finely balanced epigenetic milieu of epithelial cell can lead to inappropriate gene expression and, ultimately, carcinogenesis.

7.1.1 DNA Methylation

7.1.1.1 Historical Overview

Chronologically, DNA methylation was the first epigenetic modification to be discovered more than 50 years ago (Wyatt 1950) and linked to gene silencing and X chromosome inactivation (Holliday and Pugh 1975; Riggs 1975) in the mid 1970s. The authors of these two papers hypothesized a molecular model for switching of gene activities, and also provided explanation for the heritability of gene activity or inactivity. The first disease-associated (i.e., cancer) DNA methylation changes were described in the early 1980s when global reduction in the methylation level was detected in tumor samples compared to the corresponding normal tissues (Feinberg and Vogelstein 1983; Riggs and Jones 1983). Genome-wide mapping of cancer-associated hypomethylation revealed that it mostly occurs at various repetitive DNA sequences (Rauch et al. 2008). Soon after the first detection of global hypomethylation events, gene-specific de novo hypermethylation was also described in tumor samples (Baylin et al. 1986). Today, there are many reports documenting methylation of CpG islands associated with a large number of different genes, including almost every type of human cancer (Jones and Baylin 2007; Estecio and Issa 2011). In the last 5 years, significant developments have been made in genome-wide DNA methylation mapping methods and such hypo- and hypermethylation events have been discovered occurring in the neighborhood of CpG islands (Doi et al. 2009; Irizarry et al. 2009; Rubin and Green 2009). The function of the so called CpG island shore methylation is not obvious; it is hypothesized that it may play roles in chromatin topology and the assembly of the transcription initiation complex at the promoter region. The latest progress in the DNA methylation field is the discovery of 5-hydroxymethylcytosine (5hmC), an enzymatically oxidized version of the "classical" 5-methylcytosine (5mC) (Kriaucionis and Heintz 2009; Tahiliani et al. 2009) in mammalian cells. This finding instantly raises several basic questions, such as whether 5hmC is an intermediate of the active 5mC demethylation process or an end product with unknown function. Either way, 5hmC might play a pivotal role in reprogramming of epithelial cells during carcinogenesis (Iqbal et al. 2011; Wossidlo et al. 2011).

DNA methylation is one of the most abundant epigenetic modifications of the chromatin (Suzuki and Bird 2008). In normal cells, DNA methylation fulfills multiple obligations: (1) Silencing of transposable elements (Goodier and Kazazian 2008; Kaneko-Ishino and Ishino 2010). The two most abundant transposon families

7 Epigenetic Reprogramming in Lung Carcinomas

are constituted by the short and long interspersed nuclear elements (SINEs and LINEs) (Belancio et al. 2010); their inactivation is essential for protecting the human genome's integrity. (2) Providing stability for the terminal regions of chromosomes (e.g., centromeres and telomeres). (3) Resulting silenced genes are involved in cell- and allele-specific gene expression (Tycko 2010b). (4) Control of imprinted gene expression (Tycko 2010a). (5) Inactivation of X chromosome in females (Sharp et al. 2011). In lung carcinomas, all cellular functions of DNA methylation are damaged and may contribute to the malignant transformation of epithelial cells.

7.1.1.2 Hyper- and Hypomethylation in Lung Carcinomas

To identify the chromosomal regions that undergo de novo methylation or demethylation, a number of methods have been developed. These methods can be categorized into three groups on the basis of their principles. The *first group* of techniques is based on the sensitivity of restriction endonucleases to CpG methylation in their recognition sequence. The *second group* of methods employs bisulfite treatment of genomic DNA, which converts cytosine residues to uracil but leaves 5mC residues unaffected. This approach provides a single nucleotide resolution for methylation detection but is limited to the amplified region (Suzuki and Bird 2008). The *third group* of techniques is based on protein affinity to the methylated DNA template using anti-methyl CpG antibody (Weber et al. 2005) (MeDIP) or the MBD-protein domain-based affinity method such as MIRA (Rauch and Pfeifer 2005). There are several additional methods for analysis of DNA methylation of genes, but practically all are based upon the principles described above.

We have to emphasize that there is no a priori disease-linked epigenetic signal. These covalent modifications of the chromatin are part of the normal regulatory processes of gene expression. In the context of lung carcinogenesis, disease-linked epigenetic modification means that the given epigenetic mark is placed at the inappropriate genomic region (i.e., promoter) or the timing is incorrect. Therefore, the first step towards understanding the epigenetic background of tumor formation is to investigate the entire epigenome and determine the location of the misplaced signals. For this purpose, the proper approach is the application of genome-wide mapping methods (Heller et al. 2010). To analyze DNA methylation patterns on a *genome-wide scale*, several techniques have been developed, but none of them has reached wide acceptance since originally they were designed for single gene level analysis. It gave an impetus to the whole epigenetics when some of the original DNA methylation analysis methods were combined with tiling microarray platforms carrying shorter or longer promoter regions, whole chromosomes, or even the entire human genome (Fig. 7.1). Most of the available information regarding cancer-associated DNA methylation profile changes were gained from the coupling of the affinity-based methods and the tiling microarrays including MeDIP (Novak et al. 2006; Ruike et al. 2010; Yagi et al. 2010) and MIRA-on-chip (Tommasi et al. 2009; Dunwell et al. 2010; Wu et al. 2010). MIRA technique proved to be the most successfully used approach for detecting lung carcinoma-specific DNA methylation

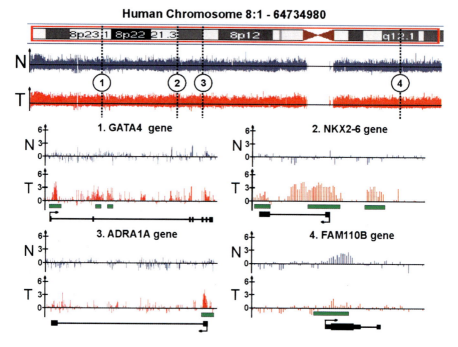

Fig. 7.1 DNA methylation profile mapping of human chromosome 8. The analysis covers 64.7 Mb-long region of the chromosome including the entire short arm and 17.8 Mb-long part of the long arm. Investigated genomic DNA was isolated from stage I lung SCC (*blue*) and matching normal lung tissue (*red*). DNA methylation pattern was mapped by MIRA-enriched methylated fractions that were amplified, labeled, and hybrized onto NimbleGen tiling microarray platforms (MIRA-on-Chip). The height of a peak is proportional with the detected DNA methylation level. *Encircled numbers* label the genomic location of four cancer-associated DNA methylation changes. *Lower four panels* show the DNA methylation profiles of the selected genes at high magnification in normal (N) and tumor (T) sample. Three genes are cancer-specifically hypermethylated (GATA4, NKX2-6, and ADRA1A). FAM110B gene is hypomethylated which may help in tumor progression. *Black boxes* (under the methylation profiles) mark exons; *black lines* are introns. The *green boxes* label CpG islands, while *arrows* indicate the direction of transcription

changes (Fig. 7.1) (Rauch et al. 2006, 2007, 2008). MIRA method is based on the observation that MBD2b protein specifically recognizes methylated DNA template, and this interaction is enhanced by MBD3L1 protein (Jiang et al. 2004). In this way, MBD2b/MBD3L1 proteins form a complex with high affinity binding to the methylated DNA, which can be used to analyze the DNA methylation status of a large number of genes by employing either microarray platforms (Rauch and Pfeifer 2005; Rauch et al. 2006).

In the near future, ChIP-sequencing (ChIP-seq) is going to replace the currently most popular epigenome analysis tools (Laird 2010). This method is the combination of chromatin immunoprecipitation (ChIP) and massive parallel DNA sequencing ("Next-Gen" technology). ChIP-seq can be used to precisely map the binding of any chromatin-associated protein or DNA methylation in the whole genome (Choi et al. 2009, 2010).

7.1.1.3 Lung Carcinoma-Associated Hypermethylations in the Genome

The first studies focusing on the discovery of de novo methylated CpG islands were mostly hypothesis driven. Particular gene or gene family members were investigated for their DNA methylation profile changes. Therefore, the early mapping studies were rather ineffective efforts for identifying a large number of disease-associated DNA methylation events. In spite of the low productivity of the gene-by-gene (i.e., CpG island by CpG island) based mapping approaches, they led to the conclusion that CpG island overlapping promoters are frequently de novo methylated in lung carcinomas but are largely methylation-free in the corresponding normal tissues. By using this research strategy, several dozen methylation-prone CpG islands were identified during the last 10–15 years. According to these studies, DNA methylation hot spots mostly overlap with promoter regions and initiate gene silencing and the targeted genes are involved in cell-cycle regulation [CDKN2A (Otterson et al. 1995), CHFR (Mizuno et al. 2002)], proliferation [CXCL12 (Suzuki et al. 2008)], various signal transduction pathways [RASSF1A (Dammann et al. 2001), NORE1A (Irimia et al. 2004), APC (Virmani et al. 2000), DKK1 (Licchesi et al. 2008)], apoptosis [CASP8 (Shivapurkar et al. 2002)], mobility/invasion [ADAMTS1 (Choi et al. 2008), TIMP3 (Dammann et al. 2005)] and DNA repair [MGMT (Esteller et al. 1999), BRCA1 (Lee et al. 2007)].

Three of the most comprehensive epigenetic studies that were focused on NSCLC samples investigated the genome-wide DNA methylation changes in stage I SCC and ADC samples (Rauch et al. 2006, 2007, 2008). The detected tumor-associated DNA methylation profiles were compared to the normal matching lung tissue profiles. MIRA method was combined with Agilent microarray platform that covered ~28,000 CpG islands of the human genome. Several hundred CpG islands showed cancer-specific hypermethylation in each of the investigated SCC samples. Five SCC samples were analyzed parallel in this study, and 36 CpG islands were hypermethylated in 5 out of 5 tumors. Interestingly, homeodomain encoding transcription factor genes (Homeobox genes), a large family of transcription regulators, were overrepresented among the methylated targets. Twelve genes (*BARHL2, EVX2, IRX2, MEIS1, MSX1, NR2E1, OC2, OSR1, OTX1, PAX6, TFAP2A,* and *ZNF577*) were further investigated in a larger series of SCC samples. The methylation frequency of these genes was 85–100% in the 20 tested primary tumors. The OTX1 and NR2E1 promoter overlapping CpG islands were methylated in all tested SCCs. The analyzed eight ADC samples carried similar numbers of cancer-specific hypermethylated CpG islands like the SCCs. Several hundred CpG islands were de novo methylated in each tumor, and 52 of them were targeted in at least 6 out of 8 samples. For further analysis, 11 genes were selected (*CHAD, DLX4, GRIK2, KCNG3, NR2E1, OSR1, OTX1, OTX2, PROX1, RUNX1,* and *VAX1*); the methylation status of these CpG islands were verified by COBRA assays that showed more than 80% methylation frequency for the tested genes.

The high methylation frequency of the identified NSCLC-related CpG islands makes them excellent candidates for DNA methylation markers. The methylation frequency of the newly discovered potential DNA methylation markers was much higher than for other previously reported DNA markers (Anglim et al. 2008).

Some of these potential SCC markers (*OTX1*, *BARHL2*, *MEIS1*, *OC2*, *TFAP2A*, and *EVX2*) proved to be highly specific for carcinogenic alterations of the lung epithelium since there was no detectable methylation in tumor-adjacent lung tissues.

DNA hypermethylation at the promoter regions leads to downregulation of gene expression and is considered to be a key mechanism for long-term silencing of tumor suppressor genes. Transcription factor encoding genes are also frequent targets of disease-associated hypermethylation in lung carcinomas (Rauch et al. 2006, 2007; Yanagawa et al. 2007; Dooley et al. 2011; Fiorentino et al. 2011). The homeobox containing HOX genes are especially favored targets of DNA methylation-mediated gene silencing (Rauch et al. 2006). Homedomain containing proteins act in concert with other transcription factors and play a crucial role in the determination of cell identity by regulating the expression of other regulators and effector molecules such as enzymes and ncRNAs. In this way, a single hit is multiplied and the expression of many other genes can be altered. This provides a plausible explanation for why they are prevalent targets of tumorigenesis-linked reprogramming. Gene-specific hypermethylation promotes gene silencing in two ways: (i) de novo DNA methylation can inhibit the binding of transactivators into the promoter region (Mancini et al. 1999) and (ii) the de novo methylated CpG dinucleotides are recognized by methyl-CpG-binding domain (MBD) proteins. Some of the MBD proteins are subunits of large chromatin remodeling complexes that promote transcriptionally incompatible chromatin formation and ultimately cancer progression (Jiang et al. 2004; Lai and Wade 2011).

The reduced DNA methylation level (i.e., hypomethylation) of cancer cell's genomes was observed prior to the gene-specific hypermethylation, but it is still less understood (Feinberg and Vogelstein 1983). Whether hypomethylation was a cause or consequence of tumorigenesis was unclear for a long time. However, using transgenic mice with reduced expression of DNA methyltransferase 1 (Dnmt1) resulted in a significant genome-wide hypomethylation in all tissues (Gaudet et al. 2003). Mutant mice were runted at birth and developed aggressive T cell lymphomas. These results indicated that DNA hypomethylation plays a causal role in tumor formation, possibly by promoting chromosomal instability. Epigenetic factors and mechanisms responsible for hypomethylation are largely unknown and mostly hypothetical. There are several conceivable scenarios for disease formation: (1) Lack of the dense DNA methylation in promoters of the transposable elements can reactivate their jumping activity and cause fatal mutations in vital genes by insertion into coding or regulatory sequences. (2) Hypomethylated telomeric, subtelomeric, and pericentromeric regions of chromosomes make the genome instable and prone to rearrangement. (3) The absence of proper DNA methylation of imprinted genes, and hiatus of silencing of tissue-specific regulated genes, contribute to aberrant gene expression profile. Recent data show that regions which are often targets of loss of heterozygosity (LOH) became frequently hypomethylated in SINEs and LINEs harboring loci (Rauch et al. 2008). In this way, the hypomethylation following reactivation of transposons may contribute to LOH and chromosome instability.

7 Epigenetic Reprogramming in Lung Carcinomas

7.1.1.4 Driver or Passenger DNA Methylation

Genome-wide studies provided a great deal of NSCLC-associated DNA methylation changes. In a single lung carcinoma sample (SCC or ADC), the average number of the detectable de novo methylation events is around 500. It is a reasonable question whether each detected epimutation plays a causative role in carcinogenesis or is just a byproduct of a reprogramming event (Kalari and Pfeifer 2010). In other words, are they drivers or passengers of carcinogenesis? The answer is relatively simple when the hypermethylation detected in a promoter region belongs to a tumor suppressor gene: this kind of epimutation most probably actively contributes, as a driver methylation, to the uncontrolled cell proliferation (Otterson et al. 1995; Dammann et al. 2001; Lee et al. 2007). Driver methylations are not exclusively hypermethylation events associated with silencing of tumor suppressors or Hox genes. Hypomethylation-mediated reactivation of oncogenes can also provoke malignant transformation and can belong to driver mutations' family. The effect of a driver methylation can be indirect on the reprogramming processes, for example, hypermethylation-mediated silencing of a DNA repair gene can lead to an increased mutation level and subsequent chromosome instability (Esteller et al. 1999). The wobbliness between driver and passenger mutations roots back to a basic problem, namely, the factors and the molecular mechanisms responsible for predisposing/exposing promoters to DNA methylation are unknown. It is very likely that posttranslational modifications (PTM) of histones, the actual chromatin structure, and short DNA sequences are also involved in the promoter selection for the subsequent DNA methylation event. Some DNA sequences (*cis* elements) have been identified that make CpG islands DNA methylation-prone or methylation-resistant sequences (Feltus et al. 2003). Recent findings show that hypermethylation of certain CpG islands is controlled in part by local DNA sequence context and *trans*-acting factors (McCabe et al. 2009). The most useful information that would help to clarify this enigma (i.e., driver vs. passenger methylation) can come from animal models with gene targeting. Although some knockout mice with the most frequently methylated genes have been created, lung carcinoma-specific hypermethylated genes have not been investigated in this model system (Tommasi et al. 2005; Ogata et al. 2006; Kansara et al. 2009). Therefore, we have to rely on data gained from other cancer-specific hypermethylation and extrapolate to lung carcinomas.

7.1.1.5 The Role of 5hmC in Lung Carcinomas

The discovery of the abundant occurrence of 5hmC in the mammalian genome is one of the most significant results of the recent 2 years (Kriaucionis and Heintz 2009). DNA methyltransferases have been extensively studied; the biochemical background of 5mC creation and maintenance is quite well-known (Suzuki and Bird 2008). Although numerous studies have been addressed, the mechanism that removes methylated cytosines from DNA (active DNA demethylation) has remained enigmatic. According to the emerging picture, oxidation of 5mC takes place in two steps catalyzed by TET enzymes. At first 5mC is converted to 5hmC and then further oxidized to 5-carboxylcytosine (5caC) (Tahiliani et al. 2009; Inoue and Zhang 2011; Iqbal et al. 2011; Wossidlo et al. 2011). The end product (5caC) of the previous process is an excellent substrate of a DNA glycosylase enzyme that is part of the base excision repair mechanism (Inoue and Zhang 2011). Note that the above-described DNA demethylation mechanism has been detected in embryonic stem cells and in the early phases of embryonic development. Although it has not been investigated in adult tissues, it is tempting to hypothesize that the same or a similar demethylation process works in somatic tissues. This hypothesis is supported by recent data that in SCCs the level of 5hmC is fivefold lower than in matching normal lung tissue (Jin et al. 2011). The reduced level of 5hmC in lung carcinomas may have significant effects on DNA methylation profile. It may counteract demethylation and promote de novo DNA methylation.

7.1.1.6 Pharmacological Aspects of DNA Methylation-Targeted Therapy

Preclinical and clinical trials have proved that DNA methylation modifiers have therapeutic potential against different types of leukemia (Szyf 2009). There have already been two FDA-approved drugs, *5-azacytidine* ("Vidaza" — Pharmion Corporation) and *5-aza-2′-deoxycytidine* ("Dacogen" — MGI Pharma), targeting the DNA methylation apparatus. By targeting DNMT1 (the maintaining methyltransferase), we can promote the gradual elimination of the aberrant methylation signal in consecutive cell divisions. Practically, Vidaza and Dacogen are reactive nucleoside analogs, and after administration, they are first phosphorylated to nucleotides and thereafter incorporated into DNA. DNMT1 is cross-linked to DNA by contact with the highly reactive nucleotide analogs. The trapped enzyme is not able to perform subsequent cytosine modification on the newly synthesized DNA strand that leads to the passive loss of the methylation signal. Inhibition of the de novo methyltransferase enzymes (DNMT3A and B) can prevent the formation of the aberrant signal on vital genes. Although these enzymes would be obvious targets of drug discovery, there are just several published efforts focusing upon the inhibition of these methyltransferases (Brueckner et al. 2005; Stresemann et al. 2006; Hagemann et al. 2011). Theoretically, selective increase of the activity of enzymes involved in DNA demethylation can be another possible way for elimination of faulty hypermethylation marks, but the DNA replication independent demethylation pathway has not been well characterized, and the acting demethylase is still unknown in adult somatic cells. It has been noticed that successful reactivation of silenced genes is frequently just a temporary effect, and repressive methylation could easily return after discontinuation of the drug (Szyf 2009). The recurrence of DNA methylation can be attributed to the retention of specific histone marks (histone K27 and H3K9). These leftover histone signals can initiate repeated DNA methylation on critical promoters. In the light of these facts, it is a suggested therapeutic strategy to apply combined drugs to eradicate both aberrant histone and DNA marks. It is well established that histone modification enzymes interact with the DNA methylation machinery and recruit DNMTs to promote DNA methylation (Li et al. 2006; Suzuki and Bird 2008). Acetylated histones are characteristic signals of transcriptionally competent chromatin, and removal of acetyl groups from histones results in the formation of such a chromatin state that is incompatible with transcriptional activity. Histone deacetylases (HDACs) are responsible for this activity and potential targets of pharmacology. It turns out that combined treatments addressing both aberrant DNA and histone marks provide the best therapeutic potential. But on the other side, combined therapies, e.g., 5-azaC/TSA treatment, can also reactivate proto-oncogenes silenced in normal cells and promote demethylation of candidate proto-oncogenes in head and neck and lung cancer (Smith et al. 2009).

Beyond the theoretical consideration, the first clinical trial of a DNA methylation-targeting drug in lung cancer treatment has been performed (Bauman et al. 2012). Combined application of 5-azacytidine and a tyrosine kinase inhibitor (Erlotinib) was studied in lung, head and neck, and ovarian cancers. Median progression-free survival was 2 months in the reported phase I study (Bauman et al. 2012). In the near future, we expect the initiation of more phase I studies focused on lung carcinomas to evaluate the application of novel DNA hypermethylation-targeting drugs in monotherapy or in combination with established anticancer drugs.

7.1.2 Histone Modifications in Lung Carcinomas

DNA methylation is not the only epigenetic mark that is involved in the reprogramming of lung epithelial cells. Besides DNA methylation, various forms of histone-tail modifications have been described including acetylation, methylation, phosphorylation, ubiquitination, sumoylation, ADP ribosylation, deimination, and proline isomerisation. DNA methylation represents a rather stable type of epigenetic modification, while the PTMs of histone molecules are flexible and more reversible covalent signals.

The structural and functional unit of chromatin is the nucleosome built up from four types of core histones (H2A, H2B, H3, and H4). Two from each core histone create the drum-like nucleosome that is wrapped around by 147-bp long DNA. Two neighboring nucleosomes are connected by a ~80-bp long "linker DNA" and a loosely associated linker histone molecule (H1). The compaction of this "beads on a string" structure determines gene activity. The tightly condensed chromatin is refractory for binding of *trans*-acting factors into the promoter regions, while the relaxed form is compatible with the formation of transcription preinitiation complex and the subsequent transcription initiation. The N-terminal parts of core histones are unstructured and the main targets of PTMs. At a given time, one histone molecule can harbor different modifications, and special combinations of these modifications possess various meanings, forming a "histone code".

Regarding lung carcinomas the two most studied forms of PTMs are acetylation and methylation of the core histones. Acetylation is catalyzed by histone acetyltransferases (HATs) that loosens up the chromatin structure and is ultimately associated with gene activation. HDACs are the enzymes which mediate the removal of acetyl groups from histones that is followed by chromatin compaction and gene silencing. Cigarette smoke induces histone H3 acetylation in lung and promotes sustained expression of proinflammatory genes (Yang et al. 2008), and in this way, it actively participates in chromatin remodeling and cell reprogramming. Global histone H4 acetylation profile changes were detected in SCC and ADC tumor samples. More specifically, aberrant hypoacetylation occurred on lysine 12 and 16, while lysine 5 and 8 were hyperacetylated in NSCLC (Van Den Broeck et al. 2008b). Many pharmacological efforts are currently being made to develop small molecular inhibitors against various subtypes of HATs (Dekker et al. 2009; Dekker and Haisma 2009; Ghizzoni et al. 2010).

Gene inactivation promoting HDACs also showed perturbed expression profiles in lung carcinomas. Aberrant expression of HDACs is implicated in tumor progression via misregulation of genes encoding proteins which play roles in cell proliferation, cell cycle regulation, and apoptosis. HDAC inhibitors can weaken the efficiency of DNA repair and promote apoptosis; this can be one plausible explanation for the observed increased cytotoxic effect in radiotherapy (Cuneo et al. 2007). HDAC inhibitors emerged as promising therapeutic agents for NSCLC (Loprevite et al. 2005; Komatsu et al. 2006; Cuneo et al. 2007; Gridelli et al. 2008). Results of the first clinical trials (phase I and II) applying HDAC inhibitors have been published (Gridelli et al. 2008) initiating new trials. Ongoing clinical trials are testing new HDAC inhibitors alone and in combination with established drugs and radiotherapy (Sacco et al. 2011; Gridelli et al. 2011a, b). Presently, HDAC inhibitors are more than promising therapeutic agents for NSCLC, but the treatment regimens including dose, schedule, and patient selection must be optimized.

Histone methylation, unlike acetylation, is not unequivocally associated with gene activation or silencing; its ultimate effect on transcriptional activity is rather complex. It depends on (i) the type of the affected residue (i.e., lysine or arginine), (ii) the position of the modified amino acid in the histone tail (e.g., K4 or K9 in histone H3), and (iii) the methylation level (mono-, di-, or tri-methylation). In addition, there is communication between histone acetylation and methylation. The combination of these factors creates a delicate language that ultimately makes the chromatin transcription permissive or repressive (Murr 2010; Oliver and Denu 2011). Histone methyltransferases (HMTs) and histone demethylases (HDMs) belong to the two enzyme families which are involved in the dynamic modifications of the methylation level of histones. It was observed that the level of histone H4 trimethylation was reduced at the lysine 20 position (i.e., H4K20me3) during the early phase of carcinogenesis (Van Den Broeck et al. 2008a). The loss of H3K20me3 methylation was correlated with the low expression of HMT SUV4-20H2 (Van Den Broeck et al. 2008c). The *HDM1* gene (or *MAPJD*) has been found to be upregulated in the majority of the investigated NSCLC samples, and RNAi-mediated inactivation of the gene resulted in a reduction of cell proliferation (Suzuki et al. 2007).

7 Epigenetic Reprogramming in Lung Carcinomas

Similar tumor-specific overexpression was observed in case of another HDM family gene (Italiano et al. 2006); the JMJD2C encoding region on chromosome 9 was amplified in tumors.

Enzymes implicated in various forms of histone methylation are potential targets of lung carcinoma therapy, but the development of HMT- and HDM-targeting drugs has somewhat lagged behind other chromatin-modifying enzyme inhibitors such as HDAC and DNA methyltransferase targeting drugs.

7.1.3 Noncoding RNAs (ncRNAs)

RNAs with no protein coding capacity can be classified into two categories (Prasanth and Spector 2007) on the basis of their size: (i) 18–200 nt-long small ncRNAs and (ii) the longer than 200 nt ncRNAs—long noncoding RNAs [lncRNAs (Gibb et al. 2011)]. The short regulatory ncRNA category encompasses functionally quite diverse ncRNA families including microRNAs [miRNAs (Bartel 2009)], small nucleolar RNAs [snoRNAs (Bachellerie et al. 2002)], small interfering RNAs [siRNAs (Tuschl 2001)], and piwi-interacting RNAs [piRNAs (Siomi et al. 2011)]. The most information is available for the members of the miRNA family in lung carcinomas (Lin et al. 2010; Liu et al. 2011). MiRNAs can act posttranscriptionally by promoting degradation of specific sets of mRNAs and/or regulating translation efficiency of protein-encoding mRNAs. Expression profiles of miRNAs are frequently disturbed in cancer cells (Chan et al. 2011). Overexpressed miRNAs can selectively eliminate the targeted mRNA(s) and significantly lower the encoded protein level. Or vice versa, low expression of a given miRNA can allow for the elevated expression of the targeted gene (i.e., mRNA). In this way, miRNAs can function as tumor suppressors or oncogenes. There is an ever growing number of miRNAs implicated in lung carcinogenesis. For example, miRNA-128b acts as a tumor suppressor; it directly regulates the expression of the EGFR gene, and its frequent loss in NSCLC contributes to the high expression of the oncogenic properties showing EGFR protein (Weiss et al. 2008). MiRNA-21 behaves the opposite way; its expression is higher in NSCLC than in the matching normal samples. The exact targets of miRNA-21 have not been identified yet, but the "TargetScanHuman" database shows a number of potential targets including transcription regulators (GATAD2B, SCLM2, BRWD1, and PHF14), receptors (GPR64, BMPR2), and receptor ligands (IL12A and PELI1).

Metastasis associated in lung adenocarcinoma transcript (MALAT1) was one of the first discovered lncRNAs but its molecular function was enigmatic for a long time (Ji et al. 2003). Its expression was higher in ADC than in normal lung tissues. According to recent data, MALAT1 plays a role in gene expression regulation; more specifically, it regulates alternative splicing via distribution of SR proteins in the nucleus (Tripathi et al. 2010). Currently MALAT1 is the only lncRNA that has been described as a contributor to epithelial reprogramming in lung carcinomas. We expect that the numbers of implicated lncRNAs in lung carcinogenesis will be increased in the near future since they play roles in every aspect of gene regulation affected in tumorigenesis.

Although DNA methylation, histone signals, and ncRNAs were discussed separately in epigenome reprogramming processes, they are tightly linked to each other (Murr 2010). Promoter-associated DNA methylation is one of the final acts in the process of gene inactivation, and before the de novo methylation occurs, replacement of the active chromatin marks (such as histone H3 lysine 4 trimethylation or the histone acetylations) takes place (Deaton and Bird 2011). HDACs and HDMs remove the transcriptionally active chromatin associated signals including the acetyl groups and the methyl groups from histone H3 and H4 molecules, while HMTs place repressive methyl groups onto histone H3 at proper positions such as lysine 9 and/or 27. So there is an intermediate state of chromatin when it is not active but still not sealed by DNA methylation. Polycomb complex-related marks (i.e., H3K27me3) can predispose promoters to DNA methylation (Margueron and Reinberg 2011). The established H3K27me3 signal is associated with silencing of genes and the targets are up to 12-fold more likely to have cancer-specific promoter DNA hypermethylation than nontargets (Widschwendter et al. 2007). It corresponds with the biochemical findings that the EZH2 enzyme, a subunit of polycomb repressing complex 2 (PRC2) that mediates the tri-methylation of lysine 27, can interact with all three DNA methyltransferases (Vire et al. 2006). In this way, EZH2 creates a recruitment platform for DNA methyltransferases. To close the regulatory circle, there is direct interaction between the EZH2 mRNA and certain miRNAs. For example, EZH2 is one of the targets of miRNA-101 and its reduced expression associated with high expression of EZH2 in NSCLC samples (Zhang et al. 2011).

Since miRNA genes are transcribed by the RNA polymerase II enzyme complex, the discussed chromatin modifying factors (i.e., DNA methyltransferases and histone modifiers) are also involved in their regulation (Schanen and Li 2011). This gives another layer onto the mutual regulatory circuits which exist among the epigenetic regulators. Promoter regions of miRNA-34b and miRNA-126 genes were silenced by DNA methylation in NSCLC cell lines and could be reactivated by 5-aza-2'-deoxycytidine treatment (Watanabe et al. 2011). Repressive histone H3 methylation signals (H3K9m3 and H3K27m3) were also detected in their promoter regions (Watanabe et al. 2011). In this intricate regulatory network, the highly organized epigenetic milieu can be damaged at many points. Genetic and epigenetic mutations can deflect a whole system from the optimal operation that ultimately leads to malignant reprogramming of epithelial cells.

7.2 Conclusions and Future Directions

Genome-wide epigenetic signal profiling studies significantly contributed to a better understanding of the reprogramming events implicated in the malignant transformation of lung carcinomas. A number of carcinoma-specifically modified promoters have been identified in the human genome. These modifications directly or indirectly affect the transcriptional activity of the targeted promoters and the gene expression level. The next challenge will be the identification of those alterations that actively contribute to tumorigenesis of the lung epithelium. Although the current picture is still not sharp enough in every detail, it offers a reasonably good picture that helps in separating driver mutations from passenger ones. Eventually, both types of mutations can be used as biomarkers for the early detection of lung carcinomas inasmuch as their frequencies are high enough. DNA methylation markers are promising biomarkers since they represent the most stable signal family among the epigenetic modifications, and the potentially applied detection method (methylation-specific PCR) is relatively simple. After identifying the genes and gene products which actively contribute to the reprogramming process, the next step will be to explore and characterize the interactions among the implicated factors. Discovery of the regulatory network will help in understanding the pathomechanisms of lung carcinomas and assign potential new enzymes for pharmacological targeting. Future studies with new approaches will provide even deeper insights into carcinoma etiology and perhaps extend the diagnostic arsenal currently used for patient selection for "personalized" therapy in lung cancer.

References

Anglim, P. P., Alonzo, T. A. & Laird-Offringa, I. A. (2008). DNA methylation-based biomarkers for early detection of non-small cell lung cancer: an update. *Mol Cancer* 7:81.

Bachellerie, J. P., Cavaille, J. & Huttenhofer, A. (2002). The expanding snoRNA world. *Biochimie* 84, 775–790.

Bartel, D. P. (2009). MicroRNAs: target recognition and regulatory functions. *Cell* 136, 215–233.

Bauman, J., Verschraegen, C., Belinsky, S., Muller, C., Rutledge, T., Fekrazad, M., Ravindranathan, M., Lee, S. J. & Jones, D. (2012). A phase I study of 5-azacytidine and erlotinib in advanced solid tumor malignancies. *Cancer Chemother Pharmacol* 69, 547–554.

Baylin, S. B., Hoppener, J. W., de Bustros, A., Steenbergh, P. H., Lips, C. J. & Nelkin, B. D. (1986). DNA methylation patterns of the calcitonin gene in human lung cancers and lymphomas. *Cancer Res* 46, 2917–2922.

Belancio, V. P., Roy-Engel, A. M. & Deininger, P. L. (2010). All y'all need to know 'bout retroelements in cancer. *Semin Cancer Biol* 20, 200–210.

Brueckner, B., Garcia, B. R., Siedlecki, P., Musch, T., Kliem, H. C., Zielenkiewicz, P., Suhai, S., Wiessler, M. & Lyko, F. (2005). Epigenetic reactivation of tumor suppressor genes by a novel small-molecule inhibitor of human DNA methyltransferases. *Cancer Res* 65, 6305–6311.

Chan, E., Prado, D. E. & Weidhaas, J. B. (2011). Cancer microRNAs: from subtype profiling to predictors of response to therapy. *Trends Mol Med* 17, 235–243.

Chi, P., Allis, C. D. & Wang, G. G. (2010). Covalent histone modifications--miswritten, misinterpreted and mis-erased in human cancers. *Nat Rev Cancer* 10, 457–469.

Choi, J. E., Kim, D. S., Kim, E. J., Chae, M. H., Cha, S. I., Kim, C. H., Jheon, S., Jung, T. H. & Park, J. Y. (2008). Aberrant methylation of ADAMTS1 in non-small cell lung cancer. *Cancer Genet Cytogenet* 187, 80–84.

Choi, J. H., Li, Y., Guo, J., Pei, L., Rauch, T. A., Kramer, R. S., Macmil, S. L., Wiley, G. B., Bennett, L. B., Schnabel, J. L., Taylor, K. H., Kim, S., Xu, D., Sreekumar, A., Pfeifer, G. P., Roe, B. A., Caldwell, C. W., Bhalla, K. N. & Shi, H. (2010). Genome-wide DNA methylation maps in follicular lymphoma cells determined by methylation-enriched bisulfite sequencing. *PLoS ONE* 5, e13020.

Choi, J. K., Bae, J. B., Lyu, J., Kim, T. Y. & Kim, Y. J. (2009). Nucleosome deposition and DNA methylation at coding region boundaries. *Genome Biol* 10, R89.

Cuneo, K. C., Fu, A., Osusky, K., Huamani, J., Hallahan, D. E. & Geng, L. (2007). Histone deacetylase inhibitor NVP-LAQ824 sensitizes human nonsmall cell lung cancer to the cytotoxic effects of ionizing radiation. *Anticancer Drugs* 18, 793–800.

Dammann, R., Strunnikova, M., Schagdarsurengin, U., Rastetter, M., Papritz, M., Hattenhorst, U. E., Hofmann, H. S., Silber, R. E., Burdach, S. & Hansen, G. (2005). CpG island methylation and expression of tumour-associated genes in lung carcinoma. *Eur J Cancer* 41, 1223–1236.

Dammann, R., Takahashi, T. & Pfeifer, G. P. (2001). The CpG island of the novel tumor suppressor gene RASSF1A is intensely methylated in primary small cell lung carcinomas. *Oncogene* 20, 3563–3567.

Deaton, A. M. & Bird, A. (2011). CpG islands and the regulation of transcription. *Genes Dev* 25, 1010–1022.

Dekker, F. J., Ghizzoni, M., van der, M. N., Wisastra, R. & Haisma, H. J. (2009). Inhibition of the PCAF histone acetyl transferase and cell proliferation by isothiazolones. *Bioorg Med Chem* 17, 460–466.

Dekker, F. J. & Haisma, H. J. (2009). Histone acetyl transferases as emerging drug targets. *Drug Discov Today* 14, 942–948.

Doi, A., Park, I. H., Wen, B., Murakami, P., Aryee, M. J., Irizarry, R., Herb, B., Ladd-Acosta, C., Rho, J., Loewer, S., Miller, J., Schlaeger, T., Daley, G. Q. & Feinberg, A. P. (2009). Differential methylation of tissue- and cancer-specific CpG island shores distinguishes human induced pluripotent stem cells, embryonic stem cells and fibroblasts. *Nat Genet* 41, 1350–1353.

Dooley, A. L., Winslow, M. M., Chiang, D. Y., Banerji, S., Stransky, N., Dayton, T. L., Snyder, E. L., Senna, S., Whittaker, C. A., Bronson, R. T., Crowley, D., Barretina, J., Garraway, L., Meyerson, M. & Jacks, T. (2011). Nuclear factor I/B is an oncogene in small cell lung cancer. *Genes Dev* 25, 1470–1475.

Dunwell, T., Hesson, L., Rauch, T. A., Wang, L., Clark, R. E., Dallol, A., Gentle, D., Catchpoole, D., Maher, E. R., Pfeifer, G. P. & Latif, F. (2010). A genome-wide screen identifies frequently methylated genes in haematological and epithelial cancers. *Mol Cancer* 9:44.

Estecio, M. R. & Issa, J. P. (2011). Dissecting DNA hypermethylation in cancer. *FEBS Lett* 585, 2078–2086.

Esteller, M., Hamilton, S. R., Burger, P. C., Baylin, S. B. & Herman, J. G. (1999). Inactivation of the DNA repair gene O6-methylguanine-DNA methyltransferase by promoter hypermethylation is a common event in primary human neoplasia. *Cancer Res* 59, 793–797.

Feinberg, A. P. & Vogelstein, B. (1983). Hypomethylation distinguishes genes of some human cancers from their normal counterparts. *Nature* 301, 89–92.

Feltus, F. A., Lee, E. K., Costello, J. F., Plass, C. & Vertino, P. M. (2003). Predicting aberrant CpG island methylation. *Proc Natl Acad Sci U S A* 100, 12253–12258.

Fiorentino, F. P., Macaluso, M., Miranda, F., Montanari, M., Russo, A., Bagella, L. & Giordano, A. (2011). CTCF and BORIS regulate Rb2/p130 gene transcription: a novel mechanism and a new paradigm for understanding the biology of lung cancer. *Mol Cancer Res* 9, 225–233.

Gaudet, F., Hodgson, J. G., Eden, A., Jackson-Grusby, L., Dausman, J., Gray, J. W., Leonhardt, H. & Jaenisch, R. (2003). Induction of tumors in mice by genomic hypomethylation. *Science* 300, 489–492.

Ghizzoni, M., Boltjes, A., Graaf, C., Haisma, H. J. & Dekker, F. J. (2010). Improved inhibition of the histone acetyltransferase PCAF by an anacardic acid derivative. *Bioorg Med Chem* 18, 5826–5834.

Gibb, E. A., Brown, C. J. & Lam, W. L. (2011). The functional role of long non-coding RNA in human carcinomas. *Mol Cancer* 10:38.

Goodier, J. L. & Kazazian, H. H., Jr. (2008). Retrotransposons revisited: the restraint and rehabilitation of parasites. *Cell* 135, 23–35.

Gridelli, C., Maione, P., Rossi, A., Bareschino, M. A., Schettino, C., Sacco, P. C. & Zeppa, R. (2011a). Pemetrexed in advanced non-small cell lung cancer. *Expert Opin Drug Saf* 10, 311–317.

Gridelli, C., Morgillo, F., Favaretto, A., de Marinis, F., Chella, A., Cerea, G., Mattioli, R., Tortora, G., Rossi, A., Fasano, M., Pasello, G., Ricciardi, S., Maione, P., Di Maio, M. & Ciardiello, F. (2011b). Sorafenib in combination with erlotinib or with gemcitabine in elderly patients with advanced non-small-cell lung cancer: a randomized phase II study. *Ann Oncol* 22, 1528–1534.

Gridelli, C., Rossi, A. & Maione, P. (2008). The potential role of histone deacetylase inhibitors in the treatment of non-small-cell lung cancer. *Crit Rev Oncol Hematol* 68, 29–36.

Hagemann, S., Heil, O., Lyko, F. & Brueckner, B. (2011). Azacytidine and decitabine induce gene-specific and non-random DNA demethylation in human cancer cell lines. *PLoS ONE* 6, e17388.

Heller, G., Zielinski, C. C. & Zochbauer-Muller, S. (2010). Lung cancer: from single-gene methylation to methylome profiling. *Cancer Metastasis Rev* 29, 95–107.

Holliday, R. & Pugh, J. E. (1975). DNA modification mechanisms and gene activity during development. *Science* 187, 226–232.

Inoue, A. & Zhang, Y. (2011). Replication-Dependent Loss of 5-Hydroxymethylcytosine in Mouse Preimplantation Embryos. *Science* 334, 194.

Iqbal, K., Jin, S. G., Pfeifer, G. P. & Szabo, P. E. (2011). Reprogramming of the paternal genome upon fertilization involves genome-wide oxidation of 5-methylcytosine. *Proc Natl Acad Sci U S A* 108, 3642–3647.

Irimia, M., Fraga, M. F., Sanchez-Cespedes, M. & Esteller, M. (2004). CpG island promoter hypermethylation of the Ras-effector gene NORE1A occurs in the context of a wild-type K-ras in lung cancer. *Oncogene* 23, 8695–8699.

Irizarry, R. A., Ladd-Acosta, C., Wen, B., Wu, Z., Montano, C., Onyango, P., Cui, H., Gabo, K., Rongione, M., Webster, M., Ji, H., Potash, J. B., Sabunciyan, S. & Feinberg, A. P. (2009). The human colon cancer methylome shows similar hypo- and hypermethylation at conserved tissue-specific CpG island shores. *Nat Genet* 41, 178–186.

Italiano, A., Attias, R., Aurias, A., Perot, G., Burel-Vandenbos, F., Otto, J., Venissac, N. & Pedeutour, F. (2006). Molecular cytogenetic characterization of a metastatic lung sarcomatoid carcinoma: 9p23 neocentromere and 9p23-p24 amplification including JAK2 and JMJD2C. *Cancer Genet Cytogenet* 167, 122–130.

Ji, P., Diederichs, S., Wang, W., Boing, S., Metzger, R., Schneider, P. M., Tidow, N., Brandt, B., Buerger, H., Bulk, E., Thomas, M., Berdel, W. E., Serve, H. & Muller-Tidow, C. (2003). MALAT-1, a novel noncoding RNA, and thymosin beta4 predict metastasis and survival in early-stage non-small cell lung cancer. *Oncogene* 22, 8031–8041.

Jiang, C. L., Jin, S. G. & Pfeifer, G. P. (2004). MBD3L1 is a transcriptional repressor that interacts with methyl-CpG-binding protein 2 (MBD2) and components of the NuRD complex. *J Biol Chem* 279, 52456–52464.

Jin, C. G., Yiang, Y., Qiu, R., Rauch, T. A., Wang, Y., Schackert, G., Krex, D., Lu, Q. & Pfeifer, G. P. (2011). 5-hydroxymethylcytosine is strongly depleted in human cancers, but its levels do not correlate with IDH1 mutations. *Cancer Res* 71, 7360–7365.

Jones, P. A. & Baylin, S. B. (2007). The epigenomics of cancer. *Cell* 128, 683–692.

Kalari, S. & Pfeifer, G. P. (2010). Identification of driver and passenger DNA methylation in cancer by epigenomic analysis. *Adv Genet* 70, 277–308.

Kaneko-Ishino, T. & Ishino, F. (2010). Retrotransposon silencing by DNA methylation contributed to the evolution of placentation and genomic imprinting in mammals. *Dev Growth Differ* 52, 533–543.

Kansara, M., Tsang, M., Kodjabachian, L., Sims, N. A., Trivett, M. K., Ehrich, M., Dobrovic, A., Slavin, J., Choong, P. F., Simmons, P. J., Dawid, I. B. & Thomas, D. M. (2009). Wnt inhibitory factor 1 is epigenetically silenced in human osteosarcoma, and targeted disruption accelerates osteosarcomagenesis in mice. *J Clin Invest* 119, 837–851.

Komatsu, N., Kawamata, N., Takeuchi, S., Yin, D., Chien, W., Miller, C. W. & Koeffler, H. P. (2006). SAHA, a HDAC inhibitor, has profound anti-growth activity against non-small cell lung cancer cells. *Oncol Rep* 15, 187–191.

Kriaucionis, S. & Heintz, N. (2009). The nuclear DNA base 5-hydroxymethylcytosine is present in Purkinje neurons and the brain. *Science* 324, 929–930.

Lai, A. Y. & Wade, P. A. (2011). Cancer biology and NuRD: a multifaceted chromatin remodelling complex. *Nat Rev Cancer* 11, 588–596.

Laird, P. W. (2010). Principles and challenges of genomewide DNA methylation analysis. *Nat Rev Genet* 11, 191–203.

Lee, M. N., Tseng, R. C., Hsu, H. S., Chen, J. Y., Tzao, C., Ho, W. L. & Wang, Y. C. (2007). Epigenetic inactivation of the chromosomal stability control genes BRCA1, BRCA2, and XRCC5 in non-small cell lung cancer. *Clin Cancer Res* 13, 832–838.

Li, H., Rauch, T., Chen, Z. X., Szabo, P. E., Riggs, A. D. & Pfeifer, G. P. (2006). The histone methyltransferase SETDB1 and the DNA methyltransferase DNMT3A interact directly and localize to promoters silenced in cancer cells. *J Biol Chem* 281, 19489–19500.

Licchesi, J. D., Westra, W. H., Hooker, C. M., Machida, E. O., Baylin, S. B. & Herman, J. G. (2008). Epigenetic alteration of Wnt pathway antagonists in progressive glandular neoplasia of the lung. *Carcinogenesis* 29, 895–904.

Lin, P. Y., Yu, S. L. & Yang, P. C. (2010). MicroRNA in lung cancer. *Br J Cancer* 103, 1144–1148.

Liu, X., Sempere, L. F., Guo, Y., Korc, M., Kauppinen, S., Freemantle, S. J. & Dmitrovsky, E. (2011). Involvement of microRNAs in lung cancer biology and therapy. *Transl Res* 157, 200–208.

Loprevite, M., Tiseo, M., Grossi, F., Scolaro, T., Semino, C., Pandolfi, A., Favoni, R. & Ardizzoni, A. (2005). In vitro study of CI-994, a histone deacetylase inhibitor, in non-small cell lung cancer cell lines. *Oncol Res* 15, 39–48.

Mancini, D. N., Singh, S. M., Archer, T. K. & Rodenhiser, D. I. (1999). Site-specific DNA methylation in the neurofibromatosis (NF1) promoter interferes with binding of CREB and SP1 transcription factors. *Oncogene* 18, 4108–4119.

Margueron, R. & Reinberg, D. (2011). The Polycomb complex PRC2 and its mark in life. *Nature* 469, 343–349.

McCabe, M. T., Lee, E. K. & Vertino, P. M. (2009). A multifactorial signature of DNA sequence and polycomb binding predicts aberrant CpG island methylation. *Cancer Res* 69, 282–291.

Mizuno, K., Osada, H., Konishi, H., Tatematsu, Y., Yatabe, Y., Mitsudomi, T., Fujii, Y. & Takahashi, T. (2002). Aberrant hypermethylation of the CHFR prophase checkpoint gene in human lung cancers. *Oncogene* 21, 2328–2333.

Murr, R. (2010). Interplay between different epigenetic modifications and mechanisms. *Adv Genet* 70, 101–141.

Novak, P., Jensen, T., Oshiro, M. M., Wozniak, R. J., Nouzova, M., Watts, G. S., Klimecki, W. T., Kim, C. & Futscher, B. W. (2006). Epigenetic inactivation of the HOXA gene cluster in breast cancer. *Cancer Res* 66, 10664–10670.

Ogata, H., Kobayashi, T., Chinen, T., Takaki, H., Sanada, T., Minoda, Y., Koga, K., Takaesu, G., Maehara, Y., Iida, M. & Yoshimura, A. (2006). Deletion of the SOCS3 gene in liver parenchymal cells promotes hepatitis-induced hepatocarcinogenesis. *Gastroenterology* 131, 179–193.

Oliver, S. S. & Denu, J. M. (2011). Dynamic interplay between histone H3 modifications and protein interpreters: emerging evidence for a "histone language". *Chembiochem* 12, 299–307.

Otterson, G. A., Khleif, S. N., Chen, W., Coxon, A. B. & Kaye, F. J. (1995). CDKN2 gene silencing in lung cancer by DNA hypermethylation and kinetics of p16INK4 protein induction by 5-aza 2'deoxycytidine. *Oncogene* 11, 1211–1216.

Prasanth, K. V. & Spector, D. L. (2007). Eukaryotic regulatory RNAs: an answer to the 'genome complexity' conundrum. *Genes Dev* 21, 11–42.

Rauch, T., Li, H., Wu, X. & Pfeifer, G. P. (2006). MIRA-assisted microarray analysis, a new technology for the determination of DNA methylation patterns, identifies frequent methylation of homeodomain-containing genes in lung cancer cells. *Cancer Res* 66, 7939–7947.

Rauch, T. & Pfeifer, G. P. (2005). Methylated-CpG island recovery assay: a new technique for the rapid detection of methylated-CpG islands in cancer. *Lab Invest* 85, 1172–1180.

Rauch, T., Wang, Z., Zhang, X., Zhong, X., Wu, X., Lau, S. K., Kernstine, K. H., Riggs, A. D. & Pfeifer, G. P. (2007). Homeobox gene methylation in lung cancer studied by genome-wide analysis with a microarray-based methylated CpG island recovery assay. *Proc Natl Acad Sci U S A* 104, 5527–5532.

Rauch, T. A., Zhong, X., Wu, X., Wang, M., Kernstine, K. H., Wang, Z., Riggs, A. D. & Pfeifer, G. P. (2008). High-resolution mapping of DNA hypermethylation and hypomethylation in lung cancer. *Proc Natl Acad Sci U S A* 105, 252–257.

Riggs, A. D. (1975). X inactivation, differentiation, and DNA methylation. *Cytogenet Cell Genet* 14, 9–25.

Riggs, A. D. & Jones, P. A. (1983). 5-methylcytosine, gene regulation, and cancer. *Adv Cancer Res* 40, 1–30.

Rubin, A. F. & Green, P. (2009). Mutation patterns in cancer genomes. *Proc Natl Acad Sci U S A* 106, 21766–21770.

Ruike, Y., Imanaka, Y., Sato, F., Shimizu, K. & Tsujimoto, G. (2010). Genome-wide analysis of aberrant methylation in human breast cancer cells using methyl-DNA immunoprecipitation combined with high-throughput sequencing. *BMC Genomics* 11:137.

Sacco, P. C., Maione, P., Rossi, A., Bareschino, M. A., Schettino, C., Guida, C., Elmo, M., Ambrosio, R., Barbato, V., Zeppa, R., Palazzolo, G. & Gridelli, C. (2011). Combination of radiotherapy and targeted therapies in the treatment of locally advanced non-small cell lung cancer. *Target Oncol* 6, 171–180.

Schanen, B. C. & Li, X. (2011). Transcriptional regulation of mammalian miRNA genes. *Genomics* 97, 1–6.

Sharp, A. J., Stathaki, E., Migliavacca, E., Brahmachary, M., Montgomery, S. B., Dupre, Y. & Antonarakis, S. E. (2011). DNA methylation profiles of human active and inactive X chromosomes. *Genome Res* 21, 1592–1600.

Shivapurkar, N., Toyooka, S., Eby, M. T., Huang, C. X., Sathyanarayana, U. G., Cunningham, H. T., Reddy, J. L., Brambilla, E., Takahashi, T., Minna, J. D., Chaudhary, P. M. & Gazdar, A. F. (2002). Differential inactivation of caspase-8 in lung cancers. *Cancer Biol Ther* 1, 65–69.

Siomi, M. C., Sato, K., Pezic, D. & Aravin, A. A. (2011). PIWI-interacting small RNAs: the vanguard of genome defence. *Nat Rev Mol Cell Biol* 12, 246–258.

Smith, I. M., Glazer, C. A., Mithani, S. K., Ochs, M. F., Sun, W., Bhan, S., Vostrov, A., Abdullaev, Z., Lobanenkov, V., Gray, A., Liu, C., Chang, S. S., Ostrow, K. L., Westra, W. H., Begum, S., Dhara, M. & Califano, J. (2009). Coordinated activation of candidate proto-oncogenes and cancer testes antigens via promoter demethylation in head and neck cancer and lung cancer. *PLoS ONE* 4, e4961.

Stresemann, C., Brueckner, B., Musch, T., Stopper, H. & Lyko, F. (2006). Functional diversity of DNA methyltransferase inhibitors in human cancer cell lines. *Cancer Res* 66, 2794–2800.

Suzuki, C., Takahashi, K., Hayama, S., Ishikawa, N., Kato, T., Ito, T., Tsuchiya, E., Nakamura, Y. & Daigo, Y. (2007). Identification of Myc-associated protein with JmjC domain as a novel therapeutic target oncogene for lung cancer. *Mol Cancer Ther* 6, 542–551.

Suzuki, M., Mohamed, S., Nakajima, T., Kubo, R., Tian, L., Fujiwara, T., Suzuki, H., Nagato, K., Chiyo, M., Motohashi, S., Yasufuku, K., Iyoda, A., Yoshida, S., Sekine, Y., Shibuya, K., Hiroshima, K., Nakatani, Y., Yoshino, I. & Fujisawa, T. (2008). Aberrant methylation of CXCL12 in non-small cell lung cancer is associated with an unfavorable prognosis. *Int J Oncol* 33, 113–119.

Suzuki, M. M. & Bird, A. (2008). DNA methylation landscapes: provocative insights from epigenomics. *Nat Rev Genet* 9, 465–476.

Szyf, M. (2009). Epigenetics, DNA methylation, and chromatin modifying drugs. *Annu Rev Pharmacol Toxicol* 49, 243–263.

Tahiliani, M., Koh, K. P., Shen, Y., Pastor, W. A., Bandukwala, H., Brudno, Y., Agarwal, S., Iyer, L. M., Liu, D. R., Aravind, L. & Rao, A. (2009). Conversion of 5-methylcytosine to 5-hydroxymethylcytosine in mammalian DNA by MLL partner TET1. *Science* 324, 930–935.

Tommasi, S., Dammann, R., Zhang, Z., Wang, Y., Liu, L., Tsark, W. M., Wilczynski, S. P., Li, J., You, M. & Pfeifer, G. P. (2005). Tumor susceptibility of Rassf1a knockout mice. *Cancer Res* 65, 92–98.

Tommasi, S., Karm, D. L., Wu, X., Yen, Y. & Pfeifer, G. P. (2009). Methylation of homeobox genes is a frequent and early epigenetic event in breast cancer. *Breast Cancer Res* 11, R14.

Tripathi, V., Ellis, J. D., Shen, Z., Song, D. Y., Pan, Q., Watt, A. T., Freier, S. M., Bennett, C. F., Sharma, A., Bubulya, P. A., Blencowe, B. J., Prasanth, S. G. & Prasanth, K. V. (2010). The nuclear-retained noncoding RNA MALAT1 regulates alternative splicing by modulating SR splicing factor phosphorylation. *Mol Cell* 39, 925–938.

Tuschl, T. (2001). RNA interference and small interfering RNAs. *Chembiochem* 2, 239–245.

Tycko, B. (2010a). Allele-specific DNA methylation: beyond imprinting. *Hum Mol Genet* 19, R210-R220.

Tycko, B. (2010b). Mapping allele-specific DNA methylation: a new tool for maximizing information from GWAS. *Am J Hum Genet* 86, 109–112.

Van Den Broeck A., Brambilla, E., Moro-Sibilot, D., Lantuejoul, S., Brambilla, C., Eymin, B., Khochbin, S. & Gazzeri, S. (2008a). Loss of histone H4K20 trimethylation occurs in preneoplasia and influences prognosis of non-small cell lung cancer. *Clin Cancer Res* 14, 7237–7245.

Van Den Broeck A., Brambilla, E., Moro-Sibilot, D., Lantuejoul, S., Brambilla, C., Eymin, B., Khochbin, S. & Gazzeri, S. (2008b). Loss of histone H4K20 trimethylation occurs in preneoplasia and influences prognosis of non-small cell lung cancer. *Clin Cancer Res* 14, 7237–7245.

Van Den Broeck A., Brambilla, E., Moro-Sibilot, D., Lantuejoul, S., Brambilla, C., Eymin, B., Khochbin, S. & Gazzeri, S. (2008c). Loss of histone H4K20 trimethylation occurs in preneoplasia and influences prognosis of non-small cell lung cancer. *Clin Cancer Res* 14, 7237–7245.

Vire, E., Brenner, C., Deplus, R., Blanchon, L., Fraga, M., Didelot, C., Morey, L., Van Eynde, A., Bernard, D., Vanderwinden, J. M., Bollen, M., Esteller, M., Di Croce, L., de Launoit, Y. & Fuks, F. (2006). The Polycomb group protein EZH2 directly controls DNA methylation. *Nature* 439, 871–874.

Virmani, A. K., Rathi, A., Zochbauer-Muller, S., Sacchi, N., Fukuyama, Y., Bryant, D., Maitra, A., Heda, S., Fong, K. M., Thunnissen, F., Minna, J. D. & Gazdar, A. F. (2000). Promoter methylation and silencing of the retinoic acid receptor-beta gene in lung carcinomas. *J Natl Cancer Inst* 92, 1303–1307.

Watanabe, K., Emoto, N., Hamano, E., Sunohara, M., Kawakami, M., Kage, H., Kitano, K., Nakajima, J., Goto, A., Fukayama, M., Nagase, T., Yatomi, Y., Ohishi, N. & Takai, D. (2011). Genome structure-based screening identified epigenetically silenced microRNA associated with invasiveness in non-small-cell lung cancer. *Int J Cancer*, 10.

Weber, M., Davies, J. J., Wittig, D., Oakeley, E. J., Haase, M., Lam, W. L. & Schubeler, D. (2005). Chromosome-wide and promoter-specific analyses identify sites of differential DNA methylation in normal and transformed human cells. *Nat Genet* 37, 853–862.

Weiss, G. J., Bemis, L. T., Nakajima, E., Sugita, M., Birks, D. K., Robinson, W. A., Varella-Garcia, M., Bunn, P. A., Jr., Haney, J., Helfrich, B. A., Kato, H., Hirsch, F. R. & Franklin, W. A. (2008). EGFR regulation by microRNA in lung cancer: correlation with clinical response and survival to gefitinib and EGFR expression in cell lines. *Ann Oncol* 19, 1053–1059.

Widschwendter, M., Fiegl, H., Egle, D., Mueller-Holzner, E., Spizzo, G., Marth, C., Weisenberger, D. J., Campan, M., Young, J., Jacobs, I. & Laird, P. W. (2007). Epigenetic stem cell signature in cancer. *Nat Genet* 39, 157–158.

Wossidlo, M., Nakamura, T., Lepikhov, K., Marques, C. J., Zakhartchenko, V., Boiani, M., Arand, J., Nakano, T., Reik, W. & Walter, J. (2011). 5-Hydroxymethylcytosine in the mammalian zygote is linked with epigenetic reprogramming. *Nat Commun* 2, 241.

Wu, X., Rauch, T. A., Zhong, X., Bennett, W. P., Latif, F., Krex, D. & Pfeifer, G. P. (2010). CpG island hypermethylation in human astrocytomas. *Cancer Res* 70, 2718–2727.

Wyatt, G. R. (1950). Occurrence of 5-methylcytosine in nucleic acids. *Nature* 166, 237–238.

Yagi, K., Akagi, K., Hayashi, H., Nagae, G., Tsuji, S., Isagawa, T., Midorikawa, Y., Nishimura, Y., Sakamoto, H., Seto, Y., Aburatani, H. & Kaneda, A. (2010). Three DNA methylation epigenotypes in human colorectal cancer. *Clin Cancer Res* 16, 21–33.

Yang, S. R., Valvo, S., Yao, H., Kode, A., Rajendrasozhan, S., Edirisinghe, I., Caito, S., Adenuga, D., Henry, R., Fromm, G., Maggirwar, S., Li, J. D., Bulger, M. & Rahman, I. (2008). IKK alpha causes chromatin modification on pro-inflammatory genes by cigarette smoke in mouse lung. *Am J Respir Cell Mol Biol* 38, 689–698.

Yanagawa, N., Tamura, G., Oizumi, H., Kanauchi, N., Endoh, M., Sadahiro, M. & Motoyama, T. (2007). Promoter hypermethylation of RASSF1A and RUNX3 genes as an independent prognostic prediction marker in surgically resected non-small cell lung cancers. *Lung Cancer* 58, 131–138.

Zhang, J. G., Guo, J. F., Liu, D. L., Liu, Q. & Wang, J. J. (2011). MicroRNA-101 exerts tumor-suppressive functions in non-small cell lung cancer through directly targeting enhancer of zeste homolog 2. *J Thorac Oncol* 6, 671–678.

Chapter 8
Epigenetic Changes in Virus-Associated Neoplasms

Hans Helmut Niller, Ferenc Banati, Eva Ay, and Janos Minarovits

Abbreviations

5-caC	5-Carboxylcytosine
5-hmC	5-Hydroxymethylcytosine
5-mC	5-Methylcytosine
AIDS-BL	AIDS-related-BL
APC	Adenomatous polyposis coli
BART	BamHI A rightward transcripts
BCBL	Body cavity-based lymphoma
BL	Burkitt's lymphoma
CBF1	C promoter-binding factor 1
CGI	CpG island
cHL	Classical Hodgkin's lymphoma
CIMP	CpG island methylator phenotype
CIN	Cervical intraepithelial neoplasm
CIN	Chromosomal instability
Cp	C promoter
CpG	Cytosine-phosphate-guanine dinucleotide
CRBP	Cellular retinol-binding protein
CTCF	CCCTC-binding factor
DLBCL	Diffuse large B-cell lymphoma
DNMT	DNA methyltransferase
DS	Dyad symmetry

H.H. Niller, MD (✉)
Institute for Medical Microbiology and Hygiene at the University
of Regensburg, Regensburg, Germany
e-mail: Hans-Helmut.Niller@klinik.uni-regensburg.de

F. Banati • E. Ay • J. Minarovits, MD, MSc
Microbiological Research Group, National Center for Epidemiology, Budapest, Hungary

J. Minarovits and H.H. Niller (eds.), *Patho-Epigenetics of Disease*,
DOI 10.1007/978-1-4614-3345-3_8, © Springer Science+Business Media New York 2012

EBER	Epstein–Barr encoded small RNA
eBL	Endemically occurring BL
EBNA	Epstein–Barr nuclear antigen
EBNA-LP	EBNA-leader protein
EBV	Epstein–Barr virus
EZH2	Enhancer of zeste homologue 2
FR	Family of repeats
GC	Gastric carcinoma
GC	Germinal center
GSTP1	Glutathione S-transferase P1
H3K27me3	Histone 3 trimethylated on lysine 27
H3K4me2	Histone 3 dimethylated on lysine 4
H3K4me3	Histone 3 trimethylated on lysine 4
H3K9me3	Histone 3 trimethylated on lysine 9
HBsAg	Hepatitis B surface antigen
HBV	Hepatitis B virus
HCC	Hepatocellular carcinoma
HCV	Hepatitis C virus
HDAC	Histone deacetylase
HHV-8	Human herpesvirus-8
HL	Hodgkin's lymphoma
HP1	Heterochromatin-associated protein 1
HPV	Human papillomavirus
HRS	Hodgkin and Reed–Sternberg cells
HTLV	Human T-lymphotropic virus
IM	Infectious mononucleosis
KDM	Lysine demethylase
KSHV	Kaposi's sarcoma herpesvirus
LANA	Latency-associated nuclear antigen
LCL	Lymphoblastoid cell line
LCR	Locus control region *or* long control region
LINE-1	Long interspersed element-1
LMP	Latent membrane protein
LMP2Ap	LMP2A promoter
lncRNA	Long noncoding RNA
MCD	Multicentric Castleman's disease
MCPyV	Merkel cell polyomavirus
MeCP2	Methylcytosine-binding protein 2
MGMT	O6-methylguanine DNA methyltransferase
miRNA	MicroRNA
NHL	Non-Hodgkin lymphoma
NPC	Nasopharyngeal carcinoma
PAN	Polyadenylated nuclear RNA
PcG	Polycomb group
PEL	Primary effusion lymphoma

PIN	Prostatic intraepithelial neoplasias
pRB	Retinoblastoma protein
PRC	Polycomb repressive complex
PTEN	Phosphatase and tensin homologue
PTLD	Posttransplant lymphoproliferative disorder
Qp	Q promoter
RARβ2	Retinoic acid receptor β2
RARRES	Retinoic acid receptor responder
RASSF1A	RAS association domain family 1 isoform A
sBL	Sporadic BL
SFRP1	Secreted frizzled-related protein 1
siRNA	Short interfering RNA
snoRNA	Small nucleolar RNA
SOCS	Suppressor of cytokine signaling
TAg	T antigen
TR	Terminal repeat
TrxG	Trithorax group
Wp	W promoter

8.1 Introduction

Viruses are associated with a significant fraction of neoplasms in mammals. Similarly to the malignant tumors elicited by other agents and the so-called spontaneous neoplasms of unknown etiology, the carcinomas, leukemias, lymphomas, and sarcomas induced by oncoviruses also frequently accumulate both genetic aberrations and epigenetic changes. It is well documented that infection by oncoviruses may introduce new information into the host cell DNA and change the structure and expression pattern of the cellular genome. DNA virus genomes or DNA copies of retrovirus (RNA tumor virus) genomes may either act as insertional mutagens that inactivate cellular genes, or may affect the regulation and expression of key cellular genes by the mechanism of promoter insertion or enhancer insertion. Furthermore, certain oncoproteins encoded by tumor viruses may directly elicit mitotic disturbances resulting in the generation of aneuploid cells or upregulate cellular enzymes with mutagenic activities.

Notwithstanding these remarkable observations, most researchers agreed that the main effect of the pleiotropic viral oncoproteins was either a direct, constitutive stimulation of cell proliferation or a continuous maintenance of cell growth in a less direct manner, via interfering with a series of tumor suppressor proteins involved in cell cycle regulation, apoptosis, and maintenance of genomic integrity. More subtle events attesting that epigenetic alterations, affecting both viral and cellular DNA sequences, also occur in transformed cells—see the review on oncogenic human adenoviruses by Walter Doerfler, Chap. 1—were not connected directly to the phenomenon of "malignant transformation" in the minds of most investigators.

Most recently, however, the focus of oncovirus research shifted, and a new concept, the epigenetic reprogramming of host cells by oncoproteins, gained momentum. Other tumor-associated pathogens, most notably *Helicobacter pylori*, a bacterium causing the majority of gastric cancer cases (see Chap. 14), and certain macroparasites, such as *Schistosoma haematobium* associated with bladder cancer (Gutierrez et al. 2004) and *Opisthorchis viverrini* (liver fluke) associated with cholangiocarcinoma (Chinnasri et al. 2009; Sriraksa et al. 2011), were also linked to the alterations of the host cell epigenome.

A recent comprehensive analysis of genome-wide DNA methylation patterns in a series of human neoplasms including colon carcinoma, lung carcinoma, breast carcinoma, thyroid carcinoma, and Wilms' tumor identified cancer-specific differentially DNA-methylated regions that lost the epigenetic stability characteristic to the corresponding normal tissues (Hansen et al. 2011). The loss of DNA methylation stability was associated with an increased gene expression variation. The major difference between normal and neoplastic tissues was the appearance of large, contiguous hypomethylated blocks in the analyzed carcinomas, although a small fraction of hypermethylated blocks were also detected. Hansen et al. suggested that the loss of epigenetic stability, i.e., increased CpG methylation variability and gene expression variability, might have a potential selective value in a changing environment. As formulated by Issa, an increased epigenetic plasticity may provide "a mechanism of Darwinian evolution at the cellular level that may underlie age-related diseases such as cancer" (Issa 2011).

The neoplasms studied by Hansen et al. are usually not regarded as virus-associated. Thus, the existence of epigenetically hypervariable regions remains to be demonstrated for virus-associated tumors. It is important to note, however, that regularly occurring stable epigenetic changes were also described both in virus-associated and non-virus-associated neoplasms. These recurrent, stable epigenetic alterations that are maintained during successive stages of tumor progression frequently affect the expression of a set of tumor suppressor genes or tumor-associated genes independently of the histological type of the neoplasm. Other stable epigenetic changes or their combinations appear to be specific for certain tumor types, or mark distinct stages of neoplastic development, and oncovirus-specific epigenetic signatures were also identified (reviewed by Niller et al. 2009). In this chapter, we focus almost exclusively on viruses associated with human neoplasms. Human viruses causing malignant tumors in experimental animals but not associated regularly with human neoplasms are either discussed in Chap. 1 (human adenoviruses) or described only briefly (human BK polyomavirus, this chapter, Sect. 8).

8.2 Epigenetic Alterations in Epstein–Barr Virus-Associated Neoplasms

Epstein–Barr virus (EBV), the first human "tumor virus" to be discovered, was observed initially by electron microscopy (Epstein et al. 1964) in cultivated cells derived from Burkitt's lymphomas (BLs, Burkitt 1958, 1962). For this historical

8 Epigenetic Changes in Virus-Associated Neoplasms

reason, and because EBV is one of the most comprehensively studied viruses with regard to the epigenetic regulation of the viral oncogenes and the pathoepigenetic consequences of viral infection, we start this chapter on oncovirus-associated dysregulation of the host cell epigenome with a brief description of the natural history and latency types of EBV (for more detailed reviews, see Niller et al. 2007, 2008).

8.2.1 Epstein–Barr Virus: Basic Facts, Natural History, and Latency Types

EBV is one of the eight known human pathogenic herpesviruses. The double-stranded linear DNA genome packaged into the virions of the prototype EBV strain B95-8 has a length of 172 kb pairs. Based on biological properties and sequence comparisons, EBV was classified as a member of the genus *Lymphocryptovirus* within the subfamily *Gammaherpesviridae* of the family *Herpesviridae*. The name *Lymphocryptovirus* refers to a fascinating property of EBV and other members of the genus: these viruses "hide" in lymphoid organs. This means that in addition to productive (lytic) replication that occurs in the epithelial cells of the oropharynx, EBV genomes can also persist for an extended period without the production of virions in latently infected host cells, notably within resting memory B lymphocytes. As we discuss it below, the expression of the EBV genome is highly restricted during latent infection in peripheral B cells, and in the virtual absence of viral protein expression (latency type 0, see Table 8.1), such EBV-infected B cells remain practically "invisible" for the immune system.

About 90% of the world's population is infected by EBV. The virus is intermittently shed into the saliva of persistently infected individuals, and the saliva is the main route of transmission to uninfected individuals. Although primary EBV infection is usually inapparent in early childhood, in adolescents or adults it may cause infectious mononucleosis (IM; also called glandular fever or kissing disease). Lymphadenopathy, an essential feature of IM, is a consequence of the proliferation of EBV-infected, activated B cells. In contrast to resting B cells, the EBV genome is not completely silenced in B lymphoblasts of IM patients, and the expressed viral oncoproteins (nuclear antigens, called EBNAs, and latent membrane proteins, LMPs; latency type III, see Table 8.1) stimulate continuous cell proliferation. Because most of the EBNAs as well as the LMPs are highly immunogenic, i.e., "visible" for the cells of the adaptive immune system, IM is usually curbed by a vigorous cellular immune response directed against latent EBV proteins. Latent, transcriptionally silent EBV genomes persist, however, lifelong in memory B cells.

EBV is associated with a series of malignant tumors including lymphomas (Burkitt's lymphoma, Hodgkin's lymphoma, T-/NK-cell lymphoma, posttransplant lymphoproliferative disease (PTLD), AIDS-associated lymphoma, X-linked lymphoproliferative syndrome), carcinomas (nasopharyngeal carcinoma (NPC), gastric carcinoma, carcinomas of the major salivary glands, thymic carcinoma, mammary carcinoma), and a sarcoma (leiomyosarcoma) (see Table 8.2). Strikingly, the incidence of Hodgkin's

Table 8.1 Latency types of Epstein–Barr vi

Latency type	Representative cell type	Active promoter	Expressed product
0	Resting B cell	Qp (?)	EBNA1 (variable?)
		LMP2Ap (?)	LMP2A (?)
I	Burkitt's lymphoma	Qp	EBNA1
		EBER1p	EBER1
		EBER2p	EBER2
		BARTp	BART, microRNAs processed from BART
II	Nasopharyngeal carcinoma	Qp	EBNA1
		EBER1p	EBER1
		EBER2p	EBER2
		BARTp	BART, microRNAs processed from BART
		LMP1p	LMP1 (variable)
		LMP2Ap	LMP2A
		LMP2Bp	LMP2B
GC	Gastric carcinoma	Qp	EBNA1
		EBER1p	EBER1
		EBER2p	EBER2
		BARTp	BART, microRNAs processed from BART
		LMP2Ap	LMP2A (variable)
		BARF1p	BARF1
III	Lymphoblastoid cell line	Cp	EBNA1-6, miR-BHRF1-3
		EBER1p	EBER1
		EBER2p	EBER2
		BARTp	BART, microRNAs processed from BART
		LMP1p	LMP1
		LMP2Ap	LMP2A
		LMP2Bp	LMP2B

Table 8.2 Epstein–Barr virus-associated neoplasms

Lymphomas
Burkitt's lymphoma
Hodgkin's disease
T-/NK-cell lymphoma
Posttransplant lymphoproliferative disease
AIDS-associated lymphoma
X-linked lymphoproliferative disease
Lymphomas in methotrexate-treated rheumatoid arthritis and polymyositis patients

Carcinomas
Nasopharyngeal carcinoma
Gastric carcinoma
Carcinomas of the major salivary glands
Thymic carcinoma
Mammary carcinoma

Sarcoma
Leiomyosarcoma

8 Epigenetic Changes in Virus-Associated Neoplasms

Table 8.3 Epigenetic regulation of latent Epstein–Barr virus promoters

		Epigenetic mark		
Promoter	Activity state	CpG methylation	Histone acetylation	H3K4me2
Qp	On	–	+	+
	Off	–	+	+
Cp	On	–	+	+
	Off	+	–	+/–
Wp	On	+		
	Off	–		
EBER1p	On	–		
	Off	?		
EBER2p	On	–		
	Off	?		
BARTp	On	(–)		
	Off	?		
LMP1p	On	–		
	Off	–		
LMP2Ap	On	–	+	+
	Off	+	–	+/–

lymphoma (HL) is increased after passing through symptomatic primary infection, i.e., IM (Niller et al. 2011), and the incidence of multiple sclerosis is increased after EBV infection by itself and even higher after IM, in addition (Niller et al. 2008).

The expression pattern of latent EBV genomes depends on the host cell phenotype. In vivo, BL cells express only a single EBV-encoded nuclear antigen, EBNA1, a transcription and replication factor (latency type I, see Table 8.1). In addition, two EBV-encoded small RNAs (EBER1 and 2) and a family of multiply spliced transcripts encoded by the *Bam*HI A fragment of the viral genome (BARTs, *Bam*HI A rightward transcripts) are also transcribed in BLs. BARTs potentially code for proteins, but they are also processed to viral microRNAs influencing the level of both viral and cellular mRNAs. In HL and NPC, latent membrane proteins (LMP1, LMP2A, LMP2B) can also be detected in addition to EBNA1, EBERs, BARTs, and viral microRNAs (latency type II, see Table 8.1). LMP1 is an oncoprotein contributing to apoptosis resistance of the infected cell, whereas LMP2A enhances metastasis formation. A unique latency type is characteristic for EBV-associated gastric carcinoma (GC) cells. Although the expression pattern is similar to that of latency type I cells, in addition to the typical latency type I EBV products, BARF1, originally characterized as a lytic cycle protein, and—variably—LMP2A are also expressed in GCs (Tables 8.1 and 8.3).

Whereas in all of the neoplasms (lymphomas and carcinomas) described above EBNA1 transcripts are initiated at Qp, a promoter located to the *Bam*HI Q fragment of the EBV genome, expanding B cells of PTLD and in vitro immortalized B lymphoblastoid cell lines (LCLs) use a B lymphoblast-specific promoter, Cp, to initiate transcripts coding for 6 nuclear antigens (EBNA1-6; Table 8.3). This expression pattern is similar to that of the proliferating B cells of IM patients (latency type III, Table 8.1). Similarly to the BARTs, viral microRNAs are also processed from the

EBNA transcripts initiated at Cp. Since EBNA2, the major transactivator protein of EBV, switches on the expression of the LMP1, LMP2A, and LMP2B genes, PTLDs and LCLs regularly express latent membrane proteins. In other LMP-expressing cell types, cellular transcription factors may substitute for EBNA2 and switch on the LMP promoters.

8.2.2 Epigenotypes of Latent Epstein–Barr Virus Genomes

The cell type-specific expression of latent EBV genes is achieved by epigenetic regulatory mechanisms controlling the activity of the alternative promoters Cp and Qp, and a third promoter for EBNA1-6 transcripts, Wp, that is less frequently used in tumor cells. The very same epigenetic regulatory mechanisms (DNA methylation, histone modifications, protein–DNA interactions) determine the activity of the promoters for LMP1, LMP2A, and LMP2B transcripts as well.

DNA methylation at position 5 of cytosine is involved in transcriptional silencing via the establishment of a "closed" chromatin structure suppressing transcription. DNA methylation patterns are *maintained* by DNA methyltransferase 1 (DNMT1) that restores the methylation pattern of the parental strands on the initially unmethylated daughter strands during DNA replication, whereas other DNMTs (DNMT3A, DNMT3B) can act on completely unmethylated DNA strands (de novo DNA methyltransferases, see Chap. 1). DNA methylation is reversible: one could distinguish active and passive mechanisms of DNA demethylation. The recently explored active pathway proceeds through conversion of 5-methylcytosine (5-mC) to 5-hydroxymethylcytosine (5-hmC) and further to 5-carboxylcytosine (5-caC) by the Tet family of dioxygenases, followed either by decarboxylation (Ito et al. 2011) or excision by thymine-DNA glycosylase that triggers the base excision repair pathway replacing thereby 5-mC with C (He et al. 2011). The passive pathway involves the inhibition of DNMT1 activity during two successive cell cycles that may result in hemimethylated and finally completely unmethylated DNA stretches.

The methylcytosine-binding protein 2 (MeCP2) can attach to hypermethylated DNA sequences with high affinity and attract histone deacetylases that remove the acetyl moieties of histone tails, thereby eliciting chromatin compaction (Nan et al. 1998b). Thus, silent promoters are frequently associated with histones H3 and H4 that are devoid of acetylation. In contrast, active promoters are frequently unmethylated, and they are associated with an "open" chromatin configuration, due to the action of histone acetyltransferases that enrich the chromatin in acetylated histones H3 and H4, thereby creating so-called acetylation islands (Roh et al. 2005). DNA methylation regulates the activity of latent EBV promoters Cp, Wp, LMP1p, LMP2Ap, and BARTp (Table 8.3). Acetylation islands were identified at the active EBV latency promoters Cp, Qp, and LMP2Ap (Table 8.3).

DNA methylation patterns are transmitted from cell generation to cell generation (epigenetic memory). An alternative system of epigenetic memory, the Polycomb–Trithorax group (TrxG) of protein complexes, also ensures a heritable regulation of gene expression. Polycomb group (PcG) protein complexes can silence promoters.

The histone lysine methyltransferase enhancer of zeste homologue 2 (EZH2), a member of the Polycomb repressive complex 2 (PRC2), trimethylates lysine 27 of histone H3, thereby producing a repressive histone mark (H3K27me3), whereas PRC1 has a histone ubiquitinase activity (reviewed by Ringrose and Paro 2007; Blomen and Boonstra 2011). The repressive mark H3K27me3 left on the chromatin by PRC2 serves as a recruitment site for PRC1. In contrast to PcG protein complexes that silence promoters, histone lysine methyltransferase members of the TrxG protein complex create an activating histone modification (H3K4me3, histone H3 trimethylated on lysine 4). It is interesting to note that certain histone lysine demethylases, which are also components of the TrxG complex, actively remove the repressive H3K27me3 mark left by PRC2. The antagonism is mutual because other histone lysine demethylases, associated with PcG complexes, remove the activating H3K4me3 mark (reviewed by Blomen and Boonstra 2011). The role of PcG protein complexes in the regulation of latent EBV promoters remains to be elucidated. The H3K4me3 mark left by the TrxG complex was identified, however, at the active promoters Cp, Qp, and LMP2Ap (Table 8.2). In summary, the latent EBV genomes carried by host cells as nuclear matrix-attached circular episomes are "decorated" with cell type-specific epigenetic marks. On this basis, one can distinguish between unique viral epigenotypes (Minarovits 2006). The epigenetic marks associated with latent EBV episomes or viral epigenotypes ensure differential expression of the identical or nearly identical viral genomes in various host cells. The best characterized epigenotypes of latent EBV genomes, that correspond to some of the major latency types, are shown in Fig. 8.1.

Whereas Fig. 8.1 depicts a linear model of the major EBV latency types based on the composition (i.e., epigenetic marks) of the chromatin fiber, recent data suggest that the 3D structure of chromatin may also influence the activity of EBV latency promoters. Tempera et al. described that distinct, alternative chromatin loops of the EBV episome could be detected in a latency type I BL cell line and a latency type III LCL (Tempera et al. 2011; Fig. 8.2). According to the chromatin conformational model of EBV latency, in the examined latency type I BL cell line, the dyad symmetry (DS) element and the family of repeats (FR) element of *oriP*, the latent origin of EBV replication, associate with Qp in concert with a binding site of the insulator protein CTCF located between *oriP* and Cp. CTCF plays an integral role in this interaction which results in the formation of a chromatin loop between Qp and *oriP* and activates the initiation of EBNA1 transcripts at Qp (Fig. 8.2a). In contrast, in a latency type III LCL, the DS element of *oriP* interacts with the CTCF-binding site situated between *oriP* and Cp, i.e., upstream of the C promoter. This interaction generates an alternative chromatin loop, in this case between *oriP* and Cp, and facilitates active transcription of the polycistronic mRNA coding for 6 EBNA proteins, at Cp (Fig. 8.2b) (Tempera et al. 2011). When independently confirmed, these data may help to build up cell type-specific 3D models of latent EBV episomes. Further studies may also help to firmly establish the relationship between alternative chromatin loops and transcriptional activity of EBV latency promoters, too.

Although Tempera et al. could successfully deplete CTCF in EBV-bacmid carrying HEK-293 cells using a short interfering RNA (siRNA) targeting CTCF (siCTCF), and thereby abolish the chromatin loop between Qp and *oriP*, this intervention had

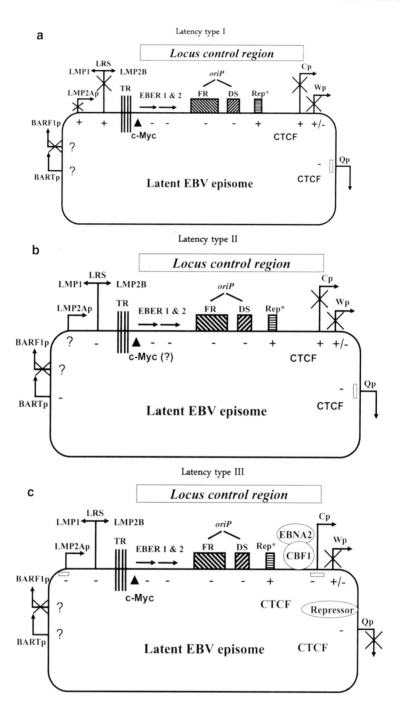

8 Epigenetic Changes in Virus-Associated Neoplasms

only a moderate impact on the activity of Qp and Cp. Accordingly, neither EBNA1 nor EBNA2 mRNA levels were significantly affected by siCTCF and the disruption of *oriP*–Qp interaction in HEK-293 cells (Tempera et al. 2011). It is also worthy to consider that Cp-reporter gene constructs lacking *oriP* but containing a binding site for the nuclear protein CBF1 (C promoter-binding factor 1/RBP-Jκ) are highly active in latency type III cells (Minarovits et al. 1994). Such constructs are apparently unable to form an intraepisomal *oriP*–Cp loop, suggesting that the presence of a large chromatin loop upstream of Cp is dispensable for Cp activity. In addition, in cell lines carrying latent EBV genomes, CTCF binds to its recognition sequence located upstream of Cp independently of Cp activity (Salamon et al. 2009). In cells actively using Cp, CTCF binding to this site does not appear to block the putative long-range enhancer activity of *oriP* (Salamon et al. 2009). It does not function as a "barrier insulator," either, because it does not prevent spreading of CpG methylation through the CTCF-bound sequence in cell lines carrying silent Cp (Salamon et al. 2009). It was also observed that CTCF binds to both silent and active Q promoters (Salamon et al. 2009). Thus, one may argue that CTCF binding to the EBV episomes does not necessarily influence the activity of the alternative promoters Cp and Qp. Based on these considerations, Takacs et al. suggested that CTCF might play a structural role in the physiology of latent EBV genomes by contributing to

Fig. 8.1 Epigenotypes of latent Epstein–Barr virus genomes. (**a**) Latency type I epigenotype. The circular episomal genome is shown with the latent viral promoters (*arrows*) and their regulatory regions (not to scale). An LCR involved in attachment to nuclear matrix includes *oriP*, the latent origin of virus replication that also acts as a long-range enhancer. *TR* terminal repeats; *c-Myc* a nuclear protein-binding upstream of EBER1 (as indicated by a *triangle*); *EBERs 1 and 2* transcription units for nontranslated viral RNA molecules; *FR, DS, Rep** elements involved in latent EBV replication; *CTCF* insulator protein. *Symbols*: "+" indicates a high level of regional CpG methylation, "−" indicates unmethylated or hypomethylated CpG dinucleotides; "crossed-out arrow," silent promoter; "open box," hyperacetylated island, a chromatin region favoring transcription. Qp, EBER1p, EBER2p, and BARTp are active. (**b**) Latency type II epigenotype. The circular episomal genome is shown with the latent viral promoters (*arrows*) and their regulatory regions (not to scale). An LCR involved in attachment to nuclear matrix includes *oriP*, the latent origin of virus replication that also acts as a long-range enhancer. *TR* terminal repeats; *c-Myc* a nuclear protein-binding upstream of EBER1 (as indicated by a *triangle*); *EBERs 1* and 2 transcription units for nontranslated viral RNA molecules; *FR, DS, Rep** elements involved in latent EBV replication; *CTCF* insulator protein. *Symbols*: "+" indicates a high level of regional CpG methylation, "−" indicates unmethylated or hypomethylated CpG dinucleotides; "crossed-out arrow," silent promoter; "open box," hyperacetylated island, a chromatin region favoring transcription. Qp, EBER1p, EBER2p, and BARTp, LMP1p, LMP2Ap, and LMP2Bp are active. (**c**) Latency type III epigenotype. The circular episomal genome is shown with the latent viral promoters (arrows) and their regulatory regions (not to scale). An LCR involved in attachment to nuclear matrix includes *oriP*, the latent origin of virus replication that also acts as a long-range enhancer. *TR* terminal repeats; *c-Myc* a nuclear protein-binding upstream of EBER1 (as indicated by a *triangle*); *EBERs 1* and 2 transcription units for nontranslated viral RNA molecules; *FR, DS, Rep** elements involved in latent EBV replication; *CTCF* insulator protein. *Symbols*: "+" indicates a high level of regional CpG methylation, "−" indicates unmethylated or hypomethylated CpG dinucleotides; "crossed-out arrow," silent promoter; "open box," hyperacetylated island, a chromatin region favoring transcription. Cp, EBER1p, EBER2p, and BARTp, LMP1p, LMP2Ap, and LMP2Bp are active

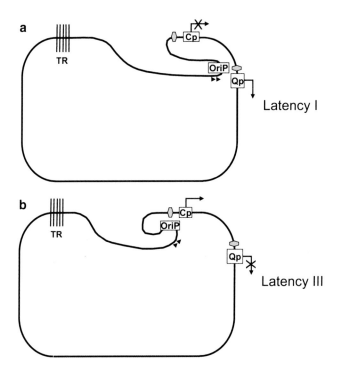

Fig. 8.2 A chromatin-looping model for the regulation of latent Epstein–Barr virus promoters (based on the work of Tempera et al. 2011). (**a**) In latency type I, *oriP*, which functions as an enhancer, associates with the chromatin insulator protein CTCF (*blue hexagonal symbol*) bound at Qp. Qp is on; Cp is off. (**b**) In latency type III, *oriP* interacts with a region upstream of Cp via CTCF (*blue hexagonal symbol*). Qp is off; Cp is on

the composition of the 3D organization of the EBV episome via recruiting additional nuclear factors and CTCF–CTCF interactions (Takacs et al. 2010). We speculate that a potential nuclear protein recruited by CTCF to EBV episomes could be cohesin, a CTCF partner known to mediate cell type-specific looping events that are unrelated to the regulation of promoter activity in lymphoid cells. Our suggestion is based on the observation of Degner et al., who studied the formation of chromatin loops during recombination of immunoglobulin gene segments. They found very similar CTCF-binding patterns throughout the *IgH* locus in different lymphoid cell types displaying very different looping patterns. In contrast, binding of the cohesin subunit Rad21 was extensive at all of the CTCF sites in pro-B cells active in V-DJ recombination but reduced in pre-B cells and thymocytes that do not show such an activity. Thus, cohesin binding to CTCF sites may facilitate multiple loop formation and V-DJ recombination (Degner et al. 2009). Recently Kang et al. described that both CTCF and cohesin were involved in the organization of chromatin loops of the Kaposi's sarcoma-associated herpesvirus (KSHV) genome (Kang et al. 2011). We conclude that the exact function of alternative 3D patterns of EBV chromatin organization remains to be explored.

8 Epigenetic Changes in Virus-Associated Neoplasms

Table 8.4 Mechanisms of epigenetic dysregulation in host cells carrying latent Epstein–Barr virus genomes

Latency product	Cellular partner	Putative outcome
EBNA1	EBNA1-binding sites	Demethylation, gene *activation*
EBNA2	Histone acetyltransferases	Gene *activation*
EBNA3C (EBNA6)	Prothymosin alpha, p300; histone deacetylases (HDAC1 and 2)	*Modulating* EBNA2-mediated transactivation
EBNA-LP (EBNA5)	Histone deacetylase 4 (HDAC4)	Displacement of HDAC4 from EBNA2-activated promoters, *coactivation*
LMP1	Upregulation of DNMT1,DNMT3A, and DNMT3B via the JNK pathway	Promoter *silencing*
	Bmi-1 (a component of PRC1)	Promoter *silencing* Promoter *activation*
	KDM6B (histone H3K27me3 demethylase)	Promoter *activation*
	Upregulation of microRNA levels via the NF-κB pathway	miR-146a and miR-155 *Modulate* cellular mRNA levels
	Downregulation of KDM3A (histone H3K9me2 demethylase) via upregulation of miR-155	Promoter *silencing*
LMP2A	Upregulation of DNMT1, DNMT3A, and DNMT3B via the NF-κB pathway	Promoter *silencing*
	Downregulation of KDM3A (histone H3K9me2 demethylase) via upregulation of miR-155	Promoter *silencing*

8.2.3 Epigenetic Alterations in Epstein–Barr Virus-Associated Neoplasms: Common Epigenetic Events and Virus-Specific Epigenetic Signatures

EBV-associated neoplastic cells differ from their normal counterparts regarding their epigenome and gene expression pattern (transcriptome). Tumor-specific epigenetic changes regularly occur, independently of the presence or absence of EBV genomes, in malignant tumors belonging to different histological types. Other tumor-specific epigenetic alterations are restricted to a certain histological type only or appear to be unique for EBV-associated neoplasms (EBV-specific epigenetic signatures).

EBV-specific epigenetic signatures may mark the epigenome of the host cell due to the interaction of certain EBV-encoded proteins with the cellular epigenetic regulatory machinery (Takacs et al. 2009, Table 8.4). The nuclear antigen EBNA1 may elicit, in principle, site-specific DNA demethylation at its cellular binding sites, similarly to the demethylation observed at *oriP*, the latent origin of EBV replication (Lin et al. 2000). The transactivator protein EBNA2, associating with cellular histone acetyltransferases (Wang et al. 2000), can switch on both viral and cellular promoters. EBNA3C (EBNA6) interacts with both histone deacetylases (HDACs) and histone acetyltransferases (Cotter and Robertson 2000; Knight et al. 2003). EBNA-LP (a leader protein, also called EBNA5) may displace HDAC4 from EBNA2-bound promoters (Portal et al. 2006; Ling et al. 2009). The transmembrane protein LMP1 upregulates DNMT1, DNMT3A, and DNMT3B as well as the PcG group protein Bmi-1 resulting in the silencing of cellular promoters (Tsai et al. 2002, 2006; Dutton et al. 2007). It is worthy to note, however, that Bmi-1 also mediates upregulation of certain LMP1 target genes implicated in lymphomagenesis. LMP1 may also activate its target genes by induction of KDM6B, a histone H3K27me3 demethylase which removes the trimethyl mark from lysine 27 of histone H3 and thereby dissociates PRC1 PRCs from their binding sites (Anderton et al. 2011). LMP1 affects cellular mRNA levels by inducing the expression of cellular microRNAs as well (Motsch et al. 2007; Cameron et al. 2008; Gatto et al. 2008; Rahadiani et al. 2008; Anastasiadou et al. 2010; Li et al. 2010). Similarly to LMP1, the transmembrane protein LMP2A may also upregulate DNMT1 transcription resulting in the inactivation of the tumor suppressor gene *PTEN* (*phosphatase and tensin homologue, deleted on chromosome ten*) by promoter methylation (Hino et al. 2009).

8.2.3.1 Burkitt's Lymphoma: Alternative Scenarios and Epigenetic Changes

BL is a high grade aggressive non-Hodgkin lymphoma (NHL) whose tumor cells resemble germinal center centroblasts. The starry-sky appearance on histological slides is due to interspersed macrophages that clear the cellular debris from the highly apoptosis-prone tumor. Sporadic BL (sBL) occurs worldwide, is EBV-infected in about 20% of cases, and mainly affects young adults. In equatorial Africa and New Guinea, BL occurs endemically (eBL) with a 100-fold higher incidence than sporadic of about 5–10 cases/100,000 children per year and is EBV-infected in more than 95% of cases. The geographical distribution of eBL coincides with areas endemic for malaria tropica. Local hotspots of high BL incidence are likely due to additional arbovirus epidemic outbreaks (van den Bosch 2004). BL occurs with intermediate incidences in other countries, e.g., in Algeria, Egypt, or Brazil where children suffer from a relatively high load of parasites other than *Plasmodium falciparum*. In those countries, the higher incidences in comparison to sBL are made up by EBV-positive cases; thus, those are in reality eBLs. BL incidence is also increased among HIV-infected, but not among otherwise severely immune-suppressed patients.

AIDS-related BL (AIDS-BL) is mostly a disease of the early lymphadenopathic phases of HIV-disease when the germinal centers are hyperactive, the CD4+ cell numbers are still within a normal range, and the immune system is still functional. Two thirds of AIDS-BL cases are EBV-negative in western countries, while they are almost entirely EBV-positive in equatorial Africa. BLs regularly carry a translocation between the c-Myc gene and one of the immunoglobulin loci as a most characteristic genetic lesion which seems to occur during the somatic hypermutation or class switch recombination in the germinal centers of the lymph nodes

8 Epigenetic Changes in Virus-Associated Neoplasms

(Goossens et al. 1998). The chromosomal translocation points of the eBL and sBL subtypes show a different local preference (Magrath et al. 1992). Myc translocation leads to the activation and constitutive overexpression of c-Myc in B cells. The pleiotropic effects of the transcription factor and chromatin modifier c-Myc on cell growth, cell cycle progression, differentiation, and apoptosis through binding to a set of cellular key promoters and the role of c-Myc overexpression for BL pathology have been extensively reviewed (e.g., Hecht and Aster 2000; Levens 2003). Expression profiling of different types of B-cell lymphomas established a molecular profile for sBL. BLs without c-Myc translocation but with an mRNA expression profile characteristic for BL do exist in about 10% of sporadic cases (Staudt and Dave 2005; Hummel et al. 2006; Dave et al. 2006). The three distinct forms of BL, endemic, sporadic, and AIDS-related, seem to arise from slightly different stages of the germinal center (GC) reaction. Therefore, sBL are supposed to originate from early centroblasts, and eBL from late germinal center cells or memory B cells (Bellan et al. 2005). Indeed, expression profiling showed that BL is a unique entity different from other B-cell malignancies and that the three BL subtypes, eBL, sBL, and AIDS-BL, with few differences in gene expression, are a highly homogeneous group which is also closely related to germinal center cells (Piccaluga et al. 2011). Also miRNA expression profiles were highly similar between the different BL subtypes with only few slight differences. Actually, the EBV and HIV status of the respective BLs had no influence on the miRNA expression profiles (Lenze et al. 2011). Even mouse models for EBV-negative and EBV-associated BL did not yield a significant difference in gene expression between both (Bieging et al. 2011). The overall small differences in the mRNA and miRNA expression profiles between the different BL subtypes support the view that they may represent only one biological entity (Lenze et al. 2011) which fits nicely to our molecular model for the origin of eBL (Niller et al. 2003, 2004a).

This BL model does not rely on the viral growth program and the accompanying expression of the major transforming viral protein of lymphoblastoid cells, EBNA2, which is not expressed in the vast majority of BLs (Kelly et al. 2002). The viral growth program, based on expression of EBNA2, seems not to be compatible with the GC reaction (Tobollik et al. 2006). This is consistent with the observation that EBV-infected cells, although physically located in GCs, did normally not participate in GC reactions which yield a clonal expansion of hundreds to thousands of cells (Araujo et al. 1999; Kurth et al. 2003). Only under conditions of immune hyperstimulation, clonal expansions of EBV-infected B cells that did not express EBNA2 were observed in GC reactions (Araujo et al. 1999). Recently, it was confirmed that EBV-infected cells, even when expressing GC markers, did not express EBNA2 and did not expand in a physiological GC reaction (Roughan et al. 2010). Obviously, LMP1 expression in the GC mostly permits only an abortive GC reaction (Uchida et al. 1999). Our model, therefore, avoids unnecessary posits and, despite converse statements on the presumptive nonexistence of a satisfactory molecular model for the origin of eBL (Thorley-Lawson 2004), remains up and standing, whereas the models relying on the EBNA2 transformation as a requirement for eBL development are unconvincing and have not been supported by later data.

We suggested that in the EBV-infected BL precursor cell, the tumorigenic and proapoptotic c-Myc protein binds to the unique Myc-binding site (Niller et al. 2004c) in the locus control region (LCR) of the viral genome. Myc binding then results both in the constitutive expression of the antiapoptotic EBER genes and supports the attachment of the viral genome to the nuclear matrix. Thereby, the viral genome has a higher likelihood to stick around in the nucleus and additional antiapoptotic viral genes of the restricted viral latency programs, e.g., EBNA1, LMP1, or LMP2A, to be expressed. Thus, the balance between apoptosis and antiapoptosis becomes permanently shifted in favor of cell survival. This allows the highly expressed c-Myc to exert its oncogenic potential and to drive lymphomagenesis in the EBV-infected B cell that happened to undergo an accidental c-Myc translocation on its way through the germinal center reaction (Niller et al. 2004b; Rossi and Bonetti 2004). This model is somewhat analogous to the pancreatic tumor model described by Pelengaris et al. (2002). However, we would like to note that from the beginning our model did not only rely on the antiapoptotic effects of the EBER RNAs but included the matrix attachment and locus control functions of c-Myc on the viral genome (Niller et al. 2003).

Tumor-specific promoter hypermethylation in BLs frequently inactivates tumor suppressor genes that are silenced, in addition to BLs, in a wide variety of carcinomas and lymphomas as well. It is also noteworthy that some of these changes could be observed only in a fraction of the EBV-positive BLs studied. As a matter of fact there is no expression of either LMP1 or LMP2A, the EBV-encoded proteins capable to induce de novo methylation, in BL cells. Thus, one may wonder how the apparently EBV-specific epigenetic signatures, e.g., hypermethylation of the activated B-cell factor (*ABF1*) gene, involved in the survival or activation of B cells, or de novo methylation of *HOXB13, CALCA, NEFL*, and *PROK2* genes marked by PRC in embryonic stem cells, are generated in BLs (Ushmorov et al. 2008; Martin-Subero et al. 2009). We speculate that EBV latency products that have not been implicated yet in the induction of cellular promoter hypermethylation may play such an unsuspected role: EBNA1, EBER1, EBER2, BART transcripts, BART encoded proteins, and microRNAs processed from BART transcripts are potential candidates. Alternatively, the observed epigenetic signatures attributed to the presence of EBV genomes could be generated by epigenetic remodeling during lymphomagenesis, independently of the putative activities of EBV latency products. In this context, one may consider that ABF1 is a transcriptional repressor interfering with the activating functions of E2A proteins that regulate early B-cell development (Massari et al. 1998). Accordingly, its expression contributes to the non-B-cell phenotype in HL, whereas the silencing of the ABF1 promoter by methylation occurs not only in BL cells carrying latent EBV genomes but also in other lymphomas of B-cell origin, including diffuse large B-cell lymphomas (DLBCLs) and follicular lymphomas that are less frequently associated with EBV (Ushmorov et al. 2008). Similarly, the stem cell chromatin pattern observed in EBV-positive BLs, i.e., the presence of a set of de novo methylated genes (*HOXB13, CALCA, NEFL, PROK2*) that are repressed by PRC2 in embryonic stem cells, could also be observed in DLBCLs in spite of the difference of BLs and DLBCLs regarding morphology, genetic background, and transcriptional pattern (Martin-Subero et al. 2009).

8 Epigenetic Changes in Virus-Associated Neoplasms

Based on these observations, Martin-Subero et al. speculated that BLs and DLBCLs either originate from cells with stem cell features or acquire such features, i.e., a stem cell-like chromatin pattern, during lymphomagenesis by remodeling their epigenotypes in a similar manner (Martin-Subero et al. 2009).

8.2.3.2 Hodgkin's Lymphoma

Like BL, classical Hodgkin's lymphoma (cHL) is a lymphoma derived from the germinal center reaction. The mononuclear Hodgkin and polynuclear Reed–Sternberg cells which comprise only about 1–2% of the total Hodgkin tumor mass are crippled GC B cells that escaped imminent apoptosis (Kuppers et al. 2002). However, cHL is an atypical B-cell lymphoma which has lost its B-cell identity (Kuppers 2009). Classical HL is grouped into four subtypes: mixed cellularity, lymphocyte depleted, nodular sclerosis, and lymphocyte predominant that are EBV-positive to varying degrees (Herbst et al. 1991; Pallesen et al. 1991, reviewed by Niller et al. 2007; Hjalgrim and Engels 2008). In Europe and the USA, the overall rate of EBV-infected tumors is between 40 and 50%, whereas in immunocompromised patients and in developing countries, the EBV-positive rate is much higher (Jarrett 2003; Hjalgrim et al. 2003; Dinand and Arya 2006; Grogg et al. 2007; Hjalgrim and Engels 2008).

Although Hodgkin and Reed–Sternberg (HRS) tumor cells are derived from B cells and carry functional rearrangements of immunoglobulin gene segments, they do not produce immunoglobulins. This can be attributed partly to mutations, partly to epigenetic alterations, and predominantly hypermethylation that suppresses the IgH promoter as well as a set of genes coding for B cell-specific transcription factors. Silencing of the genes for transcription factors OCT2, BOB.1, and PU1 as well as other B cell-specific genes including *CD19, CD20, CD78B, SYK, BCMA*, and *LCK* is attributed to the demethylation and expression of regulator genes *ABF1* and *ID2* suppressing the B cell-specific gene expression program in HRS cells. Other nuclear proteins, like GATA-3, a T and NK cell-specific transcription factor that is ectopically expressed in HL, the transcriptional repressor EAR3, the transcription factor Nrf3, and the tyrosine kinase FER, are also upregulated in HRS cells (Kuppers et al. 2003). It is interesting to note that a set of genes hypermethylated in human embryonic stem cells and CD34+ hematopoietic stem and progenitor cells were found to be demethylated during hematopoietic differentiation in a cell type-specific fashion (Calvanese et al. 2012). These included *CD19, RUNX3, CD79B*, and *TCL1* that are among the genes hypermethylated in HRS cells. Based on these observations, Calvanese et al. suggested that an unmethylated promoter landscape is established during successive stages of hematopoietic development and lymphoid differentiation. This process is disturbed, however, in neoplastic cells (Calvanese et al. 2012). Thus, the essence of epigenetic dysregulation in HL could be a defective demethylation process, rather than an anomalous de novo hypermethylation.

Although defective demethylation could certainly shape their epigenome, de novo DNA methylation induced by viral proteins is also a plausible mechanism for epigenetic dysregulation in EBV-positive HRS cells. LMP1, a transmembrane protein frequently expressed both in EBV-positive HLs and EBV-associated NPCs, is capable to upregulate the activity of DNMT1, DNMT3A, and DNMT3B, as well as the expression of the PcG protein Bmi-1 (see Sect. 2.3). Upregulation of DNMT1 by LMP1 involves the c-Jun NH2-terminal kinase/activator protein-1 (JNK/AP-1) pathway and may contribute, in principle, to CpG methylation-mediated silencing of a set of cellular promoters. In germinal center B cells infected by EBV in vitro, DNMT3A was upregulated and bound to the W promoter, where EBNA transcripts are initiated early after infection (Leonard et al. 2011). In parallel, there were non-random methylation changes at cellular genes consistent with an "instructive pattern" of methylation. DNMT3B and DNMT1 were downregulated, however, both in infected germinal center B cells and in HL cell lines (Leonard et al. 2011).

LMP1 is known to activate the NF-κB pathway as well, and this very same pathway mediates induction of the PcG protein Bmi-1. Based on this observation, Dutton et al. suggested that LMP1 may contribute to the loss of B-cell identity in EBV-positive HLs by increasing the level of Bmi-1 that inhibits the expression of a series of B-cell markers, including CD21/MS4A1, BLK, and LY9 (Dutton et al. 2007). In EBV-negative HLs, other activators of the NF-κB pathway may take over the function of LMP1. It is important to note that Bmi-1 is involved in gene activation as well, upregulating the expression of *STAT1*, *c-MET*, and *HK2*. These genes, coding for signaling molecules and hexokinase, an enzyme maintaining the high glycolytic activity of neoplastic cells, are known as the transcriptional targets of LMP1. LMP1 may upregulate its target genes in HLs by the induction of KDM6B, a H3K27me3 demethylase that removes the trimethyl mark from lysine 27 of histone H3 and thereby derepresses genes silenced by PRC1. Such a mechanism may activate a set of LMP1 target genes (*CCR7*, *NOTCH2NL*, *IER3*, *NOTCH2*, and *CD58*) in germinal center B cells, the presumptive progenitors of HRS tumor cells (Anderton et al. 2011).

LMP2A, a transmembrane protein capable to activate DNMT1 transcription in gastric carcinoma cells (Hino et al. 2009), is also expressed in EBV-positive HLs. Thus, in principle, LMP2A may also be involved in the silencing of cellular promoters in HRS tumor cells. In summary, latent EBV proteins may contribute to gene silencing events in HL by upregulating DNMTs. In addition, the LMP1-induced PcG protein Bmi-1 may inactivate cellular promoters either directly or, perhaps, by recruiting DNMTs. Bmi-1 may also mediate the activation of LMP1 target genes implicated in lymphomagenesis, in concert with the LMP1-induced histone demethylase KDM6B.

8.2.3.3 Nasopharyngeal Carcinoma

NPC is a lymphoepithelial tumor derived from the epithelium of the nasopharyngeal surface, in particular the fossa of Rosenmüller. Practically all anaplastic NPCs studied carry latent EBV episomes. This undifferentiated NPC subtype is highly endemic in South-East Asia. In some areas of South-East China, NPC has the highest incidence of all tumors among men. In Tunisians and Alaskan Inuit, NPC is endemic with an intermediate incidence. Incidences in Hong Kong, however, are receding due to changing eating habits and an improved early detection. Cantonese in the

8 Epigenetic Changes in Virus-Associated Neoplasms

Guangzhou area seem to carry a higher genetic risk for developing NPC than North Chinese in addition to the increased environmental risk due to tumor-promoting substances in local teas and fish (reviewed by Niller et al. 2007). Terminal repeat analysis of the circular viral genomes indicated that NPCs arose by monoclonal proliferation of tumor cells (Raab-Traub and Flynn 1986). This observation is consistent with a scenario envisaging EBV infection of NPC cells or their precursors already at an initial phase of carcinogenesis. The EBV oncoprotein LMP1 is frequently expressed in NPCs, and its ability to upregulate cellular DNMTs and thereby promote the formation of a transcriptional repression complex on the E-cadherin promoter resulting in hypermethylation and silencing of the promoter may directly affect tumor progression. It was observed that the frequency of *E-cadherin* (also called *CDH1*) promoter hypermethylation increased with advanced disease stage and in lymph node metastases of NPCs carrying EBV genomes (Zheng et al. 1999; Krishna et al. 2005; Niemhom et al. 2008). This indicates that LMP1-mediated silencing of a cellular gene may contribute to the progression of NPC.

By activating the NF-κB pathway, LMP1 may affect the levels of certain cellular microRNAs (miR-146a, miR-155) and thereby modulate, in a subtle way, cellular mRNA levels (Motsch et al. 2007; Cameron et al. 2008; Gatto et al. 2008; Rahadiani et al. 2008; Anastasiadou et al. 2010). NPCs express high levels of viral microRNAs as well, processed from the BART transcripts. Three of the EBV-encoded microRNAs target the LMP1 mRNA and may thereby downregulate the level of LMP1 protein in NPC cells (Lo et al. 2007; Swaminathan 2008).

A series of tumor suppressor and candidate tumor suppressor genes are regularly inactivated by promoter hypermethylation in NPCs. Their exact role in the initiation and progression of NPC remains to be established. Inactivation of the genes coding for retinoic receptor β2 (RARβ2), cellular retinol-binding proteins 1 and 4 (CRBP1, CRBP4), and retinoic acid receptor responder (RARRES/TIG1) may explain the failure of 13-*cis*-retinoic acid therapy of NPC (Kwong et al. 2002, 2005a, b; Sriuranpong et al. 2004).

8.2.3.4 EBV-Associated Gastric Carcinoma

Gastric carcinoma is among the most frequent cancers and most frequent causes of cancer-associated death worldwide, however, with large geographical differences. Although the majority of gastric carcinomas is associated with *H. pylori* infection, a subgroup of about 10% of gastric adenocarcinomas developing mostly in the upper stomach carries latent EBV genomes in the tumor cells. Adenocarcinomas are EBV-positive in about 10% of cases, the rare lymphoepithelioma subtype in more than 80% of cases. The histology of the lymphoepitheliomas resembles that of NPC. Gastric-stump carcinoma carries an intermediate frequency of EBV-positive tumors. Beyond the mucosal or in situ stage, tumors are monoclonal (reviewed by Niller et al. 2009). The methylation pattern of EBV-associated gastric carcinomas was reviewed recently (Fukayama et al. 2008; Niller et al. 2009). Most of the hypermethylated sequences were localized to so-called CpG islands (CGIs). In normal cells, CGIs that contain a higher than average level of CpG dinucleotides are typically unmethylated, and a significant fraction of cellular promoters are located on CGIs. Such CGI-associated promoters are frequently methylated, however, in certain neoplasms (CpG island methylator phenotype, CIMP). EBV-associated gastric carcinomas belong to this category, and it was observed that successive stages of neoplastic development were associated with a progressively increasing level of CGI methylation. A series of tumor suppressor genes were found to be methylated in EBV-associated gastric carcinomas. In contrast, DNA repair genes (*hMLH1* and *hMSH2*) were not inactivated by CpG methylation in EBV-positive NPCs. A fine analysis of methylation patterns showed that although the p16INK4A and p14ARF promoters were highly methylated in both EBV-negative and EBV-positive gastric carcinomas, they could be distinguished based on the methylation profiles of individual CpG motifs (Sakuma et al. 2004). How these unique methylation profiles were generated remains to be analyzed. In addition, a high level of p73 methylation could be detected in EBV-positive gastric carcinomas only (Ushiku et al. 2007), and methylation of a homeobox gene, *HOXA10*, was also associated with EBV-positive carcinomas of the stomach (Kang et al. 2008). These observations indicate that EBV-specific epigenetic signatures are left on the host cell epigenome in EBV-associated gastric carcinomas and such epigenetic marks may play a role in epigenetic reprogramming of gene expression in gastric carcinoma cells.

Although LMP1, a latent EBV protein capable to upregulate maintenance and de novo DNMT levels, could not be detected in gastric carcinomas, a fraction of EBV-positive gastric cancers express LMP2A that may contribute to epigenetic remodeling of the host cell epigenome via the activation of DNMT1 transcription (Hino et al. 2009, Table 8.4). How the upregulated DNMT is targeted to selected cellular promoters remains to be established.

8.2.3.5 Posttransplant Lymphoproliferative Disorder

A high EBV load and increased number of EBV-infected B cells could be observed in organ and bone marrow transplant patients before the onset of posttransplant lymphoproliferative disorder (PTLD) (Stevens et al. 2001; Muti et al. 2003; Baudouin et al. 2004; Davis et al. 2004). The overall PTLD frequency after allogeneic transplants is about 1%, but strongly depends on the specific organ, type and duration of immune suppression, patient age, and the EBV serostatus of donor organ and recipient (Gottschalk et al. 2005; Kamdar et al. 2011). Early lesions correspond to polyclonal proliferation of B cells carrying latent EBV genomes which express all latency genes (latency type III), similarly to the B-cell proliferations observed in IM patients (Brink et al. 1997; Tanner and Alfieri 2001, reviewed by Niller et al. 2007). In contrast to IM, where an efficient immune response curbs the B-cell proliferation in most cases, there is an uncontrolled proliferation of EBV-infected B cells in transplant patients, due to the immunosuppressive regimen. The clonality of the early-onset tumors may change towards monoclonality due to clonal evolution: More malignant late-onset PTLD may already begin with monoclonality and includes more monomorphic than polymorphic PTLD (Kremer et al. 2011). Late-onset PTLD may also include BL, large B-cell lymphoma, and HL. Accordingly, latency types I as in BL and II as in HL also occur among EBV-positive PTLDs (Brauninger et al. 2003; Timms et al. 2003). The pathogenetic mechanisms of early-onset PTLD and late-onset PTLD are different, as the latter one is due to a composite of both immune stimulation through the transplanted organ and therapeutic immune suppression (reviewed by Niller et al. 2004b).

Rossi et al. observed that *DAP-K*, a gene encoding a serine–threonine protein kinase involved in apoptosis induction, is frequently inactivated by promoter hypermethylation in monomorphic PTLD (Rossi et al. 2003). Downregulation of DAP-K may block the IFNγ-, TNFα-, and FasL-triggered apoptosis pathway and may thereby provide a selective advantage for proliferating cells carrying hypermethylated *DAP-K* alleles. In addition to *DAP-K*, the promoter for *MGMT*, a gene encoding O6-methylguanine DNMT, was also frequently inactivated by CpG methylation in monomorphic PTLD. MGMT removes DNA adducts formed by alkylating agents, and its inactivation may result in the accumulation of mutations that may facilitate lymphomagenesis or progression of the early PTLD lesions. Hypermethylation of the candidate tumor suppressor gene p73 was observed less frequently in PTLD (Rossi et al. 2003). The contribution of latent EBV proteins capable to upregulate DNMTs (LMP1, LMP2A) in PTLD-associated hypermethylation of cellular promoters remains to be clarified.

8.3 Epigenetic Changes in Neoplasms Associated with Kaposi's Sarcoma-Associated Herpesvirus

KSHV, also called human herpesvirus-8 (HHV-8), is a human gammaherpesvirus that was discovered first as a herpesvirus-like DNA sequence in AIDS-associated Kaposi's sarcoma, a malignant tumor of endothelial origin (Chang et al. 1994). The cells of two B-lymphocyte disorders, primary effusion lymphoma (PEL, also called body cavity-based lymphoma, BCBL) and multicentric Castleman's disease (MCD), also carry KSHV genomes.

8.3.1 Kaposi's Sarcoma-Associated Herpesvirus: Basic Facts and Epigenetic Landscape of the Latent Viral Genomes

After primary infection and initial productive (lytic) replication, KSHV appears to establish lifelong latency in B cells, similarly to EBV. The linear double-stranded DNA (dsDNA) genome packaged into the virions ends in terminal repetitions (TRs) that fuse with each other in latently infected cells. Thus, latent KSHV genomes persist as circular episomes and coreplicate with the cellular DNA, like the EBV episomes (reviewed by Pantry and Medveczky 2009; Tempera and Lieberman 2010).

A further similarity between the two human gammaherpesviruses is the association of the latency-associated nuclear antigen (LANA) of KSHV and EBNA1 of EBV with their latent origins of viral DNA replication that are located at the terminal repeats of KSHV and at *oriP* of EBV, respectively (reviewed by Tempera and Lieberman 2010). However, the epigenetic consequences of these protein–DNA interactions are somewhat different. LANA recruits HP1 (heterochromatin-associated protein 1) and another key component of heterochromatin formation, SUV39H1, one of the histone lysine methyltransferases acting on lysine 9 of histone H3, to the TRs (Lim et al. 2003; Sakakibara et al. 2004). Although these proteins are regularly associated with a closed chromatin structure, the TRs of KSHV are not associated with an elevated level of H3K9me3; instead, they are rich in activating histone modifications (Gunther and Grundhoff 2010). Stedman et al. also found that a combination of origin binding factors, bromodomain proteins, and histone acetyltransferases created a region of histone H3 hyperacetylation, restricted to sequences adjacent to the LANA-binding sites, at the latent origin of replication of the KSHV episomes (Stedman et al. 2004). This hyperacetylated region was flanked with hypoacetylated areas.

A major difference between the replication origins of KSHV and EBV is that EBNA1-bound *oriP* can function as a long-range transcriptional enhancer, a property not shared with the TRs of KSHV. *OriP* is enriched in H3K4me2, an activating histone mark, although in the late G1 phase of cell cycle its dyad symmetry element (DS) undergoes nucleosome remodeling in parallel with deacetylation of the associated histone H3 (Zhou et al. 2005).

According to a recent study, a most characteristic epigenetic feature of latent KSHV episomes appeared to be the deposition of both repressive (H3K27me3) and activating (H3K4me3) histone marks across the genome (Gunther and Grundhoff 2010). It is remarkable, however, that the major latency promoter located upstream of ORF73 showed very little or no H3K27 trimethylation both in BCBL1, a body cavity-based lymphoma cell line and a cell line of endothelial origin that carried strictly latent KSHV genomes. The overlapping region between nucleotides 127301 and 128901 of the KSHV genome was found to be hypomethylated in the same cell cultures and in additional lymphoma cell lines, in spite of the fact that in case of BCBL1 other regions of the genome were highly methylated. Others described binding of the histone methyltransferase EZH2 to the regions marked with H3K27me3 (Toth et al. 2010). In contrast, repressive histone marks could not be consistently detected in association with latent EBV genomes (Chau and Lieberman 2004; Day et al. 2007). Recently Kang et al. described that latent and lytic KSHV genomes are organized into chromatin loops by CTCF- and cohesin-mediated interactions. They argued that the alternative conformations of the genome may control latent and lytic gene expression programs (Kang et al. 2011).

8.3.2 Modulation of the Host Cell Epigenotype by the KSHV Nuclear Antigen LANA

As described above, interaction of LANA with the latent origin of KSHV replication results in the recruitment of nuclear proteins potentially involved in the formation of heterochromatin, although, paradoxically, the expected repressive chromatin marks were not detected at the TRs in genome-wide studies. However, a repressive chromatin structure was observed, indeed, at the LANA-binding sites of the cellular genome. LANA recruits DNMT3A, but also DNMT1 and DNMT3B to selected cellular promoters which results in promoter silencing via CpG methylation (Shamay et al. 2006). In addition, LANA interacts with the mSIN3A transcriptional corepressor complex that attracts histone deacetylases to the chromatin (Krithivas et al. 2000). Association of LANA with HP1 was also demonstrated and may target LANA and the LANA-associated KSHV episomes to heterochromatic regions within the nucleus (Lim et al. 2003; Sakakibara et al. 2004). In addition, LANA associates with MeCP2 (Matsumura et al. 2010). MeCP2, a multifunctional protein initially implicated in transcriptional silencing of methylated promoters (Nan et al. 1998a), is involved both in gene repression and gene activation (reviewed by Hansen et al. 2010; see also Chap. 3). MeCP2 facilitated both the repressor and the transactivator function of LANA, depending on the promoter analyzed (Matsumura et al. 2010). Since MeCP2 bound both methylated and unmethylated regions of the cellular genome (Yasui et al. 2007; Chahrour et al. 2008), the ability of LANA to establish an interaction with MeCP2 may direct LANA to diverse areas of the genome, resulting in gene repression or activation in a context-dependent manner.

Stuber et al. observed that in mouse cells, exogenously expressed LANA reorganized the chromatin via interaction with MeCP2 and the histone methyltransferase SUV39H1 (Stuber et al. 2007). They speculated that LANA-induced changes in the positioning of chromosomal domains in interphase nuclei may contribute to the chromosomal rearrangements and translocations observed in BCBL/PEL cell lines. They further suggested that LANA and the anchored viral episome fixed to the borders of heterochromatin may act as an epigenetic modifier, generating locally altered chromatin states. Such altered chromatin states may occur at random positions and may target genes involved in growth control by chance. Once established, however, the altered epigenetic states may be propagated during cell divisions and contribute to tumorigenesis (Stuber et al. 2007). LANA binds, induces, and relocates to nuclear heterochromatic regions Brd2/RING3, a bromodomain containing chromatin-binding protein involved in E2F-mediated activation of cell cycle regulatory genes (Platt et al. 2002; Mattsson et al. 2002; Viejo-Borbolla et al. 2005). Recent data that demonstrated a specific interaction of Brd2/RING3 with both mono- (H4K12ac) and diacetylated (H4K5ac/K12ac) histone H4 molecules (Umehara et al. 2010) are in accordance with the idea of Mattsson et al., who speculated that although the KSHV episomes are anchored to the host heterochromatin, heterochromatinization is possibly inhibited in their immediate neighborhood, due to the action of LANA and Brd2/RING3 (Mattsson et al. 2002). These LANA-induced epigenetic alterations may contribute to the gene silencing and activation events observed in cells carrying latent KSHV genomes. In endothelial cells, LANA-silenced cellular genes include *H-cadherin* (*CDH13*), *CCND2*, *LDHB*, *FOXG1B/FKHL1*, and *CREG*, whereas *p16INK4A* and *MGMT* were found to be silenced in PEL cell lines (Carbone et al. 2000; Platt et al. 2002; Shamay et al. 2006). The genes activated by LANA are mostly interferon-inducible genes encoding MxA, a protein blocking viral RNA-dependent

RNA polymerases, and transcription factors like Staf-50 (stimulated trans-acting factor of 50 kDa). LANA also upregulated the mRNAs of less well-characterized proteins (IFI 9-27, IFI 6-16, Renne et al. 2001). The role of these products in the pathobiology of KSHV infection remains to be elucidated.

8.4 Human T-Lymphotropic Virus Type I: "Hit and Run" Tumorigenesis Mediated by CpG Island Methylation?

Human T-lymphotropic virus type I (HTLV-I) infection is associated with adult T-cell lymphoma and leukemia. Similarly to other retroviruses, the RNA genome of HTLV-I is copied to dsDNA by the viral enzyme reverse transcriptase, and the DNA copy integrates into the cellular genome. The integrated HTLV-I genome (provirus) is transcribed by the cellular RNA polymerase II enzyme. Both genome-length and processed viral transcripts are transported to the cytoplasm, where the translated viral proteins assemble into virions engulfing the viral genomes and leave cells by budding through the cell membrane. Tax, the transactivator protein of HTLV-I was implicated in lymphomagenesis. Tax expression itself is downregulated, however, in a significant fraction of the tumors due to deletions or epigenetic silencing. Tax can activate a set of cellular genes, but it was also observed that this oncoprotein contributes to the silencing of *Shp1*, a gene coding for the Src homology containing protein tyrosine phosphatase. Tax-mediated silencing was attributed to the displacement of transcription factors by Tax from the *Shp1* promoter, followed by promoter hypermethylation (Nakase et al. 2009). Based on these observations, Niller et al. suggested that Tax may induce an epigenetic change at an initial phase of leukemogenesis and its presence may be dispensable, thereafter, during tumor progression (a "hit and run" scenario, Niller et al. 2011).

8.5 Epigenetic Alterations in Hepatitis B Virus-Associated Hepatocellular Carcinoma

Hepatitis B virus (HBV) is the human pathogenic representative among the *Orthohepadnavirus* genus. This virus group is characterized by a partially dsDNA genome with a DNA polymerase that still includes a reverse transcriptase function. This indicates that the replication mechanisms of hepadnaviruses are phylogenetically in between those of retroviruses and DNA viruses. HBV infects more than two billion people worldwide, and chronic HBV infection frequently results in liver cirrhosis, a condition facilitating the development of hepatocellular carcinoma (HCC) in a multistep process (reviewed by Lupberger and Hildt 2007; Tischoff and Tannapfel 2008).

8.5.1 HBV: Basic Facts and the Methylome of the Viral Genome

HBV infects the liver, thereby inducing a strong immune response against the infected cells which leads, after an incubation time of 2–6 months, to acute hepatitis in about 35% of cases. Mostly however, the immune response is not strong enough to produce icteric or other clinical symptoms. Contrary, an excessive immune response may lead to acute liver failure in less than 1% of cases. An immature immune system, like in newborns, or an insufficient immune response, in 5–10% of adults, is not able to resolve the inflammation, and chronic hepatitis may result. Infectious virus is shed from the liver to the blood and parenterally or sexually transmitted. Although integration into the cellular DNA is not a necessary step during HBV replication, most of the HBV-associated HCCs carry integrated HBV genomes (reviewed by Gatza et al. 2005; Lupberger and Hildt 2007).

Fernandez et al. constructed high-resolution CpG methylation maps of the HBV genome using samples derived from different stages of liver tumorigenesis including chronic active hepatitis, hepatic cirrhosis, and primary hepatocarcinoma. They also studied hepatic cancer cell lines. They found that the HBV genome was almost completely unmethylated in chronic active hepatitis and cirrhosis, but methylated regions could be observed in primary hepatocarcinomas and in in vitro cultivated cell lines derived from liver cancer. Methylation of the coding sequences of core (C) and surface (S) viral proteins was correlated with the lack of expression, whereas the coding sequence of the putative HBV oncoprotein, HB-X (also called pX), was unmethylated even in genomes highly methylated at other regions. Genetic changes, i.e., deletions, of the HBV genome were also recorded, more frequently in chronic active hepatitis and hepatic cirrhosis than in liver tumors (Fernandez et al. 2009).

Kaur et al. reported that three CGIs of the HBV genome were unmethylated in liver samples of chronic hepatitis patients. CGI 1, spanning the ATG start site of the surface antigen gene, was methylated, however, in 18% of cirrhotic liver samples and 30% of liver carcinomas. They argued that in HBV-infected normal hepatocytes, unlike in liver carcinoma cells, DNA methylation does not play a role in the chronic silencing of HBV surface antigen genes (Kaur et al. 2010). Thus, other epigenetic mechanisms may be responsible for the silencing of surface antigen genes in patients tested HBsAg-negative in spite of the occult HBV infection of the liver.

8.5.2 Remodeling of the Host Cell Epigenome in HBV-Associated Hepatocarcinoma: Regional Hypermethylation Versus Global Hypomethylation

Regional DNA hypermethylation that results in silencing of key cellular genes involved in cell proliferation, apoptosis, DNA repair, cell adhesion, and invasion occurs frequently in neoplastic cells (Baylin and Herman 2000). Another important phenomenon observed in neoplastic cells is a global hypomethylation of the genome, attributed mainly to a decreased methylation of repetitive sequences (reviewed by Ehrlich 2000; Hansen et al. 2011).

The viral oncoprotein HB-X (also called pX), a pleiotropic regulator expressed in hepatocarcinoma cells, not only modulates multiple signaling pathways but also interacts with the transcription factors TBP, CREB, and ATF2 directly, and activates thereby a diverse set of cellular genes coding for nuclear proteins (c-fms, c-myc), MHC class I and II proteins, interferon-β, metallothionein, and β-actin. In addition, HB-X appears to be unique among viral oncoproteins because it differentially affects the level of cellular DNA methyltransferases: HB-X upregulates DNMT1 and DNMT3A but downregulates DNMT3B. Upregulation of DNMT1 and DNMT3A1 results in local hypermethylation of selected tumor suppressor genes, including *E-cadherin* and *RASSF1A*, which occurs in parallel with a global hypomethylation of the genome, as a consequence of the HB-X-mediated downregulation of DNMT3B involved in methylation of satellite 2 repeat sequences (Park et al. 2007).

Saito et al. observed that DNMT3B4, a protein translated from a splice variant of the DNMT3B transcript, is overexpressed in chronic hepatitis, cirrhotic liver, and HCC (Saito et al. 2002). They also observed that increased levels of DNMT3B4, compared to DNMT3B3 translated from the major splice variant, correlated with hypomethylation of pericentromeric satellite regions, possibly due to an impaired DNMT activity of the DNMT3B4 variant protein that lacks the conserved methyl-transferase motifs IX and X. Transfection of DNMT3B4 cDNA into 293 cells, an adenovirus DNA-transformed human embryonic kidney cell line, induced DNA demethylation on satellite 2 sequences in pericentromeric heterochromatin DNA and resulted in an increased growth rate of the cells. Saito et al. speculated that DNA hypomethylation at these regions, especially adjacent to the centromeres of chromosomes 1 and 16, abundant in satellite 2 sequences, may lead to chromo-somal instability already in the precancerous stages of hepatocarcinogenesis. In addition, hypomethylation of certain CGI genes located to heterochromatic regions may result in aberrant expression of cancer-related genes, facilitating hepatocar-cinogenesis (Saito et al. 2002). How the HB-X-mediated downregulation of DNMT3B is related to the overexpression of DNMT3B4 remains to be clarified.

The relationship of HB-X to SALL3 (sal-like 3), a protein inhibiting the associa-tion of the de novo DNMT3A to chromatin (Shikauchi et al. 2009), also remains to be established, especially with regard to the DNA methylation-mediated silencing of *SALL3* in HCCs (Shikauchi et al. 2009). One may wonder whether HB-X could recruit DNMTs to the *SALL3* promoter. Lambert et al. detected tumor-specific hypermethylation of specific genes (*GSTP1, RASSF1A, CHRNA3, DOK1*) in hepa-tocarcinomas compared to control cirrhotic or normal liver tissues. There was a correlation between hypomethylation of *MGMT* (O6-methylguanine DNMT) which is involved in DNA repair, alcohol intake as well as hypermethylation of *GSTP1* (glutathione S-transferase P1) that is inactivating electrophilic carcinogens, and HBV infection (Lambert et al. 2011). *GSTP1* hypermethylation was observed, how-ever, in other carcinomas (prostate cancer, breast cancer, cholangiocarcinoma) as well (reviewed by Tischoff and Tannapfel 2008). In addition, Zhang et al. found a significant correlation between *GSTP1* hypermethylation and aflatoxin B1-mediated hepatocarcinogenesis (Zhang et al. 2005). One may speculate, however, that viral and chemical pathways of hepatocarcinogenesis may share certain targets, because Su et al. demonstrated an increased *GSTP1* methylation in HBV-positive hepatocarcinomas, but not in HBV-negative tumors. They also observed that *GSTP1* methylation was higher in cirrhotic versus noncirrhotic tissues (Su et al. 2007).

The promoter region of *RASSF1A*, a multifunctional tumor suppressor gene, was intensively methylated in 95% of HBV-associated HCCs (Zhong et al. 2003). Although heterogenous CpG methylation was found at a lower level also in 70% of the analyzed nontumorous tissues, hypermethylation of the *RASSF1A* promoter was diagnostic for the neoplastic samples in this study. *RASSF1A* is frequently inactivated by promoter methylation in other primary tumors as well, including lung carcinoma, breast carcinoma, prostate carcinoma, neuroblastoma, medulloblastoma, NPC, and HL (reviewed by Donninger et al. 2007). Since the RAS association domain family 1 isoform A (RASSF1A) protein, in addition to modulating multiple apoptotic and cell cycle checkpoint pathways, stabilizes the microtubules (Liu et al. 2003), its downregulation may result in genomic instability during hepatocarcinogenesis. Laurent-Puig and Zucman-Rossi defined two pathways of hepatocarcinogenesis according to the presence or absence of chromosomal instability. HBV-positive tumors fell into the chromosome instable category (Laurent-Puig and Zucman-Rossi 2006). In accordance with the above proposal, all HBV-positive HCCs showed chromosomal instability (CIN) and associated with a high level of CpG methylation of selected CGIs including the *RASSF1A* promoter in a study by Katoh et al. In contrast, all tumors with a methylator phenotype but without CIN were associated with hepatitis C virus (HCV) infection (Katoh et al. 2006). We speculate that the epigenetic downregulation of RASSF1A expression may contribute to the chromosomal instability in HBV-associated liver carcinomas.

In addition to HCCs, *CHRNA3* and *DOK1* were also found to be frequently hypermethylated in gastric carcinoma (Balassiano et al. 2011). *CHRNA3* codes for the $\alpha3$ subunit of neuronal nicotinic acetylcholine receptors (Bonati et al. 2000), and its polymorphism was found to be associated with an increased risk for squamous carcinoma of the lung (Kohno et al. 2011). The significance of CHRNA downregulation in hepatocarcinogenesis is unknown at present. *DOK1* is a putative tumor suppressor gene that is inactivated by hypermethylation in 93% of head and neck cancer, 81% of lung cancer, and 64% of Burkitt's lymphoma samples (Saulnier et al. 2011). Dok1, a docking protein that is the common substrate for activated protein-tyrosine kinases, functions as an adapter molecule anchoring these enzymes to subcellular structures (Shi et al. 2004). Dok1 appears to oppose oncogenic tyrosine kinase-mediated cell transformation (Janas and Van Aelst 2011). Thus, in principle, downregulation of Dok1 may facilitate hepatocarcinogenesis.

HBV-related HCC develops through distinct stages, and epigenetic alterations may appear already in the preneoplastic lesions. Hypermethylation of the *P16INK4A* tumor suppressor gene was observed in 62% of cirrhotic nodules that are putative preneoplastic lesions and in 70% of dysplastic nodules that surrounded HBV-positive HCC lesions (Shim et al. 2003). Um et al. described that the tumor suppressor genes *APC* (adenomatous polyposis coli) and *RASSF1A*, as well as *SOCS-1* (suppressor of cytokine signaling-1), which is a negative regulator of the JAK/STAT pathway, were methylated in a fraction of cirrhotic nodules, but the methylation levels of *APC* and *RASSF1* increased further in low-grade dysplastic nodules. *SOCS-1* methylation gradually increased during multistep carcinogenesis, peaked in early HCC, and decreased in progressed liver carcinomas (Um et al. 2011).

Feng et al. compared the epigenetic profiles of HBV-associated and HCV-associated HCCs. They found that *HOXA,* a gene coding for a transcription factor implicated in the regulation of hematopoiesis, and *SFRP1* (secreted frizzled-related protein 1) as well as *RASSF1* were preferentially methylated in HBV-positive HCCs (Feng et al. 2010). SFRP1, a putative tumor suppressor protein that acts as a Wnt signaling modulator, is frequently inactivated in gastric carcinoma and esophageal carcinoma (Kinoshita et al. 2011; Meng et al. 2011). It inhibits Wnt signaling and thereby angiogenesis and tumor growth in HCC (Hu et al. 2009). In HCV-associated HCCs, *CDKN2A (p16INK4A* and *p14ARF)* was more frequently methylated (57%) than in HBV-positive tumors (17%) (Feng et al. 2010). These data suggest that unique, HBV- or HCV-specific epigenetic signatures may mark the DNA of virus-associated HCCs.

8.6 Epigenetic Dysregulation of the Host Cell Genome in Hepatitis C Virus-Associated Hepatocellular Carcinoma

HCV belongs to the *Hepacivirus* genus of the *Flaviviridae* virus family. HCV causes inapparent infection in the majority of cases. Symptoms develop in approximately 10–15% of acute infections, but 75–80% of the infected individuals become chronic HCV carriers with an increased risk of developing HCC. Oncoviruses associated with neoplasms in humans are either DNA viruses or retroviruses that synthesize a DNA copy of their RNA genomes during replication using reverse transcriptase. Such viral genomes either integrate into the host cell genome or coreplicate with the host cell DNA. The RNA genome of HCV does not code for a reverse transcriptase, however, and in the absence of a DNA intermediate, it does not integrate into the host cell DNA, and it does not coreplicate as an extrachromosomal episome in concert with the host cell genome, either. Thus, the HCV genome is not subject of the epigenetic machinery of the host cell, and the virus persists in the absence of a latent form, by continuous productive replication in the liver. It is interesting to note that HCV also infects B cells and increases the mutation frequency of *IgH, BCL-6, TP53,* and *β-catenin* genes in in vitro infected B-cell lines, PBMCs, lymphomas, and HCCs by inducing the error-prone DNA polymerase ζ, polymerase ι, and activation-induced cytidine deaminase (Machida et al. 2004). Machida et al. suggested that HCV induces a mutator phenotype that may contribute to the development of HCV-associated oligoclonal lymphoproliferative disorders and HCCs in a "hit and run" tumorigenesis scenario. Therapeutic failure during the standard administration of pegylated interferon-alpha (IFN-α) to HCV-infected patients was traditionally attributed to viral interference with IFN-α signal transduction, i.e., blocking of the JAK–STAT pathway or disruption of type I IFN receptors (Duong et al. 2006).

Recently, however, it was demonstrated that epigenetic silencing of IFN-stimulated genes may also cause IFN resistance in cells harboring HCV replicons (Naka et al. 2006; see also Chap. 14).

A potential mechanism for HCV-mediated inactivation of cellular genes could be promoter hypermethylation because the HCV core protein is capable to activate DNMT1 and DNMT3B (Arora et al. 2008). Such a mechanism could repress the *E-cadherin* promoter, and, in principle, it could mediate the silencing of a series of tumor suppressor genes in HCV-associated HCC (Yang et al. 2003; Li et al. 2004; Narimatsu et al. 2004; Arora et al. 2008). Ko et al. observed that the *SOCS-1* gene coding for the intracellular protein Socs1 that acts as a negative regulator of the JAK/STAT signaling pathway is frequently hypermethylated in HCV-associated HCCs but not so much in HBV-associated tumors (Ko et al. 2008). Increased methylation frequencies of the tumor suppressor genes *P16INK4A, RASSF1A, APC,* and *RIZ1* as well as *GSTP1* (glutathione S-transferase P1) were also reported in HCV-positive, HBV-negative HCCs versus nontumorous tissues (Formeister et al. 2010). In parallel, a global hypomethylation of the tumor cell genome was observed, based on increased levels of hypomethylated LINE-1 repetitive sequences. Based on these data, Formeister et al. argued that aberrant CpG methylation plays a role in the pathobiology of HCV-positive HCCs. Yang et al. also observed that *APC* methylation was more frequent in HCV-positive HCCs than in virus-negative tumors (Yang et al. 2003). They also described frequent methylation of SOCS-1, in accordance with the data of Ko et al. (2008; Yang et al. 2003). In addition, Yang et al. reported a frequent methylation of *p15*, a gene encoding a cyclin-dependent kinase inhibitor in HCV-associated hepatocarcinomas (Yang et al. 2003). Decreased expression and hypermethylation of *Gadd45β*, coding for a member of the growth arrest and DNA damage family of proteins involved in stress responses, were observed in HCV-transgenic mice. This observation suggests that downregulation of Gadd45β expression in HCV-associated liver cancer may occur by a similar mechanism (Higgs et al. 2010). These data demonstrate the concurrent methylation of a set of tumor suppressor genes and other cancer-related genes in HCV-associated HCCs, reflecting the epigenetic reprogramming of the hepatocyte genome during hepatocarcinogenesis.

8.7 Epigenetic Changes in Human Papillomavirus-Associated Tumors

Human papillomaviruses (HPVs) are dsDNA viruses associated with a variety of benign epithelial proliferations. Specific high-risk HPVs (HPV-16, HPV-18, HPV-31, HPV-33, and HPV-45), however, are the causative agents of cervical carcinoma (zur Hausen 2002). Since their genome is relatively small, papillomaviruses use the host DNA synthesizing machinery to replicate their genome. The episomal viral genome frequently integrates into the host cell DNA during the carcinogenetic process. E6 and E7, the main oncoproteins encoded by high risk HPVs, are involved not only in the initiation of cervical carcinogenesis, but their persistent expression appears to be necessary for the successive steps of neoplastic development as well.

E6 proteins encoded by high-risk HPVs target the p53 tumor suppressor protein for degradation, whereas high-risk HPV E7 oncoproteins block the function of the retinoblastoma protein (pRB) involved in the regulation of the cell cycle. In addition, both E6 and E7 associate with a series of other cellular proteins as well and thereby modulate key cellular processes (reviewed by Howie et al. 2009; McLaughlin-Drubin and Munger 2009)

8.7.1 HPV: Basic Facts and Host Cell-Dependent Methylomes of the Viral Genome

Although infections with high-risk HPV strains (especially types 16 and 18) are causally related to the development of cervical cancer, only a fraction of the HPV-positive premalignant lesions appear to progress to invasive cancer (reviewed by Szalmas and Konya 2009). Genetic instability was regularly observed during HPV-induced carcinogenesis. This phenomenon is associated with the E7-initiated abnormal centrosome synthesis resulting in multipolar mitotic spindles and abnormal chromosome segregation already in an early phase of cervical carcinogenesis (Duensing and Munger 2003).

The regulatory sequences of the early genome region of HPV are located to the long control region (LCR). The transcription factor AP-1 binds to the LCR and activates the transcription of the early genes including E6 and E7. The LCR appears to be a subject of epigenetic regulation, because ectopic expression of retinoic acid receptor beta 2 (RARβ2) downregulates HPV-18 transcription by abrogating AP-1 binding and targets the LCR for de novo methylation (De Castro Arce et al. 2007). Fernandez et al. determined the DNA methylation pattern of the whole genome ("the methylome") of the high-risk HPVs HPV-16 and HPV-18 in a collection of human cervical samples corresponding to progressive disease stages, such as specimens from asymptomatic carriers, cervical intraepithelial neoplasias, and primary cervical carcinomas. They observed that the genome of HPV-16 and HPV-18 undergoes progressive de novo methylation during the successive phases of tumorigenesis (Fernandez et al. 2009). The viral genome was unmethylated or hypomethylated in samples from asymptomatic carriers, there was a low level of CpG methylation in cervical intraepithelial neoplasias considered to be premalignant lesions, and a moderate methylation level was observed in primary cervix carcinomas. The highest DNA methylation levels were found in four established cervix carcinoma cell lines cultivated in vitro.

8.7.2 Dysregulation of Cellular Epigenetic Processes by the Viral Oncoproteins E6 and E7

The E7 oncoprotein of HPV-16 may induce hypermethylation of selected cellular promoters both directly, by binding to DNMT1 and stimulating its activity

(Burgers et al. 2007), and indirectly, by releasing the so-called "activating" E2F transcription factors that are complexed with members of the retinoblastoma pocket protein family. The released E2F transcription factors not only control the coordinated transcription of genes involved in DNA replication and cell cycle progression but activate the expression of the *DNMT1* gene as well (Kimura et al. 2003; Iaquinta and Lees 2007). These mechanisms may contribute to local hypermethylation of a set of cellular promoters in HPV-16- and HPV-18-infected cells (reviewed by Szalmas and Konya 2009; Wentzensen et al. 2009).

One of the genes silenced by CpG methylation, *cyclin A1* (*CCNA1*), was considered a candidate tumor marker for the early diagnosis of HPV-associated invasive cervical cancer (Kitkumthorn et al. 2006). In a follow-up study, however, no correlation was found between the quantity of HPV and *CCNA1* promoter hypermethylation. It was observed that CCNA1 methylation was associated with the presence of integrated HPV genomes, rather than the episomal form (Yanatatsaneejit et al. 2011). Hypermethylation of the *APC* promoter was preferentially associated with a distinct histological type, adenocarcinoma (Dong et al. 2001; Wisman et al. 2006; Wentzensen et al. 2009). Methylation of *CDKN2B, RASSF1A, TIMP3,* and *TP73* was also more frequent in cervical adenocarcinomas (Henken et al. 2007). Wentzensen et al. found, based on the data of several studies, that the weighted mean methylation frequency of *TIMP3* as well as *HIC1* was considerably higher in adenocarcinoma than in squamous cell carcinoma (Wentzensen et al. 2009). In contrast, *PAX1* methylation appeared to be specific for squamous cell carcinoma (Lai et al. 2008). In addition, *DAPK1* and *CADM1* were also identified as histotype-specific markers because they were found to be significantly more frequently methylated in squamous cell carcinomas than in adenocarcinomas (Henken et al. 2007). In addition, *DAPK1* and *CADM1*, as well as *RARB*, consistently showed elevated methylations in cervical cancers across studies (Wentzensen et al. 2009). Henken et al. suggested that the *MGMT* promoter could be regarded as a *common marker* of cervical neoplasms because it was frequently methylated both in cervical adenocarcinomas and squamous cell carcinomas (Henken et al. 2007). A comparison of the reported MGMT methylation frequencies, however, showed a wide range variation among the individual studies (Wentzensen et al. 2009). Recently, Lai et al. determined the methylation levels of four genes (*SOX1, PAX1, LMX1A,* and *NKX6-1*) coding for transcription factors using a quantitative methylation polymerase chain reaction. They found a very low level of methylation in normal uterine cervix and cervical intraepithelial neoplasm types 1 and 2 (CIN1 and CIN2) (Lai et al. 2010). CIN1 and CIN2 progress to invasive cancer with a low probability. In contrast, cervical intraepithelial neoplasm 3 (CIN3) and carcinoma in situ (CIS) that progress to invasive cancer with a higher probability displayed significantly higher methylation levels of the four indicator genes, similarly to squamous cell carcinomas and adenocarcinomas of the cervix (Lai et al. 2010). Although the role of the SOX1 (sex-determining region Y, box 1), PAX1 (paired box gene 1), LMX1A (LIM homeobox transcription factor 1 α), and NKX6-1 (NK6 transcription factor-related locus 1) in the physiology of cervical epithelial cells and cervical carcinogenesis remains to be explored, assessing the DNA methylation level of their genes may be applicable in the detection of CIN3 and worse (CIN+) lesions.

The accumulation of frequent methylation events involving five candidate tumor suppressor genes (*TP73, ESR1, RARβ, DAPK1,* and *MGMT*) was also observed in an in vitro model system of cervical carcinogenesis (Henken et al. 2007). De novo methylation of *TP73* and *ESR1* started soon after primary keratinocyte cell lines carrying transfected HPV genomes became immortal. *ESR1* encodes the *estrogen receptor* that was found to be downregulated early in HPV-infected cervical dysplasia (Bekkers et al. 2005). Methylation of *RARβ* and *DAPK1* became manifest in late immortal passages, whereas methylation of *MGMT* followed the acquisition of anchorage independence (Henken et al. 2007). In contrast, no methylation was evident in preimmortal HPV-18-transfected cells characterized by extended but still finite lifespan. It is noteworthy that despite E7 expression in preimmortal cells and the known capacity of E7 to induce de novo methylation via the upregulation and activation of DNMT1, none of the 29 genes analyzed was hypermethylated at the preimmortal stage. Thus, Henken et al. speculated that the accumulation of five methylated, inactivated genes following immortalization is associated with a growth advantage of the HPV-containing keratinocytes, provided by switching off these particular genes.

In addition to regional DNA hypermethylation, a global hypomethylation of the cervical cell genome was also observed in cervix carcinomas. The level of global hypomethylation was higher in invasive cancer than in normal cervical samples or intraepithelial neoplasia (CIN1, CIN2, or CIN3) (Flatley et al. 2009). The normally methylated promoter CGIs of *CAGE* (cancer/testis antigen), a negative regulator of p53 expression, was found to be frequently hypomethylated in cervical squamous cell carcinoma (Lee et al. 2006).

E7 is a pleiotropic regulator involved not only in gene silencing but also in gene activation. In human keratinocytes, HPV-16 E7 increased histone H3 acetylation at the *E2F1* and *CDC25A* promoters (Zhang et al. 2004). In keratinocytes expressing HPV-16 E6/E7 proteins, a global decrease of H3K27me3 levels was observed, in spite of the parallel increase of the PRC2 histone lysine methyltransferase EZH2 that trimethylates lysine 27 of histone H3 (Hyland et al. 2011). This paradox phenomenon was explained by the concurrent upregulation of the KDM6A lysine demethylase, removing the methyl groups from lysine 27 in H3K27me3. In addition, the PRC1 protein Bmi-1 was downregulated. All of these changes resulted in the derepression of HOX genes. Activation of EZH2 expression by HPV-16 E7—via the E7-mediated release of E2F from pocket proteins—was also observed in HPV-positive tumor cells by Holland et al., who demonstrated that EZH2 facilitated cell cycle progression at the G1–S boundary and contributed to the apoptotic resistance of the cells, too (Holland et al. 2008).

Although the contribution of E6 versus E7 to these complex epigenetic changes remains to be elucidated, recent data by Hsu et al. suggested that E6 interacted with the arginine methyltransferases CARM1 and PRMT1 and the lysine methyltransferase SET7, downregulating their enzymatic activities (Hsu et al. 2011). E6 blocked histone methylation catalyzed by CARM1 and PRMT1 at p53-responsive promoters resulting in suppression of p53 downstream genes. McLaughlin-Drubin et al. described that HPV-16 E7 expression caused a significant reduction of the

H3K27me3 repressive mark in primary human epithelial cells. E7 induced the lysine demethylases KDM6A and KDM6B that target histone H3K27me3 and thereby disrupt Polycomb repressor complexes. In addition, KDM6B upregulated p16INK4A, a CDK4/6 inhibitor involved in oncogenic stress-induced senescence. E7 simultaneously blocks, however, pRB, a key mediator of p16INK4A-induced senescence, permitting thereby continuous cell proliferation. Epigenetic reprogramming by E7 also resulted in the activation of the homeobox genes *HOXC5* and *HOXC8* in primary human epithelial cells (McLaughlin-Drubin et al. 2011). Because *HOX* genes are involved in the regulation of epidermal differentiation, their derepression may alter the phenotype and behavior of the cells. McLaughlin-Drubin et al. speculated that due to epigenetic reprogramming, a more stem cell-like state is achieved by E7.

8.8 Tumor-Antigen-Mediated Silencing of a Master Regulator Gene in Merkel Cell Polyomavirus-Associated Carcinoma of the Skin

The association of polyomaviruses, small dsDNA viruses, with malignant tumors of humans remains to be firmly established, with the exception of Merkel cell polyomavirus, in spite of the fact that BK virus, a human polyomavirus, is capable to induce tumors in experimental animals (ter Schegget et al. 1980). The large T antigen (tumor antigen, TAg) of BK virus elicits malignant transformation of cells in vitro by inhibiting the retinoblastoma (Rb) protein family as well as the p53 tumor suppressor protein (Helt and Galloway 2003). Interaction of TAg with Rb disrupted Rb/E2F complexes resulting in the activation of a set of promoters with E2F-binding sites, including the *DNMT1* promoter in human prostate epithelial cells (McCabe et al. 2006). In a transgenic mouse model of prostate carcinogenesis (TRAMP), expression of the related simian polyomavirus SV40 TAg also elevated the expression of the murine Dnmt1 enzyme in the developing prostatic intraepithelial neoplasias (PIN) as well as prostate carcinomas and their metastases. Treatment with the Dnmt1 inhibitor 5-aza-2′-deoxycytidine prevented the progression of PIN lesions to malignant disease (McCabe et al. 2006). These data indicated that induction of DNMT activity by a polyomavirus TAg was an early event required for malignant transformation and tumorigenesis in the TRAMP model. Lytic replication of BK virus may cause nephropathy in renal transplant patients. The viral DNA appears to be unmethylated during productive viral replication (Chang et al. 2011).

Merkel cell polyomavirus (MCPyV) was discovered in Merkel cell carcinoma cells arising from specialized cells mediating mechanotransduction in touch-sensitive areas of the epidermis (Feng et al. 2008). According to lineage-tracing experiments, Merkel cells arise through the differentiation of epidermal progenitors during embryonic development, a process regulated by the basic helix-loop-helix transcription factor atonal homologue 1 (Atoh1, also called Math1 or Hath1, Van Keymeulen et al. 2009). In *Drosophila*, the corresponding atonal (Ato) protein acts as a master regulator of cell fate specification, regulating the formation and progression of retinal tumors (Bossuyt et al. 2009a). Loss of *Atoh1* promoted tumor formation in mouse models of colorectal cancer, and expression of the human *ATOH1* was reduced, due to deletion or inactivation by CpG methylation, in Merkel cell carcinomas as well as colorectal carcinomas (Bossuyt et al. 2009b). The promoter of the tumor suppressor gene *RASSF1A* was also hypermethylated in about half of the MCC samples (Helmbold et al. 2009). These observations indicate that epigenetic or genetic inactivation of a conserved master regulator gene contributes to the pathogenesis of MCPyV-associated Merkel cell carcinoma.

8.9 Conclusions and Perspectives

Alterations of the host cell epigenome regularly accompany virus-induced tumorigenesis in humans, and the epigenetic changes can be directly related to the viral oncoproteins that act as epigenetic modifiers. In addition to oncoproteins, virus-encoded or virus-induced nontranslated RNA molecules, including long noncoding RNAs (lncRNAs), may also influence, in principle, the gene expression pattern of the host cell, similarly to the cellular lncRNA HOTAIR that may retarget PRC2 to more than 800 new sites, leading to an altered pattern of histone modification and gene expression (Gupta et al. 2010, reviewed by Hung and Chang 2010). In addition to the RNA polymerase III-transcribed EBER1 and EBER2 and the microRNAs processed from EBV transcripts, EBV encodes also v-snoRNA1, a relative of small nucleolar RNAs (snoRNAs) (Hutzinger et al. 2009). The nontranslated v-snoRNA1 was expressed in EBV-positive LCLs and Burkitt's lymphoma cells, but it was absent from EBV-negative cells. Because LMP1 and LMP2A, the EBV proteins upregulating DNMTs, are not expressed in latency type I Burkitt's lymphoma cells, one may speculate that either EBNA1 or a nontranslated EBV RNA like v-snoRNA1 may mediate the change of the cellular epigenotype observed in Burkitt's lymphoma cells (see Sect. 2.3.1). In LCLs carrying latent EBV genomes, cellular noncoding RNAs upregulated by the virus (Mrazek et al. 2007) may also contribute to the epigenetic reprogramming of the host cell. It is worthy to mention that KSHV expresses a noncoding RNA, called PAN (polyadenylated nuclear RNA), during productive infection that may interfere with the immune response (Rossetto and Pari 2011), and expression of the viral noncoding RNA HSUR1 in T cells transformed by the primate gammaherpesvirus *Herpesvirus saimiri* may downregulate a cellular miRNA in a sequence-specific manner (Cazalla et al. 2010).

The data accumulated regarding the epigenotypes of human tumor-associated viruses and their host cells may allow the development of new diagnostic methods for the detection and monitoring of virus-associated neoplasms. Novel epigenetic therapies are either at the stage of preclinical studies or, as in the case of certain

8 Epigenetic Changes in Virus-Associated Neoplasms

HDAC inhibitors administered to patients with cervical carcinoma, at the stage of clinical trials already (De la Cruz-Hernandez et al. 2011; Takai et al. 2011).

References

Anastasiadou, E., Boccellato, F., Vincenti, S., Rosato, P., Bozzoni, I., Frati, L., Faggioni, A., Presutti, C. & Trivedi, P. (2010). Epstein-Barr virus encoded LMP1 downregulates TCL1 oncogene through miR-29b. *Oncogene* 29, 1316–1328.

Anderton, J. A., Bose, S., Vockerodt, M., Vrzalikova, K., Wei, W., Kuo, M., Helin, K., Christensen, J., Rowe, M., Murray, P. G. & Woodman, C. B. (2011). The H3K27me3 demethylase, KDM6B, is induced by Epstein-Barr virus and over-expressed in Hodgkin's Lymphoma. *Oncogene* 30, 2037–2043.

Araujo, I., Foss, H. D., Hummel, M., Anagnostopoulos, I., Barbosa, H. S., Bittencourt, A. & Stein, H. (1999). Frequent expansion of Epstein-Barr virus (EBV) infected cells in germinal centres of tonsils from an area with a high incidence of EBV-associated lymphoma. *J Pathol* 187, 326–330.

Arora, P., Kim, E. O., Jung, J. K. & Jang, K. L. (2008). Hepatitis C virus core protein downregulates E-cadherin expression via activation of DNA methyltransferase 1 and 3b. *Cancer Lett* 261, 244–252.

Balassiano, K., Lima, S., Jenab, M., Overvad, K., Tjonneland, A., Boutron-Ruault, M. C., Clavel-Chapelon, F., Canzian, F., Kaaks, R., Boeing, H., Meidtner, K., Trichopoulou, A., Laglou, P., Vineis, P., Panico, S., Palli, D., Grioni, S., Tumino, R., Lund, E., Bueno-de-Mesquita, H. B., Numans, M. E., Peeters, P. H., Ramon, Q. J., Sanchez, M. J., Navarro, C., Ardanaz, E., Dorronsoro, M., Hallmans, G., Stenling, R., Ehrnstrom, R., Regner, S., Allen, N. E., Travis, R. C., Khaw, K. T., Offerhaus, G. J., Sala, N., Riboli, E., Hainaut, P., Scoazec, J. Y., Sylla, B. S., Gonzalez, C. A. & Herceg, Z. (2011). Aberrant DNA methylation of cancer-associated genes in gastric cancer in the European Prospective Investigation into Cancer and Nutrition (EPIC-EURGAST). *Cancer Lett* 311, 85–95.

Baudouin, V., Dehee, A., Pedron-Grossetete, B., Ansart-Pirenne, H., Haddad, E., Maisin, A., Loirat, C. & Sterkers, G. (2004). Relationship between CD8+ T-cell phenotype and function, Epstein-Barr virus load, and clinical outcome in pediatric renal transplant recipients: a prospective study. *Transplantation* 77, 1706–1713.

Baylin, S. B. & Herman, J. G. (2000). Epigenetics and loss of gene function in cancer. In *DNA Alterations in Cancer*, pp. 293–309. Edited by M. Ehrlich. Westborough: Eaton Publishing.

Bekkers, R. L., van der Avoort, I., Melchers, W. J., Bulten, J., de Wilde, P. C. & Massuger, L. F. (2005). Down regulation of estrogen receptor expression is an early event in human papillomavirus infected cervical dysplasia. *Eur J Gynaecol Oncol* 26, 376–382.

Bellan, C., Lazzi, S., Hummel, M., Palummo, N., de Santi, M., Amato, T., Nyagol, J., Sabattini, E., Lazure, T., Pileri, S. A., Raphael, M., Stein, H., Tosi, P. & Leoncini, L. (2005). Immunoglobulin gene analysis reveals 2 distinct cells of origin for EBV-positive and EBV-negative Burkitt lymphomas. *Blood* 106, 1031–1036.

Bieging, K. T., Fish, K., Bondada, S. & Longnecker, R. (2011). A shared gene expression signature in mouse models of EBV-associated and non-EBV-associated Burkitt's lymphoma. *Blood* 118, 6849–6859.

Blomen, V. A. & Boonstra, J. (2011). Stable transmission of reversible modifications: maintenance of epigenetic information through the cell cycle. *Cell Mol Life Sci* 68, 27–44.

Bonati, M. T., Asselta, R., Duga, S., Ferini-Strambi, L., Oldani, A., Zucconi, M., Malcovati, M., Dalpra, L. & Tenchini, M. L. (2000). Refined mapping of CHRNA3/A5/B4 gene cluster and its implications in ADNFLE. *Neuroreport* 11, 2097–2101.

Bossuyt, W., De Geest, N., Aerts, S., Leenaerts, I., Marynen, P. & Hassan, B. A. (2009a). The atonal proneural transcription factor links differentiation and tumor formation in Drosophila. *PLoS Biol* 7, e40.

Bossuyt, W., Kazanjian, A., De Geest, N., Van Kelst, S., De Hertogh, G., Geboes, K., Boivin, G. P., Luciani, J., Fuks, F., Chuah, M., Van den Driessche, T., Marynen, P., Cools, J., Shroyer, N. F. & Hassan, B. A. (2009b). Atonal homolog 1 is a tumor suppressor gene. *PLoS Biol* 7, e39.

Brauninger, A., Spieker, T., Mottok, A., Baur, A. S., Kuppers, R. & Hansmann, M. L. (2003). Epstein-Barr virus (EBV)-positive lymphoproliferations in post-transplant patients show immunoglobulin V gene mutation patterns suggesting interference of EBV with normal B cell differentiation processes. *Eur J Immunol* 33, 1593–1602.

Brink, A. A., Dukers, D. F., van Den Brule, A. J., Oudejans, J. J., Middeldorp, J. M., Meijer, C. J. & Jiwa, M. (1997). Presence of Epstein-Barr virus latency type III at the single cell level in post-transplantation lymphoproliferative disorders and AIDS related lymphomas. *J Clin Pathol* 50, 911–918.

Burgers, W. A., Blanchon, L., Pradhan, S., de Launoit, Y., Kouzarides, T. & Fuks, F. (2007). Viral oncoproteins target the DNA methyltransferases. *Oncogene* 26, 1650–1655.

Burkitt, D. (1958). A sarcoma involving the jaws in African children. *Br J Surg* 45, 218–223.

Burkitt, D. (1962). A children's cancer dependent upon climatic factors. *Nature* 194, 232–234.

Calvanese, V., Fernandez, A. F., Urdinguio, R. G., Suarez-Alvarez, B., Mangas, C., Perez-Garcia, V., Bueno, C., Montes, R., Ramos-Mejia, V., Martinez-Camblor, P., Ferrero, C., Assenov, Y., Bock, C., Menendez, P., Carrera, A. C., Lopez-Larrea, C. & Fraga, M. F. (2012). A promoter DNA demethylation landscape of human hematopoietic differentiation. *Nucleic Acids Res* 40, 116–131.

Cameron, J. E., Yin, Q., Fewell, C., Lacey, M., McBride, J., Wang, X., Lin, Z., Schaefer, B. C. & Flemington, E. K. (2008). Epstein-Barr virus latent membrane protein 1 induces cellular MicroRNA miR-146a, a modulator of lymphocyte signaling pathways. *J Virol* 82, 1946–1958.

Carbone, A., Cilia, A. M., Gloghini, A., Capello, D., Fassone, L., Perin, T., Rossi, D., Canzonieri, V., De Paoli, P., Vaccher, E., Tirelli, U., Volpe, R. & Gaidano, G. (2000). Characterization of a novel HHV-8-positive cell line reveals implications for the pathogenesis and cell cycle control of primary effusion lymphoma. *Leukemia* 14, 1301–1309.

Cazalla, D., Yario, T. & Steitz, J. A. (2010). Down-regulation of a host microRNA by a Herpesvirus saimiri noncoding RNA. *Science* 328, 1563–1566.

Chahrour, M., Jung, S. Y., Shaw, C., Zhou, X., Wong, S. T., Qin, J. & Zoghbi, H. Y. (2008). MeCP2, a key contributor to neurological disease, activates and represses transcription. *Science* 320, 1224–1229.

Chang, C. F., Wang, M., Fang, C. Y., Chen, P. L., Wu, S. F., Chan, M. W. & Chang, D. (2011). Analysis of DNA methylation in human BK virus. *Virus Genes* 43, 201–207.

Chang, Y., Cesarman, E., Pessin, M. S., Lee, F., Culpepper, J., Knowles, D. M. & Moore, P. S. (1994). Identification of herpesvirus-like DNA sequences in AIDS-associated Kaposi's sarcoma. *Science* 266, 1865–1869.

Chau, C. M. & Lieberman, P. M. (2004). Dynamic chromatin boundaries delineate a latency control region of Epstein-Barr virus. *J Virol* 78, 12308–12319.

Chinnasri, P., Pairojkul, C., Jearanaikoon, P., Sripa, B., Bhudhisawasdi, V., Tantimavanich, S. & Limpaiboon, T. (2009). Preferentially different mechanisms of inactivation of 9p21 gene cluster in liver fluke-related cholangiocarcinoma. *Hum Pathol* 40, 817–826.

Cotter, M. A. & Robertson, E. S. (2000). Modulation of histone acetyltransferase activity through interaction of epstein-barr nuclear antigen 3C with prothymosin alpha. *Mol Cell Biol* 20, 5722–5735.

Dave, S. S., Fu, K., Wright, G. W., Lam, L. T., Kluin, P., Boerma, E. J., Greiner, T. C., Weisenburger, D. D., Rosenwald, A., Ott, G., Muller-Hermelink, H. K., Gascoyne, R. D., Delabie, J., Rimsza, L. M., Braziel, R. M., Grogan, T. M., Campo, E., Jaffe, E. S., Dave, B. J., Sanger, W., Bast, M., Vose, J. M., Armitage, J. O., Connors, J. M., Smeland, E. B.,

Kvaloy, S., Holte, H., Fisher, R. I., Miller, T. P., Montserrat, E., Wilson, W. H., Bahl, M., Zhao, H., Yang, L., Powell, J., Simon, R., Chan, W. C. & Staudt, L. M. (2006). Molecular diagnosis of Burkitt's lymphoma. *N Engl J Med* 354, 2431–2442.

Davis, J. E., Sherritt, M. A., Bharadwaj, M., Morrison, L. E., Elliott, S. L., Kear, L. M., Maddicks-Law, J., Kotsimbos, T., Gill, D., Malouf, M., Falk, M. C., Khanna, R. & Moss, D. J. (2004). Determining virological, serological and immunological parameters of EBV infection in the development of PTLD. *Int Immunol* 16, 983–989.

Day, L., Chau, C. M., Nebozhyn, M., Rennekamp, A. J., Showe, M. & Lieberman, P. M. (2007). Chromatin profiling of Epstein-Barr virus latency control region. *J Virol* 81, 6389–6401.

De Castro Arce, J., Gockel-Krzikalla, E. & Rosl, F. (2007). Retinoic acid receptor beta silences human papillomavirus-18 oncogene expression by induction of de novo methylation and heterochromatinization of the viral control region. *J Biol Chem* 282, 28520–28529.

De la Cruz-Hernandez, E., Perez-Plasencia, C., Perez-Cardenas, E., Gonzalez-Fierro, A., Trejo-Becerril, C., Chavez-Blanco, A., Taja-Chayeb, L., Vidal, S., Gutierrez, O., Dominguez, G. I., Trujillo, J. E. & Duenas-Gonzalez, A. (2011). Transcriptional changes induced by epigenetic therapy with hydralazine and magnesium valproate in cervical carcinoma. *Oncol Rep* 25, 399–407.

Degner, S. C., Wong, T. P., Jankevicius, G. & Feeney, A. J. (2009). Cutting edge: developmental stage-specific recruitment of cohesin to CTCF sites throughout immunoglobulin loci during B lymphocyte development. *J Immunol* 182, 44–48.

Dinand, V. & Arya, L. S. (2006). Epidemiology of childhood Hodgkin's disease: is it different in developing countries? *Indian Pediatr* 43, 141–147.

Dong, S. M., Kim, H. S., Rha, S. H. & Sidransky, D. (2001). Promoter hypermethylation of multiple genes in carcinoma of the uterine cervix. *Clin Cancer Res* 7, 1982–1986.

Donninger, H., Vos, M. D. & Clark, G. J. (2007). The RASSF1A tumor suppressor. *J Cell Sci* 120, 3163–3172.

Duensing, S. & Munger, K. (2003). Human papillomavirus type 16 E7 oncoprotein can induce abnormal centrosome duplication through a mechanism independent of inactivation of retinoblastoma protein family members. *J Virol* 77, 12331–12335.

Duong, F. H., Christen, V., Filipowicz, M. & Heim, M. H. (2006). S-Adenosylmethionine and betaine correct hepatitis C virus induced inhibition of interferon signaling in vitro. *Hepatology* 43, 796–806.

Dutton, A., Woodman, C. B., Chukwuma, M. B., Last, J. I., Wei, W., Vockerodt, M., Baumforth, K. R., Flavell, J. R., Rowe, M., Taylor, A. M., Young, L. S. & Murray, P. G. (2007). Bmi-1 is induced by the Epstein-Barr virus oncogene LMP1 and regulates the expression of viral target genes in Hodgkin lymphoma cells. *Blood* 109, 2597–2603.

Ehrlich, M. (2000). DNA hypomethylation and cancer. In *DNA Alterations in Cancer*, pp. 273–291. Edited by M. Ehrlich. Westborough: Eaton Publishing.

Epstein, M. A., Achong, B. G. & Barr, Y. M. (1964). Virus particles in cultured lymphoblasts from Burkitt's Lymphoma. *Lancet* 1, 702–703.

Feng, H., Shuda, M., Chang, Y. & Moore, P. S. (2008). Clonal integration of a polyomavirus in human Merkel cell carcinoma. *Science* 319, 1096–1100.

Feng, Q., Stern, J. E., Hawes, S. E., Lu, H., Jiang, M. & Kiviat, N. B. (2010). DNA methylation changes in normal liver tissues and hepatocellular carcinoma with different viral infection. *Exp Mol Pathol* 88, 287–292.

Fernandez, A. F., Rosales, C., Lopez-Nieva, P., Grana, O., Ballestar, E., Ropero, S., Espada, J., Melo, S. A., Lujambio, A., Fraga, M. F., Pino, I., Javierre, B., Carmona, F. J., Acquadro, F., Steenbergen, R. D., Snijders, P. J., Meijer, C. J., Pineau, P., Dejean, A., Lloveras, B., Capella, G., Quer, J., Buti, M., Esteban, J. I., Allende, H., Rodriguez-Frias, F., Castellsague, X., Minarovits, J., Ponce, J., Capello, D., Gaidano, G., Cigudosa, J. C., Gomez-Lopez, G., Pisano, D. G., Valencia, A., Piris, M. A., Bosch, F. X., Cahir-McFarland, E., Kieff, E. & Esteller, M. (2009). The dynamic DNA methylomes of double-stranded DNA viruses associated with human cancer. *Genome Res* 19, 438–451.

Flatley, J. E., McNeir, K., Balas...ani, L., Tidy, J., Stuart, E. L., Young, T. A. & Powers, H. J. (2009). Folate status and ...ant DNA methylation are associated with HPV infection and cervical pathogenesis. *Canc... idemiol Biomarkers Prev* 18, 2782–2789.

Formeister, E. J., Tsuchiya, M., Fu... H., Shpyleva, S., Pogribny, I. P. & Rusyn, I. (2010). Comparative analysis of promoter methylation and gene expression endpoints between tumorous and non-tumorous tissues from HCV-positive patients with hepatocellular carcinoma. *Mutat Res* 692, 26–33.

Fukayama, M., Hino, R. & Uozaki, H. (2008). Epstein-Barr virus and gastric carcinoma: virus-host interactions leading to carcinoma. *Cancer Sci* 99, 1726–1733.

Gatto, G., Rossi, A., Rossi, D., Kroening, S., Bonatti, S. & Mallardo, M. (2008). Epstein-Barr virus latent membrane protein 1 trans-activates miR-155 transcription through the NF-kappaB pathway. *Nucleic Acids Res* 36, 6608–6619.

Gatza, M. L., Chandhasin, C., Ducu, R. I. & Marriott, S. J. (2005). Impact of transforming viruses on cellular mutagenesis, genome stability, and cellular transformation. *Environ Mol Mutagen* 45, 304–325.

Goossens, T., Klein, U. & Kuppers, R. (1998). Frequent occurrence of deletions and duplications during somatic hypermutation: implications for oncogene translocations and heavy chain disease. *Proc Natl Acad Sci U S A* 95, 2463–2468.

Gottschalk, S., Rooney, C. M. & Heslop, H. E. (2005). Post-transplant lymphoproliferative disorders. *Annu Rev Med* 56, 29–44.

Grogg, K. L., Miller, R. F. & Dogan, A. (2007). HIV infection and lymphoma. *J Clin Pathol* 60, 1365–1372.

Gunther, T. & Grundhoff, A. (2010). The epigenetic landscape of latent Kaposi sarcoma-associated herpesvirus genomes. *PLoS Pathog* 6, e1000935.

Gupta, R. A., Shah, N., Wang, K. C., Kim, J., Horlings, H. M., Wong, D. J., Tsai, M. C., Hung, T., Argani, P., Rinn, J. L., Wang, Y., Brzoska, P., Kong, B., Li, R., West, R. B., van de Vijver, M. J., Sukumar, S. & Chang, H. Y. (2010). Long non-coding RNA HOTAIR reprograms chromatin state to promote cancer metastasis. *Nature* 464, 1071–1076.

Gutierrez, M. I., Siraj, A. K., Khaled, H., Koon, N., El Rifai, W. & Bhatia, K. (2004). CpG island methylation in Schistosoma- and non-Schistosoma-associated bladder cancer. *Mod Pathol* 17, 1268–1274.

Hansen, J. C., Ghosh, R. P. & Woodcock, C. L. (2010). Binding of the Rett syndrome protein, MeCP2, to methylated and unmethylated DNA and chromatin. *IUBMB Life* 62, 732–738.

Hansen, K. D., Timp, W., Bravo, H. C., Sabunciyan, S., Langmead, B., McDonald, O. G., Wen, B., Wu, H., Liu, Y., Diep, D., Briem, E., Zhang, K., Irizarry, R. A. & Feinberg, A. P. (2011). Increased methylation variation in epigenetic domains across cancer types. *Nat Genet* 43, 768–775.

He, Y. F., Li, B. Z., Li, Z., Liu, P., Wang, Y., Tang, Q., Ding, J., Jia, Y., Chen, Z., Li, L., Sun, Y., Li, X., Dai, Q., Song, C. X., Zhang, K., He, C. & Xu, G. L. (2011). Tet-mediated formation of 5-carboxylcytosine and its excision by TDG in mammalian DNA. *Science* 333, 1303–1307.

Hecht, J. L. & Aster, J. C. (2000). Molecular biology of Burkitt's lymphoma. *J Clin Oncol* 18, 3707–3721.

Helmbold, P., Lahtz, C., Enk, A., Herrmann-Trost, P., Marsch, W. C., Kutzner, H. & Dammann, R. H. (2009). Frequent occurrence of RASSF1A promoter hypermethylation and Merkel cell polyomavirus in Merkel cell carcinoma. *Mol Carcinog* 48, 903–909.

Helt, A. M. & Galloway, D. A. (2003). Mechanisms by which DNA tumor virus oncoproteins target the Rb family of pocket proteins. *Carcinogenesis* 24, 159–169.

Henken, F. E., Wilting, S. M., Overmeer, R. M., van Rietschoten, J. G., Nygren, A. O., Errami, A., Schouten, J. P., Meijer, C. J., Snijders, P. J. & Steenbergen, R. D. (2007). Sequential gene promoter methylation during HPV-induced cervical carcinogenesis. *Br J Cancer* 97, 1457–1464.

Herbst, H., Dallenbach, F., Hummel, M., Niedobitek, G., Pileri, S., Muller-Lantzsch, N. & Stein, H. (1991). Epstein-Barr virus latent membrane protein expression in Hodgkin and Reed-Sternberg cells. *Proc Natl Acad Sci U S A* 88, 4766–4770.

Higgs, M. R., Lerat, H. & Pawlotsky, J. M. (2010). Downregulation of Gadd45beta expression by hepatitis C virus leads to defective cell cycle arrest. *Cancer Res* 70, 4901–4911.

Hino, R., Uozaki, H., Murakami, N., Ushiku, T., Shinozaki, A., Ishikawa, S., Morikawa, T., Nakaya, T., Sakatani, T., Takada, K. & Fukayama, M. (2009). Activation of DNA methyltransferase 1 by EBV latent membrane protein 2A leads to promoter hypermethylation of PTEN gene in gastric carcinoma. *Cancer Res* 69, 2766–2774.

Hjalgrim, H., Askling, J., Rostgaard, K., Hamilton-Dutoit, S., Frisch, M., Zhang, J. S., Madsen, M., Rosdahl, N., Konradsen, H. B., Storm, H. H. & Melbye, M. (2003). Characteristics of Hodgkin's lymphoma after infectious mononucleosis. *N Engl J Med* 349, 1324–1332.

Hjalgrim, H. & Engels, E. A. (2008). Infectious aetiology of Hodgkin and non-Hodgkin lymphomas: a review of the epidemiological evidence. *J Intern Med* 264, 537–548.

Holland, D., Hoppe-Seyler, K., Schuller, B., Lohrey, C., Maroldt, J., Durst, M. & Hoppe-Seyler, F. (2008). Activation of the enhancer of zeste homologue 2 gene by the human papillomavirus E7 oncoprotein. *Cancer Res* 68, 9964–9972.

Howie, H. L., Katzenellenbogen, R. A. & Galloway, D. A. (2009). Papillomavirus E6 proteins. *Virology* 384, 324–334.

Hsu, C. H., Peng, K. L., Jhang, H. C., Lin, C. H., Wu, S. Y., Chiang, C. M., Lee, S. C., Yu, W. C. & Juan, L. J. (2011). The HPV E6 oncoprotein targets histone methyltransferases for modulating specific gene transcription. *Oncogene.* doi: 10.1038/onc.2011.415.

Hu, J., Dong, A., Fernandez-Ruiz, V., Shan, J., Kawa, M., Martinez-Anso, E., Prieto, J. & Qian, C. (2009). Blockade of Wnt signaling inhibits angiogenesis and tumor growth in hepatocellular carcinoma. *Cancer Res* 69, 6951–6959.

Hummel, M., Bentink, S., Berger, H., Klapper, W., Wessendorf, S., Barth, T. F., Bernd, H. W., Cogliatti, S. B., Dierlamm, J., Feller, A. C., Hansmann, M. L., Haralambieva, E., Harder, L., Hasenclever, D., Kuhn, M., Lenze, D., Lichter, P., Martin-Subero, J. I., Moller, P., Muller-Hermelink, H. K., Ott, G., Parwaresch, R. M., Pott, C., Rosenwald, A., Rosolowski, M., Schwaenen, C., Sturzenhofecker, B., Szczepanowski, M., Trautmann, H., Wacker, H. H., Spang, R., Loeffler, M., Trumper, L., Stein, H. & Siebert, R. (2006). A biologic definition of Burkitt's lymphoma from transcriptional and genomic profiling. *N Engl J Med* 354, 2419–2430.

Hung, T. & Chang, H. Y. (2010). Long noncoding RNA in genome regulation: prospects and mechanisms. *RNA Biol* 7, 582–585.

Hutzinger, R., Feederle, R., Mrazek, J., Schiefermeier, N., Balwierz, P. J., Zavolan, M., Polacek, N., Delecluse, H. J. & Huttenhofer, A. (2009). Expression and processing of a small nucleolar RNA from the Epstein-Barr virus genome. *PLoS Pathog* 5, e1000547.

Hyland, P. L., McDade, S. S., McCloskey, R., Dickson, G. J., Arthur, K., McCance, D. J. & Patel, D. (2011). Evidence for Alteration of EZH2, BMI1, and KDM6A and Epigenetic Reprogramming in Human Papillomavirus Type 16 E6/E7-Expressing Keratinocytes. *J Virol* 85, 10999–11006.

Iaquinta, P. J. & Lees, J. A. (2007). Life and death decisions by the E2F transcription factors. *Curr Opin Cell Biol* 19, 649–657.

Issa, J. P. (2011). Epigenetic variation and cellular Darwinism. *Nat Genet* 43, 724–726.

Ito, S., Shen, L., Dai, Q., Wu, S. C., Collins, L. B., Swenberg, J. A., He, C. & Zhang, Y. (2011). Tet proteins can convert 5-methylcytosine to 5-formylcytosine and 5-carboxylcytosine. *Science* 333, 1300–1303.

Janas, J. A. & Van Aelst, L. (2011). Oncogenic tyrosine kinases target Dok-1 for ubiquitin-mediated proteasomal degradation to promote cell transformation. *Mol Cell Biol* 31, 2552–2565.

Jarrett, R. F. (2003). Risk factors for Hodgkin's lymphoma by EBV status and significance of detection of EBV genomes in serum of patients with EBV-associated Hodgkin's lymphoma. *Leuk Lymphoma* 44 Suppl 3, S27–S32.

Kamdar, K. Y., Rooney, C. M. & Heslop, H. E. (2011). Posttransplant lymphoproliferative disease following liver transplantation. *Curr Opin Organ Transplant* 16, 274–280.

Kang, G. H., Lee, S., Cho, N. Y., Gandamihardja, T., Long, T. I., Weisenberger, D. J., Campan, M. & Laird, P. W. (2008). DNA methylation profiles of gastric carcinoma characterized by quantitative DNA methylation analysis. *Lab Invest* 88, 161–170.

Kang, H., Wiedmer, A., Yuan, Y., Robertson, E. & Lieberman, P. M. (2011). Coordination of KSHV latent and lytic gene control by CTCF-cohesin mediated chromosome conformation. *PLoS Pathog* 7, e1002140.

Katoh, H., Shibata, T., Kokubu, A., Ojima, H., Fukayama, M., Kanai, Y. & Hirohashi, S. (2006). Epigenetic instability and chromosomal instability in hepatocellular carcinoma. *Am J Pathol* 168, 1375–1384.

Kaur, P., Paliwal, A., Durantel, D., Hainaut, P., Scoazec, J. Y., Zoulim, F., Chemin, I. & Herceg, Z. (2010). DNA methylation of hepatitis B virus (HBV) genome associated with the development of hepatocellular carcinoma and occult HBV infection. *J Infect Dis* 202, 700–704.

Kelly, G., Bell, A. & Rickinson, A. (2002). Epstein-Barr virus-associated Burkitt lymphomagenesis selects for downregulation of the nuclear antigen EBNA2. *Nat Med* 8, 1098–1104.

Kimura, H., Nakamura, T., Ogawa, T., Tanaka, S. & Shiota, K. (2003). Transcription of mouse DNA methyltransferase 1 (Dnmt1) is regulated by both E2F-Rb-HDAC-dependent and -independent pathways. *Nucleic Acids Res* 31, 3101–3113.

Kinoshita, T., Nomoto, S., Kodera, Y., Koike, M., Fujiwara, M. & Nakao, A. (2011). Decreased expression and aberrant hypermethylation of the SFRP genes in human gastric cancer. *Hepatogastroenterology* 58, 1051–1056.

Kitkumthorn, N., Yanatatsanajit, P., Kiatpongsan, S., Phokaew, C., Triratanachat, S., Trivijitsilp, P., Termrungruanglert, W., Tresukosol, D., Niruthisard, S. & Mutirangura, A. (2006). Cyclin A1 promoter hypermethylation in human papillomavirus-associated cervical cancer. *BMC Cancer* 6:55.

Knight, J. S., Lan, K., Subramanian, C. & Robertson, E. S. (2003). Epstein-Barr virus nuclear antigen 3C recruits histone deacetylase activity and associates with the corepressors mSin3A and NCoR in human B-cell lines. *J Virol* 77, 4261–4272.

Ko, E., Kim, S. J., Joh, J. W., Park, C. K., Park, J. & Kim, D. H. (2008). CpG island hypermethylation of SOCS-1 gene is inversely associated with HBV infection in hepatocellular carcinoma. *Cancer Lett* 271, 240–250.

Kohno, T., Kunitoh, H., Mimaki, S., Shiraishi, K., Kuchiba, A., Yamamoto, S. & Yokota, J. (2011). Contribution of the TP53, OGG1, CHRNA3, and HLA-DQA1 genes to the risk for lung squamous cell carcinoma. *J Thorac Oncol* 6, 813–817.

Kremer, B. E., Reshef, R., Misleh, J. G., Christie, J. D., Ahya, V. N., Blumenthal, N. P., Kotloff, R. M., Hadjiliadis, D., Stadtmauer, E. A., Schuster, S. J. & Tsai, D. E. (2011). Post-transplant lymphoproliferative disorder after lung transplantation: A review of 35 cases. *J Heart Lung Transplant* 31, 296–304.

Krishna, S. M., Kattoor, J. & Balaram, P. (2005). Down regulation of adhesion protein E-cadherin in Epstein-Barr virus infected nasopharyngeal carcinomas. *Cancer Biomark* 1, 271–277.

Krithivas, A., Young, D. B., Liao, G., Greene, D. & Hayward, S. D. (2000). Human herpesvirus 8 LANA interacts with proteins of the mSin3 corepressor complex and negatively regulates Epstein-Barr virus gene expression in dually infected PEL cells. *J Virol* 74, 9637–9645.

Kuppers, R. (2009). The biology of Hodgkin's lymphoma. *Nat Rev Cancer* 9, 15–27.

Kuppers, R., Klein, U., Schwering, I., Distler, V., Brauninger, A., Cattoretti, G., Tu, Y., Stolovitzky, G. A., Califano, A., Hansmann, M. L. & Dalla-Favera, R. (2003). Identification of Hodgkin and Reed-Sternberg cell-specific genes by gene expression profiling. *J Clin Invest* 111, 529–537.

Kuppers, R., Schwering, I., Brauninger, A., Rajewsky, K. & Hansmann, M. L. (2002). Biology of Hodgkin's lymphoma. *Ann Oncol* 13 Suppl 1, 11–18.

Kurth, J., Hansmann, M. L., Rajewsky, K. & Kuppers, R. (2003). Epstein-Barr virus-infected B cells expanding in germinal centers of infectious mononucleosis patients do not participate in the germinal center reaction. *Proc Natl Acad Sci U S A* 100, 4730–4735.

Kwong, J., Lo, K. W., Chow, L. S., Chan, F. L., To, K. F. & Huang, D. P. (2005a). Silencing of the retinoid response gene TIG1 by promoter hypermethylation in nasopharyngeal carcinoma. *Int J Cancer* 113, 386–392.

Kwong, J., Lo, K. W., Chow, L. S., To, K. F., Choy, K. W., Chan, F. L., Mok, S. C. & Huang, D. P. (2005b). Epigenetic silencing of cellular retinol-binding proteins in nasopharyngeal carcinoma. *Neoplasia* 7, 67–74.

Kwong, J., Lo, K. W., To, K. F., Teo, P. M., Johnson, P. J. & Huang, D. P. (2002). Promoter hypermethylation of multiple genes in nasopharyngeal carcinoma. *Clin Cancer Res* 8, 131–137.

Lai, H. C., Lin, Y. W., Huang, R. L., Chung, M. T., Wang, H. C., Liao, Y. P., Su, P. H., Liu, Y. L. & Yu, M. H. (2010). Quantitative DNA methylation analysis detects cervical intraepithelial neoplasms type 3 and worse. *Cancer* 116, 4266–4274.

Lai, H. C., Lin, Y. W., Huang, T. H., Yan, P., Huang, R. L., Wang, H. C., Liu, J., Chan, M. W., Chu, T. Y., Sun, C. A., Chang, C. C. & Yu, M. H. (2008). Identification of novel DNA methylation markers in cervical cancer. *Int J Cancer* 123, 161–167.

Lambert, M. P., Paliwal, A., Vaissiere, T., Chemin, I., Zoulim, F., Tommasino, M., Hainaut, P., Sylla, B., Scoazec, J. Y., Tost, J. & Herceg, Z. (2011). Aberrant DNA methylation distinguishes hepatocellular carcinoma associated with HBV and HCV infection and alcohol intake. *J Hepatol* 54, 705–715.

Laurent-Puig, P. & Zucman-Rossi, J. (2006). Genetics of hepatocellular tumors. *Oncogene* 25, 3778–3786.

Lee, T. S., Kim, J. W., Kang, G. H., Park, N. H., Song, Y. S., Kang, S. B. & Lee, H. P. (2006). DNA hypomethylation of CAGE promotors in squamous cell carcinoma of uterine cervix. *Ann N Y Acad Sci* 1091, 218–224.

Lenze, D., Leoncini, L., Hummel, M., Volinia, S., Liu, C. G., Amato, T., De Falco, G., Githanga, J., Horn, H., Nyagol, J., Ott, G., Palatini, J., Pfreundschuh, M., Rogena, E., Rosenwald, A., Siebert, R., Croce, C. M. & Stein, H. (2011). The different epidemiologic subtypes of Burkitt lymphoma share a homogenous micro RNA profile distinct from diffuse large B-cell lymphoma. *Leukemia* 25, 1869–1876.

Leonard, S., Wei, W., Anderton, J., Vockerodt, M., Rowe, M., Murray, P. G. & Woodman, C. B. (2011). Epigenetic and transcriptional changes which follow Epstein-Barr virus infection of germinal center B cells and their relevance to the pathogenesis of Hodgkin's lymphoma. *J Virol* 85, 9568–9577.

Levens, D. L. (2003). Reconstructing MYC. *Genes Dev* 17, 1071–1077.

Li, G., Wu, Z., Peng, Y., Liu, X., Lu, J., Wang, L., Pan, Q., He, M. L. & Li, X. P. (2010). MicroRNA-10b induced by Epstein-Barr virus-encoded latent membrane protein-1 promotes the metastasis of human nasopharyngeal carcinoma cells. *Cancer Lett* 299, 29–36.

Li, X., Hui, A. M., Sun, L., Hasegawa, K., Torzilli, G., Minagawa, M., Takayama, T. & Makuuchi, M. (2004). p16INK4A hypermethylation is associated with hepatitis virus infection, age, and gender in hepatocellular carcinoma. *Clin Cancer Res* 10, 7484–7489.

Lim, C., Lee, D., Seo, T., Choi, C. & Choe, J. (2003). Latency-associated nuclear antigen of Kaposi's sarcoma-associated herpesvirus functionally interacts with heterochromatin protein 1. *J Biol Chem* 278, 7397–7405.

Lin, I. G., Tomzynski, T. J., Ou, Q. & Hsieh, C. L. (2000). Modulation of DNA binding protein affinity directly affects target site demethylation. *Mol Cell Biol* 20, 2343–2349.

Ling, P. D., Tan, J. & Peng, R. (2009). Nuclear-cytoplasmic shuttling is not required for the Epstein-Barr virus EBNA-LP transcriptional coactivation function. *J Virol* 83, 7109–7116.

Liu, L., Tommasi, S., Lee, D. H., Dammann, R. & Pfeifer, G. P. (2003). Control of microtubule stability by the RASSF1A tumor suppressor. *Oncogene* 22, 8125–8136.

Lo, A. K., To, K. F., Lo, K. W., Lung, R. W., Hui, J. W., Liao, G. & Hayward, S. D. (2007). Modulation of LMP1 protein expression by EBV-encoded microRNAs. *Proc Natl Acad Sci U S A* 104, 16164–16169.

Lupberger, J. & Hildt, E. (2007). Hepatitis B virus-induced oncogenesis. *World J Gastroenterol* 13, 74–81.

Machida, K., Cheng, K. T., Sung, V. M., Shimodaira, S., Lindsay, K. L., Levine, A. M., Lai, M. Y. & Lai, M. M. (2004). Hepatitis C virus induces a mutator phenotype: enhanced mutations of immunoglobulin and protooncogenes. *Proc Natl Acad Sci U S A* 101, 4262–4267.

Magrath, I., Jain, V. & Bhatia, K. (1992). Epstein-Barr virus and Burkitt's lymphoma. *Semin Cancer Biol* 3, 285–295.

Martin-Subero, J. I., Kreuz, M., Bibikova, M., Bentink, S., Ammerpohl, O., Wickham-Garcia, E., Rosolowski, M., Richter, J., Lopez-Serra, L., Ballestar, E., Berger, H., Agirre, X., Bernd, H. W., Calvanese, V., Cogliatti, S. B., Drexler, H. G., Fan, J. B., Fraga, M. F., Hansmann, M. L., Hummel, M., Klapper, W., Korn, B., Kuppers, R., Macleod, R. A., Moller, P., Ott, G., Pott, C., Prosper, F., Rosenwald, A., Schwaenen, C., Schubeler, D., Seifert, M., Sturzenhofecker, B., Weber, M., Wessendorf, S., Loeffler, M., Trumper, L., Stein, H., Spang, R., Esteller, M., Barker, D., Hasenclever, D. & Siebert, R. (2009). New insights into the biology and origin of mature aggressive B-cell lymphomas by combined epigenomic, genomic, and transcriptional profiling. *Blood* 113, 2488–2497.

Massari, M. E., Rivera, R. R., Voland, J. R., Quong, M. W., Breit, T. M., van Dongen, J. J., de Smit, O. & Murre, C. (1998). Characterization of ABF-1, a novel basic helix-loop-helix transcription factor expressed in activated B lymphocytes. *Mol Cell Biol* 18, 3130–3139.

Matsumura, S., Persson, L. M., Wong, L. & Wilson, A. C. (2010). The latency-associated nuclear antigen interacts with MeCP2 and nucleosomes through separate domains. *J Virol* 84, 2318–2330.

Mattsson, K., Kiss, C., Platt, G. M., Simpson, G. R., Kashuba, E., Klein, G., Schulz, T. F. & Szekely, L. (2002). Latent nuclear antigen of Kaposi's sarcoma herpesvirus/human herpesvirus-8 induces and relocates RING3 to nuclear heterochromatin regions. *J Gen Virol* 83, 179–188.

McCabe, M. T., Low, J. A., Daignault, S., Imperiale, M. J., Wojno, K. J. & Day, M. L. (2006). Inhibition of DNA methyltransferase activity prevents tumorigenesis in a mouse model of prostate cancer. *Cancer Res* 66, 385–392.

McLaughlin-Drubin, M. E., Crum, C. P. & Munger, K. (2011). Human papillomavirus E7 oncoprotein induces KDM6A and KDM6B histone demethylase expression and causes epigenetic reprogramming. *Proc Natl Acad Sci U S A* 108, 2130–2135.

McLaughlin-Drubin, M. E. & Munger, K. (2009). The human papillomavirus E7 oncoprotein. *Virology* 384, 335–344.

Meng, Y., Wang, Q. G., Wang, J. X., Zhu, S. T., Jiao, Y., Li, P. & Zhang, S. T. (2011). Epigenetic Inactivation of the SFRP1 Gene in Esophageal Squamous Cell Carcinoma. *Dig Dis Sci* 56, 3195–3203.

Minarovits, J. (2006). Epigenotypes of latent herpesvirus genomes. *Curr Top Microbiol Immunol* 310, 61–80.

Minarovits, J., Hu, L. F., Minarovits-Kormuta, S., Klein, G. & Ernberg, I. (1994). Sequence-specific methylation inhibits the activity of the Epstein-Barr virus LMP 1 and BCR2 enhancer-promoter regions. *Virology* 200, 661–667.

Motsch, N., Pfuhl, T., Mrazek, J., Barth, S. & Grasser, F. A. (2007). Epstein-Barr virus-encoded latent membrane protein 1 (LMP1) induces the expression of the cellular microRNA miR-146a. *RNA Biol* 4, 131–137.

Mrazek, J., Kreutmayer, S. B., Grasser, F. A., Polacek, N. & Huttenhofer, A. (2007). Subtractive hybridization identifies novel differentially expressed ncRNA species in EBV-infected human B cells. *Nucleic Acids Res* 35, e73.

Muti, G., Klersy, C., Baldanti, F., Granata, S., Oreste, P., Pezzetti, L., Gatti, M., Gargantini, L., Caramella, M., Mancini, V., Gerna, G. & Morra, E. (2003). Epstein-Barr virus (EBV) load and interleukin-10 in EBV-positive and EBV-negative post-transplant lymphoproliferative disorders. *Br J Haematol* 122, 927–933.

Naka, K., Abe, K., Takemoto, K., Dansako, H., Ikeda, M., Shimotohno, K. & Kato, N. (2006). Epigenetic silencing of interferon-inducible genes is implicated in interferon resistance of hepatitis C virus replicon-harboring cells. *J Hepatol* 44, 869–878.

8 Epigenetic Changes in Virus-Associated Neoplasms

Nakase, K., Cheng, J., Zhu, Q. & Marasco, W. A. (2009). Mechanisms of SHP-1 P2 promoter regulation in hematopoietic cells and its silencing in HTLV-1-transformed T cells. *J Leukoc Biol* 85, 165–174.

Nan, X., Cross, S. & Bird, A. (1998a). Gene silencing by methyl-CpG-binding proteins. *Novartis Found Symp* 214, 6–16.

Nan, X., Ng, H. H., Johnson, C. A., Laherty, C. D., Turner, B. M., Eisenman, R. N. & Bird, A. (1998b). Transcriptional repression by the methyl-CpG-binding protein MeCP2 involves a histone deacetylase complex. *Nature* 393, 386–389.

Narimatsu, T., Tamori, A., Koh, N., Kubo, S., Hirohashi, K., Yano, Y., Arakawa, T., Otani, S. & Nishiguchi, S. (2004). p16 promoter hypermethylation in human hepatocellular carcinoma with or without hepatitis virus infection. *Intervirology* 47, 26–31.

Niemhom, S., Kitazawa, S., Kitazawa, R., Maeda, S. & Leopairat, J. (2008). Hypermethylation of epithelial-cadherin gene promoter is associated with Epstein-Barr virus in nasopharyngeal carcinoma. *Cancer Detect Prev* 32, 127–134.

Niller, H. H., Salamon, D., Banati, F., Schwarzmann, F., Wolf, H. & Minarovits, J. (2004a). The LCR of EBV makes Burkitt's lymphoma endemic. *Trends Microbiol* 12, 495–499.

Niller, H. H., Salamon, D., Ilg, K., Koroknai, A., Banati, F., Bauml, G., Rucker, O., Schwarzmann, F., Wolf, H. & Minarovits, J. (2003). The in vivo binding site for oncoprotein c-Myc in the promoter for Epstein-Barr virus (EBV) encoding RNA (EBER) 1 suggests a specific role for EBV in lymphomagenesis. *Med Sci Monit* 9, HY1–HY9.

Niller, H. H., Salamon, D., Ilg, K., Koroknai, A., Banati, F., Schwarzmann, F., Wolf, H. & Minarovits, J. (2004b). EBV-associated neoplasms: alternative pathogenetic pathways. *Med Hypotheses* 62, 387–391.

Niller, H. H., Salamon, D., Rahmann, S., Ilg, K., Koroknai, A., Banati, F., Schwarzmann, F., Wolf, H. & Minarovits, J. (2004c). A 30 kb region of the Epstein-Barr virus genome is colinear with the rearranged human immunoglobulin gene loci: implications for a "ping-pong evolution" model for persisting viruses and their hosts. A review. *Acta Microbiol Immunol Hung* 51, 469–484.

Niller, H. H., Wolf, H. & Minarovits, J. (2007). Epstein-Barr Virus. In *Latency Strategies of Herpesviruses*, pp. 154–191. Edited by J. Minarovits, E. Gonczol & T. Valyi-Nagy. New York: Springer.

Niller, H. H., Wolf, H. & Minarovits, J. (2008). Regulation and dysregulation of Epstein-Barr virus latency: implications for the development of autoimmune diseases. *Autoimmunity* 41, 298–328.

Niller, H. H., Wolf, H. & Minarovits, J. (2009). Epigenetic dysregulation of the host cell genome in Epstein-Barr virus-associated neoplasia. *Semin Cancer Biol* 19, 158–164.

Niller, H. H., Wolf, H. & Minarovits, J. (2011). Viral hit and run-oncogenesis: Genetic and epigenetic scenarios. *Cancer Lett* 305, 200–217.

Pallesen, G., Hamilton-Dutoit, S. J., Rowe, M. & Young, L. S. (1991). Expression of Epstein-Barr virus latent gene products in tumour cells of Hodgkin's disease. *Lancet* 337, 320–322.

Pantry, S. N. & Medveczky, P. G. (2009). Epigenetic regulation of Kaposi's sarcoma-associated herpesvirus replication. *Semin Cancer Biol* 19, 153–157.

Park, I. Y., Sohn, B. H., Yu, E., Suh, D. J., Chung, Y. H., Lee, J. H., Surzycki, S. J. & Lee, Y. I. (2007). Aberrant epigenetic modifications in hepatocarcinogenesis induced by hepatitis B virus X protein. *Gastroenterology* 132, 1476–1494.

Pelengaris, S., Khan, M. & Evan, G. I. (2002). Suppression of Myc-induced apoptosis in beta cells exposes multiple oncogenic properties of Myc and triggers carcinogenic progression. *Cell* 109, 321–334.

Piccaluga, P. P., De Falco, G., Kustagi, M., Gazzola, A., Agostinelli, C., Tripodo, C., Leucci, E., Onnis, A., Astolfi, A., Sapienza, M. R., Bellan, C., Lazzi, S., Tumwine, L., Mawanda, M., Ogwang, M., Calbi, V., Formica, S., Califano, A., Pileri, S. A. & Leoncini, L. (2011). Gene expression analysis uncovers similarity and differences among Burkitt lymphoma subtypes. *Blood* 117, 3596–3608.

Platt, G., Carbone, A. & Mittnacht, S. (2002). p16INK4a loss and sensitivity in KSHV associated primary effusion lymphoma. *Oncogene* 21, 1823–1831.

Portal, D., Rosendorff, A. & Kieff, E. (2006). Epstein-Barr nuclear antigen leader protein coactivates transcription through interaction with histone deacetylase 4. *Proc Natl Acad Sci U S A* 103, 19278–19283.

Raab-Traub, N. & Flynn, K. (1986). The structure of the termini of the Epstein-Barr virus as a marker of clonal cellular proliferation. *Cell* 47, 883–889.

Rahadiani, N., Takakuwa, T., Tresnasari, K., Morii, E. & Aozasa, K. (2008). Latent membrane protein-1 of Epstein-Barr virus induces the expression of B-cell integration cluster, a precursor form of microRNA-155, in B lymphoma cell lines. *Biochem Biophys Res Commun* 377, 579–583.

Renne, R., Barry, C., Dittmer, D., Compitello, N., Brown, P. O. & Ganem, D. (2001). Modulation of cellular and viral gene expression by the latency-associated nuclear antigen of Kaposi's sarcoma-associated herpesvirus. *J Virol* 75, 458–468.

Ringrose, L. & Paro, R. (2007). Polycomb/Trithorax response elements and epigenetic memory of cell identity. *Development* 134, 223–232.

Roh, T. Y., Cuddapah, S. & Zhao, K. (2005). Active chromatin domains are defined by acetylation islands revealed by genome-wide mapping. *Genes Dev* 19, 542–552.

Rossetto, C. C. & Pari, G. S. (2011). Kaposi's Sarcoma-Associated Herpesvirus Noncoding Polyadenylated Nuclear RNA Interacts with Virus- and Host Cell-Encoded Proteins and Suppresses Expression of Genes Involved in Immune Modulation. *J Virol* 85, 13290–13297.

Rossi, D., Gaidano, G., Gloghini, A., Deambrogi, C., Franceschetti, S., Berra, E., Cerri, M., Vendramin, C., Conconi, A., Viglio, A., Muti, G., Oreste, P., Morra, E., Paulli, M., Capello, D. & Carbone, A. (2003). Frequent aberrant promoter hypermethylation of O6-methylguanine-DNA methyltransferase and death-associated protein kinase genes in immunodeficiency-related lymphomas. *Br J Haematol* 123, 475–478.

Rossi, G. & Bonetti, F. (2004). EBV and Burkitt's lymphoma. *N Engl J Med* 350, 2621.

Roughan, J. E., Torgbor, C. & Thorley-Lawson, D. A. (2010). Germinal center B cells latently infected with Epstein-Barr virus proliferate extensively but do not increase in number. *J Virol* 84, 1158–1168.

Saito, Y., Kanai, Y., Sakamoto, M., Saito, H., Ishii, H. & Hirohashi, S. (2002). Overexpression of a splice variant of DNA methyltransferase 3b, DNMT3b4, associated with DNA hypomethylation on pericentromeric satellite regions during human hepatocarcinogenesis. *Proc Natl Acad Sci U S A* 99, 10060–10065.

Sakakibara, S., Ueda, K., Nishimura, K., Do, E., Ohsaki, E., Okuno, T. & Yamanishi, K. (2004). Accumulation of heterochromatin components on the terminal repeat sequence of Kaposi's sarcoma-associated herpesvirus mediated by the latency-associated nuclear antigen. *J Virol* 78, 7299–7310.

Sakuma, K., Chong, J. M., Sudo, M., Ushiku, T., Inoue, Y., Shibahara, J., Uozaki, H., Nagai, H. & Fukayama, M. (2004). High-density methylation of p14ARF and p16INK4A in Epstein-Barr virus-associated gastric carcinoma. *Int J Cancer* 112, 273–278.

Salamon, D., Banati, F., Koroknai, A., Ravasz, M., Szenthe, K., Bathori, Z., Bakos, A., Niller, H. H., Wolf, H. & Minarovits, J. (2009). Binding of CCCTC-binding factor in vivo to the region located between Rep* and C-promoter of Epstein-Barr virus is unaffected by CpG methylation and does not correlate with Cp activity. *J Gen Virol.*

Saulnier, A., Vaissiere, T., Yue, J., Siouda, M., Malfroy, M., Accardi, R., Creveaux, M., Sebastian, S., Shahzad, N., Gheit, T., Hussain, I., Torrente, M., Maffini, F. A., Calabrese, L., Chiesa, F., Cuenin, C., Shukla, R., Fathallah, I., Matos, E., Daudt, A., Koifman, S., Wunsch-Filho, V., Menezes, A. M., Curado, M. P., Zaridze, D., Boffetta, P., Brennan, P., Tommasino, M., Herceg, Z. & Sylla, B. S. (2011). Inactivation of the putative suppressor gene DOK1 by promoter hypermethylation in primary human cancers. *Int J Cancer* 130, 2484–2894.

Shamay, M., Krithivas, A., Zhang, J. & Hayward, S. D. (2006). Recruitment of the de novo DNA methyltransferase Dnmt3a by Kaposi's sarcoma-associated herpesvirus LANA. *Proc Natl Acad Sci U S A* 103, 14554–14559.

Shi, N., Ye, S., Bartlam, M., Yang, M., Wu, J., Liu, Y., Sun, F., Han, X., Peng, X., Qiang, B., Yuan, J. & Rao, Z. (2004). Structural basis for the specific recognition of RET by the Dok1 phosphotyrosine binding domain. *J Biol Chem* 279, 4962–4969.

Shikauchi, Y., Saiura, A., Kubo, T., Niwa, Y., Yamamoto, J., Murase, Y. & Yoshikawa, H. (2009). SALL3 interacts with DNMT3A and shows the ability to inhibit CpG island methylation in hepatocellular carcinoma. *Mol Cell Biol* 29, 1944–1958.

Shim, Y. H., Yoon, G. S., Choi, H. J., Chung, Y. H. & Yu, E. (2003). p16 Hypermethylation in the early stage of hepatitis B virus-associated hepatocarcinogenesis. *Cancer Lett* 190, 213–219.

Sriraksa, R., Zeller, C., El Bahrawy, M. A., Dai, W., Daduang, J., Jearanaikoon, P., Chau-In, S., Brown, R. & Limpaiboon, T. (2011). CpG-island methylation study of liver fluke-related cholangiocarcinoma. *Br J Cancer* 104, 1313–1318.

Sriuranpong, V., Mutirangura, A., Gillespie, J. W., Patel, V., Amornphimoltham, P., Molinolo, A. A., Kerekhanjanarong, V., Supanakorn, S., Supiyaphun, P., Rangdaeng, S., Voravud, N. & Gutkind, J. S. (2004). Global gene expression profile of nasopharyngeal carcinoma by laser capture microdissection and complementary DNA microarrays. *Clin Cancer Res* 10, 4944–4958.

Staudt, L. M. & Dave, S. (2005). The biology of human lymphoid malignancies revealed by gene expression profiling. *Adv Immunol* 87, 163–208.

Stedman, W., Deng, Z., Lu, F. & Lieberman, P. M. (2004). ORC, MCM, and histone hyperacetylation at the Kaposi's sarcoma-associated herpesvirus latent replication origin. *J Virol* 78, 12566–12575.

Stevens, S. J., Verschuuren, E. A., Pronk, I., van Der Bij W., Harmsen, M. C., The, T. H., Meijer, C. J., van Den Brule, A. J. & Middeldorp, J. M. (2001). Frequent monitoring of Epstein-Barr virus DNA load in unfractionated whole blood is essential for early detection of posttransplant lymphoproliferative disease in high-risk patients. *Blood* 97, 1165–1171.

Stuber, G., Mattsson, K., Flaberg, E., Kati, E., Markasz, L., Sheldon, J. A., Klein, G., Schulz, T. F. & Szekely, L. (2007). HHV-8 encoded LANA-1 alters the higher organization of the cell nucleus. *Mol Cancer* 6:28.

Su, P. F., Lee, T. C., Lin, P. J., Lee, P. H., Jeng, Y. M., Chen, C. H., Liang, J. D., Chiou, L. L., Huang, G. T. & Lee, H. S. (2007). Differential DNA methylation associated with hepatitis B virus infection in hepatocellular carcinoma. *Int J Cancer* 121, 1257–1264.

Swaminathan, S. (2008). Noncoding RNAs produced by oncogenic human herpesviruses. *J Cell Physiol* 216, 321–326.

Szalmas, A. & Konya, J. (2009). Epigenetic alterations in cervical carcinogenesis. *Semin Cancer Biol* 19, 144–152.

Takacs, M., Banati, F., Koroknai, A., Segesdi, J., Salamon, D., Wolf, H., Niller, H. H. & Minarovits, J. (2010). Epigenetic regulation of latent Epstein-Barr virus promoters. *Biochim Biophys Acta* 1799, 228–235.

Takacs, M., Segesdi, J., Banati, F., Koroknai, A., Wolf, H., Niller, H. H. & Minarovits, J. (2009). The importance of epigenetic alterations in the development of Epstein-Barr virus-related lymphomas. *Mediterr J Hematol Infect Dis* 1, e2009012.

Takai, N., Kira, N., Ishii, T., Nishida, M., Nasu, K. & Narahara, H. (2011). Novel chemotherapy using histone deacetylase inhibitors in cervical cancer. *Asian Pac J Cancer Prev* 12, 575–580.

Tanner, J. E. & Alfieri, C. (2001). The Epstein-Barr virus and post-transplant lymphoproliferative disease: interplay of immunosuppression, EBV, and the immune system in disease pathogenesis. *Transpl Infect Dis* 3, 60–69.

Tempera, I., Klichinsky, M. & Lieberman, P. M. (2011). EBV latency types adopt alternative chromatin conformations. *PLoS Pathog* 7, e1002180.

Tempera, I. & Lieberman, P. M. (2010). Chromatin organization of gammaherpesvirus latent genomes. *Biochim Biophys Acta* 1799, 236–245.

ter Schegget, J., Voves, J., van Strien, A. & van der Noordaa J. (1980). Free viral DNA in BK virus-induced hamster tumor cells. *J Virol* 35, 331–339.

Thorley-Lawson, D. A. (2004). EBV and Burkitt's lymphoma. *N Engl J Med* 350, 2621.

Timms, J. M., Bell, A., Flavell, J. R., Murray, P. G., Rickinson, A. B., Traverse-Glehe Berger, F. & Delecluse, H. J. (2003). Target cells of Epstein-Barr-virus (EBV)-positive transplant lymphoproliferative disease: similarities to EBV-positive Hodgkin's lymph Lancet 361, 217–223.

Tischoff, I. & Tannapfel, A. (2008). DNA methylation in hepatocellular carcinoma. W J Gastroenterol 14, 1741–1748.

Tobollik, S., Meyer, L., Buettner, M., Klemmer, S., Kempkes, B., Kremmer, E., Niedobitek, G. & Jungnickel, B. (2006). Epstein-Barr virus nuclear antigen 2 inhibits AID expression during EBV-driven B-cell growth. Blood 108, 3859–3864.

Toth, Z., Maglinte, D. T., Lee, S. H., Lee, H. R., Wong, L. Y., Brulois, K. F., Lee, S., Buckley, J. D., Laird, P. W., Marquez, V. E. & Jung, J. U. (2010). Epigenetic analysis of KSHV latent and lytic genomes. PLoS Pathog 6, e1001013.

Tsai, C. L., Li, H. P., Lu, Y. J., Hsueh, C., Liang, Y., Chen, C. L., Tsao, S. W., Tse, K. P., Yu, J. S. & Chang, Y. S. (2006). Activation of DNA methyltransferase 1 by EBV LMP1 Involves c-Jun NH(2)-terminal kinase signaling. Cancer Res 66, 11668–11676.

Tsai, C. N., Tsai, C. L., Tse, K. P., Chang, H. Y. & Chang, Y. S. (2002). The Epstein-Barr virus oncogene product, latent membrane protein 1, induces the downregulation of E-cadherin gene expression via activation of DNA methyltransferases. Proc Natl Acad Sci U S A 99, 10084–10089.

Uchida, J., Yasui, T., Takaoka-Shichijo, Y., Muraoka, M., Kulwichit, W., Raab-Traub, N. & Kikutani, H. (1999). Mimicry of CD40 signals by Epstein-Barr virus LMP1 in B lymphocyte responses. Science 286, 300–303.

Um, T. H., Kim, H., Oh, B. K., Kim, M. S., Kim, K. S., Jung, G. & Park, Y. N. (2011). Aberrant CpG island hypermethylation in dysplastic nodules and early HCC of hepatitis B virus-related human multistep hepatocarcinogenesis. J Hepatol 54, 939–947.

Umehara, T., Nakamura, Y., Wakamori, M., Ozato, K., Yokoyama, S. & Padmanabhan, B. (2010). Structural implications for K5/K12-di-acetylated histone H4 recognition by the second bromodomain of BRD2. FEBS Lett 584, 3901–3908.

Ushiku, T., Chong, J. M., Uozaki, H., Hino, R., Chang, M. S., Sudo, M., Rani, B. R., Sakuma, K., Nagai, H. & Fukayama, M. (2007). p73 gene promoter methylation in Epstein-Barr virus-associated gastric carcinoma. Int J Cancer 120, 60–66.

Ushmorov, A., Leithauser, F., Ritz, O., Barth, T. F., Moller, P. & Wirth, T. (2008). ABF-1 is frequently silenced by promoter methylation in follicular lymphoma, diffuse large B-cell lymphoma and Burkitt's lymphoma. Leukemia 22, 1942–1944.

van den Bosch, C. A. (2004). Is endemic Burkitt's lymphoma an alliance between three infections and a tumour promoter? Lancet Oncol 5, 738–746.

Van Keymeulen, A., Mascre, G., Youseff, K. K., Harel, I., Michaux, C., De Geest, N., Szpalski, C., Achouri, Y., Bloch, W., Hassan, B. A. & Blanpain, C. (2009). Epidermal progenitors give rise to Merkel cells during embryonic development and adult homeostasis. J Cell Biol 187, 91–100.

Viejo-Borbolla, A., Ottinger, M., Bruning, E., Burger, A., Konig, R., Kati, E., Sheldon, J. A. & Schulz, T. F. (2005). Brd2/RING3 interacts with a chromatin-binding domain in the Kaposi's Sarcoma-associated herpesvirus latency-associated nuclear antigen 1 (LANA-1) that is required for multiple functions of LANA-1. J Virol 79, 13618–13629.

Wang, L., Grossman, S. R. & Kieff, E. (2000). Epstein-Barr virus nuclear protein 2 interacts with p300, CBP, and PCAF histone acetyltransferases in activation of the LMP1 promoter. Proc Natl Acad Sci U S A 97, 430–435.

Wentzensen, N., Sherman, M. E., Schiffman, M. & Wang, S. S. (2009). Utility of methylation markers in cervical cancer early detection: appraisal of the state-of-the-science. Gynecol Oncol 112, 293–299.

Wisman, G. B., Nijhuis, E. R., Hoque, M. O., Reesink-Peters, N., Koning, A. J., Volders, H. H., Buikema, H. J., Boezen, H. M., Hollema, H., Schuuring, E., Sidransky, D. & van der Zee, A. G. (2006). Assessment of gene promoter hypermethylation for detection of cervical neoplasia. Int J Cancer 119, 1908–1914.

Yanatatsaneejit, P., Mutirangura, A. & Kitkumthorn, N. (2011). Human papillomavirus's physical state and cyclin A1 promoter methylation in cervical cancer. *Int J Gynecol Cancer* 21, 902–906.

Yang, B., Guo, M., Herman, J. G. & Clark, D. P. (2003). Aberrant promoter methylation profiles of tumor suppressor genes in hepatocellular carcinoma. *Am J Pathol* 163, 1101–1107.

Yasui, D. H., Peddada, S., Bieda, M. C., Vallero, R. O., Hogart, A., Nagarajan, R. P., Thatcher, K. N., Farnham, P. J. & LaSalle, J. M. (2007). Integrated epigenomic analyses of neuronal MeCP2 reveal a role for long-range interaction with active genes. *Proc Natl Acad Sci U S A* 104, 19416–19421.

Zhang, B., Laribee, R. N., Klemsz, M. J. & Roman, A. (2004). Human papillomavirus type 16 E7 protein increases acetylation of histone H3 in human foreskin keratinocytes. *Virology* 329, 189–198.

Zhang, Y. J., Chen, Y., Ahsan, H., Lunn, R. M., Chen, S. Y., Lee, P. H., Chen, C. J. & Santella, R. M. (2005). Silencing of glutathione S-transferase P1 by promoter hypermethylation and its relationship to environmental chemical carcinogens in hepatocellular carcinoma. *Cancer Lett* 221, 135–143.

Zheng, Z., Pan, J., Chu, B., Wong, Y. C., Cheung, A. L. & Tsao, S. W. (1999). Downregulation and abnormal expression of E-cadherin and beta-catenin in nasopharyngeal carcinoma: close association with advanced disease stage and lymph node metastasis. *Hum Pathol* 30, 458–466.

Zhong, S., Yeo, W., Tang, M. W., Wong, N., Lai, P. B. & Johnson, P. J. (2003). Intensive hypermethylation of the CpG island of Ras association domain family 1A in hepatitis B virus-associated hepatocellular carcinomas. *Clin Cancer Res* 9, 3376–3382.

Zhou, J., Chau, C. M., Deng, Z., Shiekhattar, R., Spindler, M. P., Schepers, A. & Lieberman, P. M. (2005). Cell cycle regulation of chromatin at an origin of DNA replication. *EMBO J* 24, 1406–1417.

zur Hausen H. (2002). Papillomaviruses and cancer: from basic studies to clinical application. *Nat Rev Cancer* 2, 342–350.

Chapter 9
Genetic and Epigenetic Determinants of Aggression

Barbara Klausz, József Haller, Áron Tulogdi, and Dóra Zelena

Abbreviations

5-HT-R	Serotonin receptor
5-HTT	Serotonin transporter
ACTH	Adrenocorticotropin
ADX	Adrenalectomy
ANP	Atrial natriuretic peptide
AR	Androgen receptor
AVP	Arginine vasopressin
BDNF	Brain-derived neurotrophic factor
BNST	Bed nucleus of stria terminalis
CaMK	Calcium/calmodulin-dependent kinase
CeA	Central amygdala
CNS	Central nervous system
COMT	Catechol-O-methyltransferase
CpG	Cytosine–guanine dinucleotide
CRH	Corticotropin-releasing hormone
D2-R	Dopamine D2-receptor
DBH	Dopamine beta-hydroxylase
DNMT	DNA methyltransferase
ER	Estrogen receptor
ERα	Estrogen receptor-α
GABA	Gamma-aminobutyric acid
GR	Glucocorticoid receptor

B. Klausz • J. Haller • Á. Tulogdi • D. Zelena (✉)
Department of Behavioural Neurobiology, Laboratory of Behavioural and Stress Studies,
Institute of Experimental Medicine, Hungarian Academy of Sciences,
Szigony 43, 1083 Budapest, Hungary
e-mail: zelena.dora@koki.mta.hu

J. Minarovits and H.H. Niller (eds.), *Patho-Epigenetics of Disease*,
DOI 10.1007/978-1-4614-3345-3_9, © Springer Science+Business Media New York 2012

HAA	Hypothalamic attack area
HAT	Histone acetyltransferase
HDAC	Histone deacetylase
HPA axis	Hypothalamo–pituitary–adrenocortical axis
IL-6	Interleukin-6
KO	Knockout
L	Long variant of the 5-HTT
MAO	Monoamine oxidase
MeA	Medial amygdala
met	Methionine
MPOA	Medial preoptic area
MTHFR	Methylenetetrahydrofolate reductase
NA	Noradrenaline
NCAM	Neural cell adhesion molecule
NGF	Nerve growth factor
NK1	Neurokinin1, the receptor for substance P
NO	Nitric oxide
NOS	Nitric oxide synthase
OT	Oxytocin
PAG	Periaqueductal gray
PFC	Prefrontal cortex
POMC	Proopiomelanocortin
PVN	Nucleus paraventricularis hypothalami
S	Short variant of the 5-HTT
sc	Subcutaneous
SER	Serotonin
SON	Supraoptic nucleus
TGFα	Transforming growth factor-α
TH	Tryptophan hydroxylase
V_{1a}-R	Vasopressin V_{1a} receptor
val	Valine
VMH	Ventromedial nucleus of the hypothalamus
VPA	Valproate

9.1 Introduction

The history of genetics started with the work of Gregor Johann Mendel on pea plants, published in 1866 (Mendel 1866). Genetics deals principally with the molecular structure and function of genes, the genetic codes. However, there are heritable changes not dependent on the genomic sequence. For this kind of programming, Waddington introduced the term epigenetic in the 1940s (Waddington 1942). In the last few decades, a lot of knowledge accumulated about epigenetic modifications during development and in cancer formation; however, little is known about the role

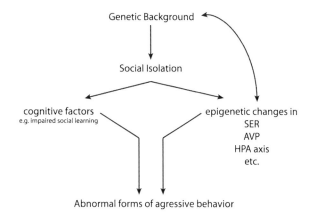

Fig. 9.1 Schematic diagram of the interaction of several factors, leading to development of pathological form of aggression. *SER* serotonin; *AVP* arginine vasopressin; *HPA axis* hypothalamo–pituitary–adrenocortical axis

of epigenetic changes in mature cells. At the moment, there is not too much direct evidence available on the connection between aggression and genetic, even more epigenetic modification. However, several facts suggest these interactions (Fig. 9.1). For example, low glucocorticoid level enhances aggression only after long-term exposure, e.g., through removal of the adrenal gland (Halasz et al. 2002), but not after an acute decrease, e.g., through metyrapone treatment (Mikics et al. 2004), raising the possibility for the development of epigenetic changes. Moreover, another model of pathologic aggression, the social isolation, is a widely used inducer of schizophrenia-like symptoms, which is—according to some theories—an epigenetic disease (Graff and Mansuy 2009). In the following, we try to summarize the present knowledge about brain areas and molecules involved in aggression, as possible targets of genetic as well as epigenetic regulation.

9.2 Why Aggression Is Important?

Every year, more than 700,000 people worldwide die because of assault (Bartolomeos et al. 2007), and many more become victims of aggressive behavior including domestic violence, terrorism, and hooliganism or get physically or psychologically injured. Besides the suffering of affected persons and their families, a large financial burden for society emerges (Neumann et al. 2010).

On the other hand, aggression is expressed by virtually all mammalian species and is of vital importance for the survival of the individual. An animal defending itself against a predator becomes aggressive in order to survive and to ensure the survival of its offspring (Gregg and Siegel 2001). Aggression against conspecifics serves to establish a dominance hierarchy for better supply of food, territory, and mating. Because winners in agonistic battles are often thought to be dominant (Bjorkqvist 2001) and aggressive behavior is often manifested in the process of hierarchy formation (de Almeida et al. 2005), many scientists equate aggressiveness with dominance. However, the dominant ones are not always the most aggressive animals in a population (Sapolsky 2005).

So, aggression is not a psychological disorder per se, but is among the symptoms of more complex psychological disorders (Vitiello et al. 1990; Scarpa and Raine 1997; Meloy 2006). It is often associated with schizophrenia, suicidal depression, and cluster B personality disorders, attention deficit/hyperactivity disorder (ADHD), which are characterized by a psychopathological complex of attentional problems, motor overactivity, and impulsivity, which is per se closely linked to behavioral problems (Retz and Rosler 2009). In addition, these disorders start early in life and, therefore, are suggested to have high impact of an individual's socialization. In fact, deficits in social behavior leading to excessive aggression may develop as a consequence of disturbed emotional regulation (Davidson et al. 2000). Accordingly, a better understanding of the link between social behaviors and emotional regulation and their neurobiological underpinnings is essential to improve the treatment of many psychopathologies (Neumann et al. 2010). When aggression occurs out of context, aggressive behaviors can become inappropriate or pathological (Veenema 2009). Interestingly, laboratory research on aggression largely ignored the existence of pathological forms of aggression and focused mainly on the mechanisms underlying natural aggression.

9.3 Models of Abnormal (Pathological) Forms of Aggression

There is no doubt that the research has to focus on pathological aggression instead of normally occurring aggressive contact during predation or defense (Blair 2001). In social species, aggression is a ritualized behavior. Therefore, it is signaled in advance to offer the weakest one the possibility to withdraw and serves for establishing a stable dominance hierarchy (Veenema 2009). It was suggested that animal aggression can be considered abnormal, if there was a mismatch between provocation and response, i.e., the aggressive response surpassed species-typical levels (de Almeida and Miczek 2002; Miczek et al. 2002); if attacks were targeted on inappropriate partners, e.g., females (de Boer et al. 2003; Natarajan et al. 2009) or inappropriate body parts, i.e., those prone to serious injury like the head, throat, and belly (Haller et al. 2001, 2005a; Haller and Kruk 2006); if attacks were not signaled by threats; or if the submissive signals of opponents were ignored (Haller et al. 2001; de Boer et al. 2003; Haller and Kruk 2006; Natarajan et al. 2009). In general, these criteria are similar to particularities of human aggressiveness that are expressed in certain aggression-related psychopathologies (Haller et al. 2005a; Haller and Kruk 2006). Four laboratory models of abnormal aggression were developed so far.

Genetic models make use of mice selected for high aggressiveness, rats selected for low anxiety, or selected subpopulations of feral rats, where abnormal features of aggression could be observed (de Boer et al. 2003; Natarajan et al. 2009; Neumann et al. 2010).

The escalated aggression model involves aggressiveness that surpasses species-typical levels and is induced by frustration or provocation (de Almeida and Miczek 2002; Miczek et al. 2002, 2004). This model is based on the attack priming phenomenon discovered by Potegal (1992).

Based upon the development of abnormal attack targeting, i.e., the ratio of attacks aimed at vulnerable targets, e.g., head, throat, and belly (Haller et al. 2001), the hypoarousal model was introduced. It involves the chronic limitation of glucocorticoid secretion, which mimics the low glucocorticoid levels seen in violent, antisocially disordered people (Haller et al. 2001; Haller and Kruk 2006). When the glucocorticoid secretion was stabilized at a low level by adrenalectomy with subcutaneous glucocorticoid pellets (ADX), a change in attack targeting can be detected (Haller et al. 2001). While control rats targeted biting attacks towards less vulnerable dorsal parts of the opponent's body, ADX rats attacked the head frequently. This was accompanied by autonomic hypoarousal and social deficit as well (Haller et al. 2004). It was also shown that the neural background and pharmacological responsiveness of attacks performed by such rats are markedly altered (Halasz et al. 2002). This suggests that mechanisms of "normal" and pathologic aggression could be different.

Aggressive behavior was increased on the long run by a variety of early adverse experiences, e.g., maternal separation (Suomi 1997; Veenema et al. 2006, 2007a) or early defeat (Delville et al. 1998; Wommack et al. 2003). However, the pathological feature of early social isolation-induced aggression was confirmed just recently (Toth et al. 2011). Namely, social isolation from weaning to the ages of 80 days not only increases the level of aggressiveness but results in abnormal attack patterns and deficits in social communication. So it models the aggression-related problems resulting from early social neglect in humans (Toth et al. 2008). Recently, it was also shown that the aggression-induced glucocorticoid and autonomic stress responses are substantially increased in these rats, suggesting that social isolation from weaning may be used as a model of aggression-related psychopathologies associated with hyperarousal (Toth et al. 2011).

These models underline that stress, thereby the hypothalamo–pituitary–adrenocortical (HPA) axis, plays a key role in the regulation of aggression.

9.4 Brain Areas Involved in Aggression

The main neuronal axis for controlling normal aggression is the medial amygdala (MeA)–hypothalamic attack area (HAA)–periaqueductal gray (PAG) axis, which is modulated by other areas such as the prefrontal cortex (PFC), lateral septum, other amygdaloid nuclei, and the brain stem monoaminergic nuclei (Gregg and Siegel 2001) (Fig. 9.2). The dysfunction of neural circuits responsible for emotional control was shown to represent an etiological factor of aggression and could be target areas for epigenetic modifications.

The so-called HAA is the only brain region from where attacks can be reliably elicited by electrical or optogenetic stimulation in all the investigated species, including cat, rat, and mouse (Lammers et al. 1988; Lin et al. 2011). This functional brain region is located in the mediobasal portion of the hypothalamus and partly overlaps with the anterior hypothalamic area, tuber cinereum area, and the ventromedial hypothalamic nucleus (Lammers et al. 1988; Hrabovszky et al. 2005).

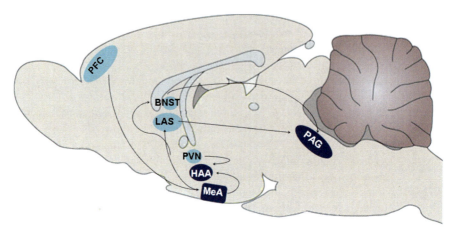

Fig. 9.2 Most important brain areas involved in aggression and their connections. *MeA* medial amygdala; *HAA* hypothalamic attack area; *PAG* periaqueductal gray; *PFC* prefrontal cortex; *LAS* lateral septum; *BNST* bed nucleus of stria terminalis; *PVN* paraventricularis nucleus of hypothalamus

The HAA stimulates the dorsal PAG via glutamatergic projections (Fuchs et al. 1985a, b). From this latter region, aggressive behavior can be elicited by electrical stimulation in the cat (Shaikh et al. 1987), but not in the rat, but still, it is an important locus for aggression control even in the rat (Lonstein and Stern 1998; Tulogdi et al. 2010).

MeA regulates the activity of the HAA through substance P neurotransmission, acting on neurokinin receptors (Shaikh et al. 1993). Antagonizing neurokinin 1 receptors systemically or by eliminating the neurons that express this receptor from the HAA results in a marked attenuation of aggressive attacks, especially that of violent attacks (Halasz et al. 2008, 2009). Moreover, stimulation of the MeA promotes intraspecific aggression, and lesion of this region attenuates aggressive behavior (Brutus et al. 1986; Vochteloo and Koolhaas 1987). Importantly, surgical lesion of the amygdala and/or the hypothalamus is a very effective method for taming even extremely violent psychiatric patients (Ramamurthi 1988; Sano and Mayanagi 1988). Structural and functional alterations of the amygdala were repeatedly shown in violent patients using different brain imaging techniques, e.g., reduced volume (Zhang et al. 2011) or asymmetrical activation pattern (Raine et al. 1997).

Together with the amygdala, the subregions of the PFC are the most frequently implicated brain regions in normal and abnormal human aggression. Some authors claim that violence is a consequence of the dysfunction of these brain regions, i.e., brain regions relevant for emotion regulation (Davidson et al. 2000). Watching angry facial expressions or thinking about personalized situations that induce anger was associated with enhanced activity in the orbitofrontal and anterior cingulate cortices (Blair et al. 1999; Dougherty et al. 1999). Impulsive personality-disordered people showed blunted metabolic responses to a serotonergic challenge in the orbitofrontal, ventromedial prefrontal, and cingulate cortices (Siever et al. 1999).

PFC lesions often lead to aggressive behavior in humans, as shown, for example, in Vietnam veterans (Grafman et al. 1996). Similar results were found in rhesus monkeys (Butter and Snyder 1972) and laboratory rodents (Rudebeck et al. 2007). These findings suggest that PFC provides the main inhibitory function over aggression. This function could be executed via direct projections to the MeA and the HAA (Toth et al. 2010).

Another important inhibitory region is the lateral septum, as lesioning or blocking this region enhances aggressiveness (Harrell and Balagura 1975; Albert and Wong 1978). The bed nucleus of stria terminalis (BNST) plays also a significant role in the expression of social preferences, affiliation, and aggression in rats, mice, and hamsters (Newman 1999; Rasia-Filho et al. 2000; Ferguson et al. 2001) and prairie voles (Wang and De Vries 1993; De Vries and Villalba 1997; Wang et al. 1997), suggesting that this region of the brain has a similar function in both highly social and less social species. The BNST is part of the accessory olfactory system, also known as the vomeronasal system, which is crucial for the detection of pheromones and influences several aspects of reproduction, including sex discrimination, attraction, and mate recognition (Wysocki 1979). In addition to having projections to the medial preoptic area (MPOA) and lateral septum, the BNST also projects to the vasopressin (arginine vasopressin, AVP) and oxytocin (OT) neurons of the nucleus paraventricularis hypothalami (PVN) and supraoptic nucleus (SON) (Sawchenko and Swanson 1983), suggesting that it can also regulate the production or release of these neuropeptides. Moreover, BNST sends dense projections to the HAA (Toth et al. 2010).

Recent data suggest that it is really important to differentiate between different forms of pathological aggression when measuring their neural background. In the hypoarousal-driven aggression, the role of the lateral hypothalamus and central amygdala increases on the expense of the roles played by the HAA and MeA (Halasz et al. 2002; Tulogdi et al. 2010). The activation patterns of the PFC (Halasz et al. 2006) and the PAG (Tulogdi et al. 2010) are also altered, while the regulatory roles of the lateral septum (Haller et al. 2006) and raphe serotonergic nuclei (Haller et al. 2005b) seem to be eliminated. Taken together, brain regions relevant for hypoarousal-driven aggression seem to modulate this form of violent behavior (Gregg and Siegel 2001; Halasz et al. 2002; Tulogdi et al. 2010). In contrast, the hyperarousal-driven aggression of socially isolated rats seems to be regulated by the hyperfunctioning of the above mentioned brain regions relevant for normal aggressive behavior, namely, the orbitofrontal cortex, MeA, and the HAA (our unpublished data).

9.5 Molecular Mechanisms of Aggression Control

The role of neurotransmitters and their receptors in preclinical studies of aggression has guided much of the development of pharmacotherapeutic interventions during the past decades (Miczek et al. 2002). The canonical aminergic transmitters such as serotonin (SER) and dopamine are still the basis for current drug treatments of violent individuals; however, they have been complemented by a better understanding

Table 9.1 Molecules involved in aggression

Group	Molecule	Knockout	Effect of KO on aggression	Polymorphism	Epigenetic changes
Sexual steroids	Testosterone	AR	↓	AR	
	Estradiol	ERα	↓		ERα
		ERβ	↑		
Catecholamines	Serotonin	$5\text{-HT}_{1A}\text{-R}$	↓	TH, 5-HTT, SER receptors	5-HTT
		$5\text{-HT}_{1B}\text{-R}$	↑		
	Dopamine	MAO-A	↑	MAO, COMT	
	Noradrenaline			DBH	
Neuropeptides	Vasopressin	$V_{1b}\text{-R}$	↓	$V_{1b}\text{-R}$	AVP, AVP– valproate
	Oxytocin	OT	↓↑		
	Substance P	NK1-R	↓		
GABA		GAT1	↓		
Neurotrophins	NGF				NGF
	BDNF	BDNF	↑		BDNF
Endocannabinoids		CB1-R			
Stress axis					CRH, POMC, GR

AR androgen receptor; *ER* estrogen receptor; *5-HT-R* serotonin receptor; *SER* serotonin; *5-HTT* serotonin transported; *MAO* monoamine oxidase; *COMT* catechol-*O*-methyltransferase; *DBH* dopamine beta-hydroxylase; $V_{1b}\text{-}R$ vasopressin 1b receptor; *AVP* vasopressin; *OT* oxytocin; *NK1-R* neurokinin 1 receptor; *GABA* gamma-aminobutyric acid; *GAT1* GABA transporter subtype 1; *NGF* nerve growth factor; *BDNF* brain-derived growth factor; *CB1-R* cannabinoid receptor 1; *CRH* corticotropin-releasing hormone; *POMC* proopiomelanocortin; *GR* glucocorticoid receptor

of modulatory influences by gamma-aminobutyric acid (GABA) as well as neuropeptides (Miczek et al. 2004; Miczek and Fish 2005). In Table 9.1, we summarized the molecules, the effect of knocking out a gene in the system and the available data on polymorphisms and epigenetic changes related to aggressive behavior.

9.5.1 Testosterone

Among all hormones, testosterone is most consistently linked to aggressive behavior (Giammanco et al. 2005). Notably, the level of aggression in males is far higher than in females (Giammanco et al. 2005), similar to the spectrum of morphological, physiological, behavioral, and psychological differences that are determined mostly by the sex hormones. This fact has brought many researchers to the conclusion that the determinants of aggression are androgens. It is supported by the well-documented fact that castration eliminates aggression (Nelson and Chiavegatto 2001).

There is evidence indicating a link between testosterone and antisocial behavior in humans as well (Dabbs and Hargrove 1997). Many researchers believe that high correlation exists between testosterone level and dominance or — in human social interaction — status-related behaviors, too (Archer 2006). A vast number of publications indicate a high level of testosterone in dominant animals, in the winners of agonistic contests (Cavigelli and Pereira 2000; Hardy et al. 2002; Oyegbile and Marler 2005), or in the winners between competing humans (McCaul et al. 1992; Bernhardt et al. 1998; Zitzmann and Nieschlag 2001; Giammanco et al. 2005; Schultheiss et al. 2005).

On the other hand, it is known that one of the main inducers of aggressive behavior is frustration (David et al. 2004), which, in turn, may be related to emotional stress. Stress suppresses androgen synthesis in rats (Andersen et al. 2004; Hardy et al. 2005; Razzoli et al. 2006), mice (Dong et al. 2004), hamsters (Castro and Matt 1997), primates (Lado-Abeal et al. 2001), and humans (Gozes et al. 1982; Ferris and Potegal 1988; Elman et al. 2001; Pavlov et al. 2012). Such suppression to a greater or lesser extent may be caused by a variety of stressors (Andersen et al. 2004; Razzoli et al. 2006). Although aggression is commonly related to a high level of testosterone, agonistic conflict is often stressogenic for both dominants and subordinates, and stress causes the inhibition of hormonal components of the reproductive system (Chichinadze and Chichinadze 2008). Thus, testosterone is not always an inducer of aggressive behavior (Aujard and Perret 1998; Morgan et al. 2000). Indeed, an association between testosterone and aggression is weak in humans except among abusers of anabolic steroids (Zitzmann and Nieschlag 2001).

In the rodent brain, aromatase converts testosterone into estradiol that is responsible for masculinization of the brain and regulates aggression and infanticide through estrogen receptor-α (ERα) (Ogawa et al. 1998; Wilson 2001).

9.5.1.1 Mechanisms of Action

According to the organizing/activating model of testosterone action suggested by vom Saal (1983), androgens during the prenatal period contribute to the formation of a neuronal network, which, in the future, participates in aggressive behavior. During pubescence, this network is activated also by testosterone, and in response to properly presented stimuli, an aggressive behavior is formed (Book et al. 2001). Thus, the primary role of testosterone may be the formation of structures that generate highly aggressive responses to external stimuli. This process might involve epigenetic mechanisms (see below). Androgens seem to promote aggressiveness at the level of the lateral septum, MPOA, amygdala, and dorsal raphe nucleus (Simon et al. 1998). Imbalance in testosterone/SER and testosterone/cortisol ratios, e.g., increased testosterone levels and reduced cortisol levels, increases the propensity toward aggression because of reduced activation of the neural circuitry of impulse control and self-regulation (Pavlov et al. 2012).

9.5.2 Serotonin

Although several neurotransmitters were connected with aggression, the SER data are the most convincing (Nelson and Chiavegatto 2001). Traditionally, many studies have shown that elevated SER levels lead to decreased aggression in many different species (Chiavegatto and Nelson 2003), including humans (Coccaro et al. 1994; Unis et al. 1997). This finding has been replicated in populations of impulsive offenders, adults, and children (Brown et al. 1979; Linnoila et al. 1983). However, the identification of genes for at least 14 SER receptor proteins, variants of the synthetic and metabolic enzymes, and transporter molecules underlines the molecular diversity of the serotonergic mechanisms of action (de Almeida et al. 2005). Moreover, genetic analyses of aggressive individuals have identified several molecules that affect the SER system directly, e.g., $5\text{-}HT_{1B}\text{-}R$, SER transporter, and monoamine oxidase-A (MAO-A), or, indirectly, e.g., neuropeptide Y, nitric oxide synthase (NOS), and brain-derived neurotrophic factor (BDNF) (Takahashi et al. 2011).

Pharmacological strategies of increasing SER levels, such as the use of SER precursors, SER reuptake inhibitors, in addition to the agonists of its receptor, $5\text{-}HT_{1A}\text{-}R$ and $5\text{-}HT_{1B}\text{-}R$, are able to reduce aggressive behavior in rodents (Olivier et al. 1995; Miczek et al. 1998; Fish et al. 1999; Lyons et al. 1999; Chiavegatto et al. 2001). When activated systemically, $5\text{-}HT_{1B}\text{-}Rs$ appear to be essential sites for the inhibition of several types of aggressive behavior. The decrease of heightened aggression was observed in studies using mice after intraperitoneal administration of $5\text{-}HT_{1B}\text{-}R$ agonists such as CP-94,253, zolmitriptan, and anpirtoline (Fish et al. 1999; de Almeida et al. 2001, 2005; de Almeida and Miczek 2002). In human samples, aggressive behavior is strongly and negatively correlated with the level of the SER metabolite 5-hydroxyindoleacetic acid, measured in the cerebrospinal fluid (Virkkunen et al. 1994). Taken together, SER was suggested to inhibit aggressive behavior in animals and violent behavior in humans (Haller et al. 2005a).

The MeA–HAA–PAG axis, as well as the PFC and lateral septum, is influenced by SER coming from the brain stem raphe nuclei. The PFC, more specifically the orbitofrontal region, has been identified to be particularly important in the inhibitory control of behavior, mainly impulsive and aggressive behavior (Blair 2001; Cardinal et al. 2004; Seguin 2004; Spinella 2004; Kheramin et al. 2005). SER facilitates prefrontal inhibition, wherefore insufficient serotonergic activity can enhance aggression (Pavlov et al. 2012) and pharmacological activation of PFC $5\text{-}HT_{1A}\text{-}R$ and $5\text{-}HT_{1B}\text{-}R$ was shown to inhibit the execution of aggressive behavior (de Almeida et al. 2005). The $5\text{-}HT_{1B}\text{-}R$ expressed in a variety of brain regions, including the basal ganglia, PAG, hippocampus, lateral septum, and raphe nuclei, either presynaptically inhibiting SER release or as a heteroreceptor modulating the release of other neurotransmitters (Nelson and Chiavegatto 2001). Although both $5\text{-}HT_{1A}\text{-}R$ and $5\text{-}HT_{1B}\text{-}R$ control the SER tone, these two receptors probably have different contributions in particular brain areas that modulate the postsynaptic SER inhibitory effects on aggression. Drugs that target the $5\text{-}HT_{1C}\text{-}R$, $5\text{-}HT_{2}\text{-}R$, or $5\text{-}HT_{3}\text{-}R$ sites have generally not influenced aggression (Valzelli 1984; Simon et al. 1998).

9 Genetic and Epigenetic Determinants of Aggression

It can be anticipated that currently developed tools for targeting the specific subtypes of SER receptors will offer new therapeutic options for reducing aggressive behavior, and the 5-HT$_{1B}$-R appears to be a promising target (Miczek et al. 2007). The modulation of GABA and GABA$_A$ receptors by SER in corticolimbic neurons promises to be particularly relevant for specific forms of escalated aggressive behavior such as alcohol-heightened aggression (Takahashi et al. 2010).

9.5.3 Other Catecholamines

9.5.3.1 Dopamine

Several studies indicate that the mesocorticolimbic dopamine (DA) system is involved in aggressive acts (Mos and van Valkenburg 1979; Louilot et al. 1986; Haney et al. 1990; Puglisi-Allegra and Cabib 1990; van Erp and Miczek 2000; Ferrari et al. 2003; de Almeida et al. 2005). Pharmacologically induced DA increases are associated with increased aggressive behavior (Senault 1968, 1971; Hasselager et al. 1972; Miczek 1974; Puech et al. 1974; Crawley et al. 1975; Ray et al. 1983).

In connection with these studies, in human aggression, the most frequent and enduring pharmacotherapeutic interventions rely on compounds that act as dopaminergic antagonists (Yudofsky et al. 1984; Gualtieri and Schroeder 1990; McDougle et al. 1998). For example, the DA D2-receptor (D2-R) antagonist haloperidol has been used for decades to treat aggressive behavior of psychotic patients (Glazer and Dickson 1998; Fitzgerald 1999). This drug also decreases violent outbursts in individuals with dementia and individuals with borderline personality disorder as well as in children and adolescents, who exhibit conduct disorder and aggression (Pies and Popli 1995; Beauchaine et al. 2000; Challman and Lipsky 2000; Kennedy et al. 2001; Diederich et al. 2003; Masi 2004). The decrease in aggression is closely linked to the sedative effects. However, the more recently developed atypical neuroleptics which are considerably less sedative are more effective and have more specific antiaggressive effects. Moreover, D2-R and D4-R gene variants and interaction between them are associated with conduct disorder and antisocial behavior (Beaver et al. 2007; Congdon et al. 2008). It was also shown that D2-Rs in the region of the MPOA area and anterior hypothalamus facilitate affective defense behavior in the cat (Sweidan et al. 1991). Aggression was decreased, and hypothalamic SER and noradrenaline (NA) were increased in birds from all strains treated with D2-R antagonist (Dennis and Cheng 2011).

The neurochemical studies link elevated DA and its metabolites in PFC and nucleus accumbens not only to the initiation of attacks and threats and its consequences but also to the defensive and submissive responses in reaction to being attacked (Puglisi-Allegra and Cabib 1990; Tidey and Miczek 1996). The lack of differentiation in mesocorticolimbic dopamine activity between attack and defensive behavior suggests that neuroleptic compounds with a high affinity for D2-R would not be specific antiaggressive treatments.

9.5.3.2 Noradrenaline

NA affects aggression on three different levels: the hormonal level, the sympathetic autonomous nervous system, and central nervous system (CNS) (Haller et al. 1998). Hormonal catecholamines (adrenaline and NA) appear to be involved in metabolic preparations for the prospective fight; the sympathetic system ensures appropriate cardiovascular reaction, while the CNS noradrenergic system prepares the animal for the prospective fight. Indirect CNS effects include the shift of attention towards socially relevant stimuli, the enhancement of olfaction (a major source of information in rodents), the decrease in pain sensitivity, and the enhancement of memory (an aggressive encounter is very relevant for the future of the animal). Concerning more aggression-specific effects, one may notice that a slight activation of the central noradrenergic system stimulates aggression, while a strong activation decreases fight readiness. This biphasic effect may allow the animal to engage or to avoid the conflict, depending on the strength of social challenge. Different receptor subtypes may influence different aspect of behavior. Namely, neurons bearing postsynaptic alpha2-adrenoceptors are responsible for the start and maintenance of aggression, while a situation-dependent fine-tuning is realized through neurons equipped with beta-adrenoceptors. The latter phenomenon may be dependent on a NA-induced glucocorticoid secretion.

9.5.3.3 Catabolism

Two major enzymes are responsible for catecholamine (SER, DA, and NA) catabolism in the brain: catechol-O-methyltransferase (COMT) and monoamine oxidase-A (MAO-A) (Shih et al. 1999). If aggressive behavior is enhanced by catecholaminergic activity, then the lower activity of COMT and MAO-A (resulting in a slower inactivation of catecholamines) should indirectly enhance aggression. This prediction has been supported by most, but not all, observations in rodents and humans. Male mice whose COMT or MAO-A genes are deleted show elevated aggression (Cases et al. 1995; Gogos et al. 1998). On the contrary, inhibition of MAO-A correlates with reduced aggression in isolated male mice (Florvall et al. 1978) and footshock-induced aggression (Datla and Bhattacharya 1990), probably as a result of increased SER levels. The COMT gene has been associated with an increased aggressive behavior, at least in several samples of psychiatric patients (Volavka et al. 2004). There is an association between MAO-A and hyperarousal-driven aggression, too (Meyer-Lindenberg et al. 2006).

9.5.4 Neuropeptides

9.5.4.1 Vasopressin

AVP is another key hormone—besides sexual steroids—that plays a crucial role in aggression (Delville et al. 1996; Bester-Meredith et al. 1999) and other social behaviors (Albers and Bamshad 1998; Ferris 2000). Our previous studies implicated that

9 Genetic and Epigenetic Determinants of Aggression

one of the most important central regulators of the stress axis is also the AVP (Zelena et al. 2009). Recently, the involvement of AVP in "normal" aggressive behavior comes to the front (Neumann et al. 2010). In pupfish, a correlation was found between vasotocin (the fish equivalent of AVP) and aggressive behavior, too (Lema 2006).

There are some putative sites where AVP might influence "normal" aggressive behavior. In male Syrian hamsters, AVP injected into the HAA stimulated aggression, while injection of a V_{1a} receptor (-R) antagonist inhibited the behavior (Ferris and Potegal 1988), although V_{1b}-Rs might be also involved (Blanchard et al. 2005). In mice, the BNST may increase AVP-Fos colocalization selectively in response to affiliation-related stimuli (Ho et al. 2010). In rats, AVP release within the lateral septum correlates positively with intermale aggression (Beiderbeck et al. 2007; Veenema and Neumann 2007; Veenema et al. 2007b), while a specific V_{1a}-R antagonist prevents an increase in aggression during a second contact (Veenema et al. 2010). The V_{1a}-R binding in the lateral septum positively correlated with maternal aggression (Caughey et al. 2011). Moreover, injection of a selective V_{1a} receptor antagonist into the CeA reduced maternal aggression in dams with high anxiety-related behavior, whereas synthetic AVP increased the low level of aggression in rats with low anxiety-related behavior. Selective aggression in pair-bonded male prairie voles was associated with increased release of AVP in the anterior hypothalamus (Gobrogge et al. 2009). Pharmacological activation of the V_{1a} receptors in the anterior hypothalamus was induced, whereas V_{1a} blockade diminished selective aggression in pair-bonded males. As this brain area does not get vasopressinergic innervation, the origin of AVP is questionable. We can hypothesize that somatic–dendritic release of AVP from the PVN might reach this area (Engelmann et al. 2004), therefore AVP might be one of the mechanisms that connect the stress response with aggressiveness.

The extrahypothalamic AVP system in the rat brain is highly sexually dimorphic and steroid responsive (De Vries et al. 1994). Adult male rats have twice more AVP-expressing cells in the BNST compared with females (Van Leeuwen et al. 1985; Miller et al. 1989b). AVP expression within this area is dependent upon gonadal hormones, as castration results in a significant decrease in AVP mRNA and protein expression, and testosterone replacement restores AVP expression (De Vries et al. 1985; Miller et al. 1989a; Brot et al. 1993). It is essentially estradiol, a metabolite of testosterone, that is the major factor regulating AVP expression mainly by acting upon neuronal ERα. Interestingly, castration is known to increase ERα mRNA together with a decrease in AVP (Brot et al. 1993; Handa et al. 1996).

There is an interaction between AVP and SER, too. Microinjections of AVP into the HAA in combination with 5-HT$_{1A}$-R or 5-HT$_{1B}$-R agonists revealed that only the 5-HT$_{1A}$-R activation inhibited AVP-facilitated aggression (Ferris et al. 1999).

9.5.4.2 Oxytocin

Oxytocin (OT) appears to act at OT receptors in the MPOA to facilitate the release of DA from neurons in the ventral tegmental nucleus, and the increased DA release then activates maternal behavior in rats (Champagne et al. 2004). OT also regulates specific forms of aggression and has differential effects depending upon the species.

OT inhibits aggressive interactions between dominant and subordinate adult female hamsters (Harmon et al. 2002). In rats, OT receptor binding positively correlated with the peak of maternal aggression, suggesting that OT may act in the lateral septum to facilitate the expression of aggressive behavior (Caughey et al. 2011).

Although many of the effects of OT are expressed primarily in females, OT also affects male behavior including social recognition (Ferguson et al. 2001), the formation of partner preferences (Cho et al. 1999), and male sexual behavior (Arletti et al. 1985; Witt 1997). In the monogamous, highly social prairie voles, a single intraperitoneal injection on the day of birth with OT or OT antagonist affected partner preference formation and aggression (Bales and Carter 2003). However, in general, OT has been shown to regulate maternal behavior (Pedersen and Prange 1979; Pedersen et al. 1982), while AVP plays a role in the expression of paternal behavior (De Vries and Villalba 1997; Bester-Meredith et al. 1999; Parker and Lee 2001). In hamsters, AVP increases aggression in males, but not in females (Cushing and Kramer 2005), and OT inhibits female aggression (Harmon et al. 2002).

Neonatal treatment with AVP increased aggression in adult male prairie voles, but not females (Stribley and Carter 1999). On the contrary, a single injection of OT on the day of birth altered the number of neurons in the PVN that expressed OT in female prairie voles, but not in males. These results support the hypothesis that while both neuropeptides may have a role in social behavior, OT may have a greater influence in the expression of prosocial behavior in females and AVP in males (Winslow and Insel 1993; De Vries and Villalba 1997; Insel et al. 1998). As the expression of OT and AVP receptors does not appear to be sexually dimorphic, other mechanisms maybe at play in establishing distinct behavioral responses to these neuropeptides.

9.5.4.3 Substance P: Neurokinin 1 Receptor

Substance P and its tachykinin NK1 receptors are highly expressed in brain regions involved in emotional control. More specifically, HAA, the only brain region in rats from which biting attacks can reliably be elicited by both electrical and neurochemical stimulation, preferentially expresses the NK1 receptors (Halasz et al. 2009). The involvement of substance P and its receptor NK1 in the induction of both defensive rage and predatory attack appears to be a consistent finding (Katsouni et al. 2009). Glucocorticoid deficiency-induced antisocial aggressiveness results from altered SER and substance P neurotransmissions (Kim and Haller 2007). Moreover, besides anxiety and depression, substance P is involved in the modulation of suicidal-related behaviors (Giegling et al. 2007).

Pharmacological studies point to a stimulatory action of substance P in aggression, as NK1 blockade lowered the development of pathologic aggression (Halasz et al. 2008). Immunohistochemical studies revealed that fos-positive, i.e., activated, neurons in the PAG of cats activated after defensive rage-inducing medial hypothalamic stimulation lie in the same region as substance-P-immunoreactive cells (Gregg and Siegel 2003). In rats, aggressive encounters activated a large number of NK1

9 Genetic and Epigenetic Determinants of Aggression

receptor-expressing neurons in HAA as well (Halasz et al. 2008). A lesion of NK1 positive neurons through the infusion of substance-P-conjugated saporin into the HAA reduced violent attacks dramatically, whereas milder forms of aggression (soft bites and offensive threats) remained unaltered (Halasz et al. 2009).

9.5.5 Gamma-Aminobutyric Acid

Earlier postmortem studies showed that brain levels of GABA and its synthesizing enzyme, glutamic acid decarboxylase, in special brain areas are decreased in mice and rats that exhibited aggressive behavior (Clement et al. 1987; Haug et al. 1987; Guillot and Chapouthier 1998). These data have been interpreted as concordant with the proposed inhibitory role of GABA on aggression.

Indeed, several other studies supported this idea. When GABA degradation is decreased through blocking the transaminase with valproate (VPA) or through inhibiting the reuptake, aggressive behavior is diminished in mice and rats (Puglisi-Allegra et al. 1979; Puglisi-Allegra and Mandel 1980; Krsiak et al. 1981; Rodgers and Depaulis 1982). Bjork et al. 2001 found a negative correlation between plasma GABA and aggressiveness in psychiatrically healthy adults with a family history of psychiatric disorders, although it is unclear how plasma levels are related to those in neural tissue. Nevertheless, benzodiazepines by enhancing the effect of GABA on $GABA_A$ receptors reduce aggression (DiMascio 1973; Jonas et al. 1992; Cherek and Lane 2001; Friedel 2004). Therefore, psychiatric patients with violent outbursts as well as those with episodic dyscontrol syndrome are often treated with benzodiazepines (Jonas et al. 1992; Cherek and Lane 2001; Gregg and Siegel 2001; Friedel 2004).

There is abundant neurochemical and behavioral evidence that GABA and SER interact in various brain regions. For example, the raphe nuclei receive a large GABAergic input and contain many GABAergic interneurons (Harandi et al. 1987). In the raphe, between 70 and 90% of serotonergic neurons also contain a subunit of the $GABA_A$ receptor (Gao et al. 1993). One of the neurobiological mechanisms for escalated aggressive behavior may involve increased activation of $GABA_A$ receptors that are modulated by the prevailing serotonergic tone in corticolimbic projection areas (de Almeida et al. 2005).

9.5.6 Neurotrophins

Nerve growth factor (NGF) was the first discovered and so far best characterized member of the neurotrophin family, which includes BDNF and neurotrophin-3 and neurotrophin-5 (Barde 1990; Huang and Reichardt 2001).

During fighting NGF is released into the bloodstream (Aloe et al. 1986; Lakshmanan 1986, 1987; Stephani et al. 1987; Alleva and Santucci 2001; Aloe et al. 2002). Serum NGF levels reflect the individual social status: mice that experienced repeated defeat and submissions, i.e., a subordinate status, show doubled serum NGF levels compared to attacking mice that achieved a dominant status (Maestripieri et al. 1990). Mice displaying a subordinate behavioral profile have high levels of NGF, while those displaying a dominant behavioral profile have high levels of BDNF in their hypothalamus (Fiore et al. 2003).

In the mouse, the largest amount of NGF is present in the submaxillary salivary glands (Aloe et al. 1994b). Sialoadenectomy, i.e., surgical removal of salivary glands, results in a minimal NGF increase after fighting, indicating that salivary glands are the main source of NGF in these conditions (Hendry and Iversen 1973; Aloe et al. 1986). Both the pronounced NGF increase found in subordinates, which bite very rarely, and the fact that NGF increases also in dominants, which do not receive bites by subordinates, suggest that NGF release is not due to the mechanics of biting (Maestripieri et al. 1990). However, NGF is also secreted from sources other than the salivary glands, including epithelial cells, fibroblasts, and lymphocytes, as well as activated macrophages (Gozes et al. 1982; Bandtlow et al. 1987; Lindholm et al. 1987; De Simone et al. 1990; Cirulli et al. 1998).

NGF is also present in the CNS, especially in the hypothalamus (Branchi et al. 2004). The relevance of the changes in NGF level in the hypothalamus following mouse intermale fighting, and the mechanisms involved, is however still a matter of discussion (Alleva and Aloe 1989). It was hypothesized that the rather rapid increase in the levels of brain NGF that follows a psychosocial stressful event may allow some phenomena of brain plasticity to take place in the adult animal. Indeed, NGF has been reported to regulate structural changes, such as formation of dendritic spines or collateral sprouting, ultimately altering the structure of neural connections in the mature brain (Diamond et al. 1987). Another possibility is that hypothalamic NGF may affect levels of other peptides or hormones present in this structure (Swanson and Sawchenko 1983; Albert et al. 1987). A series of interactive effects between NGF and thyroid hormones, adrenocorticotropin, and peptides have been reported (Aloe and Levi-Montalcini 1980; Otten et al. 1984; Wion et al. 1985).

NGF release is markedly triggered by psychosocial stressors (Weiss 1968; Henry et al. 1971; Axelrod and Reisine 1984; Alleva and Santucci 2001), i.e., those stressful conditions involving social interaction with conspecifics, including intermale fighting and at lower levels interfemale aggressive behavior, as well as precopula sexual arousal (Alleva et al. 1993; Alleva and Santucci 2001). Physical-stress conditions such as cold water swim, escapable or inescapable footshock, forced biting, or forced restraint exert a less pronounced effect on NGF release (Aloe et al. 1986; Alleva and Santucci 2001). Several in vivo and in vitro studies have shown that NGF also acts as a trophic and differentiative agent for the chromaffin tissue contained in the adrenal gland (Aloe and Levi-Montalcini 1979). It is has been widely reported that mouse adrenals change rather markedly and quickly following fighting behavior in males, while female adrenal morphology is more stable (Brain 1972). It therefore appeared likely that in mice, circulating NGF released by the salivary glands, regulated by social/aggressive behavior, controls adrenal morphology as well as adrenal functional status (Alleva et al. 1993). Indeed, exogenous NGF administrations markedly increase adrenal size (Aloe et al. 1986; Alleva and Santucci 2001).

9.5.7 Endocannabinoids

Taking into consideration the widespread occurrence of the endocannabinoid receptors, and the involvement of endocannabinoids in a wide range of physiological and pathological processes, one can hypothesize that the endocannabinoid system is involved in controlling aggression, too. Indeed, a diet deficient in polyunsaturated fatty acids, which might reduce endocannabinoid levels resulted in more aggressive rats (DeMar et al. 2006). Marijuana smoking increased aggressive responding for the first hour after smoking, which returned to placebo levels later in the day (Cherek et al. 1993). The effect could be dose dependent as the subjects in the low-dose condition tended to respond in a more aggressive manner than the subjects in the moderate- and high-dose conditions (Myerscough and Taylor 1985). Moreover, delta-tetrahydrocannabinol, the psychoactive component of marijuana, is also known to induce muricidal behavior in different rat strains (Bac et al. 2002).

9.5.8 Stress Axis

One of the best characterized pathologic aggression models is the hypoarousal model induced by ADX, i.e., removal of the glucocorticoids (Haller et al. 2004). Thereby, the stress axis seems to be a crucial component in the regulation of aggressive behavior.

Under the effect of acute and chronic stressful stimuli, the parvocellular neurons of the PVN release both corticotropin-releasing hormone (CRH) and AVP to the portal blood (Harbuz and Lightman 1989; Antoni 1993). In response to CRH and AVP, the synthesis of adrenocorticotropin (ACTH) from proopiomelanocortin (POMC) precursor as well as its release into the general circulation is increased. As a consequence, glucocorticoids are released from the adrenal cortex. Acute stress responses are blocked by an inhibitory feedback provided by adrenal glucocorticoids, which reduce the secretory activity of the endocrinomotor neurons of the hypothalamus and the POMC producing cells of the anterior pituitary (Kjaer et al. 1993).

9.6 Genetic Changes in Aggression

Despite the heterogeneity of definitions and classifications used and the difficulties regarding operationalization and assessment of antisocial behavior phenotypes, there is clear evidence from twin, adoption, and molecular genetic studies to support the notion that there are genetic influences on antisocial and aggressive behavior (Retz and Rosler 2009). A meta-analysis of 51 twin and adoption studies estimated moderate genetic (additive 32%, nonadditive 9%) and environmental influences (shared 16%, nonshared 43%) on antisocial behavior (Rhee and Waldman 2002). Nevertheless, aggression is highly heritable: by selective breeding, highly aggressive and virtually nonaggressive lines can be produced in different species. Among *Drosophila melanogaster* in only ten generations of selection, the aggressive lines became markedly more aggressive than the neutral lines (Dierick and Greenspan 2006). After 21 generations, the fighting index increased more than 30-fold. Mice genetically selected for short attack latency constitute a feasible model for hypoarousal-driven aggression (Veenema et al. 2004; Haller and Kruk 2006; Vekovischeva et al. 2007; Hood and Quigley 2008). In the rat, selecting for low anxiety resulted in a highly aggressive line, which is hyperresponsive to social stress (Veenema et al. 2007b).

Each position or rank within a group is a result of natural forms of aggression and has a distinctive brain gene expression profile that correlates with behavioral phenotype (Sneddon et al. 2011). Furthermore, transitions in rank position shift gene expression within 48 h in concurrent with the new dominance status. In the bee brain, more than 2,000 genes related to aggressive behavior were found (Chandrasekaran et al. 2011). Transcription factors played key roles including well-known regulators of neural and behavioral plasticity, e.g., CREB, as well as factors known in other biological contexts, e.g., NF-κB (immunity).

Quantitative genetic models normally partition an individual's phenotype into genetic and environmental components (Wilson et al. 2009). However, in the case of social behavior, an individual's phenotype may well be determined (at least in part) by the genotypes of interacting conspecifics. In this way, by social interaction, the "environment" is itself filled with genes and may be expected to evolve under appropriate selection (Griffing 1976; Moore et al. 1997; Wolf et al. 1998). This perspective can be accommodated using "indirect genetic effect" models (Wilson et al. 2009), in which the trait of a focal individual is potentially influenced not only by its own genotype but also by that of other individuals with which it interacts (Moore et al. 1997). In this respect, repeatable differences in aggressive behavior were present among individuals of a mice population, and the phenotypic expression also depends on the genotype of the opponent individuals (Wilson et al. 2009).

The genetic background likely stabilizes the individual's socialization by giving a frame, in which environmental factors shape the personality and behavioral styles rather than predict behavior (Retz and Rosler 2009). Allelic variation is responsible for individual differences in neural functioning, resulting in a disposition for violent aggression or delinquent behavior. The genes' effects are not deterministic, but probabilistic and leave a wide margin for self-ruled decisions.

9.6.1 Studies in Knockout Animals

9.6.1.1 Sexual Steroids

The involvement of sexual steroids was supported by androgen receptor (AR) and ER knockout (KO) mice. Male mice exhibiting a spontaneous mutation that fails to produce the long form of the AR are not aggressive (Olsen 1983; Maxson 1999). Male mice with targeted disruption of the gene encoding the ERα display reduced

9 Genetic and Epigenetic Determinants of Aggression

aggression in several testing situations due to the missing testosterone effect via estrogenic metabolites (Ogawa et al. 1997, 1998). Conversely, the ERβ-KO exhibits normal or increased aggression depending on social experience (Ogawa et al. 1999, 2000). ERα-KO females exhibit increased levels of aggression toward other female mice relative to wild-type females (Ogawa et al. 1997, 1998). Because estrogen is essential for the normal sexual differentiation of the CNS of male and possibly female mammals during development (Arnold 1996), studies of adult behavior in ER-KO are complicated by the inability to dissociate genetic from ontogenetic causes of behavior.

9.6.1.2 Serotonin

Male mice that lack functional expression of the 5-HT$_{1B}$-R attack an intruder more aggressively with a much shorter latency and higher frequency than the corresponding wild-type control (Saudou et al. 1994; Brunner and Hen 1997). Lactating female 5-HT$_{1B}$-R-KO mice also attack unfamiliar male mice more rapidly and violently (Ramboz et al. 1996). SER transporter knockout mice, where the reuptake of SER is diminished, wherefore its synaptic level is increased, show a reduction in aggressive behavior (Holmes et al. 2002).

Administration of a nonselective 5-HT$_{1B}$-R agonist (eltoprazine) significantly reduces aggressive behavior in both 5-HT$_{1B}$-R-KO and wild-type mice, suggesting the involvement of other receptor subtypes as well (Ramboz et al. 1996). Interestingly, mice lacking 5HT$_{1A}$-Rs are less reactive, and possibly less aggressive, and show more anxiety-related behavior compared with wild-type mice (Zhuang et al. 1999), a finding consistent with the observation of increased postsynaptic 5-HT$_{1A}$-R availability in limbic and cortical regions of highly aggressive mice (Korte et al. 1996). These data do not elucidate, however, the known antiaggressive effect of 5-HT$_{1A}$-R agonists in rodents (Olivier et al. 1995; Miczek et al. 1998).

9.6.1.3 Molecules Connected to Serotonin

Lifelong disruption of several molecules results in changes in SER and other systems and in this way may affect aggression. It is probable that the role attributed to a genetic deficiency in aggressive behavior is actually a consequence of secondary effects in several systems. Thus, the following results should be handled accordingly.

Calcium/calmodulin-dependent kinase II (CaMKII) is a neural-specific signaling molecule found at pre- and postsynaptic regions. CaMK-mediated phosphorylation is involved in activation of tryptophan hydroxylase (TH), the rate-limiting enzyme in SER synthesis (Ehret et al. 1989). Accordingly, SER release is reduced in the dorsal raphe of α-CaMKII mutant mice (Cases et al. 1996). α-CaMKII knockout mice display reduced aggression in resident–intruder paradigms (Cases et al. 1996). Heterozygotes, in which only one copy of the α-CaMKII gene is missing, show normal offensive aggression and elevated defensive aggression (Cases et al. 1996).

MAO-A deficiency, caused by a point mutation in its coding gene, correlates with aggression in several males from a Dutch family (Brunner et al. 1993). Ablation of the gene encoding MAO-A in mice leads to high levels of offensive aggression (Mejia et al. 2002), in spite of elevated SER concentrations in juveniles and NA concentrations in adults (Cases et al. 1995). However, the metabolic disturbance caused by chronic MAO-A deficiency induces several alterations in these mutant mice (Cases et al. 1996; Holschneider et al. 2000), including upregulation of adenosine 2A receptors, and abnormalities of SER receptor subtypes (Bou-Flores and Hilaire 2000).

Both pharmacological and genetic evidences indicate a facilitatory role for central histamine via H1-receptors in aggression in connection with SER (Noguchi et al. 1992; Yanai et al. 1998). H1-receptor KO mice exhibit less aggression and increased SER turnover in several brain areas (Yanai et al. 1998).

Neural cell adhesion molecule (NCAM) is important during development and in adult neural plasticity (Goridis and Brunet 1992; Scholey et al. 1993). Both NCAM-KO and heterozygous NCAM mice display elevated anxiety and aggression (Stork et al. 1999; Stork et al. 2000). Lower doses of 5-HT_{1A}-R agonists are necessary to reduce anxiety and presumably aggressiveness in the NCAM-KO mice compared with wild-type mice, suggesting a functional change in the 5-HT_{1A}-R (Stork et al. 1999).

Transgenic male mice that overexpress the gene encoding human transforming growth factor-α (TGFα) exhibit enhanced aggressive behavior (Hilakivi-Clarke et al. 1992) accompanied by increased plasma estradiol concentrations and reduced SER turnover in the brain. Interestingly, the heightened aggressiveness in these mice is reversed with either SER uptake inhibitors (Hilakivi-Clarke and Goldberg 1993) or by castration (Hilakivi-Clarke 1994).

Nitric oxide (NO) also serves as an aggression-modulating neurotransmitter (Nelson et al. 1997). Male neuronal NOS-KO and wild-type mice in which nNOS is pharmacologically suppressed are highly aggressive (Nelson et al. 1995; Demas et al. 1997). Castration and testosterone replacement studies in both nNOS knockout and wild-type mice exclude an activational role for gonadal steroids in the elevated aggression (Nelson et al. 1995; Kriegsfeld et al. 1997). NO also appears to affect aggressive behavior via SER. Excessive aggressiveness and impulsiveness of nNOS-KO are caused by selective decrements in SER turnover and deficient 5-HT_{1A}-R and 5-HT_{1B}-R function in brain regions regulating emotion (Chiavegatto et al. 2001). It was possible that NO from endothelial tissue could also contribute to aggression; e.g., endothelial NOS-KO displays virtually no aggression even after pharmacological normalization of blood pressure (Demas et al. 1999).

9.6.1.4 Other Genes Involved in Aggression

The disruption of the V_{1b}, but not the V_{1a}, gene reduced intermale aggression, suggesting that the former, but not the latter, is involved in the control of aggression (Wersinger et al. 2007a, b).

9   Genetic and Epigenetic Determinants of Aggression

In line with other studies on substance P and aggression, the NK1-KO mice are less aggressive (De Felipe et al. 1998). However, it could be the consequence to some secondary changes like an increase in SER function accompanied by a selective desensitization of 5-HT$_{1A}$ inhibitory autoreceptors (Santarelli et al. 2001).

The most important cannabinoid receptor in the CNS is the CB1-R. CB1-R-KO mice presented an increase in the aggressive response measured in the resident–intruder test (Martin et al. 2002).

Interleukin-6 (IL-6) is a cytokine released by activated immune cells which has been shown to affect brain function (Alleva et al. 1998). IL-6-KO mice showed a higher degree of aggressive behavior, as indicated by a higher frequency of offensive upright posture. On the contrary, IL-6 overexpressing subjects showed a tendency to be more involved in affiliative-type social interactions. As secondary change, DA levels were found to be modified in a number of brain regions in IL-6-KO mice (Alleva et al. 1998).

9.6.2   Polymorphism

9.6.2.1   Sexual Steroids

The best established, highly polymorphic and functional locus with regard to sex determination is the AR, which embraces two trinucleotide repeats (Craig and Halton 2009). Both Jonsson et al. 2001 in healthy Swedish males and Rajender et al. 2008 in Indian males found an association of shorter (and presumptively higher expressed) CAG repeats with a somewhat more dramatic phenotype, verbal aggression, or violent criminal activity.

9.6.2.2   Serotonin System

One of the best-studied genomic variations in biological psychiatry is that of the polymorphism of the SER system. Genetic polymorphism of the TH gene, which is the rate-limiting enzyme of SER biosynthesis, is associated with individual differences in aggressive disposition in the normal human population (Manuck et al. 1999). Overall, individuals with an A-allele scored significantly higher on measures of aggression and tendency to experience unprovoked anger. The covariation of the TH1 genotype with aggression and anger measures was found to be statistically robust in men, but nonsignificant among women.

The synaptic activity of SER is terminated by the reuptake into presynaptic terminals, which occurs via the SER transporter (5-HTT) protein. The gene for 5-HTT in humans shows a relatively common polymorphism characterized by a variable repeat sequence in the promoter region, resulting in two common alleles: the short (S) variant comprising 14 copies of a 20–23 base-pair repeat unit and the long (L) variant comprising 16 copies (Lesch et al. 1996). The 5-HTT promoter sequence polymorphism associates with differential transcription of the 5-HTT gene, with more efficient expression from the L-allele (Lesch et al. 1996; Greenberg et al. 1999). The 5-HTT-L variant has been shown to be associated with childhood aggression (Beitchman et al. 2006). Investigating the influence of both childhood psychosocial adversity and 5-HTT-L genotype, on the appearance of violent aggression in adult offenders, Reif et al. (2007) showed that homozygotes for the 5-HTT-L-allele were generally less likely to develop later-life violent behavior, while carriers of at least one S-allele were influenced by environmental factors.

The rhesus monkey shows a polymorphism in the 5-HTT promoter that is comparable in form and function to that in humans such that the S 5-HTT promoter variant in the monkey is also associated with decreased 5-HTT levels in brain (Lesch et al. 1997; Barr et al. 2003). Likewise, monkeys bearing the S version of the 5-HTT promoter polymorphism show reduced SER activity, increased impulsivity, and are more aggressive than animals bearing the L version of the promoter variant (Bennett et al. 2002; Champoux et al. 2002).

Regarding the SER receptors, there are a number of polymorphisms in the 5-HT$_{1B}$-R, and more than 20 association studies have been published with aggression with varying results (Sanders et al. 2002). There are several polymorphisms reported also for 5HT$_{2A}$-R; however, there is scant evidence for the functional importance of any (Craig and Halton 2009).

Polymorphism of a SER-degrading enzyme, MAO, was also discovered. Males maltreated during their youth were at higher risk of being convicted of a violent crime before 27 years of age if they had the short version of the functional polymorphism in the gene coding for MAO-A activity (Caspi et al. 2002). A study with adult males found the MAO-A–maltreatment interaction for antisocial behavior only for white subjects (Widom and Brzustowicz 2006). Another study found the same type of interaction for conduct disorder assessed during adolescence with a sample of male twins (Foley et al. 2004), and a third study with 7-year-old male twins found the significant gene–environment interaction for a composite mental health problem scale and attention deficit–hyperactivity disorder, but not for a total antisocial problem scale (Kim-Cohen et al. 2006). Thus, the MAO-A–maltreatment statistical interaction could depend on racial background and societal factors.

Several studies implicate a biallelic single nucleotide polymorphism of COMT, the other SER-degrading enzyme, with methionine (met) substituting for valine (val). This substitution increases violent behavior in a small subgroup of schizophrenic patients (Volavka et al. 2004). Recently, given the ambiguity of the data and the role of sex differences, Kulikova et al. 2008 examined the functional single nucleotid polymorphism in the manifestation of physical aggression in unselected women. They observed that the met/met homozygotes are least aggressive, while wild-type homozygotes (val/val) exhibited maximum aggression.

9.6.2.3 Other Candidates

Dopamine beta-hydroxylase (DBH) is a key enzyme in the synthesis of NA, and there is an abundance of literature describing the genetic control of DBH levels

9 Genetic and Epigenetic Determinants of Aggression

(in serum) and some indication that this may underpin aspects of antisocial behavior. Hess et al. 2009 have provided evidence that a DBH polymorphism (1021TT) was significantly associated with increased neuroticism scores and impulsive or aggressive behavior.

In pigs, a single nucleotide polymorphism of the AVP V_{1b} receptor showed a highly significant association with aggressive behavior (Murani et al. 2010). Young and colleagues examined the sequence variations in the AVP V_{1a} receptor gene (Hammock and Young 2005; Donaldson and Young 2008). Among voles, subspecies differences in social behavior are associated with a sequence variation in the V_{1a} receptor gene and differential expression of the V_{1a} receptor. Reversing the pattern of V_{1a} receptor expression eliminates the differences in selected, AVP-mediated social behavior (Robinson et al. 2008). These studies describe a causal genotype–phenotype relation (Meaney 2010).

Among these variants, functional polymorphisms in the 5-HTT and MAO-A genes may be of particular importance due to the relationship between these polymorphic variants and anatomical changes in the limbic system of aggressive people. Furthermore, functional variants of 5-HTT and MAO-A are capable of mediating the influence of environmental factors on aggression-related traits. Indeed, the above mentioned results suggest that a 5-HTT polymorphism might have a role in balancing aggressive behavior in differing societies.

9.7 Environmental Effects, Epigenetic Mechanisms

The nature–nurture debate is an essential question of the determinants of individual differences in the expression of specific traits among members of the same species. The origin of the terms nature and nurture has been credited to Richard Mulcaster, a British teacher who imagined these influences as collaborative forces that shape child development (West and King 1987). For a long time, genetic and environmental influences were considered as independent agents in the field of development (Meaney 2010). Nowadays, it is widely accepted that these two factors are highly interconnected as environmental factors may induce heritable changes, although not inside the genetic code, but on the epigenome (Murgatroyd et al. 2010). Genetic studies attempt to understand the genome that is identical in different cell types and throughout life. Epigenetic studies, however, attempt to understand the epigenome that varies between cell type and during development and could explain change and stability as well (Tremblay and Hamet 2008). It is interesting that across species, increasing complexity is associated not so much with an increase in the number of genes that actively code for proteins, but rather with the size of the noncoding region of the genome. This difference may reflect the increased complexity of the regulatory regions of the DNA that, in turn, confers enhanced capacity for tissue-specific regulation of gene expression in multiorgan animals. In addition, the increased size of the regulatory region of the genome should also correspond to an increased capacity for environmental regulation of gene expression, a process whereby an increasing range of phenotypes might emerge from a common genotype: an increased capacity for phenotypic plasticity.

In fact, the activity–inactivity of the gene expression is highly influenced by environmental factors, and these changes through epigenetic modifications, e.g., DNA methylation or histone modification, may leave marks on the genome that are transmittable to the next generations. One of the key structures connecting environment with epigenome could be the HPA axis, namely, the glucocorticoids as end hormones (Auger et al. 2011). Epigenetic modifications might underlie a wide range of stable changes in neural function following exposure to highly salient events, e.g., chronic stress, drugs of abuse, reproductive phases such as parenting, etc., and are thus logical mechanisms for environmentally induced alterations in mental health (Tsankova et al. 2007; Jiang et al. 2008; Akbarian and Huang 2009). Although epigenetic regulation was first discovered in connection with development, fully mature neurons in an adult animal also express the necessary enzymatic machinery to demethylate or remethylate DNA (Meaney 2010). It is possible that environmentally driven changes in neuronal transcriptional signals could potentially remodel the methylation state of specific regions of the DNA (Meaney and Szyf 2005). Another important epigenetic mechanism, histone protein modification, is associated with exposure to drugs of abuse and stressors in rodent models (Renthal et al. 2007; Renthal and Nestler 2008).

It is worth to mention that epigenetic changes are important determinants of development not only during the beginning of life but also in aging (Murgatroyd et al. 2010). A relationship between DNA methylation and aging was originally proposed in a pioneering study by Berdyshev et al. (1967), which showed that genomic global DNA methylation decreases with age in spawning humpbacked salmon. In support of this finding, a gradual global loss of cytosine methylation has been detected in various mouse, rat, and human tissues (Vanyushin et al. 1973; Wilson et al. 1987; Fuke et al. 2004). Aside from global hypomethylation, a number of specific loci (in cytosine–guanine islands) have been reported to become hypermethylated during aging, e.g., the ribosomal gene cluster, the estrogen receptor, c-fos, etc. (Fraga et al. 2007). Study in humans has revealed that intraindividual changes in DNA methylation show some degree of familial clustering, indicative of a genetic component (Bjornsson et al. 2008). This suggests that at least some aspects of epigenetic changes are also genetically determined.

9.7.1 Epigenetic Mechanisms

Epigenetic modifications do not alter the sequence composition of the genome. Instead, epigenetic marks on the DNA and the other features of the chromatin regulate the operation of the genome. Thus, epigenetics has been defined as a functional modification to the DNA that does not involve an alteration of sequence (Meaney 2010). Most widely studied epigenetic controls are DNA methylation and histone acetylation, as, for example, lower methylation of a gene usually leads to increased mRNA expression (Rodenhiser and Mann 2006).

9.7.1.1 Histone Modifications

The histones and DNA together are referred to as chromatin, and the nucleosome is the organization of the chromatin (Meaney 2010). Under normal conditions, there is a tight physical relation between the histone proteins and its accompanying DNA, resulting in a rather closed nucleosome configuration. This restrictive configuration is maintained, in part, by electrostatic bonds between the positively charged histones and the negatively charged DNA. The closed configuration impedes transcription factor binding and is associated with a reduced level of gene expression. The activation of gene expression commonly requires chemical modification of the chromatin that occurs on the histone proteins. Chromatin remodeling is required for increased transcription factor binding to regulatory sites on the DNA and the activation of gene expression. The dynamic alteration of chromatin structure is achieved through modifications to the histone proteins at the tail regions that protrude outside of the nucleosome. This process is achieved through a series of enzymes that bind to the histone tails and modify the local chemical properties of specific amino acids along the tail (Grunstein 1997; Jenuwein and Allis 2001; Hake and Allis 2006). The enzyme histone acetyltransferase (HAT) transfers an acetyl group onto specific lysines on the histone tails. The addition of the acetyl group diminishes the positive charge, loosening the relation between the histones and DNA, opening the chromatin, and improving the ability of transcription factors to access DNA sites. Thus, histone acetylation at specific lysine sites is commonly associated with active gene transcription. The functional antagonists of the histone acetyltransferases are a class of enzymes known as histone deacetylases (HDACs). These enzymes remove acetyl groups and prevent further acetylation, thus serving to maintain a closed chromatin structure, decreasing transcription factor and gene expression. Although several other amino acid residues and mechanisms, like phosphorylation or ubiquitination, are involved, the best studied one is lysine acetylation.

9.7.1.2 DNA Methylation

The classic epigenetic alteration is that of DNA methylation, which involves the addition of a methyl group onto cytosines in cytosine–guanine (CpG) dinucleotides in the DNA (Razin and Riggs 1980; Bird 1986; Holliday 1989; Razin and Cedar 1993). The methylation of DNA is an active biochemical modification that in mammals selectively targets cytosines and is achieved through the action of a class of enzymes, DNA methyltransferases (DNMTs), which transfer the methyl groups from methyl donors. Methylenetetrahydrofolate reductase (MTHFR) is an important enzyme in the generation of methyl groups for DNA methylation (Devlin et al. 2010). DNA methylation is a stable chemical modification, and it is associated with the silencing of gene transcription (Bestor 1998; Razin 1998; Bird and Wolffe 1999; Bird 2002). This effect appears to be mediated in one of two ways (Bird 2002). First, wide swaths of DNA can be methylated and the shear density of methylation precludes transcription factor binding to DNA sites, thus silencing gene expression. The second manner is subtler, and probably far more prevalent, in regions with more dynamic variations in gene transcription, such as the brain. In this case, selected cytosines are methylated, and the presence of the methyl group attracts a class of proteins known as methylated-DNA binding proteins (Klose and Bird 2006). These proteins, in turn, attract an entire cluster of proteins, known as repressor complexes that are the active mediators of the gene silencing. The HDACs are a critical component of the repressor complex. HDACs prevent histone acetylation and favor a closed chromatin state that constrains transcription factor binding and gene expression. Compounds that inhibit HDACs can thus increase transcription from methylated DNA.

DNA methylation-induced gene silencing mediates two of the most commonly studied examples of epigenetic silencing, namely, X-chromosome inactivation and gene imprinting. During imprinting, the expression-specific genes are determined by the parent of origin via inactivation of one allele by methylation.

Although these processes seem to have a major impact on development, recently, it was established that DNA methylation patterns are actively modified in mature, i.e., fully differentiated cells, including and perhaps especially neurons, too, and that such modifications can occur in animals in response to cellular signals driven by environmental events (Meaney and Szyf 2005; Jirtle and Skinner 2007; Bird 2007). Both mature lymphocytes (Bruniquel and Schwartz 2003; Murayama et al. 2006) and neurons (Martinowich et al. 2003; Champagne et al. 2006, 2008; Lubin et al. 2008; Sweatt 2009) show changes in the DNA methylation patterns at critical genomic regions in response to environmental stimuli that stably alter cellular function. The ability of environmental signals to actively remodel epigenetic marks that regulate gene expression is a rather radical change in our understanding of the environmental regulation of gene expression. Such epigenetic modifications are thus a candidate mechanism for the environmental "programming" of gene expression. Although DNA methylation patterns may change during the perinatal period, methylation of some genes in the brain is considered a somewhat stable event that may require maintenance throughout the life span.

9.8 Epigenetic Changes in Aggression

Despite the recommendations made over 30 years ago (Fuller and Hahn 1976; Scott 1977), studies of the genetics of variation in aggressive behaviors have generally considered aggression as a characteristic of an individual, independent of the social context in which it is expressed (Hahn and Schanz 1996). However, the behavior expressed by an individual will usually depend on the behavior of the conspecific with which it interacts (see indirect genetic effect) (Wilson et al. 2009). On the other hand, the social learning theory of aggression says that children learn to aggress from their environment, i.e., the family, peers, neighborhoods, and the media (Bandura 1973; Reiss and Roth 1993; Anderson et al. 2003; NIH 2004; Tremblay 2008). The truth seems to be somewhere in between, in the epigenome.

9 Genetic and Epigenetic Determinants of Aggression

9.8.1 Indirect Evidence on Epigenetic Mechanism in Aggression

The role of epigenetic mechanisms in the following processes could be assumed as there is a strong association between MTHFR, an enzyme required for folate metabolism, and the generation of methyl groups with global changes in DNA methylation (Frosst et al. 1995, 2002; Castro et al. 2004; Sohn et al. 2009; Devlin et al. 2010) and early intervention-induced impulsivity changes. Although most of the maternal-manipulation-induced epigenetic changes were associated with development of depression in offspring, we can assume a similar epigenetic association between early intervention and development of aggression, too.

Most of the epigenetic changes occur during early development, wherefore cross-fostering studies seem to be a good tool to dissect the effect of genetic and environmental/epigenetic changes. Manipulation of the mother–infant interaction is another widely used test to induce epigenetic changes in the offspring. Postweaning social isolation restricts environmental factors early in the development, thereby influencing gene expression most probably at epigenetic level. Indeed, postweaning social isolation induced a reduction of DNMT3b in mice.

9.8.1.1 Cross-Fostering

The nature vs. nurture question was examined in the middle of the twentieth century by cross-fostering mice with rats and studying the effects of the mother, preweaning peer group, and postweaning peer group (Denenberg et al. 1964, 1966; Hudgens et al. 1968). The results from these studies were very interesting in that they suggested that maternal interactions have a greater impact on the subsequent expression of behavior than did the genetic contribution (Hudgens et al. 1968). Indeed, mice reared by rats showed decreased aggression toward conspecifics (King and Gurney 1954; King 1957). The results suggested that the role of social context in the expression of social behavior is very important (Cushing and Kramer 2005). However, phylogenetic differences could confound the results.

Cross-fostering two mice strains, the more aggressive monogamous California mice to the less aggressive polygynous white-footed mice, the offspring showed decreased aggression as adults, and the decrease was correlated with a reduction in AVP immunoreactivity (Bester-Meredith and Marler 2001). Cross-fostering of white-footed mice to California mice resulted in the opposite pattern; aggression and AVP were increased.

9.8.1.2 Early Social Experience

"Stress diathesis" models are proposed as explanations for the relation between early experience and health (Seckl and Meaney 1993; Gorman et al. 2000; Heim and Nemeroff 2001; Meaney 2001, 2007; Repetti et al. 2002; McEwen 2003).

These models suggest that adversity in early life alters the development of neural and endocrine responses to stress in a manner that predisposes individuals to disease (Meaney 2010). Indeed, besides the widely studied depression in mood, disturbances of early mother–infant attachment relationships result in disturbance of social interaction including aggression (Veenema 2009).

The most severe early separation model consists of tactile social isolation during at least the first 6 months of life. These studies on monkeys were the first to demonstrate the devastating consequences of social isolation on normal development (Harlow et al. 1965). Isolated infants showed a total lack of exploration and social interaction, extreme high levels of fear, freezing in response to aggression of other animals, and self-directed and stereotypic motor activity (Mason and Sponholz 1963; Harlow et al. 1965, 1971; McKinney 1974; Seay and Gottfried 1975). As adolescents and adults, these monkeys exhibited excessive and inappropriate aggression toward other monkeys. Reversal of most of these early social isolation-induced behavioral deficits was observed when infants were exposed to a normal social environment within the first 6 months of life (Harlow and Suomi 1971; Cummins and Suomi 1976). In later studies, milder separation paradigms were also used. Hinde et al. (1966) demonstrated that a short separation of rhesus monkeys from their mothers continued to affect their interactions with their mothers weeks after being reunited. Female rhesus monkeys that are completely or partially deprived of maternal contact during the neonatal period, even if raised with peers, display significant changes in the expression of social behavior later in life (Seay et al. 1964; Suomi and Ripp 1983; Kraemer 1992; Kraemer and Clarke 1996). They are more aggressive, more likely to withdraw from novel social interactions and to abuse or neglect their own offspring than monkeys raised by their mothers. Nursery-reared rhesus monkeys show an increase in agonistic behaviors and stereotypy and a reduction in reciprocal social interactions relative to mother-reared monkeys (Winslow et al. 2003).

In male Wistar rats, exposure to maternal separation (3 h/day, days 1–14) induced an increase in aggressive behaviors at juvenile (play fighting) and adult (intermale aggression) age (Veenema et al. 2006; Veenema 2009). Play fighting is an essential behavior for the development of adequate adult social behaviors (Meaney and Stewart 1979; Panksepp et al. 1984; Vanderschuren et al. 1997) and consists of behavioral patterns related to adult aggressive behaviors (Panksepp et al. 1984; Pellis and Pellis 1987; Vanderschuren et al. 1997). At juvenile age, maternal separation increased the number of attacks toward the nape of the neck, decreased the number of supine behaviors, a submissive play behavior, and induced the emergence of offensive pulling and biting, a behavior hardly expressed by controls, toward an unknown age-matched play partner during the resident–intruder test (Veenema 2009). In adulthood, maternally separated rats showed significant increases in key elements of aggression, including lateral threat, offensive upright, and keep down, when being exposed as a resident to an unknown male intruder rat (Veenema et al. 2006). These data indicate that maternal separation promotes the expression of aggressive behaviors in male rats across development.

In contrast to male rats, maternal separation of C57Bl/6 mice induced a decrease in intermale aggression, as shown by longer attack latencies in maternally separated adult males compared with control males (Veenema et al. 2007a). However, maternal separation of C57Bl/6 mice induced an increase in maternal aggression towards a male CD1 intruder mouse during the first week of lactation (Veenema et al. 2007a).

9.8.1.3 Postweaning Social Isolation

In most laboratories, postweaning social isolation is performed by housing rat or mouse pups in individual cages from the first day of weaning from the dam (between postnatal days 21 and 28) for a period of 4–8 weeks. Isolated rats or mice are normally reared in a room with other isolated-reared or group-housed rats or mice. Thus, isolation-reared rats or mice have visual, auditory, and olfactory contact with other conspecifics, but they are restricted from any form of physical interaction with their conspecifics. Postweaning social isolation was shown to induce changes in a wide variety of nonsocial behaviors, including hyperreactivity to a novel environment, impaired prepulse inhibition of acoustic startle, increased ethanol intake, and increased anxiety (Lapiz et al. 2001; Fone and Porkess 2008). Moreover, postweaning social isolation altered several social behaviors. For example, postweaning socially isolated males showed reduced levels of play fighting and social grooming (Von Frijtag et al. 2002), reduced submissive behaviors toward residents (van den Berg et al. 1999), and increased aggressive behaviors toward conspecifics in dyadic interactions or when being placed in a colony (Luciano and Lore 1975; Day et al. 1982; Wongwitdecha and Marsden 1996; Vale and Montgomery 1997; Bibancos et al. 2007).

9.8.2 Epigenetic Modifications in Molecular Mechanisms Connected with Aggression

9.8.2.1 Sexual Steroids

The early testosterone treatment of the eggs increased the frequency of aggression, dominance, and sexual behavior of 1-year-old, reproductively competent house sparrows. These hormone-mediated maternal effects were supposed to be an epigenetic mechanism causing intrasexual variation in adult behavioral phenotype (Partecke and Schwabl 2008). Indeed, epigenetic modifications have been implicated in sexual differentiation of the brain (McCarthy et al. 2009; Murray et al. 2011). Many neural sex differences depend on testosterone exposure during early postnatal life (Cooke et al. 1998), and steroid hormones are assumed to work, at least in part, by orchestrating changes in the epigenome (Spencer et al. 1997; Kishimoto et al. 2006). Sexual differentiation of the BNST in mice appears to require histone acetylation, as inhibiting histone acetylation blocks the masculinization of the BNST (Murray et al. 2009; Auger et al. 2011).

Another important contributor of the masculine phenotype is the ERα receptor. The ERα gene promoter contains multiple CpG sites that are potential targets for

DNA methylation. Sexually dimorphic ERα promoter methylation and deacetylation of histones regulating the expression of the ERα appear to be partly responsible for the sexually dimorphic expression of this receptor (Pinzone et al. 2004; Kurian et al. 2010; Westberry et al. 2010). Champagne et al. (2006) found increased cytosine methylation across the ERα-gene promoter in the offspring of low-maternal care mothers in MPOA.

Castration of adult male rats, known to reduce aggression by removal of testosterone, resulted in decreased AVP mRNA expression and increased methylation within the AVP promoter in the BNST (Auger et al. 2011). Similarly, castration significantly increased ERα mRNA expression and decreased ERα promoter methylation within the BNST. These changes were prevented by testosterone replacement. This suggests that the DNA promoter methylation status of some steroid-responsive genes in the adult brain is actively maintained by the presence of circulating steroid hormones. The maintenance of methylated or demethylated states of some genes in the adult brain by the presence of steroid hormones may play a role in the homeostatic regulation of behaviorally relevant systems.

9.8.2.2 Serotonin

The 5-HTT-L variant has been shown to influence vulnerability to the impact of early stressful life events (see earlier). The methylation status of its promoter was shown to play a role in governing 5-HTT mRNA levels (Philibert et al. 2007; Devlin et al. 2010). Depressed mood during the second trimester of human pregnancy was associated with reduced methylation of the maternal and neonatal 5-HTT gene promoter region measured in the whole blood. Conceivably, such reduced methylation may lead to increased 5-HTT expression and availability of 5-HTT and as such result in increased SER reuptake and lower intrasynaptic SER. In the mature brain, this might not have a noticeable impact, but in the developing brain, such altered serotonergic tone may have long-term effects on behavior, as prior to the neurotransmitter role of SER, it plays critical roles as a trophic factor modulating neuronal differentiation and growth (Ansorge et al. 2008).

9.8.2.3 Vasopressin and the Stress Axis

Early life stress in mice (daily 3-h separation of mouse pups from their mother during postnatal days 1–10) caused persistent epigenetic marking (hypomethylation) of a key regulatory region of the AVP gene in the PVN underpinning sustained upregulation of AVP mRNA expression and increased HPA axis activity (Murgatroyd et al. 2009; Murgatroyd and Spengler 2011). Hypomethylation of the AVP promoter is achieved through phosphorylation of methyl CpG-binding protein 2 (MeCP2) (Cloud 2010). The early-life stress-induced endocrine phenotype lasted for at least 1 year following the initial adverse event. Although treatment with an AVP V_{1b} receptor antagonist reversed the mice's phenotype, but the epigenetic marking of the methylation landmarks in the AVP enhancer persisted, suggesting that early-life stress has engraved a permanent memory trace that conferred lifelong susceptibility to stress (Murgatroyd et al. 2009). Although it is clear that the epigenetic control of AVP cells in the PVN is sensitive to stress hormones during early development, it is unclear whether stress hormones alter DNA methylation patterns in adulthood within these cells.

Lower sensory input from the mother was accompanied by enhanced methylation of the CRH promoter in the hippocampus of the offspring (McClelland et al. 2011). Notably, enhanced methylation is generally associated with reduced transcription, whereas CRH gene expression was enhanced in this group that performed worse in learning and memory tests later in life.

Maternal undernutrition, as an early-life stress, may also induce epigenetic changes (Stevens et al. 2011). Not only histone acetylation but also hypomethylation of the POMC gene was detected. Parallel hypomethylation of the hypothalamic glucocorticoid receptor (GR) suggests that it might mediate regulation of a number of hypothalamic neuropeptides including POMC and neuropeptide Y.

Disturbances of early social development may alter the methylation at the GR gene promoter in the hippocampus (Weaver et al. 2004). SER, as a classic neurotransmitter responding dynamically to environmental signals, was found to regulate GR expression via epigenetic mechanisms (HAT) (Meaney 2010).This effect can be reversed by central infusion of a HDAC inhibitor, thereby normalizing the stress responses.

9.8.2.4 Neurotrophins

Neurotrophins, such as NGF and BDNF, which play a fundamental role in brain function and neuroprotection and are affected by stress, are good candidates for transducing the effects of adverse events in changes of brain function (Cirulli et al. 2009). Early adversive events may induce changes in NGF levels, which is accompanied by HPA axis dysregulation, a process, which is probably mediated by epigenetic mechanisms (Aloe et al. 1994a; Branchi et al. 2004). Indeed, the NGF pathway is able to induce histone modifications (methylation and acetylation) in cell lines, resulting in changes in opioid receptor expression (Chen et al. 2008; Park et al. 2008). During differentiation induced by in vitro NGF treatment, the mRNA and protein levels of de novo methyltransferase DNMT3b increased, whereas those of DNMT3a and DNMT1 decreased (Bai et al. 2005). Human studies also confirmed the role of NGF in diverse psychological stress, like acute alcohol, heroin, or nicotine withdrawal (Aloe et al. 1996; Lang et al. 2002). The CpG island promoter methylation of the NGF gene significantly increased in the blood of alcohol-dependent patients together with an increase in NGF serum levels (Heberlein et al. 2011).

Expression of BDNF in a rat model was shown to be sensitive to early adverse life experience, an event regulated by methylation (Roth et al. 2009). Although several epigenetic modifications of the BDNF promoter were confirmed, most of the studies were focusing on the hippocampus and depressive-like symptoms (Boulle et al. 2011). However, based upon the aforementioned role of BDNF in aggression, we might assume an epigenetic regulation of the BDNF promoter in the development of the aggressive phenotype, too.

9.8.2.5 Other Putative Mechanisms

In one study, boys from low socioeconomic background, who were found to be on a high physical aggression trajectory, were compared with boys from the same background, who followed a normal physical aggression trajectory (Broidy et al. 2003). Males on the chronic physical aggression trajectory have substantially more methylated alleles measured on T cells, more specifically on the interleukin-1B (IL-1B) cytokine gene.

The most serious self-aggression is suicide, where a significant reduction of the hippocampal GR was showed, which enhances HPA activity. This exaggerated stress response is—in the long run—the result of an increased DNA methylation of the promoter of the GR (Turner and Muller 2005).

9.8.3 Valproate Treatment

The epigenetic machinery seems to be a good target for therapeutical interventions; however, presently only few drugs are available with unspecific targets. Most of the research focused on VPA, an anticonvulsant, used in the treatment of epilepsy and bipolar disorder which, besides an effect on GABA neurotransmission, is an inhibitor of HDAC. HDAC inhibition leads to a global increase in the level of histone acetylation, which is most consistently associated with increased gene expression.

VPA administration reduces aggressive behavior in mice and rats (Sulcova et al. 1981; Oehler et al. 1985; Molina et al. 1986; Belozertseva and Andreev 1997). One of the possible background changes could be the enhanced AVP tone after VPA treatment (Murray et al. 2011).

Vehicle-treated male mice socially isolated for 2–3 weeks were very aggressive toward an untreated intruder of the same sex. Administration of met, a methyl donor enhancing the overall level of DNA methylation, to resident mice during the isolation period significantly prolonged the latency of attacks (Tremolizzo et al. 2005). This reduction of aggression was prevented by VPA coadministered with met. A VPA–met-treated resident attacked the intruder with a similar latency and duration as a vehicle-treated resident, whereas VPA alone in this experiment failed to modify aggressive behavior. The met-induced modification of aggressive behavior was dose related and persisted longer than 1 week following met withdrawal.

9.9 Conclusions

Aggression is an adaptive response to social challenges of the environment. However, pathological forms, mostly associated with other psychological disturbances, are highly destructive. Several brain regions (like HAA, MeA, PAG) and several molecules (testosterone, SER, AVP, etc.) are involved in the development of this behavior,

9 Genetic and Epigenetic Determinants of Aggression

but one of the most important determinants is the behavior of the encounter. Therefore, it is not surprising that epigenetic changes, connecting environment with gene activation, could be highly involved in fine-tuning the brain structures and molecular network taking part in aggression. Till now, most of the knowledge accumulated on the involvement of different genes and the participation of epigenetic mechanisms in the development of aggressive behavior could be only supposed based on indirect data.

References

Akbarian, S. & Huang, H. S. (2009). Epigenetic regulation in human brain-focus on histone lysine methylation. *Biol Psychiatry* 65, 198–203.

Albers, H. E. & Bamshad, M. (1998). Role of vasopressin and oxytocin in the control of social behavior in Syrian hamsters (Mesocricetus auratus). *Prog Brain Res* 119, 395–408.

Albert, D. J., Dyson, E. M., Walsh, M. L. & Gorzalka, B. B. (1987). Intermale social aggression in rats: suppression by medial hypothalamic lesions independently of enhanced defensiveness or decreased testicular testosterone. *Physiol Behav* 39, 693–698.

Albert, D. J. & Wong, R. C. (1978). Hyperreactivity, muricide, and intraspecific aggression in the rat produced by infusion of local anesthetic into the lateral septum or surrounding areas. *J Comp Physiol Psychol* 92, 1062–1073.

Alleva, E. & Aloe, L. (1989). Physiological roles of nerve growth factor in adult rodents: a biobehavioral perspective. *Int J Comp Psychol* 2, 147–163.

Alleva, E., Aloe, L. & Bigi, S. (1993). An updated role for nerve growth factor in neurobehavioural regulation of adult vertebrates. *Rev Neurosci* 4, 41–62.

Alleva, E., Cirulli, F., Bianchi, M., Bondiolotti, G. P., Chiarotti, F., De Acetis, L. & Panerai, A. E. (1998). Behavioural characterization of interleukin-6 overexpressing or deficient mice during agonistic encounters. *Eur J Neurosci* 10, 3664–3672.

Alleva, E. & Santucci, D. (2001). Psychosocial vs. "physical" stress situations in rodents and humans: role of neurotrophins. *Physiol Behav* 73, 313–320.

Aloe, L., Alleva, E., Bohm, A. & Levi-Montalcini, R. (1986). Aggressive behavior induces release of nerve growth factor from mouse salivary gland into the bloodstream. *Proc Natl Acad Sci U S A* 83, 6184–6187.

Aloe, L., Alleva, E. & Fiore, M. (2002). Stress and nerve growth factor: findings in animal models and humans. *Pharmacol Biochem Behav* 73, 159–166.

Aloe, L., Bracci-Laudiero, L., Alleva, E., Lambiase, A., Micera, A. & Tirassa, P. (1994a). Emotional stress induced by parachute jumping enhances blood nerve growth factor levels and the distribution of nerve growth factor receptors in lymphocytes. *Proc Natl Acad Sci U S A* 91, 10440–10444.

Aloe, L. & Levi-Montalcini, R. (1979). Nerve growth factor-induced transformation of immature chromaffin cells in vivo into sympathetic neurons: effect of antiserum to nerve growth factor. *Proc Natl Acad Sci U S A* 76, 1246–1250.

Aloe, L. & Levi-Montalcini, R. (1980). Comparative studies on testosterone and L-thyroxine effects on the synthesis of nerve growth factor in mouse submaxillary salivary glands. *Exp Cell Res* 125, 15–22.

Aloe, L., Skaper, S. D., Leon, A. & Levi-Montalcini, R. (1994b). Nerve growth factor and autoimmune diseases. *Autoimmunity* 19, 141–150.

Aloe, L., Tuveri, M. A., Guerra, G., Pinna, L., Tirassa, P., Micera, A. & Alleva, E. (1996). Changes in human plasma nerve growth factor level after chronic alcohol consumption and withdrawal. *Alcohol Clin Exp Res* 20, 462–465.

Andersen, M. L., Bignotto, M., Machado, R. B. & Tufik, S. (2004). Different stress modalities result in distinct steroid hormone responses by male rats. *Braz J Med Biol Res* 37, 791–797.

Anderson, C. A., Berkowitz, L., Donnerstein, E., Huesmann, L. R., Johnson, J. D., Linz, D., Malamuth, N. M. & Wartella, E. (2003). The influence of media violence on youth. *Psychol Sci Public Interest* 4, 81–110.

Ansorge, M. S., Morelli, E. & Gingrich, J. A. (2008). Inhibition of serotonin but not norepinephrine transport during development produces delayed, persistent perturbations of emotional behaviors in mice. *J Neurosci* 28, 199–207.

Antoni, F. A. (1993). Vasopressinergic control of pituitary adrenocorticotropin secretion comes of age. *Front Neuroendocrinol* 14, 76–122.

Archer, J. (2006). Testosterone and human aggression: an evaluation of the challenge hypothesis. *Neurosci Biobehav Rev* 30, 319–345.

Arletti, R., Bazzani, C., Castelli, M. & Bertolini, A. (1985). Oxytocin improves male copulatory performance in rats. *Horm Behav* 19, 14–20.

Arnold, A. P. (1996). Genetically triggered sexual differentiation of brain and behavior. *Horm Behav* 30, 495–505.

Auger, C. J., Coss, D., Auger, A. P. & Forbes-Lorman, R. M. (2011). Epigenetic control of vasopressin expression is maintained by steroid hormones in the adult male rat brain. *Proc Natl Acad Sci U S A* 108, 4242–4247.

Aujard, F. & Perret, M. (1998). Age-related effects on reproductive function and sexual competition in the male prosimian primate, Microcebus murinus. *Physiol Behav* 64, 513–519.

Axelrod, J. & Reisine, T. D. (1984). Stress hormones: their interaction and regulation. *Science* 224, 452–459.

Bac, P., Pages, N., Herrenknecht, C., Dupont, C., Maurois, P., Vamecq, J. & Durlach, J. (2002). THC aggravates rat muricide behavior induced by two levels of magnesium deficiency. *Physiol Behav* 77, 189–195.

Bai, S., Ghoshal, K., Datta, J., Majumder, S., Yoon, S. O. & Jacob, S. T. (2005). DNA methyltransferase 3b regulates nerve growth factor-induced differentiation of PC12 cells by recruiting histone deacetylase 2. *Mol Cell Biol* 25, 751–766.

Bales, K. L. & Carter, C. S. (2003). Developmental exposure to oxytocin facilitates partner preferences in male prairie voles (Microtus ochrogaster). *Behav Neurosci* 117, 854–859.

Bandtlow, C. E., Heumann, R., Schwab, M. E. & Thoenen, H. (1987). Cellular localization of nerve growth factor synthesis by in situ hybridization. *EMBO J* 6, 891–899.

Bandura, A. (1973).*Aggression: A social learning analysis*. Englewood Cliffs, NJ: Prentice Hall.

Barde, Y. A. (1990). The nerve growth factor family. *Prog Growth Factor Res* 2, 237–248.

Barr, C. S., Newman, T. K., Becker, M. L., Champoux, M., Lesch, K. P., Suomi, S. J., Goldman, D. & Higley, J. D. (2003). Serotonin transporter gene variation is associated with alcohol sensitivity in rhesus macaques exposed to early-life stress. *Alcohol Clin Exp Res* 27, 812–817.

Bartolomeos, K., Brown, D., Butchart, A., Harvey, A., Meddings, D. & Sminkey, L. (2007).*Third milestones of a global campaign for violence prevention report 2007: scaling up.* Geneva: WHO. Global campaign for violence prevention.

Beauchaine, T. P., Gartner, J. & Hagen, B. (2000). Comorbid depression and heart rate variability as predictors of aggressive and hyperactive symptom responsiveness during inpatient treatment of conduct-disordered, ADHD boys. *Aggress Behav* 26, 425–441.

Beaver, K. M., Wright, J. P., DeLisi, M., Walsh, A., Vaughn, M. G., Boisvert, D. & Vaske, J. (2007). A gene x gene interaction between DRD2 and DRD4 is associated with conduct disorder and antisocial behavior in males. *Behav Brain Funct* 3:30.

Beiderbeck, D. I., Neumann, I. D. & Veenema, A. H. (2007). Differences in intermale aggression are accompanied by opposite vasopressin release patterns within the septum in rats bred for low and high anxiety. *Eur J Neurosci* 26, 3597–3605.

Beitchman, J. H., Baldassarra, L., Mik, H., De Luca, V., King, N., Bender, D., Ehtesham, S. & Kennedy, J. L. (2006). Serotonin transporter polymorphisms and persistent, pervasive childhood aggression. *Am J Psychiatry* 163, 1103–1105.

9 Genetic and Epigenetic Determinants of Aggression

Belozertseva, I. V. & Andreev, B. V. (1997). [A pharmaco-ethological study of the GABA-ergic mechanisms regulating the depression-like behavior of mice]. *Zh Vyssh Nerv Deiat Im I P Pavlova* 47, 1024–1031.

Bennett, A. J., Lesch, K. P., Heils, A., Long, J. C., Lorenz, J. G., Shoaf, S. E., Champoux, M., Suomi, S. J., Linnoila, M. V. & Higley, J. D. (2002). Early experience and serotonin transporter gene variation interact to influence primate CNS function. *Mol Psychiatry* 7, 118–122.

Berdyshev, G. D., Korotaev, G. K., Boiarskikh, G. V. & Vanyushin, B. F. (1967). [Nucleotide composition of DNA and RNA from somatic tissues of humpback and its changes during spawning]. *Biokhimiia* 32, 988–993.

Bernhardt, P. C., Dabbs, J. M., Jr., Fielden, J. A. & Lutter, C. D. (1998). Testosterone changes during vicarious experiences of winning and losing among fans at sporting events. *Physiol Behav* 65, 59–62.

Bester-Meredith, J. K. & Marler, C. A. (2001). Vasopressin and aggression in cross-fostered California mice (Peromyscus californicus) and white-footed mice (Peromyscus leucopus). *Horm Behav* 40, 51–64.

Bester-Meredith, J. K., Young, L. J. & Marler, C. A. (1999). Species differences in paternal behavior and aggression in peromyscus and their associations with vasopressin immunoreactivity and receptors. *Horm Behav* 36, 25–38.

Bestor, T. H. (1998). Gene silencing. Methylation meets acetylation. *Nature* 393, 311–312.

Bibancos, T., Jardim, D. L., Aneas, I. & Chiavegatto, S. (2007). Social isolation and expression of serotonergic neurotransmission-related genes in several brain areas of male mice. *Genes Brain Behav* 6, 529–539.

Bird, A. (2002). DNA methylation patterns and epigenetic memory. *Genes Dev* 16, 6–21.

Bird, A. (2007). Perceptions of epigenetics. *Nature* 447, 396–398.

Bird, A. P. (1986). CpG-rich islands and the function of DNA methylation. *Nature* 321, 209–213.

Bird, A. P. & Wolffe, A. P. (1999). Methylation-induced repression--belts, braces, and chromatin. *Cell* 99, 451–454.

Bjork, J. M., Moeller, F. G., Kramer, G. L., Kram, M., Suris, A., Rush, A. J. & Petty, F. (2001). Plasma GABA levels correlate with aggressiveness in relatives of patients with unipolar depressive disorder. *Psychiatry Res* 101, 131–136.

Bjorkqvist, K. (2001). Social defeat as a stressor in humans. *Physiol Behav* 73, 435–442.

Bjornsson, H. T., Sigurdsson, M. I., Fallin, M. D., Irizarry, R. A., Aspelund, T., Cui, H., Yu, W., Rongione, M. A., Ekstrom, T. J., Harris, T. B., Launer, L. J., Eiriksdottir, G., Leppert, M. F., Sapienza, C., Gudnason, V. & Feinberg, A. P. (2008). Intra-individual change over time in DNA methylation with familial clustering. *JAMA* 299, 2877–2883.

Blair, R. J. (2001). Neurocognitive models of aggression, the antisocial personality disorders, and psychopathy. *J Neurol Neurosurg Psychiatry* 71, 727–731.

Blair, R. J., Morris, J. S., Frith, C. D., Perrett, D. I. & Dolan, R. J. (1999). Dissociable neural responses to facial expressions of sadness and anger. *Brain* 122, 883–893.

Blanchard, R. J., Griebel, G., Farrokhi, C., Markham, C., Yang, M. & Blanchard, D. C. (2005). AVP V1b selective antagonist SSR149415 blocks aggressive behaviors in hamsters. *Pharmacol Biochem Behav* 80, 189–194.

Book, A. S., Starzyk, K. B. & Quinsey, V. L. (2001). The relationship between testosterone and aggression: a meta-analysis. *Aggression and Violent Behavior* 6, 579–599.

Bou-Flores, C. & Hilaire, G. (2000). 5-Hydroxytryptamine(2A) and 5-hydroxytryptamine(1B) receptors are differently affected by the monoamine oxidase A-deficiency in the Tg8 transgenic mouse. *Neurosci Lett* 296, 141–144.

Boulle, F., van den Hove, D. L., Jakob, S. B., Rutten, B. P., Hamon, M., van Os, J., Lesch, K. P., Lanfumey, L., Steinbusch, H. W. & Kenis, G. (2011). Epigenetic regulation of the BDNF gene: implications for psychiatric disorders. *Mol Psychiatry.* doi: 10.1038/mp.2011.107.

Brain, P. F. (1972). Mammalian behavior and the adrenal cortex. A review. *Behav Biol* 7, 453–477.

Branchi, I., Francia, N. & Alleva, E. (2004). Epigenetic control of neurobehavioural plasticity: the role of neurotrophins. *Behav Pharmacol* 15, 353–362.

Broidy, L. M., Nagin, D. S., Tremblay, R. E., Bates, J. E., Brame, B., Dodge, K. A., Fergusson, D., Horwood, J. L., Loeber, R., Laird, R., Lynam, D. R., Moffitt, T. E., Pettit, G. S. & Vitaro, F. (2003). Developmental trajectories of childhood disruptive behaviors and adolescent delinquency: a six-site, cross-national study. *Dev Psychol* 39, 222–245.

Brot, M. D., De Vries, G. J. & Dorsa, D. M. (1993). Local implants of testosterone metabolites regulate vasopressin mRNA in sexually dimorphic nuclei of the rat brain. *Peptides* 14, 933–940.

Brown, G. L., Goodwin, F. K., Ballenger, J. C., Goyer, P. F. & Major, L. F. (1979). Aggression in humans correlates with cerebrospinal fluid amine metabolites. *Psychiatry Res* 1, 131–139.

Bruniquel, D. & Schwartz, R. H. (2003). Selective, stable demethylation of the interleukin-2 gene enhances transcription by an active process. *Nat Immunol* 4, 235–240.

Brunner, D. & Hen, R. (1997). Insights into the neurobiology of impulsive behavior from serotonin receptor knockout mice. *Ann N Y Acad Sci* 836, 81–105.

Brunner, H. G., Nelen, M., Breakefield, X. O., Ropers, H. H. & van Oost, B. A. (1993). Abnormal behavior associated with a point mutation in the structural gene for monoamine oxidase A. *Science* 262, 578–580.

Brutus, M., Shaikh, M. B., Edinger, H. & Siegel, A. (1986). Effects of experimental temporal lobe seizures upon hypothalamically elicited aggressive behavior in the cat. *Brain Res* 366, 53–63.

Butter, C. M. & Snyder, D. R. (1972). Alterations in aversive and aggressive behaviors following orbital frontal lesions in rhesus monkeys. *Acta Neurobiol Exp (Wars)* 32, 525–565.

Cardinal, R. N., Winstanley, C. A., Robbins, T. W. & Everitt, B. J. (2004). Limbic corticostriatal systems and delayed reinforcement. *Ann N Y Acad Sci* 1021, 33–50.

Cases, O., Seif, I., Grimsby, J., Gaspar, P., Chen, K., Pournin, S., Muller, U., Aguet, M., Babinet, C., Shih, J. C. & . (1995). Aggressive behavior and altered amounts of brain serotonin and norepinephrine in mice lacking MAOA. *Science* 268, 1763–1766.

Cases, O., Vitalis, T., Seif, I., De Maeyer, E., Sotelo, C. & Gaspar, P. (1996). Lack of barrels in the somatosensory cortex of monoamine oxidase A-deficient mice: role of a serotonin excess during the critical period. *Neuron* 16, 297–307.

Caspi, A., McClay, J., Moffitt, T. E., Mill, J., Martin, J., Craig, I. W., Taylor, A. & Poulton, R. (2002). Role of genotype in the cycle of violence in maltreated children. *Science* 297, 851–854.

Castro, R., Rivera, I., Ravasco, P., Camilo, M. E., Jakobs, C., Blom, H. J. & de Almeida, I. T. (2004). 5,10-methylenetetrahydrofolate reductase (MTHFR) 677C-->T and 1298A-->C mutations are associated with DNA hypomethylation. *J Med Genet* 41, 454–458.

Castro, W. L. & Matt, K. S. (1997). Neuroendocrine correlates of separation stress in the Siberian dwarf hamster (Phodopus sungorus). *Physiol Behav* 61, 477–484.

Caughey, S. D., Klampfl, S. M., Bishop, V. R., Pfoertsch, J., Neumann, I. D., Bosch, O. J. & Meddle, S. L. (2011). Changes in the intensity of maternal aggression and central oxytocin and vasopressin v1a receptors across the peripartum period in the rat. *J Neuroendocrinol* 23, 1113–1124.

Cavigelli, S. A. & Pereira, M. E. (2000). Mating season aggression and fecal testosterone levels in male ring-tailed lemurs (Lemur catta). *Horm Behav* 37, 246–255.

Challman, T. D. & Lipsky, J. J. (2000). Methylphenidate: its pharmacology and uses. *Mayo Clin Proc* 75, 711–721.

Champagne, D. L., Bagot, R. C., van Hasselt, F., Ramakers, G., Meaney, M. J., de Kloet, E. R., Joels, M. & Krugers, H. (2008). Maternal care and hippocampal plasticity: evidence for experience-dependent structural plasticity, altered synaptic functioning, and differential responsiveness to glucocorticoids and stress. *J Neurosci* 28, 6037–6045.

Champagne, F. A., Chretien, P., Stevenson, C. W., Zhang, T. Y., Gratton, A. & Meaney, M. J. (2004). Variations in nucleus accumbens dopamine associated with individual differences in maternal behavior in the rat. *J Neurosci* 24, 4113–4123.

Champagne, F. A., Weaver, I. C., Diorio, J., Dymov, S., Szyf, M. & Meaney, M. J. (2006). Maternal care associated with methylation of the estrogen receptor-alpha1b promoter and estrogen receptor-alpha expression in the medial preoptic area of female offspring. *Endocrinology* 147, 2909–2915.

9 Genetic and Epigenetic Determinants of Aggression

Champoux, M., Bennett, A., Shannon, C., Higley, J. D., Lesch, K. P. & Suomi, S. J. (2002). Serotonin transporter gene polymorphism, differential early rearing, and behavior in rhesus monkey neonates. *Mol Psychiatry* 7, 1058–1063.

Chandrasekaran, S., Ament, S. A., Eddy, J. A., Rodriguez-Zas, S. L., Schatz, B. R., Price, N. D. & Robinson, G. E. (2011). Behavior-specific changes in transcriptional modules lead to distinct and predictable neurogenomic states. *Proc Natl Acad Sci U S A* 108, 18020–18025.

Chen, Y. L., Law, P. Y. & Loh, H. H. (2008). NGF/PI3K signaling-mediated epigenetic regulation of delta opioid receptor gene expression. *Biochem Biophys Res Commun* 368, 755–760.

Cherek, D. R. & Lane, S. D. (2001). Acute effects of D-fenfluramine on simultaneous measures of aggressive escape and impulsive responses of adult males with and without a history of conduct disorder. *Psychopharmacology (Berl)* 157, 221–227.

Cherek, D. R., Roache, J. D., Egli, M., Davis, C., Spiga, R. & Cowan, K. (1993). Acute effects of marijuana smoking on aggressive, escape and point-maintained responding of male drug users. *Psychopharmacology (Berl)* 111, 163–168.

Chiavegatto, S., Dawson, V. L., Mamounas, L. A., Koliatsos, V. E., Dawson, T. M. & Nelson, R. J. (2001). Brain serotonin dysfunction accounts for aggression in male mice lacking neuronal nitric oxide synthase. *Proc Natl Acad Sci U S A* 98, 1277–1281.

Chiavegatto, S. & Nelson, R. J. (2003). Interaction of nitric oxide and serotonin in aggressive behavior. *Horm Behav* 44, 233–241.

Chichinadze, K. & Chichinadze, N. (2008). Stress-induced increase of testosterone: contributions of social status and sympathetic reactivity. *Physiol Behav* 94, 595–603.

Cho, M. M., DeVries, A. C., Williams, J. R. & Carter, C. S. (1999). The effects of oxytocin and vasopressin on partner preferences in male and female prairie voles (Microtus ochrogaster). *Behav Neurosci* 113, 1071–1079.

Cirulli, F., Francia, N., Berry, A., Aloe, L., Alleva, E. & Suomi, S. J. (2009). Early life stress as a risk factor for mental health: role of neurotrophins from rodents to non-human primates. *Neurosci Biobehav Rev* 33, 573–585.

Cirulli, F., Pistillo, L., De Acetis, L., Alleva, E. & Aloe, L. (1998). Increased number of mast cells in the central nervous system of adult male mice following chronic subordination stress. *Brain Behav Immun* 12, 123–133.

Clement, J., Simler, S., Ciesielski, L., Mandel, P., Cabib, S. & Puglisi-Allegra, S. (1987). Age-dependent changes of brain GABA levels, turnover rates and shock-induced aggressive behavior in inbred strains of mice. *Pharmacol Biochem Behav* 26, 83–88.

Cloud, J. (2010). Why your DNA isn't your destiny. *Time Magazine* 175, 49–53.

Coccaro, E. F., Silverman, J. M., Klar, H. M., Horvath, T. B. & Siever, L. J. (1994). Familial correlates of reduced central serotonergic system function in patients with personality disorders. *Arch Gen Psychiatry* 51, 318–324.

Congdon, E., Lesch, K. P. & Canli, T. (2008). Analysis of DRD4 and DAT polymorphisms and behavioral inhibition in healthy adults: implications for impulsivity. *Am J Med Genet B Neuropsychiatr Genet* 147B, 27–32.

Cooke, B., Hegstrom, C. D., Villeneuve, L. S. & Breedlove, S. M. (1998). Sexual differentiation of the vertebrate brain: principles and mechanisms. *Front Neuroendocrinol* 19, 323–362.

Craig, I. W. & Halton, K. E. (2009). Genetics of human aggressive behaviour. *Hum Genet* 126, 101–113.

Crawley, J. N., Schleidt, W. M. & Contrera, J. F. (1975). Does social environment decrease propensity to fight in male mice? *Behav Biol* 15, 73–83.

Cummins, M. S. & Suomi, S. J. (1976). Long-Term Effects of Social Rehabilitation in Rhesus Monkeys. *Primates* 17, 43–51.

Cushing, B. S. & Kramer, K. M. (2005). Mechanisms underlying epigenetic effects of early social experience: the role of neuropeptides and steroids. *Neurosci Biobehav Rev* 29, 1089–1105.

Dabbs, J. M. & Hargrove, M. F. (1997). Age, testosterone, and behavior among female prison inmates. *Psychosom Med* 59, 477–480.

Datla, K. P. & Bhattacharya, S. K. (1990). Effect of selective monoamine oxidase A and B inhibitors on footshock induced aggression in paired rats. *Indian J Exp Biol* 28, 742–745.

David, J. T., Cervantes, M. C., Trosky, K. A., Salinas, J. A. & Delville, Y. (2004). A neural network underlying individual differences in emotion and aggression in male golden hamsters. *Neuroscience* 126, 567–578.

Davidson, R. J., Putnam, K. M. & Larson, C. L. (2000). Dysfunction in the neural circuitry of emotion regulation--a possible prelude to violence. *Science* 289, 591–594.

Day, H. D., Seay, B. M., Hale, P. & Hendricks, D. (1982). Early social deprivation and the ontogeny of unrestricted social behavior in the laboratory rat. *Dev Psychobiol* 15, 47–59.

de Almeida, R. M., Faccidomo, S., Fish, E. & Miczek, K. A. (2001). Inhibition of alcohol-heightened aggression by action at post-synaptic 5-HT1b receptors in male mice. *Aggress Behav* 27, 234–235.

de Almeida, R. M., Ferrari, P. F., Parmigiani, S. & Miczek, K. A. (2005). Escalated aggressive behavior: dopamine, serotonin and GABA. *Eur J Pharmacol* 526, 51–64.

de Almeida, R. M. & Miczek, K. A. (2002). Aggression escalated by social instigation or by discontinuation of reinforcement ("frustration") in mice: inhibition by anpirtoline: a 5-HT1B receptor agonist. *Neuropsychopharmacology* 27, 171–181.

de Boer, S. F., van der Vegt, B. J. & Koolhaas, J. M. (2003). Individual variation in aggression of feral rodent strains: a standard for the genetics of aggression and violence? *Behav Genet* 33, 485–501.

De Felipe, C., Herrero, J. F., O'Brien, J. A., Palmer, J. A., Doyle, C. A., Smith, A. J., Laird, J. M., Belmonte, C., Cervero, F. & Hunt, S. P. (1998). Altered nociception, analgesia and aggression in mice lacking the receptor for substance P. *Nature* 392, 394–397.

De Simone, R., Alleva, E., Tirassa, P. & Aloe, L. (1990). Nerve growth factor released into the bloodstream following intraspecific fighting induces mast cell degranulation in adult male mice. *Brain Behav Immun* 4, 74–81.

De Vries, G. J., Buijs, R. M., Van Leeuwen, F. W., Caffe, A. R. & Swaab, D. F. (1985). The vasopressinergic innervation of the brain in normal and castrated rats. *J Comp Neurol* 233, 236–254.

De Vries, G. J. & Villalba, C. (1997). Brain sexual dimorphism and sex differences in parental and other social behaviors. *Ann N Y Acad Sci* 807, 273–286.

De Vries, G. J., Wang, Z., Bullock, N. A. & Numan, S. (1994). Sex differences in the effects of testosterone and its metabolites on vasopressin messenger RNA levels in the bed nucleus of the stria terminalis of rats. *J Neurosci* 14, 1789–1794.

Delville, Y., Mansour, K. M. & Ferris, C. F. (1996). Serotonin blocks vasopressin-facilitated offensive aggression: interactions within the ventrolateral hypothalamus of golden hamsters. *Physiol Behav* 59, 813–816.

Delville, Y., Melloni, R. H., Jr. & Ferris, C. F. (1998). Behavioral and neurobiological consequences of social subjugation during puberty in golden hamsters. *J Neurosci* 18, 2667–2672.

DeMar, J. C., Ma, K., Bell, J. M., Igarashi, M., Greenstein, D. & Rapoport, S. I. (2006). One generation of n-3 polyunsaturated fatty acid deprivation increases depression and aggression test scores in rats. *J Lipid Res* 47, 172–180.

Demas, G. E., Eliasson, M. J., Dawson, T. M., Dawson, V. L., Kriegsfeld, L. J., Nelson, R. J. & Snyder, S. H. (1997). Inhibition of neuronal nitric oxide synthase increases aggressive behavior in mice. *Mol Med* 3, 610–616.

Demas, G. E., Kriegsfeld, L. J., Blackshaw, S., Huang, P., Gammie, S. C., Nelson, R. J. & Snyder, S. H. (1999). Elimination of aggressive behavior in male mice lacking endothelial nitric oxide synthase. *J Neurosci* 19:RC30.

Denenberg, V. H., Hudgens, G. A. & Zarrow, M. X. (1964). Mice reared with rats: Modification of behavior by early experience with another species. *Science* 143, 380–381.

Denenberg, V. H., Schell, S. F., Karas, G. G. & Haltmeyer, G. C. (1966). Comparison of background stimulation and handling as forms of infantile stimulation. *Psychol Rep* 19, 943–948.

Dennis, R. L. & Cheng, H. W. (2011). The dopaminergic system and aggression in laying hens. *Poult Sci* 90, 2440–2448.

Devlin, A. M., Brain, U., Austin, J. & Oberlander, T. F. (2010). Prenatal exposure to maternal depressed mood and the MTHFR C677T variant affect SLC6A4 methylation in infants at birth. *PLoS ONE* 5, e12201.

Diamond, J., Coughlin, M., Macintyre, L., Holmes, M. & Visheau, B. (1987). Evidence that endogenous beta nerve growth factor is responsible for the collateral sprouting, but not the regeneration, of nociceptive axons in adult rats. *Proc Natl Acad Sci U S A* 84, 6596–6600.

Diederich, N. J., Moore, C. G., Leurgans, S. E., Chmura, T. A. & Goetz, C. G. (2003). Parkinson disease with old-age onset: a comparative study with subjects with middle-age onset. *Arch Neurol* 60, 529–533.

Dierick, H. A. & Greenspan, R. J. (2006). Molecular analysis of flies selected for aggressive behavior. *Nat Genet* 38, 1023–1031.

DiMascio, A. (1973). The effects of benzodiazepines on aggression: reduced or increased? *Psychopharmacologia* 30, 95–102.

Donaldson, Z. R. & Young, L. J. (2008). Oxytocin, vasopressin, and the neurogenetics of sociality. *Science* 322, 900–904.

Dong, Q., Salva, A., Sottas, C. M., Niu, E., Holmes, M. & Hardy, M. P. (2004). Rapid glucocorticoid mediation of suppressed testosterone biosynthesis in male mice subjected to immobilization stress. *J Androl* 25, 973–981.

Dougherty, D. D., Shin, L. M., Alpert, N. M., Pitman, R. K., Orr, S. P., Lasko, M., Macklin, M. L., Fischman, A. J. & Rauch, S. L. (1999). Anger in healthy men: a PET study using script-driven imagery. *Biol Psychiatry* 46, 466–472.

Ehret, M., Cash, C. D., Hamon, M. & Maitre, M. (1989). Formal demonstration of the phosphorylation of rat brain tryptophan hydroxylase by Ca2+/calmodulin-dependent protein kinase. *J Neurochem* 52, 1886–1891.

Elman, I., Goldstein, D. S., Adler, C. M., Shoaf, S. E. & Breier, A. (2001). Inverse relationship between plasma epinephrine and testosterone levels during acute glucoprivation in healthy men. *Life Sci* 68, 1889–1898.

Engelmann, M., Landgraf, R. & Wotjak, C. T. (2004). The hypothalamic-neurohypophysial system regulates the hypothalamic-pituitary-adrenal axis under stress: an old concept revisited. *Front Neuroendocrinol* 25, 132–149.

Ferguson, J. N., Aldag, J. M., Insel, T. R. & Young, L. J. (2001). Oxytocin in the medial amygdala is essential for social recognition in the mouse. *J Neurosci* 21, 8278–8285.

Ferrari, P. F., van Erp, A. M., Tornatzky, W. & Miczek, K. A. (2003). Accumbal dopamine and serotonin in anticipation of the next aggressive episode in rats. *Eur J Neurosci* 17, 371–378.

Ferris, C. F. (2000). Adolescent stress and neural plasticity in hamsters: a vasopressin-serotonin model of inappropriate aggressive behaviour. *Exp Physiol* 85, 85S–90S.

Ferris, C. F. & Potegal, M. (1988). Vasopressin receptor blockade in the anterior hypothalamus suppresses aggression in hamsters. *Physiol Behav* 44, 235–239.

Ferris, C. F., Stolberg, T. & Delville, Y. (1999). Serotonin regulation of aggressive behavior in male golden hamsters (Mesocricetus auratus). *Behav Neurosci* 113, 804–815.

Fiore, M., Amendola, T., Triaca, V., Tirassa, P., Alleva, E. & Aloe, L. (2003). Agonistic encounters in aged male mouse potentiate the expression of endogenous brain NGF and BDNF: possible implication for brain progenitor cells' activation. *Eur J Neurosci* 17, 1455–1464.

Fish, E. W., Faccidomo, S. & Miczek, K. A. (1999). Aggression heightened by alcohol or social instigation in mice: reduction by the 5-HT(1B) receptor agonist CP-94,253. *Psychopharmacology (Berl)* 146, 391–399.

Fitzgerald, P. (1999). Long-acting antipsychotic medication, restraint and treatment in the management of acute psychosis. *Aust N Z J Psychiatry* 33, 660–666.

Florvall, L., Ask, A. L., Ogren, S. O. & Ross, S. B. (1978). Selective monoamine oxidase inhibitors. 1. Compounds related to 4-aminophenethylamine. *J Med Chem* 21, 56–63.

Foley, D. L., Pickles, A., Maes, H. M., Silberg, J. L. & Eaves, L. J. (2004). Course and shortterm outcomes of separation anxiety disorder in a community sample of twins. *J Am Acad Child Adolesc Psychiatry* 43, 1107–1114.

Fone, K. C. & Porkess, M. V. (2008). Behavioural and neurochemical effects of post-weaning social isolation in rodents-relevance to developmental neuropsychiatric disorders. *Neurosci Biobehav Rev* 32, 1087–1102.

Fraga, M. F., Agrelo, R. & Esteller, M. (2007). Cross-talk between aging and cancer: the epigenetic language. *Ann N Y Acad Sci* 1100, 60–74.

Friedel, R. O. (2004). Dopamine dysfunction in borderline personality disorder: a hypothesis. *Neuropsychopharmacology* 29, 1029–1039.

Friso, S., Choi, S. W., Girelli, D., Mason, J. B., Dolnikowski, G. G., Bagley, P. J., Olivieri, O., Jacques, P. F., Rosenberg, I. H., Corrocher, R. & Selhub, J. (2002). A common mutation in the 5,10-methylenetetrahydrofolate reductase gene affects genomic DNA methylation through an interaction with folate status. *Proc Natl Acad Sci U S A* 99, 5606–5611.

Frosst, P., Blom, H. J., Milos, R., Goyette, P., Sheppard, C. A., Matthews, R. G., Boers, G. J., den Heijer, M., Kluijtmans, L. A., van den Heuvel, L. P. & . (1995). A candidate genetic risk factor for vascular disease: a common mutation in methylenetetrahydrofolate reductase. *Nat Genet* 10, 111–113.

Fuchs, S. A., Edinger, H. M. & Siegel, A. (1985a). The organization of the hypothalamic pathways mediating affective defense behavior in the cat. *Brain Res* 330, 77–92.

Fuchs, S. A., Edinger, H. M. & Siegel, A. (1985b). The role of the anterior hypothalamus in affective defense behavior elicited from the ventromedial hypothalamus of the cat. *Brain Res* 330, 93–107.

Fuke, C., Shimabukuro, M., Petronis, A., Sugimoto, J., Oda, T., Miura, K., Miyazaki, T., Ogura, C., Okazaki, Y. & Jinno, Y. (2004). Age related changes in 5-methylcytosine content in human peripheral leukocytes and placentas: an HPLC-based study. *Ann Hum Genet* 68, 196–204.

Fuller, J. L. & Hahn, M. E. (1976). Issues in the genetics of social behavior. *Behav Genet* 6, 391–406.

Gao, B., Fritschy, J. M., Benke, D. & Mohler, H. (1993). Neuron-specific expression of GABAA-receptor subtypes: differential association of the alpha 1- and alpha 3-subunits with serotonergic and GABAergic neurons. *Neuroscience* 54, 881–892.

Giammanco, M., Tabacchi, G., Giammanco, S., Di Majo, D. & La Guardia, M. (2005). Testosterone and aggressiveness. *Med Sci Monit* 11, RA136–RA145.

Giegling, I., Rujescu, D., Mandelli, L., Schneider, B., Hartmann, A. M., Schnabel, A., Maurer, K., De Ronchi, D., Moller, H. J. & Serretti, A. (2007). Tachykinin receptor 1 variants associated with aggression in suicidal behavior. *Am J Med Genet B Neuropsychiatr Genet* 144B, 757–761.

Glazer, W. M. & Dickson, R. A. (1998). Clozapine reduces violence and persistent aggression in schizophrenia. *J Clin Psychiatry* 59 Suppl 3, 8–14.

Gobrogge, K. L., Liu, Y., Young, L. J. & Wang, Z. (2009). Anterior hypothalamic vasopressin regulates pair-bonding and drug-induced aggression in a monogamous rodent. *Proc Natl Acad Sci U S A* 106, 19144–19149.

Gogos, J. A., Morgan, M., Luine, V., Santha, M., Ogawa, S., Pfaff, D. & Karayiorgou, M. (1998). Catechol-O-methyltransferase-deficient mice exhibit sexually dimorphic changes in catecholamine levels and behavior. *Proc Natl Acad Sci U S A* 95, 9991–9996.

Goridis, C. & Brunet, J. F. (1992). NCAM: structural diversity, function and regulation of expression. *Semin Cell Biol* 3, 189–197.

Gorman, J. M., Kent, J. M., Sullivan, G. M. & Coplan, J. D. (2000). Neuroanatomical hypothesis of panic disorder, revised. *Am J Psychiatry* 157, 493–505.

Gozes, Y., Moskowitz, M. A., Strom, T. B. & Gozes, I. (1982). Conditioned media from activated lymphocytes maintain sympathetic neurons in culture. *Brain Res* 282, 93–97.

Graff, J. & Mansuy, I. M. (2009). Epigenetic dysregulation in cognitive disorders. *Eur J Neurosci* 30, 1–8.

Grafman, J., Schwab, K., Warden, D., Pridgen, A., Brown, H. R. & Salazar, A. M. (1996). Frontal lobe injuries, violence, and aggression: a report of the Vietnam Head Injury Study. *Neurology* 46, 1231–1238.

Greenberg, B. D., Tolliver, T. J., Huang, S. J., Li, Q., Bengel, D. & Murphy, D. L. (1999). Genetic variation in the serotonin transporter promoter region affects serotonin uptake in human blood platelets. *Am J Med Genet* 88, 83–87.

Gregg, T. R. & Siegel, A. (2001). Brain structures and neurotransmitters regulating aggression in cats: implications for human aggression. *Prog Neuropsychopharmacol Biol Psychiatry* 25, 91–140.

9 Genetic and Epigenetic Determinants of Aggression 267

Gregg, T. R. & Siegel, A. (2003). Differential effects of NK1 receptors in the midbrain periaqueductal gray upon defensive rage and predatory attack in the cat. *Brain Res* 994, 55–66.

Griffing, B. (1976). Selection in reference to biological groups. VI. Use of extreme forms of nonrandom groups to increase selection efficiency. *Genetics* 82, 723–731.

Grunstein, M. (1997). Histone acetylation in chromatin structure and transcription. *Nature* 389, 349–352.

Gualtieri, C. T. & Schroeder, S. R. (1990). Pharmacotherapy for self-injurious behavior: preliminary tests of the D1 hypothesis. *Prog Neuropsychopharmacol Biol Psychiatry* 14, S81–S107.

Guillot, P. V. & Chapouthier, G. (1998). Intermale aggression, GAD activity in the olfactory bulbs and Y chromosome effect in seven inbred mouse strains. *Behav Brain Res* 90, 203–206.

Hahn, M. E. & Schanz, N. (1996). Issues in the genetics of social behavior: revisited. *Behav Genet* 26, 463–470.

Hake, S. B. & Allis, C. D. (2006). Histone H3 variants and their potential role in indexing mammalian genomes: the "H3 barcode hypothesis". *Proc Natl Acad Sci U S A* 103, 6428–6435.

Halasz, J., Liposits, Z., Kruk, M. R. & Haller, J. (2002). Neural background of glucocorticoid dysfunction-induced abnormal aggression in rats: involvement of fear- and stress-related structures. *Eur J Neurosci* 15, 561–569.

Halasz, J., Toth, M., Kallo, I., Liposits, Z. & Haller, J. (2006). The activation of prefrontal cortical neurons in aggression--a double labeling study. *Behav Brain Res* 175, 166–175.

Halasz, J., Toth, M., Mikics, E., Hrabovszky, E., Barsy, B., Barsvari, B. & Haller, J. (2008). The effect of neurokinin1 receptor blockade on territorial aggression and in a model of violent aggression. *Biol Psychiatry* 63, 271–278.

Halasz, J., Zelena, D., Toth, M., Tulogdi, A., Mikics, E. & Haller, J. (2009). Substance P neurotransmission and violent aggression: the role of tachykinin NK(1) receptors in the hypothalamic attack area. *Eur J Pharmacol* 611, 35–43.

Haller, J., Halasz, J., Mikics, E. & Kruk, M. R. (2004). Chronic glucocorticoid deficiency-induced abnormal aggression, autonomic hypoarousal, and social deficit in rats. *J Neuroendocrinol* 16, 550–557.

Haller, J. & Kruk, M. R. (2006). Normal and abnormal aggression: human disorders and novel laboratory models. *Neurosci Biobehav Rev* 30, 292–303.

Haller, J., Makara, G. B. & Kruk, M. R. (1998). Catecholaminergic involvement in the control of aggression: hormones, the peripheral sympathetic, and central noradrenergic systems. *Neurosci Biobehav Rev* 22, 85–97.

Haller, J., Mikics, E., Halasz, J. & Toth, M. (2005a). Mechanisms differentiating normal from abnormal aggression: glucocorticoids and serotonin. *Eur J Pharmacol* 526, 89–100.

Haller, J., Toth, M. & Halasz, J. (2005b). The activation of raphe serotonergic neurons in normal and hypoarousal-driven aggression: a double labeling study in rats. *Behav Brain Res* 161, 88–94.

Haller, J., Toth, M., Halasz, J. & de Boer, S. F. (2006). Patterns of violent aggression-induced brain c-fos expression in male mice selected for aggressiveness. *Physiol Behav* 88, 173–182.

Haller, J., van de Schraaf J. & Kruk, M. R. (2001). Deviant forms of aggression in glucocorticoid hyporeactive rats: a model for 'pathological' aggression? *J Neuroendocrinol* 13, 102–107.

Hammock, E. A. & Young, L. J. (2005). Microsatellite instability generates diversity in brain and sociobehavioral traits. *Science* 308, 1630–1634.

Handa, R. J., Kerr, J. E., DonCarlos, L. L., McGivern, R. F. & Hejna, G. (1996). Hormonal regulation of androgen receptor messenger RNA in the medial preoptic area of the male rat. *Brain Res Mol Brain Res* 39, 57–67.

Haney, M., Noda, K., Kream, R. & Miczek, K. A. (1990). Regional serotonin and dopamine activity: Sensitivity to amphetamine and aggressive behavior in mice. *Aggress Behav* 16, 259–270.

Harandi, M., Aguera, M., Gamrani, H., Didier, M., Maitre, M., Calas, A. & Belin, M. F. (1987). gamma-Aminobutyric acid and 5-hydroxytryptamine interrelationship in the rat nucleus raphe dorsalis: combination of radioautographic and immunocytochemical techniques at light and electron microscopy levels. *Neuroscience* 21, 237–251.

Harbuz, M. S. & Lightman, S. L. (1989). Responses of hypothalamic and pituitary mRNA to physical and psychological stress in the rat. *J Endocrinol* 122, 705–711.

Hardy, M. P., Gao, H. B., Dong, Q., Ge, R., Wang, Q., Chai, W. R., Feng, X. & Sottas, C. (2005). Stress hormone and male reproductive function. *Cell Tissue Res* 322, 147–153.

Hardy, M. P., Sottas, C. M., Ge, R., McKittrick, C. R., Tamashiro, K. L., McEwen, B. S., Haider, S. G., Markham, C. M., Blanchard, R. J., Blanchard, D. C. & Sakai, R. R. (2002). Trends of reproductive hormones in male rats during psychosocial stress: role of glucocorticoid metabolism in behavioral dominance. *Biol Reprod* 67, 1750–1755.

Harlow, H. F., Dodsworth, R. O. & Harlow, M. K. (1965). Total social isolation in monkeys. *Proc Natl Acad Sci U S A* 54, 90–97.

Harlow, H. F., Harlow, M. K. & Suomi, S. J. (1971). From thought to therapy: lessons from a primate laboratory. *Am Sci* 59, 538–549.

Harlow, H. F. & Suomi, S. J. (1971). Social recovery by isolation-reared monkeys. *Proc Natl Acad Sci U S A* 68, 1534–1538.

Harmon, A. C., Huhman, K. L., Moore, T. O. & Albers, H. E. (2002). Oxytocin inhibits aggression in female Syrian hamsters. *J Neuroendocrinol* 14, 963–969.

Harrell, L. E. & Balagura, S. (1975). Septal rage: mitigation by pre-surgical treatment with p-chlorophenylalamine. *Pharmacol Biochem Behav* 3, 157–159.

Hasselager, E., Rolinski, Z. & Randrup, A. (1972). Specific antagonism by dopamine inhibitors of items of amphetamine induced aggressive behaviour. *Psychopharmacologia* 24, 485–495.

Haug, M., Ouss-Schlegel, M. L., Spetz, J. F., Benton, D., Brain, P. F., Mandel, P., Ciesielski, L. & Simler, S. (1987). An attempt to correlate attack on lactating females and brain GABA levels in C57 and C3H strains and their reciprocal hybrids. 4 edn, pp. 83–94.

Heberlein, A., Muschler, M., Frieling, H., Behr, M., Eberlein, C., Wilhelm, J., Groschl, M., Kornhuber, J., Bleich, S. & Hillemacher, T. (2011). Epigenetic down regulation of nerve growth factor during alcohol withdrawal. *Addict Biol*. doi: 10.1111/j.1369–1600.2010.00307.x.

Heim, C. & Nemeroff, C. B. (2001). The role of childhood trauma in the neurobiology of mood and anxiety disorders: preclinical and clinical studies. *Biol Psychiatry* 49, 1023–1039.

Hendry, I. A. & Iversen, L. L. (1973). Reduction in the concentration of nerve growth factor in mice after sialectomy and castration. *Nature* 243, 550–554.

Henry, J. P., Stephens, P. M., Axelrod, J. & Mueller, R. A. (1971). Effect of psychosocial stimulation on the enzymes involved in the biosynthesis and metabolism of noradrenaline and adrenaline. *Psychosom Med* 33, 227–237.

Hess, C., Reif, A., Strobel, A., Boreatti-Hummer, A., Heine, M., Lesch, K. P. & Jacob, C. P. (2009). A functional dopamine-beta-hydroxylase gene promoter polymorphism is associated with impulsive personality styles, but not with affective disorders. *J Neural Transm* 116, 121–130.

Hilakivi-Clarke, L. (1994). Overexpression of transforming growth factor alpha in transgenic mice alters nonreproductive, sex-related behavioral differences: interaction with gonadal hormones. *Behav Neurosci* 108, 410–417.

Hilakivi-Clarke, L. A., Arora, P. K., Sabol, M. B., Clarke, R., Dickson, R. B. & Lippman, M. E. (1992). Alterations in behavior, steroid hormones and natural killer cell activity in male transgenic TGF alpha mice. *Brain Res* 588, 97–103.

Hilakivi-Clarke, L. A. & Goldberg, R. (1993). Effects of tryptophan and serotonin uptake inhibitors on behavior in male transgenic transforming growth factor alpha mice. *Eur J Pharmacol* 237, 101–108.

Hinde, R. A., Spencer-Booth, Y. & Bruce, M. (1966). Effects of 6-day maternal deprivation on rhesus monkey infants. *Nature* 210, 1021–1023.

Ho, J. M., Murray, J. H., Demas, G. E. & Goodson, J. L. (2010). Vasopressin cell groups exhibit strongly divergent responses to copulation and male-male interactions in mice. *Horm Behav* 58, 368–377.

Holliday, R. (1989). DNA methylation and epigenetic mechanisms. *Cell Biophys* 15, 15–20.

Holmes, A., Murphy, D. L. & Crawley, J. N. (2002). Reduced aggression in mice lacking the serotonin transporter. *Psychopharmacology (Berl)* 161, 160–167.

Holschneider, D. P., Scremin, O. U., Huynh, L., Chen, K., Seif, I. & Shih, J. C. (2000). Regional cerebral cortical activation in monoamine oxidase A-deficient mice: differential effects of chronic versus acute elevations in serotonin and norepinephrine. *Neuroscience* 101, 869–877.

Hood, K. E. & Quigley, K. S. (2008). Exploratory behavior in mice selectively bred for developmental differences in aggressive behavior. *Dev Psychobiol* 50, 32–47.

Hrabovszky, E., Halasz, J., Meelis, W., Kruk, M. R., Liposits, Z. & Haller, J. (2005). Neurochemical characterization of hypothalamic neurons involved in attack behavior: glutamatergic dominance and co-expression of thyrotropin-releasing hormone in a subset of glutamatergic neurons. *Neuroscience* 133, 657–666.

Huang, E. J. & Reichardt, L. F. (2001). Neurotrophins: roles in neuronal development and function. *Annu Rev Neurosci* 24, 677–736.

Hudgens, G. A., Denenberg, V. H. & Zarrow, M. X. (1968). Mice reared with rats: effects of preweaning and postweaning social interactions upon adult behaviour. *Behaviour* 30, 259–274.

Insel, T. R., Winslow, J. T., Wang, Z. & Young, L. J. (1998). Oxytocin, vasopressin, and the neuroendocrine basis of pair bond formation. *Adv Exp Med Biol* 449, 215–224.

Jenuwein, T. & Allis, C. D. (2001). Translating the histone code. *Science* 293, 1074–1080.

Jiang, Y., Jiang, J., Xiong, J., Cao, J., Li, N., Li, G. & Wang, S. (2008). Homocysteine-induced extracellular superoxide dismutase and its epigenetic mechanisms in monocytes. *J Exp Biol* 211, 911–920.

Jirtle, R. L. & Skinner, M. K. (2007). Environmental epigenomics and disease susceptibility. *Nat Rev Genet* 8, 253–262.

Jonas, J. M., Coleman, B. S., Sheridan, A. Q. & Kalinske, R. W. (1992). Comparative clinical profiles of triazolam versus other shorter-acting hypnotics. *J Clin Psychiatry* 53 Suppl, 19–31.

Jonsson, E. G., von Gertten, C., Gustavsson, J. P., Yuan, Q. P., Lindblad-Toh, K., Forslund, K., Rylander, G., Mattila-Evenden, M., Asberg, M. & Schalling, M. (2001). Androgen receptor trinucleotide repeat polymorphism and personality traits. *Psychiatr Genet* 11, 19–23.

Katsouni, E., Sakkas, P., Zarros, A., Skandali, N. & Liapi, C. (2009). The involvement of substance P in the induction of aggressive behavior. *Peptides* 30, 1586–1591.

Kennedy, J. S., Bymaster, F. P., Schuh, L., Calligaro, D. O., Nomikos, G., Felder, C. C., Bernauer, M., Kinon, B. J., Baker, R. W., Hay, D., Roth, H. J., Dossenbach, M., Kaiser, C., Beasley, C. M., Holcombe, J. H., Effron, M. B. & Breier, A. (2001). A current review of olanzapine's safety in the geriatric patient: from pre-clinical pharmacology to clinical data. *Int J Geriatr Psychiatry* 16 Suppl 1, S33–S61.

Kheramin, S., Body, S., Herrera, F. M., Bradshaw, C. M., Szabadi, E., Deakin, J. F. & Anderson, I. M. (2005). The effect of orbital prefrontal cortex lesions on performance on a progressive ratio schedule: implications for models of inter-temporal choice. *Behav Brain Res* 156, 145–152.

Kim, J. J. & Haller, J. (2007). Glucocorticoid hyper- and hypofunction: stress effects on cognition and aggression. *Ann N Y Acad Sci* 1113, 291–303.

Kim-Cohen, J., Caspi, A., Taylor, A., Williams, B., Newcombe, R., Craig, I. W. & Moffitt, T. E. (2006). MAOA, maltreatment, and gene-environment interaction predicting children's mental health: new evidence and a meta-analysis. *Mol Psychiatry* 11, 903–913.

King, J. A. (1957). Relationships between early social experience and adult aggressive behavior in inbred mice. *J Genet Psychol* 90, 151–166.

King, J. A. & Gurney, N. L. (1954). Effect of early social experience on adult aggressive behavior in C57BL/10 mice. *J Comp Physiol Psychol* 47, 326–330.

Kishimoto, M., Fujiki, R., Takezawa, S., Sasaki, Y., Nakamura, T., Yamaoka, K., Kitagawa, H. & Kato, S. (2006). Nuclear receptor mediated gene regulation through chromatin remodeling and histone modifications. *Endocr J* 53, 157–172.

Kjaer, A., Knigge, U., Bach, F. W. & Warberg, J. (1993). Impaired histamine- and stress-induced secretion of ACTH and beta-endorphin in vasopressin-deficient Brattleboro rats. *Neuroendocrinology* 57, 1035–1041.

Klose, R. J. & Bird, A. P. (2006). Genomic DNA methylation: the mark and its mediators. *Trends Biochem Sci* 31, 89–97.

Korte, S. M., Meijer, O. C., de Kloet, E. R., Buwalda, B., Keijser, J., Sluyter, F., van Oortmerssen, G. & Bohus, B. (1996). Enhanced 5-HT1A receptor expression in forebrain regions of aggressive house mice. *Brain Res* 736, 338–343.

Kraemer, G. W. (1992). Attachment and psychopathology. *Behav Brain Sci* 15, 512–541.

Kraemer, G. W. & Clarke, A. S. (1996). Social attachment, brain function, and aggression. *Ann N Y Acad Sci* 794, 121–135.

Kriegsfeld, L. J., Dawson, T. M., Dawson, V. L., Nelson, R. J. & Snyder, S. H. (1997). Aggressive behavior in male mice lacking the gene for neuronal nitric oxide synthase requires testosterone. *Brain Res* 769, 66–70.

Krsiak, M., Sulcova, A., Tomasikova, Z., Dlohozkova, N., Kosar, E. & Masek, K. (1981). Drug effects on attack defense and escape in mice. *Pharmacol Biochem Behav* 14 Suppl 1, 47–52.

Kulikova, M. A., Maluchenko, N. V., Timofeeva, M. A., Shlepzova, V. A., Schegolkova, J. V., Sysoeva, O. V., Ivanitsky, A. M. & Tonevitsky, A. G. (2008). Effect of functional catechol-O-methyltransferase Val158Met polymorphism on physical aggression. *Bull Exp Biol Med* 145, 62–64.

Kurian, J. R., Olesen, K. M. & Auger, A. P. (2010). Sex differences in epigenetic regulation of the estrogen receptor-alpha promoter within the developing preoptic area. *Endocrinology* 151, 2297–2305.

Lado-Abeal, J., Clapper, J. A. & Norman, R. L. (2001). Antagonism of central vasopressin receptors blocks hypoglycemic stress induced inhibition of luteinizing hormone release in male rhesus macaques. *J Neuroendocrinol* 13, 650–655.

Lakshmanan, J. (1986). Aggressive behavior in adult male mice elevates serum nerve growth factor levels. *Am J Physiol* 250, E386–E392.

Lakshmanan, J. (1987). Nerve growth factor levels in mouse serum: variations due to stress. *Neurochem Res* 12, 393–397.

Lammers, J. H., Kruk, M. R., Meelis, W. & van der Poel, A. M. (1988). Hypothalamic substrates for brain stimulation-induced attack, teeth-chattering and social grooming in the rat. *Brain Res* 449, 311–327.

Lang, U. E., Gallinat, J., Kuhn, S., Jockers-Scherubl, C. & Hellweg, R. (2002). Nerve growth factor and smoking cessation. *Am J Psychiatry* 159, 674–675.

Lapiz, M. D., Fulford, A., Muchimapura, S., Mason, R., Parker, T. & Marsden, C. A. (2001). Influence of postweaning social isolation in the rat on brain development, conditioned behaviour and neurotransmission. *Ross Fiziol Zh Im I M Sechenova* 87, 730–751.

Lema, S. C. (2006). Population divergence in plasticity of the AVT system and its association with aggressive behaviors in a Death Valley pupfish. *Horm Behav* 50, 183–193.

Lesch, K. P., Bengel, D., Heils, A., Sabol, S. Z., Greenberg, B. D., Petri, S., Benjamin, J., Muller, C. R., Hamer, D. H. & Murphy, D. L. (1996). Association of anxiety-related traits with a polymorphism in the serotonin transporter gene regulatory region. *Science* 274, 1527–1531.

Lesch, K. P., Meyer, J., Glatz, K., Flugge, G., Hinney, A., Hebebrand, J., Klauck, S. M., Poustka, A., Poustka, F., Bengel, D., Mossner, R., Riederer, P. & Heils, A. (1997). The 5-HT transporter gene-linked polymorphic region (5-HTTLPR) in evolutionary perspective: alternative biallelic variation in rhesus monkeys. Rapid communication. *J Neural Transm* 104, 1259–1266.

Lin, D., Boyle, M. P., Dollar, P., Lee, H., Lein, E. S., Perona, P. & Anderson, D. J. (2011). Functional identification of an aggression locus in the mouse hypothalamus. *Nature* 470, 221–226.

Lindholm, D., Heumann, R., Meyer, M. & Thoenen, H. (1987). Interleukin-1 regulates synthesis of nerve growth factor in non-neuronal cells of rat sciatic nerve. *Nature* 330, 658–659.

Linnoila, M., Virkkunen, M., Scheinin, M., Nuutila, A., Rimon, R. & Goodwin, F. K. (1983). Low cerebrospinal fluid 5-hydroxyindoleacetic acid concentration differentiates impulsive from nonimpulsive violent behavior. *Life Sci* 33, 2609–2614.

Lonstein, J. S. & Stern, J. M. (1998). Site and behavioral specificity of periaqueductal gray lesions on postpartum sexual, maternal, and aggressive behaviors in rats. *Brain Res* 804, 21–35.

9 Genetic and Epigenetic Determinants of Aggression

Louilot, A., Le Moal, M. & Simon, H. (1986). Differential reactivity of dopaminergic neurons in the nucleus accumbens in response to different behavioral situations. An in vivo voltammetric study in free moving rats. *Brain Res* 397, 395–400.

Lubin, F. D., Roth, T. L. & Sweatt, J. D. (2008). Epigenetic regulation of BDNF gene transcription in the consolidation of fear memory. *J Neurosci* 28, 10576–10586.

Luciano, D. & Lore, R. (1975). Aggression and social experience in domesticated rats. *J Comp Physiol Psychol* 88, 917–923.

Lyons, W. E., Mamounas, L. A., Ricaurte, G. A., Coppola, V., Reid, S. W., Bora, S. H., Wihler, C., Koliatsos, V. E. & Tessarollo, L. (1999). Brain-derived neurotrophic factor-deficient mice develop aggressiveness and hyperphagia in conjunction with brain serotonergic abnormalities. *Proc Natl Acad Sci U S A* 96, 15239–15244.

Maestripieri, D., De Simone, R., Aloe, L. & Alleva, E. (1990). Social status and nerve growth factor serum levels after agonistic encounters in mice. *Physiol Behav* 47, 161–164.

Manuck, S. B., Flory, J. D., Ferrell, R. E., Dent, K. M., Mann, J. J. & Muldoon, M. F. (1999). Aggression and anger-related traits associated with a polymorphism of the tryptophan hydroxylase gene. *Biol Psychiatry* 45, 603–614.

Martin, M., Ledent, C., Parmentier, M., Maldonado, R. & Valverde, O. (2002). Involvement of CB1 cannabinoid receptors in emotional behaviour. *Psychopharmacology (Berl)* 159, 379–387.

Martinowich, K., Hattori, D., Wu, H., Fouse, S., He, F., Hu, Y., Fan, G. & Sun, Y. E. (2003). DNA methylation-related chromatin remodeling in activity-dependent BDNF gene regulation. *Science* 302, 890–893.

Masi, G. (2004). Pharmacotherapy of pervasive developmental disorders in children and adolescents. *CNS Drugs* 18, 1031–1052.

Mason, W. A. & Sponholz, R. R. (1963). Behavior of rhesus monkeys raised in isolation. *J Psychiatr Res* 81, 299–306.

Maxson, S. C. (1999). Genetic influences on aggressive behavior. In *Genetic influences on neural and behavioral functions*, pp. 405–416. Edited by W. H. Berrettini, T. H. Joh, D. W. Pfaff & S. C. Maxson. Boca Raton, FL: CRC Press.

McCarthy, M. M., Auger, A. P., Bale, T. L., De Vries, G. J., Dunn, G. A., Forger, N. G., Murray, E. K., Nugent, B. M., Schwarz, J. M. & Wilson, M. E. (2009). The epigenetics of sex differences in the brain. *J Neurosci* 29, 12815–12823.

McCaul, K. D., Gladue, B. A. & Joppa, M. (1992). Winning, losing, mood, and testosterone. *Horm Behav* 26, 486–504.

McClelland, S., Korosi, A., Cope, J., Ivy, A. & Baram, T. Z. (2011). Emerging roles of epigenetic mechanisms in the enduring effects of early-life stress and experience on learning and memory. *Neurobiol Learn Mem* 96, 79–88.

McDougle, C. J., Holmes, J. P., Carlson, D. C., Pelton, G. H., Cohen, D. J. & Price, L. H. (1998). A double-blind, placebo-controlled study of risperidone in adults with autistic disorder and other pervasive developmental disorders. *Arch Gen Psychiatry* 55, 633–641.

McEwen, B. S. (2003). Early life influences on life-long patterns of behavior and health. *Ment Retard Dev Disabil Res Rev* 9, 149–154.

McKinney, W. T. (1974). Primate social isolation. Psychiatric implications. *Arch Gen Psychiatry* 31, 422–426.

Meaney, M. J. (2001). Maternal care, gene expression, and the transmission of individual differences in stress reactivity across generations. *Annu Rev Neurosci* 24, 1161–1192.

Meaney, M. J. (2007). Environmental programming of phenotypic diversity in female reproductive strategies. *Adv Genet* 59, 173–215.

Meaney, M. J. (2010). Epigenetics and the biological definition of gene x environment interactions. *Child Dev* 81, 41–79.

Meaney, M. J. & Stewart, J. (1979). Environmental factors influencing the affiliative behavior of male and female rats (Rattus norvegicus). *Anim Learn Behav* 7, 397–405.

Meaney, M. J. & Szyf, M. (2005). Maternal care as a model for experience-dependent chromatin plasticity? *Trends Neurosci* 28, 456–463.

Mejia, J. M., Ervin, F. R., Baker, G. B. & Palmour, R. M. (2002). Monoamine oxidase inhibition during brain development induces pathological aggressive behavior in mice. *Biol Psychiatry* 52, 811–821.

Meloy, J. R. (2006). Empirical basis and forensic application of affective and predatory violence. *Aust N Z J Psychiatry* 40, 539–547.

Mendel, G. (1866). Versuche über Pflanzen-Hybriden. *Verhandlungen des naturforschenden Vereines in Brünn* IV, 3–47.

Meyer-Lindenberg, A., Buckholtz, J. W., Kolachana, B., Hariri, R., Pezawas, L., Blasi, G., Wabnitz, A., Honea, R., Verchinski, B., Callicott, J. H., Egan, M., Mattay, V. & Weinberger, D. R. (2006). Neural mechanisms of genetic risk for impulsivity and violence in humans. *Proc Natl Acad Sci U S A* 103, 6269–6274.

Miczek, K. A. (1974). Intraspecies aggression in rats: effects of d-amphetamine and chlordiazepoxide. *Psychopharmacologia* 39, 275–301.

Miczek, K. A., de Almeida, R. M., Kravitz, E. A., Rissman, E. F., de Boer, S. F. & Raine, A. (2007). Neurobiology of escalated aggression and violence. *J Neurosci* 27, 11803–11806.

Miczek, K. A., Faccidomo, S., de Almeida, R. M., Bannai, M., Fish, E. W. & Debold, J. F. (2004). Escalated aggressive behavior: new pharmacotherapeutic approaches and opportunities. *Ann N Y Acad Sci* 1036, 336–355.

Miczek, K. A. & Fish, E. W. (2005). Monoamines, GABA, Glutamate and Aggression. In *Biology of Aggression*, pp. 114–149. Edited by R. J. Nelson. Oxford: Oxford University Press.

Miczek, K. A., Fish, E. W., De Bold, J. F. & de Almeida, R. M. (2002). Social and neural determinants of aggressive behavior: pharmacotherapeutic targets at serotonin, dopamine and gamma-aminobutyric acid systems. *Psychopharmacology (Berl)* 163, 434–458.

Miczek, K. A., Hussain, S. & Faccidomo, S. (1998). Alcohol-heightened aggression in mice: attenuation by 5-HT1A receptor agonists. *Psychopharmacology (Berl)* 139, 160–168.

Mikics, E., Kruk, M. R. & Haller, J. (2004). Genomic and non-genomic effects of glucocorticoids on aggressive behavior in male rats. *Psychoneuroendocrinology* 29, 618–635.

Miller, M. A., Urban, J. H. & Dorsa, D. M. (1989a). Steroid dependency of vasopressin neurons in the bed nucleus of the stria terminalis by in situ hybridization. *Endocrinology* 125, 2335–2340.

Miller, M. A., Vician, L., Clifton, D. K. & Dorsa, D. M. (1989b). Sex differences in vasopressin neurons in the bed nucleus of the stria terminalis by in situ hybridization. *Peptides* 10, 615–619.

Molina, V., Ciesielski, L., Gobaille, S. & Mandel, P. (1986). Effects of the potentiation of the GABAergic neurotransmission in the olfactory bulbs on mouse-killing behavior. *Pharmacol Biochem Behav* 24, 657–664.

Moore, A. J., Brodie, D. E. & Wolf, J. B. (1997). Interacting phenotypes and the evolutionary process: I. Direct and indirect genetic effects of social interactions. *Evolution* 51, 1352–1362.

Morgan, D., Grant, K. A., Prioleau, O. A., Nader, S. H., Kaplan, J. R. & Nader, M. A. (2000). Predictors of social status in cynomolgus monkeys (Macaca fascicularis) after group formation. *Am J Primatol* 52, 115–131.

Mos, J. & van Valkenburg, C. F. (1979). Specific effect on social stress and aggression on regional dopamine metabolism in rat brain. *Neurosci Lett* 15, 325–327.

Murani, E., Ponsuksili, S., D'Eath, R. B., Turner, S. P., Kurt, E., Evans, G., Tholking, L., Klont, R., Foury, A., Mormede, P. & Wimmers, K. (2010). Association of HPA axis-related genetic variation with stress reactivity and aggressive behaviour in pigs. *BMC Genet* 11:74.

Murayama, A., Sakura, K., Nakama, M., Yasuzawa-Tanaka, K., Fujita, E., Tateishi, Y., Wang, Y., Ushijima, T., Baba, T., Shibuya, K., Shibuya, A., Kawabe, Y. & Yanagisawa, J. (2006). A specific CpG site demethylation in the human interleukin 2 gene promoter is an epigenetic memory. *EMBO J* 25, 1081–1092.

Murgatroyd, C., Patchev, A. V., Wu, Y., Micale, V., Bockmuhl, Y., Fischer, D., Holsboer, F., Wotjak, C. T., Almeida, O. F. & Spengler, D. (2009). Dynamic DNA methylation programs persistent adverse effects of early-life stress. *Nat Neurosci* 12, 1559–1566.

Murgatroyd, C. & Spengler, D. (2011). Epigenetic programming of the HPA axis: Early life decides. *Stress* 14, 581–589.

Murgatroyd, C., Wu, Y., Bockmuhl, Y. & Spengler, D. (2010). Genes learn from stress: How infantile trauma programs us for depression. *Epigenetics* 5, 194–199.

Murray, E. K., Hien, A., De Vries, G. J. & Forger, N. G. (2009). Epigenetic control of sexual differentiation of the bed nucleus of the stria terminalis. *Endocrinology* 150, 4241–4247.

Murray, E. K., Varnum, M. M., Fernandez, J. L., De Vries, G. J. & Forger, N. G. (2011). Effects of neonatal treatment with valproic acid on vasopressin immunoreactivity and olfactory behaviour in mice. *J Neuroendocrinol* 23, 906–914.

Myerscough, R. & Taylor, S. (1985). The effects of marijuana on human physical aggression. *J Pers Soc Psychol* 49, 1541–1546.

Natarajan, D., de Vries, H., Saaltink, D. J., de Boer, S. F. & Koolhaas, J. M. (2009). Delineation of violence from functional aggression in mice: an ethological approach. *Behav Genet* 39, 73–90.

Nelson, R. J. & Chiavegatto, S. (2001). Molecular basis of aggression. *Trends Neurosci* 24, 713–719.

Nelson, R. J., Demas, G. E., Huang, P. L., Fishman, M. C., Dawson, V. L., Dawson, T. M. & Snyder, S. H. (1995). Behavioural abnormalities in male mice lacking neuronal nitric oxide synthase. *Nature* 378, 383–386.

Nelson, R. J., Kriegsfeld, L. J., Dawson, V. L. & Dawson, T. M. (1997). Effects of nitric oxide on neuroendocrine function and behavior. *Front Neuroendocrinol* 18, 463–491.

Neumann, I. D., Veenema, A. H. & Beiderbeck, D. I. (2010). Aggression and anxiety: social context and neurobiological links. *Front Behav Neurosci* 4:12.

Newman, S. W. (1999). The medial extended amygdala in male reproductive behavior. A node in the mammalian social behavior network. *Ann N Y Acad Sci* 877, 242–257.

NIH (2004). NIH State-of-the-Science Conference Statement on preventing violence and related health-risking social behaviors in adolescents. *NIH Consens State Sci Statements* 21, 1–34.

Noguchi, S., Inukai, T., Kuno, T. & Tanaka, C. (1992). The suppression of olfactory bulbectomy-induced muricide by antidepressants and antihistamines via histamine H1 receptor blocking. *Physiol Behav* 51, 1123–1127.

Oehler, J., Jahkel, M. & Schmidt, J. (1985). The influence of chronic treatment with psychotropic drugs on behavioral changes by social isolation. *Pol J Pharmacol Pharm* 37, 841–849.

Ogawa, S., Chan, J., Chester, A. E., Gustafsson, J. A., Korach, K. S. & Pfaff, D. W. (1999). Survival of reproductive behaviors in estrogen receptor beta gene-deficient (betaERKO) male and female mice. *Proc Natl Acad Sci U S A* 96, 12887–12892.

Ogawa, S., Chester, A. E., Hewitt, S. C., Walker, V. R., Gustafsson, J. A., Smithies, O., Korach, K. S. & Pfaff, D. W. (2000). Abolition of male sexual behaviors in mice lacking estrogen receptors alpha and beta (alpha beta ERKO). *Proc Natl Acad Sci U S A* 97, 14737–14741.

Ogawa, S., Eng, V., Taylor, J., Lubahn, D. B., Korach, K. S. & Pfaff, D. W. (1998). Roles of estrogen receptor-alpha gene expression in reproduction-related behaviors in female mice. *Endocrinology* 139, 5070–5081.

Ogawa, S., Lubahn, D. B., Korach, K. S. & Pfaff, D. W. (1997). Behavioral effects of estrogen receptor gene disruption in male mice. *Proc Natl Acad Sci U S A* 94, 1476–1481.

Olivier, B., Mos, J., van Oorschot, R. & Hen, R. (1995). Serotonin receptors and animal models of aggressive behavior. *Pharmacopsychiatry* 28 Suppl 2, 80–90.

Olsen, K. L. (1983). Genetic determinants of sexual differentiation. In *Hormones and Behavior in Higher Vertebrates*, pp. 138–158. Edited by J. Balthazart, E. Prove & R. Gilles. Berlin, Heidelberg: Springer.

Otten, U., Baumann, J. B. & Girard, J. (1984). Nerve growth factor induces plasma extravasation in rat skin. *Eur J Pharmacol* 106, 199–201.

Oyegbile, T. O. & Marler, C. A. (2005). Winning fights elevates testosterone levels in California mice and enhances future ability to win fights. *Horm Behav* 48, 259–267.

Panksepp, J., Siviy, S. & Normansell, L. (1984). The psychobiology of play: theoretical and methodological perspectives. *Neurosci Biobehav Rev* 8, 465–492.

Park, S. W., He, Y., Ha, S. G., Loh, H. H. & Wei, L. N. (2008). Epigenetic regulation of kappa opioid receptor gene in neuronal differentiation. *Neuroscience* 151, 1034–1041.

Parker, K. J. & Lee, T. M. (2001). Central vasopressin administration regulates the onset of facultative paternal behavior in microtus pennsylvanicus (meadow voles). *Horm Behav* 39, 285–294.

Partecke, J. & Schwabl, H. (2008). Organizational effects of maternal testosterone on reproductive behavior of adult house sparrows. *Dev Neurobiol* 68, 1538–1548.

Pavlov, K. A., Chistiakov, D. A. & Chekhonin, V. P. (2012). Genetic determinants of aggression and impulsivity in humans. *J Appl Genet* 53, 61–82.

Pedersen, C. A., Ascher, J. A., Monroe, Y. L. & Prange, A. J. (1982). Oxytocin induces maternal behavior in virgin female rats. *Science* 216, 648–650.

Pedersen, C. A. & Prange, A. J. (1979). Induction of maternal behavior in virgin rats after intracerebroventricular administration of oxytocin. *Proc Natl Acad Sci U S A* 76, 6661–6665.

Pellis, S. M. & Pellis V.C. (1987). Play-fighting differs from serious fighting in both target of attack and tactics of fighting in the laboratory rat Rattus norvegicus. *Aggress Behav* 13, 227–242.

Philibert, R., Madan, A., Andersen, A., Cadoret, R., Packer, H. & Sandhu, H. (2007). Serotonin transporter mRNA levels are associated with the methylation of an upstream CpG island. *Am J Med Genet B Neuropsychiatr Genet* 144B, 101–105.

Pies, R. W. & Popli, A. P. (1995). Self-injurious behavior: pathophysiology and implications for treatment. *J Clin Psychiatry* 56, 580–588.

Pinzone, J. J., Stevenson, H., Strobl, J. S. & Berg, P. E. (2004). Molecular and cellular determinants of estrogen receptor alpha expression. *Mol Cell Biol* 24, 4605–4612.

Potegal, M. (1992). Time course of aggressive arousal in female hamsters and male rats. *Behav Neural Biol* 58, 120–124.

Puech, A. J., Simon, P., Chermat, R. & Boissier, J. R. (1974). Profil neuropsychopharmacologique de l'apomorphine. *J Pharmacol (Paris)* 2, 241–254.

Puglisi-Allegra, S. & Cabib, S. (1990). Effects of defeat experiences on dopamine metabolism in different brain areas of the mouse. *Aggress Behav* 16, 271–284.

Puglisi-Allegra, S., Mack, G., Oliverio, A. & Mandel, P. (1979). Effects of apomorphine and sodium di-n-propylacetate on the aggressive behaviour of three strains of mice. *Prog Neuro-Psychopharmacol* 3, 491–502.

Puglisi-Allegra, S. & Mandel, P. (1980). Effects of sodium n-dipropylacetate, muscimol hydrobromide and (R,S) nipecotic acid amide on isolation-induced aggressive behavior in mice. *Psychopharmacology (Berl)* 70, 287–290.

Raine, A., Buchsbaum, M. & LaCasse, L. (1997). Brain abnormalities in murderers indicated by positron emission tomography. *Biol Psychiatry* 42, 495–508.

Rajender, S., Pandu, G., Sharma, J. D., Gandhi, K. P., Singh, L. & Thangaraj, K. (2008). Reduced CAG repeats length in androgen receptor gene is associated with violent criminal behavior. *Int J Legal Med* 122, 367–372.

Ramamurthi, B. (1988). Stereotactic operation in behaviour disorders. Amygdalotomy and hypothalamotomy. *Acta Neurochir Suppl (Wien)* 44, 152–157.

Ramboz, S., Saudou, F., Amara, D. A., Belzung, C., Segu, L., Misslin, R., Buhot, M. C. & Hen, R. (1996). 5-HT1B receptor knock out--behavioral consequences. *Behav Brain Res* 73, 305–312.

Rasia-Filho, A. A., Londero, R. G. & Achaval, M. (2000). Functional activities of the amygdala: an overview. *J Psychiatry Neurosci* 25, 14–23.

Ray, A., Sharma, K. K., Alkondon, M. & Sen, P. (1983). Possible interrelationship between the biogenic amines involved in the modulation of footshock aggression in rats. *Arch Int Pharmacodyn Ther* 265, 36–41.

Razin, A. (1998). CpG methylation, chromatin structure and gene silencing-a three-way connection. *EMBO J* 17, 4905–4908.

Razin, A. & Cedar, H. (1993). DNA methylation and embryogenesis. *EXS* 64, 343–357.

Razin, A. & Riggs, A. D. (1980). DNA methylation and gene function. *Science* 210, 604–610.

Razzoli, M., Roncari, E., Guidi, A., Carboni, L., Arban, R., Gerrard, P. & Bacchi, F. (2006). Conditioning properties of social subordination in rats: behavioral and biochemical correlates of anxiety. *Horm Behav* 50, 245–251.

Reif, A., Rosler, M., Freitag, C. M., Schneider, M., Eujen, A., Kissling, C., Wenzler, D., Jacob, C. P., Retz-Junginger, P., Thome, J., Lesch, K. P. & Retz, W. (2007). Nature and nurture

9 Genetic and Epigenetic Determinants of Aggression

predispose to violent behavior: serotonergic genes and adverse childhood environment. *Neuropsychopharmacology* 32, 2375–2383.

Reiss, A. J. & Roth, J. A. (1993). *Understanding and preventing violence.* Washington, D.C.: National Academy Press.

Renthal, W., Maze, I., Krishnan, V., Covington, H. E., Xiao, G., Kumar, A., Russo, S. J., Graham, A., Tsankova, N., Kippin, T. E., Kerstetter, K. A., Neve, R. L., Haggarty, S. J., McKinsey, T. A., Bassel-Duby, R., Olson, E. N. & Nestler, E. J. (2007). Histone deacetylase 5 epigenetically controls behavioral adaptations to chronic emotional stimuli. *Neuron* 56, 517–529.

Renthal, W. & Nestler, E. J. (2008). Epigenetic mechanisms in drug addiction. *Trends Mol Med* 14, 341–350.

Repetti, R. L., Taylor, S. E. & Seeman, T. E. (2002). Risky families: family social environments and the mental and physical health of offspring. *Psychol Bull* 128, 330–366.

Retz, W. & Rosler, M. (2009). The relation of ADHD and violent aggression: What can we learn from epidemiological and genetic studies? *Int J Law Psychiatry* 32, 235–243.

Rhee, S. H. & Waldman, I. D. (2002). Genetic and environmental influences on antisocial behavior: a meta-analysis of twin and adoption studies. *Psychol Bull* 128, 490–529.

Robinson, G. E., Fernald, R. D. & Clayton, D. F. (2008). Genes and social behavior. *Science* 322, 896–900.

Rodenhiser, D. & Mann, M. (2006). Epigenetics and human disease: translating basic biology into clinical applications. *CMAJ* 174, 341–348.

Rodgers, R. J. & Depaulis, A. (1982). GABAergic influences on defensive fighting in rats. *Pharmacol Biochem Behav* 17, 451–456.

Roth, T. L., Lubin, F. D., Funk, A. J. & Sweatt, J. D. (2009). Lasting epigenetic influence of early-life adversity on the BDNF gene. *Biol Psychiatry* 65, 760–769.

Rudebeck, P. H., Walton, M. E., Millette, B. H., Shirley, E., Rushworth, M. F. & Bannerman, D. M. (2007). Distinct contributions of frontal areas to emotion and social behaviour in the rat. *Eur J Neurosci* 26, 2315–2326.

Sanders, A. R., Duan, J. & Gejman, P. V. (2002). DNA variation and psychopharmacology of the human serotonin receptor 1B (HTR1B) gene. *Pharmacogenomics* 3, 745–762.

Sano, K. & Mayanagi, Y. (1988). Posteromedial hypothalamotomy in the treatment of violent, aggressive behaviour. *Acta Neurochir Suppl (Wien)* 44, 145–151.

Santarelli, L., Gobbi, G., Debs, P. C., Sibille, E. T., Blier, P., Hen, R. & Heath, M. J. (2001). Genetic and pharmacological disruption of neurokinin 1 receptor function decreases anxiety-related behaviors and increases serotonergic function. *Proc Natl Acad Sci U S A* 98, 1912–1917.

Sapolsky, R. M. (2005). The influence of social hierarchy on primate health. *Science* 308, 648–652.

Saudou, F., Amara, D. A., Dierich, A., LeMeur, M., Ramboz, S., Segu, L., Buhot, M. C. & Hen, R. (1994). Enhanced aggressive behavior in mice lacking 5-HT1B receptor. *Science* 265, 1875–1878.

Sawchenko, P. E. & Swanson, L. W. (1983). The organization and biochemical specificity of afferent projections to the paraventricular and supraoptic nuclei. *Prog Brain Res* 60, 19–29.

Scarpa, A. & Raine, A. (1997). Psychophysiology of anger and violent behavior. *Psychiatr Clin North Am* 20, 375–394.

Scholey, A. B., Rose, S. P., Zamani, M. R., Bock, E. & Schachner, M. (1993). A role for the neural cell adhesion molecule in a late, consolidating phase of glycoprotein synthesis six hours following passive avoidance training of the young chick. *Neuroscience* 55, 499–509.

Schultheiss, O. C., Wirth, M. M., Torges, C. M., Pang, J. S., Villacorta, M. A. & Welsh, K. M. (2005). Effects of implicit power motivation on men's and women's implicit learning and testosterone changes after social victory or defeat. *J Pers Soc Psychol* 88, 174–188.

Scott, J. P. (1977). Social genetics. *Behav Genet* 7, 327–346.

Seay, B., Alexander, B. K. & Harlow, H. F. (1964). Maternal behavior of socially deprived rhesus monkeys. *J Abnorm Psychol* 69, 345–354.

Seay, B. & Gottfried, N. W. (1975). A phylogenetic perspective for social behavior in primates. *J Gen Psychol* 92, 5–17.

Seckl, J. R. & Meaney, M. J. (1993). Early life events and later development of ischaemic heart disease. *Lancet* 342, 1236.

Seguin, J. R. (2004). Neurocognitive elements of antisocial behavior: Relevance of an orbitofrontal cortex account. *Brain Cogn* 55, 185–197.

Senault, B. (1968). Syndrome agressif induit par l'apomorphine chez le rat [Aggressive syndrome induced by apomorphine in rats]. *J Physiol (Paris)* 60 Suppl 2, 543–544.

Senault, B. (1971). Influence de l'isolement sur le comportement d'agressivité intraspécifique induit par l'apomorphine chez le rat [Influence of isolation on the aggressive behaviour induced by apomorphine in the rat]. *Psychopharmacologia* 20, 389–394.

Shaikh, M. B., Barrett, J. A. & Siegel, A. (1987). The pathways mediating affective defense and quiet biting attack behavior from the midbrain central gray of the cat: an autoradiographic study. *Brain Res* 437, 9–25.

Shaikh, M. B., Steinberg, A. & Siegel, A. (1993). Evidence that substance P is utilized in medial amygdaloid facilitation of defensive rage behavior in the cat. *Brain Res* 625, 283–294.

Shih, J. C., Chen, K. & Ridd, M. J. (1999). Monoamine oxidase: from genes to behavior. *Annu Rev Neurosci* 22, 197–217.

Siever, L. J., Buchsbaum, M. S., New, A. S., Spiegel-Cohen, J., Wei, T., Hazlett, E. A., Sevin, E., Nunn, M. & Mitropoulou, V. (1999). d,l-fenfluramine response in impulsive personality disorder assessed with [18F]fluorodeoxyglucose positron emission tomography. *Neuropsychopharmacology* 20, 413–423.

Simon, N. G., Cologer-Clifford, A., Lu, S. F., McKenna, S. E. & Hu, S. (1998). Testosterone and its metabolites modulate 5HT1A and 5HT1B agonist effects on intermale aggression. *Neurosci Biobehav Rev* 23, 325–336.

Sneddon, L. U., Schmidt, R., Fang, Y. & Cossins, A. R. (2011). Molecular correlates of social dominance: a novel role for ependymin in aggression. *PLoS ONE* 6, e18181.

Sohn, K. J., Jang, H., Campan, M., Weisenberger, D. J., Dickhout, J., Wang, Y. C., Cho, R. C., Yates, Z., Lucock, M., Chiang, E. P., Austin, R. C., Choi, S. W., Laird, P. W. & Kim, Y. I. (2009). The methylenetetrahydrofolate reductase C677T mutation induces cell-specific changes in genomic DNA methylation and uracil misincorporation: a possible molecular basis for the site-specific cancer risk modification. *Int J Cancer* 124, 1999–2005.

Spencer, T. E., Jenster, G., Burcin, M. M., Allis, C. D., Zhou, J., Mizzen, C. A., McKenna, N. J., Onate, S. A., Tsai, S. Y., Tsai, M. J. & O'Malley, B. W. (1997). Steroid receptor coactivator-1 is a histone acetyltransferase. *Nature* 389, 194–198.

Spinella, M. (2004). Neurobehavioral correlates of impulsivity: evidence of prefrontal involvement. *Int J Neurosci* 114, 95–104.

Stephani, U., Sutter, A. & Zimmermann, A. (1987). Nerve growth factor (NGF) in serum: evaluation of serum NGF levels with a sensitive bioassay employing embryonic sensory neurons. *J Neurosci Res* 17, 25–35.

Stevens, A., Begum, G. & White, A. (2011). Epigenetic changes in the hypothalamic pro-opiomelanocortin gene: a mechanism linking maternal undernutrition to obesity in the offspring? *Eur J Pharmacol* 660, 194–201.

Stork, O., Welzl, H., Wolfer, D., Schuster, T., Mantei, N., Stork, S., Hoyer, D., Lipp, H., Obata, K. & Schachner, M. (2000). Recovery of emotional behaviour in neural cell adhesion molecule (NCAM) null mutant mice through transgenic expression of NCAM180. *Eur J Neurosci* 12, 3291–3306.

Stork, O., Welzl, H., Wotjak, C. T., Hoyer, D., Delling, M., Cremer, H. & Schachner, M. (1999). Anxiety and increased 5-HT1A receptor response in NCAM null mutant mice. *J Neurobiol* 40, 343–355.

Stribley, J. M. & Carter, C. S. (1999). Developmental exposure to vasopressin increases aggression in adult prairie voles. *Proc Natl Acad Sci U S A* 96, 12601–12604.

Sulcova, A., Krsiak, M. & Masek, K. (1981). Effects of calcium valproate and aminooxyacetic acid on agonistic behaviour in mice. *Act Nerv Super (Praha)* 23, 287–289.

Suomi, S. J. (1997). Early determinants of behaviour: evidence from primate studies. *Br Med Bull* 53, 170–184.

9 Genetic and Epigenetic Determinants of Aggression

Suomi, S. J. & Ripp, C. (1983). A history of motherless mother monkeys mothering at the University of Wisconsin Primate Laboratory. In *Child Abuse: The Nonhuman Primate Data*, pp. 49–78. Edited by M. Reite & M. G. Caine. New York: Alan R Liss.

Swanson, L. W. & Sawchenko, P. E. (1983). Hypothalamic integration: organization of the paraventricular and supraoptic nuclei. *Annu Rev Neurosci* 6, 269–324.

Sweatt, J. D. (2009). Experience-dependent epigenetic modifications in the central nervous system. *Biol Psychiatry* 65, 191–197.

Sweidan, S., Edinger, H. & Siegel, A. (1991). D2 dopamine receptor-mediated mechanisms in the medial preoptic-anterior hypothalamus regulate effective defense behavior in the cat. *Brain Res* 549, 127–137.

Takahashi, A., Kwa, C., Debold, J. F. & Miczek, K. A. (2010). GABA(A) receptors in the dorsal raphe nucleus of mice: escalation of aggression after alcohol consumption. *Psychopharmacology (Berl)* 211, 467–477.

Takahashi, A., Quadros, I. M., de Almeida, R. M. & Miczek, K. A. (2011). Brain serotonin receptors and transporters: initiation vs. termination of escalated aggression. *Psychopharmacology (Berl)* 213, 183–212.

Tidey, J. W. & Miczek, K. A. (1996). Social defeat stress selectively alters mesocorticolimbic dopamine release: an in vivo microdialysis study. *Brain Res* 721, 140–149.

Toth, M., Fuzesi, T., Halasz, J., Tulogdi, A. & Haller, J. (2010). Neural inputs of the hypothalamic "aggression area" in the rat. *Behav Brain Res* 215, 7–20.

Toth, M., Halasz, J., Mikics, E., Barsy, B. & Haller, J. (2008). Early social deprivation induces disturbed social communication and violent aggression in adulthood. *Behav Neurosci* 122, 849–854.

Toth, M., Mikics, E., Tulogdi, A., Aliczki, M. & Haller, J. (2011). Post-weaning social isolation induces abnormal forms of aggression in conjunction with increased glucocorticoid and autonomic stress responses. *Horm Behav* 60, 28–36.

Tremblay, J. & Hamet, P. (2008). Impact of genetic and epigenetic factors from early life to later disease. *Metabolism* 57 Suppl 2, S27–S31.

Tremblay, R. E. (2008). Understanding development and prevention of chronic physical aggression: towards experimental epigenetic studies. *Philos Trans R Soc Lond B Biol Sci* 363, 2613–2622.

Tremolizzo, L., Doueiri, M. S., Dong, E., Grayson, D. R., Davis, J., Pinna, G., Tueting, P., Rodriguez-Menendez, V., Costa, E. & Guidotti, A. (2005). Valproate corrects the schizophrenia-like epigenetic behavioral modifications induced by methionine in mice. *Biol Psychiatry* 57, 500–509.

Tsankova, N., Renthal, W., Kumar, A. & Nestler, E. J. (2007). Epigenetic regulation in psychiatric disorders. *Nat Rev Neurosci* 8, 355–367.

Tulogdi, A., Toth, M., Halasz, J., Mikics, E., Fuzesi, T. & Haller, J. (2010). Brain mechanisms involved in predatory aggression are activated in a laboratory model of violent intra-specific aggression. *Eur J Neurosci* 32, 1744–1753.

Turner, J. D. & Muller, C. P. (2005). Structure of the glucocorticoid receptor (NR3C1) gene 5' untranslated region: identification, and tissue distribution of multiple new human exon 1. *J Mol Endocrinol* 35, 283–292.

Unis, A. S., Cook, E. H., Vincent, J. G., Gjerde, D. K., Perry, B. D., Mason, C. & Mitchell, J. (1997). Platelet serotonin measures in adolescents with conduct disorder. *Biol Psychiatry* 42, 553–559.

Vale, A. L. & Montgomery, A. M. (1997). Social interaction: responses to chlordiazepoxide and the loss of isolation-reared effects with paired-housing. *Psychopharmacology (Berl)* 133, 127–132.

Valzelli, L. (1984). Reflections on experimental and human pathology of aggression. *Prog Neuropsychopharmacol Biol Psychiatry* 8, 311–325.

van den Berg, C. L., Hol, T., Van Ree, J. M., Spruijt, B. M., Everts, H. & Koolhaas, J. M. (1999). Play is indispensable for an adequate development of coping with social challenges in the rat. *Dev Psychobiol* 34, 129–138.

van Erp, A. M. & Miczek, K. A. (2000). Aggressive behavior, increased accumbal dopamine, and decreased cortical serotonin in rats. *J Neurosci* 20, 9320–9325.

Van Leeuwen, F. W., Caffe, A. R. & De Vries, G. J. (1985). Vasopressin cells in the bed nucleus of the stria terminalis of the rat: sex differences and the influence of androgens. *Brain Res* 325, 391–394.

Vanderschuren, L. J., Niesink, R. J. & Van Ree, J. M. (1997). The neurobiology of social play behavior in rats. *Neurosci Biobehav Rev* 21, 309–326.

Vanyushin, B. F., Nemirovsky, L. E., Klimenko, V. V., Vasiliev, V. K. & Belozersky, A. N. (1973). The 5-methylcytosine in DNA of rats. Tissue and age specificity and the changes induced by hydrocortisone and other agents. *Gerontologia* 19, 138–152.

Veenema, A. H. (2009). Early life stress, the development of aggression and neuroendocrine and neurobiological correlates: what can we learn from animal models? *Front Neuroendocrinol* 30, 497–518.

Veenema, A. H., Beiderbeck, D. I., Lukas, M. & Neumann, I. D. (2010). Distinct correlations of vasopressin release within the lateral septum and the bed nucleus of the stria terminalis with the display of intermale aggression. *Horm Behav* 58, 273–281.

Veenema, A. H., Blume, A., Niederle, D., Buwalda, B. & Neumann, I. D. (2006). Effects of early life stress on adult male aggression and hypothalamic vasopressin and serotonin. *Eur J Neurosci* 24, 1711–1720.

Veenema, A. H., Bredewold, R. & Neumann, I. D. (2007a). Opposite effects of maternal separation on intermale and maternal aggression in C57BL/6 mice: link to hypothalamic vasopressin and oxytocin immunoreactivity. *Psychoneuroendocrinology* 32, 437–450.

Veenema, A. H., Koolhaas, J. M. & de Kloet, E. R. (2004). Basal and stress-induced differences in HPA axis, 5-HT responsiveness, and hippocampal cell proliferation in two mouse lines. *Ann N Y Acad Sci* 1018, 255–265.

Veenema, A. H. & Neumann, I. D. (2007). Neurobiological mechanisms of aggression and stress coping: a comparative study in mouse and rat selection lines. *Brain Behav Evol* 70, 274–285.

Veenema, A. H., Torner, L., Blume, A., Beiderbeck, D. I. & Neumann, I. D. (2007b). Low inborn anxiety correlates with high intermale aggression: link to ACTH response and neuronal activation of the hypothalamic paraventricular nucleus. *Horm Behav* 51, 11–19.

Vekovischeva, O. Y., Verbitskaya, E. V., Aitta-Aho, T., Sandnabba, K. & Korpi, E. R. (2007). Multimetric statistical analysis of behavior in mice selected for high and low levels of isolation-induced male aggression. *Behav Processes* 75, 23–32.

Virkkunen, M., Rawlings, R., Tokola, R., Poland, R. E., Guidotti, A., Nemeroff, C., Bissette, G., Kalogeras, K., Karonen, S. L. & Linnoila, M. (1994). CSF biochemistries, glucose metabolism, and diurnal activity rhythms in alcoholic, violent offenders, fire setters, and healthy volunteers. *Arch Gen Psychiatry* 51, 20–27.

Vitiello, B., Behar, D., Hunt, J., Stoff, D. & Ricciuti, A. (1990). Subtyping aggression in children and adolescents. *J Neuropsychiatry Clin Neurosci* 2, 189–192.

Vochteloo, J. D. & Koolhaas, J. M. (1987). Medial amygdala lesions in male rats reduce aggressive behavior: interference with experience. *Physiol Behav* 41, 99–102.

Volavka, J., Bilder, R. & Nolan, K. (2004). Catecholamines and aggression: the role of COMT and MAO polymorphisms. *Ann N Y Acad Sci* 1036, 393–398.

vom Saal, F. (1983). Models of early hormonal effects on intrasex aggression in mice. In *Hormones and Aggressive Behavior*. Edited by B. B. Svare. New York: Plenum Press.

Von Frijtag, J. C., Schot, M., van den Bos R. & Spruijt, B. M. (2002). Individual housing during the play period results in changed responses to and consequences of a psychosocial stress situation in rats. *Dev Psychobiol* 41, 58–69.

Waddington, C. (1942). The epigenotype. *Endeavour* 1, 18–20.

Wang, M., Haertel, G. & Walberg, H. (1997). Learning Influences. In *Psychology and Educational Practice*, pp. 199–211. Edited by H. Walberg & G. Haertel. Berkeley, CA: McCutchan.

Wang, Z. & De Vries, G. J. (1993). Testosterone effects on paternal behavior and vasopressin immunoreactive projections in prairie voles (Microtus ochrogaster). *Brain Res* 631, 156–160.

9 Genetic and Epigenetic Determinants of Aggression

Weaver, I. C., Cervoni, N., Champagne, F. A., D'Alessio, A. C., Sharma, S., Seckl, J. R., Dymov, S., Szyf, M. & Meaney, M. J. (2004). Epigenetic programming by maternal behavior. *Nat Neurosci* 7, 847–854.

Weiss, J. M. (1968). Effects of coping responses on stress. *J Comp Physiol Psychol* 65, 251–260.

Wersinger, S. R., Caldwell, H. K., Christiansen, M. & Young, W. S. (2007a). Disruption of the vasopressin 1b receptor gene impairs the attack component of aggressive behavior in mice. *Genes Brain Behav* 6, 653–660.

Wersinger, S. R., Caldwell, H. K., Martinez, L., Gold, P., Hu, S. B. & Young, W. S., III (2007b). Vasopressin 1a receptor knockout mice have a subtle olfactory deficit but normal aggression. *Genes Brain Behav* 6, 540–551.

West, M. J. & King, A. P. (1987). Settling nature and nurture into an ontogenetic niche. *Dev Psychobiol* 20, 549–562.

Westberry, J. M., Trout, A. L. & Wilson, M. E. (2010). Epigenetic regulation of estrogen receptor alpha gene expression in the mouse cortex during early postnatal development. *Endocrinology* 151, 731–740.

Widom, C. S. & Brzustowicz, L. M. (2006). MAOA and the "cycle of violence:" childhood abuse and neglect, MAOA genotype, and risk for violent and antisocial behavior. *Biol Psychiatry* 60, 684–689.

Wilson, A. J., Gelin, U., Perron, M. C. & Reale, D. (2009). Indirect genetic effects and the evolution of aggression in a vertebrate system. *Proc Biol Sci* 276, 533–541.

Wilson, J. D. (2001). Androgens, androgen receptors, and male gender role behavior. *Horm Behav* 40, 358–366.

Wilson, V. L., Smith, R. A., Ma, S. & Cutler, R. G. (1987). Genomic 5-methyldeoxycytidine decreases with age. *J Biol Chem* 262, 9948–9951.

Winslow, J. T. & Insel, T. R. (1993). Effects of central vasopressin administration to infant rats. *Eur J Pharmacol* 233, 101–107.

Winslow, J. T., Noble, P. L., Lyons, C. K., Sterk, S. M. & Insel, T. R. (2003). Rearing effects on cerebrospinal fluid oxytocin concentration and social buffering in rhesus monkeys. *Neuropsychopharmacology* 28, 910–918.

Wion, D., Barrand, P., Dicou, E., Scott, J. & Brachet, P. (1985). Serum and thyroid hormones T3 and T4 regulate nerve growth factor mRNA levels in mouse L cells. *FEBS Lett* 189, 37–41.

Witt, D. M. (1997). Regulatory mechanisms of oxytocin-mediated sociosexual behavior. *Ann N Y Acad Sci* 807, 287–301.

Wolf, J. B., Brodie, E. D., Cheverud, J. M., Moore, A. J. & Wade, M. J. (1998). Evolutionary consequences of indirect genetic effects. *Trends Ecol Evol* 13, 64–69.

Wommack, J. C., Taravosh-Lahn, K., David, J. T. & Delville, Y. (2003). Repeated exposure to social stress alters the development of agonistic behavior in male golden hamsters. *Horm Behav* 43, 229–236.

Wongwitdecha, N. & Marsden, C. A. (1996). Social isolation increases aggressive behaviour and alters the effects of diazepam in the rat social interaction test. *Behav Brain Res* 75, 27–32.

Wysocki, C. J. (1979). Neurobehavioral evidence for the involvement of the vomeronasal system in mammalian reproduction. *Neurosci Biobehav Rev* 3, 301–341.

Yanai, K., Son, L. Z., Endou, M., Sakurai, E., Nakagawasai, O., Tadano, T., Kisara, K., Inoue, I., Watanabe, T. & Watanabe, T. (1998). Behavioural characterization and amounts of brain monoamines and their metabolites in mice lacking histamine H1 receptors. *Neuroscience* 87, 479–487.

Yudofsky, S. C., Stevens, L., Silver, J., Barsa, J. & Williams, D. (1984). Propranolol in the treatment of rage and violent behavior associated with Korsakoff's psychosis. *Am J Psychiatry* 141, 114–115.

Zelena, D., Domokos, A., Jain, S. K., Jankord, R. & Filaretova, L. (2009). The stimuli-specific role of vasopressin in the hypothalamus-pituitary-adrenal axis response to stress. *J Endocrinol* 202, 263–278.

Zhang, L., Kerich, M., Schwandt, M. L., Rawlings, R. R., McKellar, J. D., Momenan, R., Hommer, D. W. & George, D. T. (2011). Smaller right amygdala in Caucasian alcohol-dependent male patients with a history of intimate partner violence: a volumetric imaging study. *Addict Biol.* doi: 10.1111/j.1369–1600.2011.00381.x.

Zhuang, X., Gross, C., Santarelli, L., Compan, V., Trillat, A. C. & Hen, R. (1999). Altered emotional states in knockout mice lacking 5-HT1A or 5-HT1B receptors. *Neuropsychopharmacology* 21, 52S–60S.

Zitzmann, M. & Nieschlag, E. (2001). Testosterone levels in healthy men and the relation to behavioural and physical characteristics: facts and constructs. *Eur J Endocrinol* 144, 183–197.

Chapter 10
Co-Regulation and Epigenetic Dysregulation in Schizophrenia and Bipolar Disorder

Dóra Zelena

Abbreviations

5HT2	Serotonin receptor 2
Ach	Acetylcholine
ACTH	Adrenocorticotropin
AMPA	2-Amino-3-(5-methyl-3-oxo-1,2-oxazol-4-yl)propanoic acid, ionotrop glutamate receptor agonist
BD	Bipolar disorders
BDNF	Brain-derived neurotrophic factor
Bp	Base pair
cAMP	Cyclic adenosine monophosphate
CaMK	Calcium/calmodulin-activated protein kinase
CBP	CREB-binding protein, a histone acetyltransferase
CNS	Central nervous system
CNV	Copy number variation
COMT	Catechol-O-methyl transferase
CpG	Cytosines next to guanine in phosphodiester bond
CREB	cAMP-response element-binding protein
CRH	Corticotropin-releasing hormone
Decitabine	5-Aza-2'-deoxycytidine
DNA	Deoxyribonucleic acid
DNMT	DNA methyl transferase
DR	Dopamine receptor
EGF	Epidermal growth factor

D. Zelena (✉)
Department of Behavioural Neurobiology, Laboratory of Behavioural and Stress Studies,
Institute of Experimental Medicine, Hungarian Academy of Sciences,
Szigony 43, 1083 Budapest, Hungary
e-mail: zelena.dora@koki.mta.hu

J. Minarovits and H.H. Niller (eds.), *Patho-Epigenetics of Disease*,
DOI 10.1007/978-1-4614-3345-3_10, © Springer Science+Business Media New York 2012

Elk-1	Ets-like protein kinase-1
ERK	Extracellular signal-regulated kinase
FDA	The US Food and Drug Administration
FST	Forced swim test
GABA	γ-Aminobutyric acid
GAD	Glutamic acid decarboxylase
GFAP	Glial fibrillary acidic protein
GR	Glucocorticoid receptor
GRIN	N-methyl-D-aspartate receptor subunit gene
GWA	Genome-wide association
H	Histone
H3K27me3	Histone 3 lysine 27 trimethylation
H3K9ac	Histone 3 lysine 9 acetylation
H3S10p	Phosphorylation of histone H3 at serine 10
HAT	Histone acetyltransferase
HDAC	Histone deacetylase
HDM	Histone demethylase
HMT	Histone methyltransferase
HPA	Hypothalamo-pituitary-adrenocortical axis
HTR	Serotonin receptor
INF	Interferon
K	Lysine
LG-ABN	Licking, grooming and arched-back nursing
lncR	Long noncoding RNA
LSD	Lysergic acid diethylamide
MAPK	Mitogen-activated protein kinase
MB-COMT	Membrane bound catechol-O-methyl transferase
MBD	Methyl-binding domain
MeCP2	Methyl-CpG-binding protein 2
Met	Methionine
miR	MicroRNA
Mll1	Mixed-lineage leukemia 1
MSK1	Mitogen- and stress-activated kinase 1
MTHFR	Methylenetetrahydrofolate reductase
nAchR	Nicotinic acetylcholine receptors
NGFI-A	Nerve growth factor inducible factor-A
NMDA	N-methyl-D-aspartate
NR	NMDA receptor subunit
NRG	Neuregulin
NVHL	Neonatal ventral hippocampal lesion
OLIG	Oligodendrocyte lineage transcription factor
PBMC	Peripheral blood mononuclear cells
PcG	Polycomb group complex

PCP	Phencyclidine
PFC	Prefrontal cortex
POMC	Proopiomelanocortin
R	Arginine
RELN	Reelin
rR	Ribosomal RNA
S	Serine
SAM	S-adenosylmethionine
SCG2	Secretogranin II
SCZ	Schizophrenia
SIRT	Sirtuin
SNP	Single nucleotide polymorphism
SOX10	Sex-determining region Y-box 10
T	Threonine
TrxG	Trithorax group complex
Val	Valine
VCFS	Velocardio facial syndrome
VGLUT	Vesicular glutamate transporter
Vorinostat	SAHA, N-hydroxy-N'-phenyl-octanediamide

10.1 Importance of the Neuropsychiatric Disorders

Psychosis is a common characteristic of neuropsychiatric conditions, including schizophrenia (SCZ) and bipolar disorder (BD) (Craddock et al. 2006). Together, these disorders are called major psychosis and constitute a considerable public health burden (Lopez et al. 2006). They disturb whole families mentally, physically, and in so many other aspects influencing the life of 10–25% of the population yearly. And the number of patients is increasing. Despite these facts, the treatment of these disorders is far from being solved.

Most neuropsychiatric disorders, like SCZ, BD or depression, share important features, including a substantial genetic predisposition and a contribution from environmental factors (Tsankova et al. 2007). Another common attribute of psychiatric conditions is long-lasting behavioral abnormalities. In most individuals, these illnesses develop gradually and show a chronic, remitting course, often over a lifetime. Likewise, the reversal of symptoms in response to treatment occurs over weeks or months. However, all reported changes in transcription factors and other nuclear regulatory proteins in animal models revert to normal within hours or days of chronic perturbation. Even the longest-lasting ones do not persist as long as the behavioral changes. Thus, the molecular basis of slowly developing, but particularly stable adaptations and maladaptations in the brain should be found somewhere else, perhaps in the epigenetic machinery.

10.1.1 Prevalence

The prevalence of SCZ is 0.3–2.7% worldwide (McGrath et al. 2008). Diagnosis is based on the appearance and duration of about 30 symptoms divided into positive, negative, and cognitive. Positive symptoms are those that most individuals do not normally experience, but are present in people with SCZ. Negative symptoms are deficits of normal emotional responses or of other thought processes, and respond less well to medication.

Estimates of the lifetime prevalence of BD, known as manic-depressive disorder, vary, with studies typically giving values in the order of 1%. BD describes a category of mood disorders defined by the presence of one or more episodes of abnormally elevated energy levels, cognition, and mood with or without one or more depressive episodes. More frequent is the unipolar disorder, also known as major depression. As per studies performed by The National Institute of Mental Health (Bethesda, MD, USA) approximately 5.3% adults and 4% adolescents suffer from serious depression annually. The term "depression" is ambiguous. It is often used to denote this syndrome, but may refer to other mood disorders or to lower mood states lacking clinical significance. The most serious consequence could be suicide, as up to 60% of people who committed suicide had depression or another mood disorder.

10.1.2 Symptoms

10.1.2.1 Schizophrenia

The term Schizophrenia was coined by Eugen Bleuler. It is a mental disorder characterized by a disintegration of thought processes and of emotional responsiveness. The onset of symptoms typically occurs in young adulthood before the age of 19, critical years in a young adult's social and vocational development (Addington et al. 2007). Diagnosis is based on observed behavior and the patient's reported experiences. A prodromal (pre-onset) phase of the illness could be detected up to 30 months before the onset of symptoms. Those who go on to develop SCZ may experience transient or self-limiting psychotic symptoms (Amminger et al. 2006) and the nonspecific symptoms of social withdrawal, irritability, dysphoria (Parnas and Jorgensen 1989), and clumsiness during the prodromal phase.

A person diagnosed with SCZ may experience hallucinations (most reported are hearing voices), delusions (often bizarre or persecutory in nature), and disorganized thinking and speech. The latter may range from loss of train of thought, to sentences only loosely connected in meaning, to incoherence known as word salad in severe cases. Social withdrawal, sloppiness of dress and hygiene, and loss of motivation and judgment are all common in SCZ (Picchioni and Murray 2007). There is often an observable pattern of emotional difficulty, for example lack of responsiveness. Impairment in social cognition is associated with SCZ, as are symptoms of paranoia; social isolation commonly occurs (Brunet-Gouet and Decety 2006).

10.1.2.2 Bipolar Disorder

Diagnosis of BD is based on the self-reported experiences of an individual, as well as abnormalities in behavior reported by family members, friends or coworkers, followed by secondary signs observed by a clinician in a clinical assessment (Ketter 2010; Loganathan et al. 2010). The cycles of BD (alternation of mania and depression) last for days, weeks, or months.

Mania is generally characterized by a distinct period of an elevated mood, which can take the form of euphoria. People commonly experience an increase in energy and a decreased need for sleep, with many often going on for days without sleeping. A person may exhibit pressured speech, with thoughts experienced as racing. Attention span is low, and a person in a manic state may be easily distracted. Judgement may become impaired, and sufferers may go on spending sprees or engage in behavior that is quite abnormal for them. They may indulge in substance abuse, particularly alcohol, cocaine, or sleeping pills. Their behavior may become aggressive, intolerant, or intrusive. People may feel out of control or unstoppable, or as if they have been chosen and are on a special mission or have other grandiose or delusional ideas.

Signs and symptoms of the *depressive phase* of BD include persistent feelings of sadness, anxiety, guilt, anger, isolation, or hopelessness; disturbances in sleep and appetite; fatigue, and loss of interest in usually enjoyable activities; problems concentrating; loneliness, self-loathing, apathy or indifference; depersonalization; loss of interest in sexual activity; shyness or social anxiety; irritability, chronic pain (with or without a known cause); lack of motivation; and morbid suicidal ideation.

10.2 Etiology

In 1977, American Psychiatrist George Engel introduced the major theory in medicine, the *Biopsychosocial Model* (Engel 1977). Engel eloquently states: "To provide a basis for understanding the determinants of disease and arriving at rational treatments and patterns of health care, a medical model must also take into account the patient, the social context in which he lives and the complementary system devised by society to deal with the disruptive effects of illness, that is, the physician role and the health care system. This requires a biopsychosocial model."

This theory was reformulated by de Kloet et al. (2007). It says that the susceptibility pathways underlying disturbed brain functions are influenced by genetic factors, early-life priming experiences and later-life events. Cortisol, the end hormone of the hypothalamo-pituitary-adrenocortical (HPA) axis (the so-called stress-axis), is an important determinant in this *three hit model*. Thus, all diseases, especially the psychiatric diseases, have three "causes": genetic predisposition, environmental factors and an actual exacerbation by stress.

10.2.1 Genetic Determination

More than a decade after the human genome sequence became available, we are still far from understanding the genetic basis of many diseases, although a significant portion of the syndromes are due to single gene perturbations, either resulting from dosage changes or mutation (Inlow and Restifo 2004).

10.2.1.1 Schizophrenia

First-degree relatives of SCZ probands have an approximately 10% probability of becoming ill (Kety et al. 1994), while about 50% of cases of SCZ are spontaneous with no other affected family member (Malaspina et al. 2002). The general belief is that approximately 50% of monozygotic twins afflicted with SCZ are discordant for the disease, although progeny of both the well and ill discordant monozygotic twin have the elevated probability typical of first degree relatives of ill individuals (Gottesman and Bertelsen 1989).

Genetic studies have linked many genes and chromosomal regions distributed throughout the genome to SCZ, but no single or small number of genes accounts for the majority of cases (Smith et al. 2010). Genes linked to SCZ do not affect a single neurobiological system and include neurotrophic factors (e.g., brain-derived neurotrophic factor (BDNF), neuregulin (NRG)), neuromodulatory receptors (dopamine receptor (DR), serotonin receptor (HTR)), members of the synaptic packaging and release machinery (SNAP25) and both inhibitory and excitatory neurotransmitter systems (glutamate, γ-aminobutyric acid (GABA)). Also, there are genes linked to folate processing and methylation (DNA methyl transferase (DNMT); catechol-O-methyl transferase (COMT)) (Pidsley and Mill 2011).

As most of the antipsychotics influence the dopamine system, this was studied in details. COMT is an S-adenosyl methionine (SAM) dependent methyltransferase enzyme which methylates catecholamines (including dopamine and norepinepherine) and catechol estrogens. COMT is solely responsible for the metabolism of dopamine in the dorsolateral prefrontal cortex (PFC), a region important for working memory performance, which is dysfunctional in SCZ. The focus of genetic studies of COMT has been the valine (Val) to methionine (Met) substitution. This polymorphism has been shown to alter the activity and thermal stability of the enzyme, so that subjects with the Met homozygous genotype are estimated to have up to 50% reduction of COMT activity (Chen et al. 2004). Dopamine signaling is therefore likely to be enhanced in subjects with the Met158 allele in comparison with subjects homozygous for the Val158 allele single nucleotide polymorphism (SNP). A groundbreaking study by Egan et al. demonstrated that COMT Val158 homozygous subjects exhibited reduced PFC cognitive performance and efficiency, in comparison with Met158 homozygous individuals (Egan et al. 2001). SCZ samples were also more likely to have Val158 SNP in COMT, and less likely to be homozygous for the Met allele than controls (Smith et al. 2010). COMT SNPs in untranslated, promoter and intronic regions are thought to impact COMT function through altered gene expression (Tunbridge et al. 2007). It became clear, however, that genetic risks do not operate in isolation; polymorphisms in the COMT gene, for example, act to moderate the influence of adolescent cannabis use on the risk of developing adult psychosis.

The advent of genome-wide association (GWA) studies has allowed a more systematic, hypothesis-free exploration of the genes associated with psychosis. Although GWA studies have identified a small number of polymorphisms associated with both SCZ (Stefansson et al. 2009) and BD (Ferreira et al. 2008), these findings are characterized by small effect sizes and await replication (Purcell et al. 2009). We can conclude that it is unlikely that common genetic variants cumulatively account for all the population variance in the risk for psychosis (O'Donovan et al. 2009).

Structural genomic alterations, such as copy number variations (CNVs), have also been implicated, but these *de novo* events are extremely rare and only found in a small number of patients (Merikangas et al. 2009).

10.2.1.2 Bipolar Disorder

Advanced paternal age has been linked to a somewhat increased chance of BD in offspring, consistent with a hypothesis of increased new genetic mutations (Frans et al. 2008). However, similar to SCZ, genetic studies have suggested many chromosomal regions and candidate genes appearing to relate to the development of BD, but the results are not consistent and often not replicable (Kato 2007). The first genetic linkage finding for mania was in 1969 (Reich et al. 1969). Later on, meta-analyses of linkage studies detected either no significant genome-wide findings or, using a different methodology, only two genome-wide significant peaks, on chromosome 6q and on 8q21 (Burmeister et al. 2008). GWA studies neither brought a consistent focus—each has identified new loci. Findings point strongly to heterogeneity, with different genes being implicated in different families (Segurado et al. 2003).

A review seeking to identify the more consistent findings suggested several genes related to serotonin, dopamine (e.g., DR4), glutamate, and cell growth and/or maintenance pathways (NRG1 and BDNF), although noting a high risk of false positives in the published literature. Individual genes were suggested to likely have only a small effect and to be involved in some aspect related to the disorder (and a broad range of "normal" human behavior) rather than the disorder per se (Serretti and Mandelli 2008).

10.2.2 Environment–Epigenetic Changes

The role of nature vs. nurture in the etiology of a disease is always questionable.

Simple organisms, such as bacteria, increase their rate of spontaneous mutations to enable the survival of the species in a changing environment. Multicellular organisms use complex mechanisms coordinated by the central nervous system (CNS) to behaviorally adapt to changing environments without paying the high price of mutating their genome (Colvis et al. 2005). Their behavioral adaptation depends on learning and long-term changes in synaptic connectivity, often mediated or supported by epigenetic mechanisms (Fischer et al. 2007; Roth and Sweatt 2009). Thus, "nature" is considered to be inherited or genetic vulnerability, and "nurture" is proposed to exert its effects through epigenetic mechanisms (Roth and Sweatt 2009). Given the essential role of the hippocampus in learning and memory, it is not surprising that the first evidence of dynamic DNA methylation in the adult CNS was found here (Miller and Sweatt 2007). The hippocampus is also one of the two regions in which a specialized form of neural plasticity, that is, the generation of new functional neurons from neural stem cells, occurs throughout adult life. This process, which is also termed adult hippocampal neurogenesis, contributes to learning and memory formation and the regulation of mood (Drapeau et al. 2003; Zhao et al. 2008).

Epigenetics is broadly defined as those heritable changes not dependent on the genomic sequence. Epigenetic programming refers to factors that are "epi," or "on top of" genetic (DNA) sequences and was coined by Waddington in the 1940s to link genes and development (Waddington 1942). Epigenetic regulation allows a single genome to code for functionally different cell types and short-term adaptation ("memory") (Abdolmaleky et al. 2004; Gavin and Sharma 2010). In contrast, DNA sequence changes are responsible for long-term adaptation and evolution. Like the DNA sequence, the epigenetic profile of somatic cells is mitotically inherited, but unlike the DNA sequence, epigenetic signals are dynamic. The epigenetic status of the genome is tissue specific, developmentally regulated, and influenced by both stochastic and environmental factors. The term "epigenetic programming" is evolving, and today refers to reversible molecular changes to DNA, RNA or proteins that regulate gene function, but do not involve DNA base changes. Epigenetic changes include DNA methylation, RNA modification (e.g., editing (addition/deletion/change to base sequence), RNA interference) and both histone (H) and nonhistone protein modifications (e.g., methylation, acetylation, phosphorylation, sumoylation, ubiquitination) (Smith et al. 2010) (see later for details). Because epigenetic processes regulate various genetic and genomic functions, epigenetic factors can have profound phenotypic effects (Mill et al. 2008).

Epigenetic modulation is able to act as interface between the environment and the genome. These processes begin shortly after DNA synthesis, although subsequent alterations may occur in response to a variety of ordinary or pathological environmental or biological factors. Epigenetic changes occur globally in early development, because they play a critical role in normal cellular differentiation during embryogenesis (Thomassin et al. 2001; Li 2002). Because epigenetic mechanisms are mitotically inherited and integral to transcription and genomic function, these dynamic epigenetic changes at key developmental periods could have a lasting influence on phenotype and disease susceptibility (Jirtle and Skinner 2007). The impact of epigenetic modification on gene expression has been extensively studied in cancer (Malik and Brown 2000; Jones and Baylin 2002). These mechanisms also regulate gene expression in neurons, but, as most neurons do not divide, chromatin modifications are instead sustained within individual cells (Tsankova et al. 2007). Although chromatin remodeling is best understood for its influence on neural development, increasing evidence suggests a role in regulating mature, fully differentiated neurons. Epigenetic changes in the brain have been associated with a range of neurobiological processes, including brain growth and development (Pidsley et al. 2010), learning and memory (Borrelli et al. 2008; Roth and Sweatt 2009), drug addiction (Renthal and Nestler 2008), neurodegeneration (Migliore and Coppede 2009), and the circadian body clock (Nakahata et al. 2007). In the last years, it has been widely speculated that epigenetic dysfunction in the brain may account for a spectrum of psychiatric disorders, including psychosis (Pidsley and Mill 2011).

Both SCZ and BD are characterized by sexual dimorphism in disease progression (Kaminsky et al. 2006). In SCZ, for example, men develop symptoms between 15 and 25 years of age, but for women the period of maximum onset is around age 30 with a smaller peak around 50 (i.e., menopausal age) (Hafner 2003). It has been shown that sex hormones, such as estrogen, often act by altering the molecular epigenetic signatures of specific chromosomal regions, modulating the access of transcription factors and producing long-lasting epigenetic effects on gene transcription (Kaminsky et al. 2006).

10.2.2.1 Inheritability and Memory Formation

Transmission of information from one generation to the next follows three routes: (1) a structural one, mainly the oocyte as the unit of cellular activity contributing cytoplasm, organelles, etc., (2) two haploid genotypes, derived from the parental ones by random segregation, linkage, and mutation, and (3) an epigenotype (de Boer et al. 2010). A first hint for a role of epigenetic transmission in genetic instability originated from the field of radiation genetics. DNA methylation is heritable via semiconservative DNA replication by the action of the DNA maintenance DNMT1 that will copy the methylation status of the template strands to the newly synthesized ones. Thus, the change in DNA methylation could serve as a memory for the acute stress endured. It was also shown that in rats, a high dose of paternal cranial irradiation led to methylation changes in the offspring (Tamminga et al. 2008).

Generally, chromatin codes (DNA and histone) are preserved through mitosis, although reprogramming may occur (Cooney et al. 2002). During meiosis and early development, complex differential global chromatin reprogramming occurs, some specific for male or female germline and others for development. As it was mentioned, some epigenetic signals, like gene-specific DNA methylation, rather than being reset and erased during gametogenesis, could be transmitted meiotically across generations (Abdolmaleky et al. 2004; Mill et al. 2008). This has to be kept in mind, as heritable phenotypic variation is assumed to result exclusively from DNA-sequence variants.

Imprinting refers to genes, though present in two copies, which are only expressed on one chromosome, dependent on the parent-of-origin (Biliya and Bulla 2010). The parental alleles are differentially methylated. Thus, imprinting is an effect brought about by epigenetic control. Epigenetic programming imprints some genes to be expressed in a parent-of-origin-dependent manner (Smith et al. 2010). Gene imprinting is proven for approximately 80 genes, and predicted for about 200 genes. Most imprinted genes are associated with growth and development. In female cells, epigenetic changes turn off all gene expression from one X chromosome randomly in each cell during early embryogenesis (Erwin and Lee 2008). This ensures that chromosome X gene expression levels are similar for female (XX) and male (XY) cells. The functional significance of DNA methylation is best established in X chromosome inactivation, and abnormal imprinting can lead to neurodevelopmental diseases.

Throughout our lives, the brain remains flexible and responsive to the outside world. In addition to receiving signals from the outside world, the brain allows us to form memories and learn from our experiences. Particularly, memories of stressful events are strong and sometimes lasting for life (Reul et al. 2009). Most likely, the reason for this is, because such memories help the organism to adapt and respond better, if similar events would reoccur in the future. Thus, usually, memories are important for survival, health, and wellbeing. However, disturbances in this cognitive process may play a role in stress-related psychiatric diseases. Many brain functions are accompanied at the cellular level by changes in gene expression. Epigenetic mechanisms stabilize gene expression, which is important for long-term storage of information. Exposure memory, expressed as epigenetic DNA modifications, allows genomic plasticity and short-term adaptation of each generation to their environment (Abdolmaleky et al. 2004).

10.2.3 Stress

Stress is believed to be one of the main triggers for the development of psychiatric disorders. Of particular importance are stressors suffered in early periods of life, the effects of which may range from structural alterations in the brain through altered stress responsiveness and an increased probability to develop psychiatric disorders (Felitti et al. 1998; Francis and Meaney 1999; Heim et al. 2001; Cavigelli and McClintock 2003; McFarlane et al. 2005; Hanson et al. 2010). Moreover, stressors suffered in infancy may be transmitted across generations both behaviorally and epigenetically (Weaver et al. 2004). It is also worth noting that the stress response is controlled at the level of the genes which strongly affect the way in which one and the same stressor is responded to endocrinologically and behaviorally (Caspi et al. 2003; Hariri et al. 2003; Szeszko et al. 2005).

According to the traditional view of Hans Selye, the father of the stress concept (Selye 1956), the main regulator of stress in our organism is the HPA axis. It consists of the corticotropin-releasing hormone (CRH) and vasopressin as the hypothalamic components, adrenocorticotropin (ACTH) as the hypophyseal component and glucocorticoids (corticosterone in rodents and cortisol in human) secreted from the adrenal cortex.

10 Co-Regulation and Epigenetic Dysregulation in Schizophrenia... 291

10.2.3.1 Schizophrenia

Patients who are in the acute phase of a psychotic disorder have an elevated HPA axis activity—similarly to depression—as shown by raised basal and stimulus-induced cortisol and ACTH levels. Dexamethasone is a potent synthetic glucocorticoid, binds primarily to the glucocorticoid receptors (GR) of the pituitary corticotrop cells and, by activating the negative feed-back mechanism, suppresses ACTH and cortisol secretion. When healthy subjects are treated with dexamethasone prior to CRH infusion, the release of ACTH is blunted. This does not always happen in patients with SCZ (Ceskova et al. 2006). However, the rate of nonsuppression in SCZ varies from 0 to 70%, with a mean rate of approximately 20%, which is much lower than that described in depression (Cotter and Pariante 2002). An increased pituitary volume was also found in patients with recent onset psychoses (Pariante 2008). Antipsychotic medications reduce HPA activation, and agents that augment cortisol release exacerbate psychotic symptoms (Pariante 2008).

In general, it is believed that the involvement of stress hormones in psychopathology is nonspecific. However, stress-related decrease of neuronal density in depression but not in SCZ, as well as stress-induced alterations in hippocampal kainate receptor function in SCZ, but not in depression, argue for disease-specific processes (Cotter and Pariante 2002; Bradley and Dinan 2010).

In line with this idea, one of the best models of SCZ is based on early stress events. Based upon the neurodevelopmental theory, several models were developed, like postweaning isolation, maternal separation, and neonatal lesions of the hippocampus (Van den Buuse et al. 2003). Rearing rat pups from weaning in isolation, to prevent social contact with conspecifics, produces reproducible, long-term changes, including neophobia, impaired sensorimotor gating, aggression, cognitive rigidity, reduced PFC volume and decreased cortical and hippocampal synaptic plasticity (Fone and Porkess 2008). These alterations are associated with hyperfunction of mesolimbic dopaminergic systems, enhanced presynaptic dopamine and serotonergic function in the nucleus accumbens, hypofunction of mesocortical dopamine, and attenuated serotonin function in the PFC and hippocampus. These behavioral, morphological, and neurochemical abnormalities strongly resemble core features of SCZ.

10.2.3.2 Bipolar Disorder

Hyperactivity of the HPA axis in major depression is one of the most consistent findings in psychiatry (Mello et al. 2003; Pariante and Lightman 2008). A significant percentage of depressed patients hypersecrete cortisol, as shown by increased 24-h urinary-free cortisol and elevated plasma and cerebrospinal fluid concentrations of cortisol, as well as by an increased volume of the pituitary gland and of the adrenal glands. In contrast to healthy people, depressed patients show enhanced ACTH responses to CRH treatments during the dexamethasone suppression test. It has even a diagnostic value, as the combined dexamethasone/CRH test is estimated to have a sensitivity of 80% in differentiating healthy subjects from patients with depression (Heuser et al. 1994). Several studies indicate an enhanced vasopressin activity in depression, too (Frank and Landgraf 2008).

There is robust evidence demonstrating abnormalities of the HPA axis in BD, as well (Daban et al. 2005). Hypercortisolism may be central to the pathogenesis of depressive symptoms and cognitive deficits, which may in turn result from neurocytotoxic effects of raised cortisol levels. Manic episodes may be preceded by increased ACTH and cortisol levels, leading to cognitive problems and functional impairments. Manipulation of the HPA axis has been shown to have therapeutic effects in preclinical and clinical studies.

10.3 Epigenetic Machinery

The epigenetic machinery includes factors that can write (covalently attach), read (differentially bind), and erase (remove) chemical moieties to chromatin thereby moderating genomic expression (Zahir and Brown 2011). These modifications are dynamic and may be controlled. It is a finely orchestrated system involving the synchronized team work of many diverse proteins, often in large multicomponent complexes that act on vast portions of the genome. Even small changes in the balance of factors may be pathogenic.

10.3.1 Possible Mechanisms

Chromatin Remodeling

Chromatin is the complex of DNA, histones, and nonhistone proteins in the cell nucleus (Yoo and Crabtree 2009). The fundamental unit of chromatin is the nucleosome, which consists of approximately 147 base pairs of DNA wrapped around a core histone octamer (about 1.65 turns) (Fig. 10.1). Each octamer contains two copies each of the histones H2A, H2B, H3, and H4. The nucleosomal structure of chromatin allows DNA to be tightly packaged into the nucleus by organized folding.

"Chromatin can't be important, otherwise bacteria would have it." said a comment, few years ago at a transcription meeting (Ordway and Curran 2002). If only genomes were important, humans would not have developed more than mice, and plants would have had the same chance of ruling the world (as human, mouse, and *Arabidopsis* genome share the same number of genes, approximately 25,000). In fact, it is the epigenome that has tremendously evolved since the appearance of the first multicellular organisms (Covic et al. 2010).

Because DNA is compacted into a highly condensed and ordered structure (chromatin), considerable interest has focused on how the transcriptional machinery gains access to the genes contained within chromatin and expresses them in an organized program (Nowak and Corces 2004). The alteration of chromatin organization—via covalent modification and/or remodeling of this structure—is

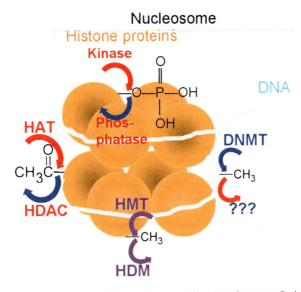

Fig. 10.1 Schematic representation of chromatin remodeling in nuleosome. *Red arrows* indicate activation, *blue arrows* indicate inhibition, *purple arrows* indicate possible activation as well as inhibition. *DNA* desoxyribonucleic acid; *DNMT* DNA methyl transferase; *HAT* histone acetyltransferase; *HDAC* histone deacetylase; *HDM* histone demethylase; *HMT* histone methyltransferase

Table 10.1 Epigenetic mechanism and their possible role in gene expression

Target	Mechanism		Effect on gene expression
DNA	Methylation		Inhibition ↓
Histone	Acetylation		Activation↑
	Methylation	Polycomb complex	Silencing↓
		Trithorax complex	Activation↑
	Phosphorylation		Activation↑
Noncoding RNA	Long noncoding RNA		Regulation↑↓
	MicroRNA		Regulation↑↓

thought to provide access to the genes for the transcription apparatus (Fig. 10.1). In simplified terms, chromatin exists in an inactivated, condensed state: heterochromatin, which does not allow transcription of genes, and in an activated, open state: euchromatin, which allows individual genes to be transcribed (Tsankova et al. 2007). In reality, chromatin can exist in many states in between these two extremes. Portions of chromatin are highly repressed, owing to DNA and histone methylation and the binding of repressor proteins, and might never be accessible for transcription. Other portions of chromatin are in repressed or permissive states; their basal activity is low owing to histone methylation and perhaps other modifications, but the genes are available for derepression and activation in response to transcription factors and transcriptional co-activators. The opening of chromatin is associated with acetylation of nearby histones. Chromatin remodeling modulates gene expression with high temporal and spatial resolution (Table 10.1).

10.3.1.1 DNA Methylation

DNA methylation occurs by transfer of a methyl group from SAM to cytosine residues in the dinucleotide sequence CpG (cytosine next to guanine in phosphodiester bond). Although CpG sequences spread throughout the genome are usually heavily methylated, those occurring in CpG islands (defined as a >500 base pair (bp) region of DNA with >55% GC content (Takai and Jones 2002)) in the promoter regions of genes are less methylated. Namely, in vertebrates, 4–8% of all cytosines, and 70% of cytosines within the CpG dinucleotide sequence, are methylated. In contrast, 70% of the cytosines at CpG dinucleotide sequences within promoter regions of active genes are unmethylated (Smith et al. 2010). The methylation of CpG sites acts to disrupt the binding of transcription factors and attract methyl-binding proteins that initiate chromatin compaction and gene silencing (Klose and Bird 2006; Illingworth and Bird 2009). Usually, dense DNA methylation is associated with irreversible silencing of gene expression, while a strong activator can overcome partial methylation. Partial promoter DNA methylation marks genes that may become unmethylated and expressed, allowing for re-adaptation to a changing micro- or macro- environment (e.g., season, ecological conditions, nutritional habits and demands of different developmental periods). An increasing amount of data implicates a role for *methylomics* in multifactorial psychiatric disorders (Abdolmaleky et al. 2004).

10.3.1.2 Histone Modifications

By far the best characterized chromatin remodeling mechanism in the brain is the posttranslational, covalent modification of histones, the core proteins of nucleosomes, at distinct amino acid residues on their amino (N)-terminal tails (Nowak and Corces 2004; Kouzarides 2007). Such modifications include acetylation, ubiquitylation or sumoylation at lysine (K) residues, methylation at K or arginine (R) residues, phosphorylation at serine (S) or threonine (T) residues, and ADP-ribosylation at glutamate residues. How tightly the DNA is wrapped around the histones, impacts the transcription. Acetylation of K residues is associated with transcriptional activation, whereas methylation may lead to both silencing and activation (Filion et al. 2006; Abdolmaleky et al. 2008; Smith et al. 2010).

The first-studied and best-understood covalent histone modification was histone acetylation by histone acetyltransferases (HATs), which neutralize the proteins' positively charged N-terminal tails (most probably histone 3 lysine 9 acetylation (H3K9ac)), making the negatively charged DNA molecule more available to the machinery of transcription (Hong et al. 1993; Nowak and Corces 2004). Conversely, the histone deacetylase (HDAC) catalyzes the removal of acetyl groups and condenses chromatin around gene promoters, generally resulting in decreased gene expression (Peterson and Laniel 2004).

Histone methylation mediated by the Polycomb (PcG, silencing) and Trithorax (TrxG, activating) group complexes. Members of the PcG complex catalyze trimethylation of lysine 27 of histone 3 (H3K27me3), which leads to transcriptional repression through local heterochromatin formation (Cedar and Bergman 2009) and/or inhibition of transcriptional elongation through stalling of RNA polymerase II (Stock et al. 2007). PcG member Bmi1 has been shown to be required for postnatal neural stem cell self-renewal (Lim et al. 2009). However, it remains elusive, how PcG complexes are recruited to chromatin in mammals (Hublitz et al. 2009).

TrxG complex is recruited by RNA polymerase II and catalyzes H3K4 trimethylation of promoter-proximal nucleosomes (Hughes et al. 2004). However, the presence of H3K4me3 does not predict efficient gene expression as the majority of inactive genes experience transcription initiation (Guenther et al. 2007). The TrxG member mixed-lineage leukemia 1 (Mll1) is required for neurogenesis in the mouse postnatal brain (Lim et al. 2009). H3K27me3 and H3K4me3 can co-occupy the same promoters and such bivalent chromatin structure is thought to allow fast activation of gene expression (Bernstein et al. 2006; Zhao et al. 2007). Typically, PcG-induced repression is easy to reverse and dynamic gain and loss of PcG- and TrxG-mediated histone modifications occur during in vitro developmental progression from embryonic stem cells to fully differentiated neurons (Mohn and Schubeler 2009).

Phosphorylation of H3 was initially linked to chromosome condensation during mitosis (Gurley et al. 1978; Wei et al. 1998). Then, evidence has accumulated that phosphorylation of H3 at serine 10 (H3S10p) has an important role in the transcriptional activation of eukaryotic genes in various organisms. It has been found that mitosis-specific phosphorylation of H3 also occurs at S28 (Goto et al. 1999) and at T11 (Preuss et al. 2003). Concomitant activation of the c-fos and c-jun immediate-early genes and H3 phosphorylation was also shown in fibroblast cells (Mahadevan et al. 1991). The observed time course of H3 phosphorylation mirrored the known expression profiles of these genes, leading the authors to postulate a link between the phosphorylation of H3 and transcriptional activation. Indeed, light pulses induce prominent phosphorylated H3 staining in the suprachiasmatic nucleus sections of the rat, and this change in the distribution of H3S10p coincides with the change in the transcriptional profile of c-fos and the circadian gene Per1 (Crosio et al. 2000). Similar changes of H3S10p and gene expression have been observed after neuronal activation by agonists of dopamine, muscarinic acetylcholine, and ionotropic glutamate receptors in hippocampal neurons (Crosio et al. 2003).

The roles of histone ubiquitylation, sumoylation, and ADP ribosylation are less well understood (Tsankova et al. 2007).

Although often investigated independently, epigenetic modifications to DNA and histones are not mutually exclusive and interact in a number of ways. Even the different histone modifications most likely act in a combinatorial or sequential fashion defining the so-called *histone code* (Grayson et al. 2010). For example, the same fraction of the H3 population is susceptible to hyperacetylation and phosphorylation (Barratt et al. 1994). Antibodies generated specifically to H3 tails that were both phosphorylated and acetylated provided the first direct proof that di-modified H3 isoforms existed in vivo (Cheung et al. 2000). However, these modifications are each deposited via independent pathways, suggesting that H3S10p is not just a signal for subsequent H3K14ac, but these processes occur via parallel-independent pathways.

10.3.1.3 Noncoding RNA

Several other general mechanisms of chromatin remodeling have been described, although they remain less well characterized in the nervous system (Tsankova et al. 2007). We might assume that these mechanism work together, although it is uncertain, what determines the locus-selectivity of the previously mentioned enzymes. However, recent evidence suggests that they are recruited to their sites of action — among others — by nonprotein-coding RNAs (Khalil et al. 2009). The genome is transcribed into a spectrum of noncoding RNAs that are implicated in a wide range of structural, regulatory, and catalytic processes. Some of these classes are well known, such as ribosomal RNAs (rRNAs) and transfer RNAs (Sproule and Kaufmann 2008). Additional classes include a wide range of long noncoding RNAs (lncRNAs) and various types of short noncoding RNAs (microRNA (miRNA), piwiRNAs, small nucleolar RNAs, promoter-associated small RNAs, and transcription initiation RNAs, among others) (Taft et al. 2010). These environmentally sensitive RNA networks are thought to efficiently couple bioenergetic properties with information storage and processing capacity and to be responsible for orchestrating a wide array of biological processes. In the CNS, these factors are implicated in mediating critical functions, including brain patterning, neural stem cell maintenance, neurogenesis and gliogenesis, stress responses, homeostasis, and synaptic and neural network connectivity and plasticity (Mehler 2008).

In the development of the vertebrate nervous system, a key step is when cells lose multipotency and begin to develop stable connections. This transition is accompanied by a switch in ATP-dependent chromatin-remodeling mechanisms that appears to coincide with the final mitotic division of neurons. It was shown that this essential transition is mediated by miRNAs (Yoo et al. 2009). miRNAs are a class of single-stranded RNAs, 19–25 nucleotides in length, generated from hairpin-shaped transcripts (Kim 2005). They control the expression levels of their target genes through an imperfect pairing with target messenger RNAs. Both the mouse and human brain express a large spectrum of distinct miRNAs compared with other organs (Babak et al. 2004; Miska et al. 2004). Therefore, the implications of dysregulation of miRNA networks in human diseases affecting the CNS are potentially enormous (Meola et al. 2009). This assumption is supported by several in vitro examples using both gain- and loss-of-function experiments. For example, the introduction of artificial miRNAs or antisense oligonucleotides induced a loss of function of primary neurons in culture (Schratt et al. 2006; Cheng et al. 2009). In an in vivo experiment in *Xenopus laevis*, miR-24a was necessary for proper neural retina development (Walker and Harland 2009).

Although lncRNAs are one of the most abundant classes of noncoding RNAs and are highly expressed in brain (Ponjavic et al. 2009), they remain poorly characterized, and their roles in the CNS have not been studied in detail. This class of noncoding RNA generally encompasses transcripts longer than 200 nucleotides, of which there are tens if not hundreds of thousands expressed from mammalian genomes (Qureshi et al. 2010). Some lncRNAs also serve as precursors for shorter regulatory noncoding RNAs (e.g., miRNAs) (Mattick and Makunin 2005).

A major function of lncRNAs appears to be to modulate the epigenetic status of proximal and distal protein-coding genes through *cis-* and *trans-*acting mechanisms that include the recruitment of chromatin remodeling complexes to specific genomic loci, thereby regulating chromatin structure.

10.3.2 Proteins Involved

In 2007, there were 197 known genes with epigenetic function (Zahir and Brown 2011), and this number is increasing rapidly. Here, we try to enumerate the most important contributors with known neuronal properties.

10.3.2.1 DNA Methyl Transferase

DNA methyl transferases (DNMT1, 3A, 3B) are the enzymes with primary epigenetic programming function (Szyf 2010). DNMT1 is considered to be the maintenance methyltransferase and DNMT3A and DNMT3B are termed *de novo* methyltransferases (Portela and Esteller 2010). DNMT1 is the most abundant methyltransferase in somatic cells. Aberrant expression of DNMT1 has been shown to result in embryonic lethality in mammals (Biniszkiewicz et al. 2002). Members of the DNMT3 family display more tissue specific expression (Rottach et al. 2009).

Today, no DNA demethylases have been identified, although their existence may be assumed. DNA methylation has traditionally been viewed as an irreversible epigenetic marker. However, some studies indicate that patterns of DNA methylation might be reversible even in adult neurons. It has been proposed that demethylation might occur through activation of second messenger signals that lead to the recruitment of HATs, which, through increased histone acetylation, would allow DNA demethylases greater access to the promoter (Meaney and Szyf 2005).

10.3.2.2 Modulators of Methylation

Transcription can be inhibited by proteins that bind directly or indirectly to methylated DNA (Smith et al. 2010). One methylated DNA-binding family, consisting of the methyl-CpG-binding protein 2 (MeCP2), MBD1, MBD2, MBD3, and MBD4 proteins, has a conserved methyl-binding domain (MBD) and binds singly methylated CpG dinucleotides (Klose and Bird 2006). MBDs are associated with large protein complexes containing HDACs and HMTs, which further repress gene transcription.

MeCP2 can act both as an activator and a repressor, affecting the expression patterns of about 85% of genes (Zahir and Brown 2011). Among others, GRIN, the *N*-methyl-D-aspartate receptor subunit gene, is also affected which is a well-documented causative gene for SCZ (Zhao et al. 2006).

The CREB-binding protein (CBP) is a transcriptional co-activator with HAT activity. CBP is a protein that is important for activating genes involved in learning and memory. Environment-induced adult neurogenesis is extrinsically regulated by CBP function in mature granule cells (Lopez-Atalaya et al. 2011).

A repressor family, all containing a zinc-finger motif, consists of Kaiso protein, which binds CGCGs, the Kaiso-binding sequence protein, and other proteins that bind lone methylated CpG dinucleotides (Filion et al. 2006). This zing-finger motif can be found also in the KAP1, a family of vertebrate-specific epigenetic repressors with largely unknown functions (Jakobsson et al. 2008). KAP1 is expressed at high levels and necessary for repression of genes in mature neurons of the mouse brain.

10.3.2.3 Chromatin Remodeling

Many HATs, which catalyze acetylation, have been identified. Several transcriptional activators (like CBP) contain intrinsic HAT activity.

HDACs catalyze deacetylation. They also associate with several transcriptional repressors to further repress chromatin activity. The superfamily of HDACs consists of at least 18 members that are divided into two main families: (1) the classic HDAC family, and (2) the silent information regulator 2-related protein (sirtuin) family. The classic or zinc-dependent HDACs include class I (HDACs 1, 2, 3, and 8), class IIa (HDACs 4, 5, 7, and 9), class IIb (HDACs 6 and 10), and class IV (HDAC11). The distribution of the HDAC1-11 enzyme is already known in the rat brain (Broide et al. 2007). The sirtuin (SIRT) family is structurally unrelated to classic HDACs. These enzymes have a unique mechanism of action requiring the cofactor NAD for enzyme activity (Mottet and Castronovo 2008). SIR2 was originally identified as a factor necessary for transcriptional silencing. Later on, it came into the limelight as an antiaging molecule, as overexpression of SIR2 was found to extend life span in yeast (Kaeberlein et al. 1999). However, recent studies confuted this idea (Burnett et al. 2011). The balance between the opposing activities of HATs and HDACs maintains acetylation on core histones and is thought to be an important determinant of transcription. Valproate, a drug used to treat epilepsy and BD, inhibits HDAC (Fukuchi et al. 2009).

Methylation at lysine (K) or arginine (R) residues is mediated by histone methyltransferases (HMTs). In general, histone K methylation is regarded as a more stable modification than other histone modifications, which seem to be more readily reversible, although the recent discovery of histone demethylases (HDMs) indicates that even methylation can be reversed.

It was demonstrated that stimulation-dependent phosphorylation of H3 is a rapid and transient event. Several different pathways include activation of kinases (e.g., MAP kinases) that—among others—could induce phosphorylation of histones. The specific kinase which is responsible for H3S10 phosphorylation is still unknown. However, a tandem kinase of the mitogen- and stress-activated kinase (MSK) family seems to be a good candidate (Jin et al. 1999). More is known about the degradation by phosphatases. The main important one could be the phosphatase type 2A (PP2A) which regulates the final phosphorylation state of the histones (Nowak et al. 2003).

10.3.3 Signaling Pathways

Several external stimuli can induce rapid changes in histone modifications in the brain, but the intracellular mediators of these signals are poorly defined.

One of the best-established scenarios is the cyclic adenosine monophosphate (cAMP) cascade. The activation of several signaling pathways involves cAMP production leading to protein kinase A activation. This leads to the phosphorylation of cAMP-response element-binding protein (CREB). CREB phosphorylation triggers the recruitment of CBP, a transcriptional co-activator, whose intrinsic HAT activity acetylates nearby histones, which loosens the chromatin and allows subsequent transcriptional activation. CBP is important for normal learning and memory, and mutations of CBP cause Rubinstein–Taybi syndrome, a form of mental retardation in humans.

The activation of cellular Ca^{2+} pathways in muscle leads to the activation of calcium/calmodulin-activated protein kinases (CaMKs), which phosphorylate class II HDACs. This phosphorylation triggers the shuttling of the enzyme out of the nucleus, and results in increased histone acetylation. This pathway has now been demonstrated in hippocampal cells and cerebellar granule neurons, indicating that chromatin signaling mechanisms in different tissues overlap substantially.

Among neural genes, BDNF is one of the most studied for its regulation by chromatin remodeling. Neuronal activity can trigger the phosphorylation of the repressor MeCP2, which regulates activity-dependent induction of BDNF expression. Phosphorylation of MeCP2 depends on CaMKII.

10.3.3.1 Histone Phosphorylation and c-Fos

Immediate-early genes, such as c-Fos transcription is induced rapidly in the brain by numerous stimuli. c-Fos induction required rapid and transient H3S10 phosphorylation and H3K14 acetylation and H4 acetylation in its promoter region, which is a brain region-specific event with a prominent appearance in the dentate gyrus (Chandramohan et al. 2008). Phosphoacetylation of H3 and resultant c-Fos expression after psychologically stressful events requires concurrent signaling via a glutamate (N-methyl-D-aspartate, NMDA) and the glucocorticoid receptor (GR) (Bilang-Bleuel et al. 2005; Chandramohan et al. 2008). Downstream of the NMDA receptor, activation of the extracellular signal-regulated kinase (ERK) mitogen-activated protein kinase (MAPK) pathway results in the phosphorylation of MSK1 and Ets-like protein kinase-1 (Elk-1) (Chandramohan et al. 2008). Subsequently, MSK1 phosphorylates H3S10 and Elk-1, through recruitment of the HAT, allows the acetylation of H3K14. Finally, these histone modifications lead to the recruitment of the gene transcription machinery resulting in the induction of c-Fos and other gene products (Chandramohan et al. 2008; Reul et al. 2009). GABA-A receptor function may affect novelty-evoked H3S10p-K14ac and c-Fos induction in dentate neurons through modulation of NMDA receptor-mediated Ca^{2+} gating. Activation of the GABA-A receptor hyperpolarizes the dentate neurons, preventing the influx of Ca^{2+} through the NMDA receptor and thereby abrogating signaling to the chromatin.

10.3.4 Dysfunction of Genes Encoding Epigenetic Regulators

Mutation in epigenetic modifiers may be particularly prone to exhibiting pleiotropy. Pleiotropy occurs when a single gene influences multiple phenotypic traits. It can arise when the dysfunctional gene's product affects a number of downstream targets, as in the case of epigenetic genes (Knight et al. 2006). Therefore, it is not a surprise that an overlap is observed between the broader clinical neurodevelopmental disorder categories, since it is not obvious, how genotypes are correlating to phenotypes. For example, patients with SCZ can display behaviors, which are part of the spectra of other disease categories, such as BD (Burmeister et al. 2008).

The loss of an epigenetically functional gene is more frequently implicated than the gain of the same gene (Lee and Scherer 2010). The indication is not that losses occur more frequently than gains, but that losses are less well tolerated than gains. It can be hypothesized that the overexpression of epigenetic regulators should not impact the overall functional outcome, because having more product would not alter the normal sequestering of these factors. However, the lack of sufficient epigenetic regulators would result in an impairment of the sum functional outcome.

There are known monogenic disorders with dysfunction of genes encoding epigenetic regulators. An example for that is the deregulation of imprinting in the Prader–Willi and Angelman syndromes (Buiting 2010). Having one faulty copy of the CBP gene causes Rubinstein–Taybi syndrome, a condition with a variety of characteristics, including mental disability. Knocking out the MBD1 protein leads to deficits in learning and social interaction (Allan et al. 2008), a typical phenotype of SCZ. Disruption of genes encoding HDACs was also analyzed for a correlation with SCZ. Kim et al. found HDAC4 genes to be associated with the disease in a large South Korean population (Kim et al. 2010). Others have found elevated brain expression levels for HDAC1 in SCZ postmortem brain samples (Sharma et al. 2008).

However, to say that a given gene may directly control a given characteristic is at best oversimplification (Hamer 2002). More prevalent is a "silent" mutation in a member of a multiunit chromatin remodeling complex, which may not be phenotypically evident in one individual. However, in another individual, who may have a variant in an additional member of the same multiunit complex, the combined effects of the mutated epigenetic regulators may manifest in a disease outcome (Zahir and Brown 2011).

Although epigenetic mechanisms have been implicated in syndromes associated with mental impairment, polymorphisms in DNMTs can determine the normal variation in human intelligence as well (Haggarty et al. 2010).

10.4 Epigenetic Changes in Schizophrenia

Besides other theories (e.g., dopamine hypothesis, glutamate hypothesis, serotonin hypothesis, neurodevelopmental hypothesis, genetic hypothesis), nowadays the epigenetic hypothesis of SCZ is also widely accepted. This was already suggested in 1982 by Gottesman et al. (1982) providing an integrative theory for the disease. The so-called multifactorial origin of SCZ may become unified under the epigenetic umbrella (Petronis et al. 1999). Since then, many mechanisms, brain areas and molecules were supposed to be involved in epigenetic regulation in SCZ patients.

10.4.1 Mechanisms

10.4.1.1 Methyl Supply

The epigenetic hypothesis is, in part, based on results from a series of clinical studies that were carried out in the late 1960s and were designed to improve the treatment of SCZ patients (Costa et al. 2002). The experiment was based on the premise that a decrease in brain dopamine levels should prove beneficial, since haloperidol, an effective antipsychotic, was able to block dopamine receptors. To this end, SCZ patients and nonpsychiatric subjects were treated with high doses of methionine (Met). Met was expected to increase the activity of enzymes that metabolize dopamine through methylation and to reduce dopamine levels in the brain. It was anticipated that the net effect would be to improve the symptoms of patients. Instead, Met was found to elicit acute psychotic episodes, if these patients had been previously symptomatic for SCZ (Berlet et al. 1965; Antun et al. 1971). Initially, it was thought that Met induced the formation of a dopamine metabolite capable of initiating a psychotic episode. However, the levels of various dopamine metabolites remained largely unchanged. Later on, it was shown that Met mediates changes in gene expression and promoter methylation in mice, both in vivo and in vitro (Dong et al. 2005; Tremolizzo et al. 2005). Interestingly, Met was shown to increase brain SAM levels, and this increase coincided with changes in promoter methylation and changes in gene expression.

In line with these findings, alterations of the one-carbon metabolism cycle were found in the brain of SCZ patients. Homocysteine is part of this cycle which generates the precursors for the molecules involved in the methylation of DNA (Frankenburg 2007). Several studies have detected abnormal levels of homocysteine and folate in the plasma of individuals with SCZ; elevated homocysteine levels in blood and urine are associated with a higher risk of the disorder (Levine et al. 2002; Haidemenos et al. 2007; Ma et al. 2009). Furthermore, high homocysteine levels in pregnancy have been shown to increase the risk of SCZ in offspring (Brown et al. 2007), and prenatal folate deficiency is known to cause neural tube defects and is linked to psychosis (Zammit et al. 2007; Krebs et al. 2009).

Methylenetetrahydrofolate reductase (encoded by the *MTHFR* gene) is a key enzyme of the one-carbon metabolic cycle and acts to catalyze the conversion of 5,10-methylenetetrahydrofolate to 5-methyltetrahydrofolate, a cosubstrate for homocysteine remethylation to Met (Roffman et al. 2008). Another enzyme, COMT is an SAM-dependent methyltransferase that methylates catechol compounds (Tunbridge et al. 2008). Research shows that functional polymorphisms in the

MTHFR and *COMT* genes interact to affect plasma total homocysteine levels. Genetic association studies also provide evidence of a link between the SNP in *MTHFR* and both SCZ and BD (Roffman et al. 2008).

10.4.1.2 Epigenetic Machinery

DNA Methylation

Mill and colleagues published the first epigenome-wide study of psychosis using postmortem tissue obtained from the PFC (Mill et al. 2008). Significant differences were found between cases and controls at a number of genes, including several that have been functionally linked to psychosis in previous work. Epigenetic differences, many sex-specific, were identified at genes involved in neuronal development and other loci implicated in both the glutamatergic and GABAergic neurotransmitter pathways. Other studies using postmortem brain samples reported DNA methylation differences between cases and controls in the *COMT* (Petronis 2000), *reelin* (*RELN*) (Schwab et al. 2000), *DR2* (Cohen et al. 2008), *and sex-determining region Y-box 10 (SOX10)* (McGowan et al. 2009) genes, although subsequent studies using quantitative methylation-profiling methods have not replicated the *COMT* and *RELN* findings (Oberlander et al. 2008; Moser et al. 2009).

As a background, DNMT1 was found to be overexpressed in the PFC of SCZ and BD patients (Veldic et al. 2005; Zhubi et al. 2009). Another DNMT, DNMT3a is also upregulated in the PFC of SCZ patients compared with control subjects (Zhubi et al. 2009).

Overall, the data suggests that DNA methylation may indeed be an epigenetic mechanism that contributes to the aberrant regulation of genes associated with SCZ.

Histone Modification

Although we know far more about DNA methylation than about posttranslational modification of histone tails in SCZ, this process seems to be important, as the functional significance of histone modifications in the regulation of genes involved in SCZ is increasingly becoming the focus of vigorous research. The special importance of these changes are that histone marks are relatively preserved in human postmortem tissues (Stadler et al. 2005; Deutsch et al. 2008; Akbarian and Huang 2009).

In 2005, Akbarian's group provided the first demonstration that there are histone modifications associated with SCZ (Akbarian et al. 2005). Specifically, in a small subset of PFC samples, they found a significantly higher open chromatin-associated H3R17 methylation in SCZ patients than controls (Akbarian et al. 2005). Later, it was also found that there is altered H3K4 and H3K27 trimethylation (Huang and Akbarian 2007; Huang et al. 2007) and increased HDAC1 expression in the PFC of SCZ patients (Sharma et al. 2008). A subsequent study by Gavin and colleagues used peripheral blood samples to show that patients with SCZ had significantly higher levels of H3K9me2 than healthy control subjects, and that among patients H3K9me2 was negatively correlated with age of onset (Gavin et al. 2009). Conversely, a study looking at different histone modifications of six lysine (K), serine (S), and arginine (R) residues of H3 and H4 found no overall significant differences in levels of histone marks in the PFC of individuals with SCZ compared with normal control subjects, but found a subset of SCZ patients with 30% higher H3R17me levels than the control group (Akbarian et al. 2005). Thus, initial findings of the role of histone modifications in psychosis are mixed, but certainly encourage further research in this area.

During recent years, an impressive catalog of data has accumulated in animal research providing strong evidence for a role of H3S10p-K14ac and associated gene expression in hippocampus-associated learning and memory processes which could be influenced by previous stress (Collins et al. 2009; Reul et al. 2009). As one of the key abnormalities of SCZ is cognitive impairment, these epigenetic modifications may also contribute to the development of symptoms. To date, specific histone modification has been studied for the *BDNF* gene. Specifically, H3 acetylation and phosphorylation is increased at *BDNF* gene promoters in association with exon-specific BDNF gene expression in an electroconvulsive seizure model of SCZ (Tsankova et al. 2004). Other forms of histone modifications at *BDNF* promoters, such as methylation, have also been described in a mouse model of depression (Tsankova et al. 2006). These data along with postmortem analyses suggest that histone modifications might also contribute to the aberrant regulation of genes associated with SCZ.

The drugs used for therapy may influence the epigenetic machinery, too. The antipsychotic drugs haloperidol and raclopride (both dopamin DR2 antagonists) rapidly induce phosphoacetylation of H3 in the mouse striatum, specifically at the c-Fos gene promoter (Li et al. 2004). Treatment with valproate that, among other actions, inhibits HDACs, increased RELN expression in vitro and in vivo, and this effect was accompanied by decreased methylation of the *RELN* promoter (Chen et al. 2002; Tremolizzo et al. 2002). Furthermore, valproate attenuated SCZ-like behavioral abnormalities in a Met-induced epigenetic mouse model of the illness. Other HDAC inhibitors produced similar results (Tremolizzo et al. 2005). Although the mechanisms by which HDAC inhibitors reduce DNA methylation are unknown, it is thought that hyperacetylation can regulate the accessibility of DNMT1 to promoter regions or that it might induce DNA demethylase activity (Grayson et al. 2006).

MicroRNAs

A link between miRNA biogenesis and SCZ has also been suggested (Maes et al. 2009; Meza-Sosa et al. 2012). For example, Dgcr8, a key protein for miRNA biosynthesis is encoded within a fragment of chromosome 22 which is commonly deleted in SCZ patients (see also "Dopamine" in Sect. 4.3.1). Recently, the genes of miRNAs encoded in the X chromosome have been sequenced in order to identify alterations that might be related to the appearance of X-linked SCZ (Bicker and

Schratt 2008). Results confirmed that ultrarare variants in the sequence of mature pre-miRNAs (miR-18b, miR-502, miR-505) or mature miRNAs (let-7f-2, miR-188, miR-325, miR-509-3, miR-510, miR-660) are frequent in men affected with this disorder, compared with healthy individuals (Feng et al. 2009). Noncoding genes of miR-198 and miR-206 were also associated with SCZ (Perkins et al. 2007) and miR-181b is up-regulated in SCZ-affected brains, where mitochondrial dysfunction has been documented (Ben Shachar and Laifenfeld 2004).

In the context of memory problems of SCZ patients, in a mouse model miR-128b was associated with formation of fear-extinction memory (Lin et al. 2011).

These results show that an altered function of miRNAs might also contribute to the development of SCZ, although further studies are required to prove this association formally.

Long Non-Coding RNAs

Although lncRNAs were first described in cancers and during development, recently their role was also implicated in many neurological disorders like neurodevelopmental or neurodegenerative diseases. More recently, lncRNAs were found to be associated with psychiatric disorders as well (Qureshi et al. 2010). More specifically, the disruption of the DISC genomic locus, which encodes both the DISC1 protein-coding gene and the DISC2 lncRNA, has been linked in a number of genetic analyses to the risk of developing SCZ, schizoaffective disorder, BD, major depression, and autistic spectrum disorders (Chubb et al. 2008). DISC2 overlaps DISC1 and is transcribed in the opposite direction. Like other antisense transcripts, DISC2 is implicated in regulating the expression of its partner, DISC1 which modulates multiple aspects of CNS structure and function including embryonic and adult neurogenesis (Brandon et al. 2009). However, DISC2 may also represent an important candidate gene for psychiatric disease, separately from its effects on DISC1 (Qureshi et al. 2010).

10.4.2 Brain Areas Involved

Functional brain imaging studies found many brain areas with possible contribution in the development of SCZ symptoms (Brown and Thompson 2010). First of all, primary and secondary sensory areas could be responsible for the hallucinations (Fig. 10.2). Negative symptoms were hypothesized to be associated with frontal brain areas. Impaired functional connections between the frontal and temporal cortex were also found. Nevertheless, epigenetic studies were mainly concentrated on the PFC.

In the PFC of SCZ subjects, abnormal DNA or histone methylation at sites of specific genes and promoters is associated with changes in RNA expression (Akbarian 2010). DNMT1 and DNMT3a were found to be overexpressed in the PFCs (Veldic et al. 2005; Zhubi et al. 2009), while levels of HDAC1 expression

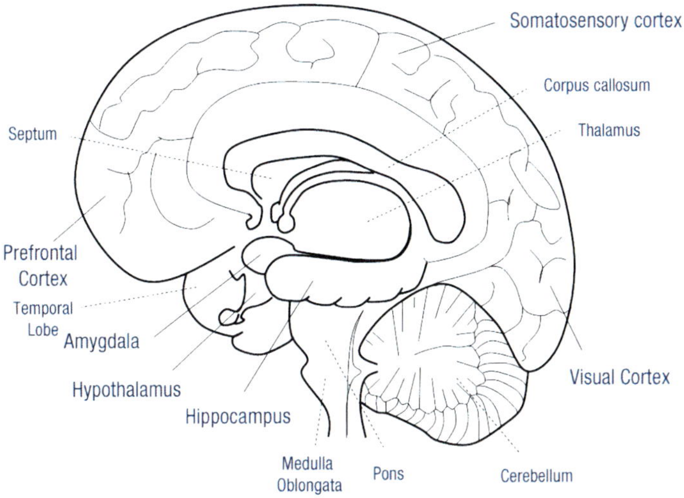

Fig. 10.2 Possible brain areas involved in the development of schizophrenia and bipolar disorders. *Dotted lines* indicate additional brain areas. Prefrontal cortex and hippocampus could play a role in both psychopathologies

were also significantly higher in the PFC of SCZ patients than in unaffected control subjects (Gavin et al. 2008). The COMT promoter was studied in the Brodmann Area 46 (dorsolateral PFC) of normal versus psychotic (SCZ, BD) individuals, and was found to be hypomethylated, accompanied by the overexpression of the COMT mRNA. Changes in the methylation of the GABA synthesizing enzyme (GABA decarboxylase 1, GAD1) and RELN gene, as well as in miRNA levels could be detected also in this brain regions (Benes and Berretta 2001; Veldic et al. 2004; Meza-Sosa et al. 2012).

Some changes in the miR-181b were found also in the temporal cortex of SCZ patients (Beveridge et al. 2008).

Similarly to the PFC, in GABAergic neurons of caudate nucleus and putamen (basal ganglia), an increased expression of DNMT1 and a decrease of RELN and GAD1 occurs in SCZ patients. However, in contrast to the PFC (Veldic et al. 2005), these changes were not detectable in BD patients, suggesting different epigenetic mechanisms in the pathogenesis of SCZ and BD (Veldic et al. 2007).

As the hippocampus is very important in memory formation, and memory disturbances are common in SCZ patients, we have to take into consideration the epigenetic changes of this area as well (Hsieh and Eisch 2010). With all the recent attention given to epigenetic modifications in mature neurons, it is easy to forget that epigenetic mechanisms were initially described for their ability to promote differentiation and drive cell fate in embryonic and early postnatal development, including neurogenesis. Given the discovery of ongoing neurogenesis in the adult brain and the intriguing links among adult hippocampal neurogenesis, hippocampal function and neuropsychiatric disorders, it is important to know that key stages and aspects of adult neurogenesis are driven by epigenetic mechanisms, too.

Table 10.2 Molecules with epigenetic modifications in schizophrenia and bipolar disorder

	Molecule	Gene involved	Epigenetic mechanism	Consequence
Schizophrenia	Dopamine	COMT	Hypomethylation	Dopamine level ↑
	Serotonin	5HT2A receptor	Hypomethylation↓	5HT2A signaling ↑
	Glutamate	NMDA receptor	Hypermethylation	Downregulation
		VGLUT1	Hypermethylation	Downregulation
		VGLUT2	Hypomethylation	Upregulation
		GLS2	Hypermethylation	Increased expression
	GABA	GAD1	Hypermethylation	GABA level ↓
	Reelin	RELN	Hypermethylation	Reelin level ↓
	BDNF		Hypermethylation	Expression and protein↓
Bipolar disorder	GABA	GAD1	Hypermethylation	GABA level ↓
	Reelin	RELN	Hypermethylation	Reelin level ↓
	BDNF		H3 acetylation	Expression ↑
	GR		Hypermethylation	Expression ↑

COMT catechol-O-methyl transferase; *5HT2A* serotonin 2A receptor; *NMDA* N-methyl-D-aspartate; *VGLUT* vesicular glutamate transporter; *GLS* glutaminase; *GABA* γ-amino butyric acid; *GAD* glutamate decarboxylase; *RELN* reelin gene; *BDNF* brain-derived neurotrophic factor; *GR* glucocorticoid receptor

10.4.3 Possible Contributing Molecules

10.4.3.1 Neurotransmitters

Dopamine

Dopamine is involved in several brain functions, including attention, executive function (e.g., working memory) and reward mechanisms. Dysfunction of the limbic circuitry in SCZ and its influence on dopamine release is considered to be the primary pathophysiological mechanism of SCZ (Weinberger et al. 2001). The dopamine hypothesis of SCZ arose, because many antipsychotic medications used for the treatment are dopamine receptor antagonists (Table 10.2).

The gene encoding the main dopamine degrading enzyme COMT has been intensively investigated in genetic studies of SCZ (Tunbridge et al. 2006). COMT is of particular interest, because of its great physiological importance, but also because of its location in the SCZ "linkage hotspot," 22q11 which is deleted in velocardio facial syndrome (VCFS, also called Di George syndrome or 22q11 deletion syndrome). (The importance of the chromosome 22 was already mentioned in 4.1.2. in connection with miRNAs.) VCFS patients have a range of symptoms, including psychosis, and together with replicated findings of linkage with SCZ in the 22q11 region, genes within this locus are strong candidates for investigation in SCZ research (Tunbridge et al. 2006). In SNP studies, there have been many reports of association between the COMT Val allele and SCZ (for details, see Sect. 2.1.1), but the data have not always been consistent. Combined genetic and epigenetic data are likely to produce more accurate predictors of psychiatric phenotypes than genetic variation alone and could be better correlated with gene-environment interactions in an integrated model. DNA methylation analyses of COMT have been conducted in postmortem brain series. The promoter of membrane bound COMT (MB-COMT) has been found to be methylated, and this isoform of COMT is predominantly responsible for the metabolism of dopamine in the brain. One study indicated that methylation of the MB-COMT promoter was reduced by approximately 50% in SCZ and BD subjects, compared with controls, especially in the left PFC (Abdolmaleky et al. 2006). Furthermore, MB-COMT gene expression was raised in SCZ and BD, compared with the controls, and subjects with the 158Val allele had lower levels of MB-COMT promoter methylation. MB-COMT hypomethylation was also correlated with promoter hypomethylation of the dopamine DR2 receptor gene in SCZ and BD, compared with controls (Abdolmaleky et al. 2006). Another study has indicated that the 158Val homozygote subjects had a greater degree of exonic DNA methylation in the PFC, but there was no association between the level of methylation and psychosis in this region (Mill et al. 2008). However, these studies did not specifically measure methylation within the dorsolateral PFC, so it has not yet been established whether methylation of COMT could be related to working memory in SCZ.

Serotonin

Altered serotonergic function is hypothesized to increase vulnerability to psychiatric diseases, including SCZ, anxiety disorders and affective disorders. To date, 14 serotonin receptors have been identified, of which the serotonin-2A (5HT2A) receptor is implicated in SCZ (Hannon and Hoyer 2008). The involvement of serotonin in SCZ first emerged from the observation that lysergic acid diethylamide (LSD), a potent 5-HT2 receptor agonist, had hallucinogenic properties (Gaddum and Hameed 1954; Woolley and Shaw 1954). The LSD-induced psychosis includes both the hallucinations and delusions observed in SCZ patients, but not the negative symptoms (such as withdrawal, blunted affect and apathy). The affinity of many hallucinogenic drugs for 5HT2 receptor sites has been found to be closely related to their potency as hallucinogens in humans (Glennon et al. 1984; Titeler et al. 1988). The atypical antipsychotic drugs and many antidepressants have high affinities for 5HT2 receptors, too (Meltzer 1999).

Genetic studies demonstrated variation of the gene encoding the 5HT2A receptor. There are several findings of association between the *HTR2A* polymorphism at position 102 and SCZ in Caucasian and Japanese studies (Inayama et al. 1996; Williams et al. 1996; Williams et al. 1997). Another study has shown that the HTR2A is paternally imprinted in human fibroblasts and transcribed from the maternal allele only (Kato et al. 1998). However, Polesskaya and Sokoloff (2002) were not able to replicate this "on or off" polymorphic imprinting, and a more recent study also found no evidence of imprinting of *HTR2A* (De Luca et al. 2007). These conflicting data suggest that these effects are not universal and could vary with brain region, ethnicity, and/or environmental influences (Kouzmenko et al. 1997, 1999; Bray et al. 2004).

Methylation events could provide explanations for the conflicting data generated by different research groups investigating *HTR2A* expression and genetic association with SCZ. Two polymorphic sites in *HTR2A* have been detected to have methylated CpG sites (Polesskaya et al. 2006). The methylation of the *HTR2A* allele at locus 102 was found to correlate with DNMT1 expression levels. Furthermore, methylation of the promoter correlated with *HTR2A* expression levels. However, in a postmortem study no differences were found in *HTR2A* methylation between controls and SCZ cases (De Luca et al. 2009). Epigenetic variation and imprinting of *HTR2A* in SCZ have yet to be extensively tested. The finding of differential DNA methylation within *HTR2A* could indicate variation in the activity of different alleles and thus confound genetic association studies of *HTR2A*, especially because the gene has not been intensively screened for methylated sites.

Glutamate

Glutamate is the most abundant fast excitatory neurotransmitter in the mammalian nervous system, with a critical role in synaptic plasticity (Mill et al. 2008). The glutamate hypothesis of SCZ was initially based on a set of clinical, neuropathological, and later, genetic findings, pointing at a hypofunction of glutamatergic signaling via the ionotropic NMDA receptors (Javitt 2010). Phencyclidine (PCP) and its analogues are non-competitive antagonists of the NMDA receptor (Javitt and Zukin 1991). PCP induces all the symptoms seen in SCZ. The so-called positive symptoms are hallucinations, particularly auditory, hostility, agitation, and paranoid delusions (Egan and Weinberger 1997). The negative symptoms are such as flattened emotional response, social withdrawal, anhedonia, and reduced initiative. PCP also induces cognitive deficits, such as impaired learning and memory (Javitt and Zukin 1991). The dopamine theory of SCZ only has strong explanatory power for the positive symptoms, which are induced by an overactivity of dopamine, as seen in amphetamine-induced psychosis. PCP psychosis gives us a better model of SCZ. Since NMDA is a glutamate receptor, and PCP is an NMDA antagonist that induces SCZ-like symptoms, it has been hypothesized that SCZ is due to glutamatergic underactivity (Kim et al. 1980).

Some authors speculated that down-regulation of the NMDA receptor subunit NR2A may stem from hypermethylation after DNMT1 upregulation (Woo et al. 2004; Costa et al. 2007). The promoters of the NMDA-receptor-subunit gene *GRIN3B* and the 2-amino-3-(5-methyl-3-oxo-1,2-oxazol-4-yl)propanoic acid (AMPA) receptor-subunit gene *GRIA2* were found to be hypomethylated in SCZ males (Gupta et al. 2005; Lau and Zukin 2007; Mill et al. 2008). Various types of glutamate transporters are present in the plasma membranes of glial cells and neurons. Two vesicular glutamate transporters (VGLUTs), which pack glutamate into synaptic vesicles, are also altered epigenetically in major psychosis. Previous studies suggested that VGLUT1 and VGLUT2 are expressed in a complementary manner in cortical neurons (Fremeau et al. 2001). The hypermethylated *VGLUT1* is downregulated in the brains of SCZ patients, while hypomethylated *VGLUT2* is upregulated (Smith et al. 2001; Eastwood and Harrison 2005). Other glutamatergic genes like *GLS2*, which encodes a glutaminase enzyme that catalyzes the hydrolysis of glutamine to glutamate, was also hypermethylated in SCZ samples. Previous studies report that the expression of this enzyme is altered in SCZ patients (Bruneau et al. 2005). The gene encoding secretogranin 2 (SCG2), a secretory protein located in neuronal vesicles that is known to stimulate the release of glutamate, was also hypomethylated in major psychosis relative to unaffected controls (Mill et al. 2008). SCG2 expression is known to be modulated by both chronic PCP exposure (Marksteiner et al. 2001), and lithium treatment (McQuillin et al. 2007).

There is a well-known interaction between stimulatory NMDA and inhibitory GABA receptors and their epigenetic regulation may also go in a parallel fashion.

γ-Aminobutyric Acid

γ-Aminobutyric acid (GABA) is synthesized from L-glutamic acid in a reaction catalyzed by glutamic acid decarboxylase (GAD) and acts as a potent inhibitory neurotransmitter (Pinal and Tobin 1998). Hypofunctioning GABAergic interneurons appear to be important in the etiology of major psychosis (Benes and Berretta 2001). The genes encoding GAD1 (or GAD67) and GAD2 (or GAD65) are downregulated in the brains of psychosis patients (Akbarian et al. 1995; Guidotti et al. 2000; Benes and Berretta 2001). A group found reduced GAD1 mRNA and H3K4me3 in female, but not male SCZ patients, with no difference in H3K4me3 or expression levels at GAD2 in comparison with control subjects (Huang and Akbarian 2007).

Altered GABA activity appears responsible for at least some of the clinical features of SCZ (Wassef et al. 2003; Daskalakis et al. 2007). Moreover, there are risk associated SNPs in GAD1, which are also associated with decreased expression in the hippocampus and dorsal PFC of patients with SCZ (Straub et al. 2007). However, this does not exclude the possibility that methylation of DNA may also contribute to the down-regulation of RELN and GAD1 activity observed in SCZ. Indeed, there is growing evidence that DNA methylation has a role in the dysfunction of GABAergic neurons in SCZ. In adult cortical GABAergic interneurons, RELN, GAD1, and DNMT1 mRNAs co-localize (Veldic et al. 2004, 2005; Kundakovic et al. 2007). Reports have shown that both DNMT1 mRNA and protein levels are significantly increased in the cortex of individuals with SCZ (Veldic et al. 2004, 2005; Ruzicka et al. 2007), and that these changes induce deficits in both RELN (for details, see "Reelin" in Sect. 4.3.2) and GAD1 (Guidotti et al. 2000; Veldic et al. 2005; Ruzicka et al. 2007). Similar results have been documented in GABAergic neurons located in the basal ganglia (Veldic et al. 2007) and hippocampus (Fatemi et al. 2000). Since these discoveries, reduction of RELN and GAD1 expression (both the mRNA and protein) have been one of the most consistent findings in the postmortem studies (Abdolmaleky et al. 2004, 2005; Dracheva et al. 2004; Akbarian and Huang 2006; Grayson et al. 2006; Straub et al. 2007; Gonzalez-Burgos and Lewis 2008).

Data suggest that the down-regulation of RELN and GAD1 transcripts is most likely due to hypermethylation of their gene promoters (Abdolmaleky et al. 2005; Grayson et al. 2005; Kundakovic et al. 2007; Costa et al. 2009). However, it is important to note that one study has documented decreased methylation of the *GAD1* promoter in SCZ patients with both repressive chromatin and lower levels of GAD1 mRNA (Huang and Akbarian 2007). A possible explanation for this apparent paradox is that down-regulation of GAD1 in the hippocampus also appears attributable to the up-regulation of an HDAC enzyme (Benes et al. 2007).

Chronic treatment of mice with Met produces an SCZ-like phenotype together with a replication of some of the molecular aspects of SCZ. These include an increase in methylation of the RELN promoter and the constituent down-regulation of RELN and GAD1 in GABAergic neurons (Tremolizzo et al. 2002; Dong et al. 2005). Using these mice, investigators have also been able to show that both *RELN* and *GAD1* promoters show increased recruitment of methyl-CpG-binding proteins, such as MeCP2 (Dong et al. 2005). Finally, using neuronal progenitor cells to further understand the epigenetic regulation of these genes, investigators have also shown that both DNMT and HDAC inhibitors activate *RELN* and *GAD1* (Kundakovic et al. 2009).

Though there is strong evidence for epigenetic changes in *RELN* and *GAD1* in SCZ, it is not likely that their epigenetic dysfunction alone confers susceptibility to SCZ. Rather, it is likely that many genes are affected, as indicated by a study utilizing GWA epigenetic approaches (Connor and Akbarian 2008; Mill et al. 2008). Indeed, GWA studies have recently found as many as 100 loci with altered CpG methylation in SCZ, including several other gene families related to the GABAergic system: glutamate receptor genes (NR3B and GRIA2), glutamate transporters (VGLUT1 and 2) and a protein that regulates production of GABA receptors (MARLIN-1) (Connor and Akbarian 2008; Mill et al. 2008).

MARLIN-1, coding for an RNA-binding protein that is widely expressed in the brain and regulates the production of functional GABA-B receptors, is hypermethylated in SCZ, BD, and major psychosis (Couve et al. 2004; Mill et al. 2008). In addition, KCNJ6, a G protein-coupled inwardly rectifying potassium channel that has been linked to the regulation of GABA neurotransmission, was found to be hypermethylated in SCZ patients, too (Koyrakh et al. 2005; Mill et al. 2008). Increasing evidence suggests that both the glutamate and GABA systems are synergistically involved in major psychosis, supporting the observation of increased *HELT* promoter methylation in SCZ and BD samples (Coyle 2004). HELT is known to determine GABAergic over glutamatergic neuronal fate in the developing mesencephalon (Nakatani et al. 2007).

Acetylcholine

Acetylcholine (Ach) is one of the main neurotransmitters both at the periphery and in the CNS. It has a variety of effects upon plasticity, arousal and reward. ACh has an important role in the enhancement of sensory perceptions as well (Himmelheber et al. 2000). Damage to the cholinergic (Ach-producing) system in the brain has been shown to be plausibly associated with memory deficits. Moreover, both muscarinic and nicotinic Ach receptors (nAchRs) have been shown to have an important role in cognition and are therefore viewed as potential therapeutic targets for drugs designed to lessen cognitive deficits (Money et al. 2010). Importantly, acetylcholinesterase inhibitors, which result in higher synaptic levels of Ach, can reduce the cognitive deficits of SCZ. In line with this finding, tobacco smoking is frequently abused by SCZ patients. The major synaptically active component inhaled from cigarettes is nicotine, which improves cognitive performance and attention in both experimental animals and in human subjects, including patients affected by neuropsychiatric disorders. Hence the smoking habit in SCZ may represent an attempt to use nicotine self-medication to correct the CNS nAchR dysfunction.

Besides the well-known functions of Ach, it may also regulate the epigenetic machinery. It was also shown, that in the PFC region of mice, repeated injections of nicotine which achieve plasma concentrations comparable to those reported in high cigarette smokers, result in an epigenetically induced increase of GAD1 expression, parallel with a decrease in the DNMT1 mRNA and protein levels (Satta et al. 2008). Down-regulation of DNMT1 expression induced by nicotine is also observed in the hippocampus, but not in striatal GABAergic neurons. The data showed that in the PFC, the same doses of nicotine that decreased DNMT1 expression also diminished the level of methylation in the GAD1 promoter and prevented the Met-induced hypermethylation of the same promoter. Pretreatment with a nAchR blocker that penetrates the blood-brain barrier, prevents the nicotine-induced decrease of PFC DNMT1 expression.

It was also confirmed that the $\alpha4\beta2$ nAChR agonists may be better suited to control the epigenetic alterations of GABAergic neurons in SCZ than the $\alpha7$ nAChR agonists (Maloku et al. 2011). A high-affinity partial agonist at $\alpha4\beta2$ nAChR elicited a 30–40% decrease of cortical DNMT1 mRNA and an increased expression of GAD1 mRNA and protein. This upregulation of GAD1 was abolished by an nAChR antagonist. Furthermore, the level of MeCP2 binding to GAD1 promoters was significantly reduced following $\alpha4\beta2$ agonist administration. In contrast, an agonist of the homomeric $\alpha7$ nAChR failed to decrease cortical DNMT1 mRNA or to induce GAD1 expression.

10.4.3.2 Neurotrophin Factors

Reelin

One hypothesis for the contribution of epigenetic malfunction to SCZ proposed that hypermethylation (regulated by DNMT1) downregulates multiple genes (such as *Reelin* (*RELN*)) in GABA-containing neurons, causing dysfunction of GABA-mediated neuronal circuitry. Higher brain function would be impaired, because synchronization with other neuronal networks would be disrupted as a consequence.

RELN encodes an extracellular matrix glycoprotein that is not only important for neural development and synapse integrity, but plays a necessary role in the long-term potentiation that supports synaptic and behavioral plasticity in the adult (Costa et al. 2002). RELN is expressed predominantly by GABAergic interneurons which regulate neighboring glutamatergic neurons (Pesold et al. 1998). Postmortem studies of patients with SCZ reveal a significant downregulation of RELN expression in several brain regions that is not associated with neuronal loss (Grayson et al. 2006). The *RELN* promoter contains a large CpG island, indicating that DNA methylation might be important for regulating its expression (Chen et al. 2002). Repeated Met administration causes hypermethylation of its promoter and results in the down-regulation of *RELN* transcription in the heterozygous RELN$^{+/-}$ mouse model of SCZ (Tremolizzo et al. 2002). Met treatment also induced MeCP2 binding to the *RELN* promoter (Dong et al. 2005). By contrast, treatment with the methylation inhibitor 5-aza-2'-deoxycytidine (decitabine) upregulated RELN expression in vitro (Chen et al. 2002).

In connection with dopamine, the promoter methylation state of the *RELN* gene was significantly linked to Val158Met genotype (for details, see Sect. 2.1.1) (Smith et al. 2010). All SCZ and control subjects possessing a Val/Val genotype had a hypermethylated *RELN* promoter and a decrease in RELN gene expression. This is consistent with the results that Val SNP is connected with SCZ and hypermethylation of the *RELN* promoter, and subsequent low expression of the RELN gene in the PFC is also correlated with SCZ (Abdolmaleky et al. 2005; Grayson et al. 2006).

RELN controls the surface expression of two NMDA receptor subunits (NR2B and NR1) contributing to a possible deficit in NMDA receptors in the inter-neurons of SCZ patients, too (Groc et al. 2007).

Brain-Derived Neurotrophic Factor

Brain-derived neurotrophic factor (BDNF) is another gene known to play an important role in cognition. Aberrant regulation of this gene has been implicated in the etiology and pathogenesis of several cognitive and mental disorders, including SCZ (Weickert et al. 2003, 2005; Angelucci et al. 2005; Tsankova et al. 2007; Lu and Martinowich 2008). The BDNF protein is synthesized from a gene that has a rather complex structure. The *BDNF* gene contains nine 5' noncoding exons (I-IX) linked to a common 3' coding exon (IX), which codes for BDNF preprotein (Liu et al. 2006; Aid et al. 2007). Despite the known importance of BDNF function and gene expression in normal neural processes and CNS disorders, there has been little investigation into the molecular mechanisms responsible for complex BDNF transcriptional readout in the brain. Recently, several studies have begun to implicate DNA methylation as a provocative molecular mechanism contributing to ongoing regulation of BDNF transcription in the CNS to mediate synaptic plasticity and memory formation (Martinowich et al. 2003; Levenson et al. 2006; Liu et al. 2006; Lubin et al. 2008; Nelson et al. 2008). It has been shown that alterations in DNA methylation levels at the *BDNF* promoter or intragenic regions occurs in response to fear learning (Lubin et al. 2008). Specifically, a Pavlovian learning paradigm (contextual fear conditioning) elicits changes in hippocampal DNA methylation across the *BDNF* gene, and this mechanism is involved in differential *BDNF* transcript read-out necessary for long-term memory formation. Furthermore, pharmacologically inhibiting DNMT activity sufficiently alters basal BDNF transcript levels in the hippocampus (Lubin et al. 2008). In addition to regulation of gene expression changes supporting synaptic plasticity and memory formation, *BDNF* DNA methylation has also been shown to play a role in altered gene expression in response to environmental influences, such as social experiences (Roth et al. 2009). In accordance with the developmental hypothesis of SCZ, stressful social experiences early in life have long-lasting effects on behavior, including increased anxiety, increased drug-seeking behavior, cognitive deficits, and altered affiliative behaviors. Depriving an infant of social interaction yields a reduction in hippocampal and cortical BDNF mRNA and protein levels that persists well into adulthood (Branchi et al. 2004; Fumagalli et al. 2007; Lippmann et al. 2007). It was also demonstrated that aversive social experiences early during the first postnatal week trigger lasting changes in DNA methylation across the *BDNF* gene that is associated with decreases in *BDNF* gene expression in the adult PFC (Roth et al. 2009). Valproate, an anticonvulsant drug with known HDAC inhibitor activity has also been shown to activate *BDNF* promoter IV in cultures of rat cortical neurons (Yasuda et al. 2009).

Together, these studies shed light on the potential role of epigenetic mechanisms in the dynamic regulation of the *BDNF* gene in the adult CNS, and highlight the fact that altered BDNF regulation could contribute to SCZ (Lu and Martinowich 2008). However, there has been little human investigations in the field. Postmortem reports have indicated that in the PFC and hippocampus of SCZ patients there is both decreased BDNF protein (Knable et al. 2004; Torrey et al. 2005; Weickert et al. 2005) and mRNA levels (Weickert et al. 2003; Angelucci et al. 2005; Weickert et al. 2005). Whether DNA methylation is a mechanism responsible for the abnormal regulation of the *BDNF* gene, is not confirmed. However, Mill and colleagues found modest evidence for an association between DNA methylation and the *BDNF* genotype at an SNP (Val66Met), which affects exonic CpG sites (Mill et al. 2008).

Neuregulin

NRG3 is a neural-enriched member of the epidermal growth factor (EGF) family and a specific ligand for the receptor tyrosine kinase ErbB4 (Zhang et al. 1997). In the mouse, NRG3 is restricted to the developing and adult CNS (Zhang et al. 1997; Longart et al. 2004) and plays a critical role in the development of the embryonic cerebral cortex via regulation of cortical cell migration and patterning (Assimacopoulos et al. 2003; Anton et al. 2004). In humans, NRG3-ErbB4 signaling promotes oligodendrocyte survival with possible participation in neurodevelopment and adult brain function (Carteron et al. 2006).

Recurrent microdeletions of chromosome 10 that involve the NRG3 gene, have been associated with a heterogeneous group of neurodevelopmental disorders (Balciuniene et al. 2007). The same locus also shows linkage to SCZ in Ashkenazi Jewish and Han Chinese populations (Fallin et al. 2003; Liu et al. 2007), and noncoding genetic variation in NRG3 has been identified as a putative risk factor for SCZ and related neuropsychiatric disorders (Wang et al. 2008; Chen et al. 2009; Morar et al. 2011). In the PFC of 400 individuals, the expression of the NRG3 isoforms is developmentally regulated and pathologically increased in SCZ (Kao et al. 2010). Genetic association studies have also identified multiple genes and epistatic locus interactions within the NRG-ErbB signaling pathway that increase the risk for SCZ, including NRG3, NRG1, and ErbB4, suggesting a pathogenic gene network (Benzel et al. 2007).

A circuit-based framework of SCZ involves glutamatergic, dopaminergic, GABAergic and cholinergic neurotransmitter systems (Lisman et al. 2008). Here, the aberrant glutamatergic or cholinergic function reduces the GABAergic control of principal neurons of the hippocampus, thereby causing disinhibition of glutamatergic outflow with increased neuronal activity in the ventral tegumental area, and subsequent generation of a hyperdopaminergic state in the hippocampus. Stimulation of NRG1-ErbB4 signaling inhibits hippocampal long-term potentiation at glutamatergic synapses of the hippocampus by rapidly increasing extracellular dopamine levels and activation of DR4 (Kwon et al. 2008). These findings position the NRG1-ErbB4 signaling pathway at the crossroads between dopaminergic and glutamatergic neurotransmission (Neddens et al. 2009).

Till now epigenetic changes of an NRG gene have been shown only for cancers. The short arm of chromosome 8 is frequently lost in epithelial cancers, and NRG1 is the most centromeric gene that is always affected. NRG1 may therefore be the major tumor suppressor gene postulated on 8p. It is in correct location, is antiproliferative and is silenced by methylation in many breast cancers (Chua et al. 2009).

10.4.3.3 *Sex-Determining Region Y-Box Containing Gene 10*

Several lines of evidence suggest the alteration of oligodendrocytes in SCZ. Electron microscopic studies revealed the ultrastructural alteration of oligodendroglias (Uranova et al. 2004). Immunohistochemical analysis revealed the downregulation of oligodendrocyte proteins in gray matter (Honer et al. 1999; Flynn et al. 2003) and alteration in number and density of oligodendrocytes (Hof et al. 2003). Gene expression analyses using DNA microarray also support the alteration of oligodendrocytes in SCZ (Bunney et al. 2003; Mirnics et al. 2004). Coordinated downregulation of a subset of oligodendrocyte-related genes (referred to as oligodendrocyte dysfunction) in the PFC (Hakak et al. 2001; Tkachev et al. 2003; Sugai et al. 2004) and temporal cortices (Aston et al. 2004) of patients with SCZ have been revealed.

Among the downregulated oligodendrocyte genes, the *SOX10* (*sex-determining region Y-box containing gene 10*) seems to be the most important one, because a combination of transcription factors, such as SOX10 and oligodendrocyte lineage transcription factor 1 (OLIG1) and OLIG2 regulates the differentiation of oligodendrocytes (Kessaris et al. 2001), and SOX10 is responsible for terminal differentiation of oligodendrocytes (Stolt et al. 2002). The *SOX10* gene, an oligodendrocyte-specific transcription factor with a large CpG promoter island, is hypermethylated and downregulated in the PFC of SCZ patients (Iwamoto et al. 2005). Interestingly, examination of SNPs in the SOX10 gene led to unequivocal results among Japanese SCZ patients (Iwamoto et al. 2006; Maeno et al. 2007). Therefore, we might assume that epigenetic changes of the *SOX10* gene is even more important than genetic variations.

10.4.3.4 Others

WNT1, an integral part of the Wnt signaling pathway which is critical for neurodevelopment and is differentially expressed in SCZ brains, was significantly hypermethylated in SCZ patients, relative to controls (Miyaoka et al. 1999; Mill et al. 2008).

The transcriptionally inducible nuclear-receptor NR4A2, downregulated in both SCZ and BD, was also found to be hypermethylated in SCZ (Mill et al. 2008).

FosB, which encodes a protein controlling cell proliferation in the brain known to be expressed following chronic antipsychotic treatment, was hypomethylated in SCZ patients, relative to controls (Kontkanen et al. 2002; Mill et al. 2008).

The genes for LIM homeobox transcription factors LMX1B and LHX5, linked to normal learning and motor functions, also showed significant methylation changes in psychosis samples, with *LMX1B* demonstrating putative hypomethylation and *LHX5* demonstrating putative hypermethylation (Paylor et al. 2001; Mill et al. 2008).

Since the phospholipid metabolism is also disturbed in SCZ (the membrane hypothesis of SCZ (Horrobin et al. 1994)), it is not a surprise that the phospholipase gene PLA2G4B was hypermethylated in SCZ samples, too (Mill et al. 2008).

As cannabis use during adolescence increases the risk of developing psychotic disorders later in life, we might assume a strong link between SCZ and the endocannabinoid system (Bossong and Niesink 2010). This system regulates other neurotransmitters, especially the glutamate and GABA release (Katona et al. 1999). Adolescent exposure to tetrahydrocannabinol, the primary psychoactive substance in cannabis, transiently disturbs physiological control of the endogenous cannabinoid system over glutamate and GABA release. This may adversely affect adolescent experience-dependent maturation of neural circuitries within PFC areas. As a result, psychosis or SCZ could develop. This idea is also supported by the fact, that early postweaning isolation, a neurodevelopmental animal model of SCZ, induces messenger RNA changes in almost all members of the endocannabinoid system (Robinson et al. 2010). Till now experimental studies did not confirm the role of epigenetic regulation in the effect of cannabinoids, although it can be strongly supposed. For example, anandamide, an endogenous cannabinoid ligand, regulates keratinocyte differentiation by inducing DNA methylation (Paradisi et al. 2008).

10.5 Epigenetic Changes in Bipolar Disorders

The significant discordance of depression between monozygotic twins (who often share the same environment as well as genes), the failure in identifying a single genetic risk factor, and depression's twofold female predominance suggest the presence of a third, nongenetic and nonenvironmental component to variability (Mill and Petronis 2007). Epigenetic modifications have been implicated as a significant contributor to this third source of variability (Krishnan and Nestler 2010).

Both clinical and preclinical epigenetic changes can be divided into different categories based upon the mechanisms involved, the targeted brain areas or molecules. However, there are remarkable overlaps between these categories, and new mechanisms suggest the involvement of other players, as well.

10.5.1 Epigenetic Mechanisms

10.5.1.1 DNA Methylation

As already mentioned, methylation changes—mostly in the promoter region of different genes—is one of the principal mechanisms of epigenetic regulation.

Although it is more typical in SCZ patients, the downregulation of GAD1 or RELN expression in GABAergic neurons of BD patients is also associated with an overexpression of DNMT1 and DNMT3a, although here the mostly involved brain area is the striatum (Veldic et al. 2007). The increased promoter methylation induced by the overexpression of DNMTs in BD patients may be the cause of the downregulation of GABAergic genes. However, the inhibitory action of DNMTs on gene expression may also occur through the formation of repressor complexes (Guidotti et al. 2011).

10.5.1.2 Histone Modification

One of the widely used tests for antidepressant efficacy is the forced swim test (FST) available both in mice and rats. It was established that behavioral changes in this model go in parallel with histone modifications (Chandramohan et al. 2008). More specifically, FST evoked a transient increase in the number of phosphoacetylated H3-positive neurons in the hippocampus. As both NMDA receptors and the ERK signaling pathway are involved in neuroplasticity processes underlying learning and memory, it was not surprising that antagonism of NMDA receptors and inhibition of ERK signaling blocked FST-induced H3 phosphoacetylation and the acquisition of the behavioral immobility response. Moreover, double knockout mice of the H3 kinase—MAPK kinase completely abolished the FST-induced increases in H3 phosphoacetylation and the behavioral immobility response. These preclinical data suggest a strong connection between histone modification and development of depression-like changes.

An indirect evidence for the involvement of histone modification in the development of BD is that valproate, an anticonvulsant drug with mood stabilizer effect, enhances GABAergic transmission through an increase of GAD1 expression via inhibition of HDACs (Guidotti et al. 2011).

10.5.1.3 MicroRNA

People who commit suicide have less-active rRNA genes than people who die of other causes. In people who commit suicide, methyl levels are higher on rRNA genes in the hippocampus, which is important for learning and memory. More methyl means less rRNA production, which means fewer ribosomes, which means less protein production.

Analyses of differentially expressed noncoding miRNAs between healthy individuals and BD patients revealed a significant reduction in the expression of miR-132 in the PFC of the patients (Miller and Wahlestedt 2010). The fact that miR-132 regulates Gap250 levels and that this protein negatively regulates BDNF-induced neurite outgrowth and synapse formation suggests that in patients with BD, its reduced level results in alteration of neurite growth and synaptic function (Vo et al. 2005). This idea is supported by the fact that lithium chloride improves BD symptoms in parallel with an increase in neuro- and synaptogenesis. Furthermore, lithium chloride and valproate altered the expression of 37 and 31 miRNAs, respectively, further regulators of neurite outgrowth and neurogenesis (reviewed by Miller and Wahlestedt 2010). Additional miRNAs may also be involved, because lithium chloride also affected the expression of miRNAs whose predicted targets are genes, such as the thyroid hormone receptor β, the dipeptidyl-peptidase 10, and the metabotropic glutamate receptor 7. All of these molecules were previously identified as risk factors for BD (Miller and Wahlestedt 2010; Meza-Sosa et al. 2012). Further, experimental data are required to validate these predictions.

10.5.2 Brain Areas Involved

Several studies suggested neural networks modulating aspects of emotional behavior to be implicated in the pathophysiology of mood disorders. These networks involve the PFC, amygdala, hippocampus, and ventromedial parts of the basal ganglia, where alterations in grey matter volume and neurophysiological activity are found in cases with recurrent depressive episodes (Rigucci et al. 2010) (Fig. 10.2). However, epigenetic changes in the hippocampus and hypothalamus were studied in detail in connection with BD.

The hippocampus is implicated in the pathophysiology and treatment of depression (Nestler et al. 2002). Chronic electroconvulsive seizures, one form of antidepressant therapy, upregulate the expression of BDNF and CREB in the hippocampus, and such upregulation has been shown to mediate antidepressant activity in animal models (Duman 2004).

The strong connection between depression and stress underlines the importance of stress-regulatory brain areas in the development of the disorder. Indeed, several epigenetic changes were found in preclinical models in the hypothalamus both in the CRH and vasopressin genes (Murgatroyd et al. 2009; McClelland et al. 2011).

As previously mentioned, miRNA expression differences were found in the PFC between BD patients and healthy controls (Miller and Wahlestedt 2010).

10.5.3 Possible Contributing Molecules

Cortical GABAergic dysfunction, a hallmark of SCZ is a substantial factor in the development of BD as well. Thus, BD pathophysiologies may relate to the hypermethylation of GABAergic gene promoters (i.e., RELN and GAD67), too (Dong et al. 2008) (for details, see "γ-Aminobutyric Acid" in Sect. 4.3.1 and "Reelin" in Sect. 4.3.2) (Table 10.2).

10.5.3.1 Brain-Derived Neurotrophic Factor

Chronic electroconvulsive seizures increased H3 acetylation at BDNF promoters 3 and 4, in correlation with increased expression of the corresponding BDNF transcripts (Tsankova et al. 2007). In contrast, acute electroconvulsive seizure-induced H4 acetylation, suggesting an epigenetically memorized aspect of the chronicity. Because of the chronic nature of neuropsychiatric disorders, chronic stimulus-induced changes seem to be more relevant for the disease.

Chronic social defeat stress, an animal model of depression that mimics many symptoms of human depression, also alters chromatin regulation of BDNF (Berton et al. 2006; Tsankova et al. 2007). At a molecular level, chronic defeat stress in mice induces sustained downregulation of the expression of two splice variants of BDNF in the hippocampus. These changes are reversed after chronic treatment with the tricyclic antidepressant imipramine. As a possible epigenetic background, chronic defeat stress induces robust and long-lasting increases in H3K27 dimethylation, a repressive modification, specifically at the promoters of the downregulated BDNF transcripts. Interestingly, it was not reversed by imipramine treatment.

Rather, chronic imipramine was shown to reverse the downregulation of BDNF transcripts III and IV in the hippocampus of animals subjected to long-term social defeat stress (Tsankova et al. 2006). Imipramine seems to reverse the repression of the BDNF gene by inducing H3 acetylation, a stimulating modification, at the same promoters. Indeed, chronic imipramine downregulates HDAC5 expression in the hippocampus, but only in animals previously subjected to chronic social defeat. Moreover, viral-mediated overexpression of HDAC5 in this region prevents imipramine's restoration of BDNF levels, as well as the drug's antidepressant effects.

10.5.3.2 Glucocorticoid Receptor

One of the earliest and most well-characterized models used to assess the effects of the environment on neurological outcomes is in relation to early postnatal maternal care. Disruption of the normal maternal behavior—among others—can lead to the development of mood disorders in the offspring (Dudley et al. 2011). In rats, some mothers naturally display high levels of nurturing behaviors, such as licking, grooming, and arched-back nursing (high-LG-ABN), whereas others have low levels of such behaviors (low-LG-ABN) (Champagne et al. 2003). Offspring of high-LG-ABN mothers are less anxious, have attenuated corticosterone responses to stress and display increased expression of glucocorticoid receptor (GR) mRNA and protein in the hippocampus when compared to pups of low-LG-ABN mothers (Meaney and Szyf 2005). The upregulation of GR mRNA specifically involves the alternatively spliced variant GR17. The promoter that drives the expression of this variant is brain specific, and contains a consensus-binding sequence for nerve growth factor inducible factor-A (NGFI-A), the expression of which is upregulated in the pups of high-LG-ABN mothers. It was shown that low-LG-ABN pups have increased methylation at the NGFI-A consensus sequence region of the GR17 promoter (Weaver et al. 2004). This difference in methylation emerged in the first week of life and persisted into adulthood. As adults, in the offspring of low-LG-ABN mothers methylation of the GR17 promoter prevented binding of the transcriptional enhancer NGFI-A, effectively disrupting the normal transcriptional regulation of the GR gene. Although this epigenetic methylation was long-lasting, it could be reversed by infusion of the class I and II HDAC inhibitor trichostatin A. Interestingly, cross-fostering also reversed the differences in methylation at that site (Weaver et al. 2004). Therefore, we might assume that the epigenetic changes were dependent on the maternal behavior of fostering, rather than the biological mother, i.e., independent of germ-line transmission (Krishnan and Nestler 2010).

Despite promising preclinical findings, subsequent studies on postmortem brains did not confirm epigenetic changes of the GR gene in depressed patients (Alt et al. 2010).

10.5.4 Possible Further Areas Based Upon New Mechanisms

10.5.4.1 Immune System

According to some concepts, depression is a disease of the dysfunction of the immune system (Leonard 2010). Both experimental and clinical evidence shows that a rise in the concentrations of proinflammatory cytokines and glucocorticoids, as occurs in chronically stressful situations and in depression, contribute to the behavioral changes associated with depression. A defect in serotonergic function is associated with hypercortisolemia, and an increase in proinflammatory cytokines that accompany depression. Glucocorticoids and proinflammatory cytokines enhance the conversion of tryptophan to kynurenine. In addition to the resulting decrease in the synthesis of brain serotonin, this leads to the formation of neurotoxins, such as the glutamate agonist quinolinic acid and contributes to the increase in apoptosis of astrocytes, oligodendroglia, and neurons. As there are several players in the immune regulation, there are many potential targets for genetic as well as epigenetic regulation.

Interferons (IFNs) and IFN receptors are present in the limbic system, where they appear to exert physiological effects, especially when levels rise during CNS infections. IFNs interact closely with cytokines and nitric oxide, signaling molecules implicated in depression. Results from knock-out mice suggest a role for IFN-gamma in moderating fear and anxiety while other lines of evidence point to a role in arousal and circadian rhythms (Kustova et al. 1998). The IFN-alpha receptor deploys an arginine methyltransferase affecting RNA editing and splicing which seem to be disrupted in SCZ and BD. The previously mentioned methyl donor SAM is an effective antidepressant, and it may influence depression—at least in part—through variations in the strength of IFN-alpha signaling by hypermethylation (Hurlock 2001).

10.5.4.2 White Matter/Glia Abnormalities

White matter is the brain region underlying the gray matter cortex, composed of neuronal fibers coated with electrical insulation called myelin (Fields 2008). Myelination continues for decades in the human brain; it is modifiable by experience, and it affects information processing by regulating the velocity and synchrony of impulse conduction between distant cortical regions. This myelin coat is composed by oligodendroglia. Glial cells were first identified as nonneuronal elements in the nineteenth century by the anatomist Virchow (Rajkowska and Miguel-Hidalgo 2007). At that time, glia was thought to be no more than silent supportive "glue" for neurons, and that glia were unable to participate in information processing. In the past 2 decades, research has changed this perception and provided evidence for glia being an important dynamic partner of neuronal cells, actively participating in brain metabolism, synaptic neurotransmission, and communication between neurons (Volterra and Meldolesi 2005). A reduction in the density and ultrastructure of oligodendrocytes was detected in the PFC and amygdala in depression. Pathological changes in oligodendrocytes may be relevant to the disruption of white matter tracts in mood disorders, as reported by diffusion tensor imaging. Myelin is involved in normal cognitive function, learning, and IQ. Cell-culture studies have identified molecular mechanisms regulating myelination by electrical activity, and myelin also limits the critical period for learning through inhibitory proteins that suppress axon sprouting and synaptogenesis. Factors, such as stress, excess of glucocorticoids, altered gene expression of neurotrophic factors and glial transporters, and changes in extracellular levels of neurotransmitters released by neurons, may modify glial cell number and affect the neurophysiology of depression.

BD and SCZ (see Sect. 4.3.3), as well as major depressive disorder may share common oligodendroglial abnormalities (Aston et al. 2005). Although methylation changes of the SOX10 gene were not confirmed in BD patients, but from SCZ studies, we might assume similar alterations in BD.

Unanticipated reductions in the density and number of glial cells were reported in fronto-limbic brain regions in major depression and BD (Rajkowska and Miguel-Hidalgo 2007). Moreover, age-dependent decreases in the density of glial fibrillary acidic protein (GFAP)-immunoreactive astrocytes were observed in the PFC of younger depressed subjects. Since astrocytes participate in the uptake, metabolism and recycling of glutamate, we hypothesize that an astrocytic deficit may account for the alterations in glutamate/GABA neurotransmission in depression.

Epigenetic modification of the GFAP promoter was described in gliomas, where its hypermethylation led to gene silencing (Restrepo et al. 2011).

In addition to findings of a stress-induced downregulation of BDNF at the epigenetic level, one of its mediating receptors in astrocytes, TrkB (tropomyosin-related kinase B), was found to be epigenetically reduced in the PFC of suicide completers (Ernst et al. 2009). Although six out of ten suicide completers with an epigenetic downregulation of TrkB suffered from depression, further analyses found no association of depression or substance abuse with hypermethylation of the TrkB promoter (Schroeder et al. 2010).

10.6 Diagnostic Possibilities

Despite the tissue-specific nature of the epigenome, there is increasing evidence that many epimutations are not limited to the affected tissue or cell type but can also be detected in other tissues (Pidsley and Mill 2011). Good examples are the epimutations at insulin-like growth factor 2 in lymphocytes (Cui et al. 2003) and mutL homologue 1 in sperm cells (Suter et al. 2004) observed in colon cancer patients; and epimutations at KCNQ1 overlapping transcript 1 in lymphocytes and skin fibroblasts in Beckwith–Wiedemann syndrome (Weksberg et al. 2002). In psychosis, studies have highlighted the skewing of X-inactivation in buccal mucosa and peripheral blood mononuclear cells (PBMCs) from females with BD (Rosa et al. 2008) and hypermethylation at a locus in lymphoblastoid cell lines derived from BD patients (Kuratomi et al. 2008). In fact, peripheral blood-based studies may be useful in revealing epigenetic changes resulting from early embryogenesis or even highlighting inherited epigenetic variation. The study of peripheral tissues may have many advantages over the use of postmortem brain tissue. Besides the fact that PBMCs are likely to accumulate fewer epigenetic changes induced by disease-related external factors, such as medications and stress, they could be used for diagnostic purposes as well as for the prediction of the treatment efficacy. Logistically, peripheral tissues are easy to obtain and could be used for prospective longitudinal epigenetic studies tracking changes associated with psychosis, as the disorder develops. Our understanding about what comprises a "normal" epigenome is still limited, and no study has attempted to systematically map epigenetic variation between specific brain regions and correlate these with DNA methylation in peripheral tissues from the same individual. Understanding the relationship between brain and peripheral epigenetic signatures could pave the way for future large-scale epigenomic studies of psychosis and other neuropsychiatric conditions.

As proof of concept, a study by Gavin and colleagues used peripheral blood samples to show that patients with SCZ had significantly higher levels of H3K9me2 than healthy control subjects, and that among patients H3K9me2 was negatively correlated with the age of onset (Gavin et al. 2009).

10.7 Treatment Possibilities

Environmental factors that affect DNA methylation include diet, proteins, drugs, and hormones (Abdolmaleky et al. 2004).

10.7.1 Diet

Diet can affect the methylation status of genes through the availability of methyl donors, thereby controlling their transcription or controlling the mode, how drugs can alter the expression levels of DNMTs (Waterland et al. 2006).

Methyl-donors, usually contained in the diet, are required for the formation of SAM (Frankenburg 2007). SAM, in turn, acts as a methyl-donor for the methylation of cytosine DNA residues. Examples of dietary factors required for the formation of SAM include folate, methionine, choline, vitamin B12, vitamin B6, and vitamin B2 (Pidsley and Mill 2011).

In one study, females of two mice strains were fed with methyl supplements prior to and during pregnancy. This treatment increased the level of DNA methylation of the studied gene (agouti) and changed the phenotype of the offspring into the healthy, longer-lived direction (Cooney et al. 2002). The results suggests that optimum dietary supplements are required for the health and longevity of the offspring.

10.7.2 Drugs Acting on the Epigenetic Machinery

Drugs targeting epigenetic mechanisms are under development for the treatment of various cancers (Acharya et al. 2005; Grayson et al. 2010). A less compelling argument can be made for the use of epigenetic drugs to modify various regulatory cascades in postmitotic cells, such as neurons. However, there has been considerable recent interest in this therapeutic approach, because, in contrast to genetic alterations, changes in epigenetic marks are potentially reversible (Egger et al. 2004). Some drugs that were already used to treat mental illness, work partially by changing gene expression. These changes in gene expression are stabilized through epigenetic mechanisms, reversing the effects of the disease.

Currently, available drugs are HDAC inhibitors and DNMT inhibitors. The majority of clinical trials that have been designed to evaluate the effects of these drugs on various types of cancers, and many have shown promising results, including favorable efficacy and safety profiles.

10.7.2.1 HDACs Inhibitors

It is noteworthy that the HDAC inhibitor N-hydroxy-N'-phenyl-octanediamide (SAHA or vorinostat) and two methylation inhibitors, 5-azacytidine and decitabine have been approved by the US Food and Drug Administration (FDA) for the treatment of refractory cutaneous T-cell lymphoma and myelodysplastic syndromes. It seems clear that many HDAC inhibitors will be potent weapons in the battle against various cancers. The effects of these drugs are likely to be much broader and more complicated than originally envisioned (Bolden et al. 2006). This is due in part to the observation that HDAC inhibitors have pleiotropic actions in different cell types. The various cellular responses induced by HDAC inhibitors include cell cycle arrest, apoptosis, angiogenesis, and immune modulation. In a wide variety of neurological and psychiatric diseases, there is enormous potential to restore patterns of gene expression and neuronal function through the use of epigenetic drugs. Several pharmacological therapies using HDAC inhibitors have been beneficial in various experimental models of brain diseases. Evidence suggests that targeting HDACs and histone acetylation might prove advantageous for seizure disorders, amyotrophic lateral sclerosis, Alzheimer's disease, Rubinstein–Taybi syndrome, spinal muscular atrophy, Rett syndrome, stroke, fragile X syndrome, and Huntington's disease, among others (Morrison et al. 2007; Abel and Zukin 2008; Kazantsev and Thompson 2008; Chuang et al. 2009). These drugs also hold promise for therapy relevant to several psychiatric disorders, including SCZ, depression, drug addiction, and anxiety disorders (Tsankova et al. 2007; Guidotti et al. 2009).

Currently, available HDAC inhibitors can be divided into four classes based on their chemical structures. This classification includes hydroxamates, short-chain fatty acids, cyclic peptides, and benzamides (Grayson et al. 2010). The majority of the currently available HDAC inhibitors blocks all classic HDACs, and the recent focus in HDAC inhibitor development has been on the improvement of specificity to overcome the nonspecific cytotoxicity of these drugs (Balasubramanian et al. 2009). In the case of treating brain disorders, an additional challenge is the permeability of the inhibitors across the blood–brain barrier (Kazantsev and Thompson 2008). So far, several drugs, including valproate, vorinostat, MS-275, sodium butyrate, and phenyl butyrate, have been shown to cross the blood–brain barrier. In addition, MS-275 has been demonstrated to be brain region selective and is 30–100 times more potent than valproate in increasing histone acetylation in vivo (Simonini et al. 2006). This drug also shows selectivity for class I HDACs, thus it might be considered a second-generation HDAC inhibitor with improved specificity.

Another possible method of controlling gene expression is the administration of therapeutics with a known mode of action on epigenetic mechanisms (Tremolizzo et al. 2005). Valproate is an HDAC inhibitor and has a long and established history of efficacy in the treatment of seizures and BD. In these conditions, oftentimes it is effective as the primary medication. However, its use in the treatment of SCZ is less straightforward. There is a sizeable body of clinical experience and published case reports in which the use of valproate as an adjunctive medication has been widespread with reported efficacy in seriously ill patients (Wassef et al. 2001). The administration of valproate in conjunction with antipsychotic medication has been shown to accelerate the onset of the antipsychotic effects in patients with SCZ (Casey et al. 2003). According to other studies, it is realistic to state that valproate is not of great benefit in randomly identified and unselected subjects with SCZ already receiving effective doses of antipsychotics (Casey et al. 2009; Citrome 2009). However, HDAC inhibitors might be effective in patients selected for epigenetic dysregulations. In an effort to link HDAC inhibitor efficacy to epigenetic parameters, it was demonstrated that lymphocyte nuclear extracts prepared from patients with SCZ have an abundance of restrictive chromatin at baseline, and that this chromatin is resistant to HDAC inhibitor treatment in vivo and in vitro (Gavin et al. 2008, 2009). Unfortunately, current clinical research in noncancer therapeutics is limited to valproate, given the safety profile of various other medications targeting the epigenetic platform. We have to note that valproate has many actions in addition to its HDAC inhibitory activity. Recent studies suggest that GAD1 and RELN expression is increased by valproate because this drug not only inhibits HDACs and modifies the histone code, but also decreases methylated sites at the RELN and GAD1 promoters (Dong et al. 2007), thus preventing the recruitment of a corepressor protein complex. It was shown that HDAC inhibitors, such as valproate, dramatically accelerate RELN and GAD1 promoter demethylation during the subsequent 24–48 h, if they are given to mice after Met withdrawal. This ability of valproate was not caused by a direct inhibitory action of this drug on DNMT1 or DNMT3a expression or DNMT activity or on an inhibitory action on SAM biosynthesis (Dong et al. 2008).

The first atypical (based upon the absence of extrapyramidal side effects) antipsychotic, clozapine, is used for the treatment of SCZ, but also for BD. Its clinically relevant doses as well as the typical antipsychotic sulpiride exhibited dose-related increases in the cortical and striatal demethylation of hypermethylated RELN and GAD1 promoters (Dong et al. 2008). Valproate potentiated these actions by increasing DNA demethylation with an unknown mechanism (Costa et al. 2009). However, another typical antipsychotic, haloperidol, or the atypical olanzapine (structurally similar to clozapine) were uneffective.

Activation of the metabotropic glutamate receptor may also influence SCZ-like symptoms (Krystal et al. 2010). At the same time, an agonist of the metabotropic glutamate receptor 2/3 was shown to influence the methylation of the GAD1, RELN, and BDNF promoters in the PFC and hippocampus (Matrisciano et al. 2011). Moreover, this treatment could also reverse the defect in social interaction seen in mice pretreated with Met. These findings suggest that many drugs could influence the epigenetic machinery, in addition to their other effects, or that this modification could be a common feature of antipsychotic drugs. Previously mentioned negative findings with haloperidol and olanzapine might suggest that these molecules affect a different epigenetic machinery or other tissues. Indeed, it was previously shown that in the rat chronic haloperidol treatment affects DNA methylation states in the brain, as well as in certain other tissues (Shimabukuro et al. 2006).

In animal models, systemically or locally administered HDAC inhibitors display antidepressant properties without obvious adverse effects on health (Tsankova et al. 2007; Covington et al. 2009), suggesting that HDAC inhibitors may function by modulating a global acetylation/deacetylation balance across several brain regions. For example, systemic administration of sodium butyrate, a nonspecific HDAC inhibitor, acts as an antidepressant in models of depression, including social defeat (Schroeder et al. 2007). Of course, histone acetylation functions in concert with several other markers of gene repression and activation, including histone methylation, phosphorylation, sumoylation, and ubiquitination (Renthal and Nestler 2008).

The majority of HDAC inhibitors that are currently either in clinical testing or on the market target multiple isoforms of the classic HDAC family, but do not inhibit SIRT family members. However, there has been growing interest in drugs that activate (resveratrol) or inhibit (sirtinol) the NAD-dependent SIRT1 protein (Kazantsev and Thompson 2008).

Future HDAC inhibitor interventions could include knowledge of the regional brain distribution of individual HDAC enzymes, similar to what has been done in the rat brain (Broide et al. 2007), and the possibility of using inhibitors targeted to a given HDAC enzyme. Treatment-resistant patients might benefit from a strategy designed first to relax regional chromatin which would then be coadministered with the optimal antipsychotic medication.

Monitoring the Treatment Efficacy

A key question is what can predict the positive outcome, i.e., what are the biomarkers (Wiedemann 2011). It is remarkable that GWA mRNA profiling studies indicate that the percentage of genes that are induced after HDAC inhibitor treatment is somewhere in the range of 2–5% (Stimson and La Thangue 2009). To test the therapeutic effectiveness in the treatment of psychiatric disorders is even more difficult as there are still no reliable tests for these diseases. The term *endophenotype* was introduced by Gottesman and Shields as a psychiatric concept and a special kind of biomarker. It divides behavioral symptoms into more stable phenotypes with clear genetic connection. Thus, endophenotypes reflect the action of sets of predisposing genes that might also be used in measuring the therapeutic efficacy of various drugs (Gottesman and Gould 2003). Recent data from twin studies have shown that the decreased amplitude of the evoked potential component (P300) might represent an endophenotype associated with SCZ (Bestelmeyer et al. 2009). However, the P300 does not reliably distinguish between patients with SCZ and those with BD and might be a better marker for psychosis.

It may be possible to measure various responses to HDAC inhibitors using lymphocytes both before and during treatment as an indirect measure of therapeutic efficacy. Preliminary studies have shown that the acetylated H3 content increased significantly when valproate is administered over a 4-week period either with or without an antipsychotic (Sharma et al. 2006). The increase was more pronounced in patients with BD who also showed higher baseline levels. In vitro studies with lymphocytes from patients with SCZ showed that baseline levels of acetylated H3K9 and H3K14 were reduced relative to the levels found in nonpsychiatric subjects (Gavin et al. 2008). More recently, baseline levels of H3K9me2 were shown to be increased in lymphocyte cultures from SCZ patients (Gavin et al. 2009). The decreased acetylated H3 and increased dimethyl H3 content suggest that lymphocyte chromatin is more restricted in patients with SCZ than in controls. The results suggest a possible means of monitoring HDAC inhibitor treatment in patients. It remains plausible that the measurement of restrictive chromatin marks and the monitoring of a neurophysiological endophenotype could be a more stringent guide to the therapeutic efficacy of new HDAC inhibitors (Grayson et al. 2010).

10.7.2.2 Methylation

Antidepressants may act through influencing the epigenetic machinery. Demethylation of H3K4 is catalyzed by BHC110/LSD1, an enzyme that has a close structural homology to monoamine oxidases. Monoamine oxidase inhibitors are a class of antidepressants, and they can increase global levels of H3K4 methylation and cause transcriptional derepression of specific genes in vitro (Tsankova et al. 2007).

10.8 Conclusions

Conventional linkage and candidate association studies revealed numerous, but also inconsistent and sometimes contradictory results (Bondy 2011). The reasons are assumed to include the complexity of the disorder with interaction of several genes of small effects, lack of a valid phenotype, and invalid statistical and methodological issues. Thus, studies were unable to identify a "susceptibility gene," therefore new approaches, as epigenetics, or gene-environment interaction, is needed in future study designs.

The previously mentioned data indicate that there are epigenetic changes associated with SCZ and BD. But do these epigenetic changes contribute to the genetic dysfunction observed in the disorder? The brief answer is that we do not know. However, basic research results help to address this question, by demonstrating that DNA methylation and the associated chromatin remodeling could indeed play a pivotal role in the development of disease-like symptoms. Epigenetic changes in the peripheral blood could be a useful diagnostic tool as well. Moreover, examination of epigenetic landmarks could form the base of a personalized therapy, the future direction of the treatment.

However, rather than examining candidate genes, the field has begun to transition toward genome-wide approaches to studying chromatin regulation (Wilkinson et al. 2009), shifting the focus from "epigenetic marks" to "epigenomic signatures." The ultimate goal would be to use transcriptional and epigenetic profiling as biomarkers to distinguish clinical categories of the illness, to determine responsivity to various drugs, and to differentiate treatment-sensitive from treatment-resistant illness. These profiles may offer new insights into subtype-specific pathophysiology and therapies and aid in the validation of our current animal models (Krishnan and Nestler 2010).

References

Abdolmaleky, H. M., Cheng, K. H., Faraone, S. V., Wilcox, M., Glatt, S. J., Gao, F., Smith, C. L., Shafa, R., Aeali, B., Carnevale, J., Pan, H., Papageorgis, P., Ponte, J. F., Sivaraman, V., Tsuang, M. T. & Thiagalingam, S. (2006). Hypomethylation of MB-COMT promoter is a major risk factor for schizophrenia and bipolar disorder. *Hum Mol Genet* 15, 3132–3145.

Abdolmaleky, H. M., Cheng, K. H., Russo, A., Smith, C. L., Faraone, S. V., Wilcox, M., Shafa, R., Glatt, S. J., Nguyen, G., Ponte, J. F., Thiagalingam, S. & Tsuang, M. T. (2005). Hypermethylation of the reelin (RELN) promoter in the brain of schizophrenic patients: a preliminary report. *Am J Med Genet B Neuropsychiatr Genet* 134B, 60–66.

Abdolmaleky, H. M., Smith, C. L., Faraone, S. V., Shafa, R., Stone, W., Glatt, S. J. & Tsuang, M. T. (2004). Methylomics in psychiatry: Modulation of gene-environment interactions may be through DNA methylation. *Am J Med Genet B Neuropsychiatr Genet* 127B, 51–59.

Abdolmaleky, H. M., Zhou, J. R., Thiagalingam, S. & Smith, C. L. (2008). Epigenetic and pharmacoepigenomic studies of major psychoses and potentials for therapeutics. *Pharmacogenomics* 9, 1809–1823.

Abel, T. & Zukin, R. S. (2008). Epigenetic targets of HDAC inhibition in neurodegenerative and psychiatric disorders. *Curr Opin Pharmacol* 8, 57–64.

Acharya, M. R., Sparreboom, A., Venitz, J. & Figg, W. D. (2005). Rational development of histone deacetylase inhibitors as anticancer agents: a review. *Mol Pharmacol* 68, 917–932.

Addington, J., Cadenhead, K. S., Cannon, T. D., Cornblatt, B., McGlashan, T. H., Perkins, D. O., Seidman, L. J., Tsuang, M., Walker, E. F., Woods, S. W. & Heinssen, R. (2007). North American Prodrome Longitudinal Study: a collaborative multisite approach to prodromal schizophrenia research. *Schizophr Bull* 33, 665–672.

Aid, T., Kazantseva, A., Piirsoo, M., Palm, K. & Timmusk, T. (2007). Mouse and rat BDNF gene structure and expression revisited. *J Neurosci Res* 85, 525–535.

Akbarian, S. (2010). Epigenetics of schizophrenia. *Curr Top Behav Neurosci* 4, 611–628.

Akbarian, S. & Huang, H. S. (2006). Molecular and cellular mechanisms of altered GAD1/GAD67 expression in schizophrenia and related disorders. *Brain Res Rev* 52, 293–304.

Akbarian, S. & Huang, H. S. (2009). Epigenetic regulation in human brain-focus on histone lysine methylation. *Biol Psychiatry* 65, 198–203.

Akbarian, S., Kim, J. J., Potkin, S. G., Hagman, J. O., Tafazzoli, A., Bunney, W. E., Jr. & Jones, E. G. (1995). Gene expression for glutamic acid decarboxylase is reduced without loss of neurons in prefrontal cortex of schizophrenics. *Arch Gen Psychiatry* 52, 258–266.

Akbarian, S., Ruehl, M. G., Bliven, E., Luiz, L. A., Peranelli, A. C., Baker, S. P., Roberts, R. C., Bunney, W. E., Jr., Conley, R. C., Jones, E. G., Tamminga, C. A. & Guo, Y. (2005). Chromatin alterations associated with down-regulated metabolic gene expression in the prefrontal cortex of subjects with schizophrenia. *Arch Gen Psychiatry* 62, 829–840.

Allan, A. M., Liang, X., Luo, Y., Pak, C., Li, X., Szulwach, K. E., Chen, D., Jin, P. & Zhao, X. (2008). The loss of methyl-CpG binding protein 1 leads to autism-like behavioral deficits. *Hum Mol Genet* 17, 2047–2057.

Alt, S. R., Turner, J. D., Klok, M. D., Meijer, O. C., Lakke, E. A., Derijk, R. H. & Muller, C. P. (2010). Differential expression of glucocorticoid receptor transcripts in major depressive disorder is not epigenetically programmed. *Psychoneuroendocrinology* 35, 544–556.

Amminger, G. P., Leicester, S., Yung, A. R., Phillips, L. J., Berger, G. E., Francey, S. M., Yuen, H. P. & McGorry, P. D. (2006). Early-onset of symptoms predicts conversion to non-affective psychosis in ultra-high risk individuals. *Schizophr Res* 84, 67–76.

Angelucci, F., Brene, S. & Mathe, A. A. (2005). BDNF in schizophrenia, depression and corresponding animal models. *Mol Psychiatry* 10, 345–352.

Anton, E. S., Ghashghaei, H. T., Weber, J. L., McCann, C., Fischer, T. M., Cheung, I. D., Gassmann, M., Messing, A., Klein, R., Schwab, M. H., Lloyd, K. C. & Lai, C. (2004). Receptor tyrosine kinase ErbB4 modulates neuroblast migration and placement in the adult forebrain. *Nat Neurosci* 7, 1319–1328.

Antun, F. T., Burnett, G. B., Cooper, A. J., Daly, R. J., Smythies, J. R. & Zealley, A. K. (1971). The effects of L-methionine (without MAOI) in schizophrenia. *J Psychiatr Res* 8, 63–71.

Assimacopoulos, S., Grove, E. A. & Ragsdale, C. W. (2003). Identification of a Pax6-dependent epidermal growth factor family signaling source at the lateral edge of the embryonic cerebral cortex. *J Neurosci* 23, 6399–6403.

Aston, C., Jiang, L. & Sokolov, B. P. (2004). Microarray analysis of postmortem temporal cortex from patients with schizophrenia. *J Neurosci Res* 77, 858–866.

Aston, C., Jiang, L. & Sokolov, B. P. (2005). Transcriptional profiling reveals evidence for signaling and oligodendroglial abnormalities in the temporal cortex from patients with major depressive disorder. *Mol Psychiatry* 10, 309–322.

Babak, T., Zhang, W., Morris, Q., Blencowe, B. J. & Hughes, T. R. (2004). Probing microRNAs with microarrays: tissue specificity and functional inference. *RNA* 10, 1813–1819.

Balasubramanian, S., Verner, E. & Buggy, J. J. (2009). Isoform-specific histone deacetylase inhibitors: the next step? *Cancer Lett* 280, 211–221.

Balciuniene, J., Feng, N., Iyadurai, K., Hirsch, B., Charnas, L., Bill, B. R., Easterday, M. C., Staaf, J., Oseth, L., Czapansky-Beilman, D., Avramopoulos, D., Thomas, G. H., Borg, A., Valle, D., Schimmenti, L. A. & Selleck, S. B. (2007). Recurrent 10q22-q23 deletions: a genomic disorder on 10q associated with cognitive and behavioral abnormalities. *Am J Hum Genet* 80, 938–947.

Barratt, M. J., Hazzalin, C. A., Cano, E. & Mahadevan, L. C. (1994). Mitogen-stimulated phosphorylation of histone H3 is targeted to a small hyperacetylation-sensitive fraction. *Proc Natl Acad Sci U S A* 91, 4781–4785.

Ben Shachar, D. & Laifenfeld, D. (2004). Mitochondria, synaptic plasticity, and schizophrenia. *Int Rev Neurobiol* 59, 273–296.

Benes, F. M. & Berretta, S. (2001). GABAergic interneurons: implications for understanding schizophrenia and bipolar disorder. *Neuropsychopharmacology* 25, 1–27.

Benes, F. M., Lim, B., Matzilevich, D., Walsh, J. P., Subburaju, S. & Minns, M. (2007). Regulation of the GABA cell phenotype in hippocampus of schizophrenics and bipolars. *Proc Natl Acad Sci U S A* 104, 10164–10169.

Benzel, I., Bansal, A., Browning, B. L., Galwey, N. W., Maycox, P. R., McGinnis, R., Smart, D., St Clair, D., Yates, P. & Purvis, I. (2007). Interactions among genes in the ErbB-Neuregulin signalling network are associated with increased susceptibility to schizophrenia. *Behav Brain Funct* 3:31.

Berlet, H. H., Matsumoto, K., Pscheidt, G. R., Spaide, J., Bull, C. & Himwich, H. E. (1965). Biochemical correlates of behavior in schizophrenic patients. Schizophrenic patients receiving tryptophan and methionine or methionine together with a monoamine oxidase inhibitor. *Arch Gen Psychiatry* 13, 521–531.

Bernstein, B. E., Mikkelsen, T. S., Xie, X., Kamal, M., Huebert, D. J., Cuff, J., Fry, B., Meissner, A., Wernig, M., Plath, K., Jaenisch, R., Wagschal, A., Feil, R., Schreiber, S. L. & Lander, E. S. (2006). A bivalent chromatin structure marks key developmental genes in embryonic stem cells. *Cell* 125, 315–326.

Berton, O., McClung, C. A., Dileone, R. J., Krishnan, V., Renthal, W., Russo, S. J., Graham, D., Tsankova, N. M., Bolanos, C. A., Rios, M., Monteggia, L. M., Self, D. W. & Nestler, E. J. (2006). Essential role of BDNF in the mesolimbic dopamine pathway in social defeat stress. *Science* 311, 864–868.

Bestelmeyer, P. E., Phillips, L. H., Crombie, C., Benson, P. & St Clair, D. (2009). The P300 as a possible endophenotype for schizophrenia and bipolar disorder: Evidence from twin and patient studies. *Psychiatry Res* 169, 212–219.

Beveridge, N. J., Tooney, P. A., Carroll, A. P., Gardiner, E., Bowden, N., Scott, R. J., Tran, N., Dedova, I. & Cairns, M. J. (2008). Dysregulation of miRNA 181b in the temporal cortex in schizophrenia. *Hum Mol Genet* 17, 1156–1168.

Bicker, S. & Schratt, G. (2008). microRNAs: tiny regulators of synapse function in development and disease. *J Cell Mol Med* 12, 1466–1476.

Bilang-Bleuel, A., Ulbricht, S., Chandramohan, Y., De Carli, S., Droste, S. K. & Reul, J. M. (2005). Psychological stress increases histone H3 phosphorylation in adult dentate gyrus granule neurons: involvement in a glucocorticoid receptor-dependent behavioural response. *Eur J Neurosci* 22, 1691–1700.

Biliya, S. & Bulla, L. A., Jr. (2010). Genomic imprinting: the influence of differential methylation in the two sexes. *Exp Biol Med (Maywood)* 235, 139–147.

Biniszkiewicz, D., Gribnau, J., Ramsahoye, B., Gaudet, F., Eggan, K., Humpherys, D., Mastrangelo, M. A., Jun, Z., Walter, J. & Jaenisch, R. (2002). Dnmt1 overexpression causes genomic hypermethylation, loss of imprinting, and embryonic lethality. *Mol Cell Biol* 22, 2124–2135.

Bolden, J. E., Peart, M. J. & Johnstone, R. W. (2006). Anticancer activities of histone deacetylase inhibitors. *Nat Rev Drug Discov* 5, 769–784.

Bondy, B. (2011). Genetics in psychiatry: are the promises met? *World J Biol Psychiatry* 12, 81–88.

Borrelli, E., Nestler, E. J., Allis, C. D. & Sassone-Corsi, P. (2008). Decoding the epigenetic language of neuronal plasticity. *Neuron* 60, 961–974.

Bossong, M. G. & Niesink, R. J. (2010). Adolescent brain maturation, the endogenous cannabinoid system and the neurobiology of cannabis-induced schizophrenia. *Prog Neurobiol* 92, 370–385.

Bradley, A. J. & Dinan, T. G. (2010). A systematic review of hypothalamic-pituitary-adrenal axis function in schizophrenia: implications for mortality. *J Psychopharmacol* 24, 91–118.

Branchi, I., Francia, N. & Alleva, E. (2004). Epigenetic control of neurobehavioural plasticity: the role of neurotrophins. *Behav Pharmacol* 15, 353–362.

Brandon, N. J., Millar, J. K., Korth, C., Sive, H., Singh, K. K. & Sawa, A. (2009). Understanding the role of DISC1 in psychiatric disease and during normal development. *J Neurosci* 29, 12768–12775.

Bray, N. J., Buckland, P. R., Hall, H., Owen, M. J. & O'Donovan, M. C. (2004). The serotonin-2A receptor gene locus does not contain common polymorphism affecting mRNA levels in adult brain. *Mol Psychiatry* 9, 109–114.

Broide, R. S., Redwine, J. M., Aftahi, N., Young, W., Bloom, F. E. & Winrow, C. J. (2007). Distribution of histone deacetylases 1-11 in the rat brain. *J Mol Neurosci* 31, 47–58.

Brown, A. S., Bottiglieri, T., Schaefer, C. A., Quesenberry, C. P., Jr., Liu, L., Bresnahan, M. & Susser, E. S. (2007). Elevated prenatal homocysteine levels as a risk factor for schizophrenia. *Arch Gen Psychiatry* 64, 31–39.

Brown, G. G. & Thompson, W. K. (2010). Functional brain imaging in schizophrenia: selected results and methods. *Curr Top Behav Neurosci* 4, 181–214.

Bruneau, E. G., McCullumsmith, R. E., Haroutunian, V., Davis, K. L. & Meador-Woodruff, J. H. (2005). Increased expression of glutaminase and glutamine synthetase mRNA in the thalamus in schizophrenia. *Schizophr Res* 75, 27–34.

Brunet-Gouet, E. & Decety, J. (2006). Social brain dysfunctions in schizophrenia: a review of neuroimaging studies. *Psychiatry Res* 148, 75–92.

Buiting, K. (2010). Prader-Willi syndrome and Angelman syndrome. *Am J Med Genet C Semin Med Genet* 154C, 365–376.

Bunney, W. E., Bunney, B. G., Vawter, M. P., Tomita, H., Li, J., Evans, S. J., Choudary, P. V., Myers, R. M., Jones, E. G., Watson, S. J. & Akil, H. (2003). Microarray technology: a review of new strategies to discover candidate vulnerability genes in psychiatric disorders. *Am J Psychiatry* 160, 657–666.

Burmeister, M., McInnis, M. G. & Zollner, S. (2008). Psychiatric genetics: progress amid controversy. *Nat Rev Genet* 9, 527–540.

Burnett, C., Valentini, S., Cabreiro, F., Goss, M., Somogyvari, M., Piper, M. D., Hoddinott, M., Sutphin, G. L., Leko, V., McElwee, J. J., Vazquez-Manrique, R. P., Orfila, A. M., Ackerman, D., Au, C., Vinti, G., Riesen, M., Howard, K., Neri, C., Bedalov, A., Kaeberlein, M., Soti, C., Partridge, L. & Gems, D. (2011). Absence of effects of Sir2 overexpression on lifespan in C. elegans and Drosophila. *Nature* 477, 482–485.

Carteron, C., Ferrer-Montiel, A. & Cabedo, H. (2006). Characterization of a neural-specific splicing form of the human neuregulin 3 gene involved in oligodendrocyte survival. *J Cell Sci* 119, 898–909.

Casey, D. E., Daniel, D. G., Tamminga, C., Kane, J. M., Tran-Johnson, T., Wozniak, P., Abi-Saab, W., Baker, J., Redden, L., Greco, N. & Saltarelli, M. (2009). Divalproex ER combined with olanzapine or risperidone for treatment of acute exacerbations of schizophrenia. *Neuropsychopharmacology* 34, 1330–1338.

Casey, D. E., Daniel, D. G., Wassef, A. A., Tracy, K. A., Wozniak, P. & Sommerville, K. W. (2003). Effect of divalproex combined with olanzapine or risperidone in patients with an acute exacerbation of schizophrenia. *Neuropsychopharmacology* 28, 182–192.

Caspi, A., Sugden, K., Moffitt, T. E., Taylor, A., Craig, I. W., Harrington, H., McClay, J., Mill, J., Martin, J., Braithwaite, A. & Poulton, R. (2003). Influence of life stress on depression: moderation by a polymorphism in the 5-HTT gene. *Science* 301, 386–389.

Cavigelli, S. A. & McClintock, M. K. (2003). Fear of novelty in infant rats predicts adult corticosterone dynamics and an early death. *Proc Natl Acad Sci U S A* 100, 16131–16136.

Cedar, H. & Bergman, Y. (2009). Linking DNA methylation and histone modification: patterns and paradigms. *Nat Rev Genet* 10, 295–304.

Ceskova, E., Kasparek, T., Zourkova, A. & Prikryl, R. (2006). Dexamethasone suppression test in first-episode schizophrenia. *Neuro Endocrinol Lett* 27, 433–437.

Champagne, F. A., Francis, D. D., Mar, A. & Meaney, M. J. (2003). Variations in maternal care in the rat as a mediating influence for the effects of environment on development. *Physiol Behav* 79, 359–371.

Chandramohan, Y., Droste, S. K., Arthur, J. S. & Reul, J. M. (2008). The forced swimming-induced behavioural immobility response involves histone H3 phospho-acetylation and c-Fos induction in dentate gyrus granule neurons via activation of the N-methyl-D-aspartate/extracellular signal-regulated kinase/mitogen- and stress-activated kinase signalling pathway. *Eur J Neurosci* 27, 2701–2713.

Chen, J., Lipska, B. K., Halim, N., Ma, Q. D., Matsumoto, M., Melhem, S., Kolachana, B. S., Hyde, T. M., Herman, M. M., Apud, J., Egan, M. F., Kleinman, J. E. & Weinberger, D. R. (2004). Functional analysis of genetic variation in catechol-O-methyltransferase (COMT): effects on mRNA, protein, and enzyme activity in postmortem human brain. *Am J Hum Genet* 75, 807–821.

Chen, P. L., Avramopoulos, D., Lasseter, V. K., McGrath, J. A., Fallin, M. D., Liang, K. Y., Nestadt, G., Feng, N., Steel, G., Cutting, A. S., Wolyniec, P., Pulver, A. E. & Valle, D. (2009). Fine mapping on chromosome 10q22-q23 implicates Neuregulin 3 in schizophrenia. *Am J Hum Genet* 84, 21–34.

Chen, Y., Sharma, R. P., Costa, R. H., Costa, E. & Grayson, D. R. (2002). On the epigenetic regulation of the human reelin promoter. *Nucleic Acids Res* 30, 2930–2939.

Cheng, L. C., Pastrana, E., Tavazoie, M. & Doetsch, F. (2009). miR-124 regulates adult neurogenesis in the subventricular zone stem cell niche. *Nat Neurosci* 12, 399–408.

Cheung, P., Tanner, K. G., Cheung, W. L., Sassone-Corsi, P., Denu, J. M. & Allis, C. D. (2000). Synergistic coupling of histone H3 phosphorylation and acetylation in response to epidermal growth factor stimulation. *Mol Cell* 5, 905–915.

Chua, Y. L., Ito, Y., Pole, J. C., Newman, S., Chin, S. F., Stein, R. C., Ellis, I. O., Caldas, C., O'Hare, M. J., Murrell, A. & Edwards, P. A. (2009). The NRG1 gene is frequently silenced by methylation in breast cancers and is a strong candidate for the 8p tumour suppressor gene. *Oncogene* 28, 4041–4052.

Chuang, D. M., Leng, Y., Marinova, Z., Kim, H. J. & Chiu, C. T. (2009). Multiple roles of HDAC inhibition in neurodegenerative conditions. *Trends Neurosci* 32, 591–601.

Chubb, J. E., Bradshaw, N. J., Soares, D. C., Porteous, D. J. & Millar, J. K. (2008). The DISC locus in psychiatric illness. *Mol Psychiatry* 13, 36–64.

Citrome, L. (2009). Adjunctive lithium and anticonvulsants for the treatment of schizophrenia: what is the evidence? *Expert Rev Neurother* 9, 55–71.

Cohen, S., Zhou, Z. & Greenberg, M. E. (2008). Medicine. Activating a repressor. *Science* 320, 1172–1173.

Collins, A., Hill, L. E., Chandramohan, Y., Whitcomb, D., Droste, S. K. & Reul, J. M. (2009). Exercise improves cognitive responses to psychological stress through enhancement of epigenetic mechanisms and gene expression in the dentate gyrus. *PLoS ONE* 4, e4330.

Colvis, C. M., Pollock, J. D., Goodman, R. H., Impey, S., Dunn, J., Mandel, G., Champagne, F. A., Mayford, M., Korzus, E., Kumar, A., Renthal, W., Theobald, D. E. & Nestler, E. J. (2005). Epigenetic mechanisms and gene networks in the nervous system. *J Neurosci* 25, 10379–10389.

Connor, C. M. & Akbarian, S. (2008). DNA methylation changes in schizophrenia and bipolar disorder. *Epigenetics* 3, 55–58.

Cooney, C. A., Dave, A. A. & Wolff, G. L. (2002). Maternal methyl supplements in mice affect epigenetic variation and DNA methylation of offspring. *J Nutr* 132, 2393S–2400S.

Costa, E., Chen, Y., Davis, J., Dong, E., Noh, J. S., Tremolizzo, L., Veldic, M., Grayson, D. R. & Guidotti, A. (2002). REELIN and schizophrenia: a disease at the interface of the genome and the epigenome. *Mol Interv* 2, 47–57.

Costa, E., Chen, Y., Dong, E., Grayson, D. R., Kundakovic, M., Maloku, E., Ruzicka, W., Satta, R., Veldic, M., Zhubi, A. & Guidotti, A. (2009). GABAergic promoter hypermethylation as a model to study the neurochemistry of schizophrenia vulnerability. *Expert Rev Neurother* 9, 87–98.

Costa, E., Dong, E., Grayson, D. R., Guidotti, A., Ruzicka, W. & Veldic, M. (2007). Reviewing the role of DNA (cytosine-5) methyltransferase overexpression in the cortical GABAergic dysfunction associated with psychosis vulnerability. *Epigenetics* 2, 29–36.

Cotter, D. & Pariante, C. M. (2002). Stress and the progression of the developmental hypothesis of schizophrenia. *Br J Psychiatry* 181, 363–365.

Couve, A., Restituito, S., Brandon, J. M., Charles, K. J., Bawagan, H., Freeman, K. B., Pangalos, M. N., Calver, A. R. & Moss, S. J. (2004). Marlin-1, a novel RNA-binding protein associates with GABA receptors. *J Biol Chem* 279, 13934–13943.

Covic, M., Karaca, E. & Lie, D. C. (2010). Epigenetic regulation of neurogenesis in the adult hippocampus. *Heredity* 105, 122–134.

Covington, H. E., Maze, I., LaPlant, Q. C., Vialou, V. F., Ohnishi, Y. N., Berton, O., Fass, D. M., Renthal, W., Rush, A. J., III, Wu, E. Y., Ghose, S., Krishnan, V., Russo, S. J., Tamminga, C., Haggarty, S. J. & Nestler, E. J. (2009). Antidepressant actions of histone deacetylase inhibitors. *J Neurosci* 29, 11451–11460.

Coyle, J. T. (2004). The GABA-glutamate connection in schizophrenia: which is the proximate cause? *Biochem Pharmacol* 68, 1507–1514.

Craddock, N., O'Donovan, M. C. & Owen, M. J. (2006). Genes for schizophrenia and bipolar disorder? Implications for psychiatric nosology. *Schizophr Bull* 32, 9–16.

Crosio, C., Cermakian, N., Allis, C. D. & Sassone-Corsi, P. (2000). Light induces chromatin modification in cells of the mammalian circadian clock. *Nat Neurosci* 3, 1241–1247.

Crosio, C., Heitz, E., Allis, C. D., Borrelli, E. & Sassone-Corsi, P. (2003). Chromatin remodeling and neuronal response: multiple signaling pathways induce specific histone H3 modifications and early gene expression in hippocampal neurons. *J Cell Sci* 116, 4905–4914.

Cui, H., Cruz-Correa, M., Giardiello, F. M., Hutcheon, D. F., Kafonek, D. R., Brandenburg, S., Wu, Y., He, X., Powe, N. R. & Feinberg, A. P. (2003). Loss of IGF2 imprinting: a potential marker of colorectal cancer risk. *Science* 299, 1753–1755.

Daban, C., Vieta, E., Mackin, P. & Young, A. H. (2005). Hypothalamic-pituitary-adrenal axis and bipolar disorder. *Psychiatr Clin North Am* 28, 469–480.

Daskalakis, Z. J., Fitzgerald, P. B. & Christensen, B. K. (2007). The role of cortical inhibition in the pathophysiology and treatment of schizophrenia. *Brain Res Rev* 56, 427–442.

de Boer, P., Ramos, L., de Vries, M. & Gochhait, S. (2010). Memoirs of an insult: sperm as a possible source of transgenerational epimutations and genetic instability. *Mol Hum Reprod* 16, 48–56.

de Kloet, E. R., Derijk, R. H. & Meijer, O. C. (2007). Therapy Insight: is there an imbalanced response of mineralocorticoid and glucocorticoid receptors in depression? *Nat Clin Pract Endocrinol Metab* 3, 168–179.

De Luca, V., Likhodi, O., Kennedy, J. L. & Wong, A. H. (2007). Parent-of-origin effect and genomic imprinting of the HTR2A receptor gene T102C polymorphism in psychosis. *Psychiatry Res* 151, 243–248.

De Luca, V., Viggiano, E., Dhoot, R., Kennedy, J. L. & Wong, A. H. (2009). Methylation and QTDT analysis of the 5-HT2A receptor 102C allele: analysis of suicidality in major psychosis. *J Psychiatr Res* 43, 532–537.

Deutsch, S. I., Rosse, R. B., Mastropaolo, J., Long, K. D. & Gaskins, B. L. (2008). Epigenetic therapeutic strategies for the treatment of neuropsychiatric disorders: ready for prime time? *Clin Neuropharmacol* 31, 104–119.

Dong, E., Agis-Balboa, R. C., Simonini, M. V., Grayson, D. R., Costa, E. & Guidotti, A. (2005). Reelin and glutamic acid decarboxylase67 promoter remodeling in an epigenetic methionine-induced mouse model of schizophrenia. *Proc Natl Acad Sci U S A* 102, 12578–12583.

Dong, E., Guidotti, A., Grayson, D. R. & Costa, E. (2007). Histone hyperacetylation induces demethylation of reelin and 67-kDa glutamic acid decarboxylase promoters. *Proc Natl Acad Sci U S A* 104, 4676–4681.

Dong, E., Nelson, M., Grayson, D. R., Costa, E. & Guidotti, A. (2008). Clozapine and sulpiride but not haloperidol or olanzapine activate brain DNA demethylation. *Proc Natl Acad Sci U S A* 105, 13614–13619.

Dracheva, S., Elhakem, S. L., McGurk, S. R., Davis, K. L. & Haroutunian, V. (2004). GAD67 and GAD65 mRNA and protein expression in cerebrocortical regions of elderly patients with schizophrenia. *J Neurosci Res* 76, 581–592.

Drapeau, E., Mayo, W., Aurousseau, C., Le Moal, M., Piazza, P. V. & Abrous, D. N. (2003). Spatial memory performances of aged rats in the water maze predict levels of hippocampal neurogenesis. *Proc Natl Acad Sci U S A* 100, 14385–14390.

Dudley, K. J., Li, X., Kobor, M. S., Kippin, T. E. & Bredy, T. W. (2011). Epigenetic mechanisms mediating vulnerability and resilience to psychiatric disorders. *Neurosci Biobehav Rev* 35, 1544–1551.

Duman, R. S. (2004). Role of neurotrophic factors in the etiology and treatment of mood disorders. *Neuromolecular Med* 5, 11–25.

Eastwood, S. L. & Harrison, P. J. (2005). Decreased expression of vesicular glutamate transporter 1 and complexin II mRNAs in schizophrenia: further evidence for a synaptic pathology affecting glutamate neurons. *Schizophr Res* 73, 159–172.

Egan, M. F., Goldberg, T. E., Kolachana, B. S., Callicott, J. H., Mazzanti, C. M., Straub, R. E., Goldman, D. & Weinberger, D. R. (2001). Effect of COMT Val108/158 Met genotype on frontal lobe function and risk for schizophrenia. *Proc Natl Acad Sci U S A* 98, 6917–6922.

Egan, M. F. & Weinberger, D. R. (1997). Neurobiology of schizophrenia. *Curr Opin Neurobiol* 7, 701–707.

Egger, G., Liang, G., Aparicio, A. & Jones, P. A. (2004). Epigenetics in human disease and prospects for epigenetic therapy. *Nature* 429, 457–463.

Engel, G. L. (1977). The need for a new medical model: a challenge for biomedicine. *Science* 196, 129–136.

Ernst, C., Deleva, V., Deng, X., Sequeira, A., Pomarenski, A., Klempan, T., Ernst, N., Quirion, R., Gratton, A., Szyf, M. & Turecki, G. (2009). Alternative splicing, methylation state, and expression profile of tropomyosin-related kinase B in the frontal cortex of suicide completers. *Arch Gen Psychiatry* 66, 22–32.

Erwin, J. A. & Lee, J. T. (2008). New twists in X-chromosome inactivation. *Curr Opin Cell Biol* 20, 349–355.

Fallin, M. D., Lasseter, V. K., Wolyniec, P. S., McGrath, J. A., Nestadt, G., Valle, D., Liang, K. Y. & Pulver, A. E. (2003). Genomewide linkage scan for schizophrenia susceptibility loci among Ashkenazi Jewish families shows evidence of linkage on chromosome 10q22. *Am J Hum Genet* 73, 601–611.

Fatemi, S. H., Earle, J. A. & McMenomy, T. (2000). Reduction in Reelin immunoreactivity in hippocampus of subjects with schizophrenia, bipolar disorder and major depression. *Mol Psychiatry* 5, 654–563.

Felitti, V. J., Anda, R. F., Nordenberg, D., Williamson, D. F., Spitz, A. M., Edwards, V., Koss, M. P. & Marks, J. S. (1998). Relationship of childhood abuse and household dysfunction to many of the leading causes of death in adults. The Adverse Childhood Experiences (ACE) Study. *Am J Prev Med* 14, 245–258.

Feng, J., Sun, G., Yan, J., Noltner, K., Li, W., Buzin, C. H., Longmate, J., Heston, L. L., Rossi, J. & Sommer, S. S. (2009). Evidence for X-chromosomal schizophrenia associated with microRNA alterations. *PLoS ONE* 4, e6121.

Ferreira, M. A., O'Donovan, M. C., Meng, Y. A., Jones, I. R., Ruderfer, D. M., Jones, L., Fan, J., Kirov, G., Perlis, R. H., Green, E. K., Smoller, J. W., Grozeva, D., Stone, J., Nikolov, I., Chambert, K., Hamshere, M. L., Nimgaonkar, V. L., Moskvina, V., Thase, M. E., Caesar, S., Sachs, G. S., Franklin, J., Gordon-Smith, K., Ardlie, K. G., Gabriel, S. B., Fraser, C., Blumenstiel, B., Defelice, M., Breen, G., Gill, M., Morris, D. W., Elkin, A., Muir, W. J., McGhee, K. A., Williamson, R., MacIntyre, D. J., MacLean, A. W., St, C. D., Robinson, M., Van Beck, M., Pereira, A. C., Kandaswamy, R., McQuillin, A., Collier, D. A., Bass, N. J., Young, A. H., Lawrence, J., Ferrier, I. N., Anjorin, A., Farmer, A., Curtis, D., Scolnick, E. M., McGuffin, P., Daly, M. J., Corvin, A. P., Holmans, P. A., Blackwood, D. H., Gurling, H. M., Owen, M. J., Purcell, S. M., Sklar, P. & Craddock, N. (2008). Collaborative genome-wide association analysis supports a role for ANK3 and CACNA1C in bipolar disorder. *Nat Genet* 40, 1056–1058.

Fields, R. D. (2008). White matter in learning, cognition and psychiatric disorders. *Trends Neurosci* 31, 361–370.

Filion, G. J., Zhenilo, S., Salozhin, S., Yamada, D., Prokhortchouk, E. & Defossez, P. A. (2006). A family of human zinc finger proteins that bind methylated DNA and repress transcription. *Mol Cell Biol* 26, 169–181.

Fischer, A., Sananbenesi, F., Wang, X., Dobbin, M. & Tsai, L. H. (2007). Recovery of learning and memory is associated with chromatin remodelling. *Nature* 447, 178–182.

Flynn, S. W., Lang, D. J., Mackay, A. L., Goghari, V., Vavasour, I. M., Whittall, K. P., Smith, G. N., Arango, V., Mann, J. J., Dwork, A. J., Falkai, P. & Honer, W. G. (2003). Abnormalities of myelination in schizophrenia detected in vivo with MRI, and post-mortem with analysis of oligodendrocyte proteins. *Mol Psychiatry* 8, 811–820.

Fone, K. C. & Porkess, M. V. (2008). Behavioural and neurochemical effects of post-weaning social isolation in rodents-relevance to developmental neuropsychiatric disorders. *Neurosci Biobehav Rev* 32, 1087–1102.

Francis, D. D. & Meaney, M. J. (1999). Maternal care and the development of stress responses. *Curr Opin Neurobiol* 9, 128–134.

Frank, E. & Landgraf, R. (2008). The vasopressin system--from antidiuresis to psychopathology. *Eur J Pharmacol* 583, 226–242.

Frankenburg, F. R. (2007). The role of one-carbon metabolism in schizophrenia and depression. *Harv Rev Psychiatry* 15, 146–160.

Frans, E. M., Sandin, S., Reichenberg, A., Lichtenstein, P., Langstrom, N. & Hultman, C. M. (2008). Advancing paternal age and bipolar disorder. *Arch Gen Psychiatry* 65, 1034–1040.

Fremeau, R. T., Jr., Troyer, M. D., Pahner, I., Nygaard, G. O., Tran, C. H., Reimer, R. J., Bellocchio, E. E., Fortin, D., Storm-Mathisen, J. & Edwards, R. H. (2001). The expression of vesicular glutamate transporters defines two classes of excitatory synapse. *Neuron* 31, 247–260.

Fukuchi, M., Nii, T., Ishimaru, N., Minamino, A., Hara, D., Takasaki, I., Tabuchi, A. & Tsuda, M. (2009). Valproic acid induces up- or down-regulation of gene expression responsible for the neuronal excitation and inhibition in rat cortical neurons through its epigenetic actions. *Neurosci Res* 65, 35–43.

Fumagalli, F., Molteni, R., Racagni, G. & Riva, M. A. (2007). Stress during development: Impact on neuroplasticity and relevance to psychopathology. *Prog Neurobiol* 81, 197–217.

Gaddum, J. H. & Hameed, K. A. (1954). Drugs which antagonize 5-hydroxytryptamine. *Br J Pharmacol Chemother* 9, 240–248.

Gavin, D. P., Kartan, S., Chase, K., Grayson, D. R. & Sharma, R. P. (2008). Reduced baseline acetylated histone 3 levels, and a blunted response to HDAC inhibition in lymphocyte cultures from schizophrenia subjects. *Schizophr Res* 103, 330–332.

Gavin, D. P., Rosen, C., Chase, K., Grayson, D. R., Tun, N. & Sharma, R. P. (2009). Dimethylated lysine 9 of histone 3 is elevated in schizophrenia and exhibits a divergent response to histone deacetylase inhibitors in lymphocyte cultures. *J Psychiatry Neurosci* 34, 232–237.

Gavin, D. P. & Sharma, R. P. (2010). Histone modifications, DNA methylation, and schizophrenia. *Neurosci Biobehav Rev* 34, 882–888.

Glennon, R. A., Titeler, M. & McKenney, J. D. (1984). Evidence for 5-HT2 involvement in the mechanism of action of hallucinogenic agents. *Life Sci* 35, 2505–2511.

Gonzalez-Burgos, G. & Lewis, D. A. (2008). GABA neurons and the mechanisms of network oscillations: implications for understanding cortical dysfunction in schizophrenia. *Schizophr Bull* 34, 944–961.

Goto, H., Tomono, Y., Ajiro, K., Kosako, H., Fujita, M., Sakurai, M., Okawa, K., Iwamatsu, A., Okigaki, T., Takahashi, T. & Inagaki, M. (1999). Identification of a novel phosphorylation site on histone H3 coupled with mitotic chromosome condensation. *J Biol Chem* 274, 25543–25549.

Gottesman, I. I. & Bertelsen, A. (1989). Confirming unexpressed genotypes for schizophrenia. Risks in the offspring of Fischer's Danish identical and fraternal discordant twins. *Arch Gen Psychiatry* 46, 867–872.

Gottesman, I. I. & Gould, T. D. (2003). The endophenotype concept in psychiatry: etymology and strategic intentions. *Am J Psychiatry* 160, 636–645.

Gottesman, I. I., Shields, J. & Hanson, D. R. (1982). *Schizophrenia: The Epigenetic Puzzle.* Cambridge, England: Cambridge University Press.

Grayson, D. R., Chen, Y., Costa, E., Dong, E., Guidotti, A., Kundakovic, M. & Sharma, R. P. (2006). The human reelin gene: transcription factors (+), repressors (-) and the methylation switch (+/-) in schizophrenia. *Pharmacol Ther* 111, 272–286.

Grayson, D. R., Jia, X., Chen, Y., Sharma, R. P., Mitchell, C. P., Guidotti, A. & Costa, E. (2005). Reelin promoter hypermethylation in schizophrenia. *Proc Natl Acad Sci U S A* 102, 9341–9346.

Grayson, D. R., Kundakovic, M. & Sharma, R. P. (2010). Is there a future for histone deacetylase inhibitors in the pharmacotherapy of psychiatric disorders? *Mol Pharmacol* 77, 126–135.

Groc, L., Choquet, D., Stephenson, F. A., Verrier, D., Manzoni, O. J. & Chavis, P. (2007). NMDA receptor surface trafficking and synaptic subunit composition are developmentally regulated by the extracellular matrix protein Reelin. *J Neurosci* 27, 10165–10175.

Guenther, M. G., Levine, S. S., Boyer, L. A., Jaenisch, R. & Young, R. A. (2007). A chromatin landmark and transcription initiation at most promoters in human cells. *Cell* 130, 77–88.

Guidotti, A., Auta, J., Chen, Y., Davis, J. M., Dong, E., Gavin, D. P., Grayson, D. R., Matrisciano, F., Pinna, G., Satta, R., Sharma, R. P., Tremolizzo, L. & Tueting, P. (2011). Epigenetic GABAergic targets in schizophrenia and bipolar disorder. *Neuropharmacology* 60, 1007–1016.

Guidotti, A., Auta, J., Davis, J. M., Giorgi-Gerevini, V., Dwivedi, Y., Grayson, D. R., Impagnatiello, F., Pandey, G., Pesold, C., Sharma, R., Uzunov, D. & Costa, E. (2000). Decrease in reelin and glutamic acid decarboxylase67 (GAD67) expression in schizophrenia and bipolar disorder: a postmortem brain study. *Arch Gen Psychiatry* 57, 1061–1069.

Guidotti, A., Dong, E., Kundakovic, M., Satta, R., Grayson, D. R. & Costa, E. (2009). Characterization of the action of antipsychotic subtypes on valproate-induced chromatin remodeling. *Trends Pharmacol Sci* 30, 55–60.

Gupta, D. S., McCullumsmith, R. E., Beneyto, M., Haroutunian, V., Davis, K. L. & Meador-Woodruff, J. H. (2005). Metabotropic glutamate receptor protein expression in the prefrontal cortex and striatum in schizophrenia. *Synapse* 57, 123–131.

Gurley, L. R., D'Anna, J. A., Barham, S. S., Deaven, L. L. & Tobey, R. A. (1978). Histone phosphorylation and chromatin structure during mitosis in Chinese hamster cells. *Eur J Biochem* 84, 1–15.

Hafner, H. (2003). Gender differences in schizophrenia. *Psychoneuroendocrinology* 28 Suppl 2, 17–54.

Haggarty, P., Hoad, G., Harris, S. E., Starr, J. M., Fox, H. C., Deary, I. J. & Whalley, L. J. (2010). Human intelligence and polymorphisms in the DNA methyltransferase genes involved in epigenetic marking. *PLoS ONE* 5, e11329.

Haidemenos, A., Kontis, D., Gazi, A., Kallai, E., Allin, M. & Lucia, B. (2007). Plasma homocysteine, folate and B12 in chronic schizophrenia. *Prog Neuropsychopharmacol Biol Psychiatry* 31, 1289–1296.

Hakak, Y., Walker, J. R., Li, C., Wong, W. H., Davis, K. L., Buxbaum, J. D., Haroutunian, V. & Fienberg, A. A. (2001). Genome-wide expression analysis reveals dysregulation of myelination-related genes in chronic schizophrenia. *Proc Natl Acad Sci U S A* 98, 4746–4751.

Hamer, D. (2002). Genetics. Rethinking behavior genetics. *Science* 298, 71–72.

Hannon, J. & Hoyer, D. (2008). Molecular biology of 5-HT receptors. *Behav Brain Res* 195, 198–213.

Hanson, J. L., Chung, M. K., Avants, B. B., Shirtcliff, E. A., Gee, J. C., Davidson, R. J. & Pollak, S. D. (2010). Early stress is associated with alterations in the orbitofrontal cortex: a tensor-based morphometry investigation of brain structure and behavioral risk. *J Neurosci* 30, 7466–7472.

Hariri, A. R., Goldberg, T. E., Mattay, V. S., Kolachana, B. S., Callicott, J. H., Egan, M. F. & Weinberger, D. R. (2003). Brain-derived neurotrophic factor val66met polymorphism affects human memory-related hippocampal activity and predicts memory performance. *J Neurosci* 23, 6690–6694.

Heim, C., Newport, D. J., Bonsall, R., Miller, A. H. & Nemeroff, C. B. (2001). Altered pituitary-adrenal axis responses to provocative challenge tests in adult survivors of childhood abuse. *Am J Psychiatry* 158, 575–581.

Heuser, I., Yassouridis, A. & Holsboer, F. (1994). The combined dexamethasone/CRH test: a refined laboratory test for psychiatric disorders. *J Psychiatr Res* 28, 341–356.

Himmelheber, A. M., Sarter, M. & Bruno, J. P. (2000). Increases in cortical acetylcholine release during sustained attention performance in rats. *Brain Res Cogn Brain Res* 9, 313–325.

Hof, P. R., Haroutunian, V., Friedrich, V. L., Jr., Byne, W., Buitron, C., Perl, D. P. & Davis, K. L. (2003). Loss and altered spatial distribution of oligodendrocytes in the superior frontal gyrus in schizophrenia. *Biol Psychiatry* 53, 1075–1085.

Honer, W. G., Falkai, P., Chen, C., Arango, V., Mann, J. J. & Dwork, A. J. (1999). Synaptic and plasticity-associated proteins in anterior frontal cortex in severe mental illness. *Neuroscience* 91, 1247–1255.

Hong, L., Schroth, G. P., Matthews, H. R., Yau, P. & Bradbury, E. M. (1993). Studies of the DNA binding properties of histone H4 amino terminus. Thermal denaturation studies reveal

336   D. Zelena

that acetylation markedly reduces the binding constant of the H4 "tail" to DNA. *J Biol Chem* 268, 305–314.

Horrobin, D. F., Glen, A. I. & Vaddadi, K. (1994). The membrane hypothesis of schizophrenia. *Schizophr Res* 13, 195–207.

Hsieh, J. & Eisch, A. J. (2010). Epigenetics, hippocampal neurogenesis, and neuropsychiatric disorders: unraveling the genome to understand the mind. *Neurobiol Dis* 39, 73–84.

Huang, H. S. & Akbarian, S. (2007). GAD1 mRNA expression and DNA methylation in prefrontal cortex of subjects with schizophrenia. *PLoS ONE* 2, e809.

Huang, H. S., Matevossian, A., Whittle, C., Kim, S. Y., Schumacher, A., Baker, S. P. & Akbarian, S. (2007). Prefrontal dysfunction in schizophrenia involves mixed-lineage leukemia 1-regulated histone methylation at GABAergic gene promoters. *J Neurosci* 27, 11254–11262.

Hublitz, P., Albert, M. & Peters, A. H. (2009). Mechanisms of transcriptional repression by histone lysine methylation. *Int J Dev Biol* 53, 335–354.

Hughes, C. M., Rozenblatt-Rosen, O., Milne, T. A., Copeland, T. D., Levine, S. S., Lee, J. C., Hayes, D. N., Shanmugam, K. S., Bhattacharjee, A., Biondi, C. A., Kay, G. F., Hayward, N. K., Hess, J. L. & Meyerson, M. (2004). Menin associates with a trithorax family histone methyltransferase complex and with the hoxc8 locus. *Mol Cell* 13, 587–597.

Hurlock, E. C. (2001). Interferons: potential roles in affect. *Med Hypotheses* 56, 558–566.

Illingworth, R. S. & Bird, A. P. (2009). CpG islands--'a rough guide'. *FEBS Lett* 583, 1713–1720.

Inayama, Y., Yoneda, H., Sakai, T., Ishida, T., Nonomura, Y., Kono, Y., Takahata, R., Koh, J., Sakai, J., Takai, A., Inada, Y. & Asaba, H. (1996). Positive association between a DNA sequence variant in the serotonin 2A receptor gene and schizophrenia. *Am J Med Genet* 67, 103–105.

Inlow, J. K. & Restifo, L. L. (2004). Molecular and comparative genetics of mental retardation. *Genetics* 166, 835–881.

Iwamoto, K., Bundo, M., Yamada, K., Takao, H., Iwayama, Y., Yoshikawa, T. & Kato, T. (2006). A family-based and case-control association study of SOX10 in schizophrenia. *Am J Med Genet B Neuropsychiatr Genet* 141B, 477–481.

Iwamoto, K., Bundo, M., Yamada, K., Takao, H., Iwayama-Shigeno, Y., Yoshikawa, T. & Kato, T. (2005). DNA methylation status of SOX10 correlates with its downregulation and oligodendrocyte dysfunction in schizophrenia. *J Neurosci* 25, 5376–5381.

Jakobsson, J., Cordero, M. I., Bisaz, R., Groner, A. C., Busskamp, V., Bensadoun, J. C., Cammas, F., Losson, R., Mansuy, I. M., Sandi, C. & Trono, D. (2008). KAP1-mediated epigenetic repression in the forebrain modulates behavioral vulnerability to stress. *Neuron* 60, 818–831.

Javitt, D. C. (2010). Glutamatergic theories of schizophrenia. *Isr J Psychiatry Relat Sci* 47, 4–16.

Javitt, D. C. & Zukin, S. R. (1991). Recent advances in the phencyclidine model of schizophrenia. *Am J Psychiatry* 148, 1301–1308.

Jin, Y., Wang, Y., Walker, D. L., Dong, H., Conley, C., Johansen, J. & Johansen, K. M. (1999). JIL-1: a novel chromosomal tandem kinase implicated in transcriptional regulation in Drosophila. *Mol Cell* 4, 129–135.

Jirtle, R. L. & Skinner, M. K. (2007). Environmental epigenomics and disease susceptibility. *Nat Rev Genet* 8, 253–262.

Jones, P. A. & Baylin, S. B. (2002). The fundamental role of epigenetic events in cancer. *Nat Rev Genet* 3, 415–428.

Kaeberlein, M., McVey, M. & Guarente, L. (1999). The SIR2/3/4 complex and SIR2 alone promote longevity in Saccharomyces cerevisiae by two different mechanisms. *Genes Dev* 13, 2570–2580.

Kaminsky, Z., Wang, S. C. & Petronis, A. (2006). Complex disease, gender and epigenetics. *Ann Med* 38, 530–544.

Kao, W. T., Wang, Y., Kleinman, J. E., Lipska, B. K., Hyde, T. M., Weinberger, D. R. & Law, A. J. (2010). Common genetic variation in Neuregulin 3 (NRG3) influences risk for schizophrenia and impacts NRG3 expression in human brain. *Proc Natl Acad Sci U S A* 107, 15619–15624.

Kato, M. V., Ikawa, Y., Hayashizaki, Y. & Shibata, H. (1998). Paternal imprinting of mouse serotonin receptor 2A gene Htr2 in embryonic eye: a conserved imprinting regulation on the RB/Rb locus. *Genomics* 47, 146–148.

Kato, T. (2007). Molecular genetics of bipolar disorder and depression. *Psychiatry Clin Neurosci* 61, 3–19.

Katona, I., Sperlagh, B., Sik, A., Kafalvi, A., Vizi, E. S., Mackie, K. & Freund, T. F. (1999). Presynaptically located CB1 cannabinoid receptors regulate GABA release from axon terminals of specific hippocampal interneurons. *J Neurosci* 19, 4544–4558.

Kazantsev, A. G. & Thompson, L. M. (2008). Therapeutic application of histone deacetylase inhibitors for central nervous system disorders. *Nat Rev Drug Discov* 7, 854–868.

Kessaris, N., Pringle, N. & Richardson, W. D. (2001). Ventral neurogenesis and the neuron-glial switch. *Neuron* 31, 677–680.

Ketter, T. A. (2010). Diagnostic features, prevalence, and impact of bipolar disorder. *J Clin Psychiatry* 71, e14.

Kety, S. S., Wender, P. H., Jacobsen, B., Ingraham, L. J., Jansson, L., Faber, B. & Kinney, D. K. (1994). Mental illness in the biological and adoptive relatives of schizophrenic adoptees. Replication of the Copenhagen Study in the rest of Denmark. *Arch Gen Psychiatry* 51, 442–455.

Khalil, A. M., Guttman, M., Huarte, M., Garber, M., Raj, A., Rivea, M. D., Thomas, K., Presser, A., Bernstein, B. E., van Oudenaarden, A., Regev, A., Lander, E. S. & Rinn, J. L. (2009). Many human large intergenic noncoding RNAs associate with chromatin-modifying complexes and affect gene expression. *Proc Natl Acad Sci U S A* 106, 11667–11672.

Kim, J. S., Kornhuber, H. H., Schmid-Burgk, W. & Holzmuller, B. (1980). Low cerebrospinal fluid glutamate in schizophrenic patients and a new hypothesis on schizophrenia. *Neurosci Lett* 20, 379–382.

Kim, T., Park, J. K., Kim, H. J., Chung, J. H. & Kim, J. W. (2010). Association of histone deacetylase genes with schizophrenia in Korean population. *Psychiatry Res* 178, 266–269.

Kim, V. N. (2005). MicroRNA biogenesis: coordinated cropping and dicing. *Nat Rev Mol Cell Biol* 6, 376–385.

Klose, R. J. & Bird, A. P. (2006). Genomic DNA methylation: the mark and its mediators. *Trends Biochem Sci* 31, 89–97.

Knable, M. B., Barci, B. M., Webster, M. J., Meador-Woodruff, J. & Torrey, E. F. (2004). Molecular abnormalities of the hippocampus in severe psychiatric illness: postmortem findings from the Stanley Neuropathology Consortium. *Mol Psychiatry* 9, 609–620.

Knight, C. G., Zitzmann, N., Prabhakar, S., Antrobus, R., Dwek, R., Hebestreit, H. & Rainey, P. B. (2006). Unraveling adaptive evolution: how a single point mutation affects the protein coregulation network. *Nat Genet* 38, 1015–1022.

Kontkanen, O., Lakso, M., Wong, G. & Castren, E. (2002). Chronic antipsychotic drug treatment induces long-lasting expression of fos and jun family genes and activator protein 1 complex in the rat prefrontal cortex. *Neuropsychopharmacology* 27, 152–162.

Kouzarides, T. (2007). Chromatin modifications and their function. *Cell* 128, 693–705.

Kouzmenko, A. P., Hayes, W. L., Pereira, A. M., Dean, B., Burnet, P. W. & Harrison, P. J. (1997). 5-HT2A receptor polymorphism and steady state receptor expression in schizophrenia. *Lancet* 349, 1815.

Kouzmenko, A. P., Scaffidi, A., Pereira, A. M., Hayes, W. L., Copolov, D. L. & Dean, B. (1999). No correlation between A(-1438)G polymorphism in 5-HT2A receptor gene promoter and the density of frontal cortical 5-HT2A receptors in schizophrenia. *Hum Hered* 49, 103–105.

Koyrakh, L., Lujan, R., Colon, J., Karschin, C., Kurachi, Y., Karschin, A. & Wickman, K. (2005). Molecular and cellular diversity of neuronal G-protein-gated potassium channels. *J Neurosci* 25, 11468–11478.

Krebs, M. O., Bellon, A., Mainguy, G., Jay, T. M. & Frieling, H. (2009). One-carbon metabolism and schizophrenia: current challenges and future directions. *Trends Mol Med* 15, 562–570.

Krishnan, V. & Nestler, E. J. (2010). Linking molecules to mood: new insight into the biology of depression. *Am J Psychiatry* 167, 1305–1320.

Krystal, J. H., Mathew, S. J., D'Souza, D. C., Garakani, A., Gunduz-Bruce, H. & Charney, D. S. (2010). Potential psychiatric applications of metabotropic glutamate receptor agonists and antagonists. *CNS Drugs* 24, 669–693.

Kundakovic, M., Chen, Y., Costa, E. & Grayson, D. R. (2007). DNA methyltransferase inhibitors coordinately induce expression of the human reelin and glutamic acid decarboxylase 67 genes. *Mol Pharmacol* 71, 644–653.

Kundakovic, M., Chen, Y., Guidotti, A. & Grayson, D. R. (2009). The reelin and GAD67 promoters are activated by epigenetic drugs that facilitate the disruption of local repressor complexes. *Mol Pharmacol* 75, 342–354.

Kuratomi, G., Iwamoto, K., Bundo, M., Kusumi, I., Kato, N., Iwata, N., Ozaki, N. & Kato, T. (2008). Aberrant DNA methylation associated with bipolar disorder identified from discordant monozygotic twins. *Mol Psychiatry* 13, 429–441.

Kustova, Y., Sei, Y., Morse, H. C., Jr. & Basile, A. S. (1998). The influence of a targeted deletion of the IFNgamma gene on emotional behaviors. *Brain Behav Immun* 12, 308–324.

Kwon, O. B., Paredes, D., Gonzalez, C. M., Neddens, J., Hernandez, L., Vullhorst, D. & Buonanno, A. (2008). Neuregulin-1 regulates LTP at CA1 hippocampal synapses through activation of dopamine D4 receptors. *Proc Natl Acad Sci U S A* 105, 15587–15592.

Lau, C. G. & Zukin, R. S. (2007). NMDA receptor trafficking in synaptic plasticity and neuropsychiatric disorders. *Nat Rev Neurosci* 8, 413–426.

Lee, C. & Scherer, S. W. (2010). The clinical context of copy number variation in the human genome. *Expert Rev Mol Med* 12, e8.

Leonard, B. E. (2010). The concept of depression as a dysfunction of the immune system. *Curr Immunol Rev* 6, 205–212.

Levenson, J. M., Roth, T. L., Lubin, F. D., Miller, C. A., Huang, I. C., Desai, P., Malone, L. M. & Sweatt, J. D. (2006). Evidence that DNA (cytosine-5) methyltransferase regulates synaptic plasticity in the hippocampus. *J Biol Chem* 281, 15763–15773.

Levine, J., Stahl, Z., Sela, B. A., Gavendo, S., Ruderman, V. & Belmaker, R. H. (2002). Elevated homocysteine levels in young male patients with schizophrenia. *Am J Psychiatry* 159, 1790–1792.

Li, E. (2002). Chromatin modification and epigenetic reprogramming in mammalian development. *Nat Rev Genet* 3, 662–673.

Li, J., Guo, Y., Schroeder, F. A., Youngs, R. M., Schmidt, T. W., Ferris, C., Konradi, C. & Akbarian, S. (2004). Dopamine D2-like antagonists induce chromatin remodeling in striatal neurons through cyclic AMP-protein kinase A and NMDA receptor signaling. *J Neurochem* 90, 1117–1131.

Lim, D. A., Huang, Y. C., Swigut, T., Mirick, A. L., Garcia-Verdugo, J. M., Wysocka, J., Ernst, P. & Alvarez-Buylla, A. (2009). Chromatin remodelling factor Mll1 is essential for neurogenesis from postnatal neural stem cells. *Nature* 458, 529–533.

Lin, Q., Wei, W., Coelho, C. M., Li, X., Baker-Andresen, D., Dudley, K., Ratnu, V. S., Boskovic, Z., Kobor, M. S., Sun, Y. E. & Bredy, T. W. (2011). The brain-specific microRNA miR-128b regulates the formation of fear-extinction memory. *Nat Neurosci* 14, 1115–1117.

Lippmann, M., Bress, A., Nemeroff, C. B., Plotsky, P. M. & Monteggia, L. M. (2007). Long-term behavioural and molecular alterations associated with maternal separation in rats. *Eur J Neurosci* 25, 3091–3098.

Lisman, J. E., Coyle, J. T., Green, R. W., Javitt, D. C., Benes, F. M., Heckers, S. & Grace, A. A. (2008). Circuit-based framework for understanding neurotransmitter and risk gene interactions in schizophrenia. *Trends Neurosci* 31, 234–242.

Liu, C. M., Liu, Y. L., Fann, C. S., Chen, W. J., Yang, W. C., Ouyang, W. C., Chen, C. Y., Jou, Y. S., Hsieh, M. H., Liu, S. K., Hwang, T. J., Faraone, S. V., Tsuang, M. T. & Hwu, H. G. (2007). Association evidence of schizophrenia with distal genomic region of NOTCH4 in Taiwanese families. *Genes Brain Behav* 6, 497–502.

Liu, Q. R., Lu, L., Zhu, X. G., Gong, J. P., Shaham, Y. & Uhl, G. R. (2006). Rodent BDNF genes, novel promoters, novel splice variants, and regulation by cocaine. *Brain Res* 1067, 1–12.

Loganathan, M., Lohano, K., Roberts, R. J., Gao, Y. & El Mallakh, R. S. (2010). When to suspect bipolar disorder. *J Fam Pract* 59, 682–688.

Longart, M., Liu, Y., Karavanova, I. & Buonanno, A. (2004). Neuregulin-2 is developmentally regulated and targeted to dendrites of central neurons. *J Comp Neurol* 472, 156–172.

Lopez, A. D., Mathers, C. D., Ezzati, M., Jamison, D. T. & Murray, C. J. L. (2006). *Global Burden of Disease and Risk Factors.* New York, Washington, D.C.: Oxford University Press, The World Bank.

Lopez-Atalaya, J. P., Ciccarelli, A., Viosca, J., Valor, L. M., Jimenez-Minchan, M., Canals, S., Giustetto, M. & Barco, A. (2011). CBP is required for environmental enrichment-induced neurogenesis and cognitive enhancement. *EMBO J* 30, 4287–4298.

Lu, B. & Martinowich, K. (2008). Cell biology of BDNF and its relevance to schizophrenia. *Novartis Found Symp* 289, 119–129.

Lubin, F. D., Roth, T. L. & Sweatt, J. D. (2008). Epigenetic regulation of BDNF gene transcription in the consolidation of fear memory. *J Neurosci* 28, 10576–10586.

Ma, Y. Y., Shek, C. C., Wong, M. C., Yip, K. C., Ng, R. M., Nguyen, D. G. & Poon, T. K. (2009). Homocysteine level in schizophrenia patients. *Aust N Z J Psychiatry* 43, 760–765.

Maeno, N., Takahashi, N., Saito, S., Ji, X., Ishihara, R., Aoyama, N., Branko, A., Miura, H., Ikeda, M., Suzuki, T., Kitajima, T., Yamanouchi, Y., Kinoshita, Y., Iwata, N., Inada, T. & Ozaki, N. (2007). Association of SOX10 with schizophrenia in the Japanese population. *Psychiatr Genet* 17, 227–231.

Maes, O. C., Chertkow, H. M., Wang, E. & Schipper, H. M. (2009). MicroRNA: Implications for Alzheimer Disease and other Human CNS Disorders. *Curr Genomics* 10, 154–168.

Mahadevan, L. C., Willis, A. C. & Barratt, M. J. (1991). Rapid histone H3 phosphorylation in response to growth factors, phorbol esters, okadaic acid, and protein synthesis inhibitors. *Cell* 65, 775–783.

Malaspina, D., Corcoran, C., Fahim, C., Berman, A., Harkavy-Friedman, J., Yale, S., Goetz, D., Goetz, R., Harlap, S. & Gorman, J. (2002). Paternal age and sporadic schizophrenia: evidence for de novo mutations. *Am J Med Genet* 114, 299–303.

Malik, K. & Brown, K. W. (2000). Epigenetic gene deregulation in cancer. *Br J Cancer* 83, 1583–1588.

Maloku, E., Kadriu, B., Zhubi, A., Dong, E., Pibiri, F., Satta, R. & Guidotti, A. (2011). Selective alpha4beta2 nicotinic acetylcholine receptor agonists target epigenetic mechanisms in cortical GABAergic neurons. *Neuropsychopharmacology* 36, 1366–1374.

Marksteiner, J., Weiss, U., Weis, C., Laslop, A., Fischer-Colbrie, R., Humpel, C., Feldon, J. & Fleischhacker, W. W. (2001). Differential regulation of chromogranin A, chromogranin B and secretogranin II in rat brain by phencyclidine treatment. *Neuroscience* 104, 325–333.

Martinowich, K., Hattori, D., Wu, H., Fouse, S., He, F., Hu, Y., Fan, G. & Sun, Y. E. (2003). DNA methylation-related chromatin remodeling in activity-dependent BDNF gene regulation. *Science* 302, 890–893.

Matrisciano, F., Dong, E., Gavin, D. P., Nicoletti, F. & Guidotti, A. (2011). Activation of group II metabotropic glutamate receptors promotes DNA demethylation in the mouse brain. *Mol Pharmacol* 80, 174–182.

Mattick, J. S. & Makunin, I. V. (2005). Small regulatory RNAs in mammals. *Hum Mol Genet* 14, R121–R132.

McClelland, S., Korosi, A., Cope, J., Ivy, A. & Baram, T. Z. (2011). Emerging roles of epigenetic mechanisms in the enduring effects of early-life stress and experience on learning and memory. *Neurobiol Learn Mem* 96, 79–88.

McFarlane, A., Clark, C. R., Bryant, R. A., Williams, L. M., Niaura, R., Paul, R. H., Hitsman, B. L., Stroud, L., Alexander, D. M. & Gordon, E. (2005). The impact of early life stress on psychophysiological, personality and behavioral measures in 740 non-clinical subjects. *J Integr Neurosci* 4, 27–40.

McGowan, P. O., Sasaki, A., D'Alessio, A. C., Dymov, S., Labonte, B., Szyf, M., Turecki, G. & Meaney, M. J. (2009). Epigenetic regulation of the glucocorticoid receptor in human brain associates with childhood abuse. *Nat Neurosci* 12, 342–348.

McGrath, J., Saha, S., Chant, D. & Welham, J. (2008). Schizophrenia: a concise overview of incidence, prevalence, and mortality. *Epidemiol Rev* 30, 67–76.

McQuillin, A., Rizig, M. & Gurling, H. M. (2007). A microarray gene expression study of the molecular pharmacology of lithium carbonate on mouse brain mRNA to understand the neurobiology of mood stabilization and treatment of bipolar affective disorder. *Pharmacogenet Genomics* 17, 605–617.

Meaney, M. J. & Szyf, M. (2005). Maternal care as a model for experience-dependent chromatin plasticity? *Trends Neurosci* 28, 456–463.

Mehler, M. F. (2008). Epigenetic principles and mechanisms underlying nervous system functions in health and disease. *Prog Neurobiol* 86, 305–341.

Mello, A. F., Mello, M. F., Carpenter, L. L. & Price, L. H. (2003). Update on stress and depression: the role of the hypothalamic-pituitary-adrenal (HPA) axis. *Rev Bras Psiquiatr* 25, 231–238.

Meltzer, H. Y. (1999). The role of serotonin in antipsychotic drug action. *Neuropsychopharmacology* 21, 106S–115S.

Meola, N., Gennarino, V. A. & Banfi, S. (2009). microRNAs and genetic diseases. *Pathogenetics* 2:7.

Merikangas, A. K., Corvin, A. P. & Gallagher, L. (2009). Copy-number variants in neurodevelopmental disorders: promises and challenges. *Trends Genet* 25, 536–544.

Meza-Sosa, K. F., Valle-Garcia, D., Pedraza-Alva, G. & Perez-Martinez, L. (2012). Role of microRNAs in central nervous system development and pathology. *J Neurosci Res* 90, 1–12.

Migliore, L. & Coppede, F. (2009). Genetics, environmental factors and the emerging role of epigenetics in neurodegenerative diseases. *Mutat Res* 667, 82–97.

Mill, J. & Petronis, A. (2007). Molecular studies of major depressive disorder: the epigenetic perspective. *Mol Psychiatry* 12, 799–814.

Mill, J., Tang, T., Kaminsky, Z., Khare, T., Yazdanpanah, S., Bouchard, L., Jia, P., Assadzadeh, A., Flanagan, J., Schumacher, A., Wang, S. C. & Petronis, A. (2008). Epigenomic profiling reveals DNA-methylation changes associated with major psychosis. *Am J Hum Genet* 82, 696–711.

Miller, B. H. & Wahlestedt, C. (2010). MicroRNA dysregulation in psychiatric disease. *Brain Res* 1338, 89–99.

Miller, C. A. & Sweatt, J. D. (2007). Covalent modification of DNA regulates memory formation. *Neuron* 53, 857–869.

Mirnics, K., Levitt, P. & Lewis, D. A. (2004). DNA microarray analysis of postmortem brain tissue. *Int Rev Neurobiol* 60, 153–181.

Miska, E. A., Alvarez-Saavedra, E., Townsend, M., Yoshii, A., Sestan, N., Rakic, P., Constantine-Paton, M. & Horvitz, H. R. (2004). Microarray analysis of microRNA expression in the developing mammalian brain. *Genome Biol* 5, R68.

Miyaoka, T., Seno, H. & Ishino, H. (1999). Increased expression of Wnt-1 in schizophrenic brains. *Schizophr Res* 38, 1–6.

Mohn, F. & Schubeler, D. (2009). Genetics and epigenetics: stability and plasticity during cellular differentiation. *Trends Genet* 25, 129–136.

Money, T. T., Scarr, E., Udawela, M., Gibbons, A. S., Jeon, W. J., Seo, M. S. & Dean, B. (2010). Treating schizophrenia: novel targets for the cholinergic system. *CNS Neurol Disord Drug Targets* 9, 241–256.

Morar, B., Dragovic, M., Waters, F. A., Chandler, D., Kalaydjieva, L. & Jablensky, A. (2011). Neuregulin 3 (NRG3) as a susceptibility gene in a schizophrenia subtype with florid delusions and relatively spared cognition. *Mol Psychiatry* 16, 860–866.

Morrison, B. E., Majdzadeh, N. & D'Mello, S. R. (2007). Histone deacetylases: focus on the nervous system. *Cell Mol Life Sci* 64, 2258–2269.

Moser, D., Ekawardhani, S., Kumsta, R., Palmason, H., Bock, C., Athanassiadou, Z., Lesch, K. P. & Meyer, J. (2009). Functional analysis of a potassium-chloride co-transporter 3 (SLC12A6) promoter polymorphism leading to an additional DNA methylation site. *Neuropsychopharmacology* 34, 458–467.

Mottet, D. & Castronovo, V. (2008). Histone deacetylases: target enzymes for cancer therapy. *Clin Exp Metastasis* 25, 183–189.

Murgatroyd, C., Patchev, A. V., Wu, Y., Micale, V., Bockmuhl, Y., Fischer, D., Holsboer, F., Wotjak, C. T., Almeida, O. F. & Spengler, D. (2009). Dynamic DNA methylation programs persistent adverse effects of early-life stress. *Nat Neurosci* 12, 1559–1566.

Nakahata, Y., Grimaldi, B., Sahar, S., Hirayama, J. & Sassone-Corsi, P. (2007). Signaling to the circadian clock: plasticity by chromatin remodeling. *Curr Opin Cell Biol* 19, 230–237.

Nakatani, T., Minaki, Y., Kumai, M. & Ono, Y. (2007). Helt determines GABAergic over glutamatergic neuronal fate by repressing Ngn genes in the developing mesencephalon. *Development* 134, 2783–2793.

Neddens, J., Vullhorst, D., Paredes, D. & Buonanno, A. (2009). Neuregulin links dopaminergic and glutamatergic neurotransmission to control hippocampal synaptic plasticity. *Commun Integr Biol* 2, 261–264.

Nelson, E. D., Kavalali, E. T. & Monteggia, L. M. (2008). Activity-dependent suppression of miniature neurotransmission through the regulation of DNA methylation. *J Neurosci* 28, 395–406.

Nestler, E. J., Barrot, M., Dileone, R. J., Eisch, A. J., Gold, S. J. & Monteggia, L. M. (2002). Neurobiology of depression. *Neuron* 34, 13–25.

Nowak, S. J. & Corces, V. G. (2004). Phosphorylation of histone H3: a balancing act between chromosome condensation and transcriptional activation. *Trends Genet* 20, 214–220.

Nowak, S. J., Pai, C. Y. & Corces, V. G. (2003). Protein phosphatase 2A activity affects histone H3 phosphorylation and transcription in Drosophila melanogaster. *Mol Cell Biol* 23, 6129–6138.

O'Donovan, M. C., Craddock, N. J. & Owen, M. J. (2009). Genetics of psychosis; insights from views across the genome. *Hum Genet* 126, 3–12.

Oberlander, T. F., Weinberg, J., Papsdorf, M., Grunau, R., Misri, S. & Devlin, A. M. (2008). Prenatal exposure to maternal depression, neonatal methylation of human glucocorticoid receptor gene (NR3C1) and infant cortisol stress responses. *Epigenetics* 3, 97–106.

Ordway, J. M. & Curran, T. (2002). Methylation matters: modeling a manageable genome. *Cell Growth Differ* 13, 149–162.

Paradisi, A., Pasquariello, N., Barcaroli, D. & Maccarrone, M. (2008). Anandamide regulates keratinocyte differentiation by inducing DNA methylation in a CB1 receptor-dependent manner. *J Biol Chem* 283, 6005–6012.

Pariante, C. M. (2008). Pituitary volume in psychosis: the first review of the evidence. *J Psychopharmacol* 22, 76–81.

Pariante, C. M. & Lightman, S. L. (2008). The HPA axis in major depression: classical theories and new developments. *Trends Neurosci* 31, 464–468.

Parnas, J. & Jorgensen, A. (1989). Pre-morbid psychopathology in schizophrenia spectrum. *Br J Psychiatry* 155, 623–627.

Paylor, R., Zhao, Y., Libbey, M., Westphal, H. & Crawley, J. N. (2001). Learning impairments and motor dysfunctions in adult Lhx5-deficient mice displaying hippocampal disorganization. *Physiol Behav* 73, 781–792.

Perkins, D. O., Jeffries, C. D., Jarskog, L. F., Thomson, J. M., Woods, K., Newman, M. A., Parker, J. S., Jin, J. & Hammond, S. M. (2007). microRNA expression in the prefrontal cortex of individuals with schizophrenia and schizoaffective disorder. *Genome Biol* 8, R27.

Pesold, C., Impagnatiello, F., Pisu, M. G., Uzunov, D. P., Costa, E., Guidotti, A. & Caruncho, H. J. (1998). Reelin is preferentially expressed in neurons synthesizing gamma-aminobutyric acid in cortex and hippocampus of adult rats. *Proc Natl Acad Sci U S A* 95, 3221–3226.

Peterson, C. L. & Laniel, M. A. (2004). Histones and histone modifications. *Curr Biol* 14, R546–R551.

Petronis, A. (2000). The genes for major psychosis: aberrant sequence or regulation? *Neuropsychopharmacology* 23, 1–12.

Petronis, A., Paterson, A. D. & Kennedy, J. L. (1999). Schizophrenia: an epigenetic puzzle? *Schizophr Bull* 25, 639–655.

Picchioni, M. M. & Murray, R. M. (2007). Schizophrenia. *BMJ* 335, 91–95.

Pidsley, R., Dempster, E. L. & Mill, J. (2010). Brain weight in males is correlated with DNA methylation at IGF2. *Mol Psychiatry* 15, 880–881.

Pidsley, R. & Mill, J. (2011). Epigenetic studies of psychosis: current findings, methodological approaches, and implications for postmortem research. *Biol Psychiatry* 69, 146–156.

Pinal, C. S. & Tobin, A. J. (1998). Uniqueness and redundancy in GABA production. *Perspect Dev Neurobiol* 5, 109–118.

Polesskaya, O. O., Aston, C. & Sokolov, B. P. (2006). Allele C-specific methylation of the 5-HT2A receptor gene: evidence for correlation with its expression and expression of DNA methylase DNMT1. *J Neurosci Res* 83, 362–373.

Polesskaya, O. O. & Sokolov, B. P. (2002). Differential expression of the "C" and "T" alleles of the 5-HT2A receptor gene in the temporal cortex of normal individuals and schizophrenics. *J Neurosci Res* 67, 812–822.

Ponjavic, J., Oliver, P. L., Lunter, G. & Ponting, C. P. (2009). Genomic and transcriptional co-localization of protein-coding and long non-coding RNA pairs in the developing brain. *PLoS Genet* 5, e1000617.

Portela, A. & Esteller, M. (2010). Epigenetic modifications and human disease. *Nat Biotechnol* 28, 1057–1068.

Preuss, U., Landsberg, G. & Scheidtmann, K. H. (2003). Novel mitosis-specific phosphorylation of histone H3 at Thr11 mediated by Dlk/ZIP kinase. *Nucleic Acids Res* 31, 878–885.

Purcell, S. M., Wray, N. R., Stone, J. L., Visscher, P. M., O'Donovan, M. C., Sullivan, P. F. & Sklar, P. (2009). Common polygenic variation contributes to risk of schizophrenia and bipolar disorder. *Nature* 460, 748–752.

Qureshi, I. A., Mattick, J. S. & Mehler, M. F. (2010). Long non-coding RNAs in nervous system function and disease. *Brain Res* 1338, 20–35.

Rajkowska, G. & Miguel-Hidalgo, J. J. (2007). Gliogenesis and glial pathology in depression. *CNS Neurol Disord Drug Targets* 6, 219–233.

Reich, T., Clayton, P. J. & Winokur, G. (1969). Family history studies: V. The genetics of mania. *Am J Psychiatry* 125, 1358–1369.

Renthal, W. & Nestler, E. J. (2008). Epigenetic mechanisms in drug addiction. *Trends Mol Med* 14, 341–350.

Restrepo, A., Smith, C. A., Agnihotri, S., Shekarforoush, M., Kongkham, P. N., Seol, H. J., Northcott, P. & Rutka, J. T. (2011). Epigenetic regulation of glial fibrillary acidic protein by DNA methylation in human malignant gliomas. *Neuro Oncol* 13, 42–50.

Reul, J. M., Hesketh, S. A., Collins, A. & Mecinas, M. G. (2009). Epigenetic mechanisms in the dentate gyrus act as a molecular switch in hippocampus-associated memory formation. *Epigenetics* 4, 434–439.

Rigucci, S., Serafini, G., Pompili, M., Kotzalidis, G. D. & Tatarelli, R. (2010). Anatomical and functional correlates in major depressive disorder: the contribution of neuroimaging studies. *World J Biol Psychiatry* 11, 165–180.

Robinson, S. A., Loiacono, R. E., Christopoulos, A., Sexton, P. M. & Malone, D. T. (2010). The effect of social isolation on rat brain expression of genes associated with endocannabinoid signaling. *Brain Res* 1343, 153–167.

Roffman, J. L., Weiss, A. P., Purcell, S., Caffalette, C. A., Freudenreich, O., Henderson, D. C., Bottiglieri, T., Wong, D. H., Halsted, C. H. & Goff, D. C. (2008). Contribution of methylenetetrahydrofolate reductase (MTHFR) polymorphisms to negative symptoms in schizophrenia. *Biol Psychiatry* 63, 42–48.

Rosa, A., Picchioni, M. M., Kalidindi, S., Loat, C. S., Knight, J., Toulopoulou, T., Vonk, R., van der Schot, A. C., Nolen, W., Kahn, R. S., McGuffin, P., Murray, R. M. & Craig, I. W. (2008). Differential methylation of the X-chromosome is a possible source of discordance for bipolar disorder female monozygotic twins. *Am J Med Genet B Neuropsychiatr Genet* 147B, 459–462.

Roth, T. L., Lubin, F. D., Funk, A. J. & Sweatt, J. D. (2009). Lasting epigenetic influence of early-life adversity on the BDNF gene. *Biol Psychiatry* 65, 760–769.

Roth, T. L. & Sweatt, J. D. (2009). Regulation of chromatin structure in memory formation. *Curr Opin Neurobiol* 19, 336–342.

Rottach, A., Leonhardt, H. & Spada, F. (2009). DNA methylation-mediated epigenetic control. *J Cell Biochem* 108, 43–51.

Ruzicka, W. B., Zhubi, A., Veldic, M., Grayson, D. R., Costa, E. & Guidotti, A. (2007). Selective epigenetic alteration of layer I GABAergic neurons isolated from prefrontal cortex of schizophrenia patients using laser-assisted microdissection. *Mol Psychiatry* 12, 385–397.

Satta, R., Maloku, E., Zhubi, A., Pibiri, F., Hajos, M., Costa, E. & Guidotti, A. (2008). Nicotine decreases DNA methyltransferase 1 expression and glutamic acid decarboxylase 67 promoter methylation in GABAergic interneurons. *Proc Natl Acad Sci U S A* 105, 16356–16361.

Schratt, G. M., Tuebing, F., Nigh, E. A., Kane, C. G., Sabatini, M. E., Kiebler, M. & Greenberg, M. E. (2006). A brain-specific microRNA regulates dendritic spine development. *Nature* 439, 283–289.

Schroeder, F. A., Lin, C. L., Crusio, W. E. & Akbarian, S. (2007). Antidepressant-like effects of the histone deacetylase inhibitor, sodium butyrate, in the mouse. *Biol Psychiatry* 62, 55–64.

Schroeder, M., Krebs, M. O., Bleich, S. & Frieling, H. (2010). Epigenetics and depression: current challenges and new therapeutic options. *Curr Opin Psychiatry* 23, 588–592.

Schwab, S. G., Hallmayer, J., Albus, M., Lerer, B., Eckstein, G. N., Borrmann, M., Segman, R. H., Hanses, C., Freymann, J., Yakir, A., Trixler, M., Falkai, P., Rietschel, M., Maier, W. & Wildenauer, D. B. (2000). A genome-wide autosomal screen for schizophrenia susceptibility loci in 71 families with affected siblings: support for loci on chromosome 10p and 6. *Mol Psychiatry* 5, 638–649.

Segurado, R., Detera-Wadleigh, S. D., Levinson, D. F., Lewis, C. M., Gill, M., Nurnberger, J. I., Jr., Craddock, N., DePaulo, J. R., Baron, M., Gershon, E. S., Ekholm, J., Cichon, S., Turecki, G., Claes, S., Kelsoe, J. R., Schofield, P. R., Badenhop, R. F., Morissette, J., Coon, H., Blackwood, D., McInnes, L. A., Foroud, T., Edenberg, H. J., Reich, T., Rice, J. P., Goate, A., McInnis, M. G., McMahon, F. J., Badner, J. A., Goldin, L. R., Bennett, P., Willour, V. L., Zandi, P. P., Liu, J., Gilliam, C., Juo, S. H., Berrettini, W. H., Yoshikawa, T., Peltonen, L., Lonnqvist, J., Nothen, M. M., Schumacher, J., Windemuth, C., Rietschel, M., Propping, P., Maier, W., Alda, M., Grof, P., Rouleau, G. A., Del Favero, J., Van Broeckhoven, C., Mendlewicz, J., Adolfsson, R., Spence, M. A., Luebbert, H., Adams, L. J., Donald, J. A., Mitchell, P. B., Barden, N., Shink, E., Byerley, W., Muir, W., Visscher, P. M., Macgregor, S., Gurling, H., Kalsi, G., McQuillin, A., Escamilla, M. A., Reus, V. I., Leon, P., Freimer, N. B., Ewald, H., Kruse, T. A., Mors, O., Radhakrishna, U., Blouin, J. L., Antonarakis, S. E. & Akarsu, N. (2003). Genome scan meta-analysis of schizophrenia and bipolar disorder, part III: Bipolar disorder. *Am J Hum Genet* 73, 49–62.

Selye, H. (1956). The stress-concept as it presents itself in 1956. *Antibiot Chemother* 3, 1–17.

Serretti, A. & Mandelli, L. (2008). The genetics of bipolar disorder: genome 'hot regions', genes, new potential candidates and future directions. *Mol Psychiatry* 13, 742–771.

Sharma, R. P., Grayson, D. R. & Gavin, D. P. (2008). Histone deactylase 1 expression is increased in the prefrontal cortex of schizophrenia subjects: analysis of the National Brain Databank microarray collection. *Schizophr Res* 98, 111–117.

Sharma, R. P., Rosen, C., Kartan, S., Guidotti, A., Costa, E., Grayson, D. R. & Chase, K. (2006). Valproic acid and chromatin remodeling in schizophrenia and bipolar disorder: preliminary results from a clinical population. *Schizophr Res* 88, 227–231.

Shimabukuro, M., Jinno, Y., Fuke, C. & Okazaki, Y. (2006). Haloperidol treatment induces tissue- and sex-specific changes in DNA methylation: a control study using rats. *Behav Brain Funct* 2:37.

Simonini, M. V., Camargo, L. M., Dong, E., Maloku, E., Veldic, M., Costa, E. & Guidotti, A. (2006). The benzamide MS-275 is a potent, long-lasting brain region-selective inhibitor of histone deacetylases. *Proc Natl Acad Sci U S A* 103, 1587–1592.

Smith, C. L., Bolton, A. & Nguyen, G. (2010). Genomic and epigenomic instability, fragile sites, schizophrenia and autism. *Curr Genomics* 11, 447–469.

Smith, R. E., Haroutunian, V., Davis, K. L. & Meador-Woodruff, J. H. (2001). Vesicular glutamate transporter transcript expression in the thalamus in schizophrenia. *Neuroreport* 12, 2885–2887.

Sproule, D. M. & Kaufmann, P. (2008). Mitochondrial encephalopathy, lactic acidosis, and strokelike episodes: basic concepts, clinical phenotype, and therapeutic management of MELAS syndrome. *Ann N Y Acad Sci* 1142, 133–158.

Stadler, F., Kolb, G., Rubusch, L., Baker, S. P., Jones, E. G. & Akbarian, S. (2005). Histone methylation at gene promoters is associated with developmental regulation and region-specific expression of ionotropic and metabotropic glutamate receptors in human brain. *J Neurochem* 94, 324–336.

Stefansson, H., Ophoff, R. A., Steinberg, S., Andreassen, O. A., Cichon, S., Rujescu, D., Werge, T., Pietilainen, O. P., Mors, O., Mortensen, P. B., Sigurdsson, E., Gustafsson, O., Nyegaard, M., Tuulio-Henriksson, A., Ingason, A., Hansen, T., Suvisaari, J., Lonnqvist, J., Paunio, T., Borglum, A. D., Hartmann, A., Fink-Jensen, A., Nordentoft, M., Hougaard, D., Norgaard-Pedersen, B., Bottcher, Y., Olesen, J., Breuer, R., Moller, H. J., Giegling, I., Rasmussen, H. B., Timm, S., Mattheisen, M., Bitter, I., Rethelyi, J. M., Magnusdottir, B. B., Sigmundsson, T., Olason, P., Masson, G., Gulcher, J. R., Haraldsson, M., Fossdal, R., Thorgeirsson, T. E., Thorsteinsdottir, U., Ruggeri, M., Tosato, S., Franke, B., Strengman, E., Kiemeney, L. A., Melle, I., Djurovic, S., Abramova, L., Kaleda, V., Sanjuan, J., de Frutos, R., Bramon, E., Vassos, E., Fraser, G., Ettinger, U., Picchioni, M., Walker, N., Toulopoulou, T., Need, A. C., Ge, D., Yoon, J. L., Shianna, K. V., Freimer, N. B., Cantor, R. M., Murray, R., Kong, A., Golimbet, V., Carracedo, A., Arango, C., Costas, J., Jonsson, E. G., Terenius, L., Agartz, I., Petursson, H., Nothen, M. M., Rietschel, M., Matthews, P. M., Muglia, P., Peltonen, L., St Clair, D., Goldstein, D. B., Stefansson, K. & Collier, D. A. (2009). Common variants conferring risk of schizophrenia. *Nature* 460, 744–747.

Stimson, L. & La Thangue, N. B. (2009). Biomarkers for predicting clinical responses to HDAC inhibitors. *Cancer Lett* 280, 177–183.

Stock, J. K., Giadrossi, S., Casanova, M., Brookes, E., Vidal, M., Koseki, H., Brockdorff, N., Fisher, A. G. & Pombo, A. (2007). Ring1-mediated ubiquitination of H2A restrains poised RNA polymerase II at bivalent genes in mouse ES cells. *Nat Cell Biol* 9, 1428–1435.

Stolt, C. C., Rehberg, S., Ader, M., Lommes, P., Riethmacher, D., Schachner, M., Bartsch, U. & Wegner, M. (2002). Terminal differentiation of myelin-forming oligodendrocytes depends on the transcription factor Sox10. *Genes Dev* 16, 165–170.

Straub, R. E., Lipska, B. K., Egan, M. F., Goldberg, T. E., Callicott, J. H., Mayhew, M. B., Vakkalanka, R. K., Kolachana, B. S., Kleinman, J. E. & Weinberger, D. R. (2007). Allelic variation in GAD1 (GAD67) is associated with schizophrenia and influences cortical function and gene expression. *Mol Psychiatry* 12, 854–869.

Sugai, T., Kawamura, M., Iritani, S., Araki, K., Makifuchi, T., Imai, C., Nakamura, R., Kakita, A., Takahashi, H. & Nawa, H. (2004). Prefrontal abnormality of schizophrenia revealed by DNA microarray: impact on glial and neurotrophic gene expression. *Ann N Y Acad Sci* 1025, 84–91.

Suter, C. M., Martin, D. I. & Ward, R. L. (2004). Germline epimutation of MLH1 in individuals with multiple cancers. *Nat Genet* 36, 497–501.

Szeszko, P. R., Lipsky, R., Mentschel, C., Robinson, D., Gunduz-Bruce, H., Sevy, S., Ashtari, M., Napolitano, B., Bilder, R. M., Kane, J. M., Goldman, D. & Malhotra, A. K. (2005). Brain-derived neurotrophic factor val66met polymorphism and volume of the hippocampal formation. *Mol Psychiatry* 10, 631–636.

Szyf, M. (2010). DNA methylation and demethylation probed by small molecules. *Biochim Biophys Acta* 1799, 750–759.

Taft, R. J., Pang, K. C., Mercer, T. R., Dinger, M. & Mattick, J. S. (2010). Non-coding RNAs: regulators of disease. *J Pathol* 220, 126–139.

Takai, D. & Jones, P. A. (2002). Comprehensive analysis of CpG islands in human chromosomes 21 and 22. *Proc Natl Acad Sci U S A* 99, 3740–3745.

Tamminga, J., Koturbash, I., Baker, M., Kutanzi, K., Kathiria, P., Pogribny, I. P., Sutherland, R. J. & Kovalchuk, O. (2008). Paternal cranial irradiation induces distant bystander DNA damage in the germline and leads to epigenetic alterations in the offspring. *Cell Cycle* 7, 1238–1245.

Thomassin, H., Flavin, M., Espinas, M. L. & Grange, T. (2001). Glucocorticoid-induced DNA demethylation and gene memory during development. *EMBO J* 20, 1974–1983.

Titeler, M., Lyon, R. A. & Glennon, R. A. (1988). Radioligand binding evidence implicates the brain 5-HT2 receptor as a site of action for LSD and phenylisopropylamine hallucinogens. *Psychopharmacology (Berl)* 94, 213–216.

Tkachev, D., Mimmack, M. L., Ryan, M. M., Wayland, M., Freeman, T., Jones, P. B., Starkey, M., Webster, M. J., Yolken, R. H. & Bahn, S. (2003). Oligodendrocyte dysfunction in schizophrenia and bipolar disorder. *Lancet* 362, 798–805.

Torrey, E. F., Barci, B. M., Webster, M. J., Bartko, J. J., Meador-Woodruff, J. H. & Knable, M. B. (2005). Neurochemical markers for schizophrenia, bipolar disorder, and major depression in postmortem brains. *Biol Psychiatry* 57, 252–260.

Tremolizzo, L., Carboni, G., Ruzicka, W. B., Mitchell, C. P., Sugaya, I., Tueting, P., Sharma, R., Grayson, D. R., Costa, E. & Guidotti, A. (2002). An epigenetic mouse model for molecular and behavioral neuropathologies related to schizophrenia vulnerability. *Proc Natl Acad Sci U S A* 99, 17095–17100.

Tremolizzo, L., Doueiri, M. S., Dong, E., Grayson, D. R., Davis, J., Pinna, G., Tueting, P., Rodriguez-Menendez, V., Costa, E. & Guidotti, A. (2005). Valproate corrects the schizophrenia-like epigenetic behavioral modifications induced by methionine in mice. *Biol Psychiatry* 57, 500–509.

Tsankova, N., Renthal, W., Kumar, A. & Nestler, E. J. (2007). Epigenetic regulation in psychiatric disorders. *Nat Rev Neurosci* 8, 355–367.

Tsankova, N. M., Berton, O., Renthal, W., Kumar, A., Neve, R. L. & Nestler, E. J. (2006). Sustained hippocampal chromatin regulation in a mouse model of depression and antidepressant action. *Nat Neurosci* 9, 519–525.

Tsankova, N. M., Kumar, A. & Nestler, E. J. (2004). Histone modifications at gene promoter regions in rat hippocampus after acute and chronic electroconvulsive seizures. *J Neurosci* 24, 5603–5610.

Tunbridge, E. M., Harrison, P. J., Warden, D. R., Johnston, C., Refsum, H. & Smith, A. D. (2008). Polymorphisms in the catechol-O-methyltransferase (COMT) gene influence plasma total homocysteine levels. *Am J Med Genet B Neuropsychiatr Genet* 147B, 996–999.

Tunbridge, E. M., Harrison, P. J. & Weinberger, D. R. (2006). Catechol-o-methyltransferase, cognition, and psychosis: Val158Met and beyond. *Biol Psychiatry* 60, 141–151.

Tunbridge, E. M., Lane, T. A. & Harrison, P. J. (2007). Expression of multiple catechol-o-methyltransferase (COMT) mRNA variants in human brain. *Am J Med Genet B Neuropsychiatr Genet* 144B, 834–839.

Uranova, N. A., Vostrikov, V. M., Orlovskaya, D. D. & Rachmanova, V. I. (2004). Oligodendroglial density in the prefrontal cortex in schizophrenia and mood disorders: a study from the Stanley Neuropathology Consortium. *Schizophr Res* 67, 269–275.

Van den Buuse, M., Garner, B. & Koch, M. (2003). Neurodevelopmental animal models of schizophrenia: effects on prepulse inhibition. *Curr Mol Med* 3, 459–471.

Veldic, M., Caruncho, H. J., Liu, W. S., Davis, J., Satta, R., Grayson, D. R., Guidotti, A. & Costa, E. (2004). DNA-methyltransferase 1 mRNA is selectively overexpressed in telencephalic GABAergic interneurons of schizophrenia brains. *Proc Natl Acad Sci U S A* 101, 348–353.

346   D. Zelena

Veldic, M., Guidotti, A., Maloku, E., Davis, J. M. & Costa, E. (2005). In psychosis, cortical interneurons overexpress DNA-methyltransferase 1. *Proc Natl Acad Sci U S A* 102, 2152–2157.

Veldic, M., Kadriu, B., Maloku, E., Agis-Balboa, R. C., Guidotti, A., Davis, J. M. & Costa, E. (2007). Epigenetic mechanisms expressed in basal ganglia GABAergic neurons differentiate schizophrenia from bipolar disorder. *Schizophr Res* 91, 51–61.

Vo, N., Klein, M. E., Varlamova, O., Keller, D. M., Yamamoto, T., Goodman, R. H. & Impey, S. (2005). A cAMP-response element binding protein-induced microRNA regulates neuronal morphogenesis. *Proc Natl Acad Sci U S A* 102, 16426–16431.

Volterra, A. & Meldolesi, J. (2005). Astrocytes, from brain glue to communication elements: the revolution continues. *Nat Rev Neurosci* 6, 626–640.

Waddington, C. (1942). The epigenotype. *Endeavour* 1, 18–20.

Walker, J. C. & Harland, R. M. (2009). microRNA-24a is required to repress apoptosis in the developing neural retina. *Genes Dev* 23, 1046–1051.

Wang, Y. C., Chen, J. Y., Chen, M. L., Chen, C. H., Lai, I. C., Chen, T. T., Hong, C. J., Tsai, S. J. & Liou, Y. J. (2008). Neuregulin 3 genetic variations and susceptibility to schizophrenia in a Chinese population. *Biol Psychiatry* 64, 1093–1096.

Wassef, A., Baker, J. & Kochan, L. D. (2003). GABA and schizophrenia: a review of basic science and clinical studies. *J Clin Psychopharmacol* 23, 601–640.

Wassef, A. A., Hafiz, N. G., Hampton, D. & Molloy, M. (2001). Divalproex sodium augmentation of haloperidol in hospitalized patients with schizophrenia: clinical and economic implications. *J Clin Psychopharmacol* 21, 21–26.

Waterland, R. A., Dolinoy, D. C., Lin, J. R., Smith, C. A., Shi, X. & Tahiliani, K. G. (2006). Maternal methyl supplements increase offspring DNA methylation at Axin Fused. *Genesis* 44, 401–406.

Weaver, I. C., Cervoni, N., Champagne, F. A., D'Alessio, A. C., Sharma, S., Seckl, J. R., Dymov, S., Szyf, M. & Meaney, M. J. (2004). Epigenetic programming by maternal behavior. *Nat Neurosci* 7, 847–854.

Wei, Y., Mizzen, C. A., Cook, R. G., Gorovsky, M. A. & Allis, C. D. (1998). Phosphorylation of histone H3 at serine 10 is correlated with chromosome condensation during mitosis and meiosis in Tetrahymena. *Proc Natl Acad Sci U S A* 95, 7480–7484.

Weickert, C. S., Hyde, T. M., Lipska, B. K., Herman, M. M., Weinberger, D. R. & Kleinman, J. E. (2003). Reduced brain-derived neurotrophic factor in prefrontal cortex of patients with schizophrenia. *Mol Psychiatry* 8, 592–610.

Weickert, C. S., Ligons, D. L., Romanczyk, T., Ungaro, G., Hyde, T. M., Herman, M. M., Weinberger, D. R. & Kleinman, J. E. (2005). Reductions in neurotrophin receptor mRNAs in the prefrontal cortex of patients with schizophrenia. *Mol Psychiatry* 10, 637–650.

Weinberger, D. R., Egan, M. F., Bertolino, A., Callicott, J. H., Mattay, V. S., Lipska, B. K., Berman, K. F. & Goldberg, T. E. (2001). Prefrontal neurons and the genetics of schizophrenia. *Biol Psychiatry* 50, 825–844.

Weksberg, R., Shuman, C., Caluseriu, O., Smith, A. C., Fei, Y. L., Nishikawa, J., Stockley, T. L., Best, L., Chitayat, D., Olney, A., Ives, E., Schneider, A., Bestor, T. H., Li, M., Sadowski, P. & Squire, J. (2002). Discordant KCNQ1OT1 imprinting in sets of monozygotic twins discordant for Beckwith-Wiedemann syndrome. *Hum Mol Genet* 11, 1317–1325.

Wiedemann, K. (2011). Biomarkers in development of psychotropic drugs. *Dialogues Clin Neurosci* 13, 225–234.

Wilkinson, M. B., Xiao, G., Kumar, A., LaPlant, Q., Renthal, W., Sikder, D., Kodadek, T. J. & Nestler, E. J. (2009). Imipramine treatment and resiliency exhibit similar chromatin regulation in the mouse nucleus accumbens in depression models. *J Neurosci* 29, 7820–7832.

Williams, J., McGuffin, P., Nothen, M. & Owen, M. J. (1997). Meta-analysis of association between the 5-HT2a receptor T102C polymorphism and schizophrenia. EMASS Collaborative Group. European Multicentre Association Study of Schizophrenia. *Lancet* 349, 1221.

Williams, J., Spurlock, G., McGuffin, P., Mallet, J., Nothen, M. M., Gill, M., Aschauer, H., Nylander, P. O., Macciardi, F. & Owen, M. J. (1996). Association between schizophrenia and T102C polymorphism of the 5-hydroxytryptamine type 2a-receptor gene. European Multicentre Association Study of Schizophrenia (EMASS) Group. *Lancet* 347, 1294–1296.

Woo, T. U., Walsh, J. P. & Benes, F. M. (2004). Density of glutamic acid decarboxylase 67 messenger RNA-containing neurons that express the N-methyl-D-aspartate receptor subunit NR2A in the anterior cingulate cortex in schizophrenia and bipolar disorder. *Arch Gen Psychiatry* 61, 649–657.

Woolley, D. W. & Shaw, E. (1954). A biochemical andpharmacological suggestion about certain mental disorders. *Proc Natl Acad Sci U S A* 40, 228–231.

Yasuda, S., Liang, M. H., Marinova, Z., Yahyavi, A. & Chuang, D. M. (2009). The mood stabilizers lithium and valproate selectively activate the promoter IV of brain-derived neurotrophic factor in neurons. *Mol Psychiatry* 14, 51–59.

Yoo, A. S. & Crabtree, G. R. (2009). ATP-dependent chromatin remodeling in neural development. *Curr Opin Neurobiol* 19, 120–126.

Yoo, A. S., Staahl, B. T., Chen, L. & Crabtree, G. R. (2009). MicroRNA-mediated switching of chromatin-remodelling complexes in neural development. *Nature* 460, 642–646.

Zahir, F. R. & Brown, C. J. (2011). Epigenetic impacts on neurodevelopment: pathophysiological mechanisms and genetic modes of action. *Pediatr Res* 69, 92R–100R.

Zammit, S., Lewis, S., Gunnell, D. & Smith, G. D. (2007). Schizophrenia and neural tube defects: comparisons from an epidemiological perspective. *Schizophr Bull* 33, 853–858.

Zhang, D., Sliwkowski, M. X., Mark, M., Frantz, G., Akita, R., Sun, Y., Hillan, K., Crowley, C., Brush, J. & Godowski, P. J. (1997). Neuregulin-3 (NRG3): a novel neural tissue-enriched protein that binds and activates ErbB4. *Proc Natl Acad Sci U S A* 94, 9562–9567.

Zhao, C., Deng, W. & Gage, F. H. (2008). Mechanisms and functional implications of adult neurogenesis. *Cell* 132, 645–660.

Zhao, X., Li, H., Shi, Y., Tang, R., Chen, W., Liu, J., Feng, G., Shi, J., Yan, L., Liu, H. & He, L. (2006). Significant association between the genetic variations in the 5′ end of the N-methyl-D-aspartate receptor subunit gene GRIN1 and schizophrenia. *Biol Psychiatry* 59, 747–753.

Zhao, X. D., Han, X., Chew, J. L., Liu, J., Chiu, K. P., Choo, A., Orlov, Y. L., Sung, W. K., Shahab, A., Kuznetsov, V. A., Bourque, G., Oh, S., Ruan, Y., Ng, H. H. & Wei, C. L. (2007). Whole-genome mapping of histone H3 Lys4 and 27 trimethylations reveals distinct genomic compartments in human embryonic stem cells. *Cell Stem Cell* 1, 286–298.

Zhubi, A., Veldic, M., Puri, N. V., Kadriu, B., Caruncho, H., Loza, I., Sershen, H., Lajtha, A., Smith, R. C., Guidotti, A., Davis, J. M. & Costa, E. (2009). An upregulation of DNA-methyltransferase 1 and 3a expressed in telencephalic GABAergic neurons of schizophrenia patients is also detected in peripheral blood lymphocytes. *Schizophr Res* 111, 115–122.

Chapter 11
Disruption of Epigenetic Mechanisms in Autoimmune Syndromes

Lorenzo de la Rica and Esteban Ballestar

11.1 Immune System, Autoimmunity, and Autoimmune Diseases

The immune system is a highly evolved system designed to protect the human body against a broad range of pathogenic agents. In vertebrates, immunity is divided into two principal categories: innate immunity and acquired or adaptive immunity. Innate immunity is present from the moment of birth and involves several types of nonspecific barriers, which may be physical (e.g., skin), chemical (e.g., lysozymes), or cellular barriers (e.g., granulocytes, monocyte-derived macrophages, and dendritic cells, as well as natural killers) (Janeway and Medzhitov 2002). Acquired immunity is a more specialized form of defense that includes the generation of lymphocytes against the antigen that has defeated innate immunity, as well as the development of memory cells in order to avoid reinfection by the same pathogen. T and B lymphocytes are the main effectors of this type of immunity, which takes longer to develop. It appeared later in evolution than innate immunity and is also known as adaptive immunity (Pancer and Cooper 2006). In order to defend ourselves, our immune systems first have to learn what is self and what is not, in other words, to limit its recognition of self-molecules to avoid harming the individual. This is achieved by a process called *immunological tolerance, which* may occur at two levels: *central* (in the thymus for T cells and in bone marrow in the case of B cells) and *peripheral* (in the case of spleen, lymph nodes, etc.) (Klinman 1996). Due to this tolerance, few self-reactive immune cells are developed, and those that survive are maintained in an inactive state (Goodnow et al. 2005).

L. de la Rica • E. Ballestar (✉)
Chromatin and Disease Group, Cancer Epigenetics and Biology Programme (PEBC),
Bellvitge Biomedical Research Institute (IDIBELL), L'Hospitalet de Llobregat,
Avda. Gran Via 199-203, Barcelona 08908, Spain
e-mail: eballestar@idibell.cat

J. Minarovits and H.H. Niller (eds.), *Patho-Epigenetics of Disease*,
DOI 10.1007/978-1-4614-3345-3_11, © Springer Science+Business Media New York 2012

However, autoreactive lymphocytes and antibodies are found in the general population (Zerrahn et al. 1997; Wardemann et al. 2003), but only in some individuals the combination of a number of factors such as genetic susceptibility and exposure to environmental risk factors leads to the dramatic disruption of tolerance, causing autoimmune disease.

Defining autoimmune diseases is a complicated matter, as they comprise a wide range of syndromes and conditions with very different outcomes. They are multifactorial illnesses caused by aberrant activation of T and/or B cells, in the absence of an ongoing infection or other discernible cause (Davidson and Diamond 2001). Autoimmunity usually refers to acquired immune system hyperactivation, as happens in rheumatoid arthritis, systemic lupus erythematosus (SLE), etc. Nevertheless, the innate immune system has also been found to be dysregulated, causing autoimmune diseases with a higher inflammatory component, which are alternatively known as autoinflammatory diseases (McGonagle and McDermott 2006). Crohn's disease and ulcerative colitis are two examples of such diseases.

In an autoimmune disorder, the immune system may react against a certain organ (organ-specific autoimmunity), or against various organs (systemic autoimmunity). The most prominent examples of systemic autoimmune diseases are SLE and rheumatoid arthritis, which affect joints (both are rheumatic diseases) as well as other organs such as skin and lungs.

In the case of organ-specific autoimmune diseases, immunity may cause reactions in the gastrointestinal tract, causing celiac, Crohn's, or Graves' disease. Other types of organ-specific autoimmune diseases affect the central nervous system in which demyelination causes multiple sclerosis. When the immune system reacts against metabolic organs, such as the pancreatic islets or liver, it causes type 1 diabetes mellitus or primary biliary cirrhosis, respectively.

Although many different outcomes of autoimmune diseases have been observed, the common characteristic of this group of diseases is that the immune system incorrectly recognizes elements as being pathogenic elements when, in fact, they are not. They share several other features such as the presence of autoantibodies against nuclear components (DNA, nucleosomes, histones, etc.) or citrullinated peptides (ACPA), hyperactivated and hyperreactive immune cells, and a stronger inflammatory response.

11.2 Autoimmune Diseases Are Complex Diseases

As previously mentioned, autoimmune diseases are complex multifactorial illnesses whose etiology remains largely unknown. Many years ago, geneticists started looking at specific loci that could help to explain the onset of such diseases. HLA genes were discovered as susceptibility loci in several autoimmune diseases, such as rheumatoid arthritis (Stastny 1978), SLE (Grumet et al. 1971) and scleroderma (Gilchrist et al. 2001) using a candidate gene approach. More recently, genome-wide association studies (GWAS) have enabled the identification of several susceptibility loci for many of these conditions, for example, *IRF5*, *BLK*, *TNFAIP3, CD40*, and *STAT4* (Delgado-Vega et al. 2010), and the confirmation of previously identified in candidate gene studies (like HLA genes, PTPN22, PAD14, CTLA4) (Gregersen 2010; Sawcer et al. 2011). Many of these genes are involved in the innate immune response (*IRF5, STAT4*), adaptive immune response (*HLA-DR, STAT4, PTPN22, BLK*), as well as immune complex clearance (*ITGAM, C1q*) (Lee and Bae 2010).

Nevertheless, genetics alone is not sufficient to explain the onset of these diseases. A classical example is the commonly observed existence of low concordance rates of these diseases in monozygotic twins (Silman et al. 1993; Grennan et al. 1997), which indicates that additional mechanisms operate on top of the genetic predisposition for autoimmune disease.

Several environmental factors are now known to contribute to the onset of autoimmune diseases, for example, infectious agents such as viruses, bacteria and parasites, drug treatment, hormone levels, and some pollutants.

Among the infectious agents, Epstein–Barr virus has been associated with rheumatoid arthritis, multiple sclerosis, and SLE. This sheds light on the possible common mechanisms driving these diseases: for example, molecular mimicry between viral antigens and self-antigens and immortalization of autoreactive clones of B lymphocytes have been proposed (Pender 2003). With respect to the common pollutant triggers of autoimmune diseases, tobacco smoke has been associated with rheumatoid arthritis, autoimmune thyroid disease (AITD), and Crohn's disease, as reviewed by Pollard et al. 2010. All these examples support the notion of the dysregulation of common immune pathways among the different types and outcomes of autoimmune diseases.

A new aspect is becoming increasingly important for understanding complex diseases: the contribution of epigenetic dysregulation. Many of the environmental factors mentioned above can dysregulate the epigenetic status of the cell. For example, the viral LMP1 protein is able to activate DNMT1, altering the DNA methylation status (Tsai et al. 2006); some hypomethylating drugs can cause a lupus-like disease (Strickland and Richardson 2008), and tobacco smoke, known to influence DNA methylation patterns in cancer (Liu et al. 2007; Hussain et al. 2009), makes a smoker twin more prone to rheumatoid arthritis (Silman et al. 1996).

Bearing all these facts in mind, it is reasonable to speculate that epigenetic regulation, which can alter the phenotype without modifying the DNA sequence, acts as an "adaptor" between the environment and the nucleus. It also regulates several developmental pathways involved in the physiological function and development of the immune system (Ballestar 2011). It is becoming increasingly clear that the study of epigenetic dysregulation events will shed light on the etiopathogenesis of these conditions. A summary of the epigenetic deregulation events taking place in autoimmune diseases can be found in Table 11.1.

The main body of evidence concerning the connections between epigenetics and autoimmune diseases has been described in rheumatic diseases such as SLE and rheumatoid arthritis. For this reason, this chapter will focus on these two well-characterized diseases. In the final section, the epigenetic research into other autoimmune diseases will be summarized.

Table 11.1 A summary of DNA methylation changes described for different autoimmune diseases

Condition	Tissue/cell type	Sequence/region	Consequences	Event	Reference
Rheumatic diseases					
SLE	CD4⁺ T lymphocytes	ITGAL/CD11a	Cell to cell adhesion	Hypomethylation	Lu et al. (2002)
		CD40LG	Stimulation of IgG overproduction in B cells	Hypomethylation	Lu et al. (2007)
		CD70	T lymphocyte proliferation	Hypomethylation	Lu et al. (2005)
		PRF1	Autoreactive killing	Hypomethylation	Kaplan et al. (2004)
		Repetitive sequences	Genomic instability	Hypomethylation	Richardson et al. (1990)
	PBMCs	IFNGR2	Receptor for interferon gamma	Hypomethylation	Javierre et al. (2010)
		MMP14	Destruction of extracellular matrix	Hypomethylation	Javierre et al. (2010)
		LCN2	Iron transporter	Hypomethylation	Javierre et al. (2010)
		rDNA (18S, 28S)	Ribosome assembly	Hypomethylation	Javierre et al. (2010)
		Repetitive sequences	Genomic instability	Hypomethylation	Javierre et al. (2010)
	CD19⁺ B lymphocytes	CD5-E1B	Interleukin production	Hypomethylation	Garaud et al. (2009)
RA	CD4⁺ T lymphocytes	Repetitive sequences	Genomic instability	Hypomethylation	Richardson et al. (1990)
	Synovial fibroblasts	LINE-1	Overexpression of MAPK13 and MET	Hypomethylation	Neidhart et al. (2000)
		miRNA-203	Overexpression of MMPs and IL-6	Hypomethylation	Stanczyk et al. (2011)
		DR3	Increased apoptosis	Hypermethylation	Takami et al. (2006)
		Repetitive sequences	Genomic instability	Hypomethylation	Karouzakis et al. (2009)
	PBMCs	IL-6	B cell response	Hypomethylation	Nile et al. (2008)
SD	CD4⁺ T lymphocytes	Repetitive sequences	Genomic instability	Hypomethylation	Lei et al. (2009)
	SD fibroblasts	FL1	Increased production of collagen	Hypermethylation	Wang et al. (2006)
	PBMCs	X chromosome	SXCIM		Ozbalkan et al. (2005)
SjS	Labial salivary glands	BP230	Involved in basal lamina anchorage	Hypermethylation	Gonzalez et al. (2011)

Nervous system

MS	Schwann cells, oligodendrocytes	*PAD2*	Increased demyelination via citrullination of MBP	Hypomethylation	Mastronardi et al. (2007)

Skin

Ps	Skin cells	*CDKN2A (p16)*	Tumor supressor	Hypermethylation	Chen et al. (2008)
		Global hypermethylation		Hypermethylation	Zhang et al. (2010a)
		PTPN6	Cell proliferation and signaling	Hypomethylation	Ruchusatsawat et al. (2006)
	Hematopoietic cells	*CDKN2B (p15)*	Cell cycle progression	Hypomethylation	Zhang et al. (2009)
		CDKN2A (p16)	Tumor supressor	Hypomethylation	Zhang et al. (2007)
		CDKN1A (p21)	Cell cycle progression	Hypomethylation	Zhang et al. (2009)
	PBMCs	Global hypermethylation		Hypermethylation	Zhang et al. (2010a, b)
		CDKN2B (p14)	Cell cycle progression	Hypermethylation.	Zhang et al. (2010a, b)

Digestive tract/endocrine glands

UC	Rectum mucosa	Global hypomethylation		Hypomethylation	Gloria et al. (1996)
		CDKN2A (p16)	Tumor supressor	Hypermethylation	Hsieh et al. (1998)
		CDKN2B (p14)	Cell cycle progression	Hypermethylation	Saito et al. (2011)
		MDR1	Transmembrane drug efflux pump	Hypermethylation	Tahara et al. (2009b)
AITDs	PBMCs	X chromosome	SXCIM		Brix et al. (2005)

11.3 Rheumatic Diseases as Evidence of the Role of Epigenetic Dysregulation in Autoimmune Disease

Most of the early studies on epigenetic alterations in autoimmune diseases have been focused on autoimmune rheumatic diseases, particularly systemic lupus erythematosus and rheumatoid arthritis.

11.3.1 Systemic Lupus Erythematosus

SLE is a systemic, chronic, autoimmune disease with a broad range of clinical presentations (D'Cruz et al. 2007). It is characterized by the production of autoantibodies against nuclear, cytoplasmic, and cell surface antigens, and the subsequent formation of immune complexes that are deposited in various tissues, from skin to central nervous system, kidneys, and joints (Rahman and Isenberg 2008). The clinical course of this disease comprises unpredictably timed relapses and remissions.

SLE affects people of all ages, races, and genders, although it is 9 times as common in women of fertile age as in men, and more so in those with African ancestry (Danchenko et al. 2006). Hormone levels are also dysregulated, indicating the importance of hormones (Cooper et al. 2002) and X-linked genes (Chagnon et al. 2006) in the etiology of this disease.

To date, the mechanisms responsible for immune tolerance dysregulation are mainly unknown, although alterations at several levels have been described: autoantibody production, apoptosis, cytokine concentrations, immune cell physiology, etc. (D'Cruz et al. 2007).

Environmental factors such as hormone levels, viral infections, and exposure to chemicals are known to influence SLE etiopathology (D'Cruz et al. 2007). Concordance rates for monozygotic and dizygotic twins are around 25% and 2%, respectively, and the heritability of SLE is more than 66%. These facts emphasize the involvement of the genetic background in the development of the disease (Sullivan 2000). Moreover, several susceptibility loci have so far been identified in large-scale GWAS (Delgado-Vega et al. 2010). These susceptibility loci could partially explain the dysregulation that takes place in SLE at the level of apoptosis, cytokine levels, and immune cell hyperreactivity.

There is an increased rate of apoptosis and an impaired clearance of apoptotic debris, probably explaining why SLE autoantibodies are directed to recognize intracellular components (Emlen et al. 1994; Kaplan et al. 2002). On the other hand, during apoptosis, cell releases nuclear material (DNA and histones) characterized by a specific epigenetic profile (Boix-Chornet et al. 2006), exposing new epitopes that could activate the immune system in SLE. Moreover, there is a loss of DNA methylation during apoptosis, and this unmethylated DNA is more likely to trigger the immune response (Goldberg et al. 2000), explaining why injection of unmethylated but not methylated DNA into healthy mice leads to a lupus-like disorder (Wen et al. 2007).

Several drugs can induce a lupus-like disease, also known as drug-induced lupus. One of these, 5-azacytidine, is a DNA methyltransferase inhibitor (DNMTi) with hypomethylating effects. This drug, approved for human use in 2004 to treat myelodysplastic syndrome (Kaminskas et al. 2005), is sufficient to induce drug-induced lupus in mice (Quddus et al. 1993). This drug, as well as other DNMTi, showed the potential involvement of DNA methylation dysregulation in this disease, and this has subsequently been demonstrated experimentally.

11.3.1.1 DNA Methylation Dysregulation in Systemic Lupus Erythematosus

T lymphocytes are the most widely studied blood cells in the context of SLE pathogenesis. The first evidence of the importance of DNA methylation in T lymphocytes came from the mid-1980s in the SLE context, where surprising effects were found when treating T cells with the hypomethylating agents (DNMTi) 5-azacytidine, hydralazine, or procainamide. DNMTi treatment was sufficient to convert $CD4^+$ T cells into autoreactive cells, so they could be activated by autologous macrophages by losing the requirement for an antigen (Richardson 1986). DNMTi were also demonstrated to influence the profile of some surface markers in $CD8^+$ T cells and result in CD4 reexpression (Richardson 1986). The observed phenotype of the $CD4^+$ T cells treated with DNMTi resembled that of $CD4^+$ T cells isolated from SLE patients (Richardson et al. 1992). The 5-methylcytosine content of $CD4^+$ T lymphocytes was found to be lower in SLE and rheumatoid arthritis (RA) (Richardson et al. 1990; Lei et al. 2009). In the case of SLE, the levels of 5-methylcytosine were correlated with the SLEDAI score (Richardson et al. 1990). Taken together, these discoveries show the importance of DNA methylation dysregulation in the pathogenesis of autoimmune disease.

Several additional lines of evidence indicate a relationship between T cell DNA hypomethylation, cell autoreactivity, and autoimmunity. For example, administering hypomethylating drugs to humans is sufficient to cause drug-induced lupus (Cornacchia et al. 1988). Moreover, treating healthy human T cells with these hypomethylating drugs and injecting them into mice are sufficient to cause a lupus-like disease because these autoreactive lymphocytes are able to lyse macrophages and secrete IL-4, IL-6, and IFN-γ (Quddus et al. 1993). Mouse Th1 and Th2 cells treated with DNA methylation inhibitors (5-azaC) also become autoreactive and induce anti-DNA antibodies if injected into syngenic mice (Yung et al. 2001). These findings suggest that macrophage killing could expose antigenic nucleosomes from the apoptotic material, and the nucleosome-reactive T cell may facilitate the production of anti-DNA antibodies. Changes in DNA methylation levels have been associated with lower levels of DNA methyltransferase 1 (DNMT1) mRNA and impaired PKC delta phosphorylation (Gorelik et al. 2007; Zhu et al. 2011). DNMT1 is the enzyme responsible for maintaining the methylation status of DNA after every cell division.

Moving from the study of global changes in DNA methylation approach to gene-specific changes, several genes that are important for the immune response and regular T lymphocyte function are known to be specifically dysregulated.

Of the genes that are important in autoimmunity, the level of surface leukocyte function-associated antigen-1 (LFA-1) (CD11a/CD18) is higher in autoreactive T cells (Richardson et al. 1994). CD11a, also known as ITGAL (integrin alpha L), is one of the components of the LFA-1 heterodimer, which is involved in T cell activation through MHCII-TCR (Hogg et al. 2003). Overexpression of CD11a with DNMTi, UV light exposure, and gene transfection in normal T cells makes them autoreactive and is sufficient to create autoimmunity (lupus-like disease) when injected into syngenic mice (adoptive transfer) (Yung et al. 1996). Moreover, the regulatory sequences of the CD11a gene promoter are specifically hypomethylated in SLE T cells (Lu et al. 2002), preventing the binding of some transcription factors such as PU.1 and Sp1.

Another gene important in autoreactivity is perforin (*PRF1*), a protein involved in cellular death through pore generation mechanisms in the cytoplasmic membrane (van den Broek and Hengartner 2000). In healthy individuals, perforin is expressed in $CD8^+$ cells, thanks to the hypomethylation of a conserved regulatory region of the gene promoter. On the other hand, this region is hypermethylated in $CD4^+$ cells and is not expressed. However, in the pathological context of SLE, this region loses its methylation, and the aberrant expression of perforin is linked with the capacity of the $CD4^+$ lupus cells to kill monocytes and macrophages (Kaplan et al. 2004; Luo et al. 2009). This situation can be reverted in DNMTi-treated lymphocytes using the perforin inhibitor concanamycin A, thereby reducing monocyte killing.

CD40L and CD70 are molecules expressed in activated $CD4^+$ T cells that belong to the tumor necrosis factor (TNF) ligand family. They are overexpressed in women with SLE (CD40L) or in SLE patients of both genders (CD70). Overexpression of CD70 can be achieved in $CD4^+$ T cells isolated from healthy donors by treating them in vitro with DNMT inhibitors (5-azacytidine and procainamide) and ERK pathway inhibitors (U0126, PD98059, and hydralazine), which are known to reduce DNMT expression. When these CD70- or CD40L-overexpressing T cells are cultured with autologous B cells, the latter are stimulated via CD27 or CD40 and overproduce IgG (Kobata et al. 1995). Inhibiting any of the costimulatory molecules (anti-CD70 or anti-CD40L) reduces antibody production to normal levels. In conclusion, CD70 overexpression in SLE T cells, along with other molecules such as CD40L, contributes to B cell stimulation (Oelke et al. 2004; Lu et al. 2005; Zhou et al. 2009; Luo et al. 2010) and may be involved in the hyperreactivity the immune system exhibits in these individuals.

As previously mentioned, CD40L is a B cell costimulatory molecule, encoded by an X chromosome gene. It is unmethylated in men, while women have one allele methylated and the other unmethylated. However, in female lupus patients, both alleles are demethylated, thus allowing CD40L overexpression in women. Demethylation of CD40L and other genes still unknown on the inactive allele of the X chromosome may explain why SLE is more common in women (Lu et al. 2007). A summary of the consequences of epigenetic deregulation in SLE at the level of DNA methylation can be found in Fig. 11.1.

A report showed that the mechanism underlying the hypomethylation and hyperacetylation of the CD11a and CD70 gene promoters acts through the lack of a transcription factor called RFX1. When this is silenced in SLE $CD4^+$ cells, it cannot

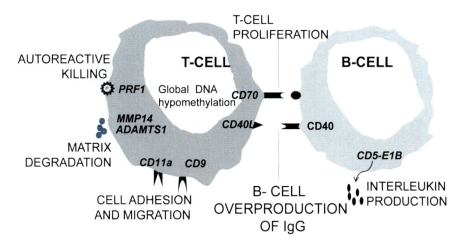

Fig. 11.1 B and T cells display DNA methylation and expression changes in SLE. Several genes important for T and B cell function are overexpressed due to hypomethylation and are involved in autoimmunity. Perforin in CD4⁺ T cells is linked with monocyte and macrophage killing. Matrix-degrading enzymes such as MMP14 and ADAMTS1 collaborate in tissue injury. Surface markers are overexpressed in T (CD11a, CD9, CD70) and B cells (CD40). These molecules are involved in cell adhesion, migration, lymphocyte proliferation, as well as production of great amounts of IgG in B cells (CD40-CD40L interaction). Production of interleukins is also deregulated in B cells due to hypomethylation of CD5-E1B locus (Richardson et al. 1990, 1994; Kaplan et al. 2004; Oelke et al. 2004; Garaud et al. 2009; Luo et al. 2009; Zhou et al. 2009; Javierre et al. 2010)

recruit DNMT1 or HDAC1, as happens in healthy individuals, thereby causing T and B cell hyperreactivity (Zhao et al. 2010a). Moreover, H3K9 trimethylation levels are also significantly reduced in these promoters, due to the impaired histone methyltransferase SUV39H1 recruitment to these promoters by smaller amounts of RFX1 (Zhao et al. 2010c).

The expression of several genes of importance in cell cycle and DNA damage is altered in SLE. Growth arrest and DNA damage-induced 45α gene (*GADD45α*) is increased in SLE CD4⁺ T cells and has also been related to DNA demethylation in *CD11a* and *CD70*, as well as T cell autoreactivity (Li et al. 2010).

Cytokine signaling is the system the immune cells use to communicate with their near or distant counterparts. This system is epigenetically altered in SLE. Specific promoter hypomethylation is found in the IL-10 and IL-13 interleukin promoters, leading to their subsequent overexpression in CD4⁺ T cells. These interleukins play important roles in Th2 differentiation and production of autoantibodies in SLE patients, and their overexpression can also be achieved by treating with DNMTi CD4⁺ T cells isolated from healthy donors (Zhao et al. 2010b). This epigenetic alteration also occurs in IL-4 and IL-6 in T cells (Mi and Zeng 2008), as well as in IFNGR2 in white blood cells (Javierre et al. 2010). Regarding IL-6 deregulation in B cell populations, its upregulation has been linked with reduced membrane CD5 levels. CD5 E1B promoter was reported to be hypomethylated in these cell types, due to the impairment of DNMT1 expression levels (Garaud et al. 2009).

There are several other examples of gene-specific promoters that undergo hypomethylation. Some are related with protein phosphatases, such as *PP2Acα* (protein phosphatase 2A) (Sunahori et al. 2011). Others are associated with the serotonin system, which is involved in B and T proliferation via *HTR1A* (5-HT1A receptor) (Xu et al. 2011) and is hypomethylated in peripheral blood lymphocytes. However, even more epigenetic information remains to be obtained.

The development of high-throughput techniques means that genome-wide analyses are displacing the candidate gene approach, not only because of the huge amount of information that can be obtained but also because of the lack of bias these techniques offer.

The first published study of SLE epigenetics using high-throughput analysis consisted in a methylation array-based method of DNA extracted from white blood cells of monozygotic twins discordant for the disease in order to avoid genetic variability (Javierre et al. 2010). In this study, a large set of gene promoters was found to be differentially methylated between healthy and SLE siblings. These were involved in several immune processes, for example, cytokine production (*STAT5A, SYK, IL-10*), immune response (*STAT5A, NOTCH4, SYK, IL-10, AIM2, LTB4R, CSF1R, CSF3*), response to external stimuli (*SYK, IL-10, CD9, LTB4R, PECAM1, GFI1, STAT5A, TNFRSF1A*), and immune synapse (*CSFR3*), as reported in Funauchi et al. 2002, as well as inflammation (*MMP14*), among others.

Using fractionated CD4$^+$ T cells from healthy and SLE donors, Sawalha's research group identified 236 hypomethylated and 105 hypermethylated CpG regions (Jeffries et al. 2011), many of which are coincident with the results of the twin study. Among the most prominent examples, it stands *CD9, BST2, RUNX3*, etc.

Further studies focusing on the DNA methylation status of fractionated blood populations would help us understand in greater depth the role of DNA methylation in SLE pathogenesis. The cutting-edge technologies under development or those already available for general use in research will continue to ensure new insights into SLE pathogenesis.

Although repetitive sequences are the main contributors of CpG dinucleotides to the genome (Wilson et al. 2007), little is known about them. The twin study identified 18 and 28S regions of the ribosomal RNA genes, which are repeated hundreds of times, as being hypomethylated in the white blood cells of SLE patients (Javierre et al. 2010). Moreover, specific LINE-1 hypomethylation was observed in specific subsets of T and B lymphocytes isolated from SLE patients, and this was correlated with SLE activity (Nakkuntod et al. 2011). The exact consequences of the presence of these repetitive sequences in the pathological phenotype are still unknown.

11.3.1.2 Histone Modification Dysregulation in Systemic Lupus Erythematosus

The expression changes observed in the various cell types isolated from SLE patients are not caused solely by alterations of the DNA methylation profile. Histone modifications contribute to gene expression (Kouzarides 2007), and in SLE, histone mark profiles have been found to be dysregulated. The analysis of the global histone H3 and H4 acetylation levels in SLE CD4+ T cells have revealed global hypoacetylation. Moreover, disease severity is inversely correlated with acetylation levels of histone H3 (Hu et al. 2008). This hypoacetylation milieu alters gene expression, specifically that of IL-10 and CD154 (CD40 ligand) overexpression and IFN-γ downregulation. These changes can be reverted after treatment with the histone deacetylase inhibitor (HDACi) trichostatin A (Mishra et al. 2001), showing the considerable degree of involvement these epigenetic marks have in the pathological phenotype.

On the other hand, in monocytes isolated from SLE patients, histone H4 has been found to be hyperacetylated in at least 179 genes. Aberrant overexpression of 225 genes was also reported, and many of them have potential IRF1 (interferon regulatory factor 1) binding sites within 5 kb of the promoter. This protein is important for monocyte development and activation (Zhang et al. 2010b).

Nevertheless, our knowledge of histone modification dysregulation in SLE is still very limited, and more in-depth analyses are required.

11.3.1.3 MicroRNA Dysregulation in Systemic Lupus Erythematosus

Epigenetic modifications are key to regulate gene expression, and therefore, deregulation of epigenetic profiles plays a major role in many diseases. The regulatory role of epigenetic modifications is amplified when this associates with microRNA (miRNA) regulation. MiRNAs are short (21–24 nucleotide), noncoding RNA species that downregulate gene expression posttranscriptionally. Although miRNAs cannot strictly be considered epigenetic factors themselves, there are complex networks that interconnect them with epigenetic modifications. Several studies have shown that epigenetic mechanisms regulate miRNA expression. Conversely, different subsets of miRNAs control the levels of important epigenetic enzymes. In the context of autoimmune disease, cross talk between miRNAs and epigenetic machinery has been reported for SLE (reviewed by Ceribelli et al. 2011), meriting a brief mention here. Several miRNAs are upregulated in SLE CD4+ T cells, specifically miR-21, miR-148a, and miR-126. Surprisingly, all of these target and therefore directly or indirectly downregulate DNMT1. This action can be prevented by inhibiting any of these miRNAs, partially alleviating their global DNA hypomethylation, as well as the aberrant overexpression of some immune-related genes, such as LFA-1 (CD11a) and CD70 (Pan et al. 2010; Zhao et al. 2011).

11.3.2 Rheumatoid Arthritis

Rheumatoid arthritis (RA) is a systemic, chronic, autoimmune disease in which joints are the main organs affected. It is characterized by synovial hyperplasia as well as joint inflammation and subsequent destruction of cartilage and bone.

The immune system is dysregulated at several levels, and many immune cells are involved directly or indirectly in the pathogenesis of this condition (Duke et al. 1982). Apart from joints, consequences of this rheumatic disease are visible in the skin (rheumatoid nodules), eyes (keratoconjunctivitis sicca), heart (pericardial effusion), nervous system (peripheral nerve entrapment and mononeuritis multiplex), and lungs (interstitial lung disease) (Young and Koduri 2007).

The inflammatory microenvironment of the affected joint is a consequence of the high concentrations of proinflammatory cytokines secreted by the immune cells that invade the joint (Choy and Panayi 2001). Moreover, synovial cells such as rheumatoid arthritis synovial fibroblasts (RASFs) and osteoclasts are hyperactivated and hyperreactive, these being the effector cells of the destruction of cartilage and bone, respectively (Scott et al. 2010).

The bulk of the information available about RA effector cells concerns RASFs. This cell type is more aggressive than normal synovial fibroblasts for several reasons. Firstly, they overexpress metalloproteinases (MMPs) and cytokines (Distler et al. 2005; Tolboom et al. 2005). Secondly, they show tumoral behavior revealed by several facts: they are more invasive in the cartilage (Muller-Ladner et al. 1996), have increased resistance to apoptosis (Baier et al. 2003), and can grow in an anchorage-independent manner (Lafyatis et al. 1989). In the literature, the behavior of RASFs is usually compared with that of osteoarthritis synovial fibroblasts (OASFs) used as a "healthy" control. This is due to the lack of an autoimmune component in the latter cell type, and because of the availability of the tissue as synovectomy is a surgical procedure commonly performed in these patients.

The triggers of the onset of RA are still unknown. Genome-wide studies performed to date indicate that, although genetics affects the susceptibility of the disease (Delgado-Vega et al. 2010), it is not sufficient nor the sole cause. The susceptibility loci identified so far include *HLA-DRB1 PTPN22, PADI4, STAT4, IL6ST, SPRED2, RBPJ, CCR6*, and *IRF5*.

Removing genetic variability from the equation, studies with monozygotic twins, who share the same genetic background, show a concordance rate for RA of around 15% (Aho et al. 1986; Silman et al. 1993). These studies also demonstrate the ability of environmental factors, such as tobacco smoke, to trigger the onset of the disease (Silman et al. 1996). Tobacco smoke can alter the methylation status of DNA (Kim et al. 2001; Liu et al. 2007; Hussain et al. 2009), so it is logical to consider epigenetic dysregulation as one of the mechanisms underlying the etiopathogenesis of RA disease.

11.3.2.1 DNA Methylation Dysregulation in Rheumatoid Arthritis

DNA methylation profile dysregulation has been studied in several cell types involved in RA pathogenesis, specifically RASFs, CD4+ T lymphocytes, and PBMCs (peripheral blood mononuclear cells). The first connections between RA and the dysregulation of DNA methylation levels were made in 1991 when lower levels of 5-methylcytosine content were measured in blood, mononuclear cells, and synovial tissue isolated from RA patients (Corvetta et al. 1991). In this context of hypomethylation, RASFs aberrantly overexpress *LINE-1* (Neidhart et al. 2000) and was associated with met proto-oncogene (*MET*), p38delta MAP kinase (*MAPK13*), and galectin-3-binding protein (*LGALS3BP*) overexpression (Kuchen et al. 2004). Functions of these genes may be involved in the aggressive phenotype this cell type exhibits. More in-depth studies of RASFs phenotype have shown global DNA hypomethylation as well as lower levels of DNMT1 than OASF. These features correlate with the aggressiveness of this cell type, as fibroblasts extracted from healthy individuals behave like RASFs after treatment with the DNA hypomethylating drug 5-aza-2-deoxycytidine (Karouzakis et al. 2009). Genes encoding miRNAs have also been found to be dysregulated at the level of methylation. The *hsa-mir-203* gene promoter is hypomethylated and thus overexpressed in RASFs in comparison to OASFs, and this enables the induction of MMP-1 and IL-6 overexpression (Stanczyk et al. 2011).

In this context of hypomethylation, some genes are specifically hypermethylated and thereby repressed, similar to the patterns displayed by tumor cells (Esteller 2008). For example, death receptor 3 (*DR3* or *TNFRSF25*), a gene involved in apoptosis, undergoes hypermethylation in RASFs, increasing the resistance to apoptosis of this cell type (Takami et al. 2006).

CD4[+] T cells are also implicated in RA pathogenesis and, as has been described for RASFs, undergo global DNA hypomethylation as well as exhibiting lower activity levels of DNMTs (Richardson et al. 1990). In blood cells isolated from RA patients, the ephrin B1 (*EFNB1*) gene promoter is hypomethylated, so higher levels of mRNA are detected in blood and synovial T cells. This gene product is a membrane protein involved in cell adhesion and inflammation signaling, and its dysregulation makes T cells migrate more actively and secrete higher levels of cytokines (Kitamura et al. 2008).

CD4[+]CD28[−◇] T cells are a subset of T lymphocytes most often found in old people, thus associated with senescence, and which are autoreactive. In RA, they are more numerous than in the normal population, and they show decreased activity of the ERK and JNK pathways, and as a consequence, DNMT1 and DNMT3 are downregulated. Genes involved in the inflammatory phenotype could help the demethylation of several gene promoters such as *CD70*, perforin (*PRF1*), and *KIR2DL4* (Liu et al. 2009; Chen et al. 2010). *CD70* and *PRF1* genes are also dysregulated in SLE CD4[+] T cells (Kaplan et al. 2004; Lu et al. 2005; Luo et al. 2010) due to aberrant hypomethylation and subsequent overexpression. These genes are important for B cell costimulation and cell-death mechanisms, respectively, and as they are shared genes that are epigenetically dysregulated in SLE and RA, the contribution to autoimmunity needs to be studied in greater depth.

IL-6 has been reported to be epigenetically dysregulated in both RA and SLE. It is hypomethylated in CD4[+] T cells isolated from SLE patients (Mi and Zeng 2008) and in PBMCs from RA patients (Nile et al. 2008). In the case of PBMCs, a single CpG controls IL-6 expression: it is hypomethylated in IL-6 overexpressing RA PMBCs. IL-6 levels are also upregulated in RASFs (Stanczyk et al. 2011), although this is achieved through miRNA dysregulation. Thus, IL-6 mRNA levels are upregulated in at least three cell types involved in both autoimmune conditions, which

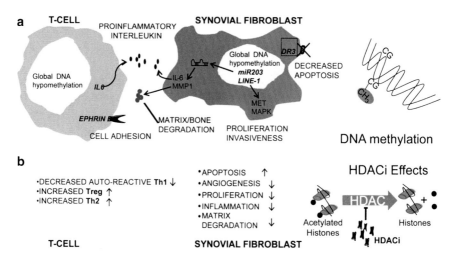

Fig. 11.2 Epigenetic deregulation in rheumatoid arthritis T cells and synovial fibroblasts. (**a**) T cells and RASFs are hyperactivated and hyperreactive. This can be in part due to aberrant methylation profiles. Both cell types undergo global DNA hypomethylation. At the gene level, ephrin B, a protein involved in cell adhesion, is upregulated in T cells. IL-6 is overexpressed in both cell types, but due to different mechanisms. In T cells, IL-6 overexpression is associated with direct hypomethylation of its promoter, while in RASFs, overexpression is indirectly related with miR-203 hypomethylation and overexpression. IL-6 is a proinflammatory cytokine that is present at higher concentrations in the affected synovia. MMP1 is also upregulated due to miR-203, and its function is to degrade the cartilage and extracellular matrix in the joints. The repetitive sequence, LINE-1, is active in RASFs due to hypomethylation, and it regulates the expression of several genes involved in proliferation and invasiveness, such as MET and MAPK. Hypermethylation of sequences is observed in DR3 promoter, inhibiting its expression, and as a consequence, conferring increased resistance to apoptosis on RASFs. (**b**) HDACi has been widely investigated in RA. Treating T cells with these compounds ameliorates their aggressive phenotype as fewer autoreactive Th1 cells are detected, and more protective Tregs and Th2 are generated. Regarding RASFs, the benefits are many: more apoptosis, less angiogenesis, decreased proliferation, and lower secretion of inflammation mediators and matrix-degrading molecules (Richardson et al. 1990; Corvetta et al. 1991; Neidhart et al. 2000; Chung et al. 2003; Kuchen et al. 2004; Distler et al. 2005; Young et al. 2005; Jungel et al. 2006; Morinobu et al. 2006; Takami et al. 2006; Tao et al. 2007; Kitamura et al. 2008; Manabe et al. 2008; Nile et al. 2008; Saouaf et al. 2009; Stanczyk et al. 2011; Zhou et al. 2011)

highlights the importance of this cytokine in immune cell hyperactivation. A scheme of the importance of DNA methylation is illustrated in Fig. 11.2a.

The balance between T lymphocyte populations in RA is shifted towards the greater representation of regulatory T and T helper cells in rheumatic joints. This has been observed as a result of using a method called *epigenetic immune lineage analysis*, in which the levels of 5-methylcytosine are determined in certain gene promoters that are important for T cell differentiation. Peripheral CD4[+] T cells have higher *FOXP3* promoter methylation, while in the synovium, *FOXP3* and *IFNG* promoters are demethylated, indicating a higher number of regulatory T and helper T cells (Janson et al. 2011).

11.3.2.2 Histone Modification Dysregulation in Rheumatoid Arthritis

Histone modifications are crucial for gene regulation (Kouzarides 2007). One of the most widely studied marks is acetylation. Histone acetyltransferases (HATs) add acetyl groups to the tail of the histone, preventing intimate union between histone and DNA, thereby relaxing the chromatin, and allowing gene transcription. Histone deacetylases (HDACs), on the other hand, eliminate this chemical group causing chromatin condensation and gene repression (Grunstein 1997). The activity of the latter type of enzymes can be inhibited by HDACi, which are molecules of very diverse chemical origin. The use of HDACi in RA has enabled a great deal of information to be obtained about the physiology of several cell types involved in the disease, although the effect of HDACi is tissue and compound dependent and their behavior is not uniform.

HDAC levels in the RA synovium have been measured by several groups, giving rise to conflicting results. Some researchers maintain that synovial fibroblasts have higher protein and activity levels of HDAC1 and that this is positively regulated by TNF-α concentrations (Kawabata et al. 2010). HDAC1 and HDAC2 depletion decreased cell proliferation, and that of HDAC2 enhanced TNF-α induced MMP-1 production (Horiuchi et al. 2009). Moreover, the promoter of this gene has been found to be hyperacetylated in RASF, with the consequence that the gene is overexpressed (Maciejewska-Rodrigues et al. 2010). On the other hand, Astrid Jüngel and coworkers showed that HDAC levels and activity in the synovial tissue of RA patients are lower than in normal individuals. Lower levels of HDACs in the synovium give rise to a hyperacetylated histone milieu and consequently gene overexpression. Intriguingly, the results observed in RA-related cell types after HDACi treatment are different (Huber et al. 2007).

Since 2003, there has been compelling evidence that HDACi treatment of RA rat or mouse models ameliorates symptoms. Using phenylbutyrate and trichostatin A (TSA), Lin-Fen Yao and colleagues reported that HDACi treatment in adjuvant arthritic rats was able to eliminate arthritic manifestations in the joints. Moreover, in vitro treatment of their RASFs prevented cell proliferation through upregulation of two cyclin-dependent kinase (CDK) inhibitors (p16^{INK4a} and p21$^{WAF1/Cip1}$) and downregulation of the levels of TNF-α, IL-1, and IL-6 in the studied joints. The effect in synovial fibroblast cell-cycle arrest was irreversible in the case of the RA rat model but fully reversible when treating "healthy" rats (Chung et al. 2003). Moreover, intravenous injection of another HDACi, FK228 (depsipeptide), in autoantibody-mediated arthritic mice was able to prevent joint swelling, synovial inflammation, and bone and cartilage destruction through the same molecular mechanisms (Nishida et al. 2004). HDACi also affect apoptosis as they are able to sensitize RASF to apoptosis via the Fas receptor and TRAIL (Jungel et al. 2006; Morinobu et al. 2006).

Several RA animal models have been treated with HDACi, and the information available about how serious symptoms are ameliorated has been added. MS-275 and suberoylanilide hydroxamic acid (SAHA) treatment in synovial fibroblasts can inhibit proinflammatory cytokines such as TNF-α, IL-1β, downregulate angiogenic factors, and MMPs, as well as results in growth arrest (Lin et al. 2007; Nasu et al. 2008; Choo et al. 2010). One of the consequences of MMP downregulation is a decrease in cartilage resorption, which helps mend the lesions that have occurred in these joints (Young et al. 2005).

Angiogenesis contributes to synovitis as well as disease progression by increasing the flow of white blood cells and nutrients to the invasive and hyperplasic synovial membrane, as reviewed by (Pap and Distler 2005). As previously mentioned, HDACi are able to inhibit angiogenesis because important factors for angiogenic development such as HIF-1α and VEGF are downregulated (Manabe et al. 2008). After HDACi treatment, synovial tissue behaves similarly to tumors, specifically reducing their size due to the lack of oxygen and nutrients through the inhibition of angiogenic factors (Kwon et al. 2002).

HDACi research has also demonstrated that T cells isolated from RA patients undergo changes in histone modification patterns. TSA and valproic acid, for example, are able to restore the function of regulatory T cells that fail to suppress CD4[+] effector T cells. This is achieved by increasing the function and number of FOXP3-expressing CD15[+]CD4[+] regulatory T cells (Tao et al. 2007), thereby ameliorating the disease in animal models (Saouaf et al. 2009). FOXP3 in T cells is present in dynamic complexes together with HATs and HDACs and is even hyperacetylated in their own lysines. Therefore, the effects of HDACi should also be addressed by looking at off-histone targets.

In mice with collagen-induced arthritis, TSA is able to inhibit T helper 1 cells that are responsive to autoantigens by promoting apoptosis and inhibiting proliferation and IFN-γ release. Moreover, it increases T helper 2 function, which is more protective, by acetylating H3 and H4 histones at the *IL-4* promoter resulting in its upregulation (Zhou et al. 2011). In this way, the balance between more aggressive and more protective T helper 1 and T helper 2 is skewed towards the latter, resulting in a better clinical outcome. A similar balancing effect has also been described in human T lymphocytes by using HDACi LAQ824, also known as dacinostat (Brogdon et al. 2007). The effects of HDACi treatment in T cells and in RASFs are summarized in Fig. 11.2b.

Bone metabolism dysregulation is extensive in RA-affected joints. Osteoblasts and osteoclasts are the main cell types responsible for de novo formation and destruction of bone, respectively. In RA, the balance is clearly tipped towards osteoclast differentiation, as there is an increase in osteoclastogenesis. This balance can be restored using HDACi, as they are able to promote osteoblast maturation by altering the expression levels of several genes, mainly growth factors, bone forming enzymes, and Wnt receptors. This accelerates matrix mineralization and restores bone anabolism (Schroeder and Westendorf 2005; Schroeder et al. 2007). On the other hand, other HDACi can inhibit osteoclastogenesis by impairing nuclear translocation of the main factor for osteoclast differentiation, NFATc1, and by increasing IFN-β, which is an inhibitor of the process. The result is that less bone is destroyed and RA symptoms are ameliorated (Nakamura et al. 2005).

There are some concerns about the effects of HDACi on nonhistone targets, and much research remains to be done into off-histone acetylation in order to clarify the potential uses of these drugs in human therapeutics. First, HDACs appeared before histones in evolution, with a role of regulating other proteins present in the cell (Gregoretti et al. 2004). Moreover, the expression of only 2–10% of genes is modified by HDACi treatment, half of them being upregulated and the other being downregulated (Peart et al. 2005). This state of affairs is not compatible with the widely held belief that the main effect of HDACi is to hyperacetylate chromatin and open it for transcription. This could explain why the effects of HDACi are so diverse and unpredictable and are not always related to the behavior of histones in nucleosomes.

11.4 Epigenetic Alterations in Other Autoimmune Diseases

There is little epigenetic information available regarding other autoimmune diseases but SLE or RA, which have received almost all the attention from chromatin researchers. Nevertheless, there is growing interest in investigating these diseases from an epigenetic point of view, as the preliminary data indicate that there is clear epigenetic dysregulation in almost all of them. Below, a brief definition of the diseases will be given, together with the currently available evidence.

11.4.1 Epigenetic Dysregulation in Inflammatory Bowel Diseases: Crohn's Disease and Ulcerative Colitis

Inflammatory bowel diseases are mainly characterized by inflammation and destruction of digestive tissue, from esophagus to colon, depending on the condition. There are two main forms: ulcerative colitis (UC) and Crohn's disease (CD).

Global hypomethylation has been reported in UC, and this is correlated with higher proliferative activity in the damaged mucosa of the rectum (Gloria et al. 1996).

As one of the main concerns for inflammatory bowel disease physicians is the high prevalence of colorectal cancer in these patients, it is important to know which genes make cells more malignant. Two principal genes, p16[INK4a] and p14[ARF] (Hsieh et al. 1998; Sato et al. 2002a), are hypermethylated in this disease and in sporadic colorectal cancer (Gonzalez-Zulueta et al. 1995; Esteller et al. 2000). Some specific genes have been proposed as severity markers for UC disease because their methylation status associated with the course of the disease. First, there is *PAR2*, which is correlated with severe clinical phenotypes of UC (Tahara et al. 2009a). *MDR1* methylation status is associated with the chronic continuous type and early onset of UC (Tahara et al. 2009b). Lastly, *CDH1* and *GDNF* methylation status is associated with the parts of the mucosa with active disease (Saito et al. 2011). Colonocytes of gastric mucosa isolated from these patients show a reduction in their telomeric sequences (Risques et al. 2008), which could be linked to the accelerated methylation of some

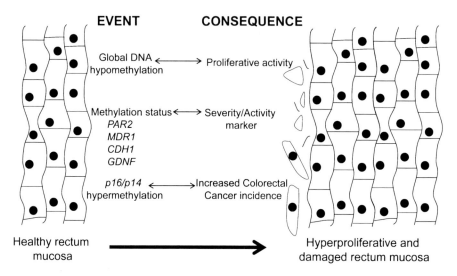

Fig. 11.3 Epigenetic events taking place in ulcerative colitis-injured mucosa. Global DNA hypomethylation has been reported in damaged UC mucosa. Specifically, the DNA methylation status of several promoters can be used to monitor the severity of the disease (*PAR2, MDR1, CDH1, GDNF*), while *p14* and *p16* promoter hypermethylation is linked with higher incidence of colorectal cancer in these patients (Gloria et al. 1996; Hsieh et al. 1998; Sato et al. 2002b; Tahara et al. 2009a, b; Saito et al. 2011)

age-related genes in UC mucosa, such as *ER*, *MYOD*, *p16*, and *CSPG2* (Issa et al. 2001). Figure 11.3 illustrates the key DNA methylation changes taking place in the affected rectum mucosa from UC patients.

Importantly, a genome-wide study performed in UC and CD has shown that there is a strong background of shared epigenetic dysregulation (Lin et al. 2011). Specifically, many immune-related genes are differentially methylated between healthy and affected individuals, for example, IL18BP, LTB4R, PECAM1, FGF2, LAT, TNFSF8, and HOXB2.

As mentioned above, there are several lines of evidence suggesting that inflammatory bowel diseases are strongly dysregulated by epigenetic means. Research in this field could help in the selection of new therapeutic pathways that could be targeted pharmacologically.

11.4.2 Epigenetic Dysregulation in Inflammatory Thyroid Diseases

AITDs are a group of autoimmune conditions characterized by immune attack on the thyroid gland. This attack may have two different clinical outcomes: Hashimoto's thyroiditis (HT) and Graves' disease (GD). HT is a hypothyroidism due to apoptosis of thyroid cells, while GD is a hyperthyroidism arising from the hyperactivation of the thyroid through TSH receptor-stimulating antibodies.

The only epigenetic information available about these diseases concerns a mechanism of skewed X chromosome inactivation mosaicism similar to what is seen in progressive systemic sclerosis (PSS) (Brix et al. 2005).

11.4.3 Epigenetic Dysregulation in Multiple Sclerosis

Multiple sclerosis is a chronic, progressive disease in which nerve cell sheaths are damaged by demyelination in the brain and spinal cord, concluding in neurodegeneration. It causes numbness, impairment of speech and muscular coordination, blurred vision, and fatigue (Compston and Coles 2008). A gene related with myelin degradation and neurodegenerative diseases, *PAD2* (peptidylarginine deiminase, type II) (Bhattacharya et al. 2006), has been found to be hypomethylated and so overexpressed in nervous tissue from multiple sclerosis patients (Mastronardi et al. 2007). More recently, CD4[+] T cells from monozygotic twins discordant for the disease were subjected to high-throughput analysis at the genetic, epigenetic, and transcriptional levels. Surprisingly, no significant differences were found in the levels of DNA methylation of T cells in this disease (Baranzini et al. 2010). However, the small number of samples included in this study limits the confidence that can be placed in the conclusions. Those results also suggest that perhaps other cell types should be studied, and emphasize the importance of selecting the tissue type for study.

11.4.4 Epigenetic Dysregulation in Primary Biliary Cirrhosis

Primary biliary cirrhosis is an organ-specific autoimmune disease in which small-to-medium bile ducts, cholestasis fibrosis, and cirrhosis take place in the liver (Poupon 2010).

There is only one report regarding epigenetic dysregulation in this disease. The expression of two X chromosomal genes, *CLIC2* and *PIN4*, is downregulated in blood cells from primary biliary cirrhosis patients relative to their healthy monozygotic twins (Mitchell et al. 2011). This suggests that mechanisms regulating X chromosomal gene silencing might be dysregulated.

11.4.5 Epigenetic Dysregulation in Progressive Systemic Sclerosis

PSS or scleroderma is a skin disease characterized by excessive collagen deposition and progressive vasculopathy. There is aberrant fibroblast activation, and as a consequence, the skin is inelastic. With respect to epigenetic dysregulation, DNMT1 has been reported to be upregulated in PSS fibroblasts (Qi et al. 2009). The *FL1* gene is hypermethylated, causing the presence of excess collagen in this cell type.

Pathological features of PSS fibroblasts can be reverted by the use of HDACi and DNA methylation inhibitors, two types of epigenetic drugs (Wang et al. 2006). On the other hand, contrary to what happens in PSS fibroblasts, T cell DNA undergoes global hypomethylation, although no biological consequences have been analyzed (Lei et al. 2009).

Lastly, a phenomenon of skewed X chromosome inactivation mosaicism has been reported in this disease (Ozbalkan et al. 2005). This mechanism for balancing gene dosage could explain the greater prevalence of this disease in women.

11.4.6 Epigenetic Dysregulation in Psoriasis

Psoriasis is a systemic autoimmune disease characterized by early keratinocyte differentiation and proliferation that cause recurring reddish patches, especially on the knees, elbow, and scalp.

In this hyperproliferative disease, *SHP-1* promoter (also known as PTPN6) 2 is hypomethylated in skin samples from psoriatic patients (Ruchusatsawat et al. 2006). Moreover, *p16* gene promoter is hypomethylated in hematopoietic cells but, conversely, hypermethylated in skin samples isolated from patients suffering psoriasis (Zhang et al. 2007; Chen et al. 2008). The *p14* gene has also been reported to be hypermethylated in skin lesions (Zhang et al. 2010a). Hematopoietic cells, on the other hand, have a lower level of promoter methylation of p15 and p21 genes (Zhang et al. 2009).

The severity of the disease, measured by the Psoriasis Area and Severity Index (PASI), is correlated with higher levels of DNA methylation in PBMCs and psoriatic skin lesions (Zhang et al. 2010a).

11.4.7 Epigenetic Dysregulation in Sjögren's Syndrome

Sjögren's syndrome (SjS) is a systemic autoimmune disease characterized by immune attack on the exocrine glands that produce saliva and tears (Mariette and Gottenberg 2010). In these tissues, hypermethylation of the *BP230* gene has been described, although higher levels of proteins are present. This protein is involved in type I hemidesmosomes (lamina anchorage), and its dysregulation could be responsible for glandule structure impairment (Gonzalez et al. 2011).

11.4.8 Epigenetic Dysregulation in Vitiligo

Vitiligo is a condition in which melanocyte cells of the skin and eye are lost, and as a consequence, patches appear in the midst of normally pigmented skin. It has autoimmune comorbidity (i.e., it appears in conjunction with other autoimmune diseases in the same individual) and affects men and women equally. The only epigenetic studies available are those of the Smyth line (SL) chicken, an animal model for human vitiligo, in which 5-azacytidine treatment induces the disease. Exposing these chickens to this epigenetic drug causes antibody production against TRP-1, a melanocyte-specific protein, and also causes skin melanocyte depletion and subsequent depigmentation (Sreekumar et al. 1996). The fact that an epigenetic drug alters the phenotype of this chicken to cause vitiligo suggests that DNA methylation alteration may drive the mechanisms of disease.

11.5 Conclusions and Future Prospects

Despite the efforts of many research groups, the molecular mechanisms contributing to the pathogenesis of autoimmune diseases are still unclear. Many of these disorders share clinical and genetic features and appear to be influenced by similar environmental factors. Epigenetic modifications are influenced by environmental factors and are known to determine gene function directly and therefore constitute an important target for investigating its participation in the etiology of these diseases. Most attempts to identify the epigenetic alterations occurring in autoimmune disease have focused on SLE and RA and have served to identify both global and sequence-specific hypomethylation and overexpression of genes that are fundamental to immune function. It is now the time to turn our attention to several important issues: to make use of high-throughput approaches, to systematically analyze all specific cell types of potential relevance to disease pathogenesis, and to find the best way of using the information obtained in the clinical setting.

References

Aho, K., Koskenvuo, M., Tuominen, J. & Kaprio, J. (1986). Occurrence of rheumatoid arthritis in a nationwide series of twins. *J Rheumatol* 13, 899–902.

Baier, A., Meineckel, I., Gay, S. & Pap, T. (2003). Apoptosis in rheumatoid arthritis. *Curr Opin Rheumatol* 15, 274–279.

Ballestar, E. (2011). Epigenetic alterations in autoimmune rheumatic diseases. *Nat Rev Rheumatol* 7, 263–271.

Baranzini, S. E., Mudge, J., van Velkinburgh, J. C., Khankhanian, P., Khrebtukova, I., Miller, N. A., Zhang, L., Farmer, A. D., Bell, C. J., Kim, R. W., May, G. D., Woodward, J. E., Caillier, S. J., McElroy, J. P., Gomez, R., Pando, M. J., Clendenen, L. E., Ganusova, E. E., Schilkey, F. D., Ramaraj, T., Khan, O. A., Huntley, J. J., Luo, S., Kwok, P. Y., Wu, T. D., Schroth, G. P., Oksenberg, J. R., Hauser, S. L. & Kingsmore, S. F. (2010). Genome, epigenome and RNA sequences of monozygotic twins discordant for multiple sclerosis. *Nature* 464, 1351–1356.

Bhattacharya, S. K., Bhat, M. B. & Takahara, H. (2006). Modulation of peptidyl arginine deiminase 2 and implication for neurodegeneration. *Curr Eye Res* 31, 1063–1071.

Boix-Chornet, M., Fraga, M. F., Villar-Garea, A., Caballero, R., Espada, J., Nunez, A., Casado, J., Largo, C., Casal, J. I., Cigudosa, J. C., Franco, L., Esteller, M. & Ballestar, E. (2006).

Release of hypoacetylated and trimethylated histone H4 is an epigenetic marker of early apoptosis. *J Biol Chem* 281, 13540–13547.

Brix, T. H., Knudsen, G. P., Kristiansen, M., Kyvik, K. O., Orstavik, K. H. & Hegedus, L. (2005). High frequency of skewed X-chromosome inactivation in females with autoimmune thyroid disease: a possible explanation for the female predisposition to thyroid autoimmunity. *J Clin Endocrinol Metab* 90, 5949–5953.

Brogdon, J. L., Xu, Y., Szabo, S. J., An, S., Buxton, F., Cohen, D. & Huang, Q. (2007). Histone deacetylase activities are required for innate immune cell control of Th1 but not Th2 effector cell function. *Blood* 109, 1123–1130.

Ceribelli, A., Yao, B., Dominguez-Gutierrez, P. R. & Chan, E. K. (2011). Lupus T cells switched on by DNA hypomethylation via microRNA? *Arthritis Rheum* 63, 1177–1181.

Chagnon, P., Schneider, R., Hebert, J., Fortin, P. R., Provost, S., Belisle, C., Gingras, M., Bolduc, V., Perreault, C., Silverman, E. & Busque, L. (2006). Identification and characterization of an Xp22.33;Yp11.2 translocation causing a triplication of several genes of the pseudoautosomal region 1 in an XX male patient with severe systemic lupus erythematosus. *Arthritis Rheum* 54, 1270–1278.

Chen, M., Chen, Z. Q., Cui, P. G., Yao, X., Li, Y. M., Li, A. S., Gong, J. Q. & Cao, Y. H. (2008). The methylation pattern of p16INK4a gene promoter in psoriatic epidermis and its clinical significance. *Br J Dermatol* 158, 987–993.

Chen, Y., Gorelik, G. J., Strickland, F. M. & Richardson, B. C. (2010). Decreased ERK and JNK signaling contribute to gene overexpression in "senescent" CD4+CD28- T cells through epigenetic mechanisms. *J Leukoc Biol* 87, 137–145.

Choo, Q. Y., Ho, P. C., Tanaka, Y. & Lin, H. S. (2010). Histone deacetylase inhibitors MS-275 and SAHA induced growth arrest and suppressed lipopolysaccharide-stimulated NF-kappaB p65 nuclear accumulation in human rheumatoid arthritis synovial fibroblastic E11 cells. *Rheumatology (Oxford)* 49, 1447–1460.

Choy, E. H. & Panayi, G. S. (2001). Cytokine pathways and joint inflammation in rheumatoid arthritis. *N Engl J Med* 344, 907–916.

Chung, Y. L., Lee, M. Y., Wang, A. J. & Yao, L. F. (2003). A therapeutic strategy uses histone deacetylase inhibitors to modulate the expression of genes involved in the pathogenesis of rheumatoid arthritis. *Mol Ther* 8, 707–717.

Compston, A. & Coles, A. (2008). Multiple sclerosis. *Lancet* 372, 1502–1517.

Cooper, G. S., Dooley, M. A., Treadwell, E. L., St Clair, E. W. & Gilkeson, G. S. (2002). Hormonal and reproductive risk factors for development of systemic lupus erythematosus: results of a population-based, case-control study. *Arthritis Rheum* 46, 1830–1839.

Cornacchia, E., Golbus, J., Maybaum, J., Strahler, J., Hanash, S. & Richardson, B. (1988). Hydralazine and procainamide inhibit T cell DNA methylation and induce autoreactivity. *J Immunol* 140, 2197–2200.

Corvetta, A., Della Bitta R., Luchetti, M. M. & Pomponio, G. (1991). 5-Methylcytosine content of DNA in blood, synovial mononuclear cells and synovial tissue from patients affected by autoimmune rheumatic diseases. *J Chromatogr* 566, 481–491.

D'Cruz, D. P., Khamashta, M. A. & Hughes, G. R. (2007). Systemic lupus erythematosus. *Lancet* 369, 587–596.

Danchenko, N., Satia, J. A. & Anthony, M. S. (2006). Epidemiology of systemic lupus erythematosus: a comparison of worldwide disease burden. *Lupus* 15, 308–318.

Davidson, A. & Diamond, B. (2001). Autoimmune diseases. *N Engl J Med* 345, 340–350.

Delgado-Vega, A., Sanchez, E., Lofgren, S., Castillejo-Lopez, C. & Alarcon-Riquelme, M. E. (2010). Recent findings on genetics of systemic autoimmune diseases. *Curr Opin Immunol* 22, 698–705.

Distler, J. H., Jungel, A., Huber, L. C., Seemayer, C. A., Reich, C. F., Gay, R. E., Michel, B. A., Fontana, A., Gay, S., Pisetsky, D. S. & Distler, O. (2005). The induction of matrix metalloproteinase and cytokine expression in synovial fibroblasts stimulated with immune cell microparticles. *Proc Natl Acad Sci U S A* 102, 2892–2897.

Duke, O., Panayi, G. S., Janossy, G. & Poulter, L. W. (1982). An immunohistological analysis of lymphocyte subpopulations and their microenvironment in the synovial membranes of patients with rheumatoid arthritis using monoclonal antibodies. *Clin Exp Immunol* 49, 22–30.

Emlen, W., Niebur, J. & Kadera, R. (1994). Accelerated in vitro apoptosis of lymphocytes from patients with systemic lupus erythematosus. *J Immunol* 152, 3685–3692.

Esteller, M. (2008). Epigenetics in cancer. *N Engl J Med* 358, 1148–1159.

Esteller, M., Tortola, S., Toyota, M., Capella, G., Peinado, M. A., Baylin, S. B. & Herman, J. G. (2000). Hypermethylation-associated inactivation of p14(ARF) is independent of p16(INK4a) methylation and p53 mutational status. *Cancer Res* 60, 129–133.

Funauchi, M., Yoo, B. S., Nozaki, Y., Sugiyama, M., Ohno, M., Kinoshita, K. & Kanamaru, A. (2002). Dysregulation of the granulocyte-macrophage colony-stimulating factor receptor is one of the causes of defective expression of CD80 antigen in systemic lupus erythematosus. *Lupus* 11, 317–321.

Garaud, S., Le Dantec, C., Jousse-Joulin, S., Hanrotel-Saliou, C., Saraux, A., Mageed, R. A., Youinou, P. & Renaudineau, Y. (2009). IL-6 modulates CD5 expression in B cells from patients with lupus by regulating DNA methylation. *J Immunol* 182, 5623–5632.

Gilchrist, F. C., Bunn, C., Foley, P. J., Lympany, P. A., Black, C. M., Welsh, K. I. & du Bois, R. M. (2001). Class II HLA associations with autoantibodies in scleroderma: a highly significant role for HLA-DP. *Genes Immun* 2, 76–81.

Gloria, L., Cravo, M., Pinto, A., de Sousa, L. S., Chaves, P., Leitao, C. N., Quina, M., Mira, F. C. & Soares, J. (1996). DNA hypomethylation and proliferative activity are increased in the rectal mucosa of patients with long-standing ulcerative colitis. *Cancer* 78, 2300–2306.

Goldberg, B., Urnovitz, H. B. & Stricker, R. B. (2000). Beyond danger: unmethylated CpG dinucleotides and the immunopathogenesis of disease. *Immunol Lett* 73, 13–18.

Gonzalez, S., Aguilera, S., Alliende, C., Urzua, U., Quest, A. F., Herrera, L., Molina, C., Hermoso, M., Ewert, P., Brito, M., Romo, R., Leyton, C., Perez, P. & Gonzalez, M. J. (2011). Alterations in type I hemidesmosome components suggestive of epigenetic control in the salivary glands of patients with Sjogren's syndrome. *Arthritis Rheum* 63, 1106–1115.

Gonzalez-Zulueta, M., Bender, C. M., Yang, A. S., Nguyen, T., Beart, R. W., Van Tornout, J. M. & Jones, P. A. (1995). Methylation of the 5' CpG island of the p16/CDKN2 tumor suppressor gene in normal and transformed human tissues correlates with gene silencing. *Cancer Res* 55, 4531–4535.

Goodnow, C. C., Sprent, J., Fazekas de St Groth, B. & Vinuesa, C. G. (2005). Cellular and genetic mechanisms of self tolerance and autoimmunity. *Nature* 435, 590–597.

Gorelik, G., Fang, J. Y., Wu, A., Sawalha, A. H. & Richardson, B. (2007). Impaired T cell protein kinase C delta activation decreases ERK pathway signaling in idiopathic and hydralazine-induced lupus. *J Immunol* 179, 5553–5563.

Gregersen, P. K. (2010). Susceptibility genes for rheumatoid arthritis - a rapidly expanding harvest. *Bull NYU Hosp Jt Dis* 68, 179–182.

Gregoretti, I. V., Lee, Y. M. & Goodson, H. V. (2004). Molecular evolution of the histone deacetylase family: functional implications of phylogenetic analysis. *J Mol Biol* 338, 17–31.

Grennan, D. M., Parfitt, A., Manolios, N., Huang, Q., Hyland, V., Dunckley, H., Doran, T., Gatenby, P. & Badcock, C. (1997). Family and twin studies in systemic lupus erythematosus. *Dis Markers* 13, 93–98.

Grumet, F. C., Coukell, A., Bodmer, J. G., Bodmer, W. F. & McDevitt, H. O. (1971). Histocompatibility (HL-A) antigens associated with systemic lupus erythematosus. A possible genetic predisposition to disease. *N Engl J Med* 285, 193–196.

Grunstein, M. (1997). Histone acetylation in chromatin structure and transcription. *Nature* 389, 349–352.

Hogg, N., Laschinger, M., Giles, K. & McDowall, A. (2003). T-cell integrins: more than just sticking points. *J Cell Sci* 116, 4695–4705.

Horiuchi, M., Morinobu, A., Chin, T., Sakai, Y., Kurosaka, M. & Kumagai, S. (2009). Expression and function of histone deacetylases in rheumatoid arthritis synovial fibroblasts. *J Rheumatol* 36, 1580–1589.

Hsieh, C. J., Klump, B., Holzmann, K., Borchard, F., Gregor, M. & Porschen, R. (1998). Hypermethylation of the p16INK4a promoter in colectomy specimens of patients with long-standing and extensive ulcerative colitis. *Cancer Res* 58, 3942–3945.

Hu, N., Qiu, X., Luo, Y., Yuan, J., Li, Y., Lei, W., Zhang, G., Zhou, Y., Su, Y. & Lu, Q. (2008). Abnormal histone modification patterns in lupus CD4+ T cells. *J Rheumatol* 35, 804–810.

Huber, L. C., Brock, M., Hemmatazad, H., Giger, O. T., Moritz, F., Trenkmann, M., Distler, J. H., Gay, R. E., Kolling, C., Moch, H., Michel, B. A., Gay, S., Distler, O. & Jungel, A. (2007). Histone deacetylase/acetylase activity in total synovial tissue derived from rheumatoid arthritis and osteoarthritis patients. *Arthritis Rheum* 56, 1087–1093.

Hussain, M., Rao, M., Humphries, A. E., Hong, J. A., Liu, F., Yang, M., Caragacianu, D. & Schrump, D. S. (2009). Tobacco smoke induces polycomb-mediated repression of Dickkopf-1 in lung cancer cells. *Cancer Res* 69, 3570–3578.

Issa, J. P., Ahuja, N., Toyota, M., Bronner, M. P. & Brentnall, T. A. (2001). Accelerated age-related CpG island methylation in ulcerative colitis. *Cancer Res* 61, 3573–3577.

Janeway, C. A. & Medzhitov, R. (2002). Innate immune recognition. *Annu Rev Immunol* 20, 197–216.

Janson, P. C., Linton, L. B., Bergman, E. A., Marits, P., Eberhardson, M., Piehl, F., Malmstrom, V. & Winqvist, O. (2011). Profiling of CD4+ T cells with epigenetic immune lineage analysis. *J Immunol* 186, 92–102.

Javierre, B. M., Fernandez, A. F., Richter, J., Al Shahrour, F., Martin-Subero, J. I., Rodriguez-Ubreva, J., Berdasco, M., Fraga, M. F., O'Hanlon, T. P., Rider, L. G., Jacinto, F. V., Lopez-Longo, F. J., Dopazo, J., Forn, M., Peinado, M. A., Carreno, L., Sawalha, A. H., Harley, J. B., Siebert, R., Esteller, M., Miller, F. W. & Ballestar, E. (2010). Changes in the pattern of DNA methylation associate with twin discordance in systemic lupus erythematosus. *Genome Res* 20, 170–179.

Jeffries, M. A., Dozmorov, M., Tang, Y., Merrill, J. T., Wren, J. D. & Sawalha, A. H. (2011). Genome-wide DNA methylation patterns in CD4+ T cells from patients with systemic lupus erythematosus. *Epigenetics* 6, 593–601.

Jungel, A., Baresova, V., Ospelt, C., Simmen, B. R., Michel, B. A., Gay, R. E., Gay, S., Seemayer, C. A. & Neidhart, M. (2006). Trichostatin A sensitises rheumatoid arthritis synovial fibroblasts for TRAIL-induced apoptosis. *Ann Rheum Dis* 65, 910–912.

Kaminskas, E., Farrell, A., Abraham, S., Baird, A., Hsieh, L. S., Lee, S. L., Leighton, J. K., Patel, H., Rahman, A., Sridhara, R., Wang, Y. C. & Pazdur, R. (2005). Approval summary: azacitidine for treatment of myelodysplastic syndrome subtypes. *Clin Cancer Res* 11, 3604–3608.

Kaplan, M. J., Lewis, E. E., Shelden, E. A., Somers, E., Pavlic, R., McCune, W. J. & Richardson, B. C. (2002). The apoptotic ligands TRAIL, TWEAK, and Fas ligand mediate monocyte death induced by autologous lupus T cells. *J Immunol* 169, 6020–6029.

Kaplan, M. J., Lu, Q., Wu, A., Attwood, J. & Richardson, B. (2004). Demethylation of promoter regulatory elements contributes to perforin overexpression in CD4+ lupus T cells. *J Immunol* 172, 3652–3661.

Karouzakis, E., Gay, R. E., Michel, B. A., Gay, S. & Neidhart, M. (2009). DNA hypomethylation in rheumatoid arthritis synovial fibroblasts. *Arthritis Rheum* 60, 3613–3622.

Kawabata, T., Nishida, K., Takasugi, K., Ogawa, H., Sada, K., Kadota, Y., Inagaki, J., Hirohata, S., Ninomiya, Y. & Makino, H. (2010). Increased activity and expression of histone deacetylase 1 in relation to tumor necrosis factor-alpha in synovial tissue of rheumatoid arthritis. *Arthritis Res Ther* 12, R133.

Kim, D. H., Nelson, H. H., Wiencke, J. K., Zheng, S., Christiani, D. C., Wain, J. C., Mark, E. J. & Kelsey, K. T. (2001). p16(INK4a) and histology-specific methylation of CpG islands by exposure to tobacco smoke in non-small cell lung cancer. *Cancer Res* 61, 3419–3424.

Kitamura, T., Kabuyama, Y., Kamataki, A., Homma, M. K., Kobayashi, H., Aota, S., Kikuchi, S. & Homma, Y. (2008). Enhancement of lymphocyte migration and cytokine production by ephrinB1 system in rheumatoid arthritis. *Am J Physiol Cell Physiol* 294, C189–C196.

Klinman, N. R. (1996). The "clonal selection hypothesis" and current concepts of B cell tolerance. *Immunity* 5, 189–195.

Kobata, T., Jacquot, S., Kozlowski, S., Agematsu, K., Schlossman, S. F. & Morimoto, C. (1995). CD27-CD70 interactions regulate B-cell activation by T cells. *Proc Natl Acad Sci U S A* 92, 11249–11253.

Kouzarides, T. (2007). Chromatin modifications and their function. *Cell* 128, 693–705.

Kuchen, S., Seemayer, C. A., Rethage, J., von Knoch, R., Kuenzler, P., Beat, A. M., Gay, R. E., Gay, S. & Neidhart, M. (2004). The L1 retroelement-related p40 protein induces p38delta MAP kinase. *Autoimmunity* 37, 57–65.

Kwon, H. J., Kim, M. S., Kim, M. J., Nakajima, H. & Kim, K. W. (2002). Histone deacetylase inhibitor FK228 inhibits tumor angiogenesis. *Int J Cancer* 97, 290–296.

Lafyatis, R., Remmers, E. F., Roberts, A. B., Yocum, D. E., Sporn, M. B. & Wilder, R. L. (1989). Anchorage-independent growth of synoviocytes from arthritic and normal joints. Stimulation by exogenous platelet-derived growth factor and inhibition by transforming growth factor-beta and retinoids. *J Clin Invest* 83, 1267–1276.

Lee, H. S. & Bae, S. C. (2010). What can we learn from genetic studies of systemic lupus erythematosus? Implications of genetic heterogeneity among populations in SLE. *Lupus* 19, 1452–1459.

Lei, W., Luo, Y., Lei, W., Luo, Y., Yan, K., Zhao, S., Li, Y., Qiu, X., Zhou, Y., Long, H., Zhao, M., Liang, Y., Su, Y. & Lu, Q. (2009). Abnormal DNA methylation in CD4+ T cells from patients with systemic lupus erythematosus, systemic sclerosis, and dermatomyositis. *Scand J Rheumatol* 38, 369–374.

Li, Y., Zhao, M., Yin, H., Gao, F., Wu, X., Luo, Y., Zhao, S., Zhang, X., Su, Y., Hu, N., Long, H., Richardson, B. & Lu, Q. (2010). Overexpression of the growth arrest and DNA damage-induced 45alpha gene contributes to autoimmunity by promoting DNA demethylation in lupus T cells. *Arthritis Rheum* 62, 1438–1447.

Lin, H. S., Hu, C. Y., Chan, H. Y., Liew, Y. Y., Huang, H. P., Lepescheux, L., Bastianelli, E., Baron, R., Rawadi, G. & Clement-Lacroix, P. (2007). Anti-rheumatic activities of histone deacetylase (HDAC) inhibitors in vivo in collagen-induced arthritis in rodents. *Br J Pharmacol* 150, 862–872.

Lin, Z., Hegarty, J. P., Cappel, J. A., Yu, W., Chen, X., Faber, P., Wang, Y., Kelly, A. A., Poritz, L. S., Peterson, B. Z., Schreiber, S., Fan, J. B. & Koltun, W. A. (2011). Identification of disease-associated DNA methylation in intestinal tissues from patients with inflammatory bowel disease. *Clin Genet* 80, 59–67.

Liu, H., Zhou, Y., Boggs, S. E., Belinsky, S. A. & Liu, J. (2007). Cigarette smoke induces demethylation of prometastatic oncogene synuclein-gamma in lung cancer cells by downregulation of DNMT3B. *Oncogene* 26, 5900–5910.

Liu, Y., Chen, Y. & Richardson, B. (2009). Decreased DNA methyltransferase levels contribute to abnormal gene expression in "senescent" CD4(+)CD28(-) T cells. *Clin Immunol* 132, 257–265.

Lu, Q., Kaplan, M., Ray, D., Ray, D., Zacharek, S., Gutsch, D. & Richardson, B. (2002). Demethylation of ITGAL (CD11a) regulatory sequences in systemic lupus erythematosus. *Arthritis Rheum* 46, 1282–1291.

Lu, Q., Wu, A. & Richardson, B. C. (2005). Demethylation of the same promoter sequence increases CD70 expression in lupus T cells and T cells treated with lupus-inducing drugs. *J Immunol* 174, 6212–6219.

Lu, Q., Wu, A., Tesmer, L., Ray, D., Yousif, N. & Richardson, B. (2007). Demethylation of CD40LG on the inactive X in T cells from women with lupus. *J Immunol* 179, 6352–6358.

Luo, Y., Zhang, X., Zhao, M. & Lu, Q. (2009). DNA demethylation of the perforin promoter in CD4(+) T cells from patients with subacute cutaneous lupus erythematosus. *J Dermatol Sci* 56, 33–36.

Luo, Y., Zhao, M. & Lu, Q. (2010). Demethylation of promoter regulatory elements contributes to CD70 overexpression in CD4+ T cells from patients with subacute cutaneous lupus erythematosus. *Clin Exp Dermatol* 35, 425–430.

Maciejewska-Rodrigues, H., Karouzakis, E., Strietholt, S., Hemmatazad, H., Neidhart, M., Ospelt, C., Gay, R. E., Michel, B. A., Pap, T., Gay, S. & Jungel, A. (2010). Epigenetics and rheumatoid arthritis: the role of SENP1 in the regulation of MMP-1 expression. *J Autoimmun* 35, 15–22.

Manabe, H., Nasu, Y., Komiyama, T., Furumatsu, T., Kitamura, A., Miyazawa, S., Ninomiya, Y., Ozaki, T., Asahara, H. & Nishida, K. (2008). Inhibition of histone deacetylase down-regulates the expression of hypoxia-induced vascular endothelial growth factor by rheumatoid synovial fibroblasts. *Inflamm Res* 57, 4–10.

Mariette, X. & Gottenberg, J. E. (2010). Pathogenesis of Sjogren's syndrome and therapeutic consequences. *Curr Opin Rheumatol* 22, 471–477.

Mastronardi, F. G., Noor, A., Wood, D. D., Paton, T. & Moscarello, M. A. (2007). Peptidyl argininedeiminase 2 CpG island in multiple sclerosis white matter is hypomethylated. *J Neurosci Res* 85, 2006–2016.

McGonagle, D. & McDermott, M. F. (2006). A proposed classification of the immunological diseases. *PLoS Med* 3, e297.

Mi, X. B. & Zeng, F. Q. (2008). Hypomethylation of interleukin-4 and -6 promoters in T cells from systemic lupus erythematosus patients. *Acta Pharmacol Sin* 29, 105–112.

Mishra, N., Brown, D. R., Olorenshaw, I. M. & Kammer, G. M. (2001). Trichostatin A reverses skewed expression of CD154, interleukin-10, and interferon-gamma gene and protein expression in lupus T cells. *Proc Natl Acad Sci U S A* 98, 2628–2633.

Mitchell, M. M., Lleo, A., Zammataro, L., Mayo, M. J., Invernizzi, P., Bach, N., Shimoda, S., Gordon, S., Podda, M., Gershwin, M. E., Selmi, C. & LaSalle, J. M. (2011). Epigenetic investigation of variably X chromosome inactivated genes in monozygotic female twins discordant for primary biliary cirrhosis. *Epigenetics* 6, 95–102.

Morinobu, A., Wang, B., Liu, J., Yoshiya, S., Kurosaka, M. & Kumagai, S. (2006). Trichostatin A cooperates with Fas-mediated signal to induce apoptosis in rheumatoid arthritis synovial fibroblasts. *J Rheumatol* 33, 1052–1060.

Muller-Ladner, U., Kriegsmann, J., Franklin, B. N., Matsumoto, S., Geiler, T., Gay, R. E. & Gay, S. (1996). Synovial fibroblasts of patients with rheumatoid arthritis attach to and invade normal human cartilage when engrafted into SCID mice. *Am J Pathol* 149, 1607–1615.

Nakamura, T., Kukita, T., Shobuike, T., Nagata, K., Wu, Z., Ogawa, K., Hotokebuchi, T., Kohashi, O. & Kukita, A. (2005). Inhibition of histone deacetylase suppresses osteoclastogenesis and bone destruction by inducing IFN-beta production. *J Immunol* 175, 5809–5816.

Nakkuntod, J., Avihingsanon, Y., Mutirangura, A. & Hirankarn, N. (2011). Hypomethylation of LINE-1 but not Alu in lymphocyte subsets of systemic lupus erythematosus patients. *Clin Chim Acta* 412, 1457–1461.

Nasu, Y., Nishida, K., Miyazawa, S., Komiyama, T., Kadota, Y., Abe, N., Yoshida, A., Hirohata, S., Ohtsuka, A. & Ozaki, T. (2008). Trichostatin A, a histone deacetylase inhibitor, suppresses synovial inflammation and subsequent cartilage destruction in a collagen antibody-induced arthritis mouse model. *Osteoarthritis Cartilage* 16, 723–732.

Neidhart, M., Rethage, J., Kuchen, S., Kunzler, P., Crowl, R. M., Billingham, M. E., Gay, R. E. & Gay, S. (2000). Retrotransposable L1 elements expressed in rheumatoid arthritis synovial tissue: association with genomic DNA hypomethylation and influence on gene expression. *Arthritis Rheum* 43, 2634–2647.

Nile, C. J., Read, R. C., Akil, M., Duff, G. W. & Wilson, A. G. (2008). Methylation status of a single CpG site in the IL6 promoter is related to IL6 messenger RNA levels and rheumatoid arthritis. *Arthritis Rheum* 58, 2686–2693.

Nishida, K., Komiyama, T., Miyazawa, S., Shen, Z. N., Furumatsu, T., Doi, H., Yoshida, A., Yamana, J., Yamamura, M., Ninomiya, Y., Inoue, H. & Asahara, H. (2004). Histone deacetylase inhibitor suppression of autoantibody-mediated arthritis in mice via regulation of p16INK4a and p21(WAF1/Cip1) expression. *Arthritis Rheum* 50, 3365–3376.

Oelke, K., Lu, Q., Richardson, D., Wu, A., Deng, C., Hanash, S. & Richardson, B. (2004). Overexpression of CD70 and overstimulation of IgG synthesis by lupus T cells and T cells treated with DNA methylation inhibitors. *Arthritis Rheum* 50, 1850–1860.

Ozbalkan, Z., Bagislar, S., Kiraz, S., Akyerli, C. B., Ozer, H. T., Yavuz, S., Birlik, A. M., Calguneri, M. & Ozcelik, T. (2005). Skewed X chromosome inactivation in blood cells of women with scleroderma. *Arthritis Rheum* 52, 1564–1570.

Pan, W., Zhu, S., Yuan, M., Cui, H., Wang, L., Luo, X., Li, J., Zhou, H., Tang, Y. & Shen, N. (2010). MicroRNA-21 and microRNA-148a contribute to DNA hypomethylation in lupus CD4+ T cells by directly and indirectly targeting DNA methyltransferase 1. *J Immunol* 184, 6773–6781.

Pancer, Z. & Cooper, M. D. (2006). The evolution of adaptive immunity. *Annu Rev Immunol* 24, 497–518.

Pap, T. & Distler, O. (2005). Linking angiogenesis to bone destruction in arthritis. *Arthritis Rheum* 52, 1346–1348.

Peart, M. J., Smyth, G. K., van Laar, R. K., Bowtell, D. D., Richon, V. M., Marks, P. A., Holloway, A. J. & Johnstone, R. W. (2005). Identification and functional significance of genes regulated by structurally different histone deacetylase inhibitors. *Proc Natl Acad Sci U S A* 102, 3697–3702.

Pender, M. P. (2003). Infection of autoreactive B lymphocytes with EBV, causing chronic autoimmune diseases. *Trends Immunol* 24, 584–588.

Pollard, K. M., Hultman, P. & Kono, D. H. (2010). Toxicology of autoimmune diseases. *Chem Res Toxicol* 23, 455–466.

Poupon, R. (2010). Primary biliary cirrhosis: a 2010 update. *J Hepatol* 52, 745–758.

Qi, Q., Guo, Q., Tan, G., Mao, Y., Tang, H., Zhou, C. & Zeng, F. (2009). Predictors of the scleroderma phenotype in fibroblasts from systemic sclerosis patients. *J Eur Acad Dermatol Venereol* 23, 160–168.

Quddus, J., Johnson, K. J., Gavalchin, J., Amento, E. P., Chrisp, C. E., Yung, R. L. & Richardson, B. C. (1993). Treating activated CD4+ T cells with either of two distinct DNA methyltransferase inhibitors, 5-azacytidine or procainamide, is sufficient to cause a lupus-like disease in syngeneic mice. *J Clin Invest* 92, 38–53.

Rahman, A. & Isenberg, D. A. (2008). Systemic lupus erythematosus. *N Engl J Med* 358, 929-939.

Richardson, B. (1986). Effect of an inhibitor of DNA methylation on T cells. II. 5-Azacytidine induces self-reactivity in antigen-specific T4+ cells. *Hum Immunol* 17, 456–470.

Richardson, B., Powers, D., Hooper, F., Yung, R. L. & O'Rourke, K. (1994). Lymphocyte function-associated antigen 1 overexpression and T cell autoreactivity. *Arthritis Rheum* 37, 1363–1372.

Richardson, B., Scheinbart, L., Strahler, J., Gross, L., Hanash, S. & Johnson, M. (1990). Evidence for impaired T cell DNA methylation in systemic lupus erythematosus and rheumatoid arthritis. *Arthritis Rheum* 33, 1665–1673.

Richardson, B. C., Strahler, J. R., Pivirotto, T. S., Quddus, J., Bayliss, G. E., Gross, L. A., O'Rourke, K. S., Powers, D., Hanash, S. M. & Johnson, M. A. (1992). Phenotypic and functional similarities between 5-azacytidine-treated T cells and a T cell subset in patients with active systemic lupus erythematosus. *Arthritis Rheum* 35, 647–662.

Risques, R. A., Lai, L. A., Brentnall, T. A., Li, L., Feng, Z., Gallaher, J., Mandelson, M. T., Potter, J. D., Bronner, M. P. & Rabinovitch, P. S. (2008). Ulcerative colitis is a disease of accelerated colon aging: evidence from telomere attrition and DNA damage. *Gastroenterology* 135, 410–418.

Ruchusatsawat, K., Wongpiyabovorn, J., Shuangshoti, S., Hirankarn, N. & Mutirangura, A. (2006). SHP-1 promoter 2 methylation in normal epithelial tissues and demethylation in psoriasis. *J Mol Med (Berl)* 84, 175–182.

Saito, S., Kato, J., Hiraoka, S., Horii, J., Suzuki, H., Higashi, R., Kaji, E., Kondo, Y. & Yamamoto, K. (2011). DNA methylation of colon mucosa in ulcerative colitis patients: correlation with inflammatory status. *Inflamm Bowel Dis* 17, 1955–1965.

Saouaf, S. J., Li, B., Zhang, G., Shen, Y., Furuuchi, N., Hancock, W. W. & Greene, M. I. (2009). Deacetylase inhibition increases regulatory T cell function and decreases incidence and severity of collagen-induced arthritis. *Exp Mol Pathol* 87, 99–104.

Sato, F., Harpaz, N., Shibata, D., Xu, Y., Yin, J., Mori, Y., Zou, T. T., Wang, S., Desai, K., Leytin, A., Selaru, F. M., Abraham, J. M. & Meltzer, S. J. (2002a). Hypermethylation of the p14(ARF) gene in ulcerative colitis-associated colorectal carcinogenesis. *Cancer Res* 62, 1148–1151.

Sato, F., Shibata, D., Harpaz, N., Xu, Y., Yin, J., Mori, Y., Wang, S., Olaru, A., Deacu, E., Selaru, F. M., Kimos, M. C., Hytiroglou, P., Young, J., Leggett, B., Gazdar, A. F., Toyooka, S., Abraham, J. M. & Meltzer, S. J. (2002b). Aberrant methylation of the HPP1 gene in ulcerative colitis-associated colorectal carcinoma. *Cancer Res* 62, 6820–6822.

Sawcer, S., Hellenthal, G., Pirinen, M., Spencer, C. C., Patsopoulos, N. A., Moutsianas, L., Dilthey, A., Su, Z., Freeman, C., Hunt, S. E., Edkins, S., Gray, E., Booth, D. R., Potter, S. C., Goris, A., Band, G., Oturai, A. B., Strange, A., Saarela, J., Bellenguez, C., Fontaine, B., Gillman, M., Hemmer, B., Gwilliam, R., Zipp, F., Jayakumar, A., Martin, R., Leslie, S., Hawkins, S., Giannoulatou, E., D'alfonso, S., Blackburn, H., Boneschi, F. M., Liddle, J., Harbo, H. F., Perez, M. L., Spurkland, A., Waller, M. J., Mycko, M. P., Ricketts, M., Comabella, M., Hammond, N., Kockum, I., McCann, O. T., Ban, M., Whittaker, P., Kemppinen, A., Weston, P., Hawkins, C., Widaa, S., Zajicek, J., Dronov, S., Robertson, N., Bumpstead, S. J., Barcellos, L. F., Ravindrarajah, R., Abraham, R., Alfredsson, L., Ardlie, K., Aubin, C., Baker, A., Baker, K., Baranzini, S. E., Bergamaschi, L., Bergamaschi, R., Bernstein, A., Berthele, A., Boggild, M., Bradfield, J. P., Brassat, D., Broadley, S. A., Buck, D., Butzkueven, H., Capra, R., Carroll, W. M., Cavalla, P., Celius, E. G., Cepok, S., Chiavacci, R., Clerget-Darpoux, F., Clysters, K., Comi, G., Cossburn, M., Cournu-Rebeix, I., Cox, M. B., Cozen, W., Cree, B. A., Cross, A. H., Cusi, D., Daly, M. J., Davis, E., de Bakker, P. I., Debouverie, M., D'hooghe, M. B., Dixon, K., Dobosi, R., Dubois, B., Ellinghaus, D., Elovaara, I., Esposito, F., Fontenille, C., Foote, S., Franke, A., Galimberti, D., Ghezzi, A., Glessner, J., Gomez, R., Gout, O., Graham, C., Grant, S. F., Guerini, F. R., Hakonarson, H., Hall, P., Hamsten, A., Hartung, H. P., Heard, R. N., Heath, S., Hobart, J., Hoshi, M., Infante-Duarte, C., Ingram, G., Ingram, W., Islam, T., Jagodic, M., Kabesch, M., Kermode, A. G., Kilpatrick, T. J., Kim, C., Klopp, N., Koivisto, K., Larsson, M., Lathrop, M., Lechner-Scott, J. S., Leone, M. A., Leppa, V., Liljedahl, U., Bomfim, I. L., Lincoln, R. R., Link, J., Liu, J., Lorentzen, A. R., Lupoli, S., Macciardi, F., Mack, T., Marriott, M., Martinelli, V., Mason, D., McCauley, J. L., Mentch, F., Mero, I. L., Mihalova, T., Montalban, X., Mottershead, J., Myhr, K. M., Naldi, P., Ollier, W., Page, A., Palotie, A., Pelletier, J., Piccio, L., Pickersgill, T., Piehl, F., Pobywajlo, S., Quach, H. L., Ramsay, P. P., Reunanen, M., Reynolds, R., Rioux, J. D., Rodegher, M., Roesner, S., Rubio, J. P., Ruckert, I. M., Salvetti, M., Salvi, E., Santaniello, A., Schaefer, C. A., Schreiber, S., Schulze, C., Scott, R. J., Sellebjerg, F., Selmaj, K. W., Sexton, D., Shen, L., Simms-Acuna, B., Skidmore, S., Sleiman, P. M., Smestad, C., Sorensen, P. S., Sondergaard, H. B., Stankovich, J., Strange, R. C., Sulonen, A. M., Sundqvist, E., Syvanen, A. C., Taddeo, F., Taylor, B., Blackwell, J. M., Tienari, P., Bramon, E., Tourbah, A., Brown, M. A., Tronczynska, E., Casas, J. P., Tubridy, N., Corvin, A., Vickery, J., Jankowski, J., Villoslada, P., Markus, H. S., Wang, K., Mathew, C. G., Wason, J., Palmer, C. N., Wichmann, H. E., Plomin, R., Willoughby, E., Rautanen, A., Winkelmann, J., Wittig, M., Trembath, R. C., Yaouanq, J., Viswanathan, A. C., Zhang, H., Wood, N. W., Zuvich, R., Deloukas, P., Langford, C., Duncanson, A., Oksenberg, J. R., Pericak-Vance, M. A., Haines, J. L., Olsson, T., Hillert, J., Ivinson, A. J., De Jager, P. L., Peltonen, L., Stewart, G. J., Hafler, D. A., Hauser, S. L., McVean, G., Donnelly, P. & Compston, A. (2011). Genetic risk and a primary role for cell-mediated immune mechanisms in multiple sclerosis. *Nature* 476, 214–219.

Schroeder, T. M., Nair, A. K., Staggs, R., Lamblin, A. F. & Westendorf, J. J. (2007). Gene profile analysis of osteoblast genes differentially regulated by histone deacetylase inhibitors. *BMC Genomics* 8, 362.

Schroeder, T. M. & Westendorf, J. J. (2005). Histone deacetylase inhibitors promote osteoblast maturation. *J Bone Miner Res* 20, 2254–2263.

Scott, D. L., Wolfe, F. & Huizinga, T. W. (2010). Rheumatoid arthritis. *Lancet* 376, 1094–1108.

Silman, A. J., MacGregor, A. J., Thomson, W., Holligan, S., Carthy, D., Farhan, A. & Ollier, W. E. (1993). Twin concordance rates for rheumatoid arthritis: results from a nationwide study. *Br J Rheumatol* 32, 903–907.

Silman, A. J., Newman, J. & MacGregor, A. J. (1996). Cigarette smoking increases the risk of rheumatoid arthritis. Results from a nationwide study of disease-discordant twins. *Arthritis Rheum* 39, 732–735.

Sreekumar, G. P., Erf, G. F. & Smyth, J. R. (1996). 5-azacytidine treatment induces autoimmune vitiligo in parental control strains of the Smyth line chicken model for autoimmune vitiligo. *Clin Immunol Immunopathol* 81, 136–144.

Stanczyk, J., Ospelt, C., Karouzakis, E., Filer, A., Raza, K., Kolling, C., Gay, R., Buckley, C. D., Tak, P. P., Gay, S. & Kyburz, D. (2011). Altered expression of microRNA-203 in rheumatoid arthritis synovial fibroblasts and its role in fibroblast activation. *Arthritis Rheum* 63, 373–381.

Stastny, P. (1978). Association of the B-cell alloantigen DRw4 with rheumatoid arthritis. *N Engl J Med* 298, 869–871.

Strickland, F. M. & Richardson, B. C. (2008). Epigenetics in human autoimmunity. Epigenetics in autoimmunity - DNA methylation in systemic lupus erythematosus and beyond. *Autoimmunity* 41, 278–286.

Sullivan, K. E. (2000). Genetics of systemic lupus erythematosus. Clinical implications. *Rheum Dis Clin North Am* 26, 229–256.

Sunahori, K., Juang, Y. T., Kyttaris, V. C. & Tsokos, G. C. (2011). Promoter hypomethylation results in increased expression of protein phosphatase 2A in T cells from patients with systemic lupus erythematosus. *J Immunol* 186, 4508–4517.

Tahara, T., Shibata, T., Nakamura, M., Yamashita, H., Yoshioka, D., Okubo, M., Maruyama, N., Kamano, T., Kamiya, Y., Fujita, H., Nakagawa, Y., Nagasaka, M., Iwata, M., Takahama, K., Watanabe, M., Nakano, H., Hirata, I. & Arisawa, T. (2009a). Promoter methylation of protease-activated receptor (PAR2) is associated with severe clinical phenotypes of ulcerative colitis (UC). *Clin Exp Med* 9, 125–130.

Tahara, T., Shibata, T., Nakamura, M., Yamashita, H., Yoshioka, D., Okubo, M., Maruyama, N., Kamano, T., Kamiya, Y., Nakagawa, Y., Fujita, H., Nagasaka, M., Iwata, M., Takahama, K., Watanabe, M., Hirata, I. & Arisawa, T. (2009b). Effect of MDR1 gene promoter methylation in patients with ulcerative colitis. *Int J Mol Med* 23, 521–527.

Takami, N., Osawa, K., Miura, Y., Komai, K., Taniguchi, M., Shiraishi, M., Sato, K., Iguchi, T., Shiozawa, K., Hashiramoto, A. & Shiozawa, S. (2006). Hypermethylated promoter region of DR3, the death receptor 3 gene, in rheumatoid arthritis synovial cells. *Arthritis Rheum* 54, 779–787.

Tao, R., de Zoeten, E. F., Ozkaynak, E., Chen, C., Wang, L., Porrett, P. M., Li, B., Turka, L. A., Olson, E. N., Greene, M. I., Wells, A. D. & Hancock, W. W. (2007). Deacetylase inhibition promotes the generation and function of regulatory T cells. *Nat Med* 13, 1299–1307.

Tolboom, T. C., van der Helm-Van Mil AH, Nelissen, R. G., Breedveld, F. C., Toes, R. E. & Huizinga, T. W. (2005). Invasiveness of fibroblast-like synoviocytes is an individual patient characteristic associated with the rate of joint destruction in patients with rheumatoid arthritis. *Arthritis Rheum* 52, 1999–2002.

Tsai, C. L., Li, H. P., Lu, Y. J., Hsueh, C., Liang, Y., Chen, C. L., Tsao, S. W., Tse, K. P., Yu, J. S. & Chang, Y. S. (2006). Activation of DNA methyltransferase 1 by EBV LMP1 Involves c-Jun NH(2)-terminal kinase signaling. *Cancer Res* 66, 11668–11676.

van den Broek, M. F. & Hengartner, H. (2000). The role of perforin in infections and tumour surveillance. *Exp Physiol* 85, 681–685.

Wang, Y., Fan, P. S. & Kahaleh, B. (2006). Association between enhanced type I collagen expression and epigenetic repression of the FLI1 gene in scleroderma fibroblasts. *Arthritis Rheum* 54, 2271–2279.

Wardemann, H., Yurasov, S., Schaefer, A., Young, J. W., Meffre, E. & Nussenzweig, M. C. (2003). Predominant autoantibody production by early human B cell precursors. *Science* 301, 1374–1377.

Wen, Z. K., Xu, W., Xu, L., Cao, Q. H., Wang, Y., Chu, Y. W. & Xiong, S. D. (2007). DNA hypomethylation is crucial for apoptotic DNA to induce systemic lupus erythematosus-like autoimmune disease in SLE-non-susceptible mice. *Rheumatology (Oxford)* 46, 1796–1803.

Wilson, A. S., Power, B. E. & Molloy, P. L. (2007). DNA hypomethylation and human diseases. *Biochim Biophys Acta* 1775, 138–162.

Xu, J., Zhang, G., Cheng, Y., Chen, B., Dong, Y., Li, L., Xu, L., Xu, X., Lu, Z. & Wen, J. (2011). Hypomethylation of the HTR1A promoter region and high expression of HTR1A in the peripheral blood lymphocytes of patients with systemic lupus erythematosus. *Lupus* 20, 678–689.

Young, A. & Koduri, G. (2007). Extra-articular manifestations and complications of rheumatoid arthritis. *Best Pract Res Clin Rheumatol* 21, 907–927.

Young, D. A., Lakey, R. L., Pennington, C. J., Jones, D., Kevorkian, L., Edwards, D. R., Cawston, T. E. & Clark, I. M. (2005). Histone deacetylase inhibitors modulate metalloproteinase gene expression in chondrocytes and block cartilage resorption. *Arthritis Res Ther* 7, R503–R512.

Yung, R., Kaplan, M., Ray, D., Schneider, K., Mo, R. R., Johnson, K. & Richardson, B. (2001). Autoreactive murine Th1 and Th2 cells kill syngeneic macrophages and induce autoantibodies. *Lupus* 10, 539–546.

Yung, R., Powers, D., Johnson, K., Amento, E., Carr, D., Laing, T., Yang, J., Chang, S., Hemati, N. & Richardson, B. (1996). Mechanisms of drug-induced lupus. II. T cells overexpressing lymphocyte function-associated antigen 1 become autoreactive and cause a lupuslike disease in syngeneic mice. *J Clin Invest* 97, 2866–2871.

Zerrahn, J., Held, W. & Raulet, D. H. (1997). The MHC reactivity of the T cell repertoire prior to positive and negative selection. *Cell* 88, 627–636.

Zhang, K., Zhang, R., Li, X., Yin, G. & Niu, X. (2009). Promoter methylation status of p15 and p21 genes in HPP-CFCs of bone marrow of patients with psoriasis. *Eur J Dermatol* 19, 141–146.

Zhang, K., Zhang, R., Li, X., Yin, G., Niu, X. & Hou, R. (2007). The mRNA expression and promoter methylation status of the p16 gene in colony-forming cells with high proliferative potential in patients with psoriasis. *Clin Exp Dermatol* 32, 702–708.

Zhang, P., Su, Y., Chen, H., Zhao, M. & Lu, Q. (2010a). Abnormal DNA methylation in skin lesions and PBMCs of patients with psoriasis vulgaris. *J Dermatol Sci* 60, 40–42.

Zhang, Z., Song, L., Maurer, K., Petri, M. A. & Sullivan, K. E. (2010b). Global H4 acetylation analysis by ChIP-chip in systemic lupus erythematosus monocytes. *Genes Immun* 11, 124–133.

Zhao, M., Sun, Y., Gao, F., Wu, X., Tang, J., Yin, H., Luo, Y., Richardson, B. & Lu, Q. (2010a). Epigenetics and SLE: RFX1 downregulation causes CD11a and CD70 overexpression by altering epigenetic modifications in lupus CD4+ T cells. *J Autoimmun* 35, 58–69.

Zhao, M., Tang, J., Gao, F., Wu, X., Liang, Y., Yin, H. & Lu, Q. (2010b). Hypomethylation of IL10 and IL13 promoters in CD4+ T cells of patients with systemic lupus erythematosus. *J Biomed Biotechnol* 2010, 931018.

Zhao, M., Wu, X., Zhang, Q., Luo, S., Liang, G., Su, Y., Tan, Y. & Lu, Q. (2010c). RFX1 regulates CD70 and CD11a expression in lupus T cells by recruiting the histone methyltransferase SUV39H1. *Arthritis Res Ther* 12, R227.

Zhao, S., Wang, Y., Liang, Y., Zhao, M., Long, H., Ding, S., Yin, H. & Lu, Q. (2011). MicroRNA-126 regulates DNA methylation in CD4+ T cells and contributes to systemic lupus erythematosus by targeting DNA methyltransferase 1. *Arthritis Rheum* 63, 1376–1386.

Zhou, X., Hua, X., Ding, X., Bian, Y. & Wang, X. (2011). Trichostatin differentially regulates Th1 and Th2 responses and alleviates rheumatoid arthritis in mice. *J Clin Immunol* 31, 395–405.

Zhou, Y., Yuan, J., Pan, Y., Fei, Y., Qiu, X., Hu, N., Luo, Y., Lei, W., Li, Y., Long, H., Sawalha, A. H., Richardson, B. & Lu, Q. (2009). T cell CD40LG gene expression and the production of IgG by autologous B cells in systemic lupus erythematosus. *Clin Immunol* 132, 362–370.

Zhu, X., Liang, J., Li, F., Yang, Y., Xiang, L. & Xu, J. (2011). Analysis of associations between the patterns of global DNA hypomethylation and expression of DNA methyltransferase in patients with systemic lupus erythematosus. *Int J Dermatol* 50, 697–704.

Chapter 12
Imprinting Disorders

Thomas Eggermann

Abbreviations

AS	Angelman syndrome
BWS	Beckwith–Wiedemann syndrome
DMR	Differentially methylated region
ID	Imprinting disorder
ICR	Imprinting control region
MLMD	Multilocus methylation defect
PWS	Prader–Willi syndrome
TNDM	Transient neonatal diabetes mellitus
UPD	Uniparental disomy
UPhD	Uniparental heterodisomy
UPiD	Uniparental isodisomy
upd(6)pat	Paternal uniparental disomy of chromosome 6
upd(7)mat	Maternal uniparental disomy of chromosome 7

12.1 Introduction

The term *genomic imprinting* describes an epigenetic marking of specific genes that allows expression from only one of the two paternal alleles (reviewed by Reik and Walter 2001). So far, more than 60 human genes are discussed to be imprinted by epigenetic mechanisms, but probably there are much more (reviewed by Horsthemke 2010). The imprinting marks are inherited from the parental gametes and are then

---

T. Eggermann (✉)

Institut für Humangenetik, RWTH Aachen, Pauwelsstrasse 30, 52074 Aachen, Germany
e-mail: teggermann@ukaachen.de

J. Minarovits and H.H. Niller (eds.), *Patho-Epigenetics of Disease*,
DOI 10.1007/978-1-4614-3345-3_12, © Springer Science+Business Media New York 2012

maintained in the somatic cells of an individual. Their programming is subject to a so-called "imprinting" cycle during life which leads to a reprogramming at each generation: during early development, methylation of the mammalian genome runs through dramatic changes and is linked to the rapid differentiation and formation of the various tissues and organs. The imprint marks are erased in the germ line and re-established according to the sex of the contributing parent for the next generation.

Genes regulated by genomic imprinting mechanisms tend to cluster; thus, the imprinting control is often not restricted to a single gene at an imprinted locus but affects the expression of several factors. On the molecular level, the expression of the involved genes within imprinted regions is influenced by specific patterns of DNA methylation, by changes in chromatin structure and by post-translational histone modifications such as acetylation, ubiquitylation, phosphorylation and methylation (reviewed by Delaval et al. 2006; Kacem and Feil 2009).

Due to the numerous factors involved in these complex mechanisms, the balanced regulation of imprinted genes is prone to different disturbances, and indeed several disorders associated with altered genomic imprinting are known, belonging to the group of congenital *imprinting disorders* (IDs).

The biological function of genomic imprinting in mammals is currently unknown. It has been suggested that they serve as mediators of the "battle of sexes" (conflict theory) in the foetal period, and indeed many paternally and maternally imprinted genes have obviously opposite functions in (early) development: whereas paternally expressed factors promote growth, maternally expressed ones suppress it (Haig and Graham 1991; Moore and Haig 1991).

This hypothesis is corroborated by the observation that the majority of the known IDs are associated with disturbed growth.

12.2 Molecular Alterations in IDs

In nearly all known IDs, the same genetic and epigenetic alterations affecting imprinted genes/gene clusters are detectable. Currently four different types of molecular disturbances are known (Fig. 12.1): (a) epimutations, i.e. aberrant methylation without alteration of the genomic DNA sequence; (b) uniparental disomy (UPD); (c) chromosomal aberrations and (d) point mutations in imprinted genes.

12.2.1 Epimutations

Epimutations, or imprinting defects, describe altered DNA methylation patterns at specific differentially methylated regions (DMRs) which regulate the expression of neighbouring genes. One of the major imprinted regions in humans is localised in the short arm of chromosome 11 (11p15) (Fig. 12.1). Several of the genes localised in

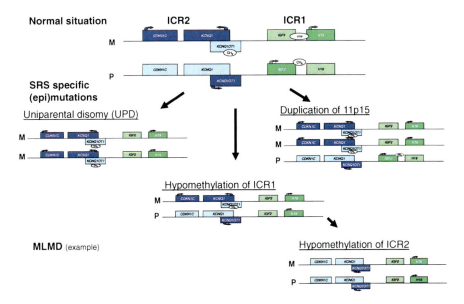

Fig. 12.1 Epigenetic/genetic alterations causing IDs, illustrated for the 11p15 region. Disturbances for this region are associated with SRS (as shown here) and with BWS. Only those genes in 11p15 are illustrated which are currently discussed to be functionally involved in the pathogenesis of SRS and BWS. As explained in the text, in some SRS patients, both ICRs are affected by hypomethylation, a situation which is frequent in MLMD carriers

11p15 are involved in human growth and development as well as in tumourigenesis. As a result, the two IDs associated with 11p15 alterations—Silver–Russell syndrome (SRS) and Beckwith–Wiedemann syndrome (BWS)—are clinically characterised by disturbed growth and, in case of BWS, associated with an increased risk for tumours.

The region 11p15 contains a number of imprinted genes, the expression of which is regulated by two different imprinting control regions (ICR1 and ICR2). In case of the telomeric ICR1, a differential chromatin architecture of the parental alleles leads to a reciprocal expression of *H19* and *IGF2* (insulin-like growth factor 2). The ICR1 contains seven CTCF target sites in the DMR 2-kb upstream of *H19* and shows allele-specific methylation. The zinc finger-binding factor CTCF binds to the maternal unmethylated ICR1 copy and thereby forms a chromatin boundary. This CTCF-binding mechanism blocks *IGF2* and promotes *H19* transcription of the maternal 11p15 copy. Whereas the paternally expressed *IGF2* is involved in foetal development and growth (DeChiara et al. 1990; Constancia et al. 2002), the function of *H19* is unknown although it was one of the first noncoding transcripts identified. The fact that *H19* is a relatively highly conserved gene among mammals suggests a profound functional relevance. A recent study indicates that *H19* functions as a primary microRNA precursor involved in the post-transcriptional downregulation of specific mRNAs during vertebrate development (Cai and Cullen 2007).

The centromeric ICR2 in 11p15 regulates the expression of *CDKN1C, KCNQ1* (*potassium channel KQT-family member 1*) and further genes and is methylated only on the maternal allele. Mutations in the paternally suppressed *CDKN1C* gene account for up to 50% of familial BWS cases and 5% of sporadic patients (Table 12.1). The gene encodes a cyclin-dependent kinase inhibitor (p57^{KIP2}) and is part of the p21^{CIP2}Cdk inhibitor family. The gene of a second non-coding RNA in 11p15, *KCNQ1OT1 (LIT1)*, is localised in intron 10 of the *KCNQ1* gene. *KCNQ1OT1* is expressed by the paternal allele and probably represses the expression of the *CDKN1C* gene. Loss of methylation (LOM) of the maternal ICR2 allele correlates with expression of *KCNQ1OT1*. In BWS, one central physiological change caused by ICR2 (epi)mutations (hypomethylation at ICR2 as well as *CDKN1C* point mutations) is the reduced expression of *CDKN1C*.

The complex regulation and interaction mechanisms in 11p15 make this region prone to epigenetic and genetic aberrations. As mentioned before, genomic and epigenetic alterations are associated with BWS and SRS (Table 12.1). In a considerable number of BWS patients (50%), a hypomethylation of the ICR2 is detectable, whereas 4% show a hypermethylation of the ICR1. The latter is predominantly hypomethylated in SRS and accounts for 40% of patients. In addition, aberrant methylation of the two ICRs has meanwhile been detected in further IDs (see 12.4).

12.2.2 Uniparental Disomy

UPD is defined as the inheritance of both copies of a chromosome or a chromosomal segment from only one parent in contrast to the regular biparental inheritance. The significance of UPD to decipher the molecular basis of congenital IDs is obvious as the genomic loci for nearly all IDs have been detected by UPD.

Several ways of UPD formation have been reported or hypothesised, the most prominent one is trisomic rescue (reviewed by Kotzot and Utermann 2005; Yamazawa et al. 2010). Here the first step is a meiotic non-disjunction resulting in a trisomic zygote. The only chance for many trisomic zygotes to survive is the subsequent loss of one of the supernumerary chromosomes. In one third of these rescues, the chromosome inherited from the parent not contributing the additional chromosome will be lost, resulting in a UPD.

In the majority of cases, UPD affects the whole chromosome, but meanwhile several UPDs affecting only parts of a chromosome have been reported ("segmental UPD"). In particular in BWS, segmental UPD of chromosome 11p15 accounts for up to 20% of cases (Table 12.1).

Dependent on the mode of its formation, two types of UPD can be distinguished: uniparental heterodisomy (UPhD) and uniparental isodisomy (UPiD). UPhD is defined as the presence of the two different homologous chromosomes from the same parent, whereas in UPiD, two copies of the identical chromosome are inherited.

Table 12.1 Overview on the currently known IDs, their estimated frequencies and the associated molecular alterations

ID	Chromosome	Frequency	Uniparental disomy (upd)	Deletion/duplication	Aberrant locus-specific methylation	Point mutations	MLMD
TNDM	6q24	1/800,000	upd(6q24)pat 40%	dup pat 40%	Hypomethylation of *ZAC1*: ~20%	ND	Yes
SRS	7	1/10,000	upd(7)mat 10%	dup(7p)mat: <1% del(7q)mat: <1%	–	ND	
	11p15		upd(11p15)mat: <1%	dup(11p)mat: <1%	Hypomethylation of ICR1: ~40%	ND	Yes
BWS	11p15	1/15,000	upd(11p15)pat: ~20%	dup(11p)pat: 2–4%	Hypermethylation of ICR1: 4% Hypomethylation of ICR2: 50%	*CDKN1C*: 5–50%[a]	Yes
Upd(14)mat	14q32	Rare	Single cases	Single cases	Hypermethylation of *DLK1/GTL2*: single cases	?	Too rare
Upd(14)pat	14q32	Rare	Single cases	Single cases	Hypomethylation of *DLK1/GTL2*: single cases	?	Too rare
PWS	15q11-q13	1/10,000 –1/25,000	upd(15)mat: <30%	del(15)pat: 70%	1%	ND	1 case
AS	15q11-q13	1/12,000 –1/20,000	upd(15)pat: 1–3%	del(15)mat: 70%	4%	*UBE3A*: 10%	no
PHPIb	20q13	Rare	ND	ND	GNAS: ?	ND	ND

ND not yet described

[a]*CDKN1C* mutations are detectable in 5% of sporadic, but in up to 50% of familial cases

Table 12.2 Overview on UPDs for the single human chromosomes reported so far

Chromosome	Maternal UPD	Paternal UPD
1	–	–
2	–	–
3	–	?
4	–	?
5	?	–
6	–	Transient diabetes mellitus
7	Silver–Russell syndrome	–
8	–	–
9	–	–
10	–	–
11	Silver–Russell syndrome	Beckwith–Wiedemann syndrome
12	–	?
13	–	–
14	UPD(14)mat syndrome	UPD(14)pat syndrome
15	Prader–Willi syndrome	Angelman syndrome
16	–	–
17	–	?
18	?	?
19	?	?
20	4 cases, specific phenotype?; (PHPIb)	–
21	–	–
22	–	–
X	–	–

So far, seven IDs are known, while numerous paternal as well as maternal UPDs have been excluded to cause specific IDs ("–"). However, some UPDs have not yet been reported ("?")

The phenotype of UPD carriers can be influenced by three mechanisms:

1. UPD can affect imprinted genes leading to their imbalanced expression. Indeed, UPD is known for nearly all human chromosomes, but due to the tendency of imprinted genes to cluster, only some UPDs are associated with specific phenotypes (Table 12.2).
2. In case a heterozygote mutation carrier of an autosomal recessive disease has a UPiD offspring, this child might be homozygous for the inherited mutated allele and will be affected by the respective disease. Meanwhile, several cases with unexpected homozygosity for a recessive allele due to UPiD have been described; therefore, a reduction to homozygosity of recessive alleles has to be considered in case of UPD.
3. As "trisomic rescue" is probably the most frequent mechanism of UPD formation and occurs post-zygotically, UPD patients might carry two cell lines with different chromosomal constitutions: a disomic but uniparental one and the original trisomic cell line. Dependent on the time of trisomic rescue and the affected tissues, such a mosaicism might remain undetected as many diagnostic tests aim only on one tissue. The clinical outcome of this mosaic constitution can therefore be influenced by the UPD as well as by the trisomy mosaicism. In this situation, a differentiation between symptoms caused by UPD or by the chromosomal mosaicism is difficult.

Due to the formation mechanism of UPD as a chromosomal disturbance, UPD testing should be considered in situations with chromosomal complements predisposing for UPD. This is the case after prenatal chorionic villus sampling where trisomy mosaicism is a relatively frequent finding and in carriers of (Robertsonian) translocations involving those chromosomes harbouring imprinted regions and genes.

12.2.3 Chromosomal Imbalances: Deletions and Duplications

The functional results of chromosomal imbalances affecting imprinted genes mirror those of UPD. The balanced expression of imprinted genes is disturbed by both disturbances, and they represent a substantial group of molecular defects in IDs as they account for up to 70% of cases (Table 12.1).

Whereas duplications or deletions affecting imprinted regions are rare in SRS, paternal duplications of the region 11p15 can be detected in nearly 20% of BWS patients. A similar frequency of paternal 6q24 duplications has been reported for transient neonatal diabetes mellitus (TNDM). In Prader–Willi and Angelman syndrome (PWS, AS), deletions affecting either the paternal or the maternal chromosome 15 are present in 70% of patients.

Similar to the profound role of chromosomal aberrations for the identification of genomic loci for monogenic disorders, deletions and duplications in IDs help to discover chromosomal regions relevant for the disease and for the understanding of imprinting and its regulation in general. A good example is the phenotypic outcome in patients with duplications of maternal chromosome 11p15 material of different sizes: whereas carriers of maternal duplications of 11p15 material including both ICR1 and ICR2 show SRS features, patients with smaller duplications harbouring only the maternal ICR1 copy are healthy (Bliek et al. 2009b). In case the maternal duplication affects only parts of the ICR1, i.e. the *H19* but not the *IGF2* gene, the patients suffer from SRS (Demars et al. 2011). It has been postulated that in this case, the functional maternal *H19* gene copy influences the *IGF2* gene expression by trans-acting mechanisms, while in case the whole ICR1 is duplicated, the *H19* copy acts in *cis*. In larger duplications including both ICRs, ICR1 and ICR2 in 11p15, there is evidence that the ICR2 is the functionally relevant region, illustrated by a patient with a duplication restricted to the ICR2 (Schonherr et al. 2007).

12.2.4 Point Mutations in Single Genes Causing IDs

A further group of alterations resulting in IDs are genomic point mutations. These mutations affect either genes underlying a regulation through imprinting centres or genes involved in the life cycle of imprinting.

In BWS and Angelman syndrome (AS), base pair mutations in genes regulated by DMRs significantly contribute to their aetiology. *CDKN1C* mutations are responsible for a large proportion of familial BWS cases (up to 50%), whereas *UBE3A1* mutations can be detected in 10–15% of AS patients (Table 12.1).

Genes responsible for the establishment or maintenance of methylation marks are good candidates to carry mutations causing aberrant methylation. Indeed, a growing number of factors involved in DNA methylation and its regulation have been identified, but so far only single cases have been published to carry point mutations in the respective genes (reviewed by Eggermann et al. 2011). In TNDM patients with multilocus hypomethylation, autosomal recessively inherited mutations in the zinc-finger protein *ZFP57* have been reported (Mackay et al. 2008). In a BWS sibship, a homozygous mutation in the *NLRP2* gene in the mother was described by (Meyer et al. 2009), providing evidence for a *trans* mechanism for the disturbed methylation pattern in 11p15 in the two children, caused by a maternal-effect mutation. However, screening studies in further enzymes involved in locus-specific methylation and regulation such as *DNMT3L* and *PLAGL1* failed to identify pathogenic variants (Jager et al. 2009; Bliek et al. 2009c).

12.3 Clinical Characteristics of IDs

Meanwhile, eight different congenital IDs have been defined (Table 12.1). As each of them is characterised by specific clinical features and appeared to be associated with specific imprinting defects, they are regarded as separate entities. However, the phenotypic transitions are fluid between several IDs; furthermore, similar (and even the same) molecular defects are detectable.

Indeed, the majority of IDs have similar clinical characteristics, i.e.:

- Pre- and/or postnatal growth retardation.
- Hypo- or hyperglycaemia.
- Failure to thrive in the newborn and early childhood period.
- Neurological abnormalities in childhood.

In particular in growth-retarded patients, it is therefore difficult to decide which ID-specific test should be applied.

Furthermore, there are ID-specific observations as the underlying mutations often occur post-zygotically:

- Asymmetry of body, head and/or limbs.
- Discordant monozygotic twins.
- Most cases are sporadic; familial cases are rare.
- A non-Mendelian inheritance.
- Genotype–phenotype correlations are difficult to delineate.

For genetic counselling of ID families, the knowledge of the nature of the molecular mutation or epimutation subtype is essential to delineate exact risk figures. Whereas the recurrence risk is generally low in case of epimutations and UPD, patients/carriers with deletions or duplications might have a 50% risk of conceiving a child with an ID, depending on the sex of the contributing patient. However, in each case, genetic counsellors are advised to continually update their knowledge for each disease.

12.3.1 Angelman and Prader–Willi Syndromes

Both AS and PWS are caused by (epi)mutations in 15q11-q13. The lack of the paternal copy of this region results in PWS, while disturbances of the maternal copy lead to AS.

AS patients as well as PWS patients are mentally retarded, but further clinical signs are different. PWS is clinically characterised by neonatal hypotonia and failure to thrive, then hyperphagia and obesity develop. Hypogonadism, short stature, behaviour problems and mild to moderate mental retardation are further characteristics (reviewed by Goldstone 2004). AS patients exhibit microcephaly, ataxia, seizures, absence of speech and sleep disorder (reviewed by Williams et al. 2006). Due to the high percentage of microdeletions in 15q11-q13 in both syndromes, AS and PWS also belong to the so-called microdeletion syndromes. In AS, ~70% of patients have de novo deletions affecting the maternal chromosome. The same frequency can be observed in PWS, but here the paternal chromosome 15 carries the deletion. Further AS-specific (epi)mutations include *UBE3A1* mutations, imprinting defects and paternal UPD(15). In PWS, maternal UPD(15) is frequent and accounts for <30% of patients, while imprinting defects are rare (~1%).

12.3.2 Silver–Russell Syndrome

SRS is mainly characterised by pre- and postnatal growth restriction. The children are relatively macrocephalic, and their face is triangular-shaped with a broad forehead and a pointed, small chin. In many cases, asymmetry of limbs and body and clinodactyly V is present. Growth failure is often accompanied by severe failure to thrive; severe feeding difficulties in early childhood are frequent. For those children without catch-up growth by the age of 2, growth hormone therapy is encouraged.

The genetic basis of SRS is very heterogeneous. In approximately 10% of SRS patients, a maternal UPD for chromosome 7 (upd(7)mat) can be found (reviewed by Eggermann et al. 2008). More than 40% of SRS patients show a hypomethylation of the ICR1 in the imprinted region 11p15 (Fig. 12.1); in single cases, maternal duplications of the whole chromosomal region in 11p15 have been reported (Gicquel et al. 2005; reviewed by Eggermann et al. 2008). Until now only one SRS patient with a maternal duplication restricted to the ICR2 has been identified (Schonherr et al. 2007). Upd(11)mat has been reported only once (Bullman et al. 2008). Interestingly, the opposite 11p15 epigenetic and genetic findings can be observed in BWS (Table 12.1, see below). Numerous (submicroscopic) chromosomal disturbances have been described in SRS patients; thus, screening for cryptic genomic imbalances is indicated after exclusion of upd(7)mat and 11p15 epimutations (Bruce et al. 2010; Spengler et al. 2010). In particular, the analysis should be focused on the chromosomal region 11p15. Generally, the recurrence risk for SRS is low because the majority of patients are sporadic. Nevertheless, the situation changes in case of familial genetic or epigenetic alterations like chromosomal rearrangements or untypical aberrant methylation patterns.

A genotype–phenotype correlation is difficult in SRS. In general, the 11p15 epimutation carriers show the more characteristic phenotype, while upd(7)mat patients are affected more mildly, but exceptions exist. Interestingly, there is evidence for a correlation between the genotype/epigenotype and endocrinological parameters (Binder et al. 2008).

12.3.3 Beckwith–Wiedemann Syndrome

BWS was initially called EMG syndrome from its three main features of exomphalos, macroglossia and (neonatal) gigantism. Additional signs include neonatal hypoglycaemia, hemihypertrophy, organomegaly, earlobe creases, polyhydramnios, haemangioma and cardiomyopathy. In 5–7% of children, embryonal tumours (most commonly Wilms' tumour) are diagnosed. The clinical diagnosis of BWS is often difficult due to its variable presentation and the phenotypic overlap with other overgrowth syndromes (reviewed by Enklaar et al. 2006).

In nearly 70% of BWS patients, an altered expression or mutations of several loci in 11p15 can be observed with a preponderance of an ICR2 hypomethylation accounting for nearly 50% of cases. UPD(11p15)pat is the second important alteration, while ICR1 hypermethylation is rare. Most BWS cases are sporadic, but familial inheritance is observed in 15% of all cases. In BWS families without aberrant 11p15 methylation, *CDKN1C* point mutations are frequent. These BWS pedigrees resemble that of an autosomal dominant inheritance with incomplete penetrance. Interestingly, an increased frequency of monozygotic twinning has been reported in BWS families with epimutations (reviewed by Bliek et al. 2009a). It was therefore hypothesised that a methylation error proceeds and possibly triggers twinning. In some monozygotic but discordant twin pairs, aberrant methylation is also detectable in leukocytes of the unaffected twin but not in other tissues of the healthy child. This discrepant finding has been explained by intrauterine sharing of hematopoietic stem cells due to the frequent vascular connections between monozygotic co-twins (Hall 2003).

A genotype/epigenotype–phenotype correlation has recently been established for BWS (Cooper et al. 2005): hemihypertrophy is strongly associated with upd(11) pat, exomphalos with ICR2 hypomethylation and *CDKN1C* mutations. Most importantly, the risk of neoplasias is significantly higher in ICR1 hypermethylation and upd(11)pat than in the other molecular subgroups. In BWS, the determination of the molecular subtype is therefore important for an individual prognosis and therapy. Nevertheless, the phenotypic transitions are fluid, and testing for all molecular subtypes should be offered in patients with BWS features.

12.3.4 Transient Neonatal Diabetes Mellitus

TNDM is a rare disease; in addition to hypoglycaemia, IUGR and abdominal wall defects are common (Temple et al. 1996). Insulin therapy is required for a an average of 3 months, afterwards the diabetes resolves, but later in life, the majority of TNDM patients develop type 2 diabetes.

TNDM is associated with an overexpression of the imprinted locus *PLAGL1/ZAC* in 6q24. Like in the other IDs, three (epi)genetic causes of TNDM have been identified: upd(6)pat, paternal duplications of 6q24 and aberrant methylation at the *PLAGL1/ZAC* locus. The *PLAGL1/ZAC* gene is a maternally imprinted gene and therefore only expressed from the paternal allele. It encodes a zinc-finger protein which binds DNA and hence influences the expression of other genes (reviewed by Abdollahi 2007).

12.3.5 Maternal and Paternal upd(14) Syndromes/upd(14)mat/pat

The upd(14)mat and upd(14)pat syndromes have firstly been described in 1991 by Temple et al. and Wang et al.; meanwhile, distinct clinical phenotypes have been defined. However, the frequencies of both syndromes are currently unknown.

Both IDs were firstly detected in patients carrying balanced Robertsonian translocations. Considering the most important formation mechanism of UPD via trisomy rescue, this observation was consequent because Robertsonian translocations are prone to trisomic offspring. Meanwhile, several cases with isolated IDs and microdeletions affecting the *DLK1/GTL2* locus in 14q32 have been described (Temple et al. 2007; Kagami et al. 2008), resulting in the same phenotypes; thus, new names for these two syndromes are necessary (Buiting et al. 2008).

The upd(14)mat phenotype is characterised by prenatal and postnatal growth retardation, muscular hypotonia, feeding difficulties, small hands and feet, recurrent otitis media, joint laxity, motor delay, truncal obesity and early onset of puberty. The facial gestalt comprises a prominent forehead, a bulbous nasal tip and a short philtrum. Patients with upd(14)mat show clinical features overlapping with PWS; thus, screening for upd(14)mat should be performed in patients with PWS-like phenotype after exclusion of the PWS-specific (epi)mutations (Mitter et al. 2006).

Upd(14)pat is associated with polyhydramnios; a typical small, bell-shaped thorax; abdominal wall defects; and a severe developmental delay. The majority of patients die in utero or in the first months of life. In addition to upd(14)pat, isolated methylation defects at the *DLK1/GTL2* locus have meanwhile been identified in upd(14)pat patients.

In all cases reported so far, the *DLK1/GTL2* locus is affected. The paternally expressed gene *DLK1* (delta, Drosophila homologue-like 1) encodes a transmembrane signalling protein; the maternally expressed *GTL2* (gene trap locus 2) is a microRNA which is involved in transcription regulation. However, the functional link between these genes and the phenotypes is currently unknown.

12.3.6 Pseudohypoparathyroidism Ib and the GNAS Locus (Chromosome 20)

Pseudohypoparathyroidism type 1 (PHPIb) is a further human ID, caused specifically by aberrant DNA methylation of a specific DMR of the GNAS locus or microdeletion in *cis* on chromosome 20q13.11 (reviewed by Kelsey 2010). *GNAS* is a complex imprinted locus that encodes several transcripts by alternative promoters and splicing. Some loci are expressed biallelically, others exclusively either from the paternal or the maternal *GNAS* allele.

To date, only three patients with a upd(20)mat have been reported (reviewed by Bastepe and Juppner 2005). All patients were characterised by severe pre- and postnatal growth retardation, but none of them showed features belonging to the *GNAS* locus mutation spectrum.

12.4 Multilocus Methylation Defects in IDs

In the last 6 years, a growing number of patients with congenital IDs have been reported exhibiting aberrant methylation at multiple imprinted loci (multilocus methylation defects, MLMD), i.e. not only the disease-specific locus was affected but also other imprinting domains showed an aberrant methylation (Table 12.1).

In 2006, the first patients with TNDM exhibiting hypomethylation at further maternally imprinted loci in addition to the disease-specific locus in 6q24 have been identified (Arima et al. 2005; Mackay et al. 2006). Two years later, Mackay et al. (2008) reported on seven TNDM/MHS pedigrees with homozygosity or compound heterozygosity for *ZFP57* mutations as the cause of the first heritable global human imprinting disorder. *ZFP57* point mutations are autosomal recessive, a finding that is important for genetic counselling of TNDM. The variable LOM mosaicism in the *ZFP57* mutation carriers suggests that it has a role in post-fertilisation maintenance of maternal and paternal DNA methylation imprints (Li et al. 2008).

Meanwhile, MLMD has been reported in numerous patients with BWS or SRS and in a single PWS patient, but in comparison to the TNDM/MLMD cases, both paternally and maternally imprinted loci are affected (multilocus methylation defect, MLMD) (Fig. 12.1) (Rossignol et al. 2006; Azzi et al. 2009; Lim et al. 2009; Bliek et al. 2009a, c; Turner et al. 2010; Baple et al. 2011; Begemann et al. 2011).

Clinically, TNDM/MLMD patients show an altered TNDM phenotype with a less severe growth retardation; omphalocoele and macroglossia are further features reminiscent to BWS (Mackay et al. 2006). In contrast, SRS and BWS patients with MLMD do not differ clinically from patients with isolated imprinting defects in the ICR1 or ICR2 in 11p15. Moreover, MLMD patients with the same aberrant methylation patterns in lymphocytes but presenting either BWS or SRS have been identified (Azzi et al. 2009). The reason for this striking observation is currently unknown; Azzi and colleagues suggested that the most severe epimutation,

12 Imprinting Disorders

in this case the lowest level of methylation, is dominant and determines the clinical outcome while it masks the expression of other phenotypes (Azzi et al. 2009). Furthermore, the studies published so far are based on lymphocytes' DNA; thus, a mosaic distribution of epimutations in other tissues influencing the phenotypic expression would not be detectable.

12.5 Molecular Genetic Testing of IDs

With the growing knowledge on IDs and the rapid development of new high-throughput technologies, genetic testing for epimutations and mutations in IDs has significantly improved. In particular, the development of bisulphite treatment approaches to differentiate between methylated and unmethylated DNA was helpful to establish fast, reliable and low-cost strategies. Locus-specific Southern-blot assays have been replaced by different PCR-based methods. With the use of PCR, only minimal amounts of patients' DNA are necessary; this aspect is particularly important for testing of neonates and deceased patients.

While techniques like methylation-specific PCRs and bisulphite sequencing allow the targeted analysis of single loci, methylation-specific *multiplex ligation probe-dependent amplification* (MS-MLPA) or methylation-specific single nucleotide primer extension (MS-SNuPE) assays now make the parallel characterisation of different loci as well as of different types of (epi)mutations (duplications, UPD, aberrant methylation) in single tube reactions possible. Thus, the identification of MLMD by simple and efficient methods becomes possible; the analysis locus by locus is not longer necessary.

The molecular confirmation of the clinical diagnosis allows a more reliable prognosis and a better directed therapy, e.g. in BWS, the risk for neoplasias can be determined more precisely and specific therapeutic options can be considered. Furthermore, the identification of genetic or epigenetic mutations is relevant not only for the patient himself but also for the family as it allows recurrence risk figures in genetic counselling.

Nevertheless, prior to genetic testing, the significance of genetic testing for the patient and his relatives should be critically discussed with the families. For each family, an individual strategy is necessary to avoid misleading and unclear results. Additionally, the putative predictive nature of a genetic test for affected as well as unaffected family members should be considered.

12.6 Conclusions

In summary, there is a growing number of conditions where genomic imprinting alterations are recognised to be associated with clinical disorders. Based on the observation that growth disturbance and behaviour abnormalities are common features of IDs, aberrant genomic imprinting should be suspected in any disorder of unknown aetiology characterised by these clinical signs. Furthermore, in disorders with unusual pattern of inheritance, genomic imprinting should also be considered. Despite the current lack of understanding of the functional basis of the known genetic/epigenetic alterations, the identification of these (epi)mutations is of major importance in terms of clinical prognosis and therapeutic options, and recurrence risks in a family.

New technologies such as microarrays and next generation sequencing together with functional analyses will contribute to the understanding of the pathophysiology of IDs.

IDs represent models for deciphering epigenetic regulation, and research on IDs will provide us with profound insights in the aetiology of many complex biological processes such as growth. Thereby we will be able to deduce the contribution of epigenetic changes to complex human disorders such as cancer and psychiatric diseases.

References

Abdollahi, A. (2007). LOT1 (ZAC1/PLAGL1) and its family members: mechanisms and functions. *J Cell Physiol* 210, 16–25.

Arima, T., Kamikihara, T., Hayashida, T., Kato, K., Inoue, T., Shirayoshi, Y., Oshimura, M., Soejima, H., Mukai, T. & Wake, N. (2005). ZAC, LIT1 (KCNQ1OT1) and p57KIP2 (CDKN1C) are in an imprinted gene network that may play a role in Beckwith-Wiedemann syndrome. *Nucleic Acids Res* 33, 2650–2660.

Azzi, S., Rossignol, S., Steunou, V., Sas, T., Thibaud, N., Danton, F., Le Jule, M., Heinrichs, C., Cabrol, S., Gicquel, C., Le Bouc, Y. & Netchine, I. (2009). Multilocus methylation analysis in a large cohort of 11p15-related foetal growth disorders (Russell Silver and Beckwith Wiedemann syndromes) reveals simultaneous loss of methylation at paternal and maternal imprinted loci. *Hum Mol Genet* 18, 4724–4733.

Baple, E. L., Poole, R. L., Mansour, S., Willoughby, C., Temple, I. K., Docherty, L. E., Taylor, R. & Mackay, D. J. (2011). An atypical case of hypomethylation at multiple imprinted loci. *Eur J Hum Genet* 19, 360–362.

Bastepe, M. & Juppner, H. (2005). GNAS locus and pseudohypoparathyroidism. *Horm Res* 63, 65–74.

Begemann, M., Spengler, S., Kanber, D., Haake, A., Baudis, M., Leisten, I., Binder, G., Markus, S., Rupprecht, T., Segerer, H., Fricke-Otto, S., Muhlenberg, R., Siebert, R., Buiting, K. & Eggermann, T. (2011). Silver-Russell patients showing a broad range of ICR1 and ICR2 hypomethylation in different tissues. *Clin Genet* 80, 83–88.

Binder, G., Seidel, A. K., Martin, D. D., Schweizer, R., Schwarze, C. P., Wollmann, H. A., Eggermann, T. & Ranke, M. B. (2008). The endocrine phenotype in Silver-Russell syndrome is defined by the underlying epigenetic alteration. *J Clin Endocrinol Metab* 93, 1402–1407.

Bliek, J., Alders, M., Maas, S. M., Oostra, R. J., Mackay, D. M., van der, L. K., Callaway, J. L., Brooks, A., van 't, P. S., Westerveld, A., Leschot, N. J. & Mannens, M. M. (2009a). Lessons from BWS twins: complex maternal and paternal hypomethylation and a common source of haematopoietic stem cells. *Eur J Hum Genet* 17, 1625–1634.

Bliek, J., Snijder, S., Maas, S. M., Polstra, A., van der, L. K., Alders, M., Knegt, A. C. & Mannens, M. M. (2009b). Phenotypic discordance upon paternal or maternal transmission of duplications of the 11p15 imprinted regions. *Eur J Med Genet* 52, 404–408.

12 Imprinting Disorders

Bliek, J., Verde, G., Callaway, J., Maas, S. M., De Crescenzo, A., Sparago, A., Cerrato, F., Russo, S., Ferraiuolo, S., Rinaldi, M. M., Fischetto, R., Lalatta, F., Giordano, L., Ferrari, P., Cubellis, M. V., Larizza, L., Temple, I. K., Mannens, M. M., Mackay, D. J. & Riccio, A. (2009c). Hypomethylation at multiple maternally methylated imprinted regions including PLAGL1 and GNAS loci in Beckwith-Wiedemann syndrome. *Eur J Hum Genet* 17, 611–619.

Bruce, S., Hannula-Jouppi, K., Puoskari, M., Fransson, I., Simola, K. O., Lipsanen-Nyman, M. & Kere, J. (2010). Submicroscopic genomic alterations in Silver-Russell syndrome and Silver-Russell-like patients. *J Med Genet* 47, 816–822.

Buiting, K., Kanber, D., Martin-Subero, J. I., Lieb, W., Terhal, P., Albrecht, B., Purmann, S., Gross, S., Lich, C., Siebert, R., Horsthemke, B. & Gillessen-Kaesbach, G. (2008). Clinical features of maternal uniparental disomy 14 in patients with an epimutation and a deletion of the imprinted DLK1/GTL2 gene cluster. *Hum Mutat* 29, 1141–1146.

Bullman, H., Lever, M., Robinson, D. O., Mackay, D. J., Holder, S. E. & Wakeling, E. L. (2008). Mosaic maternal uniparental disomy of chromosome 11 in a patient with Silver-Russell syndrome. *J Med Genet* 45, 396–399.

Cai, X. & Cullen, B. R. (2007). The imprinted H19 noncoding RNA is a primary microRNA precursor. *RNA* 13, 313–316.

Constancia, M., Hemberger, M., Hughes, J., Dean, W., Ferguson-Smith, A., Fundele, R., Stewart, F., Kelsey, G., Fowden, A., Sibley, C. & Reik, W. (2002). Placental-specific IGF-II is a major modulator of placental and fetal growth. *Nature* 417, 945–948.

Cooper, W. N., Luharia, A., Evans, G. A., Raza, H., Haire, A. C., Grundy, R., Bowdin, S. C., Riccio, A., Sebastio, G., Bliek, J., Schofield, P. N., Reik, W., Macdonald, F. & Maher, E. R. (2005). Molecular subtypes and phenotypic expression of Beckwith-Wiedemann syndrome. *Eur J Hum Genet* 13, 1025–1032.

DeChiara, T. M., Efstratiadis, A. & Robertson, E. J. (1990). A growth-deficiency phenotype in heterozygous mice carrying an insulin-like growth factor II gene disrupted by targeting. *Nature* 345, 78–80.

Delaval, K., Wagschal, A. & Feil, R. (2006). Epigenetic deregulation of imprinting in congenital diseases of aberrant growth. *Bioessays* 28, 453–459.

Demars, J., Rossignol, S., Netchine, I., Lee, K. S., Shmela, M., Faivre, L., Weill, J., Odent, S., Azzi, S., Callier, P., Lucas, J., Dubourg, C., Andrieux, J., Bouc, Y. L., El Osta, A. & Gicquel, C. (2011). New insights into the pathogenesis of Beckwith-Wiedemann and Silver-Russell syndromes: Contribution of small copy number variations to 11p15 imprinting defects. *Hum Mutat* 32, 1171–1182.

Eggermann, T., Eggermann, K. & Schonherr, N. (2008). Growth retardation versus overgrowth: Silver-Russell syndrome is genetically opposite to Beckwith-Wiedemann syndrome. *Trends Genet* 24, 195–204.

Eggermann, T., Leisten, I., Binder, G., Begemann, M. & Spengler, S. (2011). Disturbed methylation at multiple imprinted loci: an increasing observation in imprinting disorders. *Epigenomics* 3, 625–637.

Enklaar, T., Zabel, B. U. & Prawitt, D. (2006). Beckwith-Wiedemann syndrome: multiple molecular mechanisms. *Expert Rev Mol Med* 8, 1–19.

Gicquel, C., Rossignol, S., Cabrol, S., Houang, M., Steunou, V., Barbu, V., Danton, F., Thibaud, N., Le Merrer, M., Burglen, L., Bertrand, A. M., Netchine, I. & Le Bouc, Y. (2005). Epimutation of the telomeric imprinting center region on chromosome 11p15 in Silver-Russell syndrome. *Nat Genet* 37, 1003–1007.

Goldstone, A. P. (2004). Prader-Willi syndrome: advances in genetics, pathophysiology and treatment. *Trends Endocrinol Metab* 15, 12–20.

Haig, D. & Graham, C. (1991). Genomic imprinting and the strange case of the insulin-like growth factor II receptor. *Cell* 64, 1045–1046.

Hall, J. G. (2003). Twinning. *Lancet* 362, 735–743.

Horsthemke, B. (2010). Mechanisms of imprint dysregulation. *Am J Med Genet C Semin Med Genet* 154C, 321–328.

Jager, S., Schonherr, N., Spengler, S., Ranke, M. B., Wollmann, H. A., Binder, G. & Eggermann, T. (2009). LOT1 (ZAC1/PLAGL1) as member of an imprinted gene network does not harbor Silver-Russell specific variants. *J Pediatr Endocrinol Metab* 22, 555–559.

Kacem, S. & Feil, R. (2009). Chromatin mechanisms in genomic imprinting. *Mamm Genome* 20, 544–556.

Kagami, M., Sekita, Y., Nishimura, G., Irie, M., Kato, F., Okada, M., Yamamori, S., Kishimoto, H., Nakayama, M., Tanaka, Y., Matsuoka, K., Takahashi, T., Noguchi, M., Tanaka, Y., Masumoto, K., Utsunomiya, T., Kouzan, H., Komatsu, Y., Ohashi, H., Kurosawa, K., Kosaki, K., Ferguson-Smith, A. C., Ishino, F. & Ogata, T. (2008). Deletions and epimutations affecting the human 14q32.2 imprinted region in individuals with paternal and maternal upd(14)-like phenotypes. *Nat Genet* 40, 237–242.

Kelsey, G. (2010). Imprinting on chromosome 20: tissue-specific imprinting and imprinting mutations in the GNAS locus. *Am J Med Genet C Semin Med Genet* 154C, 377–386.

Kotzot, D. & Utermann, G. (2005). Uniparental disomy (UPD) other than 15: phenotypes and bibliography updated. *Am J Med Genet A* 136, 287–305.

Li, X., Ito, M., Zhou, F., Youngson, N., Zuo, X., Leder, P. & Ferguson-Smith, A. C. (2008). A maternal-zygotic effect gene, Zfp57, maintains both maternal and paternal imprints. *Dev Cell* 15, 547–557.

Lim, D., Bowdin, S. C., Tee, L., Kirby, G. A., Blair, E., Fryer, A., Lam, W., Oley, C., Cole, T., Brueton, L. A., Reik, W., Macdonald, F. & Maher, E. R. (2009). Clinical and molecular genetic features of Beckwith-Wiedemann syndrome associated with assisted reproductive technologies. *Hum Reprod* 24, 741–747.

Mackay, D. J., Boonen, S. E., Clayton-Smith, J., Goodship, J., Hahnemann, J. M., Kant, S. G., Njolstad, P. R., Robin, N. H., Robinson, D. O., Siebert, R., Shield, J. P., White, H. E. & Temple, I. K. (2006). A maternal hypomethylation syndrome presenting as transient neonatal diabetes mellitus. *Hum Genet* 120, 262–269.

Mackay, D. J., Callaway, J. L., Marks, S. M., White, H. E., Acerini, C. L., Boonen, S. E., Dayanikli, P., Firth, H. V., Goodship, J. A., Haemers, A. P., Hahnemann, J. M., Kordonouri, O., Masoud, A. F., Oestergaard, E., Storr, J., Ellard, S., Hattersley, A. T., Robinson, D. O. & Temple, I. K. (2008). Hypomethylation of multiple imprinted loci in individuals with transient neonatal diabetes is associated with mutations in ZFP57. *Nat Genet* 40, 949–951.

Meyer, E., Lim, D., Pasha, S., Tee, L. J., Rahman, F., Yates, J. R., Woods, C. G., Reik, W. & Maher, E. R. (2009). Germline mutation in NLRP2 (NALP2) in a familial imprinting disorder (Beckwith-Wiedemann Syndrome). *PLoS Genet* 5, e1000423.

Mitter, D., Buiting, K., von Eggeling, F., Kuechler, A., Liehr, T., Mau-Holzmann, U. A., Prott, E. C., Wieczorek, D. & Gillessen-Kaesbach, G. (2006). Is there a higher incidence of maternal uniparental disomy 14 [upd(14)mat]? Detection of 10 new patients by methylation-specific PCR. *Am J Med Genet A* 140, 2039–2049.

Moore, T. & Haig, D. (1991). Genomic imprinting in mammalian development: a parental tug-of-war. *Trends Genet* 7, 45–49.

Reik, W. & Walter, J. (2001). Genomic imprinting: parental influence on the genome. *Nat Rev Genet* 2, 21–32.

Rossignol, S., Steunou, V., Chalas, C., Kerjean, A., Rigolet, M., Viegas-Pequignot, E., Jouannet, P., Le Bouc, Y. & Gicquel, C. (2006). The epigenetic imprinting defect of patients with Beckwith-Wiedemann syndrome born after assisted reproductive technology is not restricted to the 11p15 region. *J Med Genet* 43, 902–907.

Schonherr, N., Meyer, E., Roos, A., Schmidt, A., Wollmann, H. A. & Eggermann, T. (2007). The centromeric 11p15 imprinting centre is also involved in Silver-Russell syndrome. *J Med Genet* 44, 59–63.

Spengler, S., Schonherr, N., Binder, G., Wollmann, H. A., Fricke-Otto, S., Muhlenberg, R., Denecke, B., Baudis, M. & Eggermann, T. (2010). Submicroscopic chromosomal imbalances in idiopathic Silver-Russell syndrome (SRS): the SRS phenotype overlaps with the 12q14 microdeletion syndrome. *J Med Genet* 47, 356–360.

Temple, I. K., Gardner, R. J., Robinson, D. O., Kibirige, M. S., Ferguson, A. W., Baum, J. D., Barber, J. C., James, R. S. & Shield, J. P. (1996). Further evidence for an imprinted gene for neonatal diabetes localised to chromosome 6q22-q23. *Hum Mol Genet* 5, 1117–1121.

Temple, I. K., Shrubb, V., Lever, M., Bullman, H. & Mackay, D. J. (2007). Isolated imprinting mutation of the DLK1/GTL2 locus associated with a clinical presentation of maternal uniparental disomy of chromosome 14. *J Med Genet* 44, 637–640.

Turner, C. L., Mackay, D. M., Callaway, J. L., Docherty, L. E., Poole, R. L., Bullman, H., Lever, M., Castle, B. M., Kivuva, E. C., Turnpenny, P. D., Mehta, S. G., Mansour, S., Wakeling, E. L., Mathew, V., Madden, J., Davies, J. H. & Temple, I. K. (2010). Methylation analysis of 79 patients with growth restriction reveals novel patterns of methylation change at imprinted loci. *Eur J Hum Genet* 18, 648–655.

Williams, C. A., Beaudet, A. L., Clayton-Smith, J., Knoll, J. H., Kyllerman, M., Laan, L. A., Magenis, R. E., Moncla, A., Schinzel, A. A., Summers, J. A. & Wagstaff, J. (2006). Angelman syndrome 2005: updated consensus for diagnostic criteria. *Am J Med Genet A* 140, 413–418.

Yamazawa, K., Ogata, T. & Ferguson-Smith, A. C. (2010). Uniparental disomy and human disease: an overview. *Am J Med Genet C Semin Med Genet* 154C, 329–334.

Chapter 13
Epigenetics and Atherosclerosis

Einari Aavik, Mikko P. Turunen, and Seppo Ylä-Herttuala

Abbreviations

ApoE   Apolipoprotein E
CHD   Coronary heart disease
CpG   Cytosine–Guanine dinucleotide
CTCF   CCCTC-binding factor
ECM   Extracellular matrix
ERα   Estrogen receptor alpha
ERβ   Estrogen receptor beta
FGF2   Fibroblast growth factor 2
HAT   Histone acetyl transferase
Hcy   Homocysteine
HDAC   Histone deacetylase
HIF1α   Hypoxia inducible factor 1α
IGF2   Insulin-like growth factor 2
KAT2B   Lysine acetyltransferase 2B
LDL   Low density lipoprotein
LDLR   LDL receptor

E. Aavik
Department of Biotechnology and Molecular Medicine, A. I. Virtanen Institute,
University of Eastern Finland, Neulaniementie 2, 70211 Kuopio, Finland

M.P. Turunen
Ark Therapeutics, Microkatu 1S, 70210 Kuopio, Finland

S. Ylä-Herttuala (✉)
Department of Biotechnology and Molecular Medicine, A. I. Virtanen Institute,
University of Eastern Finland, Neulaniementie 2, 70211 Kuopio, Finland

Gene Therapy Unit, Kuopio University Hospital, Kuopio, Finland
e-mail: Seppo.Ylaherttuala@uef.fi

J. Minarovits and H.H. Niller (eds.), *Patho-Epigenetics of Disease*,
DOI 10.1007/978-1-4614-3345-3_13, © Springer Science+Business Media New York 2012

MBD	Methyl-cytosine-binding protein
NOS	Nitric oxide synthase
PBL	Peripheral blood lymphocytes
POL2	RNA polymerase II
RISC	RNA induced silencing
SAH	S-adenosylhomocysteine
SAM	S-adenosylmethionine
siRNA	Small interfering RNA
SMC	Smooth muscle cell
SNP	Single nucleotide polymorphism
SRF	Serum response factor
TGS	Transcriptional gene silencing

13.1 Introduction

This was quite some time ago when some researchers started to doubt that not all (Rose 1964) cases of cardiovascular morbidity can be explained by common risk factors (blood lipids, hypertension, diabetes, inheritance, and lifestyle). The hundreds of risk factors for development of atherosclerosis (Hopkins and Williams 1981; Poulter 1999) can be split into two main categories—genetical and environmental risk factors, and especially the genetic component has proven difficult to assess. The genetic risk analysis is a truly complex task since the contribution of each individual single nucleotide polymorphism (SNP) is always overlaid with a personal set of atherosclerosis risk factors. Consequently, SNPs with the strongest effect only marginally increase the average risk for cardiovascular disease (Moore et al. 2010). The most forceful of all human polymorphic loci is chromosomal region 9p21, but even these polymorphisms translate into a minor increase in the risk of disease development (Palomaki et al. 2010). The inability of genetics to identify people at risk for complex diseases (including cardiovascular disease) has stimulated researchers to look for factors beyond genetic information that could "bring the phenotype into being" (Goldberg et al. 2007). Already in 1942, without any idea of the mechanisms, Conrad Waddington coined a new term—epigenetic landscape—describing proposed phenomena on top of genetic variability (Waddington 1942; Goldberg et al. 2007). Nowadays, the concept of epigenetics has been successfully exploited, and animal experiments have provided solid proof for nongenetic components (e.g., nutritional status of parents) in the increased risk for atherosclerosis development in the progeny (Collas et al. 2007). The increased risk for atherosclerosis correlates with certain alterations in chromatin structure which contribute chromatin accessibility to specific transcription factors, which in turn, leads to altered gene expression (Palomaki et al. 2010). The best studied epigenetic mechanisms regulating gene transcription are DNA methylation and histone modifications (Matouk and Marsden 2008), but other aspects of epigenetic regulation like chromatin folding, nuclear compartmentalization, and expression of noncoding RNAs are getting increasingly more attention (Goldberg et al. 2007; Lanctot et al. 2007; Matouk and Marsden 2008). Chromatin is folded a millionfold to fit into the nucleus and the actively transcribed euchromatin tends to locate centrally, while tightly packed inactive heterochromatin is found at the periphery of the nucleus (Lanctot et al. 2007). Transformation of euchromatin to heterochromatin and repositioning of chromatin segments within the nuclear compartments appear to contribute to gene regulation (Lanctot et al. 2007). However, these most fascinating findings will not be dealt with in this review as their possible role in atherosclerosis is unknown.

Although epigenetic modifications to chromatin structure are nonpermanent and they do not change the genetic code, the acquired traits still can be passed to the progeny (Napoli et al. 1999), suggesting that a particular phenotype is not entirely defined by the DNA sequence alone. This is the very essence of nongenetic (=epigenetic) regulation of gene activity (Matouk and Marsden 2008). Evidence from experimental animals (Napoli et al. 1999) allows to assume that the same holds true for humans as well, despite the fact that direct evidence in humans is still lacking (Collas et al. 2007). This is so, in part, due to the interplay of sexually dimorphic loci, which are also regulated by epigenetic mechanisms (Gabory et al. 2009). Regardless of the advancements in DNA sequencing technology which have sped up the data collection by eight orders of magnitude during the last 25 years, the need to analyze thousands of human genomes to dissect the impact of sexual dimorphism on phenotypes is still an enormous challenge (Venter 2010).

Epigenetics is an emerging field in atherosclerosis research, even though the first observation that changes in chromatin structure correlate with hypercholesterolemia was published more than 20 years ago (Lehmann et al. 1980). In this study, an intercalating fluorescent dye made it possible to distinguish between unwound actively transcribed euchromatin and tightly packed inactive heterochromatin. Subsequently, more nuclear acridine orange dye was detected in the aortic wall cells of hypercholesterolemic rabbits, suggesting chromatin modifications (Lehmann et al. 1980).

Our understanding of the pathobiology of atherosclerosis will be broadened significantly after the underlying epigenetic mechanisms have been identified. Equipped with this new knowledge, novel therapeutic approaches could be suggested to treat atherosclerosis and related diseases involving chronic inflammation.

13.2   Etiology and Pathobiology of Atherosclerosis

Accumulation of cholesterol in the walls of large- and medium-sized arteries leads to a localized chronic inflammatory response accompanied by endothelial cell dysfunction, monocyte-macrophage, and T cell infiltration (Ross 1999; Lusis 2000). Chronic inflammation, in turn, stimulates arterial smooth muscle cells (SMCs) to proliferate and to produce excessive amounts of extracellular matrix (ECM) proteins. Chronic inflammation with SMC proliferation leads to stenosis, myocardial infarction, stroke, aneurysms, and peripheral artery disease (Yla-Herttuala et al. 1989; Wissler 1991; Sanchez and Veith 1998; Lamon and Hajjar 2008).

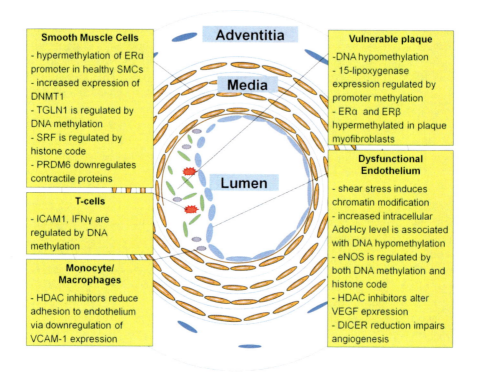

Fig. 13.1 Summary of epigenetic effects in atherosclerotic vascular wall

Atherosclerosis epidemics peaked in industrialized countries in the 1960s and 1970s, but in Far East and in third world countries, the incidence of atherosclerosis is continuing to increase. Unfavorable changes in diet and lifestyle (too much energy intake and too little physical activity) have lead to elevated blood lipid levels with concomitant enhanced uptake of cholesterol by the arterial wall SMCs. Atherosclerosis develops over decades, and despite the multiple preventative measures that have been introduced (dietary changes, effective cholesterol lowering, treatment of hypertension, diabetes, reduced smoking), it remains the main cause for premature deaths in industrialized countries (Lopez et al. 2006). The observation of continued atherosclerotic development in diabetic patients after blood sugar normalization, led to the proposal of "metabolic memory" (Nathan et al. 2005). This concept of "metabolic memory" proposes that the cells of the vascular wall are able to "remember" the past levels of high glucose (Villeneuve et al. 2008). The only known mechanism to memorize acquired traits is epigenetics. Epigenetic modifications may explain the polyclonal expansion of SMCs, and the effect of some of the dietary components (i.e., risk factors) on chromatin structure (Laukkanen et al. 1999; Hiltunen and Yla-Herttuala 2003). Some of the epigenetic effects operational in the walls of atherosclerotic arteries are summarized in Fig. 13.1.

13.2.1 Hypercholesterolemia

Genetics has provided evidence for inherited forms of cardiovascular disease, and the infliction of atherosclerosis is caused by increased levels of blood lipids, or more specifically, by hypercholesterolemia (Hansson 2005). Maternal hypercholesterolemia appears to cause more frequent incidence and faster progression of atherosclerosis in the progeny both in humans (Napoli et al. 1999), and in rodents (Napoli et al. 2000, 2002; Alkemade et al. 2007), although one study has negated this in mice (Madsen et al. 2003). Still, existing evidence supports the idea that sensitivity to atherosclerosis can be passed to the progeny, and although this has not been shown yet, perhaps epigenetic modifications of the chromatin may explain how acquired traits are inherited.

13.2.2 Homocysteinemia

The underlying pathological mechanism of homocysteinemia leading to endothelial dysfunction and, ultimately, to atherosclerosis has been widely studied but remains poorly understood. It is well known that as many as one tenth of humans can be affected by high plasma homocysteine (Hcy) levels and this goes hand-in-hand with the increased risks for heart disease and stroke (Boushey et al. 1995; Clarke et al. 2002). Hcy can be metabolized to methionine or cysteine (Finkelstein 2007). Methionine can be activated to S-adenosylmethionine (SAM), which serves as a methyl donor in biochemical pathways (Finkelstein 2007). It has been suggested that hyperhomocysteinemia mediates its effect on vascular health via S-adenosylhomocysteine (SAH), accumulation of which facilitates DNA methylation (Castro et al. 2003).

Increased plasma concentration of Hcy and SAH appear to disturb the global DNA methylation status and lead to a lowered DNA methylation level in peripheral blood lymphocytes (PBL) (Castro et al. 2003). High Hcy levels (>75 μmol/L) have also been linked to DNA hypomethylation and atherosclerosis (Lee and Wang 1999). It has been suggested that high Hcy concentrations results in competitive inhibition of SAM binding to DNA methylases and, consequently, passive loss of DNA methylation in successive rounds of replication (James et al. 2002).

Nutritional and environmental factors have been shown to be responsible for altered DNA methylation patterns both in human and experimental animal atherogenesis (Chao et al. 2000; Dayal et al. 2004). Dietary issues can have a profound effect on epigenetic modifications of the genome. It has been demonstrated that insulin-like growth factor 2 (IGF2) imprinting can be impaired by providing a diet deficient of methionine and folate; consequently, insufficient amounts of SAM are produced and imprinting fails (Dobosy et al. 2008). Colorectal cancer is a known example of even more sophisticated relations between the effective concentrations of Hcy and folate in the blood stream, as too low folate (needed for Hcy remethylation to methionine) (Kono and Chen 2005) as well as too high folate can both promote

402   E. Aavik et al.

tumor development (Pellis et al. 2008; Crott et al. 2008; Sauer et al. 2009). These examples reveal the epigenetic consequences of dietary and/or other environmental exposures, which can increase the risk for disease development.

13.3 Histone Code and Atherosclerosis

In eukaryotes, DNA is packaged into a chromatin, a higher-order structure which has repeated nucleosomes as fundamental units (Luger et al. 1997). Nucleosomes are composed of histone octamers that consist of histone (H3–H4) tetramer and two dimers of H2A and H2B. Each nucleosome consists of a nucleosome core around which 145–147 bp of DNA is wrapped around twice to create a super helix. Histone 3 (H3) and histone 4 (H4) are targets of posttranslational modifications, which include methylation, acetylation, ubiquination, and SUMOylation of lysine residues, methylation of arginine residues, and phosphorylation of serines. Histone modifications form the so-called "histone code" (Kouzarides 2007) that is read by chromatin-associated proteins and is translated into a transcriptionally active or repressed genetic state, depending on the type of modification and on the amino acid position of the modification. Histone modifications can affect the chromatin structure and gene accessibility. The epigenetic histone modifications are predominantly located at specific positions in the amino-terminal tails of histones. These protease-sensitive tails protrude from the chromatin surface and comprise approximately 25% of the mass of core histones, thus providing an exposed surface for interactions with other proteins.

The addition of a posttranslational modification to one residue can affect the subsequent modification of another residue(s) either in *cis*, on the same histone molecule, or in *trans* between histone molecules. As an example of histone modification crosstalk in *trans*, H3K4 methylation is regulated by H2B ubiquitination. This mechanism is conserved from yeast to humans; ubiquitination of histone H2BK123 in yeast or H2BK120 in humans is required for the subsequent di- and trimethylation of H3K4 at gene promoters (Dover et al. 2002).

Currently, there is only a limited amount of published data available on histone code changes associated with cardiovascular disease. Most of the studies show indirect evidence between histone code regulation and the observed phenotype. For example, trichostatin A, a specific inhibitor of histone deacetylase, has been shown to significantly increase the formation of fatty streak lesions and macrophage infiltration in LDLR−/− mice.

The transcription factor NF-κB plays an important role in regulating inflammatory gene expression. Atherosclerosis-associated inflammatory processes are also largely mediated by NF-κB activation which includes histone code modulation. Histone acetylation of lysine acetyltransferase (KAT2B) has a costimulatory effect on NF-κB-dependent transcription of inflammatory genes in PBL from diabetic humans (Miao et al. 2004; Vogel et al. 2006), whereas histone deacetylase activation suppresses inflammation (Ashburner et al. 2001). In addition to histone modifications, both KAT2B and p300 can acetylate NF-κB proteins directly and thereby increase its binding affinity to DNA and thus its transactivating activity (Chen and Greene 2003; Liu et al. 2006).

Villeneuve et al. 2008 have shown that dysregulation of epigenetic histone modifications may be a major mechanism underlying metabolic memory and sustained proinflammatory phenotype of diabetic cells. The key repressive and relatively stable epigenetic chromatin mark, H3 lysine-9 tri-methylation (H3K9me3), and Su(var) 3-9 homologue (Suv39h1) had protective roles against the preactivated state of diabetic vascular smooth muscle cells (VSMC). This work was done using VSMC derived from type 2 diabetic db/db mice, which exhibit a persistent atherogenic and inflammatory phenotype even after culture in vitro. In another study published by the same group, the levels of histone H3 lysine 4 dimethylation (H3K4me2), a key chromatin mark associated with active gene expression, were significantly elevated at the promoters of the inflammatory genes monocyte chemoattractant protein-1 and interleukin-6 in diabetic VSMCs. Furthermore, tumor necrosis factor-α-induced inflammatory gene expression, H3K4me2 levels, and recruitment of RNA polymerase II at the gene promoters were also enhanced in diabetic VSMCs, demonstrating the formation of open chromatin poised for transcriptional activation in diabetes. Histone lysine demethylase KDM1 (LSD1), a component of the CoREST2CtBP corepressor complex, was discovered in 2004 (Shi et al. 2004), and since then, a number of histone demethylases have been identified and shown to play important roles in the regulation of gene expression, as well as in cellular differentiation and development (Verrier et al. 2011). Protein levels of LSD1, which negatively regulates H3K4 methylation and its occupancy at these gene promoters, were significantly reduced in diabetic VSMCs. Therefore, dysregulation of LSD1 and H3K4 actions may be a major mechanism for vascular inflammation and metabolic memory associated with diabetic complications.

A second larger class of iron-based histone demethylases characterized by the presence of the conserved JumonjiC domain has also been identified, Jumonji histone demethylases (JHDMs) (Tsukada et al. 2006). It seems that each histone demethylase targets specific histone lysines and makes specific lysine modifications (Agger et al. 2008). JHDMs belong to a family of dioxygenases and share similarities with prolyl hydroxylases (PHDs). Interestingly, it was shown that SMC differentiation marker gene expression is regulated by H3K9 methylation and that the effects of the myocardin factors on SMC-specific transcription may involve the recruitment of JMJD1A to the SMC-specific promoters (Lockman et al. 2007). Increases in H3K9/14 acetylation and H3K4 methylation are involved in Smad2 overexpression in TAA in a cell-specific and transcription start site-specific manner (Gomez et al. 2011). H3K4 methylation and H3K9/14 acetylation are characteristic histone modifications at enhancer-promoter areas of SMC-specific genes containing CArG boxes at the SRF (serum response factor) binding site (McDonald and Owens 2007).

Hypoxia is a hallmark of many inflammatory processes including atherosclerotic plaque development. Invading macrophages adapt to hypoxic conditions by changes in gene expression (Tausendschon et al. 2011). Inhibition of JHDMs by hypoxia has been recently shown to be responsible for downregulation of chemokine signaling (CCL2, and its receptors CCR1 and CCR5) in hypoxic macrophages via increases in H2K9 di- and trimethylation and H3K36 trimethylation (Tausendschon et al. 2011). All JHDM family members JMJD1A–C and JMJD2A–D are expressed in macrophages.

The effects of in utero programming and type of postnatal diet on epigenetic histone modifications in the vasculature in relation to atherosclerosis have been studied in apolipoprotein E (ApoE) deficient animals. Differences in histone methylation profile in vascular endothelial and smooth muscle cells revealed that the offspring from apoE(−/−) mothers had significantly different responses to a high cholesterol diet when compared with offspring from wild-type mothers (Alkemade et al. 2010). Therefore, it can be concluded that both in utero programming and postnatal hypercholesterolemia affect epigenetic patterning in the vasculature.

13.4 DNA Methylation and Atherosclerosis

Alteration in DNA methylation is another epigenetic mechanism that regulates chromatin structure and gene expression (including imprinting of paternally or maternally derived genes, control on transposable element mobility and developmentally regulated genes, X chromosome inactivation). The DNA methylation pattern is copied at each mitotic cycle in embryogenesis and upon injury in tissue repair process (Goll and Bestor 2005). Methylated cytosines at regulatory areas are bound by methyl-binding domain (MBD) proteins, which effectively prevent gene activation (Matouk and Marsden 2008) and establish a long-lasting block by recruiting transcriptional corepressors introducing further modifications to the chromatin structure by histone deacetylation and methylation (Fuks et al. 2003; Ikegami et al. 2009).

The normal level of methylated cytosines in chromosomal DNA varies from 1% (Ehrlich et al. 1982) to almost 4% of all cytosines in some tissues (Hiltunen et al. 2002). The top level of cytosine methylation means that up to 90% of all cytosines in cytosine–guanine dinucleotides (CpG) are methylated (Miranda and Jones 2007). CpG dinucleotides are not uniformly distributed over the eukaryotic genome. On the contrary, CpG dinucleotides are more prominent in regulatory DNA regions; long stretches of DNA (from a few hundred to several thousand basepairs in length) enriched for CpG dinucleotides are recognized as CpG islands (Goll and Bestor 2005). The enzymatic complex responsible for DNA methylation is shared with the machinery active in histone methylation (Weaver et al. 2005; Villeneuve et al. 2008).

It is accepted that the methylation pattern of chromosomal DNA is reproduced at each mitotic cycle during embryonic development and accidental changes introduced to methylation patterns at early development can be passed to the progeny at least in mice (Collas et al. 2007). Although the methyl cytosine patterning appears to be stable enough to be passed to the progeny, there is also some evidence that pharmacological intervention in adulthood can revert the epigenetic programming introduced in early life (Napoli et al. 1999; Weaver et al. 2005, 2006; Bannister and

Kouzarides 2005). Studies of young monozygotic twins have revealed the relative stability of DNA and histone modifications in successive mitotic events (Fraga et al. 2005). This association appears to weaken upon increasing age and depended also on the twins' lifestyle (Fraga et al. 2005). Epigenetic changes identified in one third of the twins were accompanied by changes in lymphocytic gene expression profiles (Fraga et al. 2005). Furthermore, there is evidence that accumulating age-related changes in global methylation status of DNA are genetically determined. Specifically, Bjornsson et al. 2008 have been able to show that over time, some individuals gain and others lose methylated cytosines in their DNA and the arrays of paired data on intraindividual changes can be clustered family wise. These results can also explain the conclusion of another study stating that the average levels of DNA methylation in age groups do not change (Eckhardt et al. 2006).

The vast majority of vascular wall cells are SMCs, and it is safe to conclude that the epigenetic changes detected in the vasculature characterize mainly SMCs. One of the earliest developmental SMC markers is transgelin and, at least in part, the gene is regulated by DNA methylation (Yamamura et al. 1997). Estrogen receptor α (ERα) promoter is hypermethylated in rapidly proliferating SMCs, but not in the endothelial cells, which are a minor cellular constituent of the vascular wall (Post et al. 1999). Similarly, endothelial nitric oxide synthase (eNOS) is hypomethylated in endothelial cells (Chan et al. 2005) but heavily methylated in SMC (Fish et al. 2005). Upon chronic inflammation, inducible nitric oxide synthase (iNOS) is expressed in atherosclerotic plaques (Wilcox et al. 1997; Luoma et al. 1998) but repressed in the majority of other tissues via DNA methylation (Chan et al. 2005).

13.4.1 Hypermethylation

Hypermethylation of regulatory CpG islands leads to tight packaging of DNA and, consequently, to transcriptional silencing caused by inaccessibility of the transcription factors to their cognate DNA binding sites. In addition, transcription factor association and transcription activation is prevented by the five methyl-cytosine-binding proteins (MBD1, 2, 4, MeCP2, Kaiso) (Matouk and Marsden 2008).

The DNA binding sites of some transcription factors like hypoxia inducible factor 1α (HIF1α) (Wenger et al. 1998), myc (Perini et al. 2005), and CCCTC-binding factor (CTCF) (Bell and Felsenfeld 2000) have been shown to be methylated in atherosclerosis. For example, CTCF regulates the expression of maternal genes (like IGF2) when paternal CTCF target sites are methylated and inactive (Han et al. 2008). Another eukaryotic insulator YY1 regulates the expression of certain paternal genes (XIST, GNAS, PEG3), while maternal copies are fully methylated. Kaiso interacts with CTCF, but can also recognize unmethylated TNGCAGGA (Defossez et al. 2005).

Hypermethylation as a cause for transcriptional silencing has been extensively studied in cancer, but the results concerning coronary heart disease (CHD) have been controversial. Two recent studies have detected increased levels of DNA methylation in PBL isolated from CHD patients (Sharma et al. 2008; Kim et al. 2010), whereas earlier findings demonstrated the opposite (Lee and Wang 1999; Castro et al. 2006). All these studies have included relatively small numbers of human samples, and bigger studies are required for better assessment. In addition, a study by Bjornsson et al. provided evidence for both hyper- and hypomethylation, and the specific changes (either hyper- or hypomethylation) were characteristic to particular families. Consequently, studying small groups of randomly chosen people for differences in global DNA methylation may lead to opposing findings just by chance.

The complex nature of atherosclerosis becomes clearly visible when analyzing the opposing effects of DNA methylation on gene expression (Table 13.1). Homocysteinemia, as described above, has the potential to cause generalized DNA hypomethylation in PBL of atherosclerotic patients (Castro et al. 2003), but the regulatory areas of some genes like iNOS (Jiang et al. 2007) in cultured human monocytes, and fibroblast growth factor 2 (FGF2) (Chang et al. 2008) in cultured human endothelial cells remain heavily methylated.

Estrogen receptor alpha (ERα) and beta (ERβ) promoters are likewise methylated in atheromatous SMCs (Post et al. 1999; Ying et al. 2000; Kim et al. 2007). Although both estrogen receptors appear to play a role in SMC proliferation, their exact role in vivo is still unclear. Promoters of estrogen receptors show a steady age-dependent increase in DNA methylation, and in cadaveric arterial samples, the level of promoter methylation has reached 99% (Ying et al. 2000). Similarly, an age-related increase in immunomodulatory and tumor suppressor gene methylation (Thompson et al. 2010) may bring its contribution to the pathobiology of atherosclerotic lesion development. For example, the increased cardiovascular mortality of dialysis patients correlates with severity of inflammation. Inflammation, in turn, has been shown to promote DNA methylation in PBL (Stenvinkel et al. 2007). Establishing the link between inflammatory gene expression and DNA methylation is an attractive prospective for future research.

13.4.2 Hypomethylation

As discussed in the Sect. 13.2.2, loss of DNA methylation can be active (removal of methyl groups independently from DNA replication) or passive (methyl groups are not introduced during replication) (Miranda and Jones 2007). Active SMC proliferation in atherosclerotic lesions seems to support the view that passive hypomethylation could be the dominant process, but the definite answer is still missing as aging occurs simultaneously with atherosclerosis development. Global DNA hypomethylation has been linked to atherosclerosis (Hiltunen et al. 2002) as well as to aging (Wilson and Jones 1983). New findings prove that hypomethylation may affect only a fraction of the aging population and influences only 20–30% of the people (Bjornsson et al. 2008). Global hypomethylation is characteristic of hyperhomocysteinemia (Lee and Wang 1999), tumor growth (Butcher and Beck 2008), and schizophrenia (Shimabukuro et al. 2007) but also of proliferating SMCs in advanced

Table 13.1 Genes related to atherosclerotic diseases which are at least partly regulated by epigenetic mechanisms

Gene	Function	Target	Mechanism	References
Endothelial nitric oxide synthase (eNOS)	Blood pressure control	Endothelium	DNA methylation and histone code	Chan et al. (2004), Fish et al. (2005)
Inducible nitric oxide synthase (iNOS)	Blood pressure control	Inflammation, macrophages	DNA methylation and histone code	Chan et al. (2005)
Fads2	Fatty acid desaturase	Contributes to pathology of homocysteine (Hcy)	DNA methylation	Devlin et al. (2007)
c-*fos*	Transcription factor	Shear stress	Histone code	Hastings et al. (2007)
Estrogen receptor α	Transcription factor	Coronary artery	DNA methylation	Post et al. (1999)
Estrogen receptor β	Transcription factor	Coronary artery, mammary artery Carotid artery Femoral artery Saphenous vein	DNA methylation	Kim et al. (2007)
P66Shc	Docking protein in cell signaling	End-stage renal disease	DNA methylation	Geisel et al. (2007)
15-LO (ALOX15)	Lipid peroxidation	Pathogenesis of atherosclerosis	DNA methylation	Liu et al. (2004)
EC-SOD	Oxidative stress	Pathogenesis of atherosclerosis	DNA methylation	Laukkanen et al. (1999)
H19/IGF2	H19—tumor suppressor IGF2—fetal growth factor	Regulated by hyperhomocysteinemia	Imprinting, normally maternally expressed	Devlin et al. (2005, 2007)
MMP-2	Matrix degradation	Extracellular matrix	DNA methylation	Sato et al. (2003)
MMP-7	Matrix degradation	Extracellular matrix	DNA methylation	Sato et al. (2003)
MMP-9	Matrix degradation	Extracellular matrix	DNA methylation	Sato et al. (2003)
TIMP-3	Inhibitor of extracellular matrix (ECM) degradation	Extracellular matrix	DNA methylation	Wild et al. (2003)
IFN-γ	Cytokine	Inflammatory response	DNA methylation	White et al. (2002)
PDGF-A	Growth factor	Cell proliferation	DNA methylation	Lin et al. (1993)
ICAM-1	Adhesion	Inflammatory reactions	DNA methylation	Tanaka et al. (1995)
p53	Tumor suppressor	Apoptosis	DNA methylation	Schroeder and Mass (1997)

human atherosclerotic lesions, where a 9% reduction in the number of methylated cytosines was detected when compared to healthy artery (Hiltunen et al. 2002). Laboratory animals provide valuable information to dissect hypomethylation in atherosclerosis from hypomethylation related to aging. Aortas of ApoE-KO mice show hypomethylation before any histological evidence for atherosclerosis at 4 weeks (Lund et al. 2004), and in atheromas (Voo et al. 2000). Hypomethylation has been detected in the aortic neointima of New Zealand White rabbits (Laukkanen et al. 1999; Hiltunen et al. 2002).

Hypomethylation of eNOS and iNOS promoters characterizes normal human endothelial cells. In contrast, decreased methylation of superoxide dismutase (Laukkanen et al. 1999) and 15-lipoxygenase (Hiltunen et al. 2002) promoter areas characterizes human atherosclerotic lesions.

Unmethylated DNA is specifically recognized by transcription regulator CXXC1 (Voo et al. 2000), which functions in cooperation with histone 3 lysine 4 (H3K4) methylases MLL, MLL2, MLL3, and hSETI (Ansari et al. 2008). Potential consequence of global DNA hypomethylation in atherosclerosis could be transposable element activation; this has been shown to cause DNA recombination in human myeloid leukemia (Roman-Gomez et al. 2005). However, hypomethylation detected in atherosclerosis is a controversial phenomenon, as upregulation of DNA methylase DNMT1 has been confirmed in atheromas (Hiltunen et al. 2002).

It is of interest to note that global DNA hypomethylation has been shown to promote autoimmunity in mice (Richardson 2002). This is something that deserves more attention in human studies as well since altered innate and adaptive immune responses (e.g., autoantibody generation against oxLDL) are distinctive pathological features of atherosclerosis (Yla-Herttuala et al. 1994).

13.5 Noncoding RNAs

Small RNA molecules have been shown to regulate gene transcription by interacting with the promoter region and modifying the histone code (Morris et al. 2004; Li et al. 2006; Janowski et al. 2007; Turunen et al. 2009). A process known as transcriptional gene silencing (TGS) involves promoter-targeted small interfering RNA (siRNA) and leads to silent state epigenetic profile containing dimethylated histone H3 lysine 9 (H3K9me2) and trimethylated histone H3 lysine 27 (H3K27me3) (Weinberg et al. 2006). This process was first recognized in plants (Mette et al. 2000) and since has been identified in mammals (Mette et al. 2000; Morris et al. 2004). Interestingly, besides TGS, the small RNAs have also been reported to induce gene activation (Li et al. 2006; Turunen et al. 2009). This is associated with a loss of H3K9me2 at the targeted promoter sequences and seems to require a member of an RNA-induced silencing (RISC) complex, argonaute-2 protein (Li et al. 2006). TGS was earlier shown to require argonaute-1 and argonaute-2 (Janowski et al. 2006). One suggested mechanism of action for promoter-targeted small RNA-mediated gene activation and silencing is that the antisense strand of the small

RNA binds to a complementary noncoding promoter-associated antisense RNA (Han et al. 2007; Morris et al. 2008; Schwartz et al. 2008).

We have recently shown that shRNAs targeted on the VEGF-A promoter delivered by lentiviral vectors either up- or downregulated VEGF-A in mouse endothelial cell lines (Turunen et al. 2009). The delivery of downregulating shRNA led to demethylation of H3K4me2 and deacetylation of H3K9ac at the promoter and TSS but had no effect on H3K9me2 level; it also enriched nucleosome positioning both at the promoter and TSS. On the other hand, the upregulating shRNA increased H3K4me2 at TSS but not at the targeted promoter (Turunen et al. 2009). These epigenetic changes were observed already 14 h after transduction, and they further increased by 7 days. These effects were cell- and tissue-specific since the up- and downregulation and corresponding epigenetic changes were only demonstrated in some cell lines and not in others. We also evaluated the therapeutic potential of these shRNAs in vivo. Mouse hind limbs were made ischemic by ligation of the femoral artery, then they were injected with lentiviral vectors either coding for up- or downregulating shRNAs or control vector. The observed effects and epigenetic profile of muscle tissue was similar to that found in cell culture studies. As shown by ultrasound analysis, upregulating shRNA significantly increased vascularity and improved blood flow in the ischemic hind limb, demonstrating for the first time that promoter targeted shRNAs can have therapeutic effects in vivo.

Also in many cancer cell types, small RNAs have also been thoroughly studied. Small activating RNA (saRNA) was designed complementary to the p21 promoter. The upregulation of p21 in these bladder cancer cells inhibited cell proliferation and induced G1-phase arrest and apoptosis (Yang et al. 2008). Promoters of E-cadherin, p21, and VEGF were targeted in the studies by Li et al. (2006), and this resulted in induction of the indicated genes due to the epigenetic changes at the promoters. The same group also transfected African green monkey and chimpanzee cells with the same promoter targeted small RNAs; since the promoter sequence is highly conserved between humans and other primates, the activation of the genes were also induced in these cell lines. They further tested activating RNAs targeting promoters of p53, PAR4, WT1, RB1, p27, NKX3-1, VDR, IL2, and pS2 genes in primate cell lines, but induction was only achieved in p53, PAR4, WT1, and NKX3-1. Huang et al. 2010 also successfully designed saRNAs to the mouse promoter of Cyclin B1 and the rat promoter of chemokine receptor CXCR4. Since Cyclin B1 is known to promote entry into mitosis, the observation that the saRNAs did increase H3S10 phosphorylation, correlated well with chromosome condensation in mitosis.

It has been recently shown that the long intergenic noncoding RNA (lincRNA) HOTAIR acts as a molecular scaffold by binding histone methylase (polycomb repressive complex 2) and demethylase (LSD1/CoREST/REST complex) with distinct domains (Tsai et al. 2010). This directs a specific combination of histone methylations to the target gene chromatin. The report suggests a possibility that other lincRNAs could also guide distinct histone modification patterns to specific genes and therefore affect the epigenetic state of the chromatin during development and disease progression.

13.6 Conclusions

The incidence of cardiovascular diseases has undergone rapid changes within various countries, and since the human genetic code does not evolve much through classical mutations in a few decades, the transcriptional memory based on epigenetic modifications together with changes in human dietary preferences (and lifestyle, more generally) promise to provide an explanation to the epidemic of atherosclerosis. SMCs build up the bulk of cellular composition of vascular walls and their health status is of utmost importance for human well-being. Clonal expansion of lesional SMCs has been documented in human atherosclerotic plaques years ago (Benditt and Benditt 1973), and it is conceivable that a fraction of SMC population acquires the property to proliferate more rapidly and thus increase their share in the lesion area. The possible association of epigenetic modifications with clonal expansion is lacking experimental support. However, SMC-specific gene expression has been shown to depend on regulatory elements driven by epigenetic mechanisms able to alter the chromatin structure. SMC-specific gene expression is associated with histone modifications by methylation and acetylation (Qiu and Li 2002; McDonald et al. 2006). Moreover, the histone methylase PRDM6 has been shown to specifically downregulate contractile protein expression in SMCs (Davis et al. 2006), which is in part responsible for the switching to the rapidly proliferating synthetic phenotype of SMCs. These new findings suggest that underlying epigenetic mechanisms can contribute to enhanced SMC proliferation operational in cardiovascular disease and leading to rapid restenotic reocclusion after balloon dilatation or intravascular surgery of atherosclerotic arteries.

Today, we can only anticipate that a failure of an organism to resolve acute inflammation can lead to chronically inflamed atherosclerotic tissue including autoimmune responses, and we cannot exclude the epigenetic effects chronic inflammation and human dietary preferences to vascular health. Future studies linking nutrition to epigenetics will give definite answers. Currently, various drugs (e.g., histone demethylase inhibitors) are in development to target malignancies and autoimmunity. The same small molecule drugs are of great potential in atherosclerosis treatment, as chronic inflammation and autoimmunity are involved in the pathogenesis of CHD.

Expressed microRNAs derived from genes and intergenic areas constitute a large pool of effectors able to modify gene expression, and some epigenetic effects of microRNA involve changes in DNA methylation and histone modifications (Turunen et al. 2009). Epigenetic effects of double-stranded RNA molecules have been documented in several cell types by a number of research groups (Mette et al. 2000; Morris et al. 2004; Weinberg et al. 2006; Wang et al. 2008). The role of small RNAs in the fine tuning of gene expression and transcriptional memory are likely to be areas of intensive research for coming years.

To conclude, epigenetic alterations in atherosclerotic lesions contribute to the pathogenesis of arterial plaques. Deeper understanding of the operational process in normal vs. pathological tissue will provide new avenues for the development of efficient therapies.

13 Epigenetics and Atherosclerosis

Acknowledgments This work was supported by Finnish Academy, Leducq Foundation, CliniGene EU grant, and Ark Therapeutics Ltd.

References

Agger, K., Christensen, J., Cloos, P. A. & Helin, K. (2008). The emerging functions of histone demethylases. *Curr Opin Genet Dev* 18, 159–168.

Alkemade, F. E., Gittenberger-de Groot, A. C., Schiel, A. E., VanMunsteren, J. C., Hogers, B., van Vliet, L. S., Poelmann, R. E., Havekes, L. M., Willems van Dijk & DeRuiter, M. C. (2007). Intrauterine exposure to maternal atherosclerotic risk factors increases the susceptibility to atherosclerosis in adult life. *Arterioscler Thromb Vasc Biol* 27, 2228–2235.

Alkemade, F. E., van Vliet, P., Henneman, P., van Dijk, K. W., Hierck, B. P., van Munsteren, J. C., Scheerman, J. A., Goeman, J. J., Havekes, L. M., Gittenberger-de Groot, A. C., van den Elsen, P. J. & DeRuiter, M. C. (2010). Prenatal exposure to apoE deficiency and postnatal hypercholesterolemia are associated with altered cell-specific lysine methyltransferase and histone methylation patterns in the vasculature. *Am J Pathol* 176, 542–548.

Ansari, K. I., Mishra, B. P. & Mandal, S. S. (2008). Human CpG binding protein interacts with MLL1, MLL2 and hSet1 and regulates Hox gene expression. *Biochim Biophys Acta* 1779, 66–73.

Ashburner, B. P., Westerheide, S. D. & Baldwin, A. S. (2001). The p65 (RelA) subunit of NF-kappaB interacts with the histone deacetylase (HDAC) corepressors HDAC1 and HDAC2 to negatively regulate gene expression. *Mol Cell Biol* 21, 7065–7077.

Bannister, A. J. & Kouzarides, T. (2005). Reversing histone methylation. *Nature* 436, 1103–1106.

Bell, A. C. & Felsenfeld, G. (2000). Methylation of a CTCF-dependent boundary controls imprinted expression of the Igf2 gene. *Nature* 405, 482–485.

Benditt, E. P. & Benditt, J. M. (1973). Evidence for a monoclonal origin of human atherosclerotic plaques. *Proc Natl Acad Sci U S A* 70, 1753–1756.

Bjornsson, H. T., Sigurdsson, M. I., Fallin, M. D., Irizarry, R. A., Aspelund, T., Cui, H., Yu, W., Rongione, M. A., Ekstrom, T. J., Harris, T. B., Launer, L. J., Eiriksdottir, G., Leppert, M. F., Sapienza, C., Gudnason, V. & Feinberg, A. P. (2008). Intra-individual change over time in DNA methylation with familial clustering. *JAMA* 299, 2877–2883.

Boushey, C. J., Beresford, S. A., Omenn, G. S. & Motulsky, A. G. (1995). A quantitative assessment of plasma homocysteine as a risk factor for vascular disease. Probable benefits of increasing folic acid intakes. *JAMA* 274, 1049–1057.

Butcher, L. M. & Beck, S. (2008). Future impact of integrated high-throughput methylome analyses on human health and disease. *J Genet Genomics* 35, 391–401.

Castro, R., Rivera, I., Blom, H. J., Jakobs, C. & Tavares de Almeida, I. (2006). Homocysteine metabolism, hyperhomocysteinaemia and vascular disease: an overview. *J Inherit Metab Dis* 29, 3–20.

Castro, R., Rivera, I., Struys, E. A., Jansen, E. E., Ravasco, P., Camilo, M. E., Blom, H. J., Jakobs, C. & Tavares de Almeida, I. (2003). Increased homocysteine and S-adenosylhomocysteine concentrations and DNA hypomethylation in vascular disease. *Clin Chem* 49, 1292–1296.

Chan, G. C., Fish, J. E., Mawji, I. A., Leung, D. D., Rachlis, A. C. & Marsden, P. A. (2005). Epigenetic basis for the transcriptional hyporesponsiveness of the human inducible nitric oxide synthase gene in vascular endothelial cells. *J Immunol* 175, 3846–3861.

Chan, Y., Fish, J. E., D'Abreo, C., Lin, S., Robb, G. B., Teichert, A. M., Karantzoulis-Fegaras, F., Keightley, A., Steer, B. M. & Marsden, P. A. (2004). The cell-specific expression of endothelial nitric-oxide synthase: a role for DNA methylation. *J Biol Chem* 279, 35087–35100.

Chang, P. Y., Lu, S. C., Lee, C. M., Chen, Y. J., Dugan, T. A., Huang, W. H., Chang, S. F., Liao, W. S., Chen, C. H. & Lee, Y. T. (2008). Homocysteine inhibits arterial endothelial cell growth through transcriptional downregulation of fibroblast growth factor-2 involving G protein and DNA methylation. *Circ Res* 102, 933–941.

Chao, C. L., Kuo, T. L. & Lee, Y. T. (2000). Effects of methionine-induced hyperhomocysteinemia on endothelium-dependent vasodilation and oxidative status in healthy adults. *Circulation* 101, 485–490.

Chen, L. F. & Greene, W. C. (2003). Regulation of distinct biological activities of the NF-kappaB transcription factor complex by acetylation. *J Mol Med (Berl)* 81, 549–557.

Clarke, R., Collins, R., Lewington, S., Alfthan, G., Tuomilehto, J., Arnesen, E., Bonaa, K., Blacher, J., Boers, G. H., Bostom, A., Bots, M. L., Grobbee, D. E., Brattstrom, L., Breteler, M. M., Hofman, A., Chambers, J. C., Kooner, J. S., Coull, B. M., Evans, R. W., Kuller, L. H., Evers, S., Folsom, A. R., Freyburger, G., Parrot, F., Genest, J., Dalery, K., Graham, I. M., Daly, L., Hoogeveen, E. K., Kostense, P. J., Stehouwer, C. D., Hopkins, P. N., Jacques, P., Selhub, J., Luft, F. C., Jungers, P., Lindgren, A., Lolin, Y. I., Loehrer, F., Fowler, B., Mansoor, M. A., Malinow, M. R., Ducimetiere, P., Nygard, O., Refsum, H., Vollset, S. E., Ueland, P. M., Omenn, G. S., Beresford, S. A., Roseman, J. M., Parving, H. H., Gall, M. A., Perry, I. J., Ebrahim, S. B., Shaper, A. G., Robinson, K., Jacobsen, D. W., Schwartz, S. M., Siscovick, D. S., Stampfer, M. J., Hennekens, C. H., Feskens, E. J., Kromhout, D., Ubbink, J., Elwood, P., Pickering, J., Verhoef, P., von Eckardstein, A., Schulte, H., Assmann, G., Wald, N., Law, M. R., Whincup, P. H., Wilcken, D. E., Sherliker, P., Linksted, P. & Davey Smith, G. (2002). Homocysteine and risk of ischemic heart disease and stroke: a meta-analysis. *JAMA* 288, 2015–2022.

Collas, P., Noer, A. & Timoskainen, S. (2007). Programming the genome in embryonic and somatic stem cells. *J Cell Mol Med* 11, 602–620.

Crott, J. W., Liu, Z., Keyes, M. K., Choi, S. W., Jang, H., Moyer, M. P. & Mason, J. B. (2008). Moderate folate depletion modulates the expression of selected genes involved in cell cycle, intracellular signaling and folate uptake in human colonic epithelial cell lines. *J Nutr Biochem* 19, 328–335.

Davis, C. A., Haberland, M., Arnold, M. A., Sutherland, L. B., McDonald, O. G., Richardson, J. A., Childs, G., Harris, S., Owens, G. K. & Olson, E. N. (2006). PRISM/PRDM6, a transcriptional repressor that promotes the proliferative gene program in smooth muscle cells. *Mol Cell Biol* 26, 2626–2636.

Dayal, S., Arning, E., Bottiglieri, T., Boger, R. H., Sigmund, C. D., Faraci, F. M. & Lentz, S. R. (2004). Cerebral vascular dysfunction mediated by superoxide in hyperhomocysteinemic mice. *Stroke* 35, 1957–1962.

Defossez, P. A., Kelly, K. F., Filion, G. J., Perez-Torrado, R., Magdinier, F., Menoni, H., Nordgaard, C. L., Daniel, J. M. & Gilson, E. (2005). The human enhancer blocker CTC-binding factor interacts with the transcription factor Kaiso. *J Biol Chem* 280, 43017–43023.

Devlin, A. M., Bottiglieri, T., Domann, F. E. & Lentz, S. R. (2005). Tissue-specific changes in H19 methylation and expression in mice with hyperhomocysteinemia. *J Biol Chem* 280, 25506–25511.

Devlin, A. M., Singh, R., Wade, R. E., Innis, S. M., Bottiglieri, T. & Lentz, S. R. (2007). Hypermethylation of Fads2 and altered hepatic fatty acid and phospholipid metabolism in mice with hyperhomocysteinemia. *J Biol Chem* 282, 37082–37090.

Dobosy, J. R., Fu, V. X., Desotelle, J. A., Srinivasan, R., Kenowski, M. L., Almassi, N., Weindruch, R., Svaren, J. & Jarrard, D. F. (2008). A methyl-deficient diet modifies histone methylation and alters Igf2 and H19 repression in the prostate. *Prostate* 68, 1187–1195.

Dover, J., Schneider, J., Tawiah-Boateng, M. A., Wood, A., Dean, K., Johnston, M. & Shilatifard, A. (2002). Methylation of histone H3 by COMPASS requires ubiquitination of histone H2B by Rad6. *J Biol Chem* 277, 28368–28371.

Eckhardt, F., Lewin, J., Cortese, R., Rakyan, V. K., Attwood, J., Burger, M., Burton, J., Cox, T. V., Davies, R., Down, T. A., Haefliger, C., Horton, R., Howe, K., Jackson, D. K., Kunde, J., Koenig, C., Liddle, J., Niblett, D., Otto, T., Pettett, R., Seemann, S., Thompson, C., West, T., Rogers, J., Olek, A., Berlin, K. & Beck, S. (2006). DNA methylation profiling of human chromosomes 6, 20 and 22. *Nat Genet* 38, 1378–1385.

Ehrlich, M., Gama-Sosa, M. A., Huang, L. H., Midgett, R. M., Kuo, K. C., McCune, R. A. & Gehrke, C. (1982). Amount and distribution of 5-methylcytosine in human DNA from different types of tissues of cells. *Nucleic Acids Res* 10, 2709–2721.

Finkelstein, J. D. (2007). Metabolic regulatory properties of S-adenosylmethionine and S-adenosylhomocysteine. *Clin Chem Lab Med* 45, 1694–1699.

Fish, J. E., Matouk, C. C., Rachlis, A., Lin, S., Tai, S. C., D'Abreo, C. & Marsden, P. A. (2005). The expression of endothelial nitric-oxide synthase is controlled by a cell-specific histone code. *J Biol Chem* 280, 24824–24838.

Fraga, M. F., Ballestar, E., Paz, M. F., Ropero, S., Setien, F., Ballestar, M. L., Heine-Suner, D., Cigudosa, J. C., Urioste, M., Benitez, J., Boix-Chornet, M., Sanchez-Aguilera, A., Ling, C., Carlsson, E., Poulsen, P., Vaag, A., Stephan, Z., Spector, T. D., Wu, Y. Z., Plass, C. & Esteller, M. (2005). Epigenetic differences arise during the lifetime of monozygotic twins. *Proc Natl Acad Sci U S A* 102, 10604–10609.

Fuks, F., Hurd, P. J., Deplus, R. & Kouzarides, T. (2003). The DNA methyltransferases associate with HP1 and the SUV39H1 histone methyltransferase. *Nucleic Acids Res* 31, 2305–2312.

Gabory, A., Attig, L. & Junien, C. (2009). Sexual dimorphism in environmental epigenetic programming. *Mol Cell Endocrinol* 304, 8–18.

Geisel, J., Schorr, H., Heine, G. H., Bodis, M., Hubner, U., Knapp, J. P. & Herrmann, W. (2007). Decreased p66Shc promoter methylation in patients with end-stage renal disease. *Clin Chem Lab Med* 45, 1764–1770.

Goldberg, A. D., Allis, C. D. & Bernstein, E. (2007). Epigenetics: a landscape takes shape. *Cell* 128, 635–638.

Goll, M. G. & Bestor, T. H. (2005). Eukaryotic cytosine methyltransferases. *Annu Rev Biochem* 74, 481–514.

Gomez, D., Coyet, A., Ollivier, V., Jeunemaitre, X., Jondeau, G., Michel, J. B. & Vranckx, R. (2011). Epigenetic control of vascular smooth muscle cells in Marfan and non-Marfan thoracic aortic aneurysms. *Cardiovasc Res* 89, 446–456.

Han, J., Kim, D. & Morris, K. V. (2007). Promoter-associated RNA is required for RNA-directed transcriptional gene silencing in human cells. *Proc Natl Acad Sci U S A* 104, 12422–12427.

Han, L., Lee, D. H. & Szabo, P. E. (2008). CTCF is the master organizer of domain-wide allele-specific chromatin at the H19/Igf2 imprinted region. *Mol Cell Biol* 28, 1124–1135.

Hansson, G. K. (2005). Inflammation, atherosclerosis, and coronary artery disease. *N Engl J Med* 352, 1685–1695.

Hastings, N. E., Simmers, M. B., McDonald, O. G., Wamhoff, B. R. & Blackman, B. R. (2007). Atherosclerosis-prone hemodynamics differentially regulates endothelial and smooth muscle cell phenotypes and promotes pro-inflammatory priming. *Am J Physiol Cell Physiol* 293, C1824–C1833.

Hiltunen, M. O., Turunen, M. P., Hakkinen, T. P., Rutanen, J., Hedman, M., Makinen, K., Turunen, A. M., Aalto-Setala, K. & Yla-Herttuala, S. (2002). DNA hypomethylation and methyltransferase expression in atherosclerotic lesions. *Vasc Med* 7, 5–11.

Hiltunen, M. O. & Yla-Herttuala, S. (2003). DNA methylation, smooth muscle cells, and atherogenesis. *Arterioscler Thromb Vasc Biol* 23, 1750–1753.

Hopkins, P. N. & Williams, R. R. (1981). A survey of 246 suggested coronary risk factors. *Atherosclerosis* 40, 1–52.

Huang, V., Qin, Y., Wang, J., Wang, X., Place, R. F., Lin, G., Lue, T. F. & Li, L. C. (2010). RNAa is conserved in mammalian cells. *PLoS ONE* 5, e8848.

Ikegami, K., Ohgane, J., Tanaka, S., Yagi, S. & Shiota, K. (2009). Interplay between DNA methylation, histone modification and chromatin remodeling in stem cells and during development. *Int J Dev Biol* 53, 203–214.

James, S. J., Melnyk, S., Pogribna, M., Pogribny, I. P. & Caudill, M. A. (2002). Elevation in S-adenosylhomocysteine and DNA hypomethylation: potential epigenetic mechanism for homocysteine-related pathology. *J Nutr* 132, 2361S–2366S.

Janowski, B. A., Huffman, K. E., Schwartz, J. C., Ram, R., Nordsell, R., Shames, D. S., Minna, J. D. & Corey, D. R. (2006). Involvement of AGO1 and AGO2 in mammalian transcriptional silencing. *Nat Struct Mol Biol* 13, 787–792.

Janowski, B. A., Younger, S. T., Hardy, D. B., Ram, R., Huffman, K. E. & Corey, D. R. (2007). Activating gene expression in mammalian cells with promoter-targeted duplex RNAs. *Nat Chem Biol* 3, 166–173.

Jiang, Y., Zhang, J., Xiong, J., Cao, J., Li, G. & Wang, S. (2007). Ligands of peroxisome proliferator-activated receptor inhibit homocysteine-induced DNA methylation of inducible nitric oxide synthase gene. *Acta Biochim Biophys Sin (Shanghai)* 39, 366–376.

Kim, J., Kim, J. Y., Song, K. S., Lee, Y. H., Seo, J. S., Jelinek, J., Goldschmidt-Clermont, P. J. & Issa, J. P. (2007). Epigenetic changes in estrogen receptor beta gene in atherosclerotic cardiovascular tissues and in-vitro vascular senescence. *Biochim Biophys Acta* 1772, 72–80.

Kim, M., Long, T. I., Arakawa, K., Wang, R., Yu, M. C. & Laird, P. W. (2010). DNA methylation as a biomarker for cardiovascular disease risk. *PLoS ONE* 5, e9692.

Kono, S. & Chen, K. (2005). Genetic polymorphisms of methylenetetrahydrofolate reductase and colorectal cancer and adenoma. *Cancer Sci* 96, 535–542.

Kouzarides, T. (2007). Chromatin modifications and their function. *Cell* 128, 693–705.

Lamon, B. D. & Hajjar, D. P. (2008). Inflammation at the molecular interface of atherogenesis: an anthropological journey. *Am J Pathol* 173, 1253–1264.

Lanctot, C., Cheutin, T., Cremer, M., Cavalli, G. & Cremer, T. (2007). Dynamic genome architecture in the nuclear space: regulation of gene expression in three dimensions. *Nat Rev Genet* 8, 104–115.

Laukkanen, M. O., Mannermaa, S., Hiltunen, M. O., Aittomaki, S., Airenne, K., Janne, J. & Yla-Herttuala, S. (1999). Local hypomethylation in atherosclerosis found in rabbit ec-sod gene. *Arterioscler Thromb Vasc Biol* 19, 2171–2178.

Lee, M. E. & Wang, H. (1999). Homocysteine and hypomethylation. A novel link to vascular disease. *Trends Cardiovasc Med* 9, 49–54.

Lehmann, R., Denes, R., Nienhaus, R., Steinbach, T., Lusztig, G. & Sanatger, R. M. (1980). Alteration of chromatin in early experimental arteriosclerosis. *Artery* 8, 288–293.

Li, L. C., Okino, S. T., Zhao, H., Pookot, D., Place, R. F., Urakami, S., Enokida, H. & Dahiya, R. (2006). Small dsRNAs induce transcriptional activation in human cells. *Proc Natl Acad Sci U S A* 103, 17337–17342.

Lin, X. H., Guo, C., Gu, L. J. & Deuel, T. F. (1993). Site-specific methylation inhibits transcriptional activity of platelet-derived growth factor A-chain promoter. *J Biol Chem* 268, 17334–17340.

Liu, C., Xu, D., Sjoberg, J., Forsell, P., Bjorkholm, M. & Claesson, H. E. (2004). Transcriptional regulation of 15-lipoxygenase expression by promoter methylation. *Exp Cell Res* 297, 61–67.

Liu, Y., Denlinger, C. E., Rundall, B. K., Smith, P. W. & Jones, D. R. (2006). Suberoylanilide hydroxamic acid induces Akt-mediated phosphorylation of p300, which promotes acetylation and transcriptional activation of RelA/p65. *J Biol Chem* 281, 31359–31368.

Lockman, K., Taylor, J. M. & Mack, C. P. (2007). The histone demethylase, Jmjd1a, interacts with the myocardin factors to regulate SMC differentiation marker gene expression. *Circ Res* 101, e115–e123.

Lopez, A. D., Mathers, C. D., Ezzati, M., Jamison, D. T. & Murray, C. J. (2006). Global and regional burden of disease and risk factors, 2001: systematic analysis of population health data. *Lancet* 367, 1747–1757.

Luger, K., Mader, A. W., Richmond, R. K., Sargent, D. F. & Richmond, T. J. (1997). Crystal structure of the nucleosome core particle at 2.8 A resolution. *Nature* 389, 251–260.

Lund, G., Andersson, L., Lauria, M., Lindholm, M., Fraga, M. F., Villar-Garea, A., Ballestar, E., Esteller, M. & Zaina, S. (2004). DNA methylation polymorphisms precede any histological sign of atherosclerosis in mice lacking apolipoprotein E. *J Biol Chem* 279, 29147–29154.

Luoma, J. S., Stralin, P., Marklund, S. L., Hiltunen, T. P., Sarkioja, T. & Yla-Herttuala, S. (1998). Expression of extracellular SOD and iNOS in macrophages and smooth muscle cells in human and rabbit atherosclerotic lesions: colocalization with epitopes characteristic of oxidized LDL and peroxynitrite-modified proteins. *Arterioscler Thromb Vasc Biol* 18, 157–167.

13 Epigenetics and Atherosclerosis

Lusis, A. J. (2000). Atherosclerosis. *Nature* 407, 233–241.

Madsen, C., Dagnaes-Hansen, F., Moller, J. & Falk, E. (2003). Hypercholesterolemia in pregnant mice does not affect atherosclerosis in adult offspring. *Atherosclerosis* 168, 221–228.

Matouk, C. C. & Marsden, P. A. (2008). Epigenetic regulation of vascular endothelial gene expression. *Circ Res* 102, 873–887.

McDonald, O. G. & Owens, G. K. (2007). Programming smooth muscle plasticity with chromatin dynamics. *Circ Res* 100, 1428–1441.

McDonald, O. G., Wamhoff, B. R., Hoofnagle, M. H. & Owens, G. K. (2006). Control of SRF binding to CArG box chromatin regulates smooth muscle gene expression in vivo. *J Clin Invest* 116, 36–48.

Mette, M. F., Aufsatz, W., van der Winden J., Matzke, M. A. & Matzke, A. J. (2000). Transcriptional silencing and promoter methylation triggered by double-stranded RNA. *EMBO J* 19, 5194–5201.

Miao, F., Gonzalo, I. G., Lanting, L. & Natarajan, R. (2004). In vivo chromatin remodeling events leading to inflammatory gene transcription under diabetic conditions. *J Biol Chem* 279, 18091–18097.

Miranda, T. B. & Jones, P. A. (2007). DNA methylation: the nuts and bolts of repression. *J Cell Physiol* 213, 384–390.

Moore, J. H., Asselbergs, F. W. & Williams, S. M. (2010). Bioinformatics challenges for genome-wide association studies. *Bioinformatics* 26, 445–455.

Morris, K. V., Chan, S. W., Jacobsen, S. E. & Looney, D. J. (2004). Small interfering RNA-induced transcriptional gene silencing in human cells. *Science* 305, 1289–1292.

Morris, K. V., Santoso, S., Turner, A. M., Pastori, C. & Hawkins, P. G. (2008). Bidirectional transcription directs both transcriptional gene activation and suppression in human cells. *PLoS Genet* 4, e1000258.

Napoli, C., de Nigris, F., Welch, J. S., Calara, F. B., Stuart, R. O., Glass, C. K. & Palinski, W. (2002). Maternal hypercholesterolemia during pregnancy promotes early atherogenesis in LDL receptor-deficient mice and alters aortic gene expression determined by microarray. *Circulation* 105, 1360–1367.

Napoli, C., Glass, C. K., Witztum, J. L., Deutsch, R., D'Armiento, F. P. & Palinski, W. (1999). Influence of maternal hypercholesterolaemia during pregnancy on progression of early atherosclerotic lesions in childhood: Fate of Early Lesions in Children (FELIC) study. *Lancet* 354, 1234–1241.

Napoli, C., Witztum, J. L., Calara, F., de Nigris, F. & Palinski, W. (2000). Maternal hypercholesterolemia enhances atherogenesis in normocholesterolemic rabbits, which is inhibited by antioxidant or lipid-lowering intervention during pregnancy: an experimental model of atherogenic mechanisms in human fetuses. *Circ Res* 87, 946–952.

Nathan, D. M., Cleary, P. A., Backlund, J. Y., Genuth, S. M., Lachin, J. M., Orchard, T. J., Raskin, P. & Zinman, B. (2005). Intensive diabetes treatment and cardiovascular disease in patients with type 1 diabetes. *N Engl J Med* 353, 2643–2653.

Palomaki, G. E., Melillo, S. & Bradley, L. A. (2010). Association between 9p21 genomic markers and heart disease: a meta-analysis. *JAMA* 303, 648–656.

Pellis, L., Dommels, Y., Venema, D., Polanen, A., Lips, E., Baykus, H., Kok, F., Kampman, E. & Keijer, J. (2008). High folic acid increases cell turnover and lowers differentiation and iron content in human HT29 colon cancer cells. *Br J Nutr* 99, 703–708.

Perini, G., Diolaiti, D., Porro, A. & Della Valle G. (2005). In vivo transcriptional regulation of N-Myc target genes is controlled by E-box methylation. *Proc Natl Acad Sci U S A* 102, 12117–12122.

Post, W. S., Goldschmidt-Clermont, P. J., Wilhide, C. C., Heldman, A. W., Sussman, M. S., Ouyang, P., Milliken, E. E. & Issa, J. P. (1999). Methylation of the estrogen receptor gene is associated with aging and atherosclerosis in the cardiovascular system. *Cardiovasc Res* 43, 985–991.

Poulter, N. (1999). Coronary heart disease is a multifactorial disease. *Am J Hypertens* 12, 92S–95S.

Qiu, P. & Li, L. (2002). Histone acetylation and recruitment of serum responsive factor and CREB-binding protein onto SM22 promoter during SM22 gene expression. *Circ Res* 90, 858–865.

Richardson, B. C. (2002). Role of DNA methylation in the regulation of cell function: autoimmunity, aging and cancer. *J Nutr* 132, 2401S–2405S.

Roman-Gomez, J., Jimenez-Velasco, A., Agirre, X., Cervantes, F., Sanchez, J., Garate, L., Barrios, M., Castillejo, J. A., Navarro, G., Colomer, D., Prosper, F., Heiniger, A. & Torres, A. (2005). Promoter hypomethylation of the LINE-1 retrotransposable elements activates sense/antisense transcription and marks the progression of chronic myeloid leukemia. *Oncogene* 24, 7213–7223.

Rose, G. (1964). Familial patterns in ischaemic heart disease. *Br J Prev Soc Med* 18, 75–80.

Ross, R. (1999). Atherosclerosis — an inflammatory disease. *N Engl J Med* 340, 115–126.

Sanchez, L. A. & Veith, F. J. (1998). Diagnosis and treatment of chronic lower extremity ischemia. *Vasc Med* 3, 291–299.

Sato, N., Maehara, N., Su, G. H. & Goggins, M. (2003). Effects of 5-aza-2′-deoxycytidine on matrix metalloproteinase expression and pancreatic cancer cell invasiveness. *J Natl Cancer Inst* 95, 327–330.

Sauer, J., Mason, J. B. & Choi, S. W. (2009). Too much folate: a risk factor for cancer and cardiovascular disease? *Curr Opin Clin Nutr Metab Care* 12, 30–36.

Schroeder, M. & Mass, M. J. (1997). CpG methylation inactivates the transcriptional activity of the promoter of the human p53 tumor suppressor gene. *Biochem Biophys Res Commun* 235, 403–406.

Schwartz, J. C., Younger, S. T., Nguyen, N. B., Hardy, D. B., Monia, B. P., Corey, D. R. & Janowski, B. A. (2008). Antisense transcripts are targets for activating small RNAs. *Nat Struct Mol Biol* 15, 842–848.

Sharma, P., Kumar, J., Garg, G., Kumar, A., Patowary, A., Karthikeyan, G., Ramakrishnan, L., Brahmachari, V. & Sengupta, S. (2008). Detection of altered global DNA methylation in coronary artery disease patients. *DNA Cell Biol* 27, 357–365.

Shi, Y., Lan, F., Matson, C., Mulligan, P., Whetstine, J. R., Cole, P. A., Casero, R. A. & Shi, Y. (2004). Histone demethylation mediated by the nuclear amine oxidase homolog LSD1. *Cell* 119, 941–953.

Shimabukuro, M., Sasaki, T., Imamura, A., Tsujita, T., Fuke, C., Umekage, T., Tochigi, M., Hiramatsu, K., Miyazaki, T., Oda, T., Sugimoto, J., Jinno, Y. & Okazaki, Y. (2007). Global hypomethylation of peripheral leukocyte DNA in male patients with schizophrenia: a potential link between epigenetics and schizophrenia. *J Psychiatr Res* 41, 1042–1046.

Stenvinkel, P., Karimi, M., Johansson, S., Axelsson, J., Suliman, M., Lindholm, B., Heimburger, O., Barany, P., Alvestrand, A., Nordfors, L., Qureshi, A. R., Ekstrom, T. J. & Schalling, M. (2007). Impact of inflammation on epigenetic DNA methylation — a novel risk factor for cardiovascular disease? *J Intern Med* 261, 488–499.

Tanaka, Y., Fukudome, K., Hayashi, M., Takagi, S. & Yoshie, O. (1995). Induction of ICAM-1 and LFA-3 by Tax1 of human T-cell leukemia virus type 1 and mechanism of down-regulation of ICAM-1 or LFA-1 in adult-T-cell-leukemia cell lines. *Int J Cancer* 60, 554–561.

Tausendschon, M., Dehne, N. & Brune, B. (2011). Hypoxia causes epigenetic gene regulation in macrophages by attenuating Jumonji histone demethylase activity. *Cytokine* 53, 256–262.

Thompson, R. F., Atzmon, G., Gheorghe, C., Liang, H. Q., Lowes, C., Greally, J. M. & Barzilai, N. (2010). Tissue-specific dysregulation of DNA methylation in aging. *Aging Cell* 9, 506–518.

Tsai, M. C., Manor, O., Wan, Y., Mosammaparast, N., Wang, J. K., Lan, F., Shi, Y., Segal, E. & Chang, H. Y. (2010). Long noncoding RNA as modular scaffold of histone modification complexes. *Science* 329, 689–693.

Tsukada, Y., Fang, J., Erdjument-Bromage, H., Warren, M. E., Borchers, C. H., Tempst, P. & Zhang, Y. (2006). Histone demethylation by a family of JmjC domain-containing proteins. *Nature* 439, 811–816.

Turunen, M. P., Lehtola, T., Heinonen, S. E., Assefa, G. S., Korpisalo, P., Girnary, R., Glass, C. K., Vaisanen, S. & Yla-Herttuala, S. (2009). Efficient regulation of VEGF expression by promoter-targeted lentiviral shRNAs based on epigenetic mechanism: a novel example of epigenetherapy. *Circ Res* 105, 604–609.

Venter, J. C. (2010). Multiple personal genomes await. *Nature* 464, 676–677.

Verrier, L., Vandromme, M. & Trouche, D. (2011). Histone demethylases in chromatin crosstalks. *Biol Cell* 103, 381–401.

Villeneuve, L. M., Reddy, M. A., Lanting, L. L., Wang, M., Meng, L. & Natarajan, R. (2008). Epigenetic histone H3 lysine 9 methylation in metabolic memory and inflammatory phenotype of vascular smooth muscle cells in diabetes. *Proc Natl Acad Sci U S A* 105, 9047–9052.

Vogel, N. L., Boeke, M. & Ashburner, B. P. (2006). Spermidine/Spermine N1-Acetyltransferase 2 (SSAT2) functions as a coactivator for NF-kappaB and cooperates with CBP and P/CAF to enhance NF-kappaB-dependent transcription. *Biochim Biophys Acta* 1759, 470–477.

Voo, K. S., Carlone, D. L., Jacobsen, B. M., Flodin, A. & Skalnik, D. G. (2000). Cloning of a mammalian transcriptional activator that binds unmethylated CpG motifs and shares a CXXC domain with DNA methyltransferase, human trithorax, and methyl-CpG binding domain protein 1. *Mol Cell Biol* 20, 2108–2121.

Waddington, C. (1942). The epigenotype. *Endeavour* 1, 18–20.

Wang, S., Aurora, A. B., Johnson, B. A., Qi, X., McAnally, J., Hill, J. A., Richardson, J. A., Bassel-Duby, R. & Olson, E. N. (2008). The endothelial-specific microRNA miR-126 governs vascular integrity and angiogenesis. *Dev Cell* 15, 261–271.

Weaver, I. C., Champagne, F. A., Brown, S. E., Dymov, S., Sharma, S., Meaney, M. J. & Szyf, M. (2005). Reversal of maternal programming of stress responses in adult offspring through methyl supplementation: altering epigenetic marking later in life. *J Neurosci* 25, 11045–11054.

Weaver, I. C., Meaney, M. J. & Szyf, M. (2006). Maternal care effects on the hippocampal transcriptome and anxiety-mediated behaviors in the offspring that are reversible in adulthood. *Proc Natl Acad Sci U S A* 103, 3480–3485.

Weinberg, M. S., Villeneuve, L. M., Ehsani, A., Amarzguioui, M., Aagaard, L., Chen, Z. X., Riggs, A. D., Rossi, J. J. & Morris, K. V. (2006). The antisense strand of small interfering RNAs directs histone methylation and transcriptional gene silencing in human cells. *RNA* 12, 256–262.

Wenger, R. H., Kvietikova, I., Rolfs, A., Camenisch, G. & Gassmann, M. (1998). Oxygenregulated erythropoietin gene expression is dependent on a CpG methylation-free hypoxiainducible factor-1 DNA-binding site. *Eur J Biochem* 253, 771–777.

White, G. P., Watt, P. M., Holt, B. J. & Holt, P. G. (2002). Differential patterns of methylation of the IFN-gamma promoter at CpG and non-CpG sites underlie differences in IFN-gamma gene expression between human neonatal and adult CD45RO- T cells. *J Immunol* 168, 2820–2827.

Wilcox, J. N., Subramanian, R. R., Sundell, C. L., Tracey, W. R., Pollock, J. S., Harrison, D. G. & Marsden, P. A. (1997). Expression of multiple isoforms of nitric oxide synthase in normal and atherosclerotic vessels. *Arterioscler Thromb Vasc Biol* 17, 2479–2488.

Wild, A., Ramaswamy, A., Langer, P., Celik, I., Fendrich, V., Chaloupka, B., Simon, B. & Bartsch, D. K. (2003). Frequent methylation-associated silencing of the tissue inhibitor of metalloproteinase-3 gene in pancreatic endocrine tumors. *J Clin Endocrinol Metab* 88, 1367–1373.

Wilson, V. L. & Jones, P. A. (1983). DNA methylation decreases in aging but not in immortal cells. *Science* 220, 1055–1057.

Wissler, R. W. (1991). Update on the pathogenesis of atherosclerosis. *Am J Med* 91, 3S–9S.

Yamamura, H., Masuda, H., Ikeda, W., Tokuyama, T., Takagi, M., Shibata, N., Tatsuta, M. & Takahashi, K. (1997). Structure and expression of the human SM22alpha gene, assignment of the gene to chromosome 11, and repression of the promoter activity by cytosine DNA methylation. *J Biochem* 122, 157–167.

Yang, K., Zheng, X. Y., Qin, J., Wang, Y. B., Bai, Y., Mao, Q. Q., Wan, Q., Wu, Z. M. & Xie, L. P. (2008). Up-regulation of p21WAF1/Cip1 by saRNA induces G1-phase arrest and apoptosis in T24 human bladder cancer cells. *Cancer Lett* 265, 206–214.

Ying, A. K., Hassanain, H. H., Roos, C. M., Smiraglia, D. J., Issa, J. J., Michler, R. E., Caligiuri, M., Plass, C. & Goldschmidt-Clermont, P. J. (2000). Methylation of the estrogen receptor-alpha gene promoter is selectively increased in proliferating human aortic smooth muscle cells. *Cardiovasc Res* 46, 172–179.

Yla-Herttuala, S., Palinski, W., Butler, S. W., Picard, S., Steinberg, D. & Witztum, J. L. (1994). Rabbit and human atherosclerotic lesions contain IgG that recognizes epitopes of oxidized LDL. *Arterioscler Thromb* 14, 32–40.

Yla-Herttuala, S., Palinski, W., Rosenfeld, M. E., Parthasarathy, S., Carew, T. E., Butler, S., Witztum, J. L. & Steinberg, D. (1989). Evidence for the presence of oxidatively modified low density lipoprotein in atherosclerotic lesions of rabbit and man. *J Clin Invest* 84, 1086–1095.

Chapter 14
Microbe-Induced Epigenetic Alterations

Hans Helmut Niller, Ferenc Banati, Eva Ay, and Janos Minarovits

Abbreviations

actA	Gene of actin assembly-inducing protein
AID	Activation-induced cytidine deaminase
AIDS	Acquired immunodeficiency syndrome
ARC	AIDS-related complex
BL	Burkitt's lymphoma
BMDC	Bone marrow-derived cell
CagA	Cytotoxicity-associated antigen
CBF1	C promoter-binding factor 1, also termed RBP-Jκ
CGI	CpG island
CIITA	Class II transactivator
CSF1R	Colony-stimulating factor 1 receptor
CTF	CCAAT box-binding transcription factor
DNMT	DNA methyl transferase
EB	Elementary body
EhMLBP	*Entamoeba histolytica*-methylated LINE-binding protein
EMT	Epithelial–mesenchymal transition
ERF	Ets-2 repressor factor
ES	Expression site
ESB	ES body

H.H. Niller, MD
Institute for Medical Microbiology and Hygiene,
University of Regensburg, Regensburg, Germany
e-mail: Hans-Helmut.Niller@klinik.uni-regensburg.de

F. Banati • E. Ay • J. Minarovits , MD, MSc (✉)
Microbiological Research Group, National Center for Epidemiology,
Budapest, Hungary
e-mail: minimicrobi@hotmail.com

J. Minarovits and H.H. Niller (eds.), *Patho-Epigenetics of Disease*,
DOI 10.1007/978-1-4614-3345-3_14, © Springer Science+Business Media New York 2012

flaA	Flagellin A gene
GC	Gastric carcinoma
H3K27me3	Histone 3 trimethylation at lysine 27
H3K4me2	Histone 3 dimethylation at lysine 4
H3K4me3	Histone 3 trimethylation at lysine 4
H3S10	Histone 3 serine 10
H3S10pho	Histone 3 phosphorylation at serine 10
H4K16ac	Histone 4 acetylation at lysine 16
H4K20me3	Histone 4 trimethylation at lysine 20
H4R3me2	Histone 4 dimethylation at arginine 3
HAART	Highly active antiretroviral therapy
Hc1	Chlamydial histone-like protein
HCMV	Human cytomegalovirus
HCV	Hepatitis C virus
HDAC	Histone deacetylase
HIV	Human immunodeficiency virus
HL	Hodgkin's lymphoma
hly	Listeriolysin O gene
HP	Heterochromatin protein
Hp	*Helicobacter pylori*
HRF	HIV reactivating factor
IE	Immediate-early
Igf2	Insulin-like growth factor 2
IL	Interleukin
inlC	Internalin C gene
JNK	c-Jun NH_2-terminal kinase
LINE	Long interspersed nuclear element
LLO	Listeriolysin O
lncRNA	Long noncoding RNA
LTR	Long terminal repeat
MAF	Musculoaponeurotic fibrosarcoma
MBD2	Methyl-CpG-binding protein 2
MIEP	Major immediate-early promoter
MPL	Metalloproteinase
NHL	Non-Hodgkin's lymphoma
NO	Nitric oxide
NUE	Nuclear effector
OspF	Outer *Shigella* protein F
PCNSL	Primary central nervous system lymphoma
PD	Periodontal disease
PFO	Perfringolysin O
PLY	Pneumolysin
PP2A	Protein phosphatase 2A
PRC	Polycomb repressive complex

PRMT1	Protein arginine methyltransferase 1
PU.1	Purine-rich box-1
RB	Reticulate body
RT	Reverse transcriptase
SAHA	Suberoylanilide hydroxamic acid
SAM	*S*-Adenosyl-ʟ-methionine
SH2	Src homology 2
SHP2	SH2-containing tyrosine phosphatase
SPEM	Spasmolytic polypeptide-expressing metaplasia
TBP	TATA-binding protein
TLR	Toll-like receptor
TNFα	Tumor necrosis factor-alpha
TpSCOP	*Theileria parva* schizont-derived cytoskeleton-binding protein
TSA	Trichostatin A
uhpT	Glucose-6P permease gene
UPEC	Uropathogenic *Escherichia coli*
VacA	Vacuolating cytotoxin
VPA	Valproic acid
Vpr	Viral protein R
VSG	Variant surface glycoprotein

14.1 Introduction

By now, it is recognized that a broad range of pathogenic microorganisms, including both viruses and bacteria, affect the epigenetic regulatory mechanisms of their host cells, thereby inducing patho-epigenetic alterations that may result in cellular dysfunction and disease development (Minarovits 2009). One may assume that also protozoan parasites and macroparasites, including fungi, helminths, and arthropods, affect the epigenotype of their target cells in the host organism. Many of these pathogenic agents are known to use their own sophisticated epigenetic mechanisms to control the expression of their genomes (Lopez-Rubio et al. 2007b; Gissot and Kim 2008; Iyer et al. 2008; Verstrepen and Fink 2009; Merrick and Duraisingh 2010; Dixon et al. 2010; da Rosa and Kaufman 2012), and recently, unique epigenetic changes were described in neoplasms associated with macroparasite infections, namely, cholangiocarcinomas induced by the liver fluke *Opisthorchis viverrini* and bladder carcinomas caused by *Schistosoma haematobium* (Gutierrez et al. 2004; Chinnasri et al. 2009; Sriraksa et al. 2011). In this chapter, we concentrate primarily on the epigenetic consequences of host–microbe interactions causing illnesses in humans, although relevant animal models and veterinary diseases, as well as the epigenetic regulatory systems of pathogenic protozoa that are potential targets of novel, epigenetic therapeutic approaches, will also be discussed briefly, where they are relevant.

14.2 Epigenetic Changes in Virus-Infected Cells

The genomes of DNA tumor viruses and the DNA copies of RNA tumor virus genomes can either integrate into the cellular DNA or coreplicate with it once per cell cycle as extrachromosomal episomes. In Chap. 8, we discussed both the epigenetic regulation of human oncovirus genomes and the cellular epigenetic alterations elicited by the oncoproteins they are coding for. Here, we focus on the epigenetic changes observed in cells infected with human immunodeficiency virus (HIV) that is not considered to be directly involved in tumorigenesis and briefly describe how epigenetic alterations may contribute to the development of therapy resistance in hepatitis C virus (HCV)-infected, nonmalignant hepatocytes. The interactions of immediate-early proteins encoded by human cytomegalovirus (HCMV), a betaherpesvirus, with cellular histone deacetylases (HDACs) will also be delineated.

14.2.1 Epigenetic Alterations in Cells Infected with Human Immunodeficiency Virus: Background

HIV is the causative agent of the human acquired immunodeficiency syndrome (AIDS). The virus is parenterally transmitted and leads, with the exception of very rare cases of long-term nonprogressors, to the invariably lethal disease AIDS. The first phase of primary infection runs frequently without symptoms. In approximately 30% of cases, it comes along with flu- or mononucleosis-like symptoms or generalized lymphadenopathy. The initial phase is usually followed by an asymptomatic stage of latency which can last for years. During latency, the immune system can compensate for the virus-induced loss of immune cells through an even higher cell production. Towards the end of this stage, when the immune resources are shrinking, a generalized lymphadenopathy stage (LAS) follows which may finally pass via the AIDS-related complex (ARC) to full-blown AIDS. The ARC and AIDS are defined by their characteristic symptoms and by the severe course of opportunistic infections or malignant diseases which are rarely seen in immunocompetent individuals. The level of viral load, the HIV RNA copy number in blood or liquor cerebrospinalis, varies depending on the disease stage. It is an important indicator for disease progression, prognosis, infectiousness, and for therapy monitoring. Viral persistence through the integration of the viral genome into the cellular genome of long-lived resting memory CD4 T cells has been recognized as the major obstacle against curative treatment (Finzi et al. 1997; Siliciano et al. 2003, reviewed by Richman et al. 2009; Margolis 2010). The current standard of HIV therapy is HAART (highly active antiretroviral therapy), a combination of antiviral drugs with different principles of action, for patients with a measurable blood viral load after primary infection. HIV patients have a much higher risk to develop virus-associated cancers during all stages of HIV disease (Aoki and Tosato 2004; Elgui de Oliveira 2007; Grulich et al. 2007;

14 Microbe-Induced Epigenetic Alterations   423

Elgui de Oliveira 2007). Nevertheless, HIV is for the time being not regarded as a classical tumor virus, but as an indirect carcinogen. HAART has significantly decreased the incidence of AIDS-defining infectious diseases and of HIV-associated malignant tumors, like Kaposi's sarcoma, non-Hodgkin's lymphomas (NHLs), and primary central nervous system lymphoma (PCNSL). However, the incidence of Burkitt's lymphoma (BL) and Hodgkin's lymphoma (HL) has not or only slightly decreased in the era of HAART (Aoki and Tosato 2004; Bibas and Antinori 2009; Elgui de Oliveira 2007). Partially, the difference in the response rate of specific EBV-associated tumors to HAART may be explained by the different disease stages where those tumors originate and by their molecular origin (Lenoir and Bornkamm 1987; Niller et al. 2003, 2004a, b). HAART-responsive tumors depend on the severe immune suppression, i.e., they originate in the later stages of HIV disease, when the immune system is already defunct. BL and HL depend on the germinal center hyperactivity in the early phase of HIV disease when the immune system is still able to compensate for its cell losses. Certainly, a causality composed of both immune suppression and immune hyperstimulation may exist for some HIV-associated tumors.

Similarly to other retrovirus genomes, during virus replication, the single-stranded RNA genome of HIV is also converted to a double-stranded DNA (dsDNA) molecule that is integrated into the cellular genome. This integrated provirus is a potential target of epigenetic regulation, and in certain host cells, various epigenetic mechanisms may silence transcription of the proviral genome, resulting in viral latency. Infection of T-helper cells and other HIV target cells results, however, in productive infection in most cases. Expression of HIV proteins may result in cellular dysfunctions either by a direct interference with cellular effector molecules, or indirectly, via epigenetic reprogramming of the host cell.

14.2.1.1 Human Immunodeficiency Virus: Replication Cycle and Epigenetic Silencing of the Provirus in Latently Infected Cells

Like other retroviruses, HIV type 1 and type 2 also replicate with the help of reverse transcriptase (RT), a unique enzyme packaged into the virions. After receptor binding, fusion of the viral and cell membranes, nucleocapsid entry, and decapsidation, RT starts the complex process of reverse transcription that involves several macromolecular rearrangements within the preintegration complex and proceeds through minus strand DNA synthesis, degradation of the RNA genome, and synthesis of the plus strand DNA. The resulting linear dsDNA molecule contains regulatory elements (long terminal repeats, LTRs) at both ends. The viral dsDNA genome integrates into actively transcribed regions of the cellular genome at the LTRs, with the help of the viral integrase enzyme. Transcription of the integrated viral DNA by the cellular RNA polymerase II starts within the 5'-LTR and results in genome-length transcripts and spliced transcripts. The RNA genomes and the proteins translated from spliced viral mRNAs assemble into nucleocapsids that leave the cell by budding through the cell membrane associated with both viral and cellular proteins. Mature virions are formed by proteolytic processing of HIV-encoded structural proteins by the viral protease within the virus particles.

In certain target cells, the integrated HIV genomes are in a dormant state which is termed viral latency. HIV latency is due to transcriptional silencing of the viral promoter located at the 5'-LTR by the cellular epigenetic regulatory machinery. Because latent proviral genomes can be reactivated, persistence of a latent viral reservoir in long-lived cells, especially resting memory CD4+ T cells, poses a formidable barrier to curative antiretroviral therapy. The factors influencing the decision between establishment of latency through LTR silencing and virus production through active transcription of the proviral HIV genome remain to be clarified. It was suggested that infection of T cells prior to their entrance to a quiescent state or infection of naïve T lymphocytes differentiating in the thymus may favor the entry into latency (Pearson et al. 2008; Margolis 2010).

Silencing of the HIV 5'-LTR involves a series of epigenetic mechanisms. CpG methylation was shown to inhibit the activity of the HIV promoter both directly, by blocking transcription factor binding, and indirectly, by attracting the methyl-CpG-binding protein (MBD2) that established a repressive chromatin structure through recruitment of HDACs (Bednarik et al. 1987, 1990; Schulze-Forster et al. 1990; Kauder et al. 2009). Accordingly, it was observed that primary CD4+ T cells infected by HIV in vitro turned to a quiescent state after initial replication, in parallel with an increased level of CpG methylation of the LTR sequences (Kauder et al. 2009), and there was a higher level of 5'-LTR methylation in memory CD4+ T cells derived from long-time aviremic HIV-infected patients than in those of viremic patients (Blazkova et al. 2009).

Repressive chromatin structures may also contribute to the silencing of HIV LTR. The transcriptional repressors YY1 and LSF, acting in concert, and NF-κB recruited HDAC1 to the viral promoter in various in vitro models (Coull et al. 2000; Quivy et al. 2002; Williams et al. 2006). Binding of the repressor protein CTIP2 recruited HDAC1, HDAC2, and the histone methyltransferase SUV39HT to the proviral LTR in monocytes and microglial cells (Marban et al. 2007). Accordingly, enrichment of H3K9me3 and heterochromatin protein 1 (HP1) was also observed at the inactive LTR. The other repressive mark, H3K27me3, and the corresponding polycomb repressive complex 2 (PRC2) component EZH2 also marked the chromatin associated with the inactive LTR (Friedman et al. 2011). LTR binding and silencing by C promoter-binding factor 1 (CBF1), an effector of the Notch signaling pathway, in concert with HDACs and corepressor proteins, was also observed (Tyagi and Karn 2007). In vivo, in resting T cells, transcription factors c-Myc and Sp1 formed an inhibitory complex with HDAC1 (Jiang et al. 2007). Transcriptional interference by transcripts originating at upstream host genes was also implicated in silencing of HIV LTR (Han et al. 2008).

Silencing of HIV LTR by epigenetic mechanisms could be reverted by various means, including DNA methyl transferase (DNMT) inhibitors (5-aza-C and 5-aza-dC), HDAC inhibitors (suberoylanilide hydroxamic acid, SAHA; trichostatin A, TSA; valproic acid, VPA; trapoxin), NF-κB inducers (tumor necrosis factor-alpha,

14 Microbe-Induced Epigenetic Alterations

TNF-α; HIV reactivating factor, HRF, a protein derived from *Massilia timonae*), and Tat, the HIV transactivator (Bednarik et al. 1987, 1990; Schulze-Forster et al. 1990; Coull et al. 2000; Quivy et al. 2002; Williams et al. 2006; Jiang et al. 2007; Marban et al. 2007; Tyagi and Karn 2007; Blazkova et al. 2009; Kauder et al. 2009; Wolschendorf et al. 2010; Friedman et al. 2011).

These data may help to design a novel therapy that aims at the reactivation of latent HIV proviruses, thereby disrupting the latent viral reservoir. In parallel, by the application of an intensified HAART, the infection of new cells could possibly be blocked, and the elimination of cells reactivated from latency could be achieved (Richman et al. 2009).

14.2.1.2   Silencing of Cellular Genes in HIV-Infected Cells

Mikovits et al. (1998) were the first to report, as far as we know, an epigenetic change induced in host cells by a microbe pathogenic to humans. They observed that acute HIV infection of CD4+ T cells downregulated interferon-gamma (IFN-γ) expression due to increased DNA methyltransferase expression and increased methylation of the *IFN-γ* promoter (Mikovits et al. 1998). One can assume that switching off transcription of *IFN-γ* may contribute to immune evasion by HIV. The same research group also showed that upregulation of DNMT1 in HIV-infected cells increased de novo methylation and silenced the *p16INK4A* promoter (Fang et al. 2001). The significance of the latter finding, i.e., inactivation of a tumor suppressor gene by a cytopathic virus, regarding the development of AIDS and malignancies in HIV-infected patients, remains to be established.

Transcription of the HIV proviral genome results in shorter, double-spliced mRNAs in the early phase of infection. These short viral mRNAs are transported into the cytoplasm and translated into "early," accessory, or regulatory proteins such as Tat, Rev, and Nef. In a later phase of infection, Rev binds to the viral transcripts and ensures the transport of unspliced or single-spliced RNAs into the cytoplasm, followed by the translation of "late" viral proteins, mainly viral enzymes and structural proteins. Youngblood and Reich observed that upregulation of DNMT1 expression in HIV-infected cells was a function of the early HIV proteins. They also suggested that the early HIV proteins activate DNMT1 expression via the AP1 pathway (Youngblood and Reich 2008).

T cell homing and recircularization are indispensable phenomena for the adequate function of the adaptive immune system. HIV infection may disturb, however, these basic functions of T cells by blocking the sialylation of surface glycoproteins involved in these processes. Giordanengo et al. (2004) described that the promoter of the *GNE* gene became hypermethylated in HIV-infected T cell lines. *GNE* encodes UDP-*N*-acetylglucosamine 2-epimerase/*N*-acetylmannosamine kinase that generates the sialyl-donor substrate for cellular sialyltransferases. Thus, a decrease in the level of GNE may result in hyposialylation of surface glycoproteins and thereby possibly deranges the function of HIV-infected T cells, even in aviremic HIV patients who are on long-term HAART.

14.2.1.3 Induction of Premature Sister Chromatid Separation by HIV-1 Vpr, an Accessory Protein Displacing Heterochromatin Protein 1 from the Centromeres of Mitotic Chromosomes

Viral protein R (Vpr) is a pleiotropic regulator involved in the nuclear import of the HIV-1 preintegration complex that also functions as a transcriptional coactivator (Kogan and Rappaport 2011). Recently, Shimura et al. observed that Vpr aberrantly recruited the histone acetyltransferase activity of p300 and stimulated the acetylation of histone H3, resulting in the displacement of HP1-α and HP1-γ from the chromatin. In parallel, cellular proteins indispensable for proper chromosome segregation, including hRad2, hSgo-1, and hMis12, were also displaced from the centromeres of mitotic chromosomes in Vpr-expressing cells. All of these events caused premature chromatid separation. Thus, HIV-1 infection epigenetically disrupted the higher-order structure of heterochromatin thereby impairing centromere cohesion (Shimura et al. 2011).

14.2.2 Hepatitis C Virus-Induced Epigenetic Alterations May Contribute to Therapy Failure

HCV is associated with the development of hepatocellular carcinoma, and the epigenetic alterations of HCV-associated neoplasms were discussed in Chap. 8. Here, we would like to emphasize another potential consequence of HCV-induced epigenetic changes that may affect the outcome of standard therapy of patients with chronic liver disease. The success of pegylated IFN-α therapy depends on the induction of a set of cellular genes via the type I IFN receptor–Jak–STAT pathway. Activation of DNA methyltransferases by the HCV core protein (Arora et al. 2008) may result, however, in de novo methylation and epigenetic silencing of IFN-α responsive genes in HCV-infected cells and therapy failure (Naka et al. 2006). In addition, HCV infection of cell cultures upregulated protein phosphatase 2A (PP2A), an inhibitor of protein arginine methyltransferase 1 (PRMT1), thereby reducing the level of histone H4 dimethylated at arginine 3 (H4R3me2) (Duong et al. 2010). In parallel, acetylation of lysine 16 in histone 4 (H4K16ac) and trimethylation of lysine 20 in histone 4 (H4K20me3) was also significantly downregulated. The HCV-induced repressive histone modifications could be corrected by treatment of the cells with the methyl-donor S-adenosyl-L-methionine (SAM). Thus, Duong et al. suggested that co-administration of SAM with pegylated IFN and ribavirin to patients with chronic HCV infection may improve the efficiency of treatment (Duong et al. 2010).

14.2.3 Immediate-Early Proteins of Human Cytomegalovirus Block Histone Deacetylases

HCMV, a member of the betaherpesvirus family, is widespread in humans. Similarly to other herpesviruses, HCMV persists for the lifetime of its host after primary infection.

14 Microbe-Induced Epigenetic Alterations   427

The virus establishes latency in cells of the myeloid lineage. Although most infections are asymptomatic, life-threatening disease can develop in immunocompromised patients or if infection occurs in utero, due to productive, lytic replication that destroys the host cells. During the lytic cycle, more than 200 viral proteins are expressed in sequential stages.

Gönczöl et al. (1984) discovered that latent HCMV genomes carried by undifferentiated host cells can enter the lytic cycle upon cell differentiation. Since the incoming viral DNA molecules undergo chromatinization in undifferentiated host cells, in latent HCMV genomes the major immediate-early promoter (MIEP) is silenced by a repressive chromatin structure. One could speculate that chromatinization is a kind of cellular defense mechanism against viral infection (Reeves 2011). Transcriptional repression is achieved by a hypoacetylated chromatin domain associated with HP1 and Ets-2 repressor factor (ERF), a transcriptional repressor recruiting HDAC1 to the MIEP (Murphy et al. 2002; Wright et al. 2005).

The chromatin structure of the MIEP undergoes a profound change when monocytes differentiate to macrophages: the active promoter becomes associated with hypermethylated histones (Murphy et al. 2002). In addition, IE1 and IE2, the abundant protein products of the immediate-early (IE) genes expressed during the lytic cycle, interact with and block the activity of cellular HDACs (Nevels et al. 2004; Park et al. 2007). Thus, IE proteins abolish the epigenetic repression of the MIEP, thereby facilitating viral replication.

IE1 and IE2 do not bind to DNA directly, but they activate cellular promoters either by interacting with the nuclear proteins CTF1 (CCAAT box-binding transcription factor) at TATA-less promoters or TBP (TATA-binding protein) at promoters with a TATA motif, respectively (Hayhurst et al. 1995; Caswell et al. 1996). In addition, it was observed that IE1 induced acetylation of histone H3 and activated the transcription of the human telomerase reverse transcriptase gene (*hTERT*) in human fibroblasts. In parallel, there was an increased binding of the transcription factor Sp1 and a reduced level of HDAC in the *hTERT* promoter region (Straat et al. 2009). These data suggested a potential role for HCMV in the immortalization of the cells. However, this could only occur in infected cells with an incomplete lytic replication or a replication block before the egress of the virions.

14.3   Epigenetic Reprogramming of Host Cells Infected by or Interacting with Pathogenic Bacteria

Pathogenic bacteria may alter the epigenotype of host cells by producing toxins or short-chain fatty acids that alter the pattern of histone modifications in the target cells or block the activity of HDACs, respectively. In addition, effector proteins injected into target cells by bacteria or produced intracellularly are translocated into the target or host cell nuclei where they induce localized chromatin alterations, thereby inhibiting the expression of key immune defense genes. Bacterial infections may also contribute to the alterations of DNA methylation patterns in target cells either directly or by eliciting a chronic inflammatory response.

14.3.1 Alteration of Histone Modification Patterns by Bacterial Toxins: Epigenetic Dysregulation as a Tool to Evade the Innate Immune Response by Listeria monocytogenes, Streptococcus pneumoniae, Clostridium perfringens, and Aeromonas hydrophila

Listeria monocytogenes is a rod-shaped Gram-positive bacterial all-rounder with a high environmental resistance, an ability to grow even in the refrigerator, and an adaptation to a facultative intracellular life. Food-borne *Listeria* infections can cause sepsis and meningitis in immunosuppressed individuals. Sometimes unnoticed at first, intrauterine infection of the fetus is feared because it can cause abort or granulomatosis infantiseptica.

Listeria monocytogenes is phagocytosed by macrophages or granulocytes where they not only survive the phagosomal environment but even replicate. The genes for the virulence factors which are essential for the intracellular life style, viz., phospholipase C (plcA and plcB), listeriolysin O (hly), actin assembly-inducing protein (actA), zinc metalloproteinase (mpl), internalin C (inlC), and glucose-6P permease (uhpT) are all under the control of a single transcriptional activator, PrfA (Eisenreich et al. 2010). The key virulence factor listeriolysin O (LLO) which enables the bacteria to escape from the phagosome into the cytoplasm belongs to a large class of cholesterol-dependent pore-forming cytolysins which are secreted by Gram-positive bacteria (reviewed by Hamon et al. 2006). LLO strongly activates the NF-κB, MAPK, phosphatidylinositol, calcium, and protein kinase C signaling pathways, through pore formation or just by binding to the cell membrane without pore formation. Before invasion, LLO secreted by extracellular bacteria triggers a specific transcriptional response in macrophages which is revealed by transcriptional profiling (Hamon et al. 2007). Transcriptional reprogramming which leads to the repression of key immunity factors, including the chemoattractant chemokine CXCL2 which is involved in the recruitment of polymorphonuclear cells, is achieved through a modification of the cellular histone code. Specifically, *L. monocytogenes* infection decreased the overall phosphorylation level of H3S10 and the acetylation level of H3 and H4 in cell culture, while the overall methylation level at H3 remained unchanged. H3 dephosphorylation at serine 10, however, took place already at an early stage of infection, before cytoplasmic invasion had occurred. The main factor responsible for H3 dephosphorylation and H4 deacetylation was LLO. Cell membrane binding by LLO triggered the effect, while pore formation was not required. A similar effect was induced by the two related bacterial cholesterol-dependent cytolysins, pneumococcal pneumolysin (PLY), a major neurotoxin of *S. pneumoniae*, and perfringolysin O (PFO) of *C. perfringens*. The H3 dephosphorylation effect was specific to the cholesterol-dependent toxins, but not to other unrelated bacterial pore-forming cytotoxins. The exact signaling pathway by which LLO causes

14 Microbe-Induced Epigenetic Alterations

alterations of the histone code was an enigma for a while (Hamon et al. 2007). Recently, however, a pore-dependent K(+) efflux was identified as the signal required for histone H3 dephosphorylation by LLO (Hamon and Cossart 2011). Similarly to LLO, K(+) efflux and histone H3 dephosphorylation was also caused by aerolysin, an unrelated pore-forming cytolytic toxin produced by the Gram-negative bacterium *A. hydrophila* that is known as a food-borne pathogen causing diarrhea, but also wound infections (Hamon and Cossart 2011).

In later stages of *L. monocytogenes* infection, bacterial cytoplasmic invasion leads to the transcriptional activation of a set of cytokine genes (Schmeck et al. 2005). Therefore, one may speculate that signaling to establish a repressive histone code may be an immune suppressive mechanism by which bacteria keep the innate immune response at bay in expectation of its later activation. Accordingly, *L. monocytogenes* and *C. perfringens* infections are known for their initially low inflammatory activity. Furthermore, a Δ-PLY mutant of *S. pneumoniae* triggers a stronger inflammatory response than wild-type pneumococci (Benton et al. 1995). It is worthy to note that LLO, PLY, and PFO also share the capacity to impair a posttrancriptional process, protein SUMOylation by inducing the degradation of Ubc9, an essential enzyme in the SUMO-conjugation of proteins (Ribet et al. 2010).

14.3.2 *Extracellular Secretion of Butyric Acid, an Inhibitor of Histone Deacetylases, by Pathogenic Porphyromonas gingivalis Strains Implicated in Periodontal Disease*

Periodontal disease (PD) affects the majority of humans. It is caused by pathogenic Gram-negative anaerobic bacteria eliciting chronic subgingival inflammation termed periodontitis. In addition to local consequences such as the loosening and loss of teeth, PD may result in systemic diseases such as atherosclerosis, rheumatoid arthritis, and diabetes as well. Furthermore, PD is associated with adverse pregnancy outcomes and aspiration pneumonia (Darveau 2009). One of the major etiological agents of PD, *P. gingivalis*, produces short-chain fatty acids and secretes them into its environment (Tonetti et al. 1987; Kurita-Ochiai et al. 1995; Niederman et al. 1997). The dominant fatty acid species synthesized by *P. gingivalis* is butyric acid, an inhibitor of HDACs. Accumulation of butyric acid was detected in periodontal pockets, and it was observed that both butyric acid and supernatants of *P. gingivalis* cultures induced histone acetylation in a T cell line and a macrophage cell line in vitro (Tonetti et al. 1987; Niederman et al. 1997; Imai et al. 2009). One may speculate that butyric acid, produced by *P. gingivalis*, may affect the gene expression pattern of gingival epithelial cells or immune cells, thereby affecting the outcome of local immune responses.

14.3.3 Alteration of Histone Modifications at Immune Defense Gene Promoters by Bacterial Effector Proteins: Epigenetic Dysregulation as a Tool to Evade the Innate Immune Response by Shigella flexneri, Anaplasma phagocytophilum, and Mycobacterium tuberculosis

Recently, other bacteria have been shown to interfere with the immune response through advancing a repressive histone code as well. The Gram-negative bacterium *S. flexneri* causes diarrhea and injects an effector protein, OspF (outer *Shigella* protein F), into epithelial cells. OspF translocates to the host cell nucleus where it exerts its phosphatase activity that interferes with the MAPK pathway and thereby prevents histone H3S10 phosphorylation at a specific set of NF-κB responsive genes (Arbibe et al. 2007). Because *IL8* is one of the OspF-silenced genes, *S. flexneri* infection may block neutrophil recruitment, due to the downregulation of IL-8 expression.

Anaplasma phagocytophilum is a tick-transmitted rickettsial pathogen. As an obligatory intracellular bacterium, it preferentially replicates inside monocytes and granulocytes, thereby causing an acute febrile disease, human granulocytic anaplasmosis (Rikihisa 2010). *Anaplasma phagocytophilum* is capable to survive in the potentially hostile intracellular environment due to its ability to downregulate the expression of three defense gene clusters coding for antimicrobial peptides and components of a machinery generating reactive oxygen intermediates (Garcia-Garcia et al. 2009). *Anaplasma phagocytophilum* uses its effector protein, AnkA, for silencing the host genes. AnkA translocates to the nucleus and directs HDAC1 to the defense gene promoters, resulting in histone H3 deacetylation and transcriptional repression (Garcia-Garcia et al. 2009).

Mycobacteria are facultative intracellular pathogens capable to survive and multiplicate in macrophages. *Mycobacterium tuberculosis* is the causative agent of tuberculosis whereas *Mycobacterium avium* infections are common in untreated AIDS patients. Pennini et al. observed that the 19K lipoprotein of *M. tuberculosis* interfered with IFN-γ-induced expression of several immune function genes in macrophages, including class II transactivator (CIITA) (Pennini et al. 2006). *Mycobacterium avium* or *M. tuberculosis* infection as well as the 19K lipoprotein acted via MAPK and TLR2 signaling, thereby blocking IFN-γ-induced chromatin acetylation and remodeling at several immune function genes (Wang et al. 2005; Pennini et al. 2006). It seems plausible that suppression of CIITA transcription may downregulate class II MHC expression by *M. tuberculosis*-infected macrophages and facilitate immune evasion by the inhibition of antigen presentation.

14.3.4 The Nuclear Effector of Chlamydia trachomatis: A Bacterial Histone Methyltransferase Targeting Histones in Host Cell Nuclei

Chlamydia trachomatis is a sexually transmitted obligate intracellular bacterium causing infertility, ectopic pregnancy, and pelvic inflammatory disease. Ocular infection by *C. trachomatis* may cause blindness. The infectious, extracellular form of the bacteria is metabolically inactive (elementary bodies, EBs). In contrast, the intracellular form is metabolically active, but noninfectious (reticulate bodies, RBs). RBs are included within a membrane-bound vacuole in the host cell. Pennini et al. observed that the type III secretion system of *C. trachomatis* transported NUE (nuclear effector), a bacterial effector protein with histone methyltransferase activity, outside of the inclusion. NUE was detected in the host cell nuclei where it methylated histones H2B, H3, and H4, but not the chlamydial histone-like protein (Hc1), involved in compacting the bacterial DNA during RB to EB transition. Further, they speculated that NUE may alter the gene expression pattern of the host cell in the late phase of the bacterial infection (Pennini et al. 2010). Thus, the function of NUE could be similar to that of the *A. phagocytophilum* effector protein AnkA: epigenetic reprogramming of the infected cell.

14.3.5 Methylation of Histone H3 by cpnSET, a Chlamydophila pneumoniae-Encoded Protein Methyltransferase

Chlamydophila pneumoniae causes acute respiratory diseases, and it has been implicated in the pathogenesis of atherosclerosis as well. Similarly to *C. trachomatis, C. pneumoniae* is also an obligate intracellular bacterium with the alternate morphologies EB and RB (elementary and reticulate bodies). Murata et al. characterized the *C. pneumoniae*-encoded histone methyltransferase cpnSET that was capable to methylate both murine histone H3 and the chlamydial histone H1-like protein Hc1. Further, they suggested that cpnSET probably methylates Hc1 in vivo, thereby contributing to the morphological change of the bacterium, or may act on host cell histones transported into chlamydial cells (Murata et al. 2007). However, in view of the nuclear localization of NUE, a protein methyltransferase of a closely related bacterial species with a similar life cycle (see above), the modification of host cell histones within the host cell nuclei by cpnSET cannot be excluded at present.

14.3.6 Induction of Altered Gene Expression and Promoter Hypermethylation by Campylobacter rectus: A Murine Model of Infection-Mediated Intrauterine Growth Restriction

Intrauterine growth restriction causes low birth weight and thereby increases the risk for perinatal morbidity and mortality. It is a multifactorial disease due to the dysregulation of placental/fetal gene expression. One of the factors associated with intrauterine growth restriction is periodontal disease (PD) (see Sect. 14.3.2) that may have systemic effects through bacteremia. In an experimental mouse infection model, maternal infection at a site distant to the placenta with the periodontal bacteria *P. gingivalis* and *C. rectus* induced intrauterine growth restriction (Lin et al. 2003a, b; Offenbacher et al. 2005; Yeo et al. 2005). *Campylobacter rectus* infection of pregnant mice upregulated the expression of 9 genes in the placenta and down-regulated the transcription of 65 genes. The upregulated genes were related to the oxygen supply and vascular development of the fetus, whereas the suppressed genes were mostly associated with placental and fetal development. The latter gene set included a series of imprinted genes that are marked and regulated by DNA methylation in a parent-specific manner (Bobetsis et al. 2010). Although the mechanism of transcriptional repression induced by *C. rectus* has not been addressed in this particular transcriptome analysis, one of the downregulated genes, *Igf2* (insulin-like growth factor 2), was found to be silenced by promoter hypermethylation by the same research group in an earlier study (Bobetsis et al. 2007). These data suggest that maternal insult with bacteria may affect the imprinting mechanism of genes controlling fetal and placental growth.

14.3.7 An Epigenetic Mark Left by Uropathogenic Escherichia coli Infection

Urinary tract infection is associated in most cases with *E. coli* infection. The *E. coli* strains causing urinary symptoms (uropathogenic *E. coli* or UPEC strains) differ, however, from the *E. coli* strains populating the gastrointestinal tract in the same individual at the same time, as indicated by the unique expression of iron-scavenger siderophore molecules by the uropathogenic bacteria (Henderson et al. 2009). Infection of uroepithelial carcinoma cell lines by UPEC strains resulted in the formation of intracellular bacterial colonies and induced de novo DNA methyltransferase activity. DNMT1 protein levels were also upregulated. UPEC strains lacking FimH were not internalized and did not induce DNMTs. An increased methylation of the CpG island (CGI) in exon 1 of the CDKN2A gene was also observed after internalization of UPEC strains. In parallel, CDKN2A expression was downregulated, similar to MGMT transcription. Since MGMT methylation was unaltered, one may speculate that in addition to CGI methylation, other epigenetic regulatory mechanisms may also be affected by UPEC infection.

14.3.8 Helicobacter pylori: An Inducer of the CpG Island Methylator Phenotype in Gastric Carcinoma

Helicobacter pylori (Hp) infection is regarded as the most important acquired risk factor for the development of gastric carcinoma (GC). Most infections by Hp, a Gram-negative spiral-shaped microaerophilic bacterium, remain subclinical, although 10–20% of patients develop peptic ulcer, and different types of chronic gastritis such as atrophic gastritis, enlarged fold gastritis, and pangastritis also occur. Chronic inflammation can progress to gastric adenocarcinoma in about 1–2% of the infected patients through discrete steps like gastric atrophy, metaplasia, and dysplasia. Although Hp strains are diverse and the clinical outcome depends on the presence or absence of virulence factors, the risk to develop intestinal metaplasia or different types of malignant neoplasms, such as intestinal type GC, sporadic diffuse-type GC, and MALT lymphoma, is significantly higher in Hp carriers than in uninfected individuals (Uemura et al. 2001, reviewed by Nardone et al. 2007; Yamamoto et al. 2011). In addition to colonization by Hp which is highly adapted to the acidic environment of the stomach, a hereditary component appears to contribute to the development of GC as well (reviewed by El Omar et al. 2000, 2003; Machado et al. 2003; Azuma 2004; Chan et al. 2007; Kabir 2009). Several pathogenic mechanisms, including Hp virulence factors and chronic inflammation of the gastric mucosa may contribute to tumorigenesis. Hp codes for two important virulence factors, vacuolating cytotoxin (VacA) and cytotoxicity-associated antigen (CagA). VacA is a pro-apoptotic protein targeting mitochondria, and it is implicated in atrophic gastritis, a preneoplastic condition (Galmiche et al. 2000). VacA may also act as a pore-forming toxin inducing programmed necrosis of gastric epithelial cells (Radin et al. 2011). CagA, considered to be a bacterial oncoprotein, is encoded within the 40-kb bacterial cag-pathogenicity island. Both proteins are strongly associated with Hp pathogenesis. CagA is injected into gastric epithelial cells by a type IV bacterial secretion system, and after undergoing tyrosine phosphorylation at the inner surface of the plasma membrane (Stein et al. 2002), it affects multiple cellular signal transduction pathways (reviewed by Ding et al. 2010). By binding to the Src homology domain (SH2)-containing tyrosine phosphatase (SHP2), a human oncoprotein, and activating its phosphatase activity (Higashi et al. 2002), CagA may elicit uncontrolled cell proliferation as testified by the development of gastric and small intestinal adenocarcinomas, myeloid leukemias, and B cell lymphomas in CagA transgenic mice (Ohnishi et al. 2008). It is interesting to note that although chronic gastritis appears to be associated with the development of gastric carcinoma in humans, there were no signs of gastritis or systemic inflammation in CagA transgenic mice, including those with gastrointestinal carcinomas. Interaction of CagA with PAR1/MARK (partitioning defective 1/microtubule affinity-regulating kinase) may also contribute to the carcinogenic process because it induces chromosomal instability by destabilizing the microtubules during mitosis (Umeda et al. 2009).

In addition to CagA, the type IV secretion system of Hp also delivers peptidoglycan into the host cell cytoplasm, which upregulates NF-κB (Hutton et al. 2010). In gastric epithelial cells, Hp-induced NF-κB increased the expression of activation-induced cytidine deaminase (AID), an enzyme involved in DNA and RNA editing, and resulted in the accumulation of *TP53* tumor suppressor gene mutations (Matsumoto et al. 2007). Hp infection also leads to an overall increased mutation rate of the genomic and mitochondrial DNA and induces DNA repair impairment (Machado et al. 2009, reviewed by Touati 2010).

In summary, due to its pleiotropic effects, CagA is considered to be the major oncoprotein of *H. pylori* (reviewed by Suerbaum and Michetti 2002; Hatakeyama 2004; Milne et al. 2009). However, there are only indirect data as to the interaction of CagA with the epigenetic regulatory mechanisms of the host cell. Both phosphorylated and unphosphorylated CagA affects the gene expression pattern of human gastric epithelial cells, upregulating certain genes and downregulating others, thereby contributing to the process of epithelial–mesenchymal transition (EMT) (Sohn and Lee 2011). The mechanism of transcriptional repression by CagA remains to be elucidated. Promoter methylation may be involved because in gastric biopsies, *RPRM* and *MGMT* promoter hypermethylation was found to be associated with the presence of cagA (Sepulveda et al. 2010; Schneider et al. 2010), but independently also with the s1m1 allele of vacA (Schneider et al. 2010). In gastric adenocarcinomas, p16INK4A expression from the *CDKN2A* promoter was suppressed by CpG methylation in a Hp genotype-dependent manner: methylated tumors were associated with Hp bearing cagA and vacA-s1m1 genes but not the flaA gene, whereas unmethylated tumors were associated with Hp bearing flaA (Alves et al. 2010). How the *H. pylori* flaA gene encoded flagellin A, a molecule required for flagellar filament assembly and motility, affects the expression of cellular genes is unknown at present. It might not act via the Toll-like receptor 5 (TLR5), which recognizes other flagellar proteins, because the *H. pylori* flagellin escapes detection by TLR5 (Andersen-Nissen et al. 2005).

Several lines of evidence support the idea that epigenetic changes, especially hypermethylation of CGIs, and histone modifications are associated with Hp-induced inflammation, gastric carcinogenesis, and gastric MALT lymphomas (see Sects. 14.3.8.1 and 14.3.8.2). The absence of a direct link between CagA, the major oncoprotein of *H. pylori,* and the host cell epigenetic machinery suggests that either other bacterial proteins and components, or the inflammatory response of the host itself may play a decisive role in the patho-epigenetic alterations observed in Hp-associated gastritis and neoplasms. Regarding the latter possibility, a multistep mechanism may involve induction of interleukin (IL)-1β in gastric fundus mucosa by Hp infection (Wang et al. 1999), followed by the IL-1β-mediated induction of nitric oxide (NO) synthase in the target cells of IL-1β, that may finally result in the NO-mediated direct activation of DNA methyltransferase and CGI methylation (Hmadcha et al. 1999).

14 Microbe-Induced Epigenetic Alterations

14.3.8.1 Alterations of the Host Cell Methylome in *Helicobacter pylori*-Associated Diseases

CGI *hypermethylation* was significantly associated with the presence of Hp infection in gastritis patients (Maekita et al. 2006). Eradication of Hp infection resulted in a decline of CGI methylation. Hp infection is primarily acquired during childhood, and CGI hypermethylation was detected already in Hp-infected pediatric gastric mucosa samples (Shin et al. 2011b). Analysis of CGI methylation in gastric mucosa cells infected with Hp in vitro demonstrated the induction of hypermethylation at the promoter of the *IRX1* tumor suppressor gene (Guo et al. 2011). It was also observed that Hp infection increased microvillus formation and mucous secretion of the GES-1 gastric mucosa cells.

In addition to local hypermethylation of CGIs, a *global hypomethylation* — assessed by 5-methylcytosine immunostaining — was also described in Hp-positive chronic gastritis, independent of the CagA status (Compare et al. 2011). Analysis of repetitive elements by bisulfite pyrosequencing showed hypomethylation of Alu and Sat-α repeats in Hp-infected gastric mucosae of healthy volunteers and GC patients, compared to Hp-negative control samples (Yoshida et al. 2011). It is worthy to note, however, that the Alu repeats, but not the Sat-α repeats, were also hypomethylated in gastric mucosae of Hp-negative GC patients. LINE1 repeats were hypomethylated, however, in primary GC samples only (Yoshida et al. 2011).

A recent genome-wide study analyzed the methylation of 1,505 CpG sites in 807 cancer-related genes in gastric mucosae with or without Hp infection (Shin et al. 2011a). The majority of the differentially methylated genes showed increased methylation in Hp-infected tissues obtained from subjects without GC. The differentially methylated genes were related to developmental processes or signal transduction. Ten hypermethylated and three hypomethylated CpGs in the noncancerous gastric mucosa of GC patients showed an association with the presence of GC, regardless of Hp infection. These epigenetic marks may be useful predictors of the future risk of gastric carcinogenesis. A possible link between Hp infection and GC development was pinpointed by the identification of shared epigenetic alterations between Hp-infected control mucosae and Hp-negative noncancerous mucosae of GC patients: 26 hypermethylated and 13 hypomethylated CpGs fell into this category (Shin et al. 2011a).

In other studies, methylation was found at the *p16, MLH1, ECAD, DAPK,* and *MTSS1* loci in healthy mucosa of GC patients with and without Hp infection (Waki et al. 2002; Yamashita et al. 2006; Kaise et al. 2008; Alves et al. 2011). Based on the association between high methylation levels in histologically normal gastric mucosae and an increased risk of GC development, Ushijima (2007) suggested that the epigenetically altered mucosal tissue corresponded to an *epigenetic field* for cancerization. In precursor lesions of gastric cancer, traditionally thought to originate from normal mucosal progenitor cells, the level of CGI methylation increased in intestinal metaplasia compared to normal or chronic gastritis mucosa, and there was a further significant increase in flat dysplasias or polyploid adenomas unassociated with carcinoma, to a level that almost reached the methylation index of adenocarcinomas (Lee et al. 2004).

In addition to the multistep gastric carcinogenesis triggered by Hp infection that proceeds through goblet cell intestinal metaplasia consisting of cells of intestinal phenotyope (Leung and Sung 2002; Fig. 14.1), other pathways including spasmolytic polypeptide-expressing metaplasia (SPEM) composed from cells similar to antral gland cells and the direct outgrowth of signet ring form of cancer from the stem-cell zone also lead to the development of GC (Gutierrez-Gonzalez and Wright 2008). In a mouse GC model, initiated by the loss of parietal cells that play a role in luminal HCl secretion, SPEM arised from transdifferentiation of proliferating chief cells located at the base of the glands in gastric fundus mucosa (Nam et al. 2010; Fig. 14.1). Transdifferentiation resulted in the upregulation of proteins involved in DNA remodeling and expression of secreted mucosal factors involved in tumor progression (Nozaki et al. 2008). In addition to the transdifferentiation of chief cells, when chronic inflammation was present due to *Helicobacter felis* infection, engraftment of bone marrow-derived cells (BMDCs) could also be observed at the metaplastic area. BMDCs adopted the metaplastic phenotype and proliferated, and the metaplastic lesions progressed to dysplasia (Goldenring and Nomura 2006; Fig. 14.1). These observations confirmed and extended a pioneering study that demonstrated the generation of epithelial cancers from BMDCs in *Helicobacter*-infected mice (Houghton et al. 2004, reviewed by Li et al. 2006). EMT is a phenotypic change associated with the activation of genes enhancing cell motility, invasive behavior, and metastasis formation. A pathogenic Hp strain enhanced the shedding of soluble heparin-binding epithelial growth factor, thereby upregulating the EMT transcriptome (Dickson et al. 2006; Yin et al. 2010). These data suggest that there are complex changes including the alteration of cellular phenotypes and gene expression patterns, and even engraftments of new cell populations, during Hp-initiated gastric carcinogenesis. These processes are certainly controlled by epigenetic regulatory mechanisms and can be related to the alterations of the epigenome. Accordingly, Hong et al. observed a variable pattern of CpG methylation in samples from Hp-positive vs. Hp-negative gastric mucosa and GCs. All out of the eight examined CGI-containing genes were overmethylated in Hp-positive mucosa, whereas none of the six CGI-lacking, stomach-specific genes were overmethylated in the very same samples. In GC, the CGI-containing genes located distant from retro-elements were also overmethylated, but those situated close to retro-elements were not. In addition, depending on the level of loss of heterozygosity events which was estimated by using 40 microsatellite markers, there was a decreased methylation both at CGI-containing and CGI-lacking genes in GCs. These results were interpreted by comparing the data with published results of expression and methylation patterns of the stomach-specific genes in the bone marrow, assuming the fixation of bone marrow-derived stem cells in the gastric mucosa. It was concluded that the observed patterns were compatible with the engraftment of a new cell population in the gastric mucosa (Hong et al. 2010). Further, they suggested that due to the loss of heterozygosity, there was a dose-compensatory demethylation in gastric cancer. This epigenetic change possibly interrupted the nondividing terminal differentiation of newly fixed, bone marrow-derived stem cells, thereby reactivating a stem-cell intrinsic program of cell migration and proliferation (Fig. 14.1).

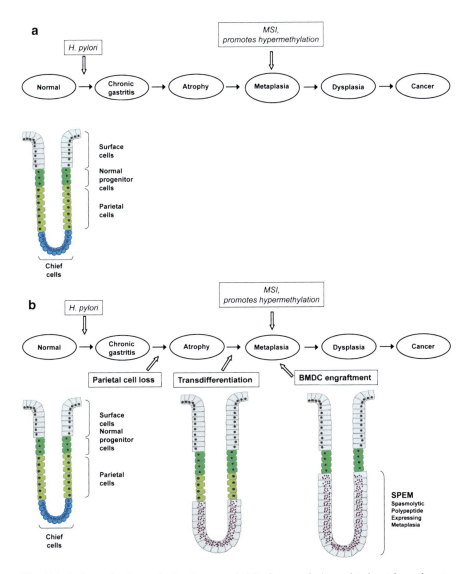

Fig. 14.1 Pathways leading to the development of *Helicobacter pylori*-associated gastric carcinoma. (**a**) The successive stages of gastric carcinogenesis. *Helicobacter pylori* infection of normal gastric mucosa causes chronic gastritis and atrophy, followed by metaplasia associated with microsatellite instability and regional hypermethylation. These changes lead to dysplasia and cancer. A normal fundic gland is depicted below the symbol for normal gastric mucosa (Leung and Sung 2002; Goldenring and Nomura 2006). (**b**) One possible pathway involving transdifferentiation of fundic gland cells after the loss of parietal cells. The developing metaplastic lesion invites bone marrow-derived stem cells that leave the circulation, engraft into the lesion, and change their phenotype, adapting to their new environment. At a later stage, they overgrow the original, local, transdifferentiated cells, contributing to the progression of SPEM (spasmolytic polypeptide-expressing metaplasia) to dysplasia (Goldenring and Nomura 2006). (**c**) The putative epigenetic and genetic events that transform bone marrow-derived stem cells into adapted and later neoplastic

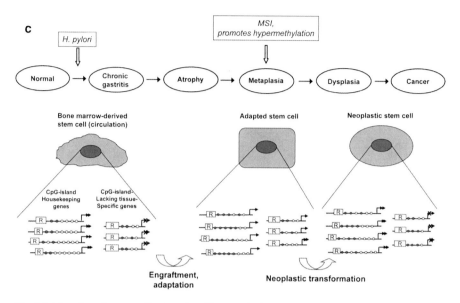

Fig. 14.1 (continued) stem cells, based on Hong et al. (2010). Bone marrow-derived stem cells in the circulation express CpG island (CGI) housekeeping genes (*double-headed arrows*) but not CGI-lacking tissue-specific genes (*crossed-out arrows*) in the vicinity of retroelements (*R*). After engraftment into a metaplastic lesion, they adapt to their new environment by switching on stomach-specific genes (*arrows* and *double-headed arrow*). In parallel, CpG methylation (*gray dots*) downregulates, but does not switch off the expression of housekeeping genes (*arrows*). Microsatellite instability results in loss of heterozygosity and a compensatory demethylation of housekeeping genes (*double arrows*) in parallel with the cessation of stomach-specific gene expression (*crossed arrows*). These changes result in the reactivation of stem-cell programs for migration and proliferation

14.3.8.2 Global Alterations of the Epigenetic Blueprint of Histone Modifications in *Helicobacter pylori*-Associated Gastric Carcinoma

Global changes in the level of certain histone modifications were observed in GC, and they were related to the prognosis of the patients. Increased levels of the repressive mark H3K9me3 correlated with the invasive behavior of the tumors and poor survival rate (Park et al. 2008). Overexpression of HDACs and a high level of H3 phosphorylation at serine 10 (H3S10pho) was also associated with an unfavorable prognosis (Takahashi et al. 2006; Weichert et al. 2008). The relationship of Hp infection to these overall shifts in the levels of modified histones remains to be established. It is interesting to note that H3K9me3 and deacetylation of histones mark heterochromatin, whereas H3S10pho plays a role both in chromatin condensation during mitosis and gene activation in interphase nuclei (Perez-Cadahia et al. 2009).

14.4 Epigenetic Modifications in Protozoan Parasites

Epigenetic modifications of histones and regulation of chromatin structure are widespread in protozoa that constitute the earliest branch of the eukaryotic lineage (Iyer et al. 2008; Fisk and Read 2011). Protozoan pathogens regularly use epigenetic mechanisms to ensure the regulated expression of virulence genes and differentiation-related gene sets. It is tempting to speculate that they may induce epigenetic alterations in the cells of their host organisms as well.

14.4.1 Epigenetic Mechanisms Control Stage Conversion and Surface Antigen Variation in Pathogenic Protozoa

The phylum *Apicomplexa* contains, among others, a series of medically important unicellular parasites that cause diseases in humans and domestic animals (Langsley et al. 2008). The apicoplast, a typical organelle characteristic for these protozoans, is probably a derivative of the chloroplast carried by an ancestral algal endosymbiont (Ralph et al. 2004).

The apicomplexan parasites *Toxoplasma gondii,* the causative agent of congenital toxoplasmosis, and *Cryptosporidium parvum,* causing persistent diarrhea, lack detectable CpG methylation in their DNA, although their genomes encode putative DNA cytosine methyltransferases (Gissot et al. 2008). The genome of the related *Plasmodium falciparum*, the causative agent of human malaria, encodes both histone lysine methyltransferases and demethylases, and methylated histone H3 and H4 could be detected in malaria parasites. These modified histones underwent dynamic changes during the asexual erythrocytic cycle of *P. falciparum* (Cui et al. 2008). In addition, antigenic variation associated with a family of surface antigens encoded by the *var* genes of *P. falciparum* was controlled by chromatin alterations: binding of the HDAC PfSir2 silenced and its removal activated a specific telomere-associated *var* gene (Duraisingh et al. 2005; Freitas-Junior et al. 2005). In addition, the activating histone marks H3K4me2 and H3K4me3 were enriched 5′ from the active *var* gene, whereas silenced *var* genes were associated with the repressive H3K9me3 mark (Lopez-Rubio et al. 2007a). Successive expression of variant surface molecules, a process regulated by epigenetic mechanisms during the proliferative phase of *P. falciparum* in red blood cells, facilitates immune evasion and thereby contributes to the establishment of a long-term chronic infection of humans. However, interfering with the epigenetic regulatory machinery of the parasite may open new therapeutic possibilities: inhibitors of HDACs have a documented antimalarial activity, and a novel inhibitor could even cure *Plasmodium berghei*-infected mice (Andrews et al. 2000, 2008; Agbor-Enoh et al. 2009).

Recently, a family of 22 telomere-associated long noncoding RNAs (lncRNAs), transcribed during intraerythrocytic development, was also described in *P. falciparum*. These lncRNAs were coordinately expressed after parasite DNA replication and may initiate post-S-phase epigenetic memory marks at nearby subtelomeric *var* loci. In addition to virulence gene regulation, lncRNAs may also play a role in telomere maintenance (Broadbent et al. 2011).

Gene sets involved in the differentiation of *T. gondii* also carried epigenetic marks: the genes expressed specifically in the rapidly replicating tachyzoite stage are marked with hyperacetylated histones H3 and H4, but after differentiation into a latent cyst form, the bradyzoite, they become hypoacetylated (Saksouk et al. 2005). In contrast, the bradyzoite-specific genes that were hypoacetylated in tachyzoite populations became acetylated during stage conversion. Gene silencing was mediated by a HDAC, whereas both histone acetylation and a *Toxoplasma* orthologue of the arginine methyltransferase CARM1 acted in concert to achieve gene activation (Saksouk et al. 2005).

It is unknown at present whether the epigenetic regulatory machinery encoded by apicomplexan parasites interferes with the epigenetic control system of their host organisms. *Toxoplasma gondii*, however, encodes a protein capable to infiltrate the host nucleus: PPC2hn, a weak protein phosphatase, may in principle modulate the histone code of the host cell chromatin. PPC2hn may either act as a phosphatase or as an inactive enzyme paralog, interfering with the activity of host cell phosphatases which modulate histone tails or other nuclear proteins (Laliberte and Carruthers 2008).

14.4.2 Theileria parva and Theileria annulata Induce Uncontrolled Proliferation of T Cells and Macrophages in Cattle: Epigenetic Dysregulation by Modulation of Signaling Pathways and Production of Nuclear Effectors?

The tick-borne apicomplexan parasites *Theileria parva* and *Theileria annulata* enter a number of different cell types in their mammalian and arthropod hosts. The presence of *Theileria* parasites immortalizes the infected bovine lymphocytes and macrophages (Heussler and Stanway 2008; Luder et al. 2009). *Theileria parva* induces primarily the transformation of T cells, although it also affects B cells, whereas *T. annulata* induces the proliferation of schizont-infected macrophages that subsequently stimulate the uncontrolled proliferation of uninfected T lymphocytes (Brown et al. 1973; Dobbelaere and Heussler 1999; Branco et al. 2010). In *T. parva*-infected cattle, the parasite replicates in the cytoplasm of T lymphocytes. The infected, transformed T cells undergo a clonal expansion and disseminate to lymphoid tissues and vital organs including the lung, kidney, and intestine causing fatal east coast fever. Remarkably, however, the uncontrolled T cell proliferation *could be reversed* by killing the parasite by a specific antitheilerian drug in vitro, which suggested that *T. parva* alone was sufficient for T cell transformation (Dobbelaere et al. 1988). *Theileria parva* activated NF-κB signaling and the c-Jun NH_2-terminal kinase (JNK) in infected T cells via its cytoskeleton-binding TpSCOP protein. These changes resulted in the autocrine growth of the cells via the induction of IL-2 and its receptor and apoptosis resistance (Dobbelaere et al. 1988; Palmer et al. 1997; Hayashida et al. 2010). Constitutive, T cell receptor-independent activation of JNK, a member of the mitogen-activated protein kinase family, caused phosphorylation of the transcription factors c-Jun, a component of the AP1 complex, and ATF2 in a parasite-dependent manner (Galley et al. 1997; Botteron and Dobbelaere 1998).

The *T. parva* genome encodes a DNA methyltransferase (Gardner et al. 2005). Whether this enzyme enters the host cell nucleus and whether *T. parva* induces epigenetic alterations in transformed bovine T cells remains to be investigated. Activation of the JNK-AP1 signaling pathway, however, may induce the expression of DNMT1, provided that DNMT1 is regulated similarly in bovine and human cells. The human parallel is provided by Epstein–Barr virus encoded LMP1 which upregulates DNMT1 via the very same AP1 signaling pathway (Tsai et al. 2006). In addition, similarly to the *T. parva*-encoded TpSCOP protein, LMP1 also activates the NF-κB pathway which resulted in the induction of another epigenetic regulator, Bmi-1, a component of polycomb repressor protein complex 1 (PRC1) in EBV-positive HLs (Dutton et al. 2007, see also Chap. 8). Based on these analogies, we suggest that the transformation of bovine lymphocytes by *T. parva* may also involve the epigenetic dysregulation of host cell genes.

Theileria annulata is the causative agent of tropical theileriosis, a cattle disease accompanied by hemorrhagic lesions due to the continuous proliferation of the parasite-infected, cytokine-producing macrophages that disseminate into nonlymphoid organs including the heart, lungs, and brain (Forsyth et al. 1999). *Theileria annulata* infects both monocytes and B cells in vitro and induces activation and proliferation of uninfected T cells both in vitro and in vivo (Spooner et al. 1989; Campbell et al. 1995; Sager et al. 1998; Branco et al. 2010). It also infects and transforms dendritic cells extracted from bovine afferent lymph (Stephens and Howard 2002). *Theileria annulata* infection of bovine monocytes and macrophages alters the gene expression pattern of the host cells (Jensen et al. 2008, 2009). The levels of c-MAF and MAFB mRNAs, coding for musculoaponeurotic fibrosarcoma oncogene (MAF) transcription factors which are involved in monocyte to macrophage differentiation, were lower in *T. annulata*-infected cell lines than in uninfected macrophages (Jensen et al. 2009). c-MAF and MAFB act by repressing the expression of other transcription factors involved in myeloid differentiation including PU.1 and RUNX1. A receptor essential for the survival of uninfected macrophages, colony-stimulating factor 1 receptor (CSF1R), was also suppressed (Jensen et al. 2009). The activity of the *PU.1* (purine-rich box-1) promoter as well as the *CSF1R* promoter is regulated by epigenetic mechanisms in human cells (Hoogenkamp et al. 2009; Lamprecht et al. 2010). The human *RUNX1* (also called *AML1*) gene has two alternative promoters (Markova et al. 2011). The epigenetic factors involved in the regulation of their activity or the activity of the corresponding bovine promoters remain to be explored. We suggest that *T. annulata* downregulates the activity of a set of host cell promoters by epigenetic mechanisms. *Theileria annulata* proteins expressed by the macro-schizont stage when a multinucleated syncytium resides free in the host cell cytoplasm are primary candidates for host cell immortalization

(Pain et al. 2005; von Schubert et al. 2010). Such proteins, released into the host cell cytoplasm or expressed on the parasite surface, could be potentially involved in the epigenetic dysregulation of bovine cells, too. As a matter of fact, a cluster of genes in the genome of *T. annulata* codes for closely related DNA-binding proteins (TashAT1–3) that carry several AT hook motifs similar to mammalian high-mobility group proteins (HMGI/Y proteins) which preferentially bind to the narrow minor groove in stretches of AT-rich sequences (Reeves and Beckerbauer 2001; Swan et al. 2001). Since TashAT1, 2, and 3 could be detected in the nuclei of infected bovine cells, Swan et al. speculated that they may alter the control of bovine leukocyte or tick cell gene expression (Swan et al. 2001).

14.4.3 Epigenetics of the Flagellated Protozoa Leishmania, Trypanosoma brucei, and Trypanosoma cruzi

Flagellated protozoa exhibit complex life cycles. *Leishmania* species alternate between their invertebrate host, the sand fly, where they grow as motile flagellated promastigotes in the gut, and their mammalian hosts, where they multiply as nonflagellated amastigotes in the phagolysosome of macrophages (Handman 1999). In humans, the disease spectrum ranges from skin ulcers to lethal infections of the internal organs in leishmaniasis patients. *Trypanosoma brucei* is the causative agent of sleeping sickness in humans and nagana in animals. *Trypanosoma brucei* adapts to the different compartments within the tsetse fly and the mammalian bloodstream, but has no intracellular form (Matthews 2005). *Trypanosoma cruzi* is the etiological agent of human Chagas' disease. Inflammatory lesions caused by the invading parasites lead to neuron destruction, fibrosis, heart failure, and visceral symptoms (Coura and Borges-Pereira 2010).

Modified histones and histone variants were identified in trypanosomes (reviewed by Martinez-Calvillo et al. 2010). Acetylated histone H3 marked the origins of polycistronic transcription in *Leishmania major* (Thomas et al. 2009). DOT1B, a histone methyltransferase, is involved both in the transcriptional silencing of the genes encoding variant surface glycoproteins (VSGs) of *T. brucei* and in rapid transcriptional VSG switching which ensures immune evasion (Figueiredo et al. 2008). VSG expression is monoallelic, and the active gene is transcribed by RNA polymerase I at a discrete nuclear expression site (ES), called the ES body (ESB). The active transcriptional state is inherited from one cell generation to the next by a cohesin-mediated, prolonged association of active *VSG* sister chromatids with each other and with the unique ESB (Landeira et al. 2009).

A modified base, β-D-glucosyl-hydroxymethyluracil, called base J, is a unique component of trypanosomal genomes. Reduction in the levels of base J in *T. cruzi* DNA resulted in an increased transcription of virulence genes followed by a global increase in the pol II transcription rate (Ekanayake et al. 2011). In parallel, histone H3 and H4 acetylation was also increased at potential promoter regions of the polycistronic transcription units (Ekanayake and Sabatini 2011). Histone H3K4me3 was enriched at transcriptional initiation sites (Respuela et al. 2008). Whether the epigenetic regulatory machinery of trypanosomatid parasites interacts with the host cell genome is currently not known.

14.4.4 *Entamoeba histolytica-Methylated LINE-Binding Protein: A Novel Target for Antiparasitic Chemotherapy?*

Entamoeba histolytica is a gastrointestinal protozoan parasite causing amebic dysentery. The amebic trophozoites residing in the large bowel occasionally invade the intestinal mucosa, causing tissue damage by apoptosis induction, although the exact mechanism of cytotoxicity remains to be explored (Ralston and Petri 2011). Furthermore, trophozoite dissemination may cause abscesses in other organs, mainly the liver. The genome of *E. histolytica* contains 5-methylcytosine and codes for the DNA methyltransferase Ehmeth, a homologue of human DNMT2. Ehmeth has a dual activity, methylating both genomic DNA prepared from trophozoites as well as tRNA(Asp) (Fisher et al. 2004; Tovy et al. 2010). Methylated regions of the *E. histolytica* genome include the ribosomal DNA region and the 5′ region of the heat shock protein 100 gene (Fisher et al. 2004; Bernes et al. 2005). Although 5-aza-cytidine-induced demethylation of the genome had limited effects on the gene expression pattern (Ali et al. 2007), it interfered with the growth of *E. histolytica* (Fisher et al. 2004). It also inhibited the cytopathic activity of trophozoites and their ability to cause liver abscesses in hamsters, although it did not affect adherence to target cells in vitro and hemolytic activity (Fisher et al. 2004). Overexpression of Ehmeth also illuminated its role in key cellular processes because it resulted in the accumulation of multinucleated cells, upregulation of heat shock protein expression, and resistance to oxidative stress (Fisher et al. 2006).

Experimental silencing of the *amebapore* (*ap-a*) gene, coding for a pore-forming cytotoxic protein, was associated with histone 3 lysine 4 demethylation in the *ap-a* gene domain and resulted in an avirulent phenotype (Mirelman et al. 2006). These studies demonstrated that epigenetic alterations may influence the virulence of *E. histolytica*. A subsequent genome-wide analysis of histone acetylation demonstrated that epigenetic mechanisms may also control the conversion of one life cycle form of the parasite to the other. It was observed that the HDAC inhibitor TSA upregulated a series of genes involved in the stage conversion, i.e., encystation of *E. histolytica*, including genes coding for heat shock proteins and signaling molecules (Ehrenkaufer et al. 2007). In parallel, TsA treatment downregulated several trophozoite-specific and virulence genes including *CP1*, encoding a cysteine protease.

Another potential target for epigenetic therapy could be the *E. histolytica*-methylated LINE-binding protein (EhMLBP), which associated with methylated long interspersed nuclear elements (LINEs) and ribosomal DNA as well (Lavi et al. 2006). EhMLBP was induced by heat shock, and its constitutive overexpression protected trophozoites against heat shock (Katz et al. 2012). Treatment of

E. histolytica with distamycin A, a drug-blocking EhMLBP binding to methylated LINE DNA in vitro, effectively inhibited the growth of the parasite (Lavi et al. 2008). In addition, *E. histolytica* transfectants containing a phage-expressed peptide specifically binding to EhMLBP also impaired the growth of trophozoites (Lavi et al. 2008). These data support the view that deciphering the unique features of epigenetic regulation may pinpoint novel targets for antiamebic chemotherapy.

14.5 Conclusions and Outlook

It looks like that certain microbial infections of great medical importance frequently leave epigenetic marks in host cells. There was an intense research effort in the past few years to describe such marks, to elucidate their patho-epigenetic consequences, and to exploit their diagnostic and therapeutic implications. In spite of these remarkable achievements, the vast majority of pathogenic microorganisms remain to be analyzed with regard to their potential interactions with the epigenetic control systems of the host cell. We expect that after analyzing mainly tumor-associated viruses and latent viral infections, epigenetic research will focus more and more on viruses not directly linked to tumorigenesis and on viruses causing acute infections. Similarly, following the studies on intracellular bacteria and chronic bacterial infections, the exploration of the epigenetic consequences of acute bacterial infections may be a logical next step. These investigations may profoundly change our ideas on the initiation and progression of microbe-induced diseases. The knowledge accumulated on the epigenetics of pathogenic protozoa is expected to be translated into new therapeutic regimes, soon. In addition, a systematic study on protozoa-induced epigenetic dysregulations may also have implications for therapy.

References

Agbor-Enoh, S., Seudieu, C., Davidson, E., Dritschilo, A. & Jung, M. (2009). Novel inhibitor of Plasmodium histone deacetylase that cures P. berghei-infected mice. *Antimicrob Agents Chemother* 53, 1727–1734.

Ali, I. K., Ehrenkaufer, G. M., Hackney, J. A. & Singh, U. (2007). Growth of the protozoan parasite Entamoeba histolytica in 5-azacytidine has limited effects on parasite gene expression. *BMC Genomics* 8:7.

Alves, M. K., Ferrasi, A. C., Lima, V. P., Ferreira, M. V., de Moura Campos Pardini MI & Rabenhorst, S. H. (2011). Inactivation of COX-2, HMLH1 and CDKN2A Gene by Promoter Methylation in Gastric Cancer: Relationship with Histological Subtype, Tumor Location and Helicobacter pylori Genotype. *Pathobiology* 78, 266–276.

Alves, M. K., Lima, V. P., Ferrasi, A. C., Rodrigues, M. A., de Moura Campos Pardini MI & Rabenhorst, S. H. (2010). CDKN2A promoter methylation is related to the tumor location and histological subtype and associated with Helicobacter pylori flaA(+) strains in gastric adenocarcinomas. *Acta Pathol Microbiol Immunol Scand* 118, 297–307.

Andersen-Nissen, E., Smith, K. D., Strobe, K. L., Barrett, S. L., Cookson, B. T., Logan, S. M. & Aderem, A. (2005). Evasion of Toll-like receptor 5 by flagellated bacteria. *Proc Natl Acad Sci U S A* 102, 9247–9252.

Andrews, K. T., Tran, T. N., Lucke, A. J., Kahnberg, P., Le, G. T., Boyle, G. M., Gardiner, D. L., Skinner-Adams, T. S. & Fairlie, D. P. (2008). Potent antimalarial activity of histone deacetylase inhibitor analogues. *Antimicrob Agents Chemother* 52, 1454–1461.

Andrews, K. T., Walduck, A., Kelso, M. J., Fairlie, D. P., Saul, A. & Parsons, P. G. (2000). Anti-malarial effect of histone deacetylation inhibitors and mammalian tumour cytodifferentiating agents. *Int J Parasitol* 30, 761–768.

Aoki, Y. & Tosato, G. (2004). Neoplastic conditions in the context of HIV-1 infection. *Curr HIV Res* 2, 343–349.

Arbibe, L., Kim, D. W., Batsche, E., Pedron, T., Mateescu, B., Muchardt, C., Parsot, C. & Sansonetti, P. J. (2007). An injected bacterial effector targets chromatin access for transcription factor NF-kappaB to alter transcription of host genes involved in immune responses. *Nat Immunol* 8, 47–56.

Arora, P., Kim, E. O., Jung, J. K. & Jang, K. L. (2008). Hepatitis C virus core protein downregulates E-cadherin expression via activation of DNA methyltransferase 1 and 3b. *Cancer Lett* 261, 244–252.

Azuma, T. (2004). Helicobacter pylori CagA protein variation associated with gastric cancer in Asia. *J Gastroenterol* 39, 97–103.

Bednarik, D. P., Cook, J. A. & Pitha, P. M. (1990). Inactivation of the HIV LTR by DNA CpG methylation: evidence for a role in latency. *EMBO J* 9, 1157–1164.

Bednarik, D. P., Mosca, J. D. & Raj, N. B. (1987). Methylation as a modulator of expression of human immunodeficiency virus. *J Virol* 61, 1253–1257.

Benton, K. A., Everson, M. P. & Briles, D. E. (1995). A pneumolysin-negative mutant of Streptococcus pneumoniae causes chronic bacteremia rather than acute sepsis in mice. *Infect Immun* 63, 448–455.

Bernes, S., Siman-Tov, R. & Ankri, S. (2005). Epigenetic and classical activation of Entamoeba histolytica heat shock protein 100 (EHsp100) expression. *FEBS Lett* 579, 6395–6402.

Bibas, M. & Antinori, A. (2009). EBV and HIV-Related Lymphoma. *Mediterr J Hematol Infect Dis* 1, e2009032.

Blazkova, J., Trejbalova, K., Gondois-Rey, F., Halfon, P., Philibert, P., Guiguen, A., Verdin, E., Olive, D., Van Lint, C., Hejnar, J. & Hirsch, I. (2009). CpG methylation controls reactivation of HIV from latency. *PLoS Pathog* 5, e1000554.

Bobetsis, Y. A., Barros, S. P., Lin, D. M., Arce, R. M. & Offenbacher, S. (2010). Altered gene expression in murine placentas in an infection-induced intrauterine growth restriction model: a microarray analysis. *J Reprod Immunol* 85, 140–148.

Bobetsis, Y. A., Barros, S. P., Lin, D. M., Weidman, J. R., Dolinoy, D. C., Jirtle, R. L., Boggess, K. A., Beck, J. D. & Offenbacher, S. (2007). Bacterial infection promotes DNA hypermethylation. *J Dent Res* 86, 169–174.

Botteron, C. & Dobbelaere, D. (1998). AP-1 and ATF-2 are constitutively activated via the JNK pathway in Theileria parva-transformed T-cells. *Biochem Biophys Res Commun* 246, 418–421.

Branco, S., Orvalho, J., Leitao, A., Pereira, I., Malta, M., Mariano, I., Carvalho, T., Baptista, R., Shiels, B. R. & Peleteiro, M. C. (2010). Fatal cases of Theileria annulata infection in calves in Portugal associated with neoplastic-like lymphoid cell proliferation. *J Vet Sci* 11, 27–34.

Broadbent, K. M., Park, D., Wolf, A. R., Van Tyne, D., Sims, J. S., Ribacke, U., Volkman, S., Duraisingh, M., Wirth, D., Sabeti, P. C. & Rinn, J. L. (2011). A global transcriptional analysis of Plasmodium falciparum malaria reveals a novel family of telomere-associated lncRNAs. *Genome Biol* 12:R56.

Brown, C. G., Stagg, D. A., Purnell, R. E., Kanhai, G. K. & Payne, R. C. (1973). Letter: Infection and transformation of bovine lymphoid cells in vitro by infective particles of Theileria parva. *Nature* 245, 101–103.

Campbell, J. D., Howie, S. E., Odling, K. A. & Glass, E. J. (1995). Theileria annulata induces abberrant T cell activation in vitro and in vivo. *Clin Exp Immunol* 99, 203–210.

Caswell, R., Bryant, L. & Sinclair, J. (1996). Human cytomegalovirus immediate-early 2 (IE2) protein can transactivate the human hsp70 promoter by alleviation of Dr1-mediated repression. *J Virol* 70, 4028–4037.

Chan, A. O., Chu, K. M., Huang, C., Lam, K. F., Leung, S. Y., Sun, Y. W., Ko, S., Xia, H. H., Cho, C. H., Hui, W. M., Lam, S. K. & Rashid, A. (2007). Association between Helicobacter pylori infection and interleukin 1beta polymorphism predispose to CpG island methylation in gastric cancer. *Gut* 56, 595–597.

Chinnasri, P., Pairojkul, C., Jearanaikoon, P., Sripa, B., Bhudhisawasdi, V., Tantimavanich, S. & Limpaiboon, T. (2009). Preferentially different mechanisms of inactivation of 9p21 gene cluster in liver fluke-related cholangiocarcinoma. *Hum Pathol* 40, 817–826.

Compare, D., Rocco, A., Liguori, E., D'Armiento, F. P., Persico, G., Masone, S., Coppola-Bottazzi, E., Suriani, R., Romano, M. & Nardone, G. (2011). Global DNA hypomethylation is an early event in Helicobacter pylori-related gastric carcinogenesis. *J Clin Pathol* 64, 677–682.

Coull, J. J., Romerio, F., Sun, J. M., Volker, J. L., Galvin, K. M., Davie, J. R., Shi, Y., Hansen, U. & Margolis, D. M. (2000). The human factors YY1 and LSF repress the human immunodeficiency virus type 1 long terminal repeat via recruitment of histone deacetylase 1. *J Virol* 74, 6790–6799.

Coura, J. R. & Borges-Pereira, J. (2010). Chagas disease: 100 years after its discovery. A systemic review. *Acta Trop* 115, 5–13.

Cui, L., Fan, Q., Cui, L. & Miao, J. (2008). Histone lysine methyltransferases and demethylases in Plasmodium falciparum. *Int J Parasitol* 38, 1083–1097.

da Rosa, J. L. & Kaufman, P. D. (2012). Chromatin-mediated Candida albicans virulence. *Biochim Biophys Acta* 1819, 349–355.

Darveau, R. P. (2009). The oral microbial consortium's interaction with the periodontal innate defense system. *DNA Cell Biol* 28, 389–395.

Dickson, J. H., Grabowska, A., El Zaatari, M., Atherton, J. & Watson, S. A. (2006). Helicobacter pylori can induce heparin-binding epidermal growth factor expression via gastrin and its receptor. *Cancer Res* 66, 7524–7531.

Ding, S. Z., Goldberg, J. B. & Hatakeyama, M. (2010). Helicobacter pylori infection, oncogenic pathways and epigenetic mechanisms in gastric carcinogenesis. *Future Oncol* 6, 851–862.

Dixon, S. E., Stilger, K. L., Elias, E. V., Naguleswaran, A. & Sullivan, W. J. (2010). A decade of epigenetic research in Toxoplasma gondii. *Mol Biochem Parasitol* 173, 1–9.

Dobbelaere, D. & Heussler, V. (1999). Transformation of leukocytes by Theileria parva and T. annulata. *Annu Rev Microbiol* 53, 1–42.

Dobbelaere, D. A., Coquerelle, T. M., Roditi, I. J., Eichhorn, M. & Williams, R. O. (1988). Theileria parva infection induces autocrine growth of bovine lymphocytes. *Proc Natl Acad Sci U S A* 85, 4730–4734.

Duong, F. H., Christen, V., Lin, S. & Heim, M. H. (2010). Hepatitis C virus-induced up-regulation of protein phosphatase 2A inhibits histone modification and DNA damage repair. *Hepatology* 51, 741–751.

Duraisingh, M. T., Voss, T. S., Marty, A. J., Duffy, M. F., Good, R. T., Thompson, J. K., Freitas-Junior, L. H., Scherf, A., Crabb, B. S. & Cowman, A. F. (2005). Heterochromatin silencing and locus repositioning linked to regulation of virulence genes in Plasmodium falciparum. *Cell* 121, 13–24.

Dutton, A., Woodman, C. B., Chukwuma, M. B., Last, J. I., Wei, W., Vockerodt, M., Baumforth, K. R., Flavell, J. R., Rowe, M., Taylor, A. M., Young, L. S. & Murray, P. G. (2007). Bmi-1 is induced by the Epstein-Barr virus oncogene LMP1 and regulates the expression of viral target genes in Hodgkin lymphoma cells. *Blood* 109, 2597–2603.

Ehrenkaufer, G. M., Eichinger, D. J. & Singh, U. (2007). Trichostatin A effects on gene expression in the protozoan parasite Entamoeba histolytica. *BMC Genomics* 8:216.

Eisenreich, W., Dandekar, T., Heesemann, J. & Goebel, W. (2010). Carbon metabolism of intracellular bacterial pathogens and possible links to virulence. *Nat Rev Microbiol* 8, 401–412.

Ekanayake, D. & Sabatini, R. (2011). Epigenetic Regulation of Polymerase II Transcription Initiation in Trypanosoma cruzi: Modulation of Nucleosome Abundance, Histone Modification, and Polymerase Occupancy by O-Linked Thymine DNA Glucosylation. *Eukaryot Cell* 10, 1465–1472.

Ekanayake, D. K., Minning, T., Weatherly, B., Gunasekera, K., Nilsson, D., Tarleton, R., Ochsenreiter, T. & Sabatini, R. (2011). Epigenetic regulation of transcription and virulence in Trypanosoma cruzi by O-linked thymine glucosylation of DNA. *Mol Cell Biol* 31, 1690–1700.

El Omar, E. M., Carrington, M., Chow, W. H., McColl, K. E., Bream, J. H., Young, H. A., Herrera, J., Lissowska, J., Yuan, C. C., Rothman, N., Lanyon, G., Martin, M., Fraumeni, J. F. & Rabkin, C. S. (2000). Interleukin-1 polymorphisms associated with increased risk of gastric cancer. *Nature* 404, 398–402.

El Omar, E. M., Rabkin, C. S., Gammon, M. D., Vaughan, T. L., Risch, H. A., Schoenberg, J. B., Stanford, J. L., Mayne, S. T., Goedert, J., Blot, W. J., Fraumeni, J. F. & Chow, W. H. (2003). Increased risk of noncardia gastric cancer associated with proinflammatory cytokine gene polymorphisms. *Gastroenterology* 124, 1193–1201.

Elgui de Oliveira, D. (2007). DNA viruses in human cancer: an integrated overview on fundamental mechanisms of viral carcinogenesis. *Cancer Lett* 247, 182–196.

Fang, J. Y., Mikovits, J. A., Bagni, R., Petrow-Sadowski, C. L. & Ruscetti, F. W. (2001). Infection of lymphoid cells by integration-defective human immunodeficiency virus type 1 increases de novo methylation. *J Virol* 75, 9753–9761.

Figueiredo, L. M., Janzen, C. J. & Cross, G. A. (2008). A histone methyltransferase modulates antigenic variation in African trypanosomes. *PLoS Biol* 6, e161.

Finzi, D., Hermankova, M., Pierson, T., Carruth, L. M., Buck, C., Chaisson, R. E., Quinn, T. C., Chadwick, K., Margolick, J., Brookmeyer, R., Gallant, J., Markowitz, M., Ho, D. D., Richman, D. D. & Siliciano, R. F. (1997). Identification of a reservoir for HIV-1 in patients on highly active antiretroviral therapy. *Science* 278, 1295–1300.

Fisher, O., Siman-Tov, R. & Ankri, S. (2004). Characterization of cytosine methylated regions and 5-cytosine DNA methyltransferase (Ehmeth) in the protozoan parasite Entamoeba histolytica. *Nucleic Acids Res* 32, 287–297.

Fisher, O., Siman-Tov, R. & Ankri, S. (2006). Pleiotropic phenotype in Entamoeba histolytica overexpressing DNA methyltransferase (Ehmeth). *Mol Biochem Parasitol* 147, 48–54.

Fisk, J. C. & Read, L. K. (2011). Protein arginine methylation in parasitic protozoa. *Eukaryot Cell* 10, 1013–1022.

Forsyth, L. M., Minns, F. C., Kirvar, E., Adamson, R. E., Hall, F. R., McOrist, S., Brown, C. G. & Preston, P. M. (1999). Tissue damage in cattle infected with Theileria annulata accompanied by metastasis of cytokine-producing, schizont-infected mononuclear phagocytes. *J Comp Pathol* 120, 39–57.

Freitas-Junior, L. H., Hernandez-Rivas, R., Ralph, S. A., Montiel-Condado, D., Ruvalcaba-Salazar, O. K., Rojas-Meza, A. P., Mancio-Silva, L., Leal-Silvestre, R. J., Gontijo, A. M., Shorte, S. & Scherf, A. (2005). Telomeric heterochromatin propagation and histone acetylation control mutually exclusive expression of antigenic variation genes in malaria parasites. *Cell* 121, 25–36.

Friedman, J., Cho, W. K., Chu, C. K., Keedy, K. S., Archin, N. M., Margolis, D. M. & Karn, J. (2011). Epigenetic Silencing of HIV-1 by the Histone H3 Lysine 27 Methyltransferase Enhancer of Zeste 2. *J Virol* 85, 9078–9089.

Galley, Y., Hagens, G., Glaser, I., Davis, W., Eichhorn, M. & Dobbelaere, D. (1997). Jun NH2-terminal kinase is constitutively activated in T cells transformed by the intracellular parasite Theileria parva. *Proc Natl Acad Sci U S A* 94, 5119–5124.

Galmiche, A., Rassow, J., Doye, A., Cagnol, S., Chambard, J. C., Contamin, S., de Thillot, V., Just, I., Ricci, V., Solcia, E., Van Obberghen, E. & Boquet, P. (2000). The N-terminal 34kDa fragment of Helicobacter pylori vacuolating cytotoxin targets mitochondria and induces cytochrome c release. *EMBO J* 19, 6361–6370.

Garcia-Garcia, J. C., Barat, N. C., Trembley, S. J. & Dumler, J. S. (2009). Epigenetic silencing of host cell defense genes enhances intracellular survival of the rickettsial pathogen Anaplasma phagocytophilum. *PLoS Pathog* 5, e1000488.

Gardner, M. J., Bishop, R., Shah, T., de Villiers, E. P., Carlton, J. M., Hall, N., Ren, Q., Paulsen, I. T., Pain, A., Berriman, M., Wilson, R. J., Sato, S., Ralph, S. A., Mann, D. J., Xiong, Z., Shallom, S. J., Weidman, J., Jiang, L., Lynn, J., Weaver, B., Shoaibi, A., Domingo, A. R., Wasawo, D., Crabtree, J., Wortman, J. R., Haas, B., Angiuoli, S. V., Creasy, T. H., Lu, C., Suh, B., Silva, J. C., Utterback, T. R., Feldblyum, T. V., Pertea, M., Allen, J., Nierman, W. C., Taracha, E. L., Salzberg, S. L., White, O. R., Fitzhugh, H. A., Morzaria, S., Venter, J. C., Fraser, C. M. & Nene, V. (2005). Genome sequence of Theileria parva, a bovine pathogen that transforms lymphocytes. *Science* 309, 134–137.

Giordanengo, V., Ollier, L., Lanteri, M., Lesimple, J., March, D., Thyss, S. & Lefebvre, J. C. (2004). Epigenetic reprogramming of UDP-N-acetylglucosamine 2-epimerase/ N-acetylmannosamine kinase (GNE) in HIV-1-infected CEM T cells. *FASEB J* 18, 1961–1963.

Gissot, M., Choi, S. W., Thompson, R. F., Greally, J. M. & Kim, K. (2008). Toxoplasma gondii and Cryptosporidium parvum lack detectable DNA cytosine methylation. *Eukaryot Cell* 7, 537–540.

Gissot, M. & Kim, K. (2008). How epigenomics contributes to the understanding of gene regulation in Toxoplasma gondii. *J Eukaryot Microbiol* 55, 476–480.

Goldenring, J. R. & Nomura, S. (2006). Differentiation of the gastric mucosa III. Animal models of oxyntic atrophy and metaplasia. *Am J Physiol Gastrointest Liver Physiol* 291, G999–G1004.

Gonczol, E., Andrews, P. W. & Plotkin, S. A. (1984). Cytomegalovirus replicates in differentiated but not in undifferentiated human embryonal carcinoma cells. *Science* 224, 159–161.

Grulich, A. E., van Leeuwen, M. T., Falster, M. O. & Vajdic, C. M. (2007). Incidence of cancers in people with HIV/AIDS compared with immunosuppressed transplant recipients: a meta-analysis. *Lancet* 370, 59–67.

Guo, X. B., Guo, L., Zhi, Q. M., Ji, J., Jiang, J. L., Zhang, R. J., Zhang, J. N., Zhang, J., Chen, X. H., Cai, Q., Li, J. F., Yan, M., Gu, Q. L., Liu, B. Y., Zhu, Z. G. & Yu, Y. Y. (2011). Helicobacter pylori induces promoter hypermethylation and downregulates gene expression of IRX1 transcription factor on human gastric mucosa. *J Gastroenterol Hepatol* 26, 1685–1690.

Gutierrez, M. I., Siraj, A. K., Khaled, H., Koon, N., El Rifai, W. & Bhatia, K. (2004). CpG island methylation in Schistosoma- and non-Schistosoma-associated bladder cancer. *Mod Pathol* 17, 1268–1274.

Gutierrez-Gonzalez, L. & Wright, N. A. (2008). Biology of intestinal metaplasia in 2008: more than a simple phenotypic alteration. *Dig Liver Dis* 40, 510–522.

Hamon, M., Bierne, H. & Cossart, P. (2006). Listeria monocytogenes: a multifaceted model. *Nat Rev Microbiol* 4, 423–434.

Hamon, M. A., Batsche, E., Regnault, B., Tham, T. N., Seveau, S., Muchardt, C. & Cossart, P. (2007). Histone modifications induced by a family of bacterial toxins. *Proc Natl Acad Sci U S A* 104, 13467–13472.

Hamon, M. A. & Cossart, P. (2011). K+ efflux is required for histone H3 dephosphorylation by Listeria monocytogenes listeriolysin O and other pore-forming toxins. *Infect Immun* 79, 2839–2846.

Han, Y., Lin, Y. B., An, W., Xu, J., Yang, H. C., O'Connell, K., Dordai, D., Boeke, J. D., Siliciano, J. D. & Siliciano, R. F. (2008). Orientation-dependent regulation of integrated HIV-1 expression by host gene transcriptional read-through. *Cell Host Microbe* 4, 134–146.

Handman, E. (1999). Cell biology of Leishmania. *Adv Parasitol* 44, 1–39.

Hatakeyama, M. (2004). Oncogenic mechanisms of the Helicobacter pylori CagA protein. *Nat Rev Cancer* 4, 688–694.

Hayashida, K., Hattori, M., Nakao, R., Tanaka, Y., Kim, J. Y., Inoue, N., Nene, V. & Sugimoto, C. (2010). A schizont-derived protein, TpSCOP, is involved in the activation of NF-kappaB in Theileria parva-infected lymphocytes. *Mol Biochem Parasitol* 174, 8–17.

Hayhurst, G. P., Bryant, L. A., Caswell, R. C., Walker, S. M. & Sinclair, J. H. (1995). CCAAT box-dependent activation of the TATA-less human DNA polymerase alpha promoter by the human cytomegalovirus 72-kilodalton major immediate-early protein. *J Virol* 69, 182–188.

Henderson, J. P., Crowley, J. R., Pinkner, J. S., Walker, J. N., Tsukayama, P., Stamm, W. E., Hooton, T. M. & Hultgren, S. J. (2009). Quantitative metabolomics reveals an epigenetic blueprint for iron acquisition in uropathogenic Escherichia coli. *PLoS Pathog* 5, e1000305.

Heussler, V. T. & Stanway, R. R. (2008). Cellular and molecular interactions between the apicomplexan parasites Plasmodium and Theileria and their host cells. *Parasite* 15, –211–218.

Higashi, H., Tsutsumi, R., Muto, S., Sugiyama, T., Azuma, T., Asaka, M. & Hatakeyama, M. (2002). SHP-2 tyrosine phosphatase as an intracellular target of Helicobacter pylori CagA protein. *Science* 295, 683–686.

Hmadcha, A., Bedoya, F. J., Sobrino, F. & Pintado, E. (1999). Methylation-dependent gene silencing induced by interleukin 1beta via nitric oxide production. *J Exp Med* 190, 1595–1604.

Hong, S. J., Oh, J. H., Jeon, E. J., Min, K. O., Kang, M. I., Choi, S. W. & Rhyu, M. G. (2010). The overmethylated genes in Helicobacter pylori-infected gastric mucosa are demethylated in gastric cancers. *BMC Gastroenterol* 10:137.

Hoogenkamp, M., Lichtinger, M., Krysinska, H., Lancrin, C., Clarke, D., Williamson, A., Mazzarella, L., Ingram, R., Jorgensen, H., Fisher, A., Tenen, D. G., Kouskoff, V., Lacaud, G. & Bonifer, C. (2009). Early chromatin unfolding by RUNX1: a molecular explanation for differential requirements during specification versus maintenance of the hematopoietic gene expression program. *Blood* 114, 299–309.

Houghton, J., Stoicov, C., Nomura, S., Rogers, A. B., Carlson, J., Li, H., Cai, X., Fox, J. G., Goldenring, J. R. & Wang, T. C. (2004). Gastric cancer originating from bone marrow-derived cells. *Science* 306, 1568–1571.

Hutton, M. L., Kaparakis-Liaskos, M., Turner, L., Cardona, A., Kwok, T. & Ferrero, R. L. (2010). Helicobacter pylori exploits cholesterol-rich microdomains for induction of NF-kappaB-dependent responses and peptidoglycan delivery in epithelial cells. *Infect Immun* 78, 4523–4531.

Imai, K., Ochiai, K. & Okamoto, T. (2009). Reactivation of latent HIV-1 infection by the periodontopathic bacterium Porphyromonas gingivalis involves histone modification. *J Immunol* 182, 3688–3695.

Iyer, L. M., Anantharaman, V., Wolf, M. Y. & Aravind, L. (2008). Comparative genomics of transcription factors and chromatin proteins in parasitic protists and other eukaryotes. *Int J Parasitol* 38, 1–31.

Jensen, K., Makins, G. D., Kaliszewska, A., Hulme, M. J., Paxton, E. & Glass, E. J. (2009). The protozoan parasite Theileria annulata alters the differentiation state of the infected macrophage and suppresses musculoaponeurotic fibrosarcoma oncogene (MAF) transcription factors. *Int J Parasitol* 39, 1099–1108.

Jensen, K., Paxton, E., Waddington, D., Talbot, R., Darghouth, M. A. & Glass, E. J. (2008). Differences in the transcriptional responses induced by Theileria annulata infection in bovine monocytes derived from resistant and susceptible cattle breeds. *Int J Parasitol* 38, 313–325.

Jiang, G., Espeseth, A., Hazuda, D. J. & Margolis, D. M. (2007). c-Myc and Sp1 contribute to proviral latency by recruiting histone deacetylase 1 to the human immunodeficiency virus type 1 promoter. *J Virol* 81, 10914–10923.

Kabir, S. (2009). Effect of Helicobacter pylori eradication on incidence of gastric cancer in human and animal models: underlying biochemical and molecular events. *Helicobacter* 14, 159–171.

Kaise, M., Yamasaki, T., Yonezawa, J., Miwa, J., Ohta, Y. & Tajiri, H. (2008). CpG island hypermethylation of tumor-suppressor genes in H. pylori-infected non-neoplastic gastric mucosa is linked with gastric cancer risk. *Helicobacter* 13, 35–41.

Katz, S., Kushnir, O., Tovy, A., Siman-Tov, R. & Ankri, S. (2012). The Entamoeba histolytica methylated LINE-binding protein EhMLBP provides protection against heat shock. *Cell Microbiol* 14, 58–70.

Kauder, S. E., Bosque, A., Lindqvist, A., Planelles, V. & Verdin, E. (2009). Epigenetic regulation of HIV-1 latency by cytosine methylation. *PLoS Pathog* 5, e1000495.

Kogan, M. & Rappaport, J. (2011). HIV-1 accessory protein Vpr: relevance in the pathogenesis of HIV and potential for therapeutic intervention. *Retrovirology* 8:25.

Kurita-Ochiai, T., Fukushima, K. & Ochiai, K. (1995). Volatile fatty acids, metabolic by-products of periodontopathic bacteria, inhibit lymphocyte proliferation and cytokine production. *J Dent Res* 74, 1367–1373.

Laliberte, J. & Carruthers, V. B. (2008). Host cell manipulation by the human pathogen Toxoplasma gondii. *Cell Mol Life Sci* 65, 1900–1915.

Lamprecht, B., Walter, K., Kreher, S., Kumar, R., Hummel, M., Lenze, D., Kochert, K., Bouhlel, M. A., Richter, J., Soler, E., Stadhouders, R., Johrens, K., Wurster, K. D., Callen, D. F., Harte, M. F., Giefing, M., Barlow, R., Stein, H., Anagnostopoulos, I., Janz, M., Cockerill, P. N., Siebert, R., Dorken, B., Bonifer, C. & Mathas, S. (2010). Derepression of an endogenous long terminal repeat activates the CSF1R proto-oncogene in human lymphoma. *Nat Med* 16, 571–579.

Landeira, D., Bart, J. M., Van Tyne, D. & Navarro, M. (2009). Cohesin regulates VSG monoallelic expression in trypanosomes. *J Cell Biol* 186, 243–254.

Langsley, G., van Noort, V., Carret, C., Meissner, M., de Villiers, E. P., Bishop, R. & Pain, A. (2008). Comparative genomics of the Rab protein family in Apicomplexan parasites. *Microbes Infect* 10, 462–470.

Lavi, T., Isakov, E., Harony, H., Fisher, O., Siman-Tov, R. & Ankri, S. (2006). Sensing DNA methylation in the protozoan parasite Entamoeba histolytica. *Mol Microbiol* 62, 1373–1386.

Lavi, T., Siman-Tov, R. & Ankri, S. (2008). EhMLBP is an essential constituent of the Entamoeba histolytica epigenetic machinery and a potential drug target. *Mol Microbiol* 69, 55–66.

Lee, J. H., Park, S. J., Abraham, S. C., Seo, J. S., Nam, J. H., Choi, C., Juhng, S. W., Rashid, A., Hamilton, S. R. & Wu, T. T. (2004). Frequent CpG island methylation in precursor lesions and early gastric adenocarcinomas. *Oncogene* 23, 4646–4654.

Lenoir, G. M. & Bornkamm, G. (1987). Burkitt's Lymphoma, a human cancer model for the study of the multistep development of cancer: proposal for a new scenario. In *Advances in Viral Oncology*, pp. 173–206. Edited by G. Klein. New York: Raven Press.

Leung, W. K. & Sung, J. J. (2002). Review article: intestinal metaplasia and gastric carcinogenesis. *Aliment Pharmacol Ther* 16, 1209–1216.

Li, H. C., Stoicov, C., Rogers, A. B. & Houghton, J. (2006). Stem cells and cancer: evidence for bone marrow stem cells in epithelial cancers. *World J Gastroenterol* 12, 363–371.

Lin, D., Smith, M. A., Champagne, C., Elter, J., Beck, J. & Offenbacher, S. (2003a). Porphyromonas gingivalis infection during pregnancy increases maternal tumor necrosis factor alpha, suppresses maternal interleukin-10, and enhances fetal growth restriction and resorption in mice. *Infect Immun* 71, 5156–5162.

Lin, D., Smith, M. A., Elter, J., Champagne, C., Downey, C. L., Beck, J. & Offenbacher, S. (2003b). Porphyromonas gingivalis infection in pregnant mice is associated with placental dissemination, an increase in the placental Th1/Th2 cytokine ratio, and fetal growth restriction. *Infect Immun* 71, 5163–5168.

Lopez-Rubio, J. J., Gontijo, A. M., Nunes, M. C., Issar, N., Hernandez Rivas R. & Scherf, A. (2007a). 5′ Flanking region of var genes nucleate histone modification patterns linked to phenotypic inheritance of virulence traits in malaria parasites. *Mol Microbiol* 66, 1296–1305.

Lopez-Rubio, J. J., Riviere, L. & Scherf, A. (2007b). Shared epigenetic mechanisms control virulence factors in protozoan parasites. *Curr Opin Microbiol* 10, 560–568.

Luder, C. G., Stanway, R. R., Chaussepied, M., Langsley, G. & Heussler, V. T. (2009). Intracellular survival of apicomplexan parasites and host cell modification. *Int J Parasitol* 39, 163–173.

Machado, A. M., Figueiredo, C., Touati, E., Maximo, V., Sousa, S., Michel, V., Carneiro, F., Nielsen, F. C., Seruca, R. & Rasmussen, L. J. (2009). Helicobacter pylori infection induces genetic instability of nuclear and mitochondrial DNA in gastric cells. *Clin Cancer Res* 15, 2995–3002.

Machado, J. C., Figueiredo, C., Canedo, P., Pharoah, P., Carvalho, R., Nabais, S., Castro, A. C., Campos, M. L., Van Doorn, L. J., Caldas, C., Seruca, R., Carneiro, F. & Sobrinho-Simoes, M. (2003). A proinflammatory genetic profile increases the risk for chronic atrophic gastritis and gastric carcinoma. *Gastroenterology* 125, 364–371.

Maekita, T., Nakazawa, K., Mihara, M., Nakajima, T., Yanaoka, K., Iguchi, M., Arii, K., Kaneda, A., Tsukamoto, T., Tatematsu, M., Tamura, G., Saito, D., Sugimura, T., Ichinose, M. & Ushijima, T. (2006). High levels of aberrant DNA methylation in Helicobacter pylori-infected gastric mucosae and its possible association with gastric cancer risk. *Clin Cancer Res* 12, 989–995.

Marban, C., Suzanne, S., Dequiedt, F., de Walque, S., Redel, L., Van Lint, C., Aunis, D. & Rohr, O. (2007). Recruitment of chromatin-modifying enzymes by CTIP2 promotes HIV-1 transcriptional silencing. *EMBO J* 26, 412–423.

Margolis, D. M. (2010). Mechanisms of HIV latency: an emerging picture of complexity. *Curr HIV/AIDS Rep* 7, 37–43.

Markova, E. N., Kantidze, O. L. & Razin, S. V. (2011). Transcriptional regulation and spatial organisation of the human AML1/RUNX1 gene. *J Cell Biochem* 112, 1997–2005.

Martinez-Calvillo, S., Vizuet-de-Rueda, J. C., Florencio-Martinez, L. E., Manning-Cela, R. G. & Figueroa-Angulo, E. E. (2010). Gene expression in trypanosomatid parasites. *J Biomed Biotechnol* 2010, 525241.

Matsumoto, Y., Marusawa, H., Kinoshita, K., Endo, Y., Kou, T., Morisawa, T., Azuma, T., Okazaki, I. M., Honjo, T. & Chiba, T. (2007). Helicobacter pylori infection triggers aberrant expression of activation-induced cytidine deaminase in gastric epithelium. *Nat Med* 13, 470–476.

Matthews, K. R. (2005). The developmental cell biology of Trypanosoma brucei. *J Cell Sci* 118, 283–290.

Merrick, C. J. & Duraisingh, M. T. (2010). Epigenetics in Plasmodium: what do we really know? *Eukaryot Cell* 9, 1150–1158.

Mikovits, J. A., Young, H. A., Vertino, P., Issa, J. P., Pitha, P. M., Turcoski-Corrales, S., Taub, D. D., Petrow, C. L., Baylin, S. B. & Ruscetti, F. W. (1998). Infection with human immunodeficiency virus type 1 upregulates DNA methyltransferase, resulting in de novo methylation of the gamma interferon (IFN-gamma) promoter and subsequent downregulation of IFN-gamma production. *Mol Cell Biol* 18, 5166–5177.

Milne, A. N., Carneiro, F., O'Morain, C. & Offerhaus, G. J. (2009). Nature meets nurture: molecular genetics of gastric cancer. *Hum Genet* 126, 615–628.

Minarovits, J. (2009). Microbe-induced epigenetic alterations in host cells: the coming era of patho-epigenetics of microbial infections. A review. *Acta Microbiol Immunol Hung* 56, 1–19.

Mirelman, D., Anbar, M., Nuchamowitz, Y. & Bracha, R. (2006). Epigenetic silencing of gene expression in Entamoeba histolytica. *Arch Med Res* 37, 226–233.

Murata, M., Azuma, Y., Miura, K., Rahman, M. A., Matsutani, M., Aoyama, M., Suzuki, H., Sugi, K. & Shirai, M. (2007). Chlamydial SET domain protein functions as a histone methyltransferase. *Microbiology* 153, 585–592.

Murphy, J. C., Fischle, W., Verdin, E. & Sinclair, J. H. (2002). Control of cytomegalovirus lytic gene expression by histone acetylation. *EMBO J* 21, 1112–1120.

Naka, K., Abe, K., Takemoto, K., Dansako, H., Ikeda, M., Shimotohno, K. & Kato, N. (2006). Epigenetic silencing of interferon-inducible genes is implicated in interferon resistance of hepatitis C virus replicon-harboring cells. *J Hepatol* 44, 869–878.

Nam, K. T., Lee, H. J., Sousa, J. F., Weis, V. G., O'Neal, R. L., Finke, P. E., Romero-Gallo, J., Shi, G., Mills, J. C., Peek, R. M., Konieczny, S. F. & Goldenring, J. R. (2010). Mature chief cells are cryptic progenitors for metaplasia in the stomach. *Gastroenterology* 139, 2028–2037.

Nardone, G., Compare, D., De Colibus, P., de Nucci, G. & Rocco, A. (2007). Helicobacter pylori and epigenetic mechanisms underlying gastric carcinogenesis. *Dig Dis* 25, 225–229.

Nevels, M., Paulus, C. & Shenk, T. (2004). Human cytomegalovirus immediate-early 1 protein facilitates viral replication by antagonizing histone deacetylation. *Proc Natl Acad Sci U S A* 101, 17234–17239.

Niederman, R., Buyle-Bodin, Y., Lu, B. Y., Robinson, P. & Naleway, C. (1997). Short-chain carboxylic acid concentration in human gingival crevicular fluid. *J Dent Res* 76, 575–579.

Niller, H. H., Salamon, D., Banati, F., Schwarzmann, F., Wolf, H. & Minarovits, J. (2004a). The LCR of EBV makes Burkitt's lymphoma endemic. *Trends Microbiol* 12, 495–499.

Niller, H. H., Salamon, D., Ilg, K., Koroknai, A., Banati, F., Bauml, G., Rucker, O., Schwarzmann, F., Wolf, H. & Minarovits, J. (2003). The in vivo binding site for oncoprotein c-Myc in the promoter for Epstein-Barr virus (EBV) encoding RNA (EBER) 1 suggests a specific role for EBV in lymphomagenesis. *Med Sci Monit* 9, HY1-HY9.

Niller, H. H., Salamon, D., Ilg, K., Koroknai, A., Banati, F., Schwarzmann, F., Wolf, H. & Minarovits, J. (2004b). EBV-associated neoplasms: alternative pathogenetic pathways. *Med Hypotheses* 62, 387–391.

Nozaki, K., Ogawa, M., Williams, J. A., Lafleur, B. J., Ng, V., Drapkin, R. I., Mills, J. C., Konieczny, S. F., Nomura, S. & Goldenring, J. R. (2008). A molecular signature of gastric metaplasia arising in response to acute parietal cell loss. *Gastroenterology* 134, 511–522.

Offenbacher, S., Riche, E. L., Barros, S. P., Bobetsis, Y. A., Lin, D. & Beck, J. D. (2005). Effects of maternal Campylobacter rectus infection on murine placenta, fetal and neonatal survival, and brain development. *J Periodontol* 76, 2133–2143.

Ohnishi, N., Yuasa, H., Tanaka, S., Sawa, H., Miura, M., Matsui, A., Higashi, H., Musashi, M., Iwabuchi, K., Suzuki, M., Yamada, G., Azuma, T. & Hatakeyama, M. (2008). Transgenic expression of Helicobacter pylori CagA induces gastrointestinal and hematopoietic neoplasms in mouse. *Proc Natl Acad Sci U S A* 105, 1003–1008.

Pain, A., Renauld, H., Berriman, M., Murphy, L., Yeats, C. A., Weir, W., Kerhornou, A., Aslett, M., Bishop, R., Bouchier, C., Cochet, M., Coulson, R. M., Cronin, A., de Villiers, E. P., Fraser, A., Fosker, N., Gardner, M., Goble, A., Griffiths-Jones, S., Harris, D. E., Katzer, F., Larke, N., Lord, A., Maser, P., McKellar, S., Mooney, P., Morton, F., Nene, V., O'Neil, S., Price, C., Quail, M. A., Rabbinowitsch, E., Rawlings, N. D., Rutter, S., Saunders, D., Seeger, K., Shah, T., Squares, R., Squares, S., Tivey, A., Walker, A. R., Woodward, J., Dobbelaere, D. A., Langsley, G., Rajandream, M. A., McKeever, D., Shiels, B., Tait, A., Barrell, B. & Hall, N. (2005). Genome of the host-cell transforming parasite Theileria annulata compared with T. parva. *Science* 309, 131–133.

Palmer, G. H., Machado, J., Fernandez, P., Heussler, V., Perinat, T. & Dobbelaere, D. A. (1997). Parasite-mediated nuclear factor kappaB regulation in lymphoproliferation caused by Theileria parva infection. *Proc Natl Acad Sci U S A* 94, 12527–12532.

Park, J. J., Kim, Y. E., Pham, H. T., Kim, E. T., Chung, Y. H. & Ahn, J. H. (2007). Functional interaction of the human cytomegalovirus IE2 protein with histone deacetylase 2 in infected human fibroblasts. *J Gen Virol* 88, 3214–3223.

Park, Y. S., Jin, M. Y., Kim, Y. J., Yook, J. H., Kim, B. S. & Jang, S. J. (2008). The global histone modification pattern correlates with cancer recurrence and overall survival in gastric adenocarcinoma. *Ann Surg Oncol* 15, 1968–1976.

Pearson, R., Kim, Y. K., Hokello, J., Lassen, K., Friedman, J., Tyagi, M. & Karn, J. (2008). Epigenetic silencing of human immunodeficiency virus (HIV) transcription by formation of restrictive chromatin structures at the viral long terminal repeat drives the progressive entry of HIV into latency. *J Virol* 82, 12291–12303.

Pennini, M. E., Pai, R. K., Schultz, D. C., Boom, W. H. & Harding, C. V. (2006). Mycobacterium tuberculosis 19-kDa lipoprotein inhibits IFN-gamma-induced chromatin remodeling of MHC2TA by TLR2 and MAPK signaling. *J Immunol* 176, 4323–4330.

Pennini, M. E., Perrinet, S., Dautry-Varsat, A. & Subtil, A. (2010). Histone methylation by NUE, a novel nuclear effector of the intracellular pathogen Chlamydia trachomatis. *PLoS Pathog* 6, e1000995.

14 Microbe-Induced Epigenetic Alterations 453

Perez-Cadahia, B., Drobic, B. & Davie, J. R. (2009). H3 phosphorylation: dual role in mitosis and interphase. *Biochem Cell Biol* 87, 695–709.

Quivy, V., Adam, E., Collette, Y., Demonte, D., Chariot, A., Vanhulle, C., Berkhout, B., Castellano, R., de Launoit, Y., Burny, A., Piette, J., Bours, V. & Van Lint, C. (2002). Synergistic activation of human immunodeficiency virus type 1 promoter activity by NF-kappaB and inhibitors of deacetylases: potential perspectives for the development of therapeutic strategies. *J Virol* 76, 11091–11103.

Radin, J. N., Gonzalez-Rivera, C., Ivie, S. E., McClain, M. S. & Cover, T. L. (2011). Helicobacter pylori VacA induces programmed necrosis in gastric epithelial cells. *Infect Immun* 79, 2535–2543.

Ralph, S. A., Foth, B. J., Hall, N. & McFadden, G. I. (2004). Evolutionary pressures on apicoplast transit peptides. *Mol Biol Evol* 21, 2183–2194.

Ralston, K. S. & Petri, W. A. (2011). Tissue destruction and invasion by Entamoeba histolytica. *Trends Parasitol* 27, 254–263.

Reeves, M. B. (2011). Chromatin-mediated regulation of cytomegalovirus gene expression. *Virus Res* 157, 134–143.

Reeves, R. & Beckerbauer, L. (2001). HMGI/Y proteins: flexible regulators of transcription and chromatin structure. *Biochim Biophys Acta* 1519, 13–29.

Respuela, P., Ferella, M., Rada-Iglesias, A. & Aslund, L. (2008). Histone acetylation and methylation at sites initiating divergent polycistronic transcription in Trypanosoma cruzi. *J Biol Chem* 283, 15884–15892.

Ribet, D., Hamon, M., Gouin, E., Nahori, M. A., Impens, F., Neyret-Kahn, H., Gevaert, K., Vandekerckhove, J., Dejean, A. & Cossart, P. (2010). Listeria monocytogenes impairs SUMOylation for efficient infection. *Nature* 464, 1192–1195.

Richman, D. D., Margolis, D. M., Delaney, M., Greene, W. C., Hazuda, D. & Pomerantz, R. J. (2009). The challenge of finding a cure for HIV infection. *Science* 323, 1304–1307.

Rikihisa, Y. (2010). Anaplasma phagocytophilum and Ehrlichia chaffeensis: subversive manipulators of host cells. *Nat Rev Microbiol* 8, 328–339.

Sager, H., Bertoni, G. & Jungi, T. W. (1998). Differences between B cell and macrophage transformation by the bovine parasite, Theileria annulata: a clonal approach. *J Immunol* 161, 335–341.

Saksouk, N., Bhatti, M. M., Kieffer, S., Smith, A. T., Musset, K., Garin, J., Sullivan, W. J., Cesbron-Delauw, M. F. & Hakimi, M. A. (2005). Histone-modifying complexes regulate gene expression pertinent to the differentiation of the protozoan parasite Toxoplasma gondii. *Mol Cell Biol* 25, 10301–10314.

Schmeck, B., Beermann, W., van, L., V, Zahlten, J., Opitz, B., Witzenrath, M., Hocke, A. C., Chakraborty, T., Kracht, M., Rosseau, S., Suttorp, N. & Hippenstiel, S. (2005). Intracellular bacteria differentially regulated endothelial cytokine release by MAPK-dependent histone modification. *J Immunol* 175, 2843–2850.

Schneider, B. G., Peng, D. F., Camargo, M. C., Piazuelo, M. B., Sicinschi, L. A., Mera, R., Romero-Gallo, J., Delgado, A. G., Bravo, L. E., Wilson, K. T., Peek, R. M., Correa, P. & El Rifai, W. (2010). Promoter DNA hypermethylation in gastric biopsies from subjects at high and low risk for gastric cancer. *Int J Cancer* 127, 2588–2597.

Schulze-Forster, K., Gotz, F., Wagner, H., Kroger, H. & Simon, D. (1990). Transcription of HIV1 is inhibited by DNA methylation. *Biochem Biophys Res Commun* 168, 141–147.

Sepulveda, A. R., Yao, Y., Yan, W., Park, D. I., Kim, J. J., Gooding, W., Abudayyeh, S. & Graham, D. Y. (2010). CpG methylation and reduced expression of O6-methylguanine DNA methyltransferase is associated with Helicobacter pylori infection. *Gastroenterology* 138, 1836–1844.

Shimura, M., Toyoda, Y., Iijima, K., Kinomoto, M., Tokunaga, K., Yoda, K., Yanagida, M., Sata, T. & Ishizaka, Y. (2011). Epigenetic displacement of HP1 from heterochromatin by HIV-1 Vpr causes premature sister chromatid separation. *J Cell Biol* 194, 721–735.

Shin, C. M., Kim, N., Jung, Y., Park, J. H., Kang, G. H., Park, W. Y., Kim, J. S., Jung, H. C. & Song, I. S. (2011a). Genome-wide DNA methylation profiles in noncancerous gastric mucosae with regard to Helicobacter pylori infection and the presence of gastric cancer. *Helicobacter* 16, 179–188.

Shin, S. H., Park, S. Y., Ko, J. S., Kim, N. & Kang, G. H. (2011b). Aberrant CpG island hypermethylation in pediatric gastric mucosa in association with Helicobacter pylori infection. *Arch Pathol Lab Med* 135, 759–765.

Siliciano, J. D., Kajdas, J., Finzi, D., Quinn, T. C., Chadwick, K., Margolick, J. B., Kovacs, C., Gange, S. J. & Siliciano, R. F. (2003). Long-term follow-up studies confirm the stability of the latent reservoir for HIV-1 in resting CD4+ T cells. *Nat Med* 9, 727–728.

Sohn, S. H. & Lee, Y. C. (2011). The genome-wide expression profile of gastric epithelial cells infected by naturally occurring cagA isogenic strains of Helicobacter pylori. *Environ Toxicol Pharmacol* 32, 382–389.

Spooner, R. L., Innes, E. A., Glass, E. J. & Brown, C. G. (1989). Theileria annulata and T. parva infect and transform different bovine mononuclear cells. *Immunology* 66, 284–288.

Sriraksa, R., Zeller, C., El Bahrawy, M. A., Dai, W., Daduang, J., Jearanaikoon, P., Chau-In, S., Brown, R. & Limpaiboon, T. (2011). CpG-island methylation study of liver fluke-related cholangiocarcinoma. *Br J Cancer* 104, 1313–1318.

Stein, M., Bagnoli, F., Halenbeck, R., Rappuoli, R., Fantl, W. J. & Covacci, A. (2002). c-Src/Lyn kinases activate Helicobacter pylori CagA through tyrosine phosphorylation of the EPIYA motifs. *Mol Microbiol* 43, 971–980.

Stephens, S. A. & Howard, C. J. (2002). Infection and transformation of dendritic cells from bovine afferent lymph by Theileria annulata. *Parasitology* 124, 485–493.

Straat, K., Liu, C., Rahbar, A., Zhu, Q., Liu, L., Wolmer-Solberg, N., Lou, F., Liu, Z., Shen, J., Jia, J., Kyo, S., Bjorkholm, M., Sjoberg, J., Soderberg-Naucler, C. & Xu, D. (2009). Activation of telomerase by human cytomegalovirus. *J Natl Cancer Inst* 101, 488–497.

Suerbaum, S. & Michetti, P. (2002). Helicobacter pylori infection. *N Engl J Med* 347, 1175–1186.

Swan, D. G., Stern, R., McKellar, S., Phillips, K., Oura, C. A., Karagenc, T. I., Stadler, L. & Shiels, B. R. (2001). Characterisation of a cluster of genes encoding Theileria annulata AT hook DNA-binding proteins and evidence for localisation to the host cell nucleus. *J Cell Sci* 114, 2747–2754.

Takahashi, H., Murai, Y., Tsuneyama, K., Nomoto, K., Okada, E., Fujita, H. & Takano, Y. (2006). Overexpression of phosphorylated histone H3 is an indicator of poor prognosis in gastric adenocarcinoma patients. *Appl Immunohistochem Mol Morphol* 14, 296–302.

Thomas, S., Green, A., Sturm, N. R., Campbell, D. A. & Myler, P. J. (2009). Histone acetylations mark origins of polycistronic transcription in Leishmania major. *BMC Genomics* 10, 152.

Tonetti, M., Eftimiadi, C., Damiani, G., Buffa, P., Buffa, D. & Botta, G. A. (1987). Short chain fatty acids present in periodontal pockets may play a role in human periodontal diseases. *J Periodontal Res* 22, 190–191.

Touati, E. (2010). When bacteria become mutagenic and carcinogenic: lessons from H. pylori. *Mutat Res* 703, 66–70.

Tovy, A., Hofmann, B., Helm, M. & Ankri, S. (2010). In vitro tRNA methylation assay with the Entamoeba histolytica DNA and tRNA methyltransferase Dnmt2 (Ehmeth) enzyme. *J Vis Exp* pii: 2390.

Tsai, C. L., Li, H. P., Lu, Y. J., Hsueh, C., Liang, Y., Chen, C. L., Tsao, S. W., Tse, K. P., Yu, J. S. & Chang, Y. S. (2006). Activation of DNA methyltransferase 1 by EBV LMP1 Involves c-Jun NH(2)-terminal kinase signaling. *Cancer Res* 66, 11668–11676.

Tyagi, M. & Karn, J. (2007). CBF-1 promotes transcriptional silencing during the establishment of HIV-1 latency. *EMBO J* 26, 4985–4995.

Uemura, N., Okamoto, S., Yamamoto, S., Matsumura, N., Yamaguchi, S., Yamakido, M., Taniyama, K., Sasaki, N. & Schlemper, R. J. (2001). Helicobacter pylori infection and the development of gastric cancer. *N Engl J Med* 345, 784–789.

Umeda, M., Murata-Kamiya, N., Saito, Y., Ohba, Y., Takahashi, M. & Hatakeyama, M. (2009). Helicobacter pylori CagA causes mitotic impairment and induces chromosomal instability. *J Biol Chem* 284, 22166–22172.

Ushijima, T. (2007). Epigenetic field for cancerization. *J Biochem Mol Biol* 40, 142–150.

Verstrepen, K. J. & Fink, G. R. (2009). Genetic and epigenetic mechanisms underlying cell-surface variability in protozoa and fungi. *Annu Rev Genet* 43, 1–24.

14 Microbe-Induced Epigenetic Alterations

von Schubert, C., Xue, G., Schmuckli-Maurer, J., Woods, K. L., Nigg, E. A. & Dobbelaere, D. A. (2010). The transforming parasite Theileria co-opts host cell mitotic and central spindles to persist in continuously dividing cells. *PLoS Biol* 8, e1000499.

Waki, T., Tamura, G., Tsuchiya, T., Sato, K., Nishizuka, S. & Motoyama, T. (2002). Promoter methylation status of E-cadherin, hMLH1, and p16 genes in nonneoplastic gastric epithelia. *Am J Pathol* 161, 399–403.

Wang, M., Furuta, T., Takashima, M., Futami, H., Shirai, N., Hanai, H. & Kaneko, E. (1999). Relation between interleukin-1beta messenger RNA in gastric fundic mucosa and gastric juice pH in patients infected with Helicobacter pylori. *J Gastroenterol* 34 Suppl 11, 10–17.

Wang, Y., Curry, H. M., Zwilling, B. S. & Lafuse, W. P. (2005). Mycobacteria inhibition of IFN-gamma induced HLA-DR gene expression by up-regulating histone deacetylation at the promoter region in human THP-1 monocytic cells. *J Immunol* 174, 5687–5694.

Weichert, W., Roske, A., Gekeler, V., Beckers, T., Ebert, M. P., Pross, M., Dietel, M., Denkert, C. & Rocken, C. (2008). Association of patterns of class I histone deacetylase expression with patient prognosis in gastric cancer: a retrospective analysis. *Lancet Oncol* 9, 139–148.

Williams, S. A., Chen, L. F., Kwon, H., Ruiz-Jarabo, C. M., Verdin, E. & Greene, W. C. (2006). NF-kappaB p50 promotes HIV latency through HDAC recruitment and repression of transcriptional initiation. *EMBO J* 25, 139–149.

Wolschendorf, F., Duverger, A., Jones, J., Wagner, F. H., Huff, J., Benjamin, W. H., Saag, M. S., Niederweis, M. & Kutsch, O. (2010). Hit-and-run stimulation: a novel concept to reactivate latent HIV-1 infection without cytokine gene induction. *J Virol* 84, 8712–8720.

Wright, E., Bain, M., Teague, L., Murphy, J. & Sinclair, J. (2005). Ets-2 repressor factor recruits histone deacetylase to silence human cytomegalovirus immediate-early gene expression in non-permissive cells. *J Gen Virol* 86, 535–544.

Yamamoto, E., Suzuki, H., Takamaru, H., Yamamoto, H., Toyota, M. & Shinomura, Y. (2011). Role of DNA methylation in the development of diffuse-type gastric cancer. *Digestion* 83, 241–249.

Yamashita, S., Tsujino, Y., Moriguchi, K., Tatematsu, M. & Ushijima, T. (2006). Chemical genomic screening for methylation-silenced genes in gastric cancer cell lines using 5-aza-2'-deoxycytidine treatment and oligonucleotide microarray. *Cancer Sci* 97, 64–71.

Yeo, A., Smith, M. A., Lin, D., Riche, E. L., Moore, A., Elter, J. & Offenbacher, S. (2005). Campylobacter rectus mediates growth restriction in pregnant mice. *J Periodontol* 76, 551–557.

Yin, Y., Grabowska, A. M., Clarke, P. A., Whelband, E., Robinson, K., Argent, R. H., Tobias, A., Kumari, R., Atherton, J. C. & Watson, S. A. (2010). Helicobacter pylori potentiates epithelial:mesenchymal transition in gastric cancer: links to soluble HB-EGF, gastrin and matrix metalloproteinase-7. *Gut* 59, 1037–1045.

Yoshida, T., Yamashita, S., Takamura-Enya, T., Niwa, T., Ando, T., Enomoto, S., Maekita, T., Nakazawa, K., Tatematsu, M., Ichinose, M. & Ushijima, T. (2011). Alu and Satalpha hypomethylation in Helicobacter pylori-infected gastric mucosae. *Int J Cancer* 128, 33–39.

Youngblood, B. & Reich, N. O. (2008). The early expressed HIV-1 genes regulate DNMT1 expression. *Epigenetics* 3, 149–156.

Index

A

Acetylcholine, 310–311
ADHD. *See* Attention deficit/hyperactivity disorder
Aeromonas hydrophila, 429
Aggression, genetic and epigenetic determinants
  brain areas
    BNST, 232, 233
    glutamatergic projections, 232
    HAA, 231–233
    PFC, 231–233
  catecholamines
    catabolism, 238
    dopamine, 237
    noradrenaline, 238
  endocannabinoids, 234, 243
  environmental effects, 249–250
  epigenetic mechanisms
    C57Bl/6 mice, maternal separation, 254
    cross-fostering, 253
    DNA methylation, 251–252
    histone modifications, 251
    postweaning social isolation, 255
    social isolation, 254
    stress diathesis models, 253–254
  epigenetic modifications
    neurotrophins, 257
    putative mechanisms, 258
    serotonin, 256
    sexual steroids, 255–256
    vasopressin and stress-axis, 256–257
  epigenetic regulation, 229, 250, 257
  GABA, 234, 237, 241, 258
  genetic changes
    antisocial and aggressive behavior, 243
    factors, 244
    gene disruption, 246–247
    hypoarousal driven aggression, 244
    molecules disruption, 245–246
    polymorphism, 247–249
    serotonin, 245
    sexual steroids, 244–245
  mammalian species, 229
  neuropeptides
    oxytocin, 239–240
    substance P - neurokinin 1 receptor, 240–241
    vasopressin, 238–239
  neurotrophins, 234, 241–242
  pathological form models, 230–231
  psychological disorders, 230
  serotonin, 236–237
  stress axis, 234, 243
  testosterone, 234–235
  valproate treatment, 258
Anaplasma phagocytophilum, 430
Angelman syndromes (AS), 56, 387
Atherosclerosis
  DNA methylation
    cytosine methylation, 404
    estrogen receptor, 405
    hypermethylation, 405–407
    hypomethylation, 406, 408
    methyl-binding domain protein, 404
    transgelin, 405
  etiology and pathobiology
    epigenetic effects, 400
    extracellular matrix protein, 399
    homocysteinemia, 401–402
    hypercholesterolemia, 401
    metabolic memory, 400
    preventative measures, 400

J. Minarovits and H.H. Niller (eds.), *Patho-Epigenetics of Disease*,
DOI 10.1007/978-1-4614-3345-3, © Springer Science+Business Media New York 2012

458   Index

Atherosclerosis (*cont.*)
  histone code, 402–404
  noncoding RNA, 408–409
Attention deficit/hyperactivity disorder
    (ADHD), 230
Autoimmune syndromes
  DNA methylation, 351–353
  environmental factors, 351
  epigenetic dysregulation
    Crohn's disease and ulcerative
      colitis, 365–366
    inflammatory thyroid diseases,
      366–367
    multiple sclerosis, 367
    primary biliary cirrhosis, 367
    progressive systemic sclerosis, 367–368
    psoriasis, 368
    Sjögren's syndrome, 368
    vitiligo, 368–369
  HLA genes, 350
  immunological tolerance, 349
  innate and adaptive immunity, 349
  organ-specific autoimmunity, 350
  rheumatoid arthritis
    CD70 and PRF1 genes, 361
    CD4+ T cells, 361
    cytokines, 360
    death receptor 3, 361
    epigenetic immune lineage analysis,
      361–362
    histone modification dysregulation,
      363–365
    hsa-mir-203 gene, 361
    IL-6 mRNA levels, 361–362
    metalloproteinases, 360
    OASF, 360
    osteoclasts, 360
    RASF, 360, 361
    synovial hyperplasia, 359
  SLE
    B and T cells, 356, 357
    CD11a gene promoter, 356
    CD40L and CD70, 356
    cutting-edge technologies, 358
    cytokine signaling, 357
    DNA methylation inhibitors, 355
    DNMTi, 355–356
    drug-induced lupus, 355
    environmental factors, 354
    histone modification dysregulation,
      358–359
    H3K9 trimethylation levels, 357
    microRNA dysregulation, 359

    perforin, 356
    susceptibility loci, 354
  systemic autoimmunity, 350
5-Aza-2'-deoxycytidine, 112, 137,
    138, 170, 211

B

BDNF. *See* Brain-derived neurotrophic
    factor (BDNF)
Beckwith–Wiedemann syndrome (BWS), 388
Bed nucleus of stria terminalis
    (BNST), 232, 233, 239, 255, 256
Belinostat, 142
Bipolar disorder
  diagnosis, 321–322
  environment-epigenetic changes
    epigenetic programming, 288
    inheritability and memory
      formation, 289–290
    nature *vs.* nurture, 287–288
    sexual dimorphism, 289
  epigenetic machinery
    chromatin remodeling, 292–293, 298
    DNA methylation, 294, 316
    DNA methyl transferase, 297
    genes encoding epigenetic
      regulators, 300
    histone modifications, 294–295,
      316–317
    methylation, modulators of, 297–298
    noncoding RNA, 296–297
    signaling pathway, 299
  genetic determination, 287
  glucocorticoid receptor, 319
  immune system, 319–320
  microRNA, 317
  neural networks, 317–318
  neurotrophic factor, 318
  prevalence, 284
  stress, 291–292
  symptoms, 285
  treatment requirement
    diet, 322
    HDACs inhibitor, 323–326
    methylation, 326
    postmitotic cell, 322
  white matter/glia abnormalities, 320–321
BNST. *See* Bed nucleus of stria terminalis
Brain-derived neurotrophic factor (BDNF),
    312–313
BRCA1. *See* Germline mutations
    of breast cancer 1

Index

Breast cancer
aberrant methylation patterns
BRCA1, 95
candidate gene approach, 94
estrogen receptor alpha, 94–95
genome-wide methylation
patterning, 96–97
hypomethylation, 93
RASSF1A gene, 95
biomarkers, 92, 108
chromatin, 97, 98, 101
CpG methylation, 91
DNA methyltransferases, 93
environmental epigenetic influences
alcohol, 107
endocrine-disrupting chemicals, 107
lifetime, 106
transgenerational effects, 107–108
xenoestrogens, 106
epigenetic regulation, 91, 92
field cancerization, 111
histone modifications measuring
methods, 98–100
mammary gland development, 104
mammary stem cells, 104–105
metastasis, 110–111
miRNA
gene expression regulation, 101
lncRNA, 103
miR-9, 103
miR-34, 103
miR-205, 102
oncomirs, 102
piRNA, 102
tumor suppressors, 102
nuclear organization, 100–101
oncogenes, 91
prognosis and progression, 109
stem cells, 105
subtype classification, 108–109
therapeutic interventions, 111–113
tumor suppressor genes, 91
Burkitt's lymphoma, 192–195
BWS. *See* Beckwith–Wiedemann syndrome

C
Campylobacter rectus, 432
Catecholamines
catabolism, 238
dopamine, 237
noradrenaline, 238
Chlamydia trachomatis, 431
Chlamydophila pneumoniae, 431

Chromokinesin homolog (KIF4A), 32
Clostridium perfringens, 428–429
Constitutive centromere protein
(CENP-C), 32
CpG islands, 160, 162, 163, 165
Crohn's disease and ulcerative colitis,
365–366
Cryptosporidium parvum, 439

D
Decitabine. *See* 5-Aza-2'-deoxycytidine
DNA methyltransferase 3B gene (DNMT3B)
mutations
catalytic mechanism, 28
chromatin remodeling, 24
crystal structure identification, 29
de novo methylation, 22
DNA methylation, mammals
CHG and CHH methylation, 19
CpG dinucleotides, 19, 20
cytosine methylation, 19
gene body methylation, 20
genomic imprinting, 21
5-hydroxymethylcytosine, 19–20
RING finger domain–containing
protein 1, 21
ubiquitin-like plant homeodomain, 21
up-to-date model, 21
enzymatic activity reduction, 30
epigenetic network
CENP-C, 32
chromatin-associated enzymatic
activities, 33
HP1 protein, 33
juxtacentromeric chromatin, 31, 34
KIF4A, 32
SYBL1 and G6PD genes, 32
functional domains and enzymatic
properties, 26–28
heterochromatin protein 1, 23
histone deacetylases, 23
H3K4 methylation, 23
H3K36 methylation, 23
H814R variant proteins, 28, 29
loss-of-function mutations, 30
mouse models, 24–26
RdDM, 24
R823G variant, 29
SAM binding–defective mutants, 28, 29
targeting models, 23
DNA methyltransferase inhibitors, 112
Dopamine, 306–307
Doxorubicin, 141

460   Index

E

EBV. *See* Epstein–Barr virus
Endocannabinoids, 234, 243
Entamoeba histolytica, 443–444
Entinostat, 142
Epigenetic immune lineage analysis, 361–362
Epstein–Barr virus (EBV), 184
  acetylation islands, 186
  antagonism, 187
  Burkitt's lymphoma, 182, 192–195
  chromatin-looping model, 187, 190
  Cp-reporter gene, 189
  CTCF binding, 189–190
  DNA methylation, 186
  EBNA1, 185
  epigenetic dysregulation mechanisms, 191
  epigenetic memory, 186
  epigenetic regulation, 185
  gastric carcinoma, 197–198
  Hodgkin's lymphoma, 195–196
  human pathogenic herpesviruses, 183
  infectious mononucleosis, 183
  latency types, 183, 184, 187–189
  lymphocryptovirus, 183
  MeCP2, 186
  nasopharyngeal carcinoma, 196–197
  PcG protein, 187
  PTLD, 198–199

F

Foreign DNA integration
  Ad12-transformed hamster cell lines, 6
  de novo methylation, 2–4
  epigenetic alterations, 5
  eukaryotic genomes, 10
  hamster cells transgenic, 6–7
  HIV infections and AIDS, 10
  human cell line HCT116, 7–8
  lincRNA, 8
  mammalian cells transgenic, 5
  retroviral/retrotransposon sequences, 8
  site-directed mutagenesis, 4
  subtractive hybridization methods, 7
  tumor disease, 9
  viral genomes, 4
Fragile X syndrome (FRAXA), 3, 56

G

Gamma-aminobutyric acid (GABA), 234, 237,
    241, 258
Gastric carcinoma, 197–198

GBM. *See* Glioblastoma multiforme
Genomic imprinting, 379, 380
Germline mutations of breast cancer 1
    (BRCA1), 95
Glandular fever/kissing disease, 183
Glioblastoma multiforme (GBM)
  advantage, 80
  De Novo GBM genome, 79, 80
  description and incidence, 73
  epigenome and CNS malignancy, 72–73
  HDAC inhibitors, 80–82
  histone deacetylase inhibitor, 80
  H3K27me3, 81
  5-Hydroxymethylcytosine, 77–78
  IDH mutation and DNA hypermethylator
      phenotype, 75–77
  lysine demethylase 1, 81
  phenylbutyrate, 82
  primary and secondary, 73–74
  SAHA, 81
Glucocorticoid receptor, 319

H

HAA. *See* Hypothalamic attack area
HDAC inhibitors. *See* Histone deacetylases
    inhibitors
Hepatitis B virus (HBV)
  methylome, 203
  Orthohepadnavirus genus, 202
  regional hypermethylation *vs.* global
      hypomethylation, 203–206
Hepatitis C virus (HCV), 206–207, 426
Highly active antiretroviral therapy
    (HAART), 422, 423
Histone deacetylases (HDAC) inhibitors,
    112–113
  advantages, 323
  antidepressant property, 325
  antitumor activity, 142
  belinostat, 142
  classes, 139
  clinical trials, 139, 140
  corepressor protein complex, 324
  doxorubicin, 141
  entinostat, 142
  metabotropic glutamate receptor
      activation, 324–325
  N-hydroxy-N'-phenyl-octanediamide, 323
  OSU-HDAC42, 142
  panobinostat, 141
  romidepsin, 142
  sirtuins, 139

Index

suberoylanilide hydroxamic acid, 139
treatment efficacy, 325–326
US Food and Drug Administration, 323
valproic acid, 142
vorinostat, 141
Hodgkin's lymphoma, 195–196
Homocysteinemia, 401–402
HPV. *See* Human papillomaviruses
Human cytomegalovirus block histone
deacetylases, 426–427
Human papillomaviruses (HPV)
cellular epigenetic processes
dysregulation, 208–211
E6 proteins, 208
high-risk, 207
host cell-dependent methylomes, 208
Human T-lymphotropic virus type I
(HTLV-I), 202
Hypercholesterolemia, 401
Hypothalamic attack area (HAA), 231–233

I

Immunodeficiency, centromere instability,
facial abnormalities (ICF) syndrome
clinical manifestation and diagnosis, 16–18
DNMT3B mutations (*see* DNA
methyltransferase 3B gene
mutations)
ICF Type 1 and ICF Type 2 genetics,
18–19
Imprinting disorders (ID)
clinical characteristics
Angelman syndromes, 387
Beckwith–Wiedemann syndrome, 388
molecular defects, 386
Prader–Willi Syndromes, 387
pseudohypoparathyroidism type 1
(PHPIb), 390
Silver–Russell syndrome, 387–388
transient neonatal diabetes mellitus,
388–389
upd(14)mat and upd(14)pat
syndromes, 389
genomic imprinting, 379, 380
molecular alterations
chromosomal imbalances, 385
epimutations, 380–382
genomic point mutations, 385–386
UPD, 382–385
molecular genetic testing, 391
multilocus methylation defects, 390–391
Inflammatory thyroid diseases, 366–367

K

Kaposi's sarcoma-associated herpesvirus
(KSHV)
genome, 190
heterochromatin-associated protein 1, 200
H3K27me3, 200
LANA, 200–202
lytic replication, 199
OriP, 200

L

Latency-associated nuclear antigen
(LANA), 200–202
Leishmania, 442
LIM homeobox, 315
Listeria monocytogenes, 428–429
Lung carcinomas, epigenetic
reprogramming
anti-methyl CpG antibody, 161
CpG islands, 160, 162, 163, 165
driver/passenger DNA methylation, 165
genomic DNA, bisulfite treatment, 161
histone modifications
aberrant hypoacetylation, 168
beads on a string, 167
H4 tri-methylation, 168
JMJD2C, 169
linker DNA, 167
phase I and II, 168
posttranslational
modifications, 167, 168
5hmC role, 160, 165–166
hypermethylations in genome
cell-cycle regulation, 163
COBRA assays, 163
epigenetic studies, 163
mechanism, 164
SINEs and LINEs harboring loci, 164
MIRA-on-chip, 161–162
molecular model, 160
multifactorial process, 159
non-coding RNAs
5-aza-2'-deoxycytidine treatment, 170
MALAT1, 169
signals, 170
TargetScanHuman database, 169
tumor suppressors/oncogenes, 169
NSCLC, 159
pharmacological aspects of DNA
methylation, 166–167
SCLC, 159
Lymphocryptovirus, 182

M

Merkel cell polyomavirus (MCPyV), 211–212
Metastamir, 110
Methyl-CpG-binding protein (MeCP2), 186
  AMPAkines, 59
  Angelman syndrome, 56
  Bdnf gene coding, 58–59
  "BDNF-mimetic" trophic factors, 59
  CamkII promoter, 57, 58
  Cre-ER protein, 57
  Cre/loxP recombination system, 54
  desipramine, 59
  dietary choline supplementation, 60
  Down syndrome, 56
  fiber-Purkinje cell synapses, 60
  Fragile X syndrome, 56
  gene function, 51–52
  knockout mice, 53
  lentiviruses, intraventricular injections, 55
  loxP sites, 54
  loxP-STOP-loxP cassette, 58
  MeCP2 A140V mouse, 55
  neurofibromatosis-1, mouse model, 56
  neuron-specific transgene expression, 57
  P1-derived artificial chromosome, 55
  postmitotic neurons, 57
  Rubinstein–Taybi syndrome, 56
  R168X mouse, 55
  short hairpin RNA, 55
  Tau-MeCP2 fusion protein, 57
  truncated protein, 53–54
Microbe-induced epigenetic alterations
  bacterial effector proteins, 430
  bacterial infections, 427
  bacterial toxins, 428–429
  Campylobacter rectus, 432
  Chlamydia trachomatis, 431
  Chlamydophila pneumoniae, 431
  helicobacter pylori infection
    cytotoxicity-associated
      antigen, 433–434
    epithelial–mesenchymal transition, 434
    histone modifications, 438
    host cell methylome
      alterations, 435–438
    pleiotropic effects, 434
    TP53 tumor suppressor gene
      mutations, 434
    transcriptional repression
      mechanism, 434
    vacuolating cytotoxin, 433
    virulence factors, 433
  hepatitis C virus, 426

HIV infected cell
  cellular genes silencing, 425
  HAART, 422, 423
  lymphadenopathy, 422
  replication cycle and epigenetic
    silencing, 423–425
  viral load, 422
  viral protein R, 426
human cytomegalovirus block histone
  deacetylases, 426–427
periodontal disease, 429
protozoan parasites
  congenital toxoplasmosis, 439
  Cryptosporidium parvum, 439
  Entamoeba histolytica, 443–444
  Leishmania, 442
  Plasmodium falciparum, 439
  Theileria annulata, 440–442
  Theileria parva, 440–442
  Toxoplasma gondii, 439, 440
  Trypanosoma brucei, 442
  Trypanosoma cruzi, 442
uropathogenic E.coli infection, 432
Microcephaly, 44
Multiple sclerosis, 367
Mycobacterium tuberculosis, 430

N

Nasopharyngeal carcinoma, 196–197
Neuregulin, 313–314
Neuropeptides
  oxytocin, 239–240
  substance P - neurokinin 1 receptor,
    240–241
  vasopressin, 238–239
Neurotrophins, 16–17
Non-small cell lung carcinomas
  (NSCLC), 159

O

Osteoarthritis synovial fibroblasts
  (OASF), 360
Oxytocin, 14

P

Panobinostat, 141
Periodontal disease, 429
PFC. *See* Prefrontal cortex (PFC)
Plasmodium falciparum, 439
Polycomb group proteins (PcG), 99–100

Polymorphism
    AVP V1b receptor, 249
    serotonin system, 247–248
    sexual steroids, 247
Postmitotic cell, 322
Posttransplant lymphoproliferative disorder
    (PTLD), 198–199
Prader–Willi syndromes (PWS), 387
Prefrontal cortex (PFC), 231–233, 236, 237
Primary biliary cirrhosis, 367
Progressive systemic sclerosis, 367–368
Prostate cancer
    biomarkers
        carcinoma tissue analysis, 131
        epigenetic catastrophe, 131
        GSTP1 methylation, 128–130
        hypermethylated genes, 126, 127
        meta-analysis, 129
        PIN lesions, 128
        RASSF1A gene, 130, 131
        serum PSA concentration
           measurement, 126
    clinically significant cancer, 123
    genetic alterations, 124–125
    lethal tumors, 124
    prognostic biomarkers
        biochemical recurrence, 134, 135
        DNA hypermethylation events, 133
        global hypomethylation, 132
        histone modifications and histone-
           modifying enzymes, 135–136
        MCAM hypermethylation, 133
        MGMT hypermethylation, 132
        MLH1 hypermethylation, 132
        multigene testing, 135
        PITX2 promoter hypermethylation, 134
        PTGS2 hypermethylation, 134
        SOCS3, 134
    prostate-specific antigen, 123
    therapeutic targets
        DNA methylation, 137–139
        HDAC inhibitors (*see* Histone
           deacetylases inhibitors)
        histone demethylase inhibitors, 143–144
        histone methyltransferase
           inhibitors, 143
Prostatic intraepithelial neoplasia
    (PIN) lesions, 128
Pseudohypoparathyroidism type 1
    (PHPIb), 390
Psoriasis, 368
PTLD. *See* Posttransplant lymphoproliferative
    disorder
PWS. *See* Prader–Willi syndromes

R

Ras-associated domain family member 1 gene
    (RASSF1A), 95
RASF. *See* Rheumatoid arthritis synovial
    fibroblasts
Reelin, 311–312
Rett syndrome (RTT)
    autism spectrum disorders, 49–51
    clinical features
        breathing irregularities, 47
        cardiac conduction abnormalities, 47
        consensus criteria, 44–46
        gallbladder dysfunction, 48
        medical issues, 46, 47
        microcephaly, 44
        paroxysmal/epileptic episodes, 48
        regression phase, 44
        scoliosis, 47
        temporal profile, 46
        video-EEG monitoring, 48
    MeCP2
        AMPAkines, 59
        Angelman syndrome, 56
        Bdnf gene coding, 58–59
        "BDNF-mimetic" trophic factors, 59
        CamkII promoter, 57, 58
        Cre-ER protein, 57
        Cre/loxP recombination system, 54
        desipramine, 59
        dietary choline supplementation, 60
        Down syndrome, 56
        fiber-Purkinje cell synapses, 60
        Fragile X syndrome, 56
        gene function, 51–52
        knockout mice, 53
        lentiviruses, intraventricular
           injections, 55
        loxP sites, 54
        loxP-STOP-loxP cassette, 58
        MeCP2 A140V mouse, 55
        neurofibromatosis-1, mouse model, 56
        neuron-specific transgene
           expression, 57
        P1-derived artificial chromosome, 55
        postmitotic neurons, 57
        Rubinstein–Taybi syndrome, 56
        R168X mouse, 55
        short hairpin RNA, 55
        Tau-MeCP2 fusion protein, 57
        truncated protein, 53–54
    neuropathology, 48–49
Rheumatoid arthritis
    CD70 and PRF1 genes, 361
    CD4⁺ T cells, 361

Rheumatoid arthritis (*cont.*)
  cytokines, 360
  death receptor 3, 361
  epigenetic immune lineage
      analysis, 361–362
  histone modification
      dysregulation, 363–365
  hsa-mir-203 gene, 361
  IL-6 mRNA levels, 361–362
  metalloproteinases, 360
  OASF, 360
  osteoclasts, 360
  RASF, 360, 361
  synovial hyperplasia, 359
Rheumatoid arthritis synovial fibroblasts
    (RASF), 360
RNA-directed DNA methylation (RdDM), 24
Romidepsin, 142
RTT. *See* Rett syndrome
Rubinstein–Taybi syndrome, 56

S

Schizophrenia (SCZ)
  Brodmann area, 305
  cannabis, 315
  diagnosis, 321–322
  environment–epigenetic changes
      epigenetic programming, 288
      inheritability and memory
          formation, 289–290
      nature *vs.* nurture, 287–288
      sexual dimorphism, 289
  epigenetic machinery
      chromatin remodeling, 292–293, 298
      DNA methylation, 294, 302
      DNA methyl transferase, 297
      genes encoding epigenetic
          regulators, 300
      histone modifications, 294–295,
          302–303
      methylation, modulators of, 297–298
      microRNA, 292–293, 298, 303–304
      noncoding RNA, 296–297, 304
      signaling pathway, 299
  FosB, 315
  GABA synthesizing enzyme, 305
  genetic determination, 285–287
  LIM homeobox, 315
  methyl supply, 301–302
  neurotransmitters
      acetylcholine, 310–311
      γ-aminobutyric acid, 309–310
      dopamine, 306–307

glutamate, 308–309
      serotonin, 307–308
  neurotrophin factors
      BDNF, 312–313
      neuregulin, 313–314
      reelin, 311–312
  nuclear-receptor NR4A2, 315
  prevalence, 284
  primary and secondary sensory
      areas, 304
  psychopathology, 304, 305
  stress, 285, 290–291
  symptoms, 284
  treatment requirement
      diet, 322
      HDACs inhibitors, 323–326
      methylation, 326
      postmitotic cell, 322
  Wnt signaling pathway, 315
  Y-Box containing gene 10, 314–315
SCLC. *See* Small cell lung carcinomas
SCZ. *See* Schizophrenia
Shigella flexneri, 430
Silver–Russell syndrome (SRS), 387–388
Sjögren's syndrome, 368
SLE. *See* Systemic lupus erythematosus
Small cell lung carcinomas (SCLC), 159
SRS. *See* Silver–Russell syndrome
Streptococcus pneumoniae, 428–429
Systemic lupus erythematosus (SLE)
  B and T cells, 356, 357
  CD11a gene promoter, 356
  CD40L and CD70, 356
  cutting-edge technologies, 358
  cytokine signaling, 357
  DNA methylation inhibitors, 355
  DNMTi, 355–356
  drug-induced lupus, 355
  environmental factors, 354
  histone modification dysregulation,
      358–359
  H3K9 trimethylation levels, 357
  microRNA dysregulation, 359
  perforin, 356
  susceptibility loci, 354

T

Testosterone, 234–235
Theileria annulata, 440–442
Theileria parva, 440–442
Toxoplasma gondii, 439, 440
Transient neonatal diabetes mellitus
    (TNDM), 388–389

Index

Trypanosoma brucei, 442
Trypanosoma cruzi, 442
Tumor biology. *See* Foreign DNA integration

U
Uniparental disomy, 382–385
Upd(14)mat and upd(14)pat syndromes, 389

V
Valproate treatment, 258
Vasopressin, 238–239
Vidaza™, 112
Virus-associated neoplasms
  EBV, 184
    acetylation islands, 186
    antagonism, 187
    Burkitt's lymphoma, 182, 192–195
    chromatin-looping model, 187, 190
    Cp-reporter gene, 189
    CTCF binding, 189–190
    DNA methylation, 186
    EBNA1, 185
    epigenetic dysregulation
      mechanisms, 191
    epigenetic memory, 186
    epigenetic regulation, 185
    gastric carcinoma, 197–198
    Hodgkin's lymphoma, 195–196
    human pathogenic herpesviruses, 183
    infectious mononucleosis, 183

  latency types, 183, 184, 187–189
  lymphocryptovirus, 183
  MeCP2, 186
  nasopharyngeal carcinoma, 196–197
  PcG protein, 187
  PTLD, 198–199
HBV
  methylome, 203
  Orthohepadnavirus genus, 202
  regional hypermethylation *vs.* global
    hypomethylation, 203–206
HCV, 206–207
HPV
  cellular epigenetic processes
    dysregulation, 208–211
  E6 proteins, 208
  high-risk, 207
  host cell-dependent methylomes, 208
HTLV-I, 202
KSHV (*see* Kaposi's sarcoma-associated
  herpesvirus)
MCPyV, 211–212
Vitiligo, 368–369
Vorinostat, 141

W
Wilms' tumor, 182

Y
Y-Box containing gene 10, 314–315

Printed by Publishers' Graphics LLC
BT20121030.19.18.61